CHEMISTRY

IN THE MARKETPLACE

Fourth Edition

Illustration by Mary Fuller

CHEMISTRY
IN THE MARKETPLACE
Fourth Edition

BEN SELINGER

Department of Chemistry
Faculty of Science
Australian National University, Canberra

HARCOURT
BRACE

Sydney   Fort Worth   London   Orlando   Toronto

Harcourt Brace & Company, Australia
30–52 Smidmore Street, Marrickville, NSW 2204

Harcourt Brace & Company
24–28 Oval Road, London NW1 7DX

Harcourt Brace & Company
Orlando, Florida 32887

First edition 1975
Reprinted 1976, 1977
Second edition 1978
Reprinted 1981
Third edition 1986
Reprinted 1987, 1988
Fourth edition 1989
Reprinted 1991, 1994, 1996

National Library of Australia Cataloguing-in-Publication Data

Selinger, Ben, date-.
Chemistry in the marketplace.

4th ed.
Bibliography
Includes index.
ISBN 0 7295 0334 8.

1. Chemistry. 2. Chemistry — Popular works. 1. Title.

540

Printed in Australia by Star Printery

Contents

Preface

Science teaching internationally has caught up with the idea that teaching must be made more relevant to society to better motivate our children and to give adults an informed opinion on technological issues. In 1988 the national guidelines that came into effect for chemistry teaching at General Certificate of Secondary Education (GCSE) level in the UK took into account the approach used in *Chemistry in the marketplace*. These guidelines specify that social, economic, environmental and technological issues of chemistry must represent at least 30% of the syllabus. Across the Atlantic in the same year, the American Chemical Society publication, *ChemCom* (Chemistry in the Community), was produced in its final form under the project management of Sylvia A. Ware, with whom I have had many interesting discussions.

The first edition of *Chemistry in the marketplace* was published in 1975, and it has been widely used as a text around which teachers and lecturers have organised their own courses. It also seems to have fulfilled its major aim of providing the general reader with a reference book to the broader community issues in which chemistry is involved. The examples chosen are mostly Australian, although this fourth edition also caters for international readers. The problems are much the same the world over. How much has this edition changed since the third? Enough new material is included to fill a double-sided Macintosh disk (800K).

Chemistry in the marketplace is written for the general reader. Such 'academic' chemistry as it contains can be skipped without loss of continuity or understanding. It is included for those who can benefit from the additional information. On the other hand, the general reader will hopefully gain some insight into the way chemists think.

Chapter 1 does a tourist tour of basic concepts. I am under no illusion on its limited usefulness. This edition includes a section on the naming of organic chemicals, so that the reader can appreciate that there is a logic in those horrible chemical names that is no worse than in other languages such as Latin or Basic.

We start in Chapter 2 with the laundry, where considerable changes have occurred in the market and thus also in the book. Chapter 3, in the kitchen, still emphasises fat chemistry, but also deals with technology of cooking. Chapter 4, in the boudoir, covers cosmetics, sunscreens and surface chemistry, the latter providing a cross-link in ideas and concepts with many other topics dealt with elsewhere in the book. When we move into the garden we are confronted with soils and pesticides. For the summer, there is swimming pool chemistry, important to our health and providing another cross-link, not so much this time with other sections of the book, but with topics in senior high school and junior tertiary curricula.

In the shopping centre, we take Chapter 6 into the hardware store, and first take a look at the science of materials such as plastics and metals. This is expanded in Chapter 7 to cater for fabrics, including carpets and leather, while Chapter 8 considers paints and glues, concrete and fun materials.

Chapter 9 on drugs has been completely rewritten and updated and we take a much closer look at pharmacies and illicit drugs.

The chemistry of energy has expanded Chapter 10 again, in spite of the deletion of older material. Related to this chapter is Appendix VIII, 'The entropy game', which should provide both enjoyment and insight.

Chapter 11, in the dining room, has been updated to include changes in legislation for food additives and Appendix XII gives the decoder for the additive numbers you find on your packaged food (introduced by the author as a member of the Food Standards Committee of the National Health and Medical Research Council (NH&MRC) in 1974).

In Chapter 12 an explanation and discussion of the consumer implications of radiation has been added to the information on heavy metals.

Chapter 13 gives a large selection of consumer experiments in considerable detail for people wishing to set them up in a laboratory. Elementary experiments and simple lecture demonstrations (for professional teachers only) are placed where relevant throughout the book. A large selection of consumer information sources and library references is included in Appendix XIV, while an ever expanding glossary acts as a finale.

I have tried to maintain the philosophy of emphasising the product, followed by the chemistry necessary to understand it, rather than the usual inverted approach. To keep this edition up to date, I found myself making more use of articles in the 'professional-popular' science digests such as *Ecos* from Australia and *New Scientist* from the UK, which provide reference to primary sources.

The eclectic approach to chemistry that characterises this book has meant diverse sources of information and consequent difficulty in validation. As in the past, I welcome correspondence on errors of omission and commission, and updates where things have changed. Questions to which readers believe the book should have addressed itself are particularly welcome, although in a limited volume (at whose size the publishers already baulk) the problem of compromise between breadth and depth is not readily resolvable.

I hope you enjoy reading this book as much as I have enjoyed writing it.

USER FRIENDLY CHEMISTRY

The author is Head of the Department of Chemistry, Faculty of Science, the Australian National University, Canberra, Australia. The department is part of the Faculty of Science and sits astride the physical and biological sciences. The first year offers a traditional chemistry course in conjunction with an innovative practical section, which emphasises consumer aspects of chemical analysis. The theory course is backed by computer-aided learning, which has proved popular with students.

In the second year, semester units are offered in biological/organic, materials chemistry (physical/inorganic) and instrumental analysis, units which cut across the traditional boundaries. In the practical work there is emphasis on learning to use modern instruments and to interpret the results, as well as on developing experimental accuracy, skills that are very marketable.

In the third year there are traditional units in inorganic, organic and physical chemistry as well as more specialist programs. The Research School of Chemistry, which is physically linked to the Department of Chemistry, also provides a wide range of input, and its world-class facilities are available to senior undergraduate and graduate students. Related sections of the CSIRO as well as the National Science and Technology Centre are nearby.

Research at the Department of Chemistry varies with staff interest, but includes the chemistry of surfaces and colloidal solutions, natural products and organic reaction mechanisms, spectroscopy, chemistry and the law, coordination chemistry, pharmacology and biophysical chemistry, and, of course consumer chemistry.

WARNING

While all care has been taken to check the experiments and provide safe instructions, readers are warned that chemistry experiments can be dangerous and should be carried out with safety always in mind. The author does not warrant that any instruction, recipe or formula in the book is free of possible danger to the user.

Acknowledgments

While the problem with the first edition was in obtaining information, the climate generated by organised consumer activity has in the meantime resulted in so great an availability of relevant sources as to make selection the major concern.

Chapter 1. May and Baker Ltd (Essex, UK) provided Figure 1.1, the periodic table of the elements (one of the few that gives each element its full name). For help with the section on nomenclature and much else, I thank Dr Tom Bellas and Dr Malcolm Rasmussen. *Song of the elements* is reproduced with the permission of Tom Lehrer. I had fallen in love with Terry Sedgwick's ceramic goannas and I am pleased he has allowed me to reproduce a photograph of the chemical goanna he made for me (Fig. 1.13).

Chapter 2. Mr Peter Strasser is thanked both as author and editor for the use of his published lecture on detergents, as well as for his continuing advice on updating this material, while New Science Publications allowed Ariadne's photograph of the royal statue attacked by detergent to appear (Fig. 2.8). The *Canberra Times* provided the photograph of the foaming Molonglo River (Fig. 2.12).

Chapter 3. Hugh Sinclair's magnificent poetical parody *Hiawatha's lipid* was brought to my attention by Sir Henry Somerset and I am grateful for Dr Sinclair's permission to publish it here. Professor Rod Rickards' article on prostaglandins is reproduced with his permission. I am also indebted to the CSIRO Division of Food Processing for its permission to use material from *Food Research Quarterly* and a large section of its Information Sheet no. 17-1, *Nutritional value of processed food*, compiled by CSIRO editor, Gordon Walker.

Chapter 4. A stimulating talk by Ms Marilyn Elfverson on market research (presented to the Australian Cosmetics Association in Canberra, April 1975) is reproduced with permission, to contrast with the report of the Industries Assistance Commission on cosmetics. The section on hair is reproduced from 'A hair piece for Christmas', by J. Chorfas, *New Scientist* 19(26), December 1985, with permission. The photograph of *in vivo* testing of sunscreens (Fig. 4.10) is courtesy of Greiter, A.G. The two diagrams on the ultraviolet penetration of skin (Fig. 4.7) and eye (Fig. 4.11) were provided by Dr Keith Lokan, Director of the Australian Radiation Laboratory, Melbourne.

Chapter 5. The Total Environment Centre in Sydney has allowed me to reproduce the table on pesticides and alternatives (Table 5.3) and I hope that this will publicise the efforts of the centre in this important area. The editor of

Chemistry in Australia has allowed me to reuse material I had published in the journal, including sections of my Nyholm Memorial Lecture and Bayliss Youth Lecture. Standards Australia is thanked for granting permission to reproduce sections from Australian standards and drafts in this and other chapters. Figures 5.1 and 5.2 are reproduced from *Soils: an outline of their properties and management* (CSIRO, Melbourne, 1977), with permission of CSIRO publications department.

Chapter 6. The Plastics Institute of Australia is thanked for providing Tables 6.3 and 6.4 on plastic films, and for material from the seminar, Plastics and Food Packaging in Perspective. Dr Steve Glover allowed me to reproduce his article on the design of sailboards from *RSC News*.

Chapter 7. The late Dr Tom Pressley provided guidance in the area of fabric flammability and granted permission to make extensive use of his research material. The United Kingdom Consumers' Association, through Mrs Gillian Clegg, allowed reproduction of a section of their monumental project on carpets.

Chapter 8. Mr Timothy Crick at Canberra CAE made me aware of the importance of the design of consumer products, and their chemical basis. The American Chemical Society granted permission to reprint Figure 8.6 on the chalking of paints.

Chapter 9. Alan Foley Pty Ltd gave permission for the 'Wizard of ID' cartoon (Fig. 9.2). I am grateful to the editor of *Chemistry in Britain* (Chemical Society, London) for allowing the generous use of material on the chemistry of drugs. Emeritus Professor Adrien Albert provided much help and support on the topic of drugs and I am pleased to recommend his book, *Xenobiosis* (Chapman and Hall, 1988), finished on his eightieth birthday, as an excellent reference. Dr Sam Wong (Commonwealth Department of Community Services and Health) helped update aspects of the pharmaceutical sections. Mr Peter Crothers of the Pharmaceutical Society of Australia provided a helpful insight into the pharmacy industry. Plate X, showing the Reflotron diagnostic system, is reproduced courtesy of Boehringer-Mannheim.

Chapter 10. The Australian Consumers' Association allowed reproduction of material from *Choice* and generous use has been made of this. CSIRO encouraged the use of material from its publications, such as *Rural Research* (courtesy B.J. Woodruff), for the section 'Energy down on the farm'. Dr K.G. Neill of ICIANZ provided the information for the section on brake fluids and Figure 10.16. Figure 10.5 is reproduced courtesy of the NSW Department of Energy. The photograph of the Solar Marsupial (Fig. 10.11) was provided by Dick Smith and *Australian Geographic*. Automobiles Peugeot provided the diagram of the electric Peugeot (Fig. 10.12). Figure 10.17 showing a comparison of the capacity and power of storage batteries with those of a cyclist is included by courtesy of MIT Press. The diagrams depicting atmospheric reactions (Fig. 10.21) are from G. Pimentel *et al.*, *Opportunities in chemistry* (National Academy Press, Washington DC 1985), pp. 199–200.

Chapter 11. In 1984, the last edition of the *Monash Science Magazine* was produced. The magazine's editor, Associate Professor Ian D. Rae, has allowed me to use some material on liquid extraction from that fateful final edition. Mr Michael Tracey (Director, CSIRO Division of Food Processing) gave permission for the extract from his paper 'The price of making our foods safe and suitable'. The section on food law in Australia incorporates comments from Val Johanson of the federal Bureau of Consumer Affairs (Attorney General's Department). Mr Bill McCray (Director of Biochemistry, Animal Research Institute, Queensland) allowed me to use a section of his talk on 'The myth of pure natural foods'. Emeritus Professor F.H. Reuter is thanked for allowing the generous use of many articles from the journal *Food Technology in Australia*. Dr Frank Peters (analyst for the Australian government) allowed me to use his talk on 'Legal aspects of alcoholic products' and provided an interesting insight into aspects of food irradiation. Mr Richard Boston allowed the use of his article in the *Guardian*, 'More than barley and hops'. Gerald Duckworth Co. Ltd allowed reproduction of an extract of the poem 'On food' by Hilaire Belloc. The front cover of *Consumer Reports*, March 1978 (Plate XIII), is included courtesy of the Consumers Union, USA. Plate XV appears courtesy of Fritz Sondereggar, Urambi Hills Wholemeal Bakery, ACT. Permission to reproduce the following figures is also acknowledged: Fig. 11.1, United Features Syndicate Inc.; Fig. 11.2, Alan Foley Pty Ltd; Fig. 11.14, *New Scientist* and Searle; Fig. 11.19, Dr Simon Brooke-Taylor, Eagle Hawk Hill Brewery Co., Sutton, NSW; Figs. 11.22 and 11.23, Bread Research Institute of Australia.

Chapter 12. Mr Anthony Tucker, science correspondent for the *Guardian*, gave permission to use material from his book *The toxic metals*. A major source was provided by the article 'Trace elements in food' by Mr J.W.C. Neuhaus (NSW Health Commission Analytical Laboratories). To the Shire Clerk of the City of Broken Hill go my thanks for supplying a city street map in which the chemical names say it all (Fig. 12.2). Dr Keith H. Lokan, Director of the Australian Radiation Laboratory is thanked for very helpful comments on the section on radioactivity and for permission to use the diagrams on sampling radon levels (Fig. 12.14). Figure 12.5 appears courtesy of the *Newcastle Morning Herald*. Figures 12.3, 12.11 and 12.12 are from the *New Scientist*, 9 January 1975 and 11 February 1988.

Chapter 13. Permission was granted by Canberra Consumers Inc. to use data for Table 13.1. The Royal Society of Chemistry allowed reproduction of material from my own paper published in *Education in Chemistry*. Some experiments from the consumer course on which this book was based were incorporated into our general first-year practical teaching program at the ANU, where improvements took place. These have now been incorporated in this edition. The wine analysis experiment (13.16), however, was initiated in the first-year practical and has been included in this book. This is gratefully acknowledged. Dr Dennis Mulcahy, South Australian Institute of Technology, has been a constant and continuous source of inspiration on many experimental aspects (e.g. 13.10).

Appendixes. Mrs Yvonne Preston and the *Financial Review* are thanked for permission to reproduce a moving and succinct review of the book, *Thalidomide and the power of the drug companies* (Appendix IV). The Child and Home Safety Advisory Committee of the National Safety Council of South Australia granted permission to reproduce part of its comprehensive statistics on accidental poisoning in children (Appendix V). The Information Section, Pharmaceutical Benefits Branch, Australian Department of Health, provided the statistics on the prescription frequency of pharmaceutical benefits (Appendix VI). Mr R.R. Couper (CSIRO Division of Building Research) gave permission to reprint the article on space heaters from *Rebuild*, June 1977 (Appendix IX). Appendix X on octane values is based on articles in *Chemistry and Engineering News*. The National Health and Medical Research Council gave permission for reproduction from their standards and booklet, *Lead glazes in pottery* (Appendix XIII). Mr Mike Vernon AM, veteran chairman of the ACT Consumer Affairs Council, continued to make available his encyclopaedic knowledge of product safety (Appendix XIV). His initiative in establishing Consumer Interpol is a landmark in the move towards making consumer protection international. He provided most of the sources of consumer information and assisted with the updating of consumer standards information.

General. I am indebted to Dr David Hill (Chemistry Department, University of Queensland), who edited and largely produced the booklet *Chemistry experiments for high school students*, 2nd edition, and to the Royal Australian Chemical Institute Chemical Education Branch, Queensland and Northern Territory, for permission to use the following home experiments in various chapters: Separation by solubility (Exp. 1.1); Rusting of iron (Exp. 1.2); Clarification of river water (Exp. 2.1); Hydrolysis of starch with saliva (Exp. 3.1); Test for starch (Exp. 3.2); Preparation of ferric tannate (Exp. 8.1); Electrolysis of acetic acid solution (Exp. 10.1); Tarnishing and corrosion (Exp. 12.1); Exchanging metals (Exp. 12.2). Dr Tom Bellas (Division of Entomology, CSIRO) applied his considerable editorial skills to checking details in the manuscript. The original acknowledgment to Dr Malcolm Rasmussen and Dr Dereham Scott in the first edition needs to be repeated. Without their help and encouragement in presenting the course in consumer chemistry, this book would never have appeared. With the encouragement of ANU Press and its experienced editor, Patricia Croft, the course notes were turned into a book. In 1984, at the time when ANU Press was closed down, the publication of the third edition was taken over by Harcourt Brace Jovanovich. I would like to thank Dallas Cox for coping with what turned out to be a major reworking of that edition. Although the fourth edition was meant to be a minor revision, this was not to be. HBJ Senior editor Carol Natsis has my gratitude for ensuring that considerable rewriting was incorporated smoothly. A check for consistency of the use of SI units was initiated by Angus Henry of the Northern Territory Department of Education, Curriculum and Assessment Branch. For this I am particularly grateful. Versel Scientific Consultants (PO Box 71, Garran, ACT 2605) allowed me to use ideas that had resulted from commissioned projects (with the consent of clients).

My service on the following bodies has given me an invaluable insight into the role of chemistry in society:

- Food Standards Committee, NH & MRC, Commonwealth Department of Health (1973–75)
- Executive of Canberra Consumers Inc. (1974–79)
- Council of the Australian Consumers' Association (1980–86)
- Editorial panel, Australian Academy of Science School Chemistry Project, *Elements of chemistry: earth, air, fire and water* (1980–81)
- Advisory Council of CSIRO, observer (1983–84)

- ACT Asbestos Advisory Committee, Chair (1983–86)
- National Health and Safety Commission, now Worksafe Australia, commissioner for the ACT (1984–85)
- Advisory Committee to the Australian Government Analytical Laboratories (1986)

While every attempt has been made to find the original copyright owners of all sources of information, the material, having been collected over many years without thought to publication, may sometimes not be acknowledged. In that event I would be grateful for information so that due acknowledgment can be given.

It is to my late father, Herbert Selinger, that I am particularly indebted. His long involvement with the consumer movement (he was also on the council of the ACA), as well as his firm belief that scientists should do useful things (at least some of the time), has had a profound influence on me.

My 16-year-old son Michael is thanked for providing some of the new drawings for this edition, and 18-year-old Adam helped me test some of the experiments.

Finally, I thank my wife Veronique for her patience and understanding, in spite of my broken promise never to write another edition.

Some basic chemical ideas

1

INTRODUCTION

I shall open this chapter with some lightly edited extracts from a polemic 'What is chemistry, that I may teach it?' by Peter Nelson of the University of Hull, UK.

The meaning of the word 'chemistry' has changed many times over the centuries. In the third century of our common era, it referred to the fraudulent practice of imitating precious metals and stones (Greek *khemeia*). By the Middle Ages, this practice had evolved into the quest for a substance that would actually transmute base metals into gold, and into the study of matter associated with this process (Arabic *alkimia*) to give alchemy. In the sixteenth century, following the redirection of alchemy towards medicine by Paracelsus, the word came to mean the making of medicines. A dealer in medicinal drugs became a chemist. With Boyle's critique of alchemical theory in his *The Sceptical Chymist* of 1661, 'chemistry' came to mean also the study of matter in a scientific manner.

There is ample precedent for changing the meaning of the word 'chemistry' and for including 'applied' activities within its scope . . . Today's academics have been brought up to think of chemistry as a science and the more empirical aspects of 'applied' chemistry as not really chemistry . . . Chemistry occupies a special position in science as being one area in which explanations of the kind used in physics 'rub shoulders with' explanations of the kind used in biology . . . There can be no doubt that the general tendency in schools and universities is to lay much greater emphasis on the more *rational* aspects of chemistry than on the more *intuitive*. [This leads to important topics such as catalysis and colloid and surface chemistry remaining untaught because they are not yet wholly rational] . . .

Applied chemistry is a very wide-ranging activity comprising the chemical industry (i.e. manufacture of materials) and service industry (purification of water supplies, toxicology, forensic chemistry etc.). The materials used by the chemical industry and the equipment that is used to produce them vary very widely—from pig iron made in blast furnaces and petrol in refineries to pharmaceuticals made in glassware, and precious metal salts at the bench. Because of its dependence on raw materials and its provision of materials that are essential to other industries, the chemical industry plays a major part in world trade *and a significant part in world politics* (e.g. Middle East oil, Southern African minerals). Applied chemistry ranges from its use within the public service sector of society to the manufacture of pottery, paper and glass . . .

The nomenclature of chemistry is difficult and is made confusing when pure chemists say 'ethyne' but welders use 'acetylene'. I know how grateful I am to my botanist colleagues when they speak to me about daisies, but not *Bellis perennis*.

Matter under investigation

Matter comes in different forms. Traditionally these are *solids, liquids and gases*. These three categories are not exclusive; for example liquid crystals are materials that have ordered structure in one or two dimensions. This contrasts with solids, which are structured in three dimensions, and liquids, in none. Actually this last statement is not strictly true. Liquids do show some ordering and solids invariably have some disorder. Solids resist change in shape and volume, liquids resist change in volume but not shape, and gases resist change neither in shape nor volume. Glass is a liquid; therefore resistance is a flexible concept.

We can't operate without categories. They are a necessary part of efficient thinking and transfer of information. However, they are our own constructs and not those of nature. They are for our convenience and they are tools, not experimental or theoretical results.

First we must explore why we feel that matter falls into such apparently obvious categories. When two categories of matter are in contact, the thing which separates them is a surface. The shape of a solid is defined by the solid-air surface. Oil floating on water is separated by a surface. This surface can be apparently disrupted by a detergent, which allows a form of mixing of the two liquids, the formation of an emulsion, with droplets of one liquid suspended in the other. The total surface separating the two liquids is now enormous. Categories of matter that are separated by surfaces are called *phases*. The word 'phase' is heavily overworked. The Oxford dictionary gives five distinct meanings, of which we are using one. Even then we can still run into problems. We need to confront questions about when a solid *mixture* of two *components* is one or two phases. Mixture? Component? The Oxford dictionary will not help us for long in our quest for understanding.

The fundamental question is, 'What is matter and what holds it together?' There are things we call atoms and molecules, elements and compounds. The atoms and elements are like letters. Just as letters form words, the atoms form molecules. Only certain orderings of letters form sensible words and only certain orderings of atoms form sensible molecules. What are the rules of sensibleness? Words are those combinations of letters that stay in the language long enough to be given a defined meaning. Molecules are those combinations of atoms that stay around long enough to be worth defining a meaning for. Most molecules, like most words, are around for a long time. But aren't molecules real? Are they just categories for sensibly dividing the world of matter into chunks suitable for human discourse, like words? If you have a molecule isolated in space, as a gas molecule say, then it is a very real category. But when it is part of a liquid or a solid, its isolated existence is more a convenience than a reality.

One of the great difficulties of translating from one language into another is that each language chunks the world in a slightly different way. There is no neutral abstract way of assigning words to meaning. A word means what most people using it want it to mean. This consensus is not static within a language and even less so between languages. Even within chemistry, there are many different meanings given to a word. As an example, chemistry teachers might like to consider the word *equivalent* (of all words!). How many different ways is 'equivalent' used in chemistry alone?

The arbitrary manner in which we define our molecular chunks results from the hierarchy of bonding. What is a hierarchy of bonding? If you were to define the social structure of a community, you would start by categorising the social interactions. Starting with the strong interactions you find that they are fewer in number than the weak ones. A nuclear family unit has strong interactions, in the sense that the members of a family interact over long periods of time and the individuals influence each other mutually to a large degree. Families are relatively small units (2–8 members) and, initially at least, are highly localised. The next step up gives you a choice of categories. For example, a circle of close family friends or the people at work. Then you have the club or the church group. The larger the group, the weaker the interactions. The categories are arbitrary but some are more obvious and useful than others. They are *never* clear-cut and there are always questions at the edges. Is Uncle Joe, three times removed, family or close friend? Anyone making out a wedding invitation list knows the problem.

With chemical bonding the same problem arises. With extreme examples (i.e. the ones always chosen to illustrate the point) everything is clear-cut. An isolated gaseous molecule of hydrogen chloride is like a couple on their honeymoon—a unique, an unperturbed, bond between hydrogen and chlorine, oblivious to all other interaction. But when the molecule dissolves in water, the honeymoon is over and all the interactions change. Systematic chemistry deals with the family life of the 90-plus individual elements. The most basic categorisation in chemistry is the periodic table of the chemical elements (Fig. 1.1).

The chemical elements, in order of increasing relative atomic mass (later corrected to atomic number), were found to exhibit periodic behaviour. This was one of chemistry's major achievements. In the periodic table these systematic variations in the properties of the elements change slowly down a column and also along a row. We are no longer dealing with 90-plus individuals but with a system that allows extrapolation of the properties of one element from those of others. Missing elements were defined according to their predicted properties, which were later confirmed when they were found. Additional elements at the end of the table had their properties extrapolated from within the table before their artificial discovery and production. It was a change as dramatic as switching from doing arithmetic with Roman numerals to doing it with Arabic numbers. Only much later was a rational basis for the periodic table provided by an understanding of the substructure of the atom in terms of protons, neutrons and electrons.

LANGUAGE AND LABELS

To talk about chemistry, or any other subject for that matter, you need to know some of the jargon. The existence of a specialised language peculiar to a particular field can be the most formidable obstacle in approaching an unfamiliar subject. Let us try to overcome this hurdle first.

Atoms

All matter is composed of atoms, of which there are about 90 distinctly different kinds occurring naturally on the earth. In very simple terms, each atom consists of a *positively-charged nucleus* containing protons and neutrons surrounded by a cloud of *negatively-charged electrons*, so that, overall, the atom has no net electrical charge.

Elements

Elements are those substances that contain only atoms with the same number of protons. Consequently, there are only about 90 natural elements. Familiar examples are copper, tin, iron, aluminium, oxygen, nitrogen, hydrogen and carbon. Each element is given a symbol—O for oxygen, H for hydrogen, C for carbon, and so on. These symbols may be regarded as the letters of the chemical alphabet.

Isotopes

Elements with the same number of protons but a different number of neutrons are isotopes of each other.

Compound substances

The remainder of the millions of other substances around us are either *compounds*, formed by specific combinations of different elements, or *mixtures* of a number of compounds. The basic combination of atoms characteristic of each compound is called a *molecule*, for which a formula can be written using the symbols of the elements making up the compound. The formula also includes a number to indicate the number of each type of atom present (but we don't bother to use '1' if only one atom is present). For example, carbon, C, forms two compounds with oxygen, O: carbon (mon)oxide, CO, and carbon (di)oxide, CO_2. Water is a compound of hydrogen and oxygen, H_2O. Most molecules are *polyatomic*; that is, they have many atoms, up to many thousands. Table sugar, for example, has the molecular formula $C_{12}H_{22}O_{11}$. Since these formulae are combinations of the letters of the chemical alphabet, we can consider them chemical words.

Valency

Just as there are spelling rules for putting letters together to form words, so there are the rules of *valency* (or *combining power*) for constructing the formulae of molecules from the atomic symbols. In water, H_2O, for example, hydrogen has a

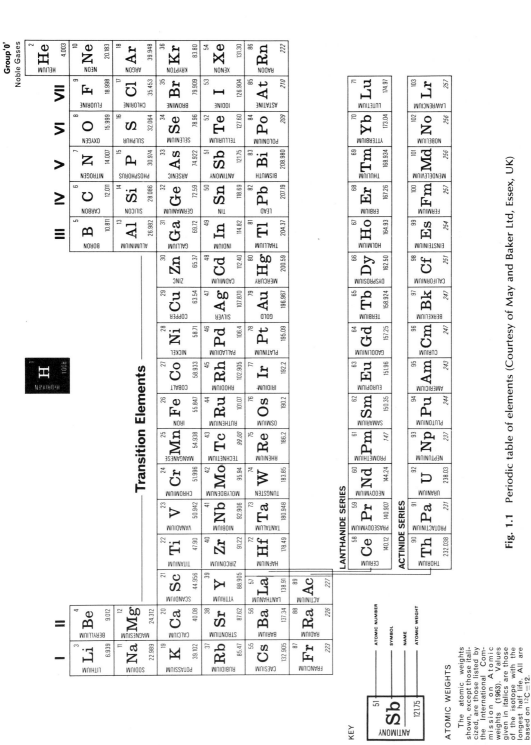

Fig. 1.1   Periodic table of elements (Courtesy of May and Baker Ltd, Essex, UK)

SONG OF THE ELEMENTS
by Tom Lehrer

There's antimony[1], arsenic[2], aluminium[3], selenium[4]
And hydrogen[5] and oxygen[6] and nitrogen[7] and rhenium[8]
And nickel[9], neodymium[10], neptunium[11], germanium[12]
And iron[13], americium[14], ruthenium[15], uranium[16]
Europium[17], zirconium[18], lutetium[19], vanadium[20]
And lanthanum[21] and osmium[22] and astatine[23] and radium[24]
And gold[25] and protactinium[26] and indium and gallium[27]
And iodine[28] and thorium[29] and thulium[30] and thallium.
There's yttrium[31], ytterbium[32], actinium[33], rubidium[34]
And boron[35], gadolinium[36], niobium[37], iridium[38]
And strontium[39] and silicon[40] and silver[41] and samarium[42]
And bismuth[43], bromine[44], lithium[45], beryllium[46] and barium[47].
There's holmium[48], and helium[49] and hafnium[50] and erbium[51]
And phosphorus[52] and francium[53] and fluorine[54] and terbium[55]
And manganese[56] and mercury[57], molybdenum[58], magnesium[59]
Dysprosium[60] and scandium[61] and cerium[62] and caesium[63]
And lead[64], praseodymium[65] and platinum[66], plutonium[67]
Palladium[68], promethium[69], potassium[70], polonium[71]
And tantalum[72], technetium[73], titanium[74], tellurium[75]
And cadmium[76] and calcium[77] and chromium[78] and curium[79]
There's sulfur[80], californium[81], and fermium[82], berkelium[83]
And also mendelevium[84], einsteinium[85], nobelium[86]
And argon[87], krypton[88], neon[89], radon[90], xenon[91], zinc[92] and rhodium[93]
And chlorine[94], carbon[95], cobalt[96], copper[97], tungsten[98], tin[99] and sodium[100].
These are the only ones of which the news has come to Harvard
And there may be many others, but they haven't been discarvard.

Notes
The origins of the elements are taken from diverse contradictory sources, including *The Handbook of Chemistry and Physics* (51st edn, 1970–1971). Abbreviations: Arab., Arabic; Ger., German; Gk, Greek; Icel., Icelandic; It., Italian; L, Latin; OE, Old English; ON, Old Norse; Sp., Spanish; Swed., Swedish.

1. L *stibium*, mark. It marks paper like lead.

2. Gk *arsenikon*, orpiment—artist's pigment, As_2S_3.

3. L *alumen*, alum—an astringent crystalline substance.

4. Gk *selene*, the moon.

5. Gk *hydros* + *genes*, water-forming.

6. Gk *oxys* + *genes*, acid forming. The theory that oxygen was essential to acids is wrong.

7. L *nitrium*; Gk *nitron*, native soda.

8. L *Rhenus*, the Rhine.

9. Ger. *kupfernickel*, copper demon.

10. Gk *neo* + *didymos*, new twin.

11. The planet Neptune.

12. Germany.

13. OE *iren*, *isen*; Ger. *eisen*. Symbol Fe from L *ferrum*.

14. The Americas.

15. L *Ruthenia*, Russia.

16. The planet Uranus.

17. Europe.

18. Pers. *zargun*, gold-coloured (mineral) zircon (other varieties of which

are called *argon, hya-cinth, jacinth,* or *ligure.*

19. Lutetia, an ancient name for Paris.
20. Icel. *Vanad(is),* epithet of Norse goddess Freya, discovered in Sweden.
21. Gk *lanthanein,* to escape notice.
22. Gk *osme,* smell or odour. The most dense element forms pongy compounds.
23. Gk *astatos,* unstable.
24. L *radius,* ray.
25. ON *gull,* Gothic *gulth* (*ghel,* yellow; Ger. *gelb*), Sanskrit *Jval.* Symbol from L *aurum.*
26. Gk *protos,* first (of a series), + actinium.
27. L *Gallia,* France.
28. Gk *iodes,* rust-coloured (assigned by a colour-blind chemist?).
29. Scandinavian god of thunder.
30. Thule, an early name for Scandinavia.
31. From Ytterby, a town in Sweden.
32. From Ytterby, a town in Sweden.
33. Gk *aktis,* beam or ray.
34. L *rubidus,* red (in allusion to the two red lines in its flame spectral emission).
35. Pers. *borah*; Arab. *buraq* for borax.
36. Finnish chemist J. Gadolin (1760–1852).
37. Gk *Niobe,* mythological daughter of Tantalus (niobium found with tantalum), originally called Columbium.
38. L *iris,* rainbow—iridescence in solution.
39. Strontian, a town in Scotland.
40. L *silex,* flint.
41. Anglo-Saxon *seolfor, siolfur.* Symbol from L *argentum.*
42. Samarskite (a mineral named after a Russian mine official).
43. Ger. *Wismut,* from Ger. *Weisse Masse,* white mass.
44. Gk *bromos,* stench.

45. Gk *lithos,* stone.
46. Gk *beryl,* beryl (a gem), originally (1798) called *glucinium* from Gk *glykos,* sweet.
47. Gk *barys,* heavy.
48. L *Holmia,* Stockholm.
49. Gk *helios,* the sun.
50. L *Hafnia,* Copenhagen.
51. Ytterby, a town in Sweden.
52. Gk *phosphorus,* light-bearing; a name applied to the planet Venus when appearing as a morning star.
53. France.
54. L *fluere,* to flow—the mineral fluorspar, CaF_2, used as a *flux* for metalwork.
55. Ytterby, a town in Sweden.
56. L *magnes,* magnet, from the magnetic properties of pyrolusite. It. *manganese,* corrupt form of magnesia.
57. The planet Mercury.
58. Gk *molybdaina,* galena—lead sulfide.
59. Magnesia, a district in Thessaly.
60. Gk *dysprositos,* hard to get at.
61. L *Scandis,* Scandinavia.
62. The asteroid Ceres.
63. L *caesium,* bluish-grey.
64. Icel. *leidha.* Symbol from L *plumbum.*
65. L *prasius,* a leek-green stone, + Gk *didymos* double, twin.
66. Sp. *plata,* silver; *platina,* diminutive.
67. The planet Pluto.
68. Named after the asteroid Pallas, then recently discovered (1803).
69. Gk *Prometheus,* one of the gods.
70. Pot-ashes from early Dutch *potasschen* L *kalium* from Arab. *gali,* alkali, gives symbol K.
71. Poland, birthplace of Marie Curie, who discovered polonium in 1898.
72. Gk Tantalus, Greek god.
73. Gk *technetos,* artificial, the first element to be made that does not occur

in nature; once called masurium (Masuria region in N.E. Poland).
74. Gk Titan, gods of great strength and size.
75. L *tellus,* the earth.
76. L *cadmia,* calamine (a zinc ore containing small amounts of cadmium); calamine lotion, suspension of zinc oxide + 0.5% iron oxide.
77. L *calx,* lime (calcium hydroxide).
78. Gk *chroma,* colour.
79. Marie Curie, Polish chemist who worked in France.
80. Sanskrit *sulvere*; L *sulphurium.*
81. California.
82. Enrico Fermi, Italian nuclear physicist who fled from the fascists to the US.
83. Berkeley, California.
84. Dimitri Mendeleev (1834–1907), Russian discoverer of the periodic table of the elements.
85. Albert Einstein, German physicist who fled from the fascists to the US.
86. Alfred Nobel (of prize fame).
87. Gk *argos,* idle.
88. Gk *kryptos,* hidden.
89. Gk *neos,* new.
90. From radium.
91. Gk *xenon,* stranger.
92. Ger. *Zink,* origin uncertain.
93. Gk *rhodios,* roselike.
94. Gk *chloritis,* kind of greenstone.
95. L *carbo,* coal or charcoal.
96. Ger., *Kobalt,* evil spirit or goblin.
97. L *cuprum,* from the island of Cyprus.
98. Swed. *tung* + *sten,* heavy stone.
99. Germanic *tinam,* of unknown origin.
100. Arab. *suwwad* or *suda* headache; glasswort plant (L *sodanum*) used as a headache remedy. L *natrium* gives symbol Na.

combining power of one, while oxygen has a combining power of two. Thus one oxygen atom may combine with two hydrogen atoms. Consequently, the arrangement of atoms in the water molecule must be something of the form H . . . O . . . H, with the oxygen atom in the middle.

Elements often show different valencies in different compounds. For example, lead, Pb, forms two compounds with chlorine, $PbCl_2$ and $PbCl_4$, showing valencies of two and four, while chlorine in both compounds has a valency of one. Some of the elements you will meet are shown in Table 1.1.

TABLE 1.1
Some common elements

	Hydrogen	Oxygen	Carbon	Nitrogen	Phosphorus	Lead	Mercury
Symbol	H	O	C	N	P	Pb	Hg
Valencies	1	2	4	3, 5	3, 5	2, 4	1, 2

Bonding

Since all molecules of any one compound are composed of the same atoms arranged in essentially the same way to satisfy the rules of valency, there must be some definite forces holding the atoms together in that particular way. These forces are the *chemical bonds* and, broadly, they arise from the sharing of electrons between the atoms. Thus, in the molecule of hydrogen, H_2, the two atoms are bonded together by the sharing of two electrons, one from each atom. We write it H : H, where the two dots represent the electrons, or, more commonly, H–H, where the dash represents the electron pair or the chemical bond between the atoms. A bond of this sharing type is called a *covalent bond*, which is by far the most common kind.

Again, for hydrogen chloride, we can write H : Cl or H–Cl, but here, because the two atoms of the molecule are different, the bonding pair is not shared equally; rather, the chlorine has more than its fair share. We could show this as H :Cl, or, because the chlorine not only has its own electron but a share of the electron from the hydrogen as well, and hence has gained a slight additional negative charge, as

$$\overset{\delta^+ \quad \delta^-}{H—Cl}$$

Because this molecule has positive and negative ends, or *poles*, it is said to be a polar molecule and the bond is called a *polar bond*. As you will see, this uneven distribution of electrons can have most important consequences for the behaviour (or *properties*) of the molecule, in comparison with the behaviour of *non-polar* molecules.

The bond in hydrogen chloride is still a covalent (or sharing) bond, albeit a polar one, but when the molecule dissolves in water, the chlorine atom takes the bonding electron completely from the hydrogen atom, assuming in the process a unit negative charge (the charge of one electron) and becoming what is called a *negative ion* or *anion*. Similarly, the hydrogen atom has become a *positive ion* or *cation*. Compounds composed of ions are called *ionic compounds* and are said to be held together by *ionic bonds*. As a general rule, ionic compounds are water-soluble and fat-insoluble while the reverse is true of covalent compounds. Where a molecule has both types of bonds, intermediate behaviour is to be expected. Sodium chloride, common salt, is another example of an ionic compound. We show this as Na^+Cl^-, even in the solid state.

Chemical equations

Equations are chemical sentences, composed of words (molecules) and conveying information about the transformation of chemical substances; that is, they describe *chemical reactions*. Thus

$$2H_2 + O_2 \rightarrow 2H_2O$$

tells us that two molecules of hydrogen combine with one molecule of oxygen to give two molecules of water. These chemical equations are a statement of the over-all change and make no attempt to indicate *how* the change occurs.

Shapes of molecules

Earlier, we concluded that the arrangement of the atoms in the water molecule was something of the form H—O—H, where one oxygen atom with a valency of two forms two covalent bonds with two hydrogen atoms. These bonds are polar, and we could show this as

$$\delta^+ \quad \delta^- \quad \delta^+$$
$$H—O—H$$

But do the three atoms of the water molecule lie in a straight line? If not, what is the shape of the water molecule? Consideration of molecular shapes and their consequences is known as *stereochemistry*, a most important aspect of the properties of the molecule.

Consider these three compounds: methane CH_4; ammonia, NH_3; and water, H_2O. Carbon has a valency of four, and in methane the four hydrogen atoms lie at the corners of a tetrahedron with the carbon atom at the centre. Ammonia, containing the central atom of nitrogen with a valency of three, has a similar shape, except that one of the corners of the tetrahedron does not have a hydrogen atom; the ammonia molecule is pyramidal in shape. The water molecule, with only two hydrogen atoms, is bent (see Plate I).

Names of chemicals

In order to communicate information about chemistry and chemicals there has to be a way of identifying and naming them. A trade or brand name can be given to a chemical or a formulation of chemicals and registered by a firm for its exclusive use. Such a trade name gives no information, and the composition of the product to which it refers can vary. In many cases a trade name can be so popular that it becomes, by common usage, a common name (e.g., aspirin, biro, cellophane, bakelite). Manufacturers must keep on insisting that their trade names be spelt with a capital letter, because once a trade name has become a common name it may be legally defined as the *generic* or non-proprietary term for a particular material. Today 'aspirin' is not the Bayer trade mark but the generic name for a particular chemical substance. Whereas there can be an enormous number of trade names for a chemical, there are only one or two generic names. The generic name can then be qualified to provide further generic names for related compounds (e.g., penicillin and penicillin G).

In order to have an unambiguous precise name for a chemical substance a system has been devised to provide a *systematic name* for any substance. These names are built up according to strict rules. Each compound has only one correct systematic name and that name conveys the complete information about the detailed structure of the compound: it is a stylised written description of the chemical structure. For example, 'Aspro' is a trade name; 'aspirin' or 'acetylsalicylic acid', the generic term; and '2-acetyloxybenzoic acid', the systematic name.

Chemical formulae and diagrams

Chemical formulae and diagrams are symbolic representations of the composition and structure of molecules and compounds. There is a variety of conventions for them, depending on the sort of information to be conveyed.

At the lowest level of information is the *empirical formula*, which lists only the numbers and types of atoms present in the molecule and tells us nothing about the structure. For example, for aspirin it is $C_9H_8O_4$. Many different chemicals can have the same empirical formula.

The *group formula* places atoms together in groups that correspond to the grouping in the actual molecule, and gives some indication (using prefix symbols) of how the groups fit together. For example, the group formula for aspirin is 2-CH_3COO—C_6H_4COOH. As you become more familiar with them, these group formulae can give a fairly complete (marginally ambiguous) description of the structure.

The *condensed structural diagram*, the most common form of line diagram, gives a two-dimensional representation of a three-dimensional structure. It leaves out a lot of the atoms, but their presence is implied by the shorthand conventions. For aspirin the condensed structural diagram is as shown in Figure 1.2.

The hexagon is a ring of six carbon atoms joined by alternating double and single bonds. This ring is called the benzene ring. Carbon has a valency of four, so

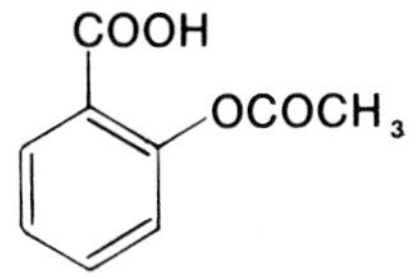

Fig. 1.2 Condensed structural diagram for aspirin

that any missing bonds are taken up by a bond to a hydrogen atom. There are other apparent ambiguities (although convention clarifies them). In fact this diagram is a condensation of the *full structural diagram*, which is rarely used, in which the three-dimensional structure of the compound and the bond angles are indicated. The full structural diagram for aspirin is shown in Figure 1.3.

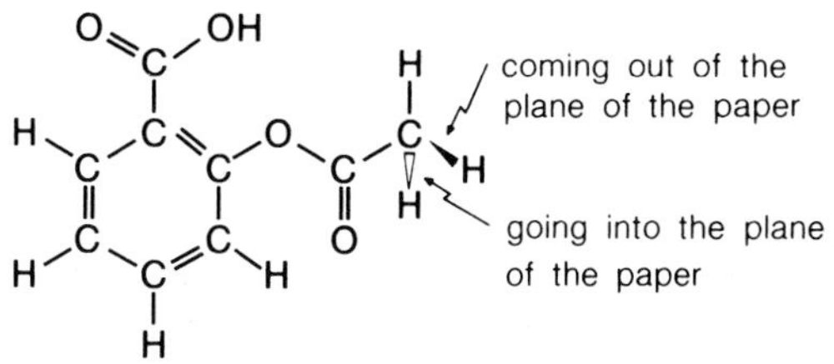

Fig. 1.3 Full structural diagram for aspirin

Note that each carbon atom, C, and hydrogen atom, H, is specifically depicted, whereas in the more usual, condensed version the carbon atoms are implied to be at the intersections of the bond lines and the hydrogen atoms are implied to fill positions so that the valence of four for carbon is satisfied. Throughout the text condensed structural diagrams will be used, but extra information will be included when required to emphasise a particular feature.

The stick and space-filling model is useful to indicate the actual geometry of a molecule. This may be essential for explaining its biological activity (see also section on DDT in Chapter 5, under 'Pesticides and alternatives—Organochlorine compounds' and Plate IV). The stick and space-filling model for aspirin is shown in Plate II.

Organic chemistry

As its name implies, organic chemistry originally dealt with chemical substances produced by living organisms. The key element in all organic molecules is carbon, which forms many more compounds than any other element (with the possible exception of hydrogen). In its stable compounds, carbon always has a valency of four, and when it is bonded to four other atoms, they are arranged tetrahedrally about the carbon atom (see Plate I). Carbon also shows, to a remarkable extent, the property of *catenation* (chain forming); that is, many carbon atoms can bond together to form chains or rings.

The simplest group of these chain molecules are the *paraffins*, or *alkanes*, which are found in natural gas and petroleum. The first members of the series are methane, CH_4, ethane, C_2H_6, propane, C_3H_8, and butane, C_4H_{10} (Fig. 1.4). Propane and butane are the main constituents of bottled fuel gas, whereas petrol contains principally the members with seven (heptane) or eight (octane) carbon atoms. The lubricating oils have much longer chains.

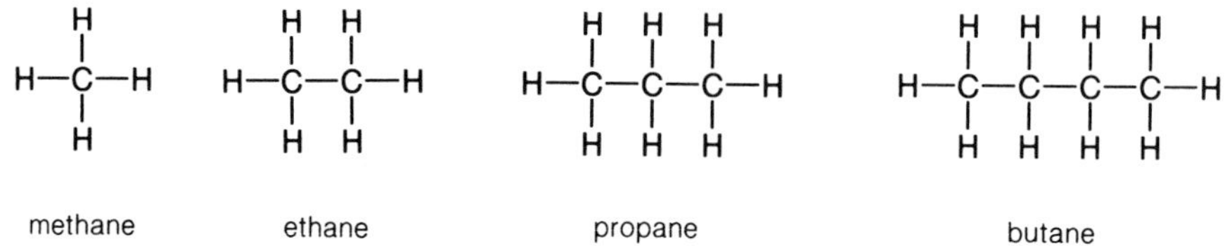

methane   ethane   propane   butane

Fig. 1.4   Formulae for some paraffins

In all of the paraffin compounds, the carbon atom shows its constant valency of four and the atoms bonded to it are arranged tetrahedrally. However, in many cases, two or three of the four valencies may be used up in forming a *double* or even a *triple* covalent bond with another atom. Carbon dioxide, $O{=}C{=}O$, is a familiar example.

There is a series of compounds, called the *olefins* or *alkenes*, which are closely related to the alkane series, but each member has two less hydrogen atoms than the corresponding alkane, which means that each has one double bond. The olefins are the simplest unsaturated compounds. Figure 1.5 gives the structural formulae for ethylene, C_2H_4 (compare ethane), the basic molecule from which polythene (polyethylene) is made, and for propylene, C_3H_6, which is used in the new plastic polypropylene. Note that the ending *-ene* is indicative of the alk*enes*, as *-ane* is of the alk*anes*.

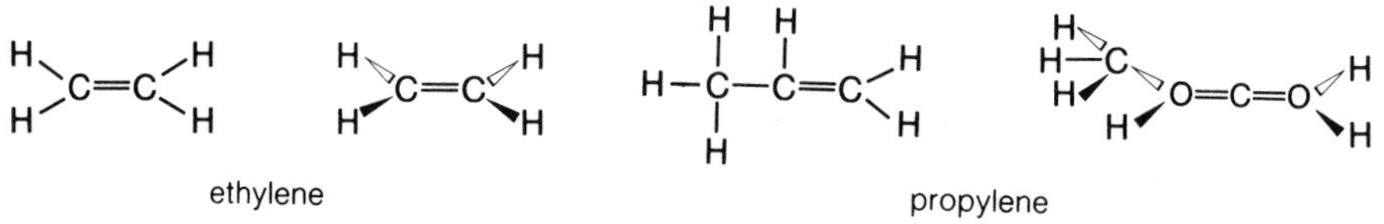

ethylene   propylene

Fig. 1.5   Structural formulae for ethylene and propylene

Another type of unsaturated hydrocarbon is butadiene, C_4H_8, which contains two double bonds (Fig. 1.6). It is the basic constituent of the earliest synthetic rubbers.

Fig. 1.6 Butadiene

The simplest member of the triple-bonded series, the *alkynes*, is acetylene, H—C≡C—H, the gas used in welding. Such carbon compounds containing *multiple* bonds are said to be *unsaturated*—as in polyunsaturated margarine, which we shall discuss in Chapter 3—because it is possible to add more atoms to their molecules, thereby *saturating* them. For example, ethylene can be converted simply to ethane by the addition of two hydrogen atoms (Fig. 1.7).

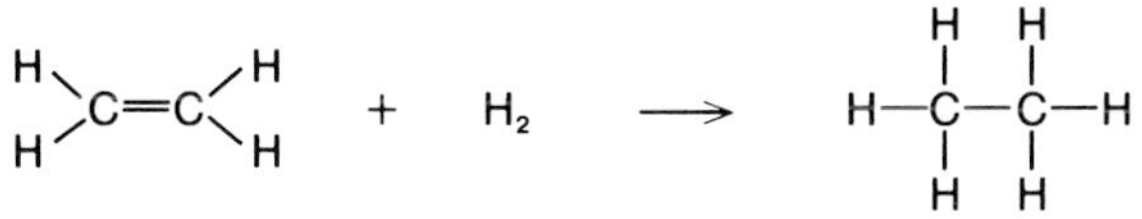

Fig. 1.7 Hydrogenation of ethylene

Another important group of carbon compounds are those containing ring molecules, such as benzene, toluene and naphthalene (Fig. 1.8).

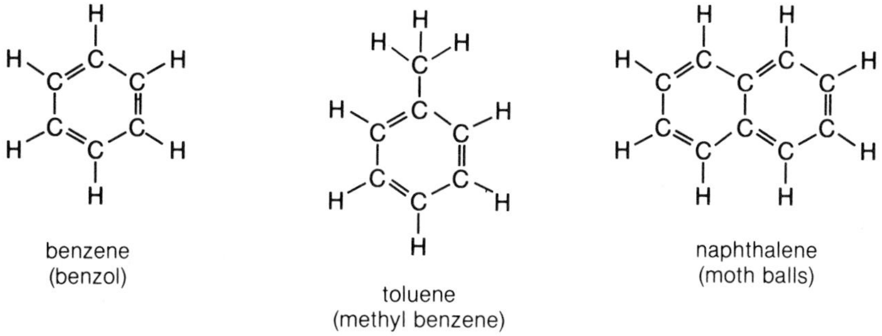

benzene
(benzol)

toluene
(methyl benzene)

naphthalene
(moth balls)

Fig. 1.8 Some aromatic hydrocarbons

So far we have considered only organic molecules containing carbon and hydrogen—the hydrocarbons—but it is possible, by replacing some of the hydrogen atoms with other atoms, to make whole new series of compounds, with different properties, depending upon the nature of the *substituting* atoms. It is important to note at this stage that apparently minor changes to a molecule can produce major changes in physical, chemical and physiological properties. Thus, from methane, CH_4, we can prepare CH_3Cl (Methyl chloride), CH_2Cl_2 (dichloromethane), used in paint strippers, $CHCl_3$ (chloroform), the first anaesthetic other than alcohol, and CCl_4 (carbon tetrachloride), once widely used in dry-cleaning until its toxic properties were admitted (see Table 1.2).

TABLE 1.2
Chlorinated hydrocarbons

Methane	Methyl chloride	Dichloromethane	Chloroform	Carbon tetrachloride
CH_4 non-toxic flammable gas	CH_3Cl toxic	CH_2Cl_2 paint stripper; relatively non-toxic	$CHCl_3$ first anaesthetic; damages liver	CCl_4 former dry-cleaning fluid; more toxic than $CHCl_3$; fire extinguisher (electrical) (no longer)

Some other examples of property changes resulting from small changes in a molecule can be seen in the different structures of ethane, C_2H_6, ethanol (formerly ethyl alcohol) C_2H_5OH, and acetic acid, CH_2COOH; and of benzene, C_6H_6, benzoic acid, C_6H_5COOH, and acetylsalicylic acid, $CH_3COOC_6H_4COOH$ (Fig. 1.9).

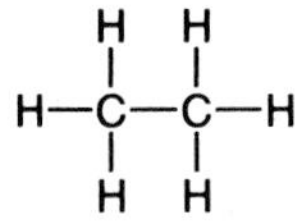

ethane
(colourless,
non-toxic gas)

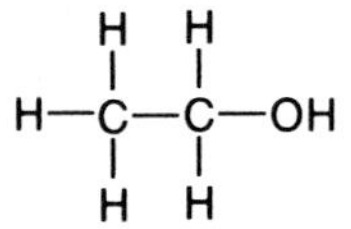

ethanol
(the common alcohol
in drinks)

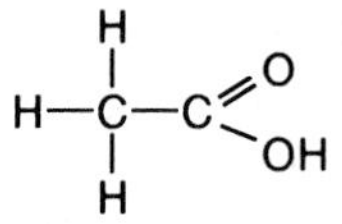

acetic or ethanoic
acid (gives vinegar
its bite)

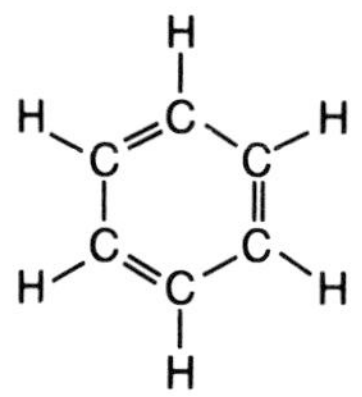

benzene (liquid)
(toxic)

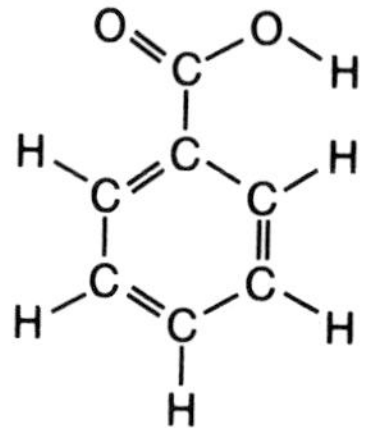

benzoic acid (solid)
(food preservative)

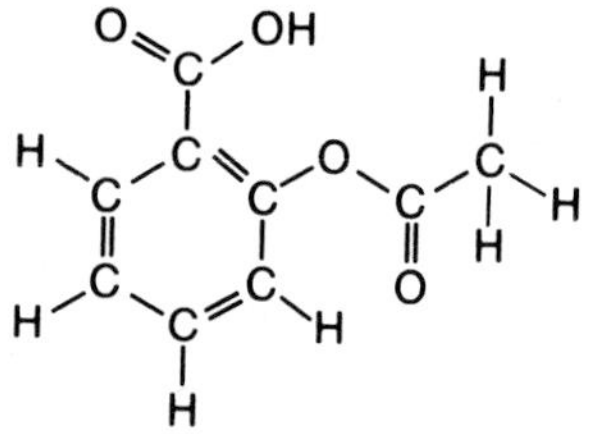

acetylsalicylic acid (solid)
(aspirin, analgesic)

Fig. 1.9 Property changes due to molecular changes

Yet another example is given by the successive replacement of the hydrogen atoms of acetic acid with fluorine (Fig. 1.10). Acetic acid, in vinegar, is relatively non-toxic, whereas fluoroacetic acid is used as the sodium salt in '1080' rabbit poison and is highly toxic to humans. It is also the poisonous constituent in some poisonous plants. Difluoroacetic acid is not primarily toxic, whereas trifluoro-acetic acid forms the basis of some weed killers, such as Dalapon.

Many of the changes in physical and chemical properties accompanying substitutions of this sort can be predicted, but changes in physiological properties

acetic acid   fluoroacetic acid   difluoroacetic acid   trifluoroacetic acid

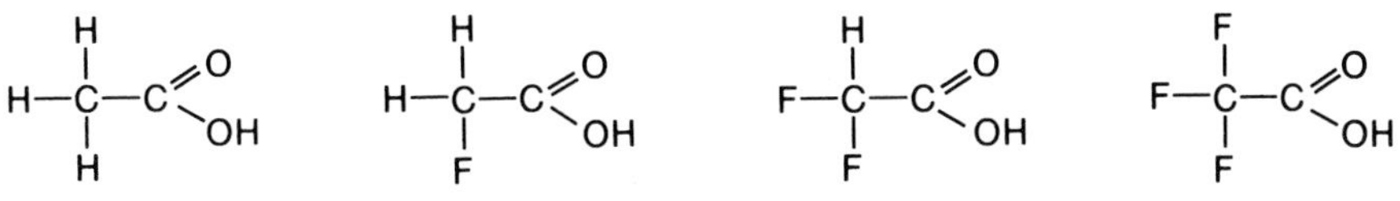

Fig. 1.10   Replacement of hydrogen atoms of acetic acid with fluorine

are often unexpected. However, by the same token, certain geometric groupings of atoms within a molecule are known to produce certain physiological effects, and much drug research is devoted to modifying these basic structures in the hope of enhancing their pharmacological activity and, at the same time, reducing undesirable side-effects. It is safest to assume that all chemicals are toxic.

Organic compounds may be classified according to the nature of the *functional groups* that have replaced hydrogen in the basic molecule. Three important functional groups are alcohols, organic acids and esters.

Alcohols have the general formula R—OH, where R stands for the rest of the hydrocarbon skeleton. The simplest alcohol, derived from methane, CH_4, is methyl alcohol, CH_3—OH. This substance, often called 'wood alcohol' because it is obtained by heating wood, is more toxic than the familiar ethanol, CH_3CH_2—OH.

Organic acids. Most contain the characteristic carboxyl group COOH. Any molecule containing this group is called a *carboxylic acid*. Some examples are illustrated in Figure 1.11.

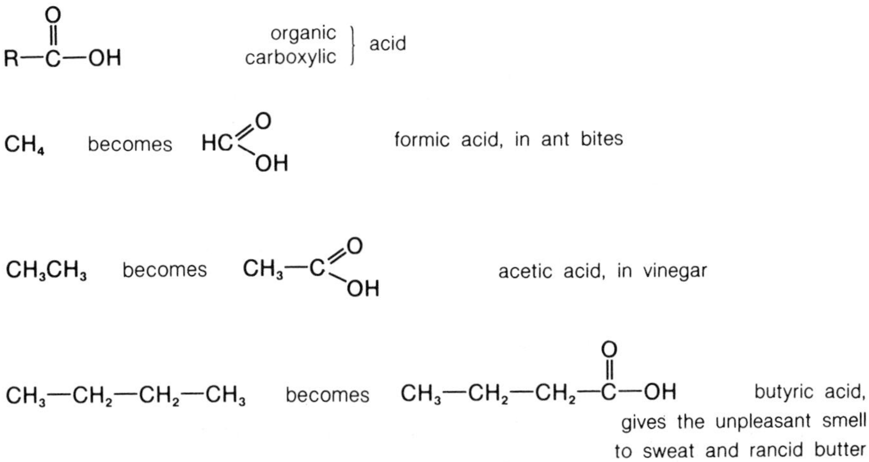

Fig. 1.11   Organic acids

Esters are produced by the combination of alcohols and acids in an *esterification* reaction (Fig. 1.12). Ethyl acetate is an important solvent in the paint and adhesive industries. Many of the natural flavours and smells result from the presence of volatile esters in flowers and fruits, whereas meat fat consists of solid esters, and oils are liquid esters.

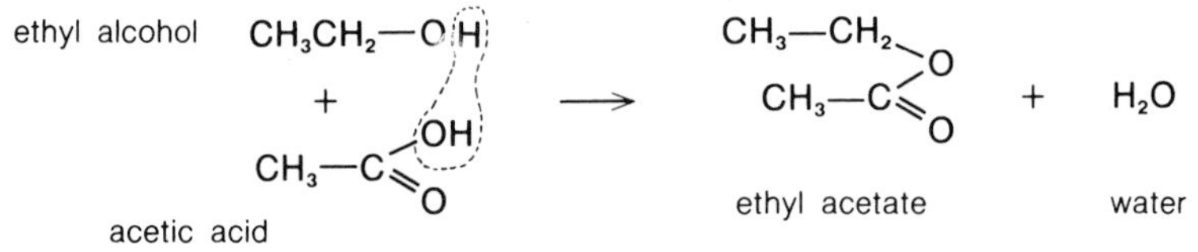

ethyl alcohol $CH_3CH_2-O{:}H$

$+$

$CH_3-C{\lessapprox}^{OH}_{O}$

acetic acid

$\longrightarrow$

$CH_3-CH_2{\searrow}O$
$CH_3-C{\lessapprox}^{O}_{O}$

ethyl acetate

$+$ H_2O

water

Fig. 1.12   Esterification reaction

Acids and bases

You will have heard of sulfuric acid and hydrochloric acid, and of caustic soda, which is a member of a group of compounds called bases. You may know that if caustic soda is mixed with hydrochloric acid, the corrosive properties of both solutions are overcome, or *neutralised*. The chemical basis for the concepts of acidity and basicity is rather complex and we must begin with a simplified approach. For our purposes, all substances that we will consider as acids, such as hydrochloric acid, HCl, sulfuric acid, H_2SO_4, nitric acid, HNO_3, and acetic acid, CH_3COOH, are characterised by the presence, in the molecule, of one or more reactive hydrogen atoms that can be displaced by a base in the process called neutralisation. The reactive or *acidic* hydrogen atoms in the molecules shown in Figure 1.14 are marked by an asterisk. Notice that the methyl hydrogens of acetic acid, the ones attached directly to the carbon atom, are non-acidic. Among the

Fig. 1.13   The chemical goanna (made by Terry Sedgwick)

common bases are sodium hydroxide (caustic soda), NaOH; calcium hydroxide (slaked lime), $Ca(OH)_2$; and ammonium hydroxide ('cleaning' ammonia), NH_4OH.

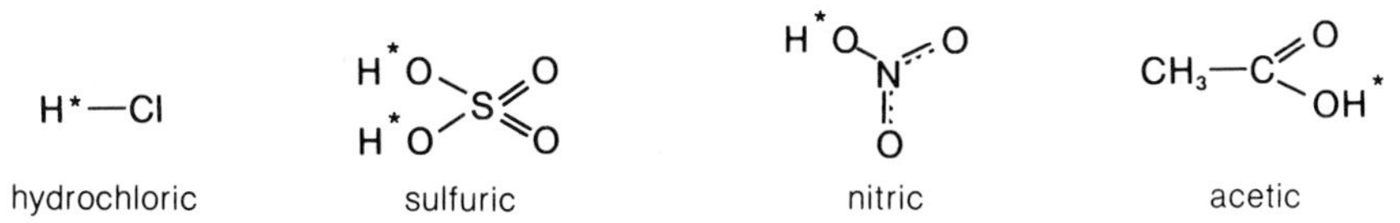

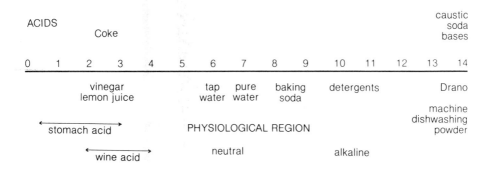

hydrochloric   sulfuric   nitric   acetic

Fig. 1.14 Reactive or acidic hydrogen atoms in acids

The neutralisation reaction between an acid and a base follows the general pattern

an acid + a base ⇌ a salt + water

e.g.   H—Cl + Na^+OH^- → Na^+Cl^- + H—OH

which is clearly very closely related to the esterification reaction between organic acids and alcohols already discussed.

Simple acids have one other important property: when dissolved in water, to a greater or lesser extent, the reactive hydrogen atom can transfer to the water molecule and thus form an ion. Thus:

$$H—Cl + H_2O \rightarrow H_3O^+ + Cl^-$$
$$CH_3COOH + H_2O \rightarrow H_3O^+ + CH_3COO^-$$

On solution in water then, acids produce the H_3O^+ ion. Strong acids such as HCl transfer the active hydrogen completely; only some weak acids, such as CH_3COOH, transfer their hydrogen.

The amount, or rather the concentration, of the species H_3O^+ produced when a given acid is dissolved in water may be expressed on a scale known as the *pH scale* (see Fig. 1.15). Pure water has a pH of 7 (neutral pH). A solution of a strong acid in water at unit concentration (1 mole/litre) has a pH of 0, whereas a solution of a strong base at unit concentration has a pH of 14.

ACIDS   caustic soda bases

Coke

0 1 2 3 4 5 6 7 8 9 10 11 12 13 14

vinegar   tap   pure   baking   detergents   Drano
lemon juice   water water   soda

                                               machine
←————→   dishwashing
 stomach acid   PHYSIOLOGICAL REGION   powder

←—→   neutral   alkaline
 wine acid

Fig. 1.15 pH scale

AN UNEXPECTED REACTION

The infamous Russian peasant monk Grigori Rasputin, who was the power behind Tsar Nicholas II of Russia, was assassinated during the First World War (1914–1918) when Russia was doing very badly. He proved difficult to kill and was finally shot, although some initial attempts involved some botched chemistry. At the palace of Prince Yusupov (the Tsar's nephew) a Dr Lazovert mixed into some cakes offered to Rasputin generous quantities of a white powder from a box allegedly containing potassium cyanide. Rasputin appeared to suffer no ill effects. What had presumably occurred was a reaction between carbon dioxide from the air with the potassium cyanide. Carbonic acid ($pK_a=6.35$) is a stronger acid than hydrocyanic acid ($pK_a=9.22$) and so readily displaces it. The poisonous potassium cyanide had probably been converted to harmless potassium carbonate.

Nomenclature

At school you would have studied some *in*organic chemistry, for example the halogens, including fluorine. Your teacher would have emphasised their reactivity and the fact that fluorine has seven electrons in its outer shell. In the world outside the classroom you are more likely to hear about fluoride in drinking water. How can a violently reactive element be good for teeth? And as for hydrofluoric acid, which can etch through glass like no other acid . . . Fluor*ine*, fluor*ide*: chemists are finicky about naming (nomenclature). A one-letter misprint can be the difference between life and death! You will have heard less about the naming of organic compounds, so we shall explain it in more detail.

Visita interiora terrae rectificando invenies occultum lapidem. Probably the names of chemicals you see mean as much to you as this Latin expression. Chemists are interested in molecular structure and in conveying that information rather than in the names for their own sake. That has not always been the case and the ancient alchemists in fact tried to obscure their naming with acronyms and other devices. The Latin expression above suggests the preparation of a material: 'Visit the interior of the earth, by rectifying you will find the hidden stone', and the first letters spell it out—*vitriol*—from the Latin *vitrium*, glass. From this we have obtained vitreous (glassy) and vitriolic, and later oil of vitriol, i.e. sulfuric acid.

In what follows, I will lead you as gently as possible through the minefield of the various nomenclature practices used in the world today. Even this will not be sufficient for you to read chemical names, let alone write them, but the underlying logic will be exposed.

The aim of any systematic nomenclature is to provide the equivalent in words of a chemical structure. The verbal building blocks can be transferred from one substance to another while preserving intact the chunk of chemical information. There are a number of naming conventions, each with its own role. The

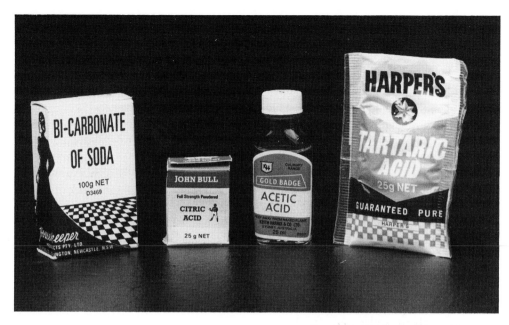

Fig. 1.16 Common acids and bases

International Union of Pure and Applied Chemistry (IUPAC) has established a systematic naming procedure (which fills several books with its rules; blue for organic etc.).

DDT (see Chapter 5) is shorthand for *di*chloro*di*phenyl*tri*chloroethane.

$$(C_6H_4Cl)_2CH\text{—}CCl_3$$

The formula is written so as to follow this naming. However, this name is ambiguous and is no longer acceptable. The IUPAC name is:

1,1,1-trichloro-2,2-bis (4-chlorophenyl)ethane

On the right, the root word *ethane* establishes a chain of two carbon atoms $H_3C\text{—}CH_3$, on which the other structural parts are built.

On the left, *1,1,1-trichloro* says there are three chlorine atoms (for atoms the Latin *tri* is used to mean thrice) replacing the three hydrogen atoms on carbon atom 1 of the ethane, i.e. $Cl_3C\text{—}CH_3$.

Then, *2,2-bis* says there are two identical groups (for groups the Greek *bis* is used to mean twice) replacing two hydrogen atoms on the carbon atom 2.

Then, *(4-chlorophenyl)* says what the two groups are. Each group is a benzene ring with a bond to the carbon on the ethane (an attached benzene ring is called a phenyl group), and with one chlorine atom replacing one hydrogen in position 4, i.e. diametrically opposite the point of attachment. The third hydrogen at C2 on the ethane is not replaced.

IUPAC names are not always unique, because you can chunk some molecules in different ways, with different rules. For example, the most common method involves a *substitution* sequence, which was used for DDT. Let us try another example, the drug barbital (Fig. 1.17).

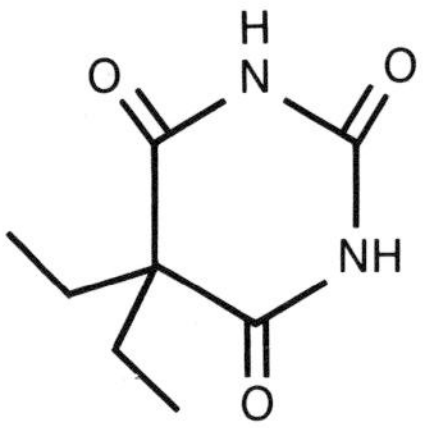

IUPAC 5,5–diethylbarbituric acid
IUPAC 5,5–diethyl(1H, 3H, 5H)-pyrimidine-2,4,6-trione

Fig. 1.17   Barbital

As soon as a compound appears in the chemical literature, it needs to be classified. This work is done by the Chemical Abstracts Service (CAS). CAS abstracts the chemical literature of the whole world and insists on a unique name for every one of the eight million or so chemicals it has currently indexed. Whereas IUPAC names begin at the beginning and proceed to the end, CAS needs a system which lends itself to indexing, and so a name often appears in the form of a parent compound followed by bracketed phrases denoting the nature of a particular derivative. For barbital:

> CAS (index form) 2,4,6(1*H*, 3*H*, 5*H*)-Pyrimidinetrione, 5,5-diethyl-CAS reg. number [57–44–3] (These numbers are unique but a compound may have more than one number.)

We can now list derivatives other than the 5,5-diethyl. This inverted presentation is in keeping with the concept of an index. The alphabetical listing of IUPAC names does not collect together compounds of the same chemical class. For example, bromobenzene would be found under 'b', while dibromobenzene would be under 'd'. Because barbitone is a drug it will appear in the *British Pharmacopoeia* with a British Approved Name (BAN). The World Health Organisation (WHO) agrees on a list of International Non-Proprietary Names (INNs).

e.g. WHO   barbital (INN)

These names consist of a few syllables, devised to be distinct from names already in use. At the same time, the name should tell the doctor or nurse something about the drug. Thus the various barbiturates will be 'generically' related in name, like phenobarbital, pentobarbital etc. (see Fig. 9.15). Australia follows the *British Pharmacopoeia* and uses BANs in the first instance, in preference to the INNs, where these differ. Thus our barbiturates are barbitones rather than barbitals. These names are given in a publication called *Australian Approved Names and other Names for Therapeutic Substances* (AGPS, 4th edn,

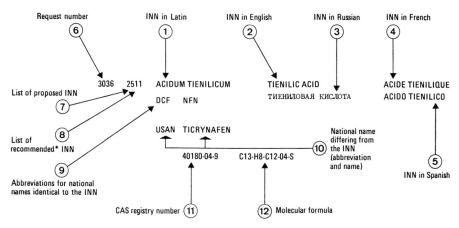

Fig. 1.18 Layout of information for INNS

1986). From the third edition onwards, this book has adopted the international nomenclature wherever possible.

Sometimes a name comes from a condensation of a systematic name:

e.g. 2-ph*thal*im*id*oglutar*imide*

INNs are published in Latin, English, French, Spanish and Russian, along with the names used in some national pharmacopoeias. While INNs may have arisen from lapsed trade names, they must be distinct from and not used as current trade marks (see Fig. 1.18).

A similar system for naming pesticides in Australia is found in a booklet *Pesticides Synonyms and Chemical Names* (7th edn, 1985) issued by the Commonwealth Department of Health. The International Standards Organisation (ISO) sets the common names recommended/approved for international use and SAA does the same for local use. The IUPAC name is given as an alternative.

For dyestuffs and colourants, the internationally recognised source is the *Colour Index* (CI), a joint publication of the UK Society of Dyers and Colourists and the American Association of Textile Chemists and Colorists (with the UK spelling winning out in the name!). The CI numbers are used in the table of food colours in Appendix XII. The index contains the CI common names, commercial name(s) and a CI number. The *Colour Index* gives the structure, where available, or otherwise a description of the method of manufacture. Food colours have additional names given to them by national regulating bodies which are so indicated (e.g. the US uses the FD & C—Food, Drug and Cosmetics Act and the European Economic Community uses E.)

For food chemicals and additives, the *US Food Chemicals Codex* of the US National Academy of Sciences gives systematic names (and/or structures). *McCutcheon's Detergents and Emulsifiers* (Allured Publishing Corporation) is an index of trade names, with the name of manufacturer, class and formula, a type classification and further remarks.

In common use and industry, names are used which relate to properties (e.g. *caustic* soda, *slaked* lime, *killed* spirits), use (e.g. *bleaching* powder, *heating* oil, *battery* acid), or source (e.g. muriatic acid—how is your Latin?). It becomes more complicated when only initials are used. Try your hand at DDT, MEK, RDX, BHT, KIOP, TDI, DNA, PVC, QUAT, ABS, SAN, PCB. TCP can mean trichlorophenol, trichloracetophenone, tricresyl phosphate etc. BHC stands for benzene hexachloride, which it is not. In fact, it is hexachlorocyclohexane, a completely different compound (see Fig. 1.19).

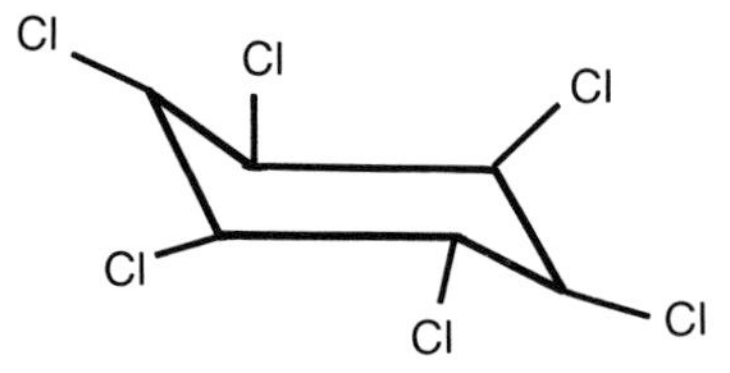

Fig. 1.19   Hexachlorocyclohexane (BHC)

Once you know the logic in some code names, they can be very useful. For example, in discussing the depletion of the ozone layer by 'freon' fluorocarbons note that:

freon 11	methane, trichlorofluoro-	CCl_3F	[75-69-4]
freon 12	methane, dichlorodifluoro-	CCl_2F_2	[75-71-8]
freon 22	methane, chlorodifluoro-	$CHClF_2$	[75-45-6]
freon 115	ethane, chloropentafluoro-	$CClF_2CF_3$	[76-15-3]
freon 132	ethane,1,2-dichloro-1,2-difluoro-	$CHClFCHClF$	[413-06-1]

The last digit is the number of fluorine atoms. The next to last digit is one plus the number of hydrogen atoms. The next digit forward is the number of carbon atoms minus one. The balance is the number of chlorine atoms. The designation freon 11 etc. is used to avoid brand names.

Some rules for IUPAC naming

We have tried a few simple examples of interpreting chemical names, now we shall see how they are assembled. Naming compounds fills several books, so that all we can do here is to give a few of the simpler rules, first for linear (non-cyclic) compounds and then for cyclic compounds.

Linear compounds

Step 1. Identify all the functional groups present. Select the *principal* functional group (PFG) according to the order of priority given in Table 1.3.

TABLE 1.3
Order of priority of functional groups

Order of priority	Functional group	Formula	Order of priority	Functional group	Formula
1	Carboxylic acid	—COOH	8	Ketone	—CO
2	Sulfonic acid	—SO$_3$H	9	Alcohol	—OH
3	Ester	—COOR	10	Phenol	—OH
4	Acid chloride	—COCl	11	Thiol	—SH
5	Amide	—CONH$_2$	12	Amine	—NH$_2$
6	Nitrile	—CN	13	Ether[a]	—OR
7	Aldehyde	—CHO	14	Sulfide[a]	—SR

[a]Simple ethers and sulfides are often found named using the radico-functional system, ethyl methyl ether, instead of methoxymethane; but not more complex ones, e.g. 2-ethoxypropane.

Step 2. Check for unsaturation of the carbon skeleton, i.e. the presence of double (C═C) or triple (C≡C) bonds.

Step 3. Select the longest continuous carbon chain that contains the principal functional group and the maximum number of unsaturated bonds. This is called the *principal chain*. This supplies the stem name (see Table 1.4).

Step 4. Number the carbon chain from one end to the other so that the PFG has the lowest number possible.

Step 5. The PFG now provides the suffix for the chain name. Any other functional groups present are cited as prefixes.

TABLE 1.4
Simple stem names

Hydrocarbon chain	Stem name	Alkane/radical	Alkane/radical name
1 carbon atom	meth-	CH$_4$/CH$_3$-	methane/methyl
2 carbon atoms	eth-	C$_2$H$_6$/C$_2$H$_5$-	ethane/ethyl
3 carbon atoms	prop-	C$_3$H$_8$/C$_3$H$_7$-	propane/propyl
4 carbon atoms	but-	C$_4$H$_{10}$/C$_4$H$_9$-	butane/butyl
5 carbon atoms	pent-	C$_5$H$_{12}$/C$_5$H$_{11}$-	pentane/pentyl
6 carbon atoms	hex-	C$_6$H$_{14}$/C$_6$H$_{13}$-	hexane/hexyl
7 carbon atoms	hept-	C$_7$H$_{16}$/C$_7$H$_{15}$-	heptane/heptyl
8 carbon atoms	oct-	C$_8$H$_{18}$/C$_8$H$_{17}$-	octane/octyl

The list in Table 1.5 is in order of priority for principal functional groups. The alternative radico-functional names for some of these are given later.

Step 6. Insert all the lesser priority functional groups and branching alkyl groups (radicals) as prefixes, using the names from Table 1.5. These prefix names are listed alphabetically ignoring, for ordering purposes, any multiplicity modifiers. (Multiplicity terms are *di* or *bis* (×2), *tri* or *tris* (×3), *tetra* (×4), *penta* (×5), *hexa* (×6) etc.; e.g. bis(dimethylamino)). The numbers, indicating the location

TABLE 1.5
Suffixes and prefixes

	Group structure	Prefix	Suffix
1. Carboxylic acid	X—COOH	carboxy-	-oic acid
2. Sulfonic acid	X—SO$_2$—OH	sulfo-	-sulfonic acid
3. Ester	X—CO—OR	R-oxycarbonyl-	-alkyl -oate
4. Acid chloride	X—CO—Cl	halo-formyl-	oyl halide
5. Amide	X—CONRR	carbamoyl-	-carboxamide
	X—NR—COR	acylamido-	
6. Nitrile	X—CN	cyano-	-nitrile
7. Aldehyde	X—CH=O	formyl-	-al
8. Ketone	X—COR	oxo-	-one
9. Alcohol	X—OH	hydroxy-	-ol
10. Phenol	X—C$_6$H$_4$—OH	hydroxy-	-ol
11. Thiol	X—SH	mercapto-	-thiol
12. Amine	X—NRR	amino-	-amine
13. Alkene[a]	X—C=CRR	alkenyl-	-ene
14. Alkyne[a]	X—C≡CR	alkylnyl-	-yne
Groups cited only as prefixes			
15. Fluoride	X—F	fluoro-	
16. Chloride	X—Cl	chloro-	
17. Bromide	X—Br	bromo-	
18. Iodide	X—I	iodo-	
19. Azide	X—N$_3$	azido-	
20. Nitroso	X—NO	nitroso-	
21. Nitro	X—NO$_2$	nitro-	
22. Ether	X—OR	R-oxy-	
23. Sulfide	X—SR	R-thio-	

[a] 13,14 as suffix only if no PFG present.

along the chain, are placed in front of the functional or branching alkyl groups they locate.

Step 7. Assemble the complete name as a single word (there are some exceptions), using commas to separate numbers from numbers and hyphens to separate numbers from words:

2,2-dichlorohept-3-en-1-ol

Breakdown of molecule for nomenclature purposes
The following steps are followed when naming an organic compound. Table 1.3 gives the order of priority for choosing a suffix.

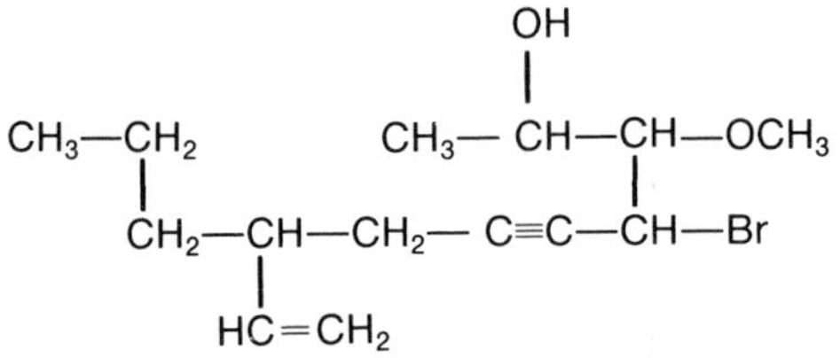

Fig. 1.20

Step 1. Functional groups are determined.
OH—hydroxyl will determine the suffix (see Table 1.3).
OCH_3—methoxy; Br—bromo.

Step 2. Unsaturated? Yes, it contains both $C=C$ and $C\equiv C$.

Step 3. Selecting principal chain.
The stem is chosen to include the PFG and maximum number of double and triple bonds.

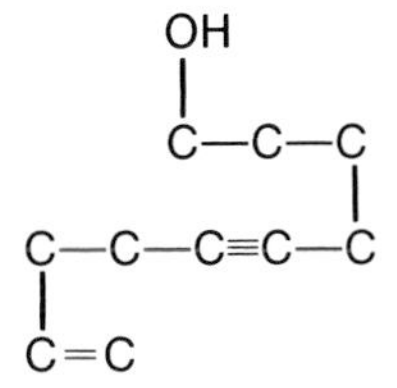

$$
\begin{array}{c}
OH \\
| \\
C-C-C \\
\qquad | \\
C-C-C\equiv C-C \\
| \\
C=C
\end{array}
$$

Fig. 1.21

Principal chain (stem) name: dec-en-yn. This order is maintained.

Steps 4–5. Suffix for PFG and numbers added.

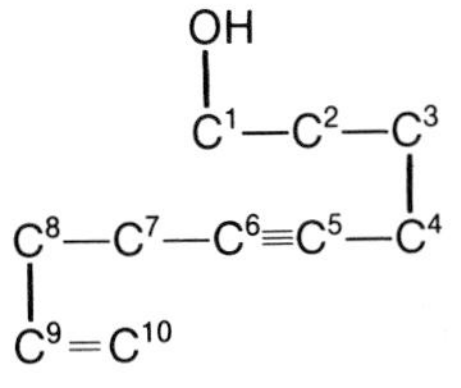

$$
\begin{array}{c}
OH \\
| \\
C^1-C^2-C^3 \\
\qquad\qquad | \\
C^8-C^7-C^6\equiv C^5-C^4 \\
| \\
C^9=C^{10}
\end{array}
$$

Fig. 1.22

Basic name: dec-9-en-5-yn-2-ol.
   If a suffix begins with a vowel (or y), when adding the suffix to a hydrocarbon name the final e is dropped. Thus, from pentane comes pentanol, pentanoic acid, pentanal etc. When assembling the stem name, if both a double and a triple bond is present the e is elided between -ene and -yne: thus pentenyne.

Step 6. Inclusion of lower priority functional groups and branching chains as substituents.

$$
\begin{array}{ll}
CH_3CH_2CH_2- & \text{propyl} \\
CH_3O- & \text{methoxy} \\
Br- & \text{bromo}
\end{array}
$$

These substituents are listed alphabetically as prefixes.

Step 7
IUPAC name: 4-bromo-3-methoxy-8-propyldec-9-en-5-yn-2-ol.

The rules for Chemical Abstracts are not quite the same. The CAS name for the same compound is 4-bromo-7-ethenyl-3-methoxyundec-5-yn-2-ol. The CAS sequence of application of principles (neglecting heteroatoms and rings) is as follows:

1. Greatest number of principal chemical functional group (PFG).
2. Largest index heading parent (i.e. longest chain).
3. Greatest number of multiple bonds (but double bonds preferred over triple bonds—inferred from example).
4. Lowest numbers for PFGs.
5. Lowest number for all multiple bonds.
6. Lowest number for double bonds.

In the IUPAC system rule 3 takes precedence over rule 2.

Cyclic compounds
Aromatic compounds are related to benzene either by replacing a hydrogen atom with a substituent, or by fusing the rings together to give naphthalene or anthracene etc., as shown in Figure 1.23.

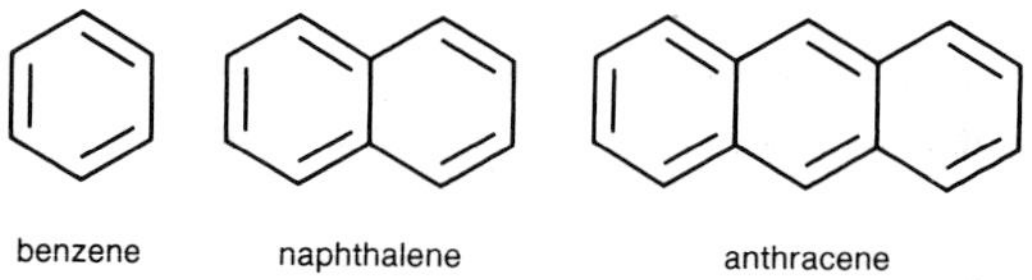

benzene   naphthalene   anthracene

Fig. 1.23 Aromatic hydrocarbons

Note: the radical derived from benzene, C_6H_5—, is called *phenyl* (not benzenyl); other related radicals are $C_6H_5CH_2$—, called *benzyl*, and C_6H_5CO—, called *benzoyl*. Most simple aromatic compounds are known by trivial names: e.g. C_6H_5OH, called phenol; $C_6H_5NH_2$, aniline; $C_6H_5CH_3$, toluene; $C_6H_5CH{=}CH_2$, styrene; $C_6H_5(CH_3)_2$, xylene; $(C_6H_5)_2CO$ benzophenone.
Semi-systematic names are formed as shown in Figure 1.24.

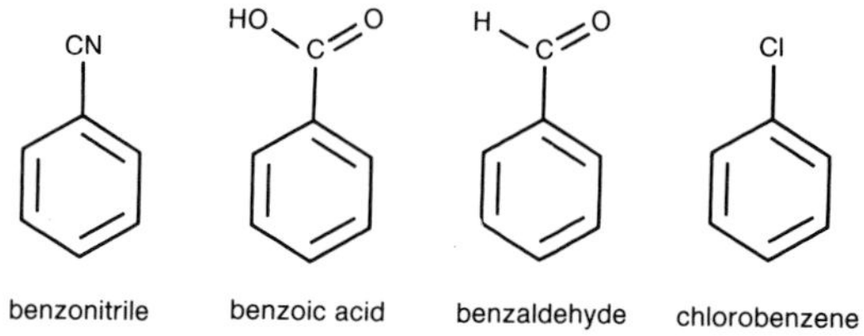

benzonitrile   benzoic acid   benzaldehyde   chlorobenzene

Fig. 1.24 Aromatic compounds

Two substituents in the benzene ring on adjacent positions 1,2 are labelled ortho or *o*-; on positions 1,3 are labelled meta or *m*-; on opposite positions 1,4 are labelled para or *p*-. IUPAC prefers the use of the numbers rather than these names, which apply only to benzene.

Again the PFG attracts the lowest number, as shown in Figure 1.25.

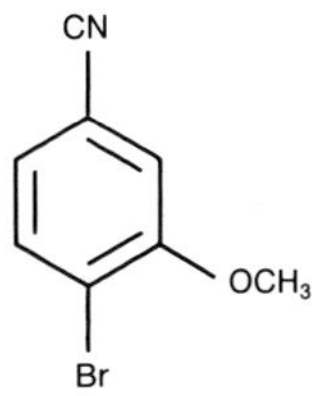

Fig. 1.25   4-bromo-3-methoxybenzonitrile

Non-aromatic cyclic systems follow similar rules. Thus cyclohexane is a C_6 ring and the numbering is the same as for benzene.

Nomenclature exercise
Try to name the structures of the compounds listed on the label of a solar block sunscreen as follows:

octyl methoxycinnamate 7.5%
2 hydroxy-4-methoxy-benzophenone 3%
4-t-butyl-4′-methoxydibenzoylmethane 2%

Unfortunately, you are still not in a position to entangle this completely. For that you will need to consult some references (see the bibliography at the end of the chapter for details). A good learning resource, and the reference used for this section, is *Chemical nomenclature* by J.D. Coyle and E.W. Godly. The standard reference, 'the Blue Book', is *Nomenclature of organic chemistry*, in particular 'Characteristic groups containing carbon, hydrogen, oxygen, nitrogen, halogens, sulfur, selenium and/or tellurium', issued by the IUPAC Commission on Nomenclature in Organic Chemistry. As a reference for particular compounds, be they trade, trivial or systematic, you cannot better the *Chemical index guide* which comes with each volume of *Chemical abstracts*, the major chemistry reference series.

Solubility

Ordinarily, when we say a substance is soluble, we mean that it dissolves to an appreciable extent in water. However, we do make every day use of other solvents—dry-cleaning spirit to dissolve grease stains, turpentine to dissolve paints, and so on. The question of solubility in oils and fats is of great importance to the physiological action of, for example, pesticides and drugs, and the cleaning action of detergents. There is a simple rule which works well in predicting solubilities of substances in various solvents—*like dissolves like*. This means that ionic and highly polar substances are usually soluble in polar solvents such as water, whereas covalent non-polar substances are soluble in non-polar solvents such as benzene, petrol and carbon tetrachloride.

Although many substances do not dissolve in water or other solvents, it is often possible to produce a stable to semi-stable mixture or *dispersion* of solute and solvent. This process is called *solubilisation*. (The *solute* is what is dissolved; the *solvent* is what does the dissolving.) The resulting dispersion may be stabilised by the addition of another substance, such as a detergent. *Emulsions* are an important group of these dispersions. Milk and cream, for example, are emulsions of fat in water stabilised by the emulsifying agent casein, a protein. When producing butter from cream the emulsion changes to one of water in oil (20% water). In egg mayonnaise, the oil-water emulsion is stabilised by the egg white, also a protein (see also 'Emulsions' in Chapter 4). These principles will be illustrated in the discussions on detergents and cosmetics (Chapters 2 and 4).

STANDARDS

The word *standard* has a variety of meanings—from a flag, to a weight or measure, to a connotation of average quality (e.g. we say something is of a high or low standard). Thus, standards are very much a part of our everyday lives. After all, language itself is standardised, although its standard is constantly changed by common usage or agreement.

The main body concerned with this is Standards Australia (SA), which derives its legal status from a Royal Charter and its finance from government grants, subscriptions and the sale of documents detailing SA Standards. There is a Consumer Standards Advisory Committee, whose task is to define areas in need of consumer standards.

Not surprisingly these consumer standards are directed in the first instance at safety; for example, the safety of children's toys, furniture and playground equipment, of protective helmets for pedal cyclists, of swimming pool covers and swimming aids, and of plastics that are used by the food industry or to store food. Secondly, they are directed towards providing information; for example, the care labelling of textiles, contracts for consumer transactions, and the burning behaviour of textiles and textile products. Safety and information are the two categories that can be regulated under Section 63 of the Trade Practices Act in Australia.

Although there can be considerable argument on SA committees between suppliers and customers in reaching a consensus on these points, there is at least a reasonably well-defined aim. A safety standard sets a level of *maximum acceptable dangerousness*, and this can be fairly arbitrary. An example is the regulatory speed limit aspect of road safety. An arbitrary maximum speed is chosen that is deemed to give a reasonable compromise between convenience and safety. The extent to which an electrical appliance is shielded against giving its user a fatal shock is another example.

The same arbitrariness is true for a third classification, that of performance. Defining standard performance is the aim of committees involved in dealing with electrical appliances, school and college wear, contraceptive devices and household detergents. However, a little contemplation reveals that, for many consumer products at least, the performance, like beauty, is in the eye of the beholder.

Having mentioned beauty, a pertinent example is that of cosmetics. On the one hand is the consumer, who is buying a fervent hope; on the other is the manufacturer, who is selling an image (see 'Market research' in Chapter 4). Setting of performance standards here would require the skill of a catcher of mirages.

On the other hand, for contraceptives, performance is well defined but cost, convenience and, possibly, unwanted side-effects are other important parameters. Consumers are not going to agree on the combination that represents an optimum—even the same consumer will have different views in different circumstances.

Standard setting does for industry and commerce what public health measures do for individuals. They each provide cost-effective, preventative measures for ensuring a healthy environment. They yield security in return for a small limitation in freedom of action. Standards Australia has acted as a quiet, moderately efficient, industrial and commercial arbitration commission that has benefited all sections of the community.

OLD SCIENCE

Figure 1.26 shows a time series for major chemical innovations from 1840 to 1980. The old technologies on which industrial growth in the period 1940-70 has depended have reached their zenith, are mature and are declining. The period 1930-60, in particular, was filled with chemical innovations based on earlier science.

Since 1960, there has been no major chemical innovation.[1] New science as a basis for new innovation has occurred in molecular biology, microelectronics, lasers, optical fibres and new materials. These will be the areas about which consumers will need to be concerned in the year 2000 and after. Meanwhile, it is the 'old' science that needs to be understood properly to cope with our current concerns.

EXPERIMENT 1.1

Separation by solubility

Equipment

A glass container, non-rechargeable, dry cell battery (new or used, but *not* alkaline), spoon or spatula, funnel, cotton wool, bowl, hack-saw.

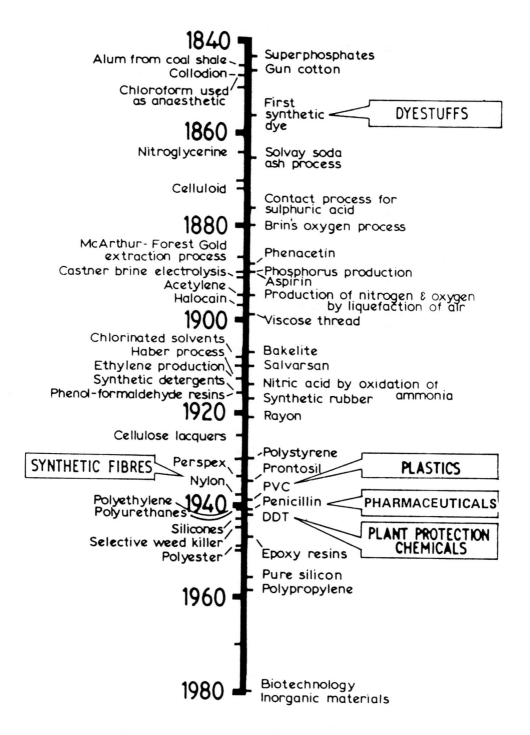

Fig. 1.26 Major chemical innovations 1840–1980 (Reproduced with permission from Sharp and West, *The chemical industry*, Ellis Horwood Ltd, Chichester, UK, 1982)

WARNING

A letter in *New Scientist* (20 November 1986) correctly warns about the danger of choosing the wrong batteries. While the Leclanche cell's components are mildly poisonous, other cells have very dangerous constituents such as nickel, cadmium, lithium, silver and mercury salts, metallic mercury, potassium hydroxide and thionyl chloride! Do not examine any other type of battery.

Procedure

Have a dry cell battery cut open with a hack-saw (cut down one side of the battery). Take care not to damage the carbon rod central electrode of the battery, which is useful for electrolysis experiments. Also, retain the zinc electrode, which constitutes the metal outer case of the battery.

Remove the contents of the battery with a spoon and put it in a glass container half full of water. Stir the contents of the glass thoroughly.

Take a funnel and put a small wad of cotton wool in the neck as shown in Figure 1.27. Filter the solution through the funnel into a glass bottle.

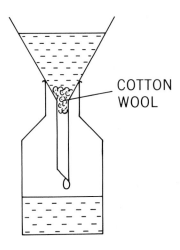

COTTON WOOL

Fig. 1.27 Filtering apparatus

The black material remaining in the filter funnel is mainly manganese dioxide (MnO_2). If you allow the water to evaporate from the clear solution (this is best done using a wide-neck glass or plastic bowl placed in the sun), the white material that remains is chiefly ammonium chloride (NH_4Cl) with some zinc chloride ($ZnCl_2$). You can test for ammonia by adding solid sodium bicarbonate and warming. This should release ammonia gas, which you can smell. You may also notice a fine white precipitate, zinc hydroxide ($Zn(OH)_2$), which may redissolve to form the zincate ion ($Zn(OH)_4^{2-}$).

Note

If you have young brothers or sisters or animals, make sure you don't leave the clear solution (which looks like pure water) where they can reach it. Although not highly toxic, it has a mild diuretic effect.

EXPERIMENT 1.2

Rusting of iron

Equipment

A glass container, a saucer, a steel wool pad (without any soap), water.

Procedure

Put some water in the saucer. Place the steel wool in the centre of the saucer and cover the steel wool with the upturned glass as shown in Figure 1.28. Mark the water level on the outside of the glass.

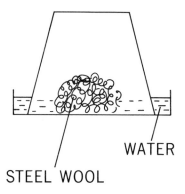

WATER

STEEL WOOL

Fig. 1.28   Rusting of iron

Allow the experiment to stand for several days, topping the water level up to the mark when necessary.

Water enters the glass as the oxygen inside is used up. After a week, estimate the fraction of the air in the glass that has been used in the reaction with the iron.

Reactions

$$Fe + \tfrac{1}{2}O_2 + H_2 \rightarrow Fe(OH)_2\downarrow$$

iron   oxygen   water   ferrous hydroxide (white)

$$4Fe(OH)_2 + O_2 \rightarrow 2Fe_2O_3\cdot3H_2O + H_2O$$

rust (ferric oxide)
approximate formula

Notes

1.   The rusting process is more rapid in the presence of salts (e.g., common salt, NaCl), which make the water more conducting, or sulfur dioxide, (SO_2), which is an air pollutant in cities.

2. You can test the effect of common salt. Note, however, that the salt does not enter into the rusting reaction above.

REFERENCE

1. A. Baklien, 'Chemistry in the 1990s', *Chem. Aust.* **52**(1), January 1985, 22.

BIBLIOGRAPHY

Asimov, Isaac. *Asimov on chemistry*, Scientific Book Club, London, 1976.
Bennett, H. *The chemical formulary*, 24 vols. Chemical Publishing Co., New York, 1933-1982. Commerical recipes for a magnificent array of products, updated each volume.
Bucat, R. (ed.). *The elements of chemistry*. Australian Academy of Science, Canberra, 1984.
Coyle, J.D., and Godly, E.W. *Chemical nomenclature*. Open University and Laboratory of the Government Chemist, Milton Keynes, UK, 1984.
Grayson, M. (ed.). *Kirk-Othmer encyclopedia of chemical technology*, 3rd edn. Wiley, New York, 1983. The major reference series for the fields to be discussed in the following chapters.
ICI. *Inorganic chemicals*. ICI Educational Publication, 1978.
International Union of Pure and Applied Chemistry (IUPAC). *Nomenclature of organic chemistry*. Butterworths, London, 1969.
Selinger, Ben. 'Nyholm Memorial Lecture: Chemistry in the marketplace'. *Chem. Aust.*, **45**(1), 1978, 11.
Selinger, Ben. 'Bayliss Youth Lecture: chemistry—the best buy in the scientific supermarket'. *Chem. Aust.* **47**(12), 1980, 491.

Chemistry in the laundry

2

INTRODUCTION

In primitive societies, even today, clothes are cleaned by beating them on rocks near a stream. Certain plants, such as soapworts, have leaves that produce *saponins*, chemical compounds that give a soapy lather. These were probably the first detergents people used.

If you look up the word *detergent* in a dictionary it is simply defined as *cleaning agent*. During the past two to three decades, however, the word detergent has tended to imply synthetic detergent, or *syndet* for short, rather than the older *soap*. In fact, commercial formulations consist of a number of components, and we shall use the term *surface-active agent*, or its abbreviation *surfactant*, to describe the special active ingredients that give detergents their unusual properties.

Soap, by this definition, is a surfactant. In fact, it is the oldest one and has been in use for more than 4500 years. Some soap manufacture took place in Venice and Savona in the fifteenth century, and in Marseilles in the seventeenth century. By the eighteenth century, manufacture was widespread throughout Europe and North America, and by the nineteenth century the making of soap had become a major industry. As a matter of fact, soap became a detergent in 1907 when the German firm Henkel & Cie put the product Persil on the market. In addition to the carboxylic acid soap, Persil contained sodium perborate, sodium silicate and sodium carbonate. Hence *per*borate + *sil*icate = *Persil*.

SOAPS

Ordinary soaps are the sodium salts of long-chain fatty acids. They have the general formula $RCOO^-Na^+$, where R is a long hydrocarbon chain, $CH_3(CH_2)_{10-16}$. These salts can be made by the simple neutralisation reaction

$$
\underset{\text{acid}}{R-\overset{\overset{\displaystyle O}{\|}}{C}-OH} \;+\; \underset{\text{base}}{NaOH} \;\longrightarrow\; \underset{\text{salt}}{R-\overset{\overset{\displaystyle O}{\|}}{C}-O^-Na^+} \;+\; \underset{\text{water}}{H_2O}
$$

However the cheapest sources of the fatty acids are animal fats and certain vegetable oils, which are largely *esters*. In practice, therefore, soaps are made by the *saponification* reaction

34

$$R-\overset{\overset{\displaystyle O}{\|}}{C}-O-R' \quad + \quad NaOH \quad \longrightarrow \quad R-\overset{\overset{\displaystyle O}{\|}}{C}-O^-Na^+ \quad + \quad R'OH$$

| ester | base | salt of fatty acid | alcohol |
| (fat) | (caustic soda) | (soap) | (e.g. glycerine) |

which is essentially the reverse of the esterification reaction. Beef tallow gives principally sodium stearate, $CH_3(CH_2)_{16}COO^-Na^+$, the most common soap. Palm oil gives sodium palmitate, $CH_3(CH_2)_{14}COO^-Na^+$, a component in more expensive soaps.

The standard for personal soap in Australia provides for not less than 70% of fatty matter—actual soap plus so-called *superfatting* agents, which can be fats, fatty acids, wool wax etc., but these agents are limited to a maximum of 10%. The total amount of water allowed is 17%. In addition, there will be some sodium chloride and glycerol left from the production process. Also added are preservatives, antioxidants, perfume and colouring matter (titanium dioxide in the case of white soaps). In laundry soap the amount of fatty material allowed is lower (60%) and the amount of water is higher (34%).

The combination of frequent showering and soft water means that the use of toilet soap is high in Australia. The consumption per person (kg/inhabitant) rose from 1.9 kg in 1977 to 2.25 kg in 1982, compared to a rise in West Germany from 0.90 kg to 0.97 kg. During the same period, consumption *fell* in the UK from 1.48 kg/person to 1.25 kg; and in France, from 0.67 kg to 0.63 kg.

If the sodium ion of ordinary soap is replaced by other metal ions, soaps with different properties are produced. When potassium hydroxide is used instead of sodium hyroxide in the manufacturing process, *soft soaps* are formed. These are semi-solid soaps, once used in shampoos and special-purpose soaps. However they are more expensive than the ordinary soaps. Most other metals give soaps that are insoluble in water. Clearly these are not much use for washing, but they do find applications as additives for greases and heavy lubricating oils, where their principal function is still as detergents. Copper stearate has been used as a waterproofing colour agent; not only is it water-repellent, but the copper ion is poisonous to mildew. The heavy metal stearates are also used as stabilisers or release agents in plastics such as poly(vinyl chloride) (PVC) and polythene (see 'Chemistry of plastics—Vinyl polymers' in Chapter 6).

Synthetic surfactant or soap?

You may well ask why soap, which served well for so many years, was eventually displaced. In some ways, ordinary soaps are better than the newer detergents. They are cheap and they are manufactured from a renewable source, whereas many of the synthetic detergents are made from petrochemicals. Soaps are also biodegradable; that is, they are readily broken down by bacteria, and thus they do not pollute rivers. However, because of their gelling properties, soaps do have a greater tendency to clog sewerage reticulation systems than synthetic detergents. The grease trap of a non-sewered house was often laden with soap. But the most important reason for the displacement of soap is the fact that, when a carboxylic acid soap is used in hard water, precipitation occurs. The calcium and magnesium ions, which give hardness to the water, form insoluble salts with the fatty acid in the soap and a curd-like precipitate occurs and settles, of course, on whatever is being washed. By using a large excess of soap, it is possible to redisperse the precipitate, but it is extremely sticky and difficult to move. This problem with soap can be demonstrated by a simple experiment in which a concentrated solution of hard-water salts is added to a 0.1% solution of soap and also to a 0.1% solution of synthetic surfactant. The soap precipitates, but the synthetic surfactant remains clear because its salts are water soluble.

You may live in an area (such as Melbourne) where the water is extremely soft. But calcium and magnesium ions are present in the dirt that you wash out of your clothes, so that some precipitation still occurs if soap is used, and gradually deposits are built up in the fabric.

There are other disadvantages with soap; it deteriorates on storage, and it lacks cleaning power when compared with the modern synthetic surfactants, which can be designed to perform specialised cleaning tasks. Finally, and very importantly from a domestic laundry point of view, soap does not rinse out; it tends to leave residues behind in the fabric that is being washed. These residues gradually build up and cause bad odour, deterioration of the fabric and other problems.

What is the difference between a surfactant and soap? In general terms the difference can be likened to the difference between cotton and nylon. On the one hand, soap and cotton are produced from natural products by a relatively small modification. On the other hand, synthetic surfactants and nylon are produced entirely in a chemical factory. Synthetic surfactants are not very new, either. Back in 1834 the first forerunner of today's synthetic surfactants was produced in the form of a sulfated castor oil, which was used in the textile industry.

The development of the first detergents in an effort to overcome the reaction of soaps with hard water provides a good illustration of one of the standard chemical approaches. If a useful substance has some undesirable property, an attempt is made to prepare an analogue, a near chemical relation, which will prove more satisfactory.

The petroleum industry had, as a waste product, the compound propylene, CH_3—CH=CH_2, which used to be burnt off. By joining four of these propylene molecules together propylene tetramer is obtained (Fig 2.1). If benzene is attached at the double bond, the resulting compound reacts with sulfuric acid, H_2SO_4. Then sodium hydroxide is added to neutralise the sulfonic acid, and the sodium salt shown in Figure 2.2 is obtained, which is a branched-chain alkylbenzene sulfonate (ABS).

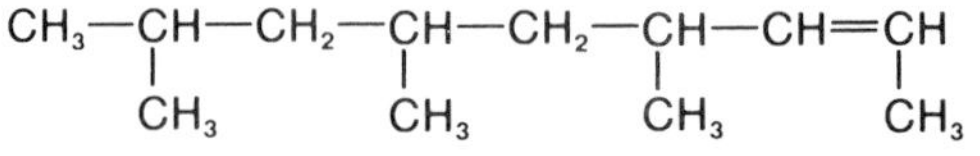

$$CH_3—CH—CH_2—CH—CH_2—CH—CH=CH$$
$$\qquad |\qquad\qquad |\qquad\qquad |\qquad\quad |$$
$$\qquad CH_3\qquad\; CH_3\qquad\; CH_3\qquad CH_3$$

Fig. 2.1 Propylene tetramer

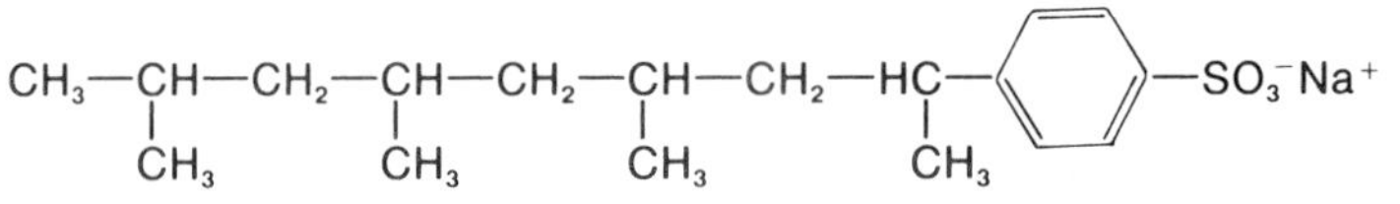

$$CH_3—CH—CH_2—CH—CH_2—CH—CH_2—HC—\langle\!\!\!\bigcirc\!\!\!\rangle—SO_3^-Na^+$$
$$\qquad |\qquad\qquad |\qquad\qquad |\qquad\qquad |$$
$$\qquad CH_3\qquad\; CH_3\qquad\; CH_3\qquad\; CH_3$$

Fig. 2.2 Alkylbenzene sulfonate

Clearly, the new substance is closely related to an ordinary soap (Fig 2.3) and is an excellent detergent.

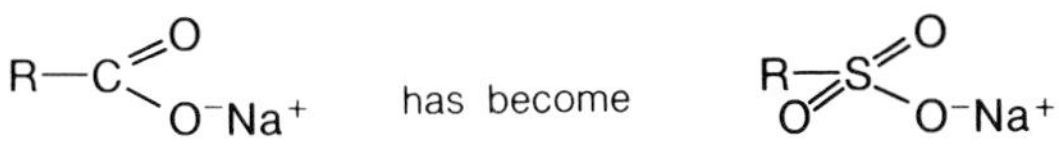

$$R-C\overset{\displaystyle O}{\underset{\displaystyle O^- Na^+}{\big<}} \qquad \text{has become} \qquad R-S\overset{\displaystyle O}{\underset{\displaystyle O^- Na^+}{\underset{O}{\big<}}}$$

Fig. 2.3 Soap and detergent

The detergents produced in this way are much more soluble than soap, and their calcium and magnesium salts are soluble, so that a scum is not formed with hard water. However, they are more stable than soaps and persist in the waste water long after use. The consequence of this was the fouling of sewerage works and rivers with tremendous masses of froth. The increased stability of the detergents resulted from both the greater stability of the sulfonate grouping and the fact that the raw material hydrocarbon chain molecules contained large proportions of carbon chains that were *branched*, in contrast to the straight-chain hydrocarbons from animal fats. As bacteria break down the branched chains more slowly, detergents were once considered not to be biodegradable at all (see 'Domestic laundry detergents—Biodegradability' later in this chapter).

How do surfactants work?

The surfactant molecule is often described as tadpole-like because it has a fairly long fatty tail, which is water-insoluble or *hydrophobic*, and a small, often electrically charged head, which is water-soluble or *hydrophilic*. There are four possible combinations (Fig. 2.4):

1. The *anionic* surface-active agents, in which the surfactant is an anion (that is, it carries a negative charge) and the charge is concentrated in the hydrophilic or water-soluble head.
2. The *cationic* products (the opposite of anionic), in which the head carries a positive charge.
3. The so-called *non-ionic* detergents. These do not have a specific charge, but the hydrophilic or water-soluble portion of the molecule is usually achieved by incorporating a polyethylene oxide group into the molecule (see below in the section on surfactants in domestic laundry detergents). You can see that, because it is less polar than an ion, the hydrophilic portion of these molecules is usually rather bigger than in the case of the ionic surfactants.
4. Finally, there are some specialised products that carry both a positive and a negative charge in the same molecule. These are called *amphoteric*, and they are particularly useful for very specialised applications (such as hair shampoos). Because they carry both an anionic and a cationic centre, they behave as either an anion or a cation, depending on the pH of the solution in which they are used.

In domestic detergents, anionic surfactants are predominant. Non-ionics are increasingly used but cationic surfactants are not. Cationic surfactants have two interesting and useful properties. First, they are mildly antiseptic and may be used (in combination with non-ionic surfactants) in nappy washes, hair conditioners and throat lozenges, and as algaecides in swimming pools. Second, the positive

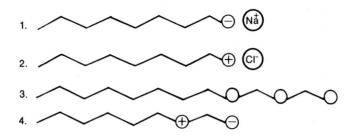

Fig. 2.4   Diagrammatic representation of the shapes and electric charges of surfactant molecules

charge on the chain makes them useful for washing plastic articles but not glass. Glass normally acquires a surface negative charge, which to a certain extent attracts dirt. Anionic detergents can remove this dirt, but cationic surfactants are attracted to the glass so strongly that a thin layer adheres to the glass, with the long hydrophobic (fatty) chain outwards, thus making the glass non-wettable and apparently greasy. The reverse is true for plastic articles, which normally have a positive surface charge. Have you ever noticed how dirt clings to plastic and is hard to remove by ordinary washing?

The positive charge also makes cationic surfactants useful as fabric softeners (Comfort, Cuddly, Huggie). These are liquids and are added to the rinse cycle. The cationic charge has a strong affinity for wet, negatively charged fabric and forms a uniform layer on the surface of the fibres, thus lubricating them and reducing friction and static. Double chain (di-C_{18}) cationic surfactants are much less water soluble and are used in solid packs in clothes dryers. They are transferred as the solid to the clothes (e.g. the sheets of Breeze or Fluffy). The shorter chain version of these molecules (di-C_{12}) appears to have subtle effects on cell membranes and may act in suppressing body immunity.

Because their chains have opposite charges, anionic and cationic surfactants are generally incompatible. On mixing, they give a scum, so that, except in very special cases, it is not possible to formulate a detergent that contains both types of surfactant and hence combines their advantages. (See 'Domestic laundry detergents—Multicomponent detergents' later in this chapter.)

The cleaning action of surfactants

The molecules of surfactants tend to concentrate in the surface layers of the water (because the water-insoluble portion wants to get out of the water), lowering the *surface tension* of the water and allowing the water to wet non-wettable surfaces. Some of the cleaning power of surfactants thus results from the enhanced ability of the water to wet the normally hydrophobic surface and lift off the dirt. If you place a strip of cotton fabric, weighted at one end to pull it into the liquid, in a 0.1% solution of surfactant and another in pure water, you will see a difference in behaviour. Because the pure water alone does not wet the cotton, the fabric strip remains 'upright', whereas the fabric in the surfactant solution wets out and sinks immediately.

The long hydrocarbon tails of the detergent molecules are soluble in non-polar substances such as oil, whereas the polar carboxyl or sulfonate groups are soluble in water. Thus the molecules promote solubilisation of oil in water by lying across the oil-water interface. When the concentration of the detergent molecules in the water reaches a certain value, called the *critical micellar concentration*, the molecules aggregate into communes called *micelles*, which contain roughly 40 to 100 molecules. In these aggregates, which at high enough concentrations give soap solutions their cloudy appearance (because they scatter light just as dust does in the air), the hydrocarbon tails lie towards the centre, while the surface of the micelle contains the water-soluble polar ends (see Fig. 2.5).

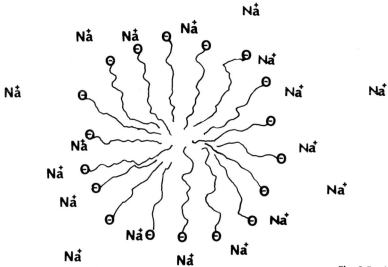

Fig. 2.5 A surfactant micelle

How does a surfactant remove the oily soil and, with that, frequently a lot of the particulate soil? This is a complex process. The insides of the micelles are virtually small oil droplets and so can dissolve oily materials, but the main action of the surfactant is to stimulate emulsification. If some olive oil is carefully poured into water and also into a solution of surfactant and each vessel is vigorously shaken and then put down again, the oil immediately rises to the surface of the water, but remains emulsified and dispersed in the detergent solution and can therefore be rinsed out. The process of emulsification (e.g. the removal of fat from a plate) is illustrated in Figure 2.6.

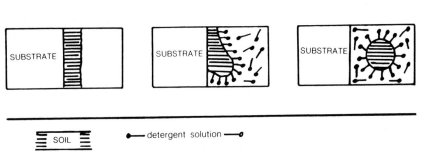

Fig. 2.6 Emulsification

DOMESTIC LAUNDRY DETERGENTS

The ingredients in any commercially produced detergent fall into five groups: the surfactant system; the builders; a filler; bleaches; and fluorescers.

Surfactant

The starting material for the synthetic surfactant that forms the major active ingredient is a material known as alkylbenzene. In appearance it is a kerosene-like liquid with a slightly oily odour. In fact, it is a product of the petroleum industry and is made by the condensation of an α-olefin with benzene. As such it is completely insoluble in water. By treating this material in a process called sulfonation, it is converted to the corresponding sulfonic acid, which is a viscous, dark-brown material (see Fig. 2.2). Chemically, sulfonation is achieved by treating alkylbenzene with an excess of sulfuric acid, giving the sulfonic acid plus water.

Inorganic builders

Turning now from surfactants let us look at the builders. Builders are included in modern domestic laundry detergents in order to assist the surfactant system in its action. Builders are both organic and inorganic. Among the inorganic builders, *sodium tripolyphosphate* is the major one used. It is a polyphosphate equivalent to having been formed from two molecules of disodium monohydrogen phosphate and one molecule of monosodium dihydrogen phosphate to give what is correctly called pentasodium triphosphate, by the elimination of water, as shown in Figure 2.7.

$$2Na_2HPO_4 \quad + \quad NaH_2PO_4 \quad \longrightarrow \quad Na_5P_3O_{10} \quad + \quad 2H_2O$$

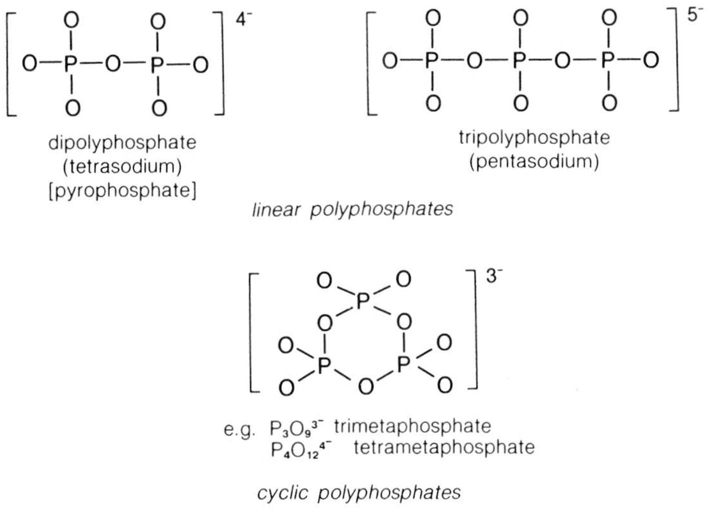

dipolyphosphate
(tetrasodium)
[pyrophosphate]

tripolyphosphate
(pentasodium)

linear polyphosphates

e.g. $P_3O_9{}^{3-}$ trimetaphosphate
  $P_4O_{12}{}^{4-}$ tetrametaphosphate

cyclic polyphosphates

Fig. 2.7 Polyphosphates

Why should tripolyphosphate be put into a detergent? It has a number of advantages from the point of view of safety. It is non-toxic (toxicological information indicates that it is comparable to common table salt) and non-irritant. It does three things in the detergent. First, it buffers the washing water to a milder pH than would otherwise be obtained. You may recall that the first detergent was Persil, in which sodium carbonate and sodium silicate had been added to soap. One of the drawbacks of the early versions of Persil was high alkalinity and the damage that this could do to fabrics. On the other hand, a certain amount of alkalinity is needed to wash fabrics, particularly cotton, successfully. Second, sodium tripolyphosphate sequesters hard-water ions. Although synthetic surfactants do not precipitate with the ions in hard water (such as calcium and magnesium), the presence of these ions does tend to decrease detergency to some degree. The full cleaning power of the surfactant is preserved if the ions are sequestered by tripolyphosphate. Finally, tripolyphosphate is important in its deflocculating action; that is to say, it helps to keep a clay-type dirt in suspension.

The contribution of hardness salts from mains water is readily calculated from data supplied by a water supply authority. The distribution for Australia is shown in Table 2.1.

TABLE 2.1
Distribution of water hardness in Australia

Water hardness (mg/kg as calcium carbonate)	Population (%)
10–20	26
21–40	11
41–60	37
61–120	8
121–180	14
181–240	1
Over 240	3

To this base level of hardness must be added the hardness contributed by the soiled fabrics to be washed. The equivalent of 20–30 mg/kg as calcium carbonate is used. You can measure water hardness with a dipstick test (see 'Dipstick chemistry' in Chapter 13) and calculate how much polyphosphate a detergent suitable for your local conditions should contain.

The height of controversy over phosphates in detergents came in the early 1970s in the USA and Australia, and later in Europe. In Canada there is a limit on the use of phosphates in household detergents of 2.2% phosphorus (equivalent to 9% of polyphosphate). In West Germany polyphosphate levels dropped from 40% in 1975 (before legislation limits went into effect) to 30%. This meant a drop from 270 000 tonnes to 160 000 tonnes. In 1984 the level dropped to 20–24%. In the USA, Desoto went to sodium carbonate, Lever Bros reformulated Wisk liquid detergent with sodium citrate and Henkel began to capitalise on its nearly worldwide patents based on sodium aluminosilicate—zeolites. In Japan, for example, zeolites are the sole builder in more than 90% of all detergents sold. In August 1983, zeolite production capacity in Europe was at least 210 000 tonnes per year. They are used only in problem areas and are unlikely to be used in British products.

In Australia the topic of phosphate reduction is under discussion, but there is a feeling that the problem here is more political than actual. There has been no move in Australia towards the use of citrates, polyacrylates or zeolites and Wisk here has a phosphate builder, as do the other Australian laundry liquids of comparable performance.

Sodium nitrilotriacetate (NTA) has never quite cleared itself of early fears about carcinogenicity, or about heavy metals carried into the drinking water supply by this superefficient sequestering agent. It appears to be used freely only in Canada and with strict limits in Germany and the Netherlands. Canada's practice of combining sewage and storm water provides a major dilutent effect on the NTA, which then appears to be less of a problem.

The second inorganic builder is a *sodium silicate*, a material that is better known by the name 'water-glass'. It is silicate obtained by having the ratio of sodium oxide to silicon dioxide different from that of metasilicate (see 'Machine dishwashing detergents' in this chapter). In fact the ratio is 1:2. This product in the detergent further controls the alkalinity. It also acts as a corrosion inhibitor, protecting die cast washing-machine parts especially, and it plays an important role in strengthening the physical part of the detergent powder.

The organic builder

The organic builder used in detergents is a product called *sodium carboxymethylcellulose*, which is produced by treating pure cellulose with caustic soda and chloracetic acid. It is used at a concentration of less than 1%, and its major function is to act as an anti-redeposition agent: it *increases the negative charge* in fabrics, which then repel the dirt particles because they are themselves negatively charged. Imagine that you have a white handkerchief with a black sooty spot in one corner. If you washed it in a detergent that did not contain an anti-redeposition agent, the soot would be dislodged but would tend to redeposit all over the handkerchief so that it would emerge a uniform grey.

Unfortunately this material is active only on cellulosic fabrics (cotton, rayon etc.) and on fabric blends with a cellulosic component. Nevertheless these fabrics do represent around 70% of Australian washloads. Polyacrylic acids and polyacrylates (1-6%) in a formulation can handle synthetics and synthetic blends.

An inert filler—sodium sulfate

Some *sodium sulfate* ($\sim$ 6%) is produced when the excess sulfuric acid used for sulfonation is neutralised, but extra is added and up to 50% may be found in some products. A certain amount is needed to form a crisp powder. Although it is marginally useful in lowering the critical micelle concentration of ionic surfactants and thus possibly reducing the amount of detergent needed, its main purpose would appear to be to produce a free-flowing powder and to add bulk to the product.

Fluorescers—'whiter than white'

In the 'good old days' we added washing blue so that cotton ageing naturally to yellow would look white. Today, very small amounts of *fluorescers* are added to detergent powders. They are in fact already in the new, gleaming business shirt when you buy it, but wear and washing remove them. These compounds absorb ultraviolet light (which is invisible) and re-emit blue light (which yellow fabrics do not reflect fully from sunlight), and so restore the mixture of colours reflected to that which a white fabric would reflect. The exact nature of the brighteners differs with geographic location. We are conditioned to blue-white as the accepted hue for cleanliness, whereas in South America a red-white colouration is the culturally accepted colour for clean. Fluorescers do not clean but they do whiten the fabric, which some consumers consider to be aesthetically desirable. On the Australian market, the main reason for using a combination of fluorescers is to cope with the variation in washing conditions—notably wash temperatures, wash time (rate of adsorption of the fluorescer), and the comparatively high incidence of re-use of wash solution for two or more loads. Also, different fabrics carry different charges. Nylon carries a positive charge and cotton a negative one; thus oppositely charged fluorescers are needed.

Foam

The relationship between foaming power and detergency has always been of interest, and foaming power has become associated in many consumers' minds with high detergent power. The first liquid detergent on the Australian market was Trix. It was non-foaming and was soon replaced because of consumer resistance. However, it is generally conceded by detergent technologists that foam has no direct relationship to detergency in ordinary fabric-washing systems.

But in systems where the amount of washing fluid is low, foam may play an important role. The individual foam films tend to take up and hold particles of soil that have been removed from the item, preventing them from being redeposited and allowing them to be washed or scraped away. This effect is very important in the on-location shampooing of carpets, and to a certain extent in the cosmetic shampooing of hair. Front-loading washing machines work by bashing clothes against the side of the tub—the high-tech version of beating clothes on rocks. Front-loaders clean clothes better than top loaders, but only if a low-suds detergent is used, because the suds cushion the impact and reduce the cleaning action. The suds can also cause electrical short circuits in time switches, etc.

In some detergent formulations a small amount of soap is included to serve a number of functions. Depending on the hardness of the wash water and the balance of the formulation, any soap added will act either as a water softener or as a surface active agent. However, the primary aim for including soap is to bring about a rapid collapse of foam during rinsing after the wash.

Because washing conditions and habits vary around the world, detergent formulations vary greatly also (Table 2.2).

Sodium perborate bleach

In water, *sodium perborate* (5–15%) releases hydrogen peroxide, which is a powerful *oxidising* agent. The oxidation removes a lot of the stains while generally not affecting fast colours. It is particularly effective when the material is left to soak, but requires a fairly high temperature to be effective during a wash.

Unlike chlorine, this bleach has virtually no adverse effect on textile fibres or on most dyes.

The most striking change currently occurring worldwide that is affecting the soap and detergent industry is a move towards lower washing temperatures. In Europe a 'hot' wash is 90–95°C, whereas in the USA and Australia a 'hot' wash may be a much lower temperature (50–60°C). In Japan the usual wash temperature is 30°C or below, and in Mexico, historically, cold water is used for washing.

As a bleach, perborate fits in nicely with European washing habits. Cotton and linens can be washed at the boil at 95°C, at which temperature perborate releases 90% of its oxygen and bleaches heavy soil and wine stains. At 55°C, perborate is only about 60% effective, but as clothes are worn for much shorter times between washing in Australia, separate bleaches are sold to be used only when an actual need for bleaching is perceived. These bleaches are often chlorine based, but the rapid rise in synthetic fibres, which are adversely affected by chlorine, has boosted the use of perborate.

The Europeans were hit hard by the energy crisis of 1973 and began dropping their wash temperature and wearing their shirts for less than a week. European manufacturers stayed with perborate but developed bleach activators such as pentaacetyl glucose and tetraacetylethylenediamine (TAED), which decompose to the unstable peracetic acid, which acts as a bleach at low temperature. They have been slow to accept sodium percarbonate. An interesting speciality chemical is magnesium monoperoxyphthalate hexahydrate, which releases the peroxyacid anion in solution (used, for example, in Fewa from Henkel).

Another problem at the lower temperatures is that an enzyme found in many biological strains, called catalase, decomposes hydrogen peroxide rapidly (at a rate of a million molecules a minute), and so can destroy the perborate bleach at room temperature in a few minutes (the enzyme is deactivated at higher temperature). Conversely, the high concentration of a related enzyme in blood stains is used in a presumptive forensic test for dried blood.

WARNING
Perborate and chlorine should not be used together.

IS IT LIKELY TO BE BLOOD?

Presumptive blood tests are generally catalytic tests that involve the use of hydrogen peroxide and an indicator that changes colour (or luminesces) when oxidised. The release of oxygen by the peroxidase activity of the haemoglobin causes the indicator to change. Common indicators used include o-tolidine (see 'Swimming pools' in Chapter 5), and luminol. Another indicator used, phenolphthalin (made from phenolphthalein), also reacts with other oxidisers. Unfortunately other enzymes (such as vegetable peroxidases) also give positive results.

Fig. 2.8 A royal statue attacked by detergent
(Courtesy of New Scientist Publications)

Washing in machines

The type of washing machine used in Australia tends to control the standard of wash, since the percentage of separate dwellings in Australia that have washing machines has risen from 86% in 1976 to virtually 100% in 1984. Automation has brought with it daily washing (Monday wash-day is a myth!), small loads, little soaking, little sorting of fabric types, and less care about the end result. Whereas the English and Europeans wash most loads for at least an hour in water up to 80°C (today closer to 60°C) and use plenty of detergent, Australians wash a load for an average of 10 minutes in water around 40°C and use one-third of the detergent Europeans do. (Median dose: Australia (Sydney), 1.33 g/L; USA, 1.5–2.0; UK, 2.6–3.5; France 3.5–4.0; Germany, 6.0.) However, Australians don't wear clothes as long before washing them, and so we wash much more often. It all seems to even out.

A typical Australian synthetic domestic laundry powder would have the average composition shown in Table 2.3.

Liquid laundry detergents

First marketed in the late 1970s, domestic liquid laundry detergents have been the growth product in the mid-1980s. Starting from virtually zero, by 1986 they had reached 25% of the Australian domestic laundry detergent market, while in the USA the level exceeds 30%. In contrast, usage in Europe is still low but rising rapidly, 5.5% in the UK and 9.5% in France. Local brands include Dynamo, Liquid Surf and Liquid Fab.

You may wonder about the term *surfactant system*. In most liquid detergents, and increasingly in the powders, more than one surfactant is used. There are a number of reasons for this, the most important of which is the property of

Table 2.2
Detergent formulations (percentage of ingredient in total formulation)

Ingredient	Heavy-duty powders				Heavy-duty liquids	Light-duty powders (primarily Europe)	Light-duty liquids	Detergent laundry bars
	USA Canada Australia	South America Middle East Africa	Europe	Japan				
Active[a]	8–20	17–32	8–18	19–23	15–50[b]	14–25	15–37	18–30
Foam boosters[c]	0–2	—	0–3	—	0–4	—	0–5	—
Foam depressants[d]	—	—	0.3–5	1–2	—	—	—	—
Builders[e]								
Sodium tripolyphosphate	25–35	20–30	20–35	0–15	10–30	10–40	—	—
Mixed or nonphosphate	15–30	25–30	20–45	0–20	15–20	—	—	25–45
Sodium carbonate	0–50	0–60	—	—	—	—	—	15–30
Antiredeposition agents[f]	0.1–0.9	0.2–1	0.4–1.5	—	0–2	0–0.9	—	0.3–2.0
Anticorrosion agents[g]	5–10	5–12	5–9	15	2–5	0–8	—	3–8
Optical brighteners	0.1–0.75	0.08–0.5	0.1–0.75	—	0.1–0.5	0–0.3	—	0.05–0.3
Bleach[h]	—[i]	—	15–30	—	—	—	—	—
Enzymes[j]	—[i]	—	0–0.75	—	0–1	—	—	—
Moisture	6–20	6–13	4–20	—	Balance	6–14	Balance	3–14
Fillers[k]	20–45	10–35	5–45	—	—	15–45	—	see sodium carbonate
Other ingredients[l]	—	—	—	—	—	—	6–9[m]	15–20[n]

[a] Mostly alkylsulfonates (linear and branched), fatty alcohol ethoxylates and fatty alcohol sulfates. [b] Includes up to 10% soap in European formulations. [c] Ethanolamides, such as coconut monoethanolamide. [d] In small levels, silicones; at higher levels, soap. [e] Some formulations use only sodium tripolyphosphate; others have mixtures of tripoly with other phosphates (for example, sodium orthophosphate), aluminosilicate zeolites, the sodium salt of nitrilotriacetic acid, sodium citrate and sodium carbonate; still others use the other builders alone or in various combinations. [f] Sodium carboxymethylcellulose, other cellulose-based polymers or synthetic polymers. [g] Sodium silicates (as purchased solutions). [h] In USA, sodium perborate, when used. In Europe, formulations for high-temperature have sodium perborate, whereas those for use at lower temperatures include boosters such as tetraacetylethylenediamine. [i] A few formulations include 20–25% bleach and up to 0.75% enzymes. [j] As purchased granules. [k] Predominantly sodium sulfate. [l] Usually also include small amounts of colouring agents, opacifiers if desired, fragrance etc. [m] Primarily viscosity builders. [n] Predominantly sodium bicarbonate, but can also include other materials such as kaolin and talc.

Source: C & EN, 23 January 1984.

TABLE 2.3
Composition of a typical Australian laundry powder

Ingredient	Percentage of total formulation 1977	1986
Surfactants	10–20	10–20
Sodium tripolyphosphate	20–30	15–25
Sodium silicate	7–10	7–10
Sodium carbonate	—	5–15
Sodium carboxymethylcellulose	0.3–1	0.3–1
Fluorescers	0.3–1	0.3–1
Perfume	0.1–0.2	0.1–0.2
Water	8–12	8–12
Sodium sulfate	to 100	to 100

potentiation. By this is meant a mutual reinforcement of the cleaning actions of two surfactants used together. Figure 2.9 illustrates this. The horizontal axis represents the compositions of mixtures of two surfactants A and B, grading from 100% A (no B) on the left to no A (100% B) on the right. If we measure the detergency or cleaning power of these blends by any convenient method, the intermediate detergencies frequently do not lie on the expected straight line. On the contrary, there is a peak that clearly shows that there is an optimum composition of two surfactants.

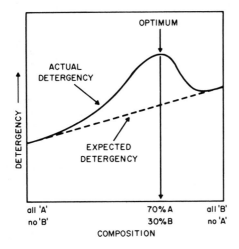

Fig. 2.9 Potentiation

In liquid laundry detergents, the second surfactant is frequently a non-ionic one, and it may be either a *coconut diethanolamide* (an alkylolamide) or a *synthetic fatty alcohol ethoxylate.* Both of these are rather waxy products, substantially all active material. The alkylolamide is prepared by making the fatty acids obtained from coconut oil react with an ethylene oxide derivative called monoethanolamine. A condensation reaction takes place with the elimination of water and a waxy product is formed (Fig. 2.10).

On the other hand, the formation of a fatty alcohol ethoxylate illustrates the method of manufacture of an ever-increasing class of surfactants, the ethylene

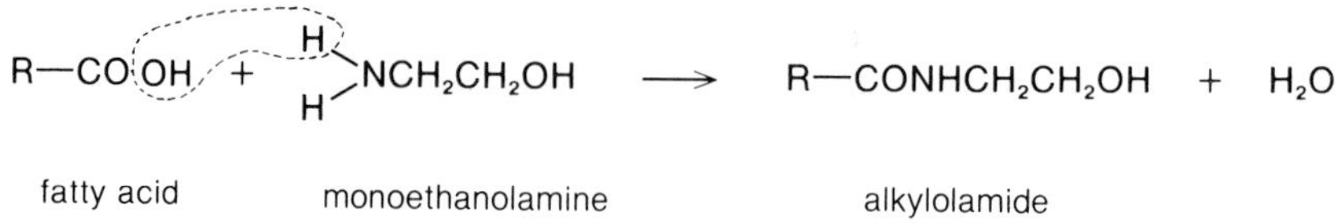

$$R{-}CO\,OH \; + \quad \overset{H}{\underset{H}{\diagdown}}NCH_2CH_2OH \quad \longrightarrow \quad R{-}CONHCH_2CH_2OH \; + \; H_2O$$

fatty acid   monoethanolamine   alkylolamide

Fig. 2.10   Condensation reaction

oxide condensates. They are now made in very large quantities all over the world (including Australia). Ethylene oxide is a toxic gas, it is flammable and it forms explosive mixtures with air in any proportion from 3% upwards. Whereas in the sulfonation reaction the chemical nature of the product is relatively definite (as is also the case for the coconut monoethanolamide), this is not so for the ethylene oxide condensation products, because we are now dealing with a type of *polymerisation* (see 'Chemistry of plastics—Condensation polymerisation' in Chapter 6). It is usual for these condensation reactions to take place by starting with a fatty alcohol, which is also a waxy material and chemically is the long fatty chain forming the hydrophobic part of our surfactant. It has a terminal hydroxyl group to which a molecule of ethylene oxide can be added thus:

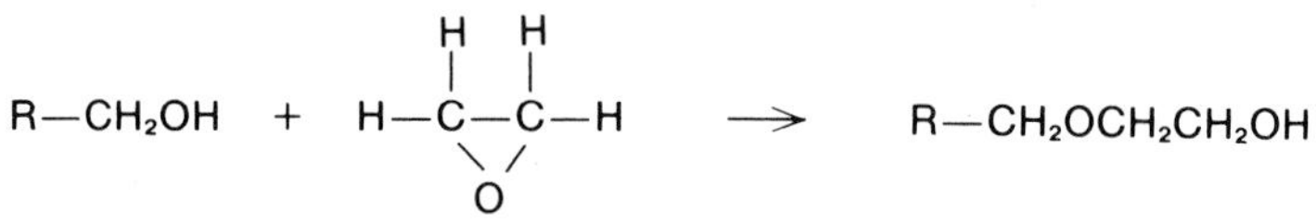

$$R{-}CH_2OH \; + \; H{-}\overset{\overset{H}{|}}{C}{-}\overset{\overset{H}{|}}{\underset{\diagdown\;O\;\diagup}{C}}{-}H \quad \longrightarrow \quad R{-}CH_2OCH_2CH_2OH$$

Further molecules of ethylene oxide can then add on to the terminal hydroxyl group which is formed at each step of the addition:

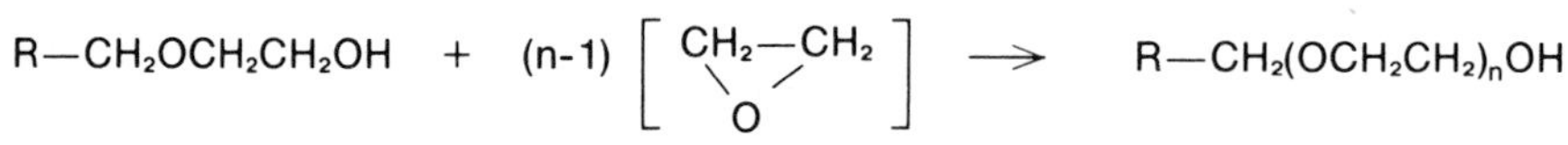

$$R{-}CH_2OCH_2CH_2OH \; + \; (n{-}1)\left[\underset{O}{CH_2{-}CH_2} \right] \quad \longrightarrow \quad R{-}CH_2(OCH_2CH_2)_nOH$$

If the ethylene oxide and tallow fatty alcohol are reacted together in the proportion of ten molecules of ethylene oxide to one molecule of tallow fatty alcohol, not all of the ethylene oxide molecules will combine with the tallow alcohol molecules in the proportion of ten to one. You will in fact get some molecules with fewer ethylene oxide groups and some molecules with considerably more. The graph in Figure 2.11 illustrates the kind of distribution curves that are obtained when different ethylene oxide condensates are manufactured.

In May 1984, at the World Congress on Surfactants, Bosch-Siemens unveiled the washing machine of the future. A microprocessor controlled not only the optimum time and temperature cycles for the load entered, but also made up the correct detergent from components stored in solenoid controlled containers in the side of the machine—surfactant blends, builders, bleach, bleach activators and softeners etc—all as liquids. A top-up every three to six months or so would be required!

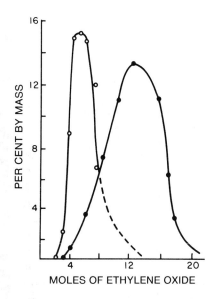

Fig. 2.11 The distribution of the number of ethylene oxide molecules in the surfactant molecule for two different mean values of ethylene oxide reacted

Biodegradability

A sight such as that shown in Figure 2.12 was once relatively common in the 1960s under normal weather conditions. That was in the days before detergents were what is called *biodegradable*. 'Biodegradability' is a term that requires careful definition. In essence, it means the process of decomposition of an organic material by naturally occurring micro-organisms—and note that it applies only to organic materials. Such a process is one that obviously depends on time, concentration and temperature. The surfactants that were used previously were biodegradable but only very slowly. They did not degrade quickly enough for the surfactants to be destroyed in the conventional sewage treatment plant, or to be decomposed reasonably rapidly in flowing rivers.

Enzymes

The use of enzymes for washing has a long history, starting with a patent in 1913 for a trypsin enzyme marketed as Burnus. This product was an impure proteolytic (protein digesting) enzyme of low activity. There was some sporadic activity in Switzerland around 1935, but it was not until 1945 that Bio 38 was launched. However it had limited commercial success. In 1958, Novo, an enzyme soak stage, was introduced in Denmark. The first commercial success occurred in 1963 in Holland, where Biotex captured 20% of the detergent market. In the mid-1960s enzymes were introduced in Britain and later in Australia.

Fig. 2.12   Detergent foam in the Molonglo River below the Scrivener Dam, which holds back Lake Burley Griffin in Canberra, during one of the overflows of the (upstream) Queanbeyan sewerage system (Courtesy of *Canberra Times*, 25 June 1975)

Enzymes are proteins that act as biological catalysts and, as such, can alter the rate of particular chemical reactions but remain unchanged once the reaction is over. Of the 2000 enzymes classified, about 150 have commercial uses. They operate in tepid water in conditions of mild acidity or alkalinity (pH 4 to 8.5). However enzymes are easily inactivated by a range of chemicals or adverse conditions.

Although enzymes can digest proteins in stains, they can also cause severe allergic reactions in people, just like the proteins in bee stings. However, generally the allergic response does not occur until after a sensitisation period (see 'Nutrition—Allergies' in Chapter 11). Asthma is now recognised as an industrial disease that can be caused by exposure to the enzymes used in detergents. Some consumers suffer from allergies when they wear clothes washed in enzyme detergents. Enzymes were withdrawn from detergents in the mid-1970s by Lever Bros, but other manufacturers continued to use them in modified form, encapsulated in wax, which melts and releases them in the wash. The manufacturing process has been greatly improved to protect the workforce and sensitive (atopic) people are screened out of such work.

The most common enzyme is alkaline protease, which digests protein in alkaline conditions. It is produced from the bacterium *Bacillus licheniformis* or *B. subtilis* in large fermenters. By working at pH 7, the enzyme produced is inactivated and prevented from digesting the bacteria producing it! At pH 9–10, it becomes active again and will work in the presence of polyphosphate and perborate. In Novo's process the enzyme is granulated with salt as a preservative and bound with carboxymethyl cellulose and coated with wax.

Amylases are enzymes used in detergents to degrade starches to water-soluble sugars (see also 'Alcoholic products—Beer: more than barley and hops' in Chapter 11). Cellulases remove cellulose microfibrils, which are released in cotton after repeated washing and cause stiffness and greying. Lipases are enzymes that hydrolyse fats and are being developed for dealing with fatty soil in clothes. There is an interesting biochemical reason why enzymes are not used in liquid formulations with builders. The builders are basically metal-complexing agents and tie up, *inter alia*, magnesium which is an essential cofactor for the operation of enzymes. In powder formulations, the time that the two components come into contact in solution is apparently too short to cause deactivation.

In Britain, Lever Bros quietly reintroduced enzymes into Persil Automatic. The reintroduction of enzymes has been generally done with little publicity and with the manufacturer using words like 'protease' or 'amylase', which appear to have no connection with the earlier products. In the USA, enzyme activated products are used in about 50% of liquid laundry detergents, 25% of powders and essentially all powdered bleach laundry additives; in Europe they comprise over 75% of the premium-quality heavy-duty products including automatic dishwashing powders. In 1978, in Japan, all major brands were reformulated with enzymes.

There were enzyme-containing powders on the Australian market from the late 1960s until 1972. Apart from a short period in 1987 when an enzyme-containing product from a small company was on the market, enzyme products are not in use in Australia. Technical advances in the industrial use of enzymes have meant that there are now few potential occupational health problems yet to be resolved.

Multicomponent detergents

These are nonwoven rayon sheets on which the different components are each deposited in separate small piles. A second sheet is then laid on top and laminated to the first to isolate the cleaning components. The consumer tosses one or more sheets into the washing machine. The detergent and organic peracid bleach dissolve during washing, but the waxy spots in which the softeners are embedded remain. The consumer then transfers the whole load to the dryer without removing the sheet(s). Heat from the dryer sublimes the fabric softener.

The lower performance of these products is balanced by their convenience, particularly for lightly soiled loads.

Should we re-use our water?

We are very fortunate in Australia that we have not had to face the problem of water re-use. After all, most of the population lives in a relatively narrow coastal strip and its effluent, particularly that from the main cities, goes into the sea more or less untreated. On the other hand, the regular droughts that have caused water supply problems in Canberra, Sydney and Melbourne over the years clearly show that a case could be made for the re-use of at least some of our water. In Australia we do not have long rivers along which cities are located, taking their water from, and returning it to, the river. It has been said, only half jokingly, that, if you live in a city near the mouth of the Mississippi River, the water you drink has probably passed through eleven stomachs before yours.

The fact that sewage treatment was incomplete, as shown by the appearance of foam in what was regarded as 'treated' water, caused chemists to look at what might be done to improve the biodegradability of surfactants. In the case of alkylbenzene, this was achieved by changing the structure of the alkyl part of the alkylbenzene. In Australia, before 1971, the alkyl part came from a propylene tetramer and was what the organic chemists called 'branched in its chain'. It was shown that this branching inhibited attack by the micro-organisms and, as a result, slowed down biodegradation. The remedies for this problem are again an interesting illustration of how chemistry works.

With the advent of plastics, propylene was suddenly useful again for making polypropylene. When the long-chain, branched hydrocarbons became a desirable component in automobile petrol, because they contributed a high-octane (anti-knock) rating, it left the straight-chain hydrocarbons available for making detergents. They are converted into the soft (biodegradable), linear alkylbenzene sulfonates (LABS). However, instead of simply sulfonating the straight-chain hydrocarbons directly, they can first be converted to the alcohol, and then sulfated (Fig. 2.13).

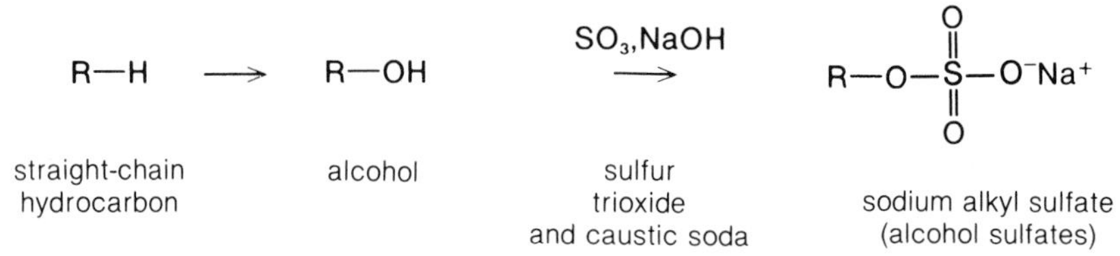

$$R-H \longrightarrow R-OH \xrightarrow{SO_3, NaOH} R-O-\overset{\overset{O}{\|}}{\underset{\underset{O}{\|}}{S}}-O^-Na^+$$

straight-chain hydrocarbon   alcohol   sulfur trioxide and caustic soda   sodium alkyl sulfate (alcohol sulfates)

Fig. 2.13 Production of sodium alkyl sulfate

Notice the slight difference between the sulfonate, in which the alkyl group, R, is attached directly to the sulfur atom (see Fig. 2.3), and the sulfate (Fig. 2.13), which has an oxygen atom between the sulfur atom and the alkyl group. The sulfate is readily *hydrolysed* under acid conditions; that is to say it is attacked by water to give the original alcohol, which has no detergent action. Toothpastes contain alkyl sulfate detergents because the abrasive contains minerals that would react with the soap and make it unusable.

Figure 2.14 illustrates the differences in the biodegradability of the different types of detergent. As you can see, soap degrades completely, and there are surfactants available which will essentially do the same.

OTHER HOUSEHOLD CLEANING AGENTS

Manual dishwashing detergents

Other detergents used in the home are liquid ones and are mainly used for dishwashing. Much the same combination of ingredients is used in these liquid detergents, except that inorganic builders are usually omitted. In order to provide

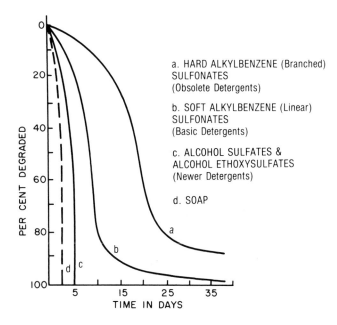

a. HARD ALKYLBENZENE (Branched)
SULFONATES
(Obsolete Detergents)

b. SOFT ALKYLBENZENE (Linear)
SULFONATES
(Basic Detergents)

c. ALCOHOL SULFATES &
ALCOHOL ETHOXYSULFATES
(Newer Detergents)

d. SOAP

Fig. 2.14 Biodegradation using Standards Australia (SA) standard for biodegradability, AS1792 (Courtesy of Surfactants Association of Victoria)

greater solubility, the alkylbenzene sulfonic acid is neutralised, not only with caustic soda, but also with, for example, triethanolamine, $(HOCH_2CH_2)_3N$. The salts formed are soluble in both water and hydrocarbons and so are good emulsifying agents. (The triethanolamine is produced from ethylene oxide and ammonia.)

The 1970s saw a market rush to produce the cheapest and lowest quality manual dishwashing detergents (Experiment 13.1 in Chapter 13 shows you how to measure the water content of products such as detergents). Consumers lugged home from the supermarket multilitre bottles containing 90% to 95% H_2O. At the same time some manufacturers introduced concentrates to which consumers added their own water at home. The quality of some of these also quickly deteriorated without any indication of change of composition on the label. Salt was added as a thickening agent, to give the impression of a more concentrated product, while urea and alcohol were added to aid the solubility of magnesium sulfate. (The magnesium ions improve the detergency of LABS in soft water.) The third edition of this book gives some results obtained at that time.

Consumer organisations agitated for standards, and an Australian Standard AS1999–1977 was finally agreed to. I chaired that committee and I remember the aggravation involved in the consensus process required for setting the levels for active ingredients. This level for regular manual dishwashing detergents was eventually pushed down to 6%. (Consensus = If at first you don't agree with me— try, try again!) Not one manufacturer adopted the standard, but in this instance the market enforced what consumers and standards never achieved.

A new player entered the kitchen—Morning Fresh by Cussons. It weighed in at about 36% active ingredients double the strength of the market leaders, and soon grabbed 25% of the market. Effective advertising and an effective fragrance were the keys to its success. The fragrance appeared to cover over the 'greasy' washing-up odours. Imitation is the sincerest form of flattery and we soon saw the 'me-too' products of Double-Strength Sunlight and Double-Strength Palmolive Green. Cussons upped the ante with Morning Fresh currently containing 42% active ingredients. But do the experiments, because it may all crash down again!

Machine dishwashing detergents

Generally, machine dishwashing detergents contain only about 2% of low-foaming, non-ionic surfactant (usually the block co-polymers or propylene oxide and ethylene oxide). Their efficiency depends more on their physical character-istics. A typical formulation of dishwashing detergent is shown in Table 2.4.

TABLE 2.4
Typical formulation of a machine dishwashing detergent

Ingredient	Percentage of total formulation
Anhydrous sodium tripolyphosphate	30
Anhydrous sodium metasilicate	30
Anhydrous sodium carbonate	37.5
Low-foam surfactant	0.5
Sodium dichloroisocyanurate (56–64% available chlorine)	2
Corrosion inhibitors	9.5

Sodium metasilicate, in particular, is quite caustic and very dangerous if swallowed. Some commercial powders contain 65% hydrated metasilicate, and consumer agitation (along with comment in earlier editions of this book) has ensured that these products are now adequately labelled.

Preparations with greater than 5% sodium carbonate can cause sheet erosion of glassware, which gradually thins the glass. Chlorine-containing powders can cause deterioration of plastic kitchenware, but without chlorine the tannin stains from tea are not removed. Aluminium-ware is also attacked quite strongly; therefore your saucepans are probably losing weight at an appreciable rate. Corrosion inhibitors consisting of aluminium salts are added to protect aluminium com-ponents—a case of shifting the equilibrium back again!

Liquid formulations have been introduced for machine dishwashing as well. Examples are Liquid Finish and Palmolive Automatic. In order to keep the liquid from seeping from a standard powder dispenser, it has an additive such as bentonite clay, to give it thixotropic properties (see box 'Fluid flow' in Chapter 8). Liquids overcome some of the problems in powder formulation such as residues, resistance of certain soils, and difficult product solubility due to low-temperature dishwasher water (a lot of water heat is used in the first cycle to heat the machine).

The standard rinse agent consists of 60% low-foaming wetting agent (Teric 164 or Triton DX12) plus 20% propanol and 20% water.

Scouring powders

Scouring powders consist mainly of abrasive powder (~80%), which can be screened silica, felspar, calcite or limestone, with the size of most of the material 44 micrometres or smaller. The rest of the powder is sodium carbonate or similar alkaline salts, with about 2% surfactant and in some cases chlorine bleach. The blue dye which appears on wetting and which is for appearance only can be finely divided copper phthalocyanine, and this sometimes bleaches while in use. (See AS1962, 1976 Hard Surface (Scouring) Powder.) (General purpose liquid cleaners are simply a suspension of scouring powder.)

Drain cleaners

Drain cleaners are a mixture of caustic soda (NaOH) and aluminium filings, which react in and with water to provide heat to melt fat and to saponify the fat.

Bleach

Hypochlorite salts (e.g. sodium, potassium, calcium, magnesium) serve as disinfectants, bleaches and deodorisers. Specifically they are used to disinfect contaminated utensils, nappies, and water for drinking or swimming. Liquid household bleach is usually 5% sodium hypochlorite (NaOCl). Commercial solutions are made by adding gaseous chlorine to 12–15% sodium hydroxide solution (NaOH) until alkalinity is just neutralised, and the resulting solution is diluted to 5%. Even a little too much free chlorine results in an acidic solution that is unstable. Manufacturers with inadequate production controls use too little chlorine and market a product with a free alkalinity that may run as high as 1% NaOH, whereas a well-controlled product has much lower alkalinity. (This chemistry is discussed in more detail in the section on swimming pools in Chapter 5.) The thickened stabilised bleaches such as Domestos contain an amine oxide surfactant as the main thickening agent.

Dangers

Mixing household bleach with preparations containing ammonia leads to the formation of chloramines (NH_2Cl, $NHCl_2$), which form acrid fumes and cause respiratory distress but have no permanent consequences. If swallowed, household bleach causes immediate vomiting but with no serious consequences except those associated with vomiting in general. The treatment for the ingestion of bleach is to swallow immediately one or more of milk, egg white, starch paste or milk of magnesia. Avoid sodium bicarbonate because it releases carbon dioxide. *Do not* use acidic antidotes.

Mould removers

The domestic moulds found in kitchens and bathrooms in Australia belong to a small range of common fungal types. The moulds are normally found growing between the tiles. Some, especially *Phoma*, can cause allergies, but the main household problem is their unsightliness. A survey by the Australian Consumers' Association (ACA) found the common species around the country to be: Adelaide (*Phoma, Penicillium*); Brisbane (*Phoma, Rhizopus*); Canberra (*Phoma, Penicillium*); Darwin (*Phoma, Phialophora, Rhizopus*); Hobart (*Phoma, Phialophora*); Melbourne (*Phoma, Fusarium*); Perth (*Phoma*); Sydney (*Phoma, Fusarium*).

Mould removers are generally only thickened, stabilised bleach solutions in bottles with spray caps. The same safety instructions as for bleaches apply.

Cloudy ammonia

When ammonia (NH_3) was first made from coal tar, the solutions were very murky. Later the Haber process for 'fixing' the nitrogen of the air gave a very pure product, but by this time people were used to 'cloudy ammonia'. For this reason soap is added to keep pure, clear ammonia cloudy. Fresh household ammonia ranges in concentration up to 10% actual NH_3. Ammonia vapour is extremely irritating and the solution is very alkaline and acts as a caustic. A little ammonia in a dish of hot water placed in a cold oven overnight will saponify the fat to a certain extent and make oven cleaning easier.

If ammonia is swallowed, the general rules to follow are:

> Do not use emetics—*do not* induce vomiting.
> Drink large quantities of water or weak acids such as diluted vinegar or lemon juice.
> Drink egg white, milk or olive oil, which act as soothing agents.

Contaminated eyes or skin should be washed with copious amounts of tap water.

Stain removal

To a large extent, stain-removal procedures are based on solubility patterns (like dissolves like) or on chemical reactions. Stains caused by fatty substances such as chocolate, butter or grease can be removed by treatment with typical dry-cleaning solvents such as tetrachloroethylene, $CCl_2 = CCl_2$. Removal of iron stains (e.g. rust and some inks) involves a chemical reaction: treatment with oxalic acid (poisonous!) forms a soluble complex with the iron. Much stain removal is carried out by oxidation, with oxidising bleaches such as hydrogen peroxide, sodium perborate (which forms hydrogen peroxide in water), or laundry bleach (sodium hypochlorite). These bleaches work on mildews and blood, for example, but bleach should not be used on wool because it attacks the linkages that hold the wool together.

The working of some methods is not completely understood. For example, the use of copious amounts of salt on a red wine stain on a table cloth probably operates by *osmosis*; that is, the salt draws out the water from the fibres of the cloth and takes the red stain along with it. This method only works on a fresh stain before the red dye becomes firmly attached to the cloth.

Dry-cleaning

The term *dry-cleaning* was introduced generally to cover the cleaning of textiles with organic solvents rather than water. When cleaning in water, the water-soluble components of the soil are not taken into account, because they spontaneously dissolve in the water and so their removal is not a problem. In dry-cleaning the situation is quite similar with regard to oily and greasy dirt. The necessity for also removing water-soluble substances such as salt and sugar is one of the reasons why *water* is usually added to the dry-cleaning bath.

The organic solvents most frequently used in dry-cleaning can dissolve only extremely low quantities of water. Generally, chlorinated hydrocarbons such as chlorinated ethylene are used. However, the surfactants form 'inverse' micelles, where the polar groups are *inside* and the hydrocarbon tails *outside* in the solvent phase (compare Fig 2.5). The interior of these micelles can dissolve additional water (about 1.5 molecules of water per surfactant molecule). The removal of solid particles is promoted by surfactants in dry-cleaning as well as in normal washing. Sufficient water is solubilised to maintain a reasonably high relative humidity of water vapour above the solvent; this is a convenient measure of the 'activity' of water *in* the solvent. (An explanation of the Biblical phenomenon of Gideon's fleece in Judges 6 given by C.M. Giles is vaguely relevant to this discussion.[1]) In practical dry-cleaning, however, an upper limit to the relative humidity is determined by the fact that textile material shrinks, wrinkles or felts if the relative humidity is too high.

ORGANIC POLLUTION

Organic pollution is the most widespread type of river pollution. 'Organic' materials are those complex carbon-based compounds that are such important parts of living matter—proteins, carbohydrates and fats. But many other substances, including detergents and pesticides, are chemically very similar. Untreated sewage, the discharges from sewage works, and waste from paper factories and food factories all add organic matter to our rivers.

If the discharge of this organic waste is small compared to the amount of water in the river, it is broken down to simple inorganic substances by bacteria and fungi in the river water. It is a natural process and is known as self-purification. The main chemical elements in organic matter are carbon (C), hydrogen (H), some oxygen (O), nitrogen (N), sulfur (S) and phosphorus (P). The large organic molecules are broken down and the elements form carbon dioxide, water, nitrates, sulfates and phosphates, which are harmless in small amounts.

If you look at Figure 2.15 you will notice that oxygen is needed for self-purification. This oxygen must come from the oxygen dissolved in the river water. The dissolved oxygen removed in this way by the bacteria during self-purification is replaced by oxygen seeping through the surface of the water and, during daylight hours, by oxygen given off by the submerged green plants. However, if the pollution is heavy—if there is a lot of organic matter—oxygen will be removed from the water faster than it can be replaced, and self-purification stops. When

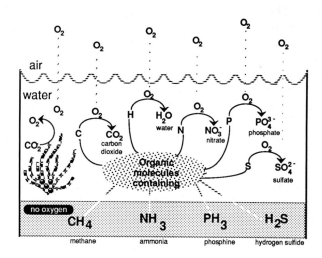

Fig. 2.15 Movement of molecules between air and water

there is no oxygen present at all, some types of bacteria can still break down organic matter, but marsh gas (methane) and offensive and poisonous substances such as ammonia and hydrogen sulfide (rotten egg gas) are formed instead of the mild products of self-purification. The smells and colour of the bottom of a pond or stagnant water come from this airless breakdown of organic matter. Fortunately, few rivers in industrialised countries today are as badly polluted with organic matter as this, except sometimes by accident—and accidents do happen.

EXPERIMENT 2.1

Clarification of river water

Equipment
Jam jars, potash alum, clay soil or river water.

Procedure
Muddy river water contains soil held as a colloid in the water. Colloidal particles are generally charged, either with positive or negative charges, and it is the repulsive forces between charges of the same sign that keep the colloid particles suspended in the solution.

If compounds that form ions are added to the colloidal solution, the positive ions will gather around any particles that are negatively charged (and negative ions around particles that are positively charged), thereby effectively screening the repulsive forces between the colloidal particles. The reduction in mutual repulsion allows the particles to clump together and thus settle out of the water (flocculation).

Many colloidal suspensions are found to carry negative charges; thus ionic compounds that yield multicharged positive ions are particularly effective flocculating agents. The most commonly used is potash alum, $KAl(SO_4)_2 \cdot 12H_2O$, which can be purchased at many pharmacies, photo shops or swimming pool supply shops. It is sometimes used to clarify city water supplies, but a CSIRO process, SIROFLOC, is now more frequently used.

Fill a jam jar with water and add a small quantity of clay soil to the jar. Shake the jar and then allow the heavy solid material to settle out by leaving it to stand for a few minutes. (You could use creek or river water instead of the clay suspension.)

Pour some of the muddy water you have prepared into two similar jam jars.

To one of the jars add a small quantity (about a teaspoonful) of potash alum and shake the jar.

Allow the two jars to stand. Check the clarity of the water in the jars from time to time. You should notice the colloid suspension in the jar to which you have added the alum begin to clarify as the colloidal particles flocculate and settle to the bottom of the jar.

Test other ionic compounds for their effectiveness as flocculating agents. Consider the relative rates at which given quantities of the compounds clarify the water.

Note

Do not try to drink the clarified water. Special care is taken at drinking-water treatment works to ensure the purity of the reticulated water supply. The clarified water you have prepared is *not* suitable for drinking.

DISCUSSION TOPICS

1. What would be the problems associated with a return to using only soap? [Competition with fat used for food—e.g. margarine.]

2. The production of soap required a lot of caustic soda, which is made by electrolysing salt. The other product of the electrolysis is chlorine. What is it used for? [PVC] Balancing the need for and production of caustic soda and chlorine is the basic problem of the chemical industry in any emerging chemical technology.

3. Are phosphates a real environmental problem in Australia?

4. How much more does it cost to wash with hot water rather than cold water? [Consider the water capacity of the washing machine; number of cycles requiring fresh water; cost per unit of electricity. For example, if you normally use 55 L of hot water in your average-sized machine, and cold water for rinsing, then using the fact that one unit of electricity will raise the temperature of 15.6 L of water by 55°C (i.e. from 15°C to 70°C) and that a unit costs say 7.5 cents (including heat losses), the cost per wash will be 27 cents.

5. What advantages does a good liquid laundry detergent have? [1. Convenience! 2. The variety of products could be greater because there are fewer formulating difficulties. 3. The use of detergent can be automated, by having a reservoir in the washing machine to automatically dispense liquid laundry detergent—and save on wastage resulting from lazy measuring of solids. 4. The type of surfactant used in liquid detergents is generally less affected by hard water.]

6. How much phosphate should a laundry detergent contain?

The water-softening reaction requires one molecule of tripolyphosphate for each molecule of hardness salt. From this equivalence it is possible to calculate the amount of tripolyphosphate required to compensate for each level of hardness. However, it is necessary to know the concentration of washing powder in the wash solution. A survey by Lever and Kitchen in Sydney revealed the wide distribution shown in Table 2.5.

TABLE 2.5
Distribution of the amount of detergent used by consumers in doing their washing

Product dosage used by consumers (g/L)	Consumers (%)
0–0.4	8
0.5–0.9	21
1.0–1.4	25
1.5–1.9	17
2.0–2.4	10
2.5–2.9	7
3.0–3.4	5
3.5–3.9	1
4.0 and above	6

The median concentration is 1.3 g/L. The median is the value with equal percentage of greater and lesser usage. Why is the median preferred to the mean as a measure of average? What is meant by the 'robustness' of a measure of average?

For a wash solution of hardness 80 mg/kg as calcium carbonate and a dosage of detergent at 1.3 g/L, what is the adequate percentage of tripolyphosphate (TPP)? [22%]. Remember, at this level, half the population will be underdosing. What percentage TPP would cover 75% of the Sydney population? [36%] Of course, at this level three-quarters of the population will be overdosing and polluting unnecessarily!

7. How much surfactant should a laundry detergent contain? [Well, the amount of surfactant in the final wash should (according to the manufacturers) be at the critical micelle concentration. For anionic surfactants this is relatively high, although it is lowered by all the other ingredients present ($\sim 5 \times 10^{-2}$ mole/L). A powder designed for an expected dose of 2 g of product per litre of water might contain 20% surfactant. The molecular mass of sodium dodecylbenzene sulfonate is 340, so this dose represents $\sim 10^{-3}$ mole/L. The amount of tripolyphosphate etc. will also be adjusted accordingly. 2 g/L is equivalent to 4 oz/12 gal (100 g/50 L); and 1 gal (British) = 1.2 gal (US) = 4.54 L.]

Note that detergent powders differ quite considerably in density so that measuring by volume, which the manufacturers specify in their instructions on the packet, can be misleading as a basis for comparison. With a cost for detergent powder of say $2.50 per kilogram or about 20 cents per wash, the cost of the detergent is almost as much as the cost of the hot water.

8. Have you ever had medical problems with a detergent or soap, such as asthma, dermatitis, photodermatitis (caused by sunlight on a skin sensitised by a chemical)? Why do you buy the brand that you do—advertisement on television; well-known company; washes better; cheaper; on special at the shop?

9. Have you ever wondered why packets of laundry detergent from the two major companies cost approximately the same in all the eastern capital cities? Both companies manufacture in Sydney and the freight costs are quite large. [Ask the companies.]

10. How would you go about dividing the cost of the packet of detergent up into:
 ingredients
 production costs
 packaging
 advertising and sales
 research
Would it be easy to bias these figures depending on the case you were trying to present?

11. Soap bubbles are blown on a T-piece with three taps so that one bubble is bigger than the other (Fig. 2.16). The input tap A is closed and the other two taps are opened so that air can pass from one bubble to the other. The question is—does the big bubble blow the little bubble up till they are equal (socialist) or does the big bubble get bigger and the little get littler (capitalist)? [You're wrong—they are capitalists! Why? See 'Chemistry of surfaces' in Chapter 4 for an answer.] A high stability soap bubble solution may be made from ammonium lauryl sulfate surfactant with glycerol (glycerine) and table salt (NaCl).

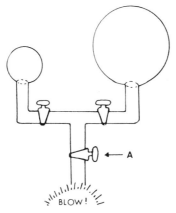

BLOW!

Fig. 2.16 Blowing soap bubbles with a T-piece

REFERENCE

1. C.M. Giles, *J. Chem. Ed.* **39**, 1962, 584.

BIBLIOGRAPHY

General

Falbe, J., *Surfactants in consumer products: theory, technology and application*, Springer Verlag, Heidelberg, 1987.

Gostelow, J., and Dean, P. *Experiments in detergency*. Shell International, London.

Grayson, M. (ed.). *Kirk-Othmer encyclopedia of chemical technology*, 3rd edn. Wiley, New York, 1983.

Isenberg, C. *The science of soap films and soap bubbles*. Tieto Ltd, Clevedon, UK, 1978.

Layman, P. L. 'Detergent report' *C & EN*, 23 January 1984, 17.

Stinson, S. C. 'Consumer preferences spur innovation in detergents'. *C & EN*, 26 January 1987, 21.

Strasser, P. H. A. 'Surface-activity in the service of women' (the Hartung Youth Lecture 1973). Published in *Proceedings of the Royal Australian Chemical Institute* **41**(April), 1974. To a large extent this lecture is the basis for this chapter. I am grateful to Peter Strasser for continuing to supply further information on this topic for each new edition.

Winton, J. M. *et al*, 'Soaps and detergents: a basketful of high-tech products' *Chemical Week*, 21 January 1987, 22.

Dishwashing detergents

'Automatic dishwashers'. *Choice*, Australian Consumers Association (ACA), August 1982; May 1987; October 1987.

'Dishwashing detergents'. *Choice*, ACA, October 1983.

Laundry detergents

'Fabric softeners for tumble dryers'. *Choice*, ACA, January 1985.

'Hand laundry detergents'. *Consumer Reports*, Consumers' Union (USA), February 1982.

'Formulation of laundry detergents'. Lever and Kitchen, August 1977. Submission to the Industries Assistance Commission 'Enquiry into Tariffs on Soaps and Detergents etc.'

'Laundry detergents'. *Choice*, ACA, June 1986; October 1987.

'Washing machines'. *Choice*, ACA, May, October 1982. See also *Consumer*, NZ Consumer Council, May 1981; *Which*, Consumers' Association (UK), May 1982; *Consumer Reports*, Consumers' Union (USA), October 1982.

Nappy treatments

'Nappy treatments'. *Choice*, ACA, September 1985.

Other cleansers

'All-purpose cleaners'. *Consumer Reports*, Consumers' Union (USA), October 1982.

'Mould removers'. *Choice*, ACA, June 1983.

Shampoos

'Shampoo'. *Choice*, ACA, January 1983. See also *Consumer*, NZ Consumer Council, June 1982.

Soap

'How do you choose a soap?' *Choice*, ACA, October 1986.

Chemistry in the kitchen

3

BUTTER, MARGARINE AND OTHER FATS, OILS AND WAXES

Fats, oils and some waxes are the naturally occurring *esters* of long, straight-chain carboxylic acids. These esters are the materials from which soaps are made (see 'Soaps' in Chapter 2). We have seen that an ester is produced when an organic acid combines with an alcohol (Fig. 1.12). Whenever the alcohol is *glycerol*—glycerine, a by-product in the manufacture of soap from fat—the esters are fats or oils (see Fig. 3.1). The difference between fats and oils is merely one of melting point: fats are solid or semi-solid at room temperature, whereas oils are liquids.

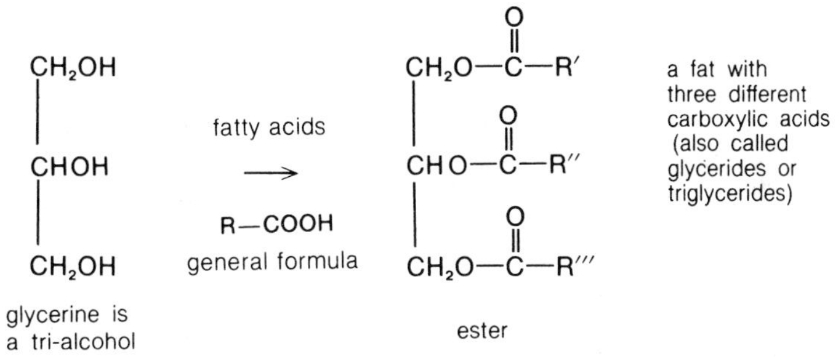

Fig. 3.1 Esterification of glycerine to form fats

Since glycerol is common to all fats, whether animal or vegetable, it is the fatty acid part of the fat that is of interest. The differences between fats depend on the nature of the acid groups—the length of the chain (which controls the molecular mass), and the number and position of double bonds (unsaturation). Before proceeding further I will briefly describe the acids. There are three groups important to this discussion: the saturated fatty acids; the straight-chain unsaturated fatty acids; and the polyunsaturated fatty acids.

Normal saturated fatty acids

Normal saturated fatty acids have the general formula $CH_3(CH_2)_n COOH$, where n is usually even and varies from 2 to 24. The building block for producing fatty acids is the acetate ion, CH_3COO^-; hence the predominance of even carbon chains. The most common examples of normal saturated fatty acids are palmitic acid ($n = 14$), and stearic acid ($n = 16$), which is illustrated in Figure 3.3. Others are lauric acid ($n = 10$), which is dominant in coconut and palm kernel, and myristic acid ($n = 12$), dominant in nutmeg.

Even shorter-chain fatty acids (n less than 10) form a large proportion of the fats in milk fats, especially those of ruminants. Odd-numbered acids do occur, but only in traces, and then over a wide range up to $n = 23$, generally in ruminants. The unusual composition occurs because it is bacteria in the rumen that carry out the reactions for the animal. Geographic variations in products such as milk can cause temporary diarrhoea until readjustments are made by the victim's body.

Normal straight-chain unsaturated fatty acids

The most important unsaturated acids have 18 carbon atoms, usually with one double bond at the middle of the chain. If other double bonds are present they lie closer to the carboxyl group, COOH. The double bond cannot be rotated and so there are two distinct geometries possible, which are called *cis* and *trans* (Fig. 3.2).

cis   trans

Fig. 3.2   *Cis* and *trans* isomers

Oleic acid, a compoent of olive oil and the most abundant of all fatty acids, is

$$CH_3(CH_2)_7CH\overset{cis}{=}CH(CH_2)_7COOH$$

(see also Fig. 3.3). The vast majority of olefinic (unsaturated) linkages in fats and oils are *cis*.

Polyunsaturated fatty acids

The polyunsaturated acids are those that have more than one *cis*-methylene interrupted double bond, and they have the general formula:

$$CH_3(CH_2)_x(CH=CHCH_2)_y-(CH_2)_zCOOH$$

where x and z range between 3 and 20, and y is usually from 1 to 4. Several polyunsaturated acids are illustrated in Figure 3.3.

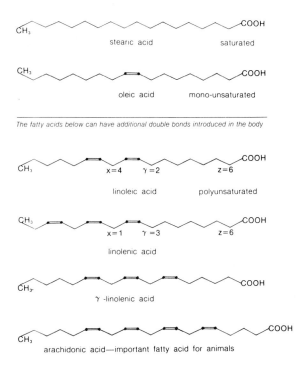

Fig. 3.3 The structure of some fatty acids

PROSTAGLANDINS

The arachidonic acid cascade exerts control on almost every part of your body. It may provide the contraceptives of the future.

Fatty acids used to be uninteresting components to all but a few dedicated specialists, who were generally considered to be way behind the frontiers of science. They were mentioned only briefly in chemistry courses: students were told that they were straight chains of carbon atoms ending in an acid group, varying only in length and the number of hydrogen atoms attached to the chain. Unsaturated fatty acids have less hydrogen atoms than saturated ones. Polyunsaturated fatty acids have less still. Arachidonic acid was just another polyunsaturated fatty acid, with 20 carbon atoms.

Times change. Housewives buy margarine containing a high ratio of poly-unsaturated to saturated fatty acids. Arachidonic acid itself has come of age with a vengence. It is recognised as the source in the body of a range of compounds that are critical for the health and illness of humans and animals—the arachidonic acid cascade. These compounds are formed from arachidonic acid by most tissues, sometimes in response to appropriate stimuli. They are present in minute quantities, about one part per million and often have lifetimes of only minutes in the body; yet their functions and effects are dramatic.

Prostaglandins, thromboxanes, prostacyclins and leukotrienes are all members of the arachidonic acid cascade. The prostaglandins have been known longest. The name is misleading, since they are not solely produced by the prostate gland. Their effects in human seminal plasma were observed as long ago as 1930, although they were not isolated and studied closely until after the Second World War. The first clinical use of a prostaglandin ($F_{2\alpha}$) was in 1968, to induce labour in childbirth. This application, together with its use in nonsurgical abortion and the control of the oestrus cycle, has been the most studied area of prostaglandin research. They also control aspects of respiration, nervous system activity, muscle function, blood pressure and immune response, among other things. Significantly, aspirin blocks the conversion of arachidonic acid to prostaglandins.

The potency of prostaglandins is remarkable. One kilogram of a synthetic prostaglandin (an analogue of $F_{2\alpha}$) is sufficient to synchronise heat in one million cows. Thus, if each cow was injected with one thousandth of a gram, and there was enough bull around, they would all calve at the same time.

In 1975 a new member of the arachidonic acid cascade was reported. This was thromboxane A_2, a crucial factor of unprecedented potency in causing aggregation of blood platelets and hence thrombosis in humans. It is extremely unstable and short-lived—half of it is decomposed in 30 seconds in water. Thromboxanes are largely undesirable, but it was only a year later in 1976 when the natural counter which controlled blood platelet aggregation was isolated and named prostacyclin. Prostacyclin is more stable, lasting up to 10 minutes in water. It again is very powerful stuff. Platelet aggregation in humans is completely but temporarily suppressed by infusing about one millionth of a gram per minute into the blood. The normal flow of blood depends largely upon a delicate balance between prostacyclins and thromboxanes.

Recently the cascade showed its next facet—leukotriene C, involved in the body's allergic responses and anaphylactic shock, was identified as a derivative of arachidonic acid. In an asthma attack its production is triggered by pollen or other allergens, and constriction of the air passages results.

These compounds open a new frontier in medicine, potentially as important as steroid hormones or penicillin. The frontier is expanding fast; some nine publications on them appeared in the scientific literature each day in 1977, including Sundays. They provide new insight into the basic chemistry that controls the human body. For pharmaceutical firms they also provide visions of gold at the ends of rainbows. However, they are difficult to use effectively in clinical practice because of their instability and rapid degradation by enzymes in the body. In some cases their range of effects is too wide, and their potency almost too high. Their medical future may lie not with the natural compounds themselves, although many of these can now be synthesised in the laboratory. Modified versions of the natural compounds, tailor-made to achieve greater stability and specificity of action, may have advantages. An attractive alternative approach is the design and synthesis of other compounds which selectively prevent production by the body of undesirable species, such as the thromboxanes.

Source: Professor R. W. Rickards, Research School of Chemistry, ANU, in *RSC News*, August 1979.

Fats and oils

Polyunsaturated fats are essential to all animals; they are the building blocks for the important prostaglandins, and they are also alleged to be effective in lowering the cholesterol content of the blood.

The double bond, especially when *cis* (see Fig. 3.2), means that the molecules do not pack together easily, which is seen in the low melting point of double bond containing material (i.e. oils). Substances made up of shorter chains also melt at lower temperatures.

Why do animals produce fats, mainly saturated, while plants produce oils, mainly unsaturated? Plants suffer extremes of temperature and require their fats or oils to be semi-fluid even at low overnight or winter temperatures, whereas some animals can maintain a high temperature through internal heating and insulation, and in the case of mammals can even regulate it. In mammals a higher melting compound is preferable because the fats also have a structural part to play and must not be too fluid. Kidney fat is solid so as to provide mechanical support for the organ, although fat circulating in the cells is fluid even at lower temperatures—otherwise you might get a solid casing in a cold shower.

FATS AND OILS IN ANIMALS AND PLANTS

The distribution of fats and oils varies in different plants and animals. Fats and oils appear to be a biological solution to the problem of storing, transporting and utilising those fatty acids that an organism requires for its metabolic processes. The ester bond is quite stable but is easily split, when required, by a specific *enzyme*. Lipids (a term used in biochemistry) includes fats and fat-like materials of biological origin. They are the major energy storage in animals (38 kJ/g compared with 17 for carbohydrates and 23 for protein). When fats are 'burnt' in the body to produce energy, they also produce water, and more than from burning sugar. The hump of the camel is, in fact, a fat-storage unit that provides the camel with both energy and water. Hence the snide camel-selling trick of pumping the hump with air!

Plants, including moulds, yeasts and bacteria, synthesise both fats and their component fatty acids. Animals can synthesise much, although not all, of their fatty acid requirements, but they prefer to ingest plant foods and modify them to their own needs. Only plants are known to synthesise linoleic and linolenic acids, but animals can increase the chain length and further increase unsaturation, giving, for example, acids characteristic of fish oils, which are particularly rich in unsaturated acids and have up to six double bonds.

Fats are traditionally classified by two chemical tests that measure two important characteristics of fats and oils and that lead to the first two index numbers.

1. *Saponification value.* The test for this index number involves the hydrolysis of the fats into their component fatty acids (as their anions or soaps) and glycerol. The saponification value is defined as the number of milligrams of potassium hydroxide required to saponify one gram of fat (or oil). It gives a measure of the average chain length—i.e. the molecular mass of the fatty acids. Coconut 'oil' has a saponification value of 250–260, whereas for butter it is 245–255. (Actually coconut oil is solid outside the tropics—i.e. a fat. It occurs in a tropical plant and can afford to have a higher melting point.) These high values are the result of a large percentage of short chains. Some other saponification values of interest in the kitchen are lard, 193–200, and peanut oil, 185–195; and outside the kitchen, linseed oil, 189–196.

2. *Iodine value.* This second index number is a measure of the number of double bonds present. Iodine reacts with the double bond:

$$\underset{R}{\overset{H}{>}}C=C\underset{R'}{\overset{H}{<}} \quad + \quad I_2 \quad \longrightarrow \quad \underset{R}{\overset{I}{\underset{|}{H-C}}}-\underset{I}{\overset{H}{\underset{|}{C-R'}}}$$

The number of grams of iodine that react with 100 g of fat or oil is known as the iodine value. Some iodine values are:

coconut oil	8–10	} low—predominantly saturated
butter	26–45	
lard	46–66	
peanut oil	83–98	} high because of polyunsaturated component
linseed oil	170–204	

3. *Acid value.* This is a measure of the extent to which glycerides in the fat or oil have been decomposed to free acid. The Australian Food Standards Code sets values for this parameter.

4. *Peroxide value.* This measures the extent to which oxygen has been taken up by the oil (to form peroxides) and is a measure of freshness of the oil. An accelerated (60°C) test of stability of oils to oxygen uptake is given below (5). The Model Food Legislation sets a limit of 10 millimoles per kilogram (mmol/kg) of oil for rapeseed, safflower, sunflower and soyabean oils. Rats have shown harmful effects from oil at 50 mmol/kg.

    The peroxide formation is a *positive* feature of polyunsaturated oils when used as drying oils in paints (see 'Oil-drying paint' in Chapter 8).

5. *Oxygen uptake.* If samples of oil are incubated at 60°C for an extended period and weighed at daily intervals, a weight gain is noted for polyunsaturated samples. The induction time is taken as the point at which sample weight begins to rise rapidly (0–25 days). Alternatively, the time taken to reach a 1% uptake of oxygen can be established. The index being measured here is the effectiveness of natural or added antioxidants in the oil (see 'Antioxidants', below).

Hydrogenation

In general, vegetable oils are a good source of polyunsaturates and animal fats are a poor source, although there are exceptions to this rule (e.g. coconut oil). However unsaturated fats can be saturated by adding hydrogen to the double bonds in the presence of a nickel catalyst. This hydrogenation converts a substance with the properties of a vegetable oil into one with the properties of an animal fat; it changes a liquid into a solid. Thus linoleic and oleic acids would turn into stearic acid (see Fig. 3.3). When margarines are said to be made *from* pure vegetable oils the emphasis could be very much on the 'from'.

$$\underset{R}{\overset{H}{\diagdown}}C=C\underset{R'}{\overset{H}{\diagup}} \quad + \quad H_2 \quad \xrightarrow[\text{catalyst}]{Ni} \quad \underset{R}{\overset{H}{\diagdown}}H-C-C\underset{R'}{\overset{H}{\diagup}}H$$

There has been increasing concern that during hydrogenation other reactions occur in which the natural *cis*-methylene interrupted double bond configuration of the polyunsaturated acid has been altered by a double bond shifting one place closer to the adjacent double bond. Such isomers are said to have a conjugated double bond system. The shifted double bond usually alters from *cis* to the more stable *trans* arrangement.

The presence of such fatty acid artefacts in processed vegetable oils has been the subject of some controversy. Although it is true that fat in products derived from ruminant animals (beef, lamb and dairy products) also contains conjugated and *trans* unsaturated fatty acids (arising from biohydrogenation of unsaturated fatty acids in the rumen), these are present only in small amounts. *Trans* fatty acids in processed oils such as shortenings may reach levels of over 50%. Tub margarines vary considerably in composition. Surveys in 1974 and 1978 showed values of nil to 18%, the lower values occurring in polyunsaturated products. These 'unnatural' fatty acids are reported to appear to have no deleterious effect, provided an adequate intake of essential fatty acids such as linoleic is maintained (FAO/WHO 1977).

Rancidity

The effect of the oxygen in the air on unsaturated fats is to cause oxidation, which is basically the addition of oxygen. This addition occurs adjacent to the double bond and further reactions that produce a complex mixture of volatile rancid-smelling products are then possible. (The chain is split at the double bond giving smaller and, hence, more easily evaporated compounds.) Although two-thirds of the fatty acids of butter are C_{16-18}, a significant proportion are shorter. The milk of sheep and goats has an even greater concentration of short-chain fatty acids. The cheeses are correspondingly stronger. Merely producing the fatty acids from butter fat gives these products, whereas with margarine the longer chains must first be 'broken' before the short-chain, rancid-smelling substances can be produced. Hence margarine rarely becomes rancid. Commercial fats and oils have *antioxidant* added to them (see p. 71).

Heating of oils and fats

There are three important points in the heating spectrum of an oil or a fat. The first is the *smoke point*, the temperature at which a fat breaks down into visible gaseous products and thin wisps of bluish smoke begin to rise from the surface.

THE CHEMICALS OUR BODIES EMIT

The same short-chain acids as those present in rancid butter are produced by bitches on heat to make them attractive to the opposite sex. They are also present in human perspiration, although the sex attractant properties in this case are still under dispute.

An intriguing study carried out by the US armed forces over the years has to do with 'close' situations in submarines and spacecraft. It involved collecting and identifying the chemicals given off by humans as vapours. The list is enormous and obviously will give air-pollution authorities a fascinating task. (Perhaps public meetings will be banned because safe levels of mercaptans have been exceeded!) Quantitative data were reported for five compounds (Table 3.1). The presence of isoprene as one of the most common effluents was unexpected.

TABLE 3.1
Rates of emission of five human effluents

Compound	Rate of emission (μg/hour)		
	Subject 1	Subject 2	Subject 3
ethanol	25	58	100
isoprene	425	251	270
acetone	360	240	470
butanol	16	26	41
toluene	0.6	14	13

Source: Journal of Chromatography **100**, 1974, 137.

The *flash point* (See 'Chemistry of the car—Safety aspects', Chapter 10) occurs when brief but sustained bursts of flame start to shoot up. Higher still is the *ignition temperature*, a temperature at which the entire surface of the frying medium becomes covered with a continuous sheet of flame.

The temperature of the smoke point does not stay constant: it tends to fall with the continued use of the oil or fat (because the oil or fat decomposes and the free fatty acids lower the smoke point). It drops about 20°C each time the fat is re-used for 30 minutes at 270°C. Thus the higher the initial smoke point, the longer the fat is usable before it starts to smoke. The US Consumers' Union and the Australian Consumers' Association (ACA) have both carried out tests on fats and oils. They agree that the smoke point of an oil or fat is an important piece of information for consumers and should be listed on the label. Table 3.2 details some results published in *Choice*, the ACA's journal.

Heating does not significantly change the P/S ratio of polyunsaturated oils, but it causes the formation of oxidised compounds, which tend to destroy the vitamin E content and can also cause the oils to have a tendency to polymerise, which makes them unpalatable (see 'Oil-drying paint' in Chapter 8). Big changes to the

TABLE 3.2
Smoke points of some oils

Oil type	Av. smoke point range in °C	P/S ratio[a]
Safflower	246–258	6.0–7.4
Sunflower	229–252	4.7–5.2
Maize	229–268	3.1–4.2
Peanut	246–251	1.9–3.5
Soyabean	256	3.7–3.9
Olive	204	0.5–0.7

[a]The P/S ratio is the ratio of polyunsaturated fatty acids in the fat or oil to the saturated fatty acids present. Mono-unsaturated acids are ignored. The range of P/S values can be affected by such factors as where the seeds are grown.

Source: Choice, April 1975.

peroxide value of oils after heating reveal how heating oxidises oils. Values up to 10 times the initial value have been reported for standard tests, which often bring them over the legal limit for unused oils. Olive oil (which is mainly mono-unsaturated oleic acid) appears to be the most stable cooking oil because it also contains a steroid stabiliser. This is probably why it has been used so long in Mediterranean countries. It needs no refining, preservatives or refrigeration.

Antioxidants

The two most common antioxidants added to consumer products during processing are butylated hydroxyanisole (BHA) and butylated hydroxytoluene (BHT). Their structural formulae are given in Figure 3.4.

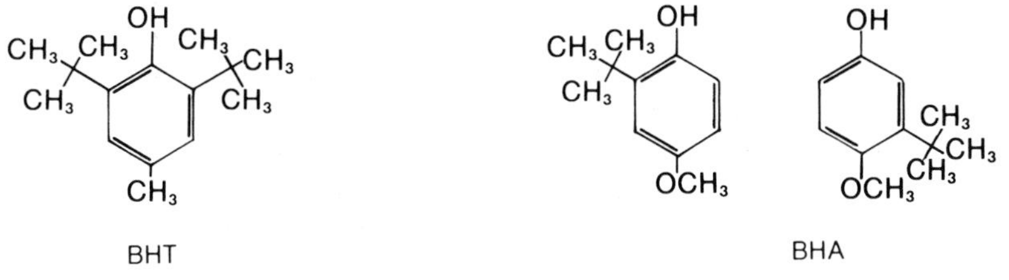

Fig. 3.4 The antioxidants BHT and BHA

Other antioxidants allowed are propyl gallate and mono-*tert*-butylhydroquinone (TBHQ). Their structural formulae are illustrated in Figure 3.5.

Butylated hydroxyanisole (BHA) is legally allowed in edible oil and fat products at levels of 0.01–0.02% (NH & MRC). It is also allowed in dried mashed potatoes, edible oils and fats, margarine and essential oils. *Butylated hydroxytoluene* (BHT) is used in petrol, lubricating oils and rubber, but not generally in foods, although there are exceptions. As it is used in polythene film, a legal standard allowing an

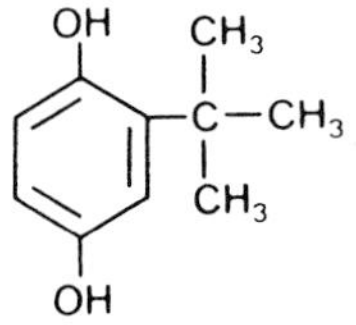

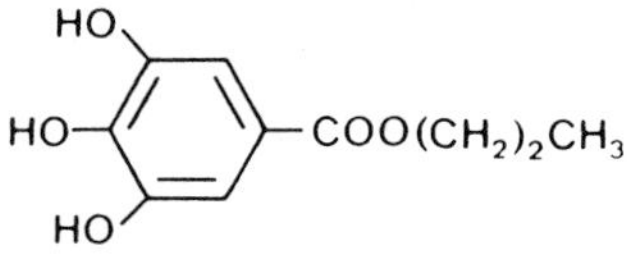

TBHQ   Propyl gallate

Fig. 3.5   The antioxidants TBHQ and propyl gallate

amount of up to 2 mg/kg to be absorbed by the food has been set (see also 'Physics of plastics—Plastics for packaging', Chapter 6; 'Packaging and other accidental additives', Chapter 11). Ironically, BHT appears to have antiviral properties and is now actually sold as a health food.

Antioxidants are preferentially oxidised; that is, they, and not the fat, are attacked by oxygen. They are oil-soluble, easy and cheap to produce, and are related to the 'natural' oil-soluble antioxidant α-tocopherol (vitamin E; see Fig. 3.6). Vitamin E occurs in vegetable oils—the most important source is wheatgerm oil—and protects them against oxidation. It has been found to be essential for the growth and reproduction of rats, but these results cannot be extrapolated to humans. It is the subject of controversy as an anti-ageing agent.

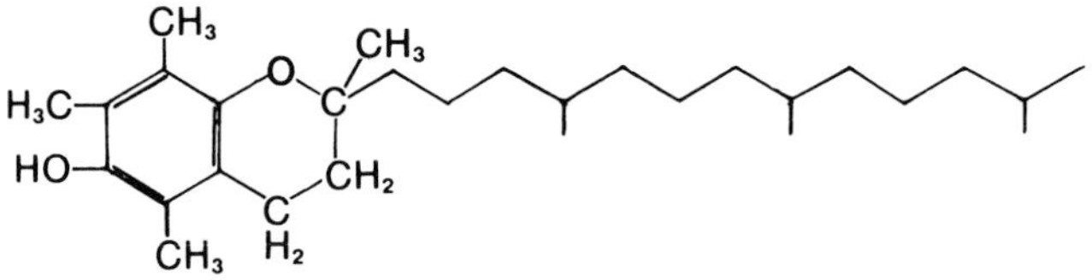

Fig. 3.6   Vitamin E

Humans have a need for vitamin E in an amount proportional to the amount of polyunsaturated fat in the diet: it is apparently required when the fats are laid down in the body, and presumably it is also an antioxidant. But note that excessive amounts of any oil-soluble vitamin can be dangerous.

Margarine

The origin of margarine dates from 1869 when Napoleon III proposed a competition with the aim of discovering 'for the use of the working class and, incidentally the Navy, a clean fat, cheap and with good keeping qualities, suitable to replace butter'. The prize was won by a chemist, Hippolyte Mège-Mouriés.

The legal definition of margarine recommended by the NH & MRC Food Standards is: margarine is a mixture of edible fats, oils and water prepared in the form of a solid or semi-solid emulsion (water in oil). It includes all substances made in imitation or semblance of butter and all preparations resembling butter, the fat contents of which are not derived exclusively from milk, and it should not contain more than 16% water nor more than 4% salt (i.e., in terms of energy, margarine is equivalent to butter). Table margarine must contain at least 8.5 mg of vitamin A and at least 55 μg of vitamin D per kilogram.

The use of the term *polyunsaturated* is permitted where

> the proportion of *cis*-methylene interrupted polyunsaturated fatty acids [see Fig. 3.7] present in the margarine is at least 49% [this requirement is to prevent the inclusion of synthetic materials]; the proportion of saturated fatty acids does not exceed 20% of the total fatty acids; and the ratio of polyunsaturated to saturated fatty acids is at least 2:1. The total cholesterol content expressed in mg/100 g must appear with equal prominence.

The other 40% of the fatty acids can be *mono*-unsaturated (e.g. oleic acid), which neither increases nor decreases the problem of cholesterol deposition. These recommendations have been partially adopted by some of the State governments.

cis-methylene interrupted

polyene is easier
to make synthetically

Fig. 3.7 The distinction between polyunsaturated and polyene

Manufacturers also produce a margarine with a polyunsaturation (P/S) ratio of 3:1. It is softer than the other margarines and requires constant refrigeration.

Antioxidants, flavouring (3-hydroxy-2-butanone and diacetyl, which give butter its characteristic flavour), and vegetable colouring (usually carotene, a source of vitamin A, which gives the colour to butter) may be added to table margarine (but not to cooking margarine in Victoria and Tasmania). In the USA, prior to 1950, margarine manufacturers could not colour their products unless they paid an additional 22 cents per kilogram Federal tax, the result of an effective lobby by the dairy industry. By changing from coconut oil to soyabean oil as a raw material, the makers of oleomargarine were able to enlist the aid of soyabean growers to combat the dairy lobby, and the law was repealed in 1950.

Quotas on the production of table margarine were imposed during the Second World War (1940) to protect the butter industry at a time when table margarines were highly saturated because they were made from imported coconut oil.

Cholesterol

Cholesterol is not technically a fat but a steroid related chemically to the bile acids, cortisone, the sex hormones and vitamin D—a motley collection. Cholesterol is a necessary substance that is found in all the cells of the body. It is produced in the liver and it may also be taken in directly from foods of animal origin, so that the level in the blood comes from two sources, both of which can vary.

About 93% of the body's cholesterol exists in cells, especially the cell membranes that encase them, where it is vital for structural support and certain biochemical reactions. The remaining 7% circulates in the blood, where problems can occur. Because cholesterol and fats are insoluble in water they are coated with a phospholipid-protein envelope called lipoprotein. This comes in a high-density

(HDL) and low-density (LDL) form. LDL carries cholesterol. Special receptors on the cell recognise LDL and let it in. If your genes code for too few LDL receptors, then you are likely to have high levels of cholesterol in your blood, with the consequent health risks.

In countries like Australia and the USA, where people eat large amounts of meat and dairy products—that is, their diets are high in cholesterol and saturated fats—the mortality rate from heart disease is high. In countries like Japan, where diets are traditionally low in cholesterol and rich in the polyunsaturated fats found in vegetable oils and fish, the death rate from coronary disease is lower. But Japanese who migrate to the USA soon follow the US pattern.

A major prospective study at the National Heart, Lung, and Blood Institute in the USA covering a 10 year period involved 3806 middle-aged men with no signs of heart disease but with raised cholesterol levels (av. 2.9, where the range is 1.0–3.0 g/L plasma.). All the men were first put on a diet that reduced their cholesterol level by 3.5%. Half of the men, selected at random, were given daily doses of a cholesterol-lowering drug, cholestyramine, and the other half received a placebo. Neither the doctors nor the subjects knew who was receiving which. (This is an example of a double-blind study.) Not surprisingly, the study showed that the drug regime caused a modest lowering of cholesterol levels, which was significantly more than in the controls. More importantly, the group of men on the drug had fewer symptoms of incipient heart disease and significantly fewer non-fatal heart attacks or coronary heart-disease deaths.

However, the study does not recommend the general use of the drug and recommends it be restricted to clinical use for severe cases. The study also emphasises the other factors, such as smoking and high blood-pressure. Medical authorities emphasise that people should not radically alter their diets without being advised by their doctor.

Vegetable oils contain the phytosterols instead of cholesterol. Isolation of ergosterol used to be employed as evidence proving the addition of vegetable oil to animal products.

HIAWATHA'S LIPID

by H.M. Sinclair

From his brief case Hiawatha
Took his paper for the meeting,
Typed in triple-spacing and in
Triplicate on foolscap paper;
Glanced upon the crowd before him,
Critical and very hostile,
Like the lions in the arena
Waiting for a Christian victim,
As the surgeons in some theater
Wait impatiently the patient;
Saw them with their notebooks
  waiting,
Saw the tape recorder ready
Gleaming in its chromium plating

By the Ampex Corporation
And preserving all the nonsense
Spoken by the previous speakers,
As the snow upon the prairies
Uselessly records the footprints,
So preserving all the nonsense
Spoken by the previous speakers.

Introduction

Hiawatha, taking courage
Started on the Introduction
Giving first a brief description
Of the Proto-Keysian period
When all fats in equal measure

Raised cholesterol in serum:[1]
Butter, sardines, walrus liver,
Margarine, or safflower seed oil,
Or arachidonic acid,
Or the body fat of quokkas,[2]
Or adrenals of the muskrat,
Or the milk of female reindeer—
As these fats in equal measure
Raise cholesterol in serum,
As the rain in Minneapolis
Fills the ditches in the roadways
(So at least thought Hiawatha)
So these fats in equal measure
Raised cholesterol in serum.

Then the Meso-Keysian period
When it's known from work of others
Quantitative variations
Do occur when different lipids
Are included in our diets,
Plentifully in our diets:
Butter, sardines, walrus liver,
Margarine, or safflower seed oil,
Or arachidonic acid,
Or the body fat of quokkas,
Or adrenals of the muskrat,
Or the milk of female reindeer—
Do not in an equal measure
Raise cholesterol in serum;
As the smoke is wind-swept upward
Randomly with Brownian movement
Wandering above the wigwam,
So these fats in different measure
Raise cholesterol in serum.

Then the Neo-Keysian period
When arithmetic will tell us
By an intricate equation[3]
What cholesterol in serum
We will have when we have eaten
Butter, sardines, walrus liver,
Margarine, or safflower seed oil,
Or arachidonic acid,
Or the body fat of quokkas,
Or the adrenals of the muskrat,
Or the milk of the female reindeer;
Count the double bonds and add by[4]
Electronic automation
On a digital computer
From the Ampex Corporation
There's no need to estimate it—
All cholesterol in serum
Follows now the Keys equation;
As the caribou in summer
Migrate by accustomed pathways
And predictably are herded,
Dietetic computation
Of the double bonds in lipids
With a slide-rule calculation
Gives you now a neat prognosis
Whether you will die tomorrow

From a thrombus in your vessels—
Myocardial infarction
Or ischaemic heart diseases—
As a cork pushed in a bottle
Stops the wine from flowing freely
(Vin rosé of California);
Atheromatosis also
Is predicted by this method,
By this skilful Keysian method.

Others are not quite so lucky:
Larry's lowered lipid levels
After vegetable seed oils—
Polyethenoic acids
Or essential fatty acids
From the vegetable seed oils—
Follow a more simple pattern,
So at least thinks Hiawatha
In unpublished observations:
As the sun comes up in the morning,
As the sun goes down in the evening,
So the laws of lipid levels
Are predictably determined—
Saturated fatty acids
Raise cholesterol in serum,
Polyethenoic acids
Lower serum lipid levels—
So at least thinks Hiawatha
In unpublished observations.

Methodology

After this review of others
Hiawatha turned to methods
(Methodology, he called it
Making it more scientific—
Longer words are scientific);
Talked about silicic acid,
Mead's silicic acid column,
How he trapped the different lipids
As he used to trap the beaver.
Then he pushed them back and
    forward
Countercurrent distribution—
As the frightened hare or reindeer
Runs at random back and forward,
When he boiled them up with potash
Alcoholic potash mixture,
Following the rules established—
Reimenschneider's 'skilful witchcraft'
So politely called by Mattson
(Personal communication)—
As the dinner in the stewpot
Is boiled up by Minnehaha,
So he boiled them up with potash
And the double bonds determined
Spectrophotometrically.
Then he used the latest method,
Gas-chromatographic method
Introduced by James and Martin
Showing peaks upon the paper[5]
Like the Rockies at the sunset
Like the mole hills in the prairies,

Thus he estimated lipids
As he wondered if it mattered,
Wondered secretly about it
With unpublished wond'rings.

Results

Thus supplied with diverse methods
Hiawatha took some serum
From his arm by venipuncture
And cholesterol determined;
Why he had no clear conception
But there's wild enthusiasm
For cholesterol in serum;
As the children round the camp fire
Dance and shout in exultation,
So there's wild enthusiasm
For cholesterol in serum:
Why it rises on infusion
Of suspended phospholipids
(Ethanolamine and choline
Joined to phosphatidic acid
With unsaturated acids—
Polyethenoic acids—
Also saturated acids,
From the glycerol projecting
Like the branches of a cactus),
Coming out from unknown tissues—
Red cells, liver, spleen and kidneys,
Atheromatous aortas,
Hepatectomized adrenals;
Why it falls when you have eaten
Polyethenoic acids
Or essential fatty acids.[6]

Thinking that this single value
For the level in his serum
Might not be sufficient data
To establish without question
What the normal value should be,
Hiawatha with his cunning
Took a logarithmic table,
Photographed a page at random
For a lantern slide of figures,
Showed it very confidentially
With his back toward the audience
Talking fast and very softly
At the figures thus projected
Which were very small and many
Like the sands upon the seashore;
And the audience, not hearing
What he spoke toward the blackboard
Very softly, very swiftly
Like the gentle brook in springtime,
Thought him wise and very clever
To have got so many figures
And their standard deviations,
Arithmetical progressions,
Geometrical regressions
And regression coefficients;
Praised his industry, his brilliance,
And applauded his statistics,
For they had not understood him
Nor could read his logarithms.[7]

Having thus established clearly
What the normal value should be,
Hiawatha took a patient
Who had grave thrombotic symptoms,
Used his methods on the serum
(Methodology he called it),
Found a curious lipid in it—
Ante-iso-*trans*-oleic;
Recognized it by the usual
Gas-chromatographic method,
By the bumps upon the paper
Like the Rockies at the sunset,
By the bumps upon the paper
Like the mole hills in the prairies,
By a very curious spicule
Like the tower of Hotel Raddison
Coming in a new position
Which unquestionably proved it
Ante-iso-*trans*-oleic;
Called it Hiawatha's UFA[8]
'Hiawathianic acid';
No one else had found this lipid
In the serum of a patient;
Called it Hiawatha's syndrome,
Hiawatha's lipidosis,[9]
But he did not know his patient
Had been bitten by a viper—
Viperus Russelianus—
And in Russell's viper venom
There is but one type of UFA—
Ante-iso-*trans*-oleic,
Hiawathianic acid.

Therapy

So he started quick to treat him;
Gave him safflower oil and corn oil,
Gave him pints and quarts of corn oil,
Gave it by infusion, also
Gave it by inunction, also
Poured it down, *per os*, his pharynx,
(As the beaver in the flood time
Being drowned in swirling waters
Soon becomes a bloated carcass).
Every orifice was needed
For administ'ring the doses
Of essential fatty acids;
But the patient still had in him
Ante-iso-*trans*-oleic,
Hiawathianic acid.
So he tried specific treatment;
Gave some linoleic acid
(Octadecadienoic),
Gave arachidonic acid,
Named you might suppose from
    peanuts,
But it is not found in peanuts
And is plentiful in spiders,
So perhaps he spelt it wrongly—
So 'arach*ni*donic acid',[10]
Like arachnoidea mater,
Which, as everyone remembers,
Is the inmost spidery mother

Which ensheaths and wraps the
cortex;
But the patient still had in him
Ante-iso-*trans*-oleic,
Hiawathianic acid,
Which had come, if he had known it,
From the Russell's viper venom.

Summary

The moral of this story is then
Take some care when you have eaten
Butter, sardines, walrus liver,
Margarine or safflower seed oil,
Or arachidonic acid,
Or the body fat of quokkas,
Or adrenals of the muskrat,
Or the milk of female reindeer;
To avoid thrombosis don't get
Bitten by a Russell's viper
Which has but one type of UFA—
Ante-iso-*trans*-oleic,
Hiawathianic acid.

Source: Hiawatha's lipid was originally produced for after dinner at the Sixth Annual Symposium on Lipids, San Francisco, February 1958.

Notes
1. Before Keys it was believed all fats equally raised cholesterol levels in the blood.
2. A marsupial (that is, it carries its nipples in a bag) which has a ruminant's digestion like a cow (which doesn't).
3. Different fats have different effects on cholesterol levels.
4. Depending on their degree of polyunsaturation.
5. The degree of polyunsaturation of the fats is determined by releasing the component fatty acids and separating them.
6. Cholesterol in the blood serum drops when polyunsaturated (polyethenoic) fats are eaten.
7. Standard conference lecture technique!
8. UFA—unsaturated fatty acid.
9. -*osis* denotes a process or condition (makes anything sound more medical and hence serious). Compare -*itis*, which inflames whatever it is attached to.
10. The author shows his Magdalen College (Oxford) background. It would seem that either the original chemistry or the etymology of the word was wrong.

Fish oils

Supplements of fish oils are sometimes taken because they are high in polyunsaturated fats. Eskimos have long been known to bleed and bruise easily. As fish oils affect platelet function, they can increase bleeding time, severely in those with a pre-existing subclinical abnormality. For most people, however, 10 g day has less effect than one soluble aspirin.

Ice-cream

Ice-cream is a foam that is preserved by freezing. Four different phases can be seen under a microscope: solid globules of milk fat; air cells, which should be no bigger than 0.1 mm; tiny ice crystals, formed from freezing out pure water; leaving behind a fourth phase of water with concentrated sugars, salts and suspended milk proteins. The ice freezes out to a point where the lowering of the freezing point caused by the concentration of the remaining solutes in the water corresponds to the freezer temperature.

Legally the mix can be expanded with air to double its volume (called maximum 100% overrun in the trade). Ice-cream with more air feels fluffier and warmer to taste. Less milk fat means bigger ice crystals and a coarser texture and colder taste. Emulsifiers and stabilisers can mask the lower fat properties and can impart a gummy, sticky quality to the product.

If ice-cream is not stored properly, then the partial thawing causes the *smaller* crystals to melt and refreezing causes the *larger* crystals to grow (this is related to the demonstration of small bubbles blowing up large ones in Figure 2.16; see also Appendix VIII). This coarsening of texture from partial thawing can also occur by the crystallising out of the lactose (milk sugar), which tends to persist after the ice has melted either in a dish or on the tongue. (Lactose occurs only in milk and is one-tenth as soluble as sucrose.)

Chocolate

Chocolate and related products begin with the beans of the cacao tree (*Theobroma cacao*, 'food of the Gods'). These grow in elongated melon-shaped seed pods, each of which contains about 25 to 40 white or pale purple beans, each slightly larger than a coffee bean. The beans are collected, heaped, covered with leaves and allowed to ferment through the action of microbes and enzymes naturally present. This process kills the germ of the bean, removes adhering pulp and modifies the flavour and colour (now brown). After drying, the beans are ready for export.

For chocolate manufacture the beans are roasted and passed through a complex set of milling processes. The heat of grinding melts the fat and produces chocolate liquor, which is composed of about 55% fat, 17% carbohydrate, 11% protein, tannins, ash. Theobromine, the stimulant alkaloid related to caffeine, is found in amounts ranging from 0.8% to 1.7%, depending on its source. Somewhat less caffeine is also found. The solidified liquor forms the bitter cooking or baking chocolate. The fat removed from the chocolate liquor is cocoa butter. This consists mainly of triglycerides in which the middle fatty acid is oleic and the two outside fatty acids are saturated, generally stearic or palmitic. In beef tallow the opposite is true: the dominant triglyceride has saturated fatty acid on the inside and unsaturated on the outer positions. Although these two fats have rather similar overall fatty acid compositions (albeit that cocoa butter is more saturated than any animal fat!), their triglyceride composition is strikingly different, and so are their physical characteristics. The simple composition of cocoa butter leads to a relatively sharp melting point, 30°C to 35°C, which makes it attractive in chocolate. However the solid is polymorphic: it can crystallise in at least three, possibly six, different crystal forms, with melting points varying from 17.3°C to 36.4°C. Only the fifth of these (a so-called β-3 type) with a melting point of 33.8°C, is suitable). If the fat crystallises in an unstable form it will cause problems.

Question: Could one make a polyunsaturated chocolate?

A high-quality sweet chocolate might consist of 32% chocolate liquor, 16% additional cocoa butter, 50% sugar, plus flavouring. The mixture is ground to about 25 micrometres (μm) or less and then 'conched' (kneaded) for 96 to 120 hours for a high-quality product. The critical operation of controlled crystallisation follows. The liquid chocolate is cooled to initiate crystallisation and reheated to just below the melting point of the desired crystal structure so as to melt undesirable lower-melting types of crystals. The chocolate is then stirred at this temperature for a while (tempered) and then crystallised quickly to produce fine crystals. Different sources describe the procedure differently—chocolate making is still a dark art!

For milk chocolate, the Australian Model Food Act specifies a minimum of 45 g/kg milk fat, 105 g/kg non-fat milk solids (milk sugars mainly) and 30 g/kg water-free, fat-free cocoa paste. While cocoa paste is defined as the product prepared by grinding solidified chocolate liquor containing not less than 480 g/kg of cocoa butter, the fat-free specification means there is *no minimum requirement* for chocolate to contain cocoa butter (since October 1983). There is an obvious incentive to replace the expensive (and often variable in quality) cocoa butter with a cheaper fat. White chocolate, on the other hand, does have a minimum content of 200 g/kg of cocoa butter specified and also must contain not more than 550 g/kg of sugar, i.e. it can be over half sugar.[1]

Fat bloom is the development of a new phase in a chocolate fat, causing surface disruption with large clusters (5 μm), to give the grey mould-like coating inevitably blamed on poor consumer storage. Fat bloom in chocolate is distinguished from loss of gloss which occurs when small crystals (0.5 μm) on the surface grow into large ones and scatter light. The growth of large at the expense of small is thermodynamically favoured. The use of emulsifiers and stabilisers can greatly affect the rate at which crystal changes occur in the solid state. Various additives, particularly sorbitan fatty acid esters, are used to control crystallisation and phase change in substitute chocolate. Partial glycerides are also good stabilisers. They occur naturally in high levels in palm oil, which is used in margarines to stabilise the β' crystal, which, if it transforms to the β crystal, causes the margarine to become grainy.

Waxes

Fats and oils are fatty acid esters of the trialcohol glycerol. Waxes are esters of long-chain (C_{16} and above) alcohols with one hydroxyl group and long-chain (C_{16} and above) fatty acids. The natural waxes are mixtures of esters and contain hydrocarbons as well.

Spermaceti crystallises and separates when the oil from the head of the sperm whale (a member of the Cetacea) is chilled. It is mainly the ester cetyl palmitate, $C_{15}H_{31}COO—C_{16}H_{33}$, which has a melting point of 42°C to 47°C, and is used in skin lotions.

Beeswax. The cells of the honeycomb contain esters of C_{26} and C_{28} acids with the C_{30} and C_{32} alcohols plus 14% hydrocarbons (mainly C_{31}). Beeswax is used in furniture polishes.

Carnauba wax, the most valuable of the natural waxes, is obtained from the coating on the leaves of a Brazilian palm. Hard and impervious, it is used in car and floor polishes. It has a melting point of 80°C to 87°C, and consists of the esters of the C_{24} and C_{28} acids with the C_{32} and C_{34} alcohols. There is also a considerable amount of ω-hydroxy fatty acids, $HO—(CH_2)_xCOOH$, where $x = 17$–29 (ω, omega, is the last letter of the Greek alphabet and means that the —OH is at the end of the chain). These ω-hydroxy fatty acids can form long polymer esters, which give this wax its unique properties of being hard and impervious.

Wool wax (wool grease, degras) is recovered from the scouring of wool and is unusual because it forms a stable, semi-solid emulsion containing up to 80% water—a purified product known as lanolin is used as a base for salves and ointments in which it is desired to incorporate both water-soluble and fat-soluble substances. This 'wax' consists mainly of fatty acid esters of cholesterol, lanosterol and fatty alcohols.

BODY FAT

A discussion of fats and chemistry in the kitchen would not be complete without some comments on body fat and dieting. Today, many Australians are obsessed with their body weight. At the extremes are the anorexics, who starve themselves to skin and bone, and the bulimics, who have an unhealthy hunger. In between, there are many who expend a lot of energy thinking about food and struggling to avoid eating it. Although weight loss is easy for most people, maintaining the lower weight is achieved by only a fraction of all dieters. Once overweight, many people don't actually eat more than thin people. In fact, because of their better fat

Fig. 3.8   To lose 1 kg of fat — 1200 km at 30 km/h.

insulation and because they get less exercise, their energy needs decrease and this may mean that they eat less. After all, lack of sufficient food was the norm for all the human race in earlier times and so the efficient storage and use of fat was valuable for survival. There is a theory that suggests that our bodies defend against change. This set-point or 'natural weight' depends on heredity factors.

There is another key biochemical element, called the metabolic rate, which is the fuel the body burns when it is 'idling'. Just as cars differ in the petrol they use when idling, so do people when they are not active. In contrast to cars, this idling energy is the major part of the total energy we use each day and so is critical in determining our weight balance. Even with strenous exercise, it still contributes a significant amount.

Research suggests that taking regular exercise *increases* the idling metabolic rate for up to 15 hours after the activity. You can apparently make your body less biochemically efficient by being in better trim! On the other hand, dieting can be counterproductive. The metabolic rate slows down and you become more energy efficient when the energy intake is reduced. The body protects itself against what it believes is famine and not a voluntary diet. When you give up the diet, the body is now more energy efficient and a return to the 'normal' rate of eating results in putting on weight at a faster rate than before.

Human nutrition is a Cinderella science and a great deal more needs to be learnt. One of the areas of 'inefficiency' needing study is the way we store food energy. There is a belief that lean people store carbohydrate as fats and lose about 20% of the energy value of the food in the process, whereas fat people store carbohydrate as glycogen, a form efficiently converted to glucose. What is clear is that a simple conversion of fat energy to work by exercising does not give the right result.

Given the wide variety and contradictory nature of diets (e.g. high protein/low carbohydrate, low protein, very low energy, reduced energy), the long-term answer for most people is probably not to be found in dieting alone.

CHEMISTRY OF COOKING[2]

For an experimental chemist, cooking can be divided into four basic methods:
dry heating (baking, roasting etc.) < 250°C
wet heating (boiling, steaming etc.) 100–120°C
hot oil frying (frying etc.) < 300°C
microwave

The changes brought about fall into three classes: physical changes; chemical changes; and microbiological changes.

Physical changes

A very simple experiment is to place some peas in a measuring cylinder and measure the volume of water needed to fill the cylinder to a mark. Do this before and after 15 minutes boiling. Is water gained or lost? Calculate the percentage.

When starch grains are suspended in a drop of water and viewed under a low-power microscope ($\times$ 100), the discrete grains have a characteristic appearance, depending on the source. Compare potato, rice, wheat, corn etc. When starch grains are boiled the starch swells and the cellulose envelope bursts, allowing the starch to form a gelatinous mass and leaving empty cellulose envelopes behind as 'ghosts'. There is a characteristic temperature range at which this occurs for different starches (Table 3.3).

TABLE 3.3
Gelating action of some starches

Starch	Barley	Tapioca	Pea	Potato	Wheat	Corn	Rice
Gel temperature (°C)	51–61	52–64	57–70	58–66	59–64	62–70	68–70

Source: D. J. Smith, 'Chemistry of Cooking', *School Science Review* **58** (202), 1976.

Chemical changes

Cooking and digestion

Although we are told that cooking aids digestion, it would be nice to see some (simple) experimental evidence for this. A test based on the colouring of starch by iodine works well. When starches are heated, the long polymer chain is broken into shorter lengths called dextrins, which react differently with a solution of iodine (iodine, 1 g; potassium iodide, 2 g; water to 100 mL). The large polymers give the characteristic blue colour, the intermediate ones (of size equivalent to glycogen) give a brown to red colour, and smaller molecules give no colour.

QUANTUM MECHANICS IN THE KITCHEN

If you want to discuss quantum mechanics in the kitchen, then the explanation for this colour progression is that one iodide ion (I^-) is linked to many iodine molecules (I_2) in a chain, the length of which depends on the length of the spiral dextrin in which it lies. The single extra electron is shared by all the iodine atoms and 'moves' along the chain as a particle in a one-dimensional (quantum) box. The longer the box, the lower the frequency of the absorption bands. The large starch molecule allows absorption of long wavelength red light to give a blue colour, medium dextrins allow absorption of blue to give reddish brown colours and shorter dextrins shift the absorption into the ultraviolet range and hence no visible absorption is noted.

Dextrins form from starches during toasting, and samples of bread, before and after toasting, can be compared. An electric coffee grinder can be used to grind up the bread (toast), and the grind is shaken with water and filtered with a Buchner funnel. Dextrins are detected with the iodine solution. Saturation of the solution with crystals of ammonium sulfate will precipitate starch and other large molecules leaving smaller dextrins in solution. Confirm that thin slices of toast produce more dextrins than thick slices.

The effect of the salivary enzyme amylase on raw and boiled starch can be measured using dipsticks specific for glucose. (A suitable brand is Ames Clinistix or those recommended by your local doctor or pharmacist. See Chapter 13 for more on dipsticks and further experiments, including testing for vitamin C.) In our Western cuisine we carefully boil superb quality vegetables, throw away the water containing most of the vitamins and minerals and eat the almost indigestible remains. When vegetables such as spinach, brussels sprouts and spring greens have been cooked and filtered, you can see the colouring of beta-carotene (vitamin A precursor) in the water. You can also test for vitamin C. The presence of iron in the water can be demonstrated by reaction with potassium ferrocyanide (1% potassium ferrocyanide plus 1% hydrochloric acid mixed just before use). A deep-blue precipitate indicates iron. Semi-quantitative dipsticks can also be used. Care must be taken to avoid iron pots for this experiment. However, in passing, it should be noted that, in Third-World countries today, and in Australia until recently, much of the human dietary need for iron came from iron cooking pots or from iron salts in clay vessels. Aluminium pots and quality enamelware possibly contribute to anaemia!

The inactivation of enzymes can be demonstrated in many ways. A very simple but effective experiment consists in dropping small pieces of raw and cooked liver into test tubes containing 10 vol (3%) hydrogen peroxide and comparing the rate of effervescence. The enzyme peroxidase catalase is responsible for the reaction. Fresh pineapple contains a powerful proteolytic enzyme called bromelase, but in tinned pineapple it is inactivated. A pineapple garnish for ham (gammon) makes digestive sense if it is fresh, but not when it is canned. On the other hand, fresh pineapple plays havoc with setting gelatine and whipped egg whites because the enzyme is active.

Blanching inactivates enzymes. Potatoes cut for chips, boiled for two minutes and then cooled in a refrigerator for a further two minutes do not brown anywhere near as fast as potatoes that are only cut and left.

Drying and microbiology

Changing the 'availability' of water is the most common method of preserving food. The ability of micro-organisms to use water (and thus multiply) depends not on how much water is present in the food but on its activity. The activity of pure water is 1, and any dissolved material reduces the activity. The higher the concentration of dissolved species (molecules, dissociated ions etc.), the lower the activity. This activity can be measured directly by measuring the equilibrium vapour pressure of the water (or, less conveniently in a food, by its freezing or boiling point). For this reason we 'dry' food, which removes most water and increases the concentration of natural solutes. Alternatively, we add solutes such as salt or sugar, or in the case of preserved meats such as ham, sodium nitrate. A plot of the water activity versus percentage of water is shown in Figure 3.9 for a number of foods and also, for calibration, a number of saturated solutions of inorganic solutes. You can determine the percentage of water by using the Dean and Starke distillation (Chapter 13, Experiment 13.1) for most of the products shown on the plot. The lowering in activity needed to stop various micro-organisms from growing is also shown. (The pH is also very important: an acid medium stops many 'nasties', for example *Clostridium. botulinum*, from growing.)

Micro-organisms themselves are also used to preserve foods—for example, fermented raw meats such as salami, cheese and sauerkraut.

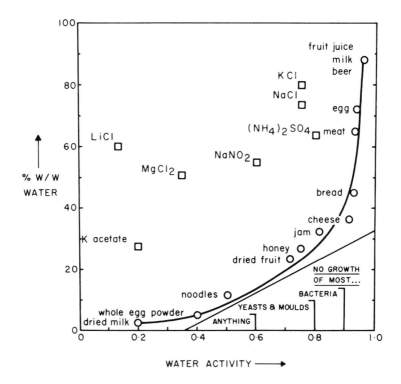

Fig. 3.9 Water activity versus water concentration for selected foods and selected standard saturated salt solutions

MOLECULES ON THE MOVE

When toast burns in the kitchen, you can smell it 'instantaneously' in the bedroom at the other end of the house. How fast do gas molecules move? A truck with a load of chlorine turns over and releases its cargo. You smell it immediately but you can run away without coming to grief (provided you don't run downwind!). Why is bulk diffusion so very slow?

There is obviously a difference between the speed of a select fraction of molecules that by chance have few collisions with other atmospheric molecules (root mean speed) and the mean free path of the bulk of the gas, which determines mass diffusion. (Reference topic: Maxwell Boltzmann Distribution and the Kinetic Theory of Gases.)

How are nutrients lost in processing food?[3]

The nutritional value of a processed food is rarely better than that of the raw ingredients from which it is produced, unless nutrients are added in purified or concentrated form. Nevertheless, there are some beneficial effects of processing (e.g. the destruction of trypsin inhibitor in legumes, and the liberation of bound niacin in cereals).

Although any loss of nutritional value cannot be judged by the consumer, a reduction in nutritional value is usually accompanied by a reduction in other food qualities, such as colour, flavour and texture, which *can* be judged by the consumer. During processing, nutrients are lost because they react with other constituents of the food, oxygen, light or heat, or because they are leached by water at some stage of the process. Trace elements and enzymes may also catalyse the destruction of the nutrients. Loss of nutrients may occur at any or all of the following steps between garden and gullet: harvesting, handling and transport; preparation and processing; storage; distribution; and handling and preparation in the home, restaurant or institution. Of these, the greatest loss of nutrients in many processed foods occurs during handling and preparation in the home, restaurant or institution. The reasons for this are described below.

How stable are nutrients?

In addition to the major food constituents such as carbohydrates, fats and proteins, about 50 other nutrients (amino acids, minerals and vitamins) are essential for adequate nutrition by humans.

Foods from plants and animals contain nutrients in varying amounts, and they are unequally sensitive to temperature, light, oxygen and leaching. Consequently, the nutrient loss resulting from processing varies according to the type of food, the length of time that processing takes, and the particular nutrient. Table 3.4 gives an indication of the stability of some vitamins, amino acids and minerals under various conditions.

Loss of nutrients during processing

The importance of nutrient losses during processing depends on the total diet. If we ate a balanced diet there would be no cause for concern. Inevitably, great differences exist between the diets of people of different ages and social groups. For example, the diet of babies normally lacks variety; and often so does the diet of the aged living alone. The diets of some social groups are quite unbalanced. Over indulgence in alcohol or dedication to macrobiotic diets brings other problems. In these groups, where one or more essential dietary components may be in limited supply, any loss of nutrients may be critically important.

Foods of animal origin

The successful canning of meats for human consumption depends on the application of sufficient heat to a sealed container to render the contents 'commercially sterile'.

TABLE 3.4
Stability of nutrients

Nutrient	Effect of solutions that are:			Effect of exposure to:			Cooking losses (%)
	acid	neutral	alkaline	air	light	heat	
Vitamins							
Vitamin A	U	S	S	U	U	U	0–40
Vitamin C	S	U	U	U	U	U	0–100
Biotin	S	S	S	S	S	U	0–60
Vitamin D	—	S	U	U	U	U	0–40
Folic acid	U	U	S	U	U	U	0–100
Vitamin K	U	S	U	S	U	S	0–5
Niacin	S	S	S	S	S	S	0–75
Riboflavin	S	S	U	S	U	U	0–75
Thiamin	S	U	U	U	S	U	0–80
Essential amino acids							
Isoleucine	S	S	S	S	S	S	0–10
Leucine	S	S	S	S	S	S	0–10
Lysine	S	S	S	S	S	U	0–40
Methionine	S	S	S	S	S	S	0–10
Phenylalanine	S	S	S	S	S	S	0–5
Threonine	U	S	U	S	S	U	0–20
Tryptophan	U	S	S	S	U	S	0–15
Valine	S	S	S	S	S	S	0–10
Mineral salts	S	S	S	S	S	S	0–3

S: Stable (No important destruction)
U: Unstable (Significant destruction)

Source: G. J. Walker, *Nutritional value of processed food*, CSIRO, Information Service, 1979.

In general, canned meats contain less thiamin than fresh meats cooked by roasting, braising or boiling because some thiamin is destroyed by the extended heat treatment used to sterilise them.

Foods of plant origin
In their preservation by canning, freezing or dehydration, most vegetables are blanched by immersion in steam or hot water. During the preparation of vegetables for freezing, blanching inactivates enzymes that otherwise would cause deterioration of the food during storage and subsequent thawing.

Values for the loss of vitamins resulting from blanching are given in Table 3.5. Steam blanching results in greater retention of water-soluble nutrients than water blanching. A comparison of microwave, steam and water blanching shows that microwave blanching results in better ascorbic acid retention in brussels sprouts. However, the best product was obtained with combined microwave and water blanching.

Pasteurisation of milk
The introduction of compulsory pasteurisation of milk gave rise to more controversy than has attended the introduction of any other food-processing

TABLE 3.5
Vitamin losses resulting from blanching

Food	Loss of vitamin C (%)	Loss of thiamin (%)
Asparagus	10	—[a]
Green beans	20–25	10
Lima beans	20–25	35
Sprouts	20–25	—
Cauliflower	20–25	—
Peas	20–25	10
Broccoli	35	—
Spinach	50	60

[a]The dash indicates not available.

Source: G. J. Walker, *Nutritional value of processed food,* CSIRO Information Service, 1979.

procedure. What effect does pasteurisation have on the nutrients in milk?

- Minerals, vitamins A and D, pyridoxine, niacin, pantothenic acid and biotin levels are unaffected.
- Riboflavin and vitamins E and K are practically unaffected.
- Thiamin is reduced by 3% to 20%.
- There may be a coagulation of up to 10% of the proteins albumin and globulin.

Since milk is not a rich source of vitamin B_1 (thiamin) and vitamin C (ascorbic acid), the moderate loss from pasteurisation is not significant. On the other hand, pasteurisation destroys all pathogenic organisms in milk, and most other bacteria, so that the great advance in reducing the risk of milk-borne infection gained by processing far outweighs the insignificant losses of nutrients.

Loss of nutrients during commercial storage

The amounts of nutrients lost from a processed food depend on the temperature experienced, the time that elapses between processing and consumption, the nature of the food and sometimes the nature of the container. From production in the factory until delivery to the consumer, processed food may be held in a warehouse at the factory, in railway vans, trucks or ships during transport, in distribution warehouses, and in retail stores. The length of time that foods remain in any part of this distribution chain and the temperatures that they experience are extremely variable. Table 3.6 presents figures that show how vitamins are retained in canned foods stored at 10°, 18° and 27°C.

Loss of nutrients during preparation in the home

Foods of animal origin

The major foods of animal origin consumed by Australians are beef, pork, veal, lamb, poultry, fish, eggs, milk and milk products. The nutrients in these foods that cooking and processing affect to an important extent are the B vitamins, proteins and fat.

TABLE 3.6
Percentage retention of vitamins in canned foods

	Temp. (°C)	Peas			Orange juice		Tomatoes	
		Vitamin C	Thiamin	Carotene	Vitamin C	Thiamin	Vitamin C	Thiamin
12	10	93	92	98	97	100	95	94
months	18	91	87	94	92	98	94	93
storage	27	86	74	91	77	89	82	82
24	10	91	90	94	95	101	89	91
months	18	89	85	91	80	89	87	87
storage	27	81	70	90	50	83	70	70

Source: G. J. Walker, *Nutritional value of processed food*, CSIRO Information Service, 1979.

Freezing, thawing and cold storage

In frozen storage of a few months duration, muscle meats and liver lose 20% to 40% of the thiamin, 0% to 30% of the riboflavin and up to 18% of the niacin and pantothenic acid. In refrigerator storage for two weeks, losses are 8% to 10% for thiamin and pantothenic acid, 10% to 15% for riboflavin, and less than 10% for niacin.

The method of thawing usually has little effect on the nutrient levels, except that thawing in running water or by warming results in a loss of thiamin and pantothenic acid. Up to 33% of pantothenic acid is lost in the 'drip' from thawed beef, much more than other B vitamins.

To control bacteria, temperatures should either be very high or very low (see Fig. 3.10).

Thermal processing, cooking and canning

Pyridoxine (vitamin B_6), folic acid and pantothenic acid are the most vulnerable of the B vitamins in foods of animal origin. From 50% to 90% of folic acid and pyridoxine, and 10% to 50% of pantothenic acid may be lost during cooking. Choline is stable and biotin relatively so.

Foods of plant origin

A few years ago, a survey of the vitamin content of some 2000 foods consumed by a selected group of Australians showed that, particularly in institutions, cooked fruits and vegetables contained virtually no vitamin C. The most disturbing finding of the survey showed low concentrations of vitamin C in the plasma of many of these people. These low concentrations probably resulted from diets being deficient in citrus fruits or juices and other rich sources of vitamin C. The destruction of the vitamin resulting from institutional cooking practices and poor handling of the food in some of the domestic kitchens also led to low concentrations of vitamin C in the foods consumed by some people.

Boiling

Vegetables with large surface-to-mass ratios (e.g. spinach) are especially sensitive to loss of vitamins. Vitamin C is a most unstable nutrient in neutral and alkaline

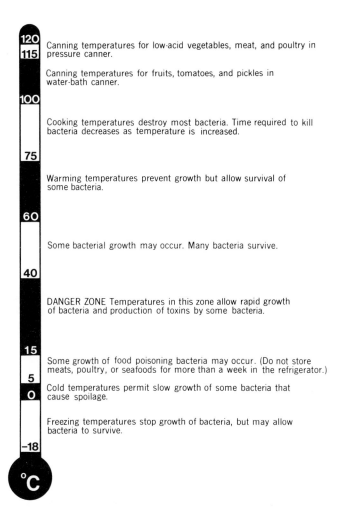

Canning temperatures for low-acid vegetables, meat, and poultry in pressure canner.

Canning temperatures for fruits, tomatoes, and pickles in water-bath canner.

Cooking temperatures destroy most bacteria. Time required to kill bacteria decreases as temperature is increased.

Warming temperatures prevent growth but allow survival of some bacteria.

Some bacterial growth may occur. Many bacteria survive.

DANGER ZONE Temperatures in this zone allow rapid growth of bacteria and production of toxins by some bacteria.

Some growth of food poisoning bacteria may occur. (Do not store meats, poultry, or seafoods for more than a week in the refrigerator.)

Cold temperatures permit slow growth of some bacteria that cause spoilage.

Freezing temperatures stop growth of bacteria, but may allow bacteria to survive.

Fig. 3.10   Temperature of food for control of bacteria

conditions, and in the presence of oxygen; folic acid may be even more unstable; and thiamin, riboflavin, carotene and niacin are sensitive under certain conditions. Nutrient losses increase if large amounts of cooking water are used. Cooking for longer times also results in a greater loss of nutrients.

Baking

Retention of nutrients may be low during baking; for example, the thiamin retention in potatoes has been reported to be 41% to 48%. Apples retained 40% of their vitamin C when baked for 60 to 90 minutes at 200°C in an open dish. When baked in a covered dish, retention was 25%, and only 20% when baked as a pie.

Baking powders destroy thiamin; slightly acid conditions are desirable to keep thiamin stable during baking.

Bread loses little of its nutritional value during baking, but toasting results in a greater loss of protein and vitamins, depending on the degree of drying out and browning.

Manufactured breakfast cereals are prepared in a variety of ways. The heating, rolling and flaking processes do not affect the protein, but the more severe process of explosion puffing has been shown to reduce the nutritive value. However, it is necessary to consider these foods in perspective as they are almost invariably eaten with milk.

Retention of nutrients

Table 3.7 gives approximate figures for losses of nutrients from some foods during cooking.

TABLE 3.7
Percentage of original vitamin content lost from foods after commercial, followed by home, cooking

	Thiamin	Riboflavin	Niacin	Pantothenate
Bacon				
Crisped at 175°C	5	—	—	—
Fried at 290°C	5	—	—	—
Beef				
Boiled 55 min,	30	0	10	25
then freeze dried, then reheated	90	0	50	—
Canned and fried	80	0	10	20
Chicken				
Simmered 50 min,	50	30	40	50
then freeze dried, rehydrated,				
boiled 30 min	55	70	60	—
Boiled 35 min, canned,	98	15	20	30
then reheated 30 min,				
contents drained	98	—	30	—

	Vitamin C			

Carrots		Green beans	
Boiled 5 min, dried,	30	Steamed 4 min, dried	70
then cooked 12 min	70	Simmered 10 min	70
Steamed 15 min, dried,	20	Steamed 9 min, freeze-dried,	70
then boiled 20 min	70	then reconstituted	80
Steamed 5 min, canned,	50	Steamed 4 min, canned,	75
then reheated	80	then reheated	75

Source: G. J. Walker, *Nutritional value of processed food*, CSIRO Information Service, 1979.

How to retain nutrients in food

The following steps should be followed if the maximum amounts of nutrients are to be retained during the preparation and cooking of food in the home.

1. Take frozen food home from the supermarket in an insulated container and place it in the home freezer immediately.
2. Store foods at the lowest appropriate temperature and for the shortest possible time.
3. If possible, minimise the amount of chopping and cutting of vegetables for cooking.

4. Do not use sodium bicarbonate in cooking as it increases the rate of loss of thiamin and vitamin C.
5. Try to avoid storing cooked foods for lengthy periods in the freezer or refrigerator.

Microwave cooking

Microwaves, like radiowaves and light waves, are part of the electromagnetic spectrum. The region of 300 MHz (megahertz) to 300 000 MHz (corresponding to wavelengths of 1 m to 1 mm) is the microwave region, and microwave ovens operate at the fixed frequency of 2450 MHz (12 cm) in Australia. The microwave radiation is generated in an electronic tube called a magnetron and passes along a waveguide into the metal oven cavity.

Other uses of microwave include radar and TV microwave links; medical diathermy operates at around 27 MHz; the longer wavelength penetrates more deeply.

Why are microwave ovens so efficient for cooking? After all, in a conventional oven, the electricity is converted with 100% efficiency into heat, whereas, in a microwave oven, electricity is converted with only about 50% efficiency into microwaves. The reason is that microwave energy is all strongly absorbed by some food components, particularly *mobile polar molecules* such as water. The electric field of the passing radiowave can force rotations and vibrations in the water molecule, which is a form of heat. Collisions pass this heat on to other molecules in the food.

Because water is a very strong absorber of microwaves, cooking occurs through the water in the food boiling. Because the same process could occur in our body, it is important that we are properly shielded from the microwaves (particularly the eyes). For this reason, microwave ovens are designed with at least two safety interlock switches that cut the power when the door is opened.

In a *conventional* oven the heat generated consists of radiation spread over an enormous range of frequencies (Planck or Black Body distribution, for physics students), from radiowaves to infra-red, with a little bit of visible light if the oven is hot enough. Only limited sections of this radiation are absorbed by the food and the rest is absorbed by the oven walls and conducted to the room. The longer-wavelength microwave radiation in the microwave oven is *reflected* by the oven walls and is later absorbed preferentially by the food.

Even when electromagnetic radiation is reflected from a material it penetrates the surface to a distance of about one wavelength. The infra-red heat radiation from a conventional oven is less than a tenth of a millimetre in wavelength and so heats only the outside of the food, from where slow conduction takes it into the interior. The 12 cm wavelength microwaves, however, penetrate many centimetres into the food.

The NH & MRC standard for microwave ovens requires that the power flux density of microwave radiation shall not exceed 50 W/m^2 at any point 5 cm or more from the external surface of the oven (any closer than that and the simple testing devices that measure the electric field power are not accurate because both the electronic and magnetic fields have to be measured when closer than 6 cm, half

SUPERHEATING IN A MICROWAVE

When tap water is heated on a conventional stove top, the maximum superheating is 100.75°C. In a microwave oven, however, superheating temperatures of 105° to 106°C can be sustained, with periods at 110°C. (A thermocouple was placed in a glass well that contained cooking oil. The leads of the thermocouple were not attached to the voltmeter while the microwave was running, because the radiation would cause spurious readings. Extremely fine, insulated thermocouple wires were used so that the microwave door could be closed to prevent leakage.)

The microwave oven heats the water *directly* and the beaker *indirectly*, the opposite of what happens when heating by conduction with a burner. The water is not heated near the beaker surface where nucleation is most likely to occur. Evaporation occurs only at the water surface and is not fast enough to cool the bulk water. There is much less superheating in plastic cups because water does not wet plastic and many more nucleation sites are available for boiling.

Source: R.E. Apfel and R. L. Day, 'Superheating water in a microwave', *Nature* **321**, 12 June 1986, 65.

a microwave wavelength, to the cavity). If a microwave oven door or door frame becomes distorted through mechanical abuse or the grille in the window is damaged, there is a danger of radiation leakage. The cheaper monitoring detectors consist of a piece of wire of length just under half a wavelength (5.7 cm) with a diode in the centre across which a light emitting diode (LED) is placed to indicate a power reading. These devices should be brought in from a distance so that the diode does not saturate and give a zero reading when in fact the level is high.

TABLE 3.8
Dielectric properties of plastics

	Dielectric constant[a] (ε at 1 MHz)	Dissipation factor (1 MHz)
PVC/PVA	3.0–4.0	0.006–0.02
PVC/PVDC		
Melamine-formaldehyde	7.2–8.4	0.03
Polyamide (Nylon)	3.4–3.6	0.04
Polycarbonate	2.96	0.01
Polyethylene (Polythene)	1.3	<0.0005
Polypropylene	3.2–2.6	0.0005–0.002
Polystyrene	2.4–2.65	0.0001–0.0004
Polysulfone	—	—
Polytetrafluoroethylene (Teflon)	2	0.0002

[a]Values at 1 MHz only are available.

Source: Handbook of chemistry and physics, The Chemical Rubber Company, Cleveland, Ohio.

Materials suitable for use as containers or covers in a microwave cooker should not absorb the radiation strongly. Therefore, they should be made from material with low polarity or, in other words, low dielectric constant and low dissipation factor (at 2450 MHz); for example, some plastics and glass. Metals reflect microwaves and are unsuitable (except for special local shielding as with the use of foil strips). The first counter to microwave radar developed during the Second World War was the dropping of metallic strips called 'chaff' or 'window'.

Look at Table 3.8 and see which materials you think would be suitable for use in a microwave oven. The plastic used must also be stable at high temperature and have good resistance to oil vapour.

EXPERIMENT 3.1

Hydrolysis of starch with saliva

Equipment
Two small glass containers, cooking pot, and a source of starch (such as noodles, or potato).

Procedure
Prepare a solution of starch by boiling a few noodles (or a piece of potato) in a small quantity of water. Pour about 10–20 mL of this aqueous solution into each of two small glass containers.

Add about 5 mL of saliva to one of the containers of starch solution.

After about 10 minutes add a small drop of iodine solution to each container to test for the presence of starch (see Experiment 3.2).

Starch and cellulose are both polymers made up of glucose units, but the units are linked together in a slightly different way in the two polymers. In cellulose the polymer chains are essentially linear, whereas in starch the chains have a helical structure. These differences give the two polymers vastly different properties; for example cellulose is the water-insoluble, fibrous building material of plants, but starch has no fibrous structure and has different solubility properties.

Reaction
Saliva contains an enzyme (ptyalin) that is capable of breaking the links between the glucose units in starch, so that the solution treated with saliva no longer gives a positive reaction for starch following the action of the enzyme.

$$starch + ptyalin \rightarrow maltose\ (a\ sugar)$$

Maltose is a compound made up of two units of glucose joined together. The enzyme is very selective. It breaks the bonds in starch, but not those in cellulose.

Note
You can test the effect of temperature on the rate of your hydrolysis reaction by carrying out the hydrolysis in the refrigerator or at temperatures above room temperature.

EXPERIMENT 3.2

Test for starch

Equipment
Tincture of iodine, potato, saucepan, cotton wool, funnel.

Procedure
Cut up your potato and boil it with water in a saucepan for a few minutes. Allow the water to cool. Decant or filter the solution to separate the soluble amylase from the insoluble amylopectin of the starch. (A cotton wool pad in the neck of a funnel will provide an ideal filter.)

Add a drop of tincture of iodine to your filtered starch solution. An intense blue colour will appear if any starch is present. If your test doesn't work, try adding some iodised salt to your potatoes when you boil them.

Reactions
The soluble part of the starch contains the substance β-amylase, which has the empirical formula $C_6H_{10}O_5$.

The blue starch colour is believed to be caused by the formation of a complex between iodine and β-amylase of the general composition:

$$(\beta\text{-amylase})_p(I^-)(I_2)_r\ (H_2O)_s$$

in which r is very much smaller than p and s is very much greater than p, i.e. gelation as described earlier. (The small value of r means that this reaction can also be used as a very sensitive test for the presence of iodine). See also box, 'Quantum mechanics in the kitchen', in this chapter.

Notes
1.   Try your test on other plants and foods (e.g. beans, peas, bread, spaghetti).
2.   Tincture of iodine is used to treat wounds and may be in your home medicine cabinet. If not, it is available from pharmacies.

REFERENCES

1.   US regulations are given in M. Grayson (ed.), *Kirk-Othmer Encyclopedia of chemical technology*, 3rd edn, Wiley, New York 1983, in the section on chocolate.
2.   This section is based on D. J. Smith, 'Chemistry of cooking', *School Science Review* **58** (202), 1976, 25.
3.   The sections on nutrients are based on G. J. Walker, *Nutritional value of processed food*, CSIRO Information Service, 1979 (CSIRO Division of Food Processing, PO Box 52, North Ryde 2113, Australia), and references therein.
4.   This section was written with the assistance of Vic Burgess, CSIRO Radiophysics.

BIBLIOGRAPHY

General
Beeton, Isabella. *The book of household management*. Originally published by S.O. Beeton, 1859–61. Reproduced in facsimile, Jonathan Cape, London, 1968,
Noller, W. B. *Chemistry of organic compounds*, 3rd edn. Saunders, Philadelphia, Pa., 1966. An excellent organic chemistry textbook of the more traditional type. It gives background information on industrial and agricultural aspects of chemistry.

Cooking
Coultate, T. P. *Food: the chemistry of its components*. Royal Society of Chemistry Paperbacks, London, 1984.
McGee, H. *On food and cooking: the science and lore in the kitchen*. Unwin Hyman, London, 1987.

Diet
Gurr, M. I. and Kirtland, J. 'Obesity'. *Chemistry and Industry*, 18 September 1976, 766.
Smith, D. 'The great diet myth'. *National Times*, 9–15 January 1983.

Nutrition
Wahlqvist, M. L. (ed.). *Food and nutrition in Australia*, Cassell, Sydney, 1981.

Oils and fats
Berger, K. G. 'Catalysis and inhibition of oxidation processes, Part II. The use of anti-oxidants in food'. *Chemistry and Industry*, 1 March 1975, 194.
The chemistry of glycerides. Unilever Educational Booklet, Advanced Series no. 4. Unilever, London, 1968.
Consumers' Union (USA). 'Cooking oils and fats'. *Consumer Reports*, September 1973; 'Margarines', January 1983.
'Cooking oils and fats', *Choice*, ACA, April 1975.
Harwood J. L. 'Nutritional aspects of oils and fats'. *Chemistry and Industry*, 16 September 1978, 687.
Opfer, W. B. 'Margarine production: a review of modern manufacturing processes'. *Chemistry and Industry*, 16 September 1978, 681.
'Salad and cooking oils'. *Choice*, ACA, March 1983.
Thomas, T. 'Chemistry and biology: an interface of oils'. *Chemistry and Industry*, 17 July 1982, 484.
Vegetables oils and fats. Unilever Educational Booklet, Ordinary Series no. 2. Unilever, London, 1968. A descriptive pamphlet on oil-producing plants and their seeds; colour photos, statistics and a discussion of oil making; very readable.
Young, V. 'Processing of oils and fats'. *Chemistry and Industry*, 16 September 1978, 692.

Polyunsaturated
'Fats in the diet. Why and where?' Institute of Food Technologists, 221 N. La Salle Street, Chicago, USA, 1981.
Fogerty, A. C., Ford G. L. and Pearson, J. A. 'Composition of some Australian table margarines'. *Food Res. Q.* **39**, 1979, 38.
'Polyunsaturated dairy products'. *Rural Research* **79**, March 1973. CSIRO, Canberra. An excellent discussion on the production and properties of polyunsaturated meat and dairy products; well illustrated and very readable.

Chemistry in the boudoir

4

INTRODUCTION

In 1770 a Bill was allegedly introduced into the British Parliament which read:

> ... that all women, of whatever age, rank, profession or degree, whether virgins, maids or widows, that shall impose upon, seduce and betray into matrimony, any of His Majesty's subjects by the scents, sprays, cosmetic washes, artificial teeth, false hair, Spanish wool, iron stays, hoops, high heeled shoes and bolstered hips, shall incur the penalty of the law in force against witchcraft and like misdemeanours and that the marriage upon conviction shall stand null and void.

However, after a reader's enquiry, I searched Great Britain's *Statutes at Large* right back to Magna Carta and found no such Act recorded.

Although cosmetics have been used—by men as well as women—since prehistory, the popular general use of cosmetics is a modern phenomenon. Many new and improved cosmetics have been made available by scientific research, and extensive advertising has been an important factor in bringing them to the attention of consumers and in increasing the number of people who use cosmetics. Consumer products range from those where the composition and performance of the product are of fundamental importance to the consumer, to those where the image and personality of the product virtually is the product. With cosmetics, the consumer is buying a hope. Because the chemistry of a hope might be best described by a market researcher, the first part of this chapter is an edited version of a talk given by Ms Marilyn Elfverson to the Australian Cosmetics Association in Canberra in April 1975, followed by a brief description of the Australian cosmetics industry. This will give you some idea of how manufacturers develop consumer products, assess the potential market for their goods, design their packaging and devise their advertising strategies.

The second part of this chapter deals with the chemistry of cosmetics and related products and takes a more detailed look at surface chemistry.

COSMETICS TIME-SCALE

BC

5000	Eye shadow of green copper ore.
3500	Eye-lids darkened and lustre of eyes increased by Egyptian women using kohl (a fine-powdered form of antimony).
3000	Balsam and perfume was a Cypriot industry of the Bronze Age.
1600	Red hair-dye in use.
200	Dye extracted from a root called rizion used for rouging the cheeks (mentioned by Theophrastus).
68–30	Asses-milk bath used by Cleopatra to improve and whiten the skin. Henna used to dye fingernails, palms of hands and soles of feet. Egyptian beauty shops sold cosmetics and perfumes.

AD

800–1100	Perfume distilled by Arabs.
1100	Gifts of cosmetics brought back from the crusades.
1558–1603	Reign of Queen Elizabeth I. Toilet preparations in use. For example, perfume boxes ('sweet coffers') were a necessary item of bedroom furniture. Face powder made from powdered marble, borax and starch. Women of the court washed their faces in wine to achieve a ruddy complexion.
1570	Mary, Queen of Scots, bathed in wine.
1610–43	Use of cosmetics encouraged by Louis XIII (France).
1649–58	Use of cosmetics discouraged by Cromwell.
1660–85	The Restoration. Cosmetics played vital role in improving physical appearance. Charles II encouraged use of cosmetics.
1700	Poisonous white lead (lead carbonate) was used to coat the face by many women, many of whom died as a result.
1770	Widespread use of cosmetics allegedly stimulated the introduction of a Bill into the British Parliament to protect men from being falsely attracted into marriage (see opposite).
1780	Lavoisier delivered a paper on cosmetics to the French Academy of Science. He distinguished between those rouges made from plant and animal extracts (plant rouge is decolourised by spirits of wine; animal rouge is decolourised by alkali) and those made from minerals (not decolourised by either chemical).
1800	A great revival of interest in cosmetics early in the century during Napoleon's time.
1850	Synthetic perfumes and synthetic dye stuffs developed.
1939	World War II caused a 75% reduction in British cosmetic manufacture. This resulted in a fall in morale and a decline in the number of factory employees. When cosmetic rooms were set up in factories, women flocked to them, and production efficiency improved—proving a direct relationship between the sense of well-being imparted by cosmetics and excellence of performance (industry advertisement).

MARKET RESEARCH[1]

What is market research? Essentially, market research operates as a channel of communication between the marketer and his market. It is a *dialogue* with the consumer, a means of keeping the marketer in touch with the consumer. Market research provides the marketer with vital feedback about consumers' needs, attitudes and behaviour. This *understanding* of the consumer is an essential basis for successful marketing. It is essential because, in many product areas, consumers can satisfy their basic needs in a variety of ways: they have a *choice* and it is this choice which has led to the marketing philosophy of consumer orientation. Take the example of a woman who has just purchased a bottle of skin cleanser. This purchase involved a *choice* between different *brands* of cleanser, e.g. Revlon, Helena Rubinstein, and between different *types* of cleanser, e.g. cream or lotion, but also a choice about whether to use a special cleanser product or to use soap and water—or just water. Marketers are in a better position to channel consumers' generalised needs and wants into specific product and brand purchases if they have a full understanding of the consumer and of her attitudes towards competing alternatives.

Market research would not be needed if management were in direct contact with its market. As an example, imagine the case of a very small company which makes hand lotion. The company consists of one person who spends his mornings making hand lotion and his afternoons selling it directly to consumers. This manufacturer has a first-hand knowledge of each and every customer. He receives instant feedback on his product and on his new market offerings and, because he knows each customer personally, he can develop products and a selling approach to meet individual needs. For example, he may know that Mrs Jones has six children and does a lot of washing and a lot of washing up. She has a problem with her skin drying out and cracking. She wants a product with healing properties and our manufacturer can develop a product to meet that need. Mrs Smith does a lot of gardening and spends a lot of time outdoors; she wants something to protect her hands and he can sell her a hand-care product with this benefit. Miss Brown is eighteen. She doesn't spend much time with her hands in detergent, but she wants her hands to feel soft to touch. He can sell a product to her that promises to make her hands nice to hold. His thorough understanding of his market leads to success and soon he outgrows his manufacturing capacity. He puts on an assistant to help manufacture his hand lotion, and then another. Soon he needs a second car and then a third to sell his product and he is beginning to lose that personal contact with his customers. *As his firm grows, so does the need for research.* Research is a substitute for personal contact with the customer. As the firm grows, the personal communication to and from customers has to be replaced by other media; advertising and promotion on the one hand, which communicates to the consumer, and *research* on the other hand, to obtain feedback from the consumer about his market offering.

Market research covers a lot of areas—it is used for a variety of purposes. For example, it is used to provide feedback about *what* is happening in the marketplace: where does a specific brand stand in relation to competition? Where is the industry as a whole going—is it expanding, is it contracting? This type of research is often described as 'market intelligence'. It helps the marketer to define his current position, predict his future position, identify strengths and weaknesses. Research is also used to help the marketer develop and present a product that will sell. Some people do that very effectively without ever doing any research. But research can help to reduce the risks of introducing a new product, or of modifying the mix of an existing product; it can help to identify opportunities for new products, or new product positionings; and it can help to maximise the effectiveness of the product's marketing mix.

Brand personality

When we talk of a *product* we are talking about more than a set of physical properties. We are talking about the *overall impression* that a product makes on the consumer through the *collective* impact of its physical characteristics, its performance characteristics and its presentational elements, including its brand name, packaging, price, distribution and advertising. *In most product categories, we are not just selling a product, we are selling a proposition.* Toilet soaps share a number of physical properties and they all clean—but what determines whether I choose to pamper with Palmolive, luxuriate with Lux, or freshen up with Lime Fresh? These different brands do much the same thing. And if one of them was to go off the market, consumers could quite happily make do with another one. The consumer's relationship with a product goes beyond what the product is or what that product does. A product has a *personality* and this personality may be the strongest element in attracting purchase. Research is often used to help in the development of a brand personality, and to ensure that this personality is being effectively communicated to the target market.

The way a *product* dresses tells you something about that product. If the pack design tells you something you don't like about that product, maybe you won't give it a chance. Impressions count with products as well as with people. If I take two identical hand lotions and put one in an Estee Lauder pack and one in a Woolworths pack, and ask women which one they want, I would predict that the majority would go for the Estee Lauder—as long as they didn't have to pay for it! But more than that, the impression they get from these packs will probably affect their perceptions of the product itself. If they are asked to try the lotions, it is likely that they will say that the Estee Lauder product smells nicer, makes their hands feel softer, has a finer texture, and is more effective. And they will actually believe that. Of course, not all women will be attracted to the Estee Lauder personality—for some it may be threatening, for some it just may not be the kind of personality they can relate to, or aspire to.

What people say and *how* they say it influences how you feel about them. Products talk too. They talk through advertising, they talk through their pack copy. It is important for a product to say the right thing to the right people. Where a person lives, how much money they earn, whether they are successful, influences how you feel about them. The price of a product, where you buy it, the advertising medium chosen for it—*Vogue* or the *Women's Weekly* or the particular radio station—the types of presenters used in its commercials, all communicate in a similar way. A higher price *tends* to suggest better quality, a headache remedy retailed only through pharmacies tends to have more ethical overtones than one sold through a supermarket.

A very useful way of establishing how consumers really feel about a product is to ask them to talk about it as if it were a person. Often when you ask a person why they buy one brand rather than another, they can't really tell you 'why', or they give you superficial answers such as 'it gets clothes whiter'; yet when you ask them to talk about the two brands as if they were people, they may have very different images and feelings about these two brands that come out in these descriptions. Try it for yourself; try to think about some products in personality terms. Let's take Brut—aftershave or perfume. Think about the Brut brand—think about Brut as if it were a person. Let's say that suddenly Brut came to life as a person. What kind of person would Brut be? For starters, is Brut male or female? Is Brut a young person or an old person? How does she/he get on with the opposite sex?

OK, now let's take Lux—Lux toilet soap. If Lux were to come to life as a person, what kind of person would Lux be? Male/female? What would Lux's favourite colour be? What kind of clothes would Lux wear?

People are more likely to buy a product if they like its personality. People buy products that they feel comfortable with, products that they trust. People buy a product or brand that feels 'more me' than others. The product personality can act as an affirmation of their own self-image. For example, when a woman is buying soup for her family, she may be looking for a product to match her image of herself as someone concerned to provide wholesome, nutritious food. When she buys cosmetics, she may look for a product to express her more glamorous sensual potential.

The importance of the personality varies to some extent depending on the product, but it is probably true to say that, for most products, a strong, distinctive personality that consumers can relate to improves their chances of success. The product personality may not be a major consideration for a product such as a scouring pad, but it is a definite consideration in the choice of many every-day food and toiletry products, and it is at work in full force in the purchase of cigarettes, cars, *cosmetics* and alcohol. It is particularly important in these categories because they are *social* products. In these categories, brand choice says something to other people about 'me'. We buy the brand that will create the impression of ourselves that we want to create.

Now, let's look at how research is used in the development of a brand personality. Research needs to take into account all the elements that contribute to the personality, that is, it needs to cover the whole spectrum of consumers' contact with the brand: the product's attributes—its appearance, smell, texture and performance characteristics; its packaging; its pricing; its advertising; its retail outlet; and so on. The interaction between these elements is important. If the personality is to be strong, these elements need to be consistent; they need to be mutually reinforcing.

The research question differs a little depending on whether we are dealing with an existing brand or introducing a new brand or a new product. With an existing brand, the first step is to find out how consumers see its current personality and how they see the personalities of its competitors. With an existing brand, it is often difficult to completely change the personality: it is difficult to change people's attitudes to a brand overnight, and while they may respond to being shown a different side of the brand's personality, they may not be able to relate to a complete change of face. However, when the positive and the negative aspects of the brand's personality are known, it is possible to develop strategies which will reinforce its positive aspects and play down its negative aspects.

When a completely new personality is to be developed for a new product or a new brand, the first step is to find out how consumers feel about the product category and about available brands. Brand personality must be consistent with product personality. For example, some products are seen as 'goodies' and some are seen as 'baddies'. Lemonade and milk chocolate are products which tend to be seen as virtue products; cola and coffee tend to be regarded a bit as vice products. If you try to infuse a tough product with a tender personality, this incompatibility is likely to result in a weakened impact. In the same way, product characteristics must be consistent with expectations for the product category. For example, if you take a mouthwash product, consumers have certain expectations about it: they expect it to work, they relate to it as a medicinal product. Medicinal products are expected to have certain attributes, e.g. a bad taste; that is, these products almost need to punish you before you feel that they are doing you any good. Listerine has capitalised on this expectation; rather than changing the taste of the product, or trying to tell people that it doesn't taste too bad, they have played up its bad taste in their advertising strategy. There would be no point in giving a product which is seen as having an almost medicinal function a personality which is bright, happy, easygoing. The product is essentially serious, and although the creative approach can be humorous, the product itself must be seen seriously.

Research can help to provide this understanding of what the product means to consumers and this understanding can help in setting parameters for the product or brand personality. If a personality is being developed for a new brand, then it is important to understand the personalities of existing brands. A personality will have more impact if it is unique; there is no point in being a 'me too' personality. The final personality probably will not come out of research but from creative inspiration; however, research can help in defining the parameters for the personality.

Understanding the consumer

A typical research program might start with some basic attitude and motivation research. The technique that is mainly used for this type of research is the group discussion technique. This technique involves getting together small groups of consumers, usually about six to eight, and just having them talk about the product category. Group discussions are usually conducted by a psychologist and the approach taken is relatively unstructured. At the beginning of the discussion consumers are just invited to talk together about the product category in any way they want to; this is called the non-directive approach. And it is a good way of getting to understand the consumer's point of view. It provides an opportunity for ideas to come up that the marketer and researcher might not have thought about. A typical group discussion might last for about two hours—this allows time for a relaxed atmosphere to develop, in which the consumers feel free to express their thoughts and feelings about the topic and to express them in their own way and at their own pace. The group discussion technique also allows a range of probing techniques to be used to elicit in-depth understanding of consumers' attitudes, needs, and motivations.

From such research we find out what motivates consumers to use the type of product under discussion. We find out what they are seeking from the product both in terms of performance benefits and in terms of more subjective benefits. We find out whether they have any dissatisfactions with available products which could be capitalised on by a new brand. We find out how consumers perceive existing brands, what image they hold of each brand, and how they relate to these different images. We also find out how they actually use the products—the way they use a product may suggest opportunities for new positionings in terms of usage, or may suggest a creative approach to capitalise upon usage factors.

There are some problems in obtaining an understanding of what really influences consumers' behaviour. For example, if we are trying to find out attitudes to toilet paper, consumers are likely to say things like, 'I want it to be soft'. If the softness dimension in toilet paper is to be fully understood, we need to know a little more than that. We need to know what softness *means* in consumer terms, whether this positive dimension has any implicit negatives, e.g. in relation to strength, and how important softness is in relation to other product attributes and benefits. Let's just look at the question of what softness means to the consumer. When a consumer says she wants a soft toilet paper, it is possible that she doesn't have a very strong feeling about it, it's just a word that springs to mind when she's asked to say what she likes in a toilet paper. Consumers have learned a vocabulary from advertising and they will often use words that seem meaningful, but which really may just be a play-back of an advertising claim. Examples of such words are: soft, pure, mild, natural, reliable.

The research problem is to break through the superficial response and find out whether these words do signify something important to consumers and what it is that they signify. One way of doing this is to ask people the *opposite* of the word. For example, if consumers are talking about a skin care product, and they describe it as 'pure', we might ask them, 'What's the opposite of *pure*?', and this procedure often helps to reveal the dimensions of the word *pure*. The opposites they give might be *irritating, contaminated, perfumed, artificial, harsh, cheap*. Consumers often find it easier to talk about negatives, what they don't want, rather than what they do want, and the emotional implications of a product attribute or benefit are often clearer when examined from the opposite point of view.

Another problem in research is that often people's behaviour is influenced by emotional factors of which they are not fully conscious. For example, if you ask a woman why she likes a particular brand of detergent, she's not going to tell you that it is because she can identify with the attractive, competent, efficient, modern, interesting image of the brand—she is more likely to say that it is because it gets clothes really clean, that it is economical. She is more likely to say that she doesn't use another brand because it doesn't get clothes as clean or bright rather than because it connotes to her a rather frumpy, fairly sloppy, harassed mother who leads a fairly housebound existence. So, how do we find out the underlying *emotional* relationship with the brand which is probably the most important factor determining behaviour? One way is the one already discussed—ask consumers to personify the product, to describe it as if it were a person. Another technique is what we call the *projective technique*. With this technique, we might show to respondents two pictures of women. These pictures will be fairly unstructured; they will be more outlines than concrete pictures of two women. We then say, 'this women uses brand A and this woman uses brand B', and we ask the respondents to just tell us about these women. What are they like? How old are they? What kind of life do they lead? How do they approach the washing? This technique often brings out well-defined and very different images of the users of these different brands.

Another technique for getting at consumers' emotional relationship with a product or brand is to use what we call a *fantasy*. For instance, if we are trying to find out what deodorant means to a woman, we may ask her to sit back, close her eyes and have a fantasy about a time when she's run out of deodorant. We then ask her to talk about what was happening, where she was, how she felt. This fantasy approach helps to bring consumers into contact with their feelings about the product. For example, they might play-back a fantasy of being alone, sitting all scrunched up, avoiding people, feeling dirty, and so on.

Group discussions are the most useful technique when we are trying to get an in-depth understanding of consumers' relationships with a product or brand. However, the technique is based on small sample sizes and for this reason the findings often cannot be extrapolated to the whole population. Consumers often differ in their attitudes to a product and in their motivations for using it. Similarly, a brand personality which has strong impact is unlikely to appeal to all consumers because consumers themselves differ in terms of personality, life-style, income, and so on. To provide an indication of the size of different market segments, a large-scale survey needs to be conducted. The group discussion technique is often used in an exploratory fashion to identify the factors that are important in the market and to identify the range of consumer attitudes and images; this information can then be used to structure a questionnaire for use in a large-scale survey of the market.

Concept and product testing

Depending on the marketing problem—e.g. are we looking for opportunities for introducing a new brand? are we looking for ways of revitalising or repositioning an existing brand whose sales' performance is dropping off? are we simply checking whether our market offering is as effective as it could be?—the research findings will be analysed in various ways. For a *new* product or brand, we will probably look very carefully at any consumer needs that are not being satisfied, any dissatisfactions with available products, and at the strengths of available brands. If we already have a brand in this market, it is possible that it is appealing to a particular segment of people and not touching others, and this may indicate an opportunity for a line extension or a new brand. With an existing brand, we will probably be looking carefully at its strengths and weaknesses relative to competing brands and at whether its consumer appeal could be strengthened in any way.

Let's take as an example the problem of introducing a new brand into an existing market. Our exploratory research has indicated that there is an opportunity for a new brand and has provided broad guidelines in terms of desirable product attributes, the benefits that should be promoted and the kind of personality that this brand should project. The next stage of research will probably be concept testing. The product or advertising concept usually embodies the key claims that will be made for the product and some elements of the product personality. It might take the form of some advertising copy, a product name, a pack design, and it might be presented in the form of a fairly rough press layout. Often at this stage, we will develop a number of alternative concepts to see which is the strongest. These are then researched to check that the intended product benefits are being clearly communicated, that they are meaningful to consumers and that they generate interest in the product. We can also check whether the name and pack design elements are consistent with the expectations generated by the advertising copy. This stage of the research may identify problems in communication or in the mix, which can then be modified before going into expensive creative production. We can also probe the kind of product expectations generated by this concept, in terms of attributes such as texture, colour, perfume and so on. If a product has already been developed, it can be shown to consumers to see if the actual product matches their expectations.

Another important research stage is *product testing*. Once a product has been developed it is usually tested on a sample of consumers in a real-life situation, that is, in the home. The product is left with consumers for a typical usage period and then they are interviewed to determine their reactions to it. Their reactions to the product are often assessed in comparison with their reactions to the brand they usually use. This stage of the research is very important; the ultimate test of the consumer/product relationship is whether the product matches their expectations. A brilliant brand personality might achieve a first sale, but if the product itself disappoints the consumer, it will not obtain a repeat sale. If the product test indicates a high level of acceptance of the product, the program will probably move into the final stage. If the product does not achieve a high level of acceptance, the research will usually indicate where the weaknesses are. The

Fig. 4.1 Chemistry of a hope

product can then be modified and re-tested. Before reaching this stage, *elements* of the product may have been tested; for example, taste characteristics and perfume characteristics may be examined in separate research projects to ensure consumer acceptability.

When we know we have the right concept, the right pack, and the right product, it is time to develop the final program. Again, this should be researched to check that it is an effective expression of the intended strategy. The advertising is tested in the target market to find the answers to the following questions. Does it have impact? Does it communicate effectively? Does it generate purchase interest? The final packaging is tested both in terms of design and function. In terms of design we ask: does it stand out—will it be noticed on the shelf? is it consistent with the product personality? If there are no problems, we are ready to go.

THE AUSTRALIAN COSMETICS INDUSTRY

In May 1975 the Industries Assistance Commission (IAC) presented its report *Cosmetics and toilet preparations*. It contains some interesting information about the Australian cosmetics industry.

> The local manufacturing industry has a turnover of over $100 million and employs directly about 5000 persons, the majority of whom are women. The industry is mainly located in Sydney [and]... is largely foreign owned and controlled . . . In addition to companies which manufacture and sell their own products, there are contract manufacturers which produce cosmetics . . . for marketing companies, other manufacturers and retailers . . .
>
> Local production was encouraged in 1929 when the Australian Government raised the customs duties . . . to 45% Preferential [most favoured nation agreements] and 60% General.

The report goes on to say that, when an additional special duty of 50% of the Customs Tariff duty was imposed in 1930, this represented such a virtual embargo on imports that many Australian importers and, later, large multinationals commenced production in Australia. It continues:

> Avon, an American owned company . . . commenced local manufacturing operations in 1968 and is now believed to be the largest local manufacturer and the largest marketer of cosmetics and toilet preparations in Australia.
>
> The production of aerosol deodorants and antiperspirants has risen significantly from approximately 6 million units in 1969, to around 17 million units in 1973. Other products which have shown considerable growth since 1966–67 are facial creams and lotions, hair tonics, suntan oils and shampoo. Products which showed a decrease . . . were face powder, hand cream, barrier creams and certain hair preparations . . .
>
> The number of individual ingredients used in a product varies according to the formulation used . . . the cost of ingredients ranged from about 3% to 40% of the cost to manufacture, although the percentage for most of the products was below 25% . . .
>
> Packaging accounted for between 20% to 60% of cost to manufacture, exceeding 40% in most instances . . . This expenditure on packaging materials represented 58% of the total cost . . . compared with a figure of only 5% for all manufacturing industry . . .

Avon estimated that the total Australian cosmetics and toilet preparations market at the retail level was valued at about $230 million in 1973 . . . [and] that the medium-priced segment accounted for about 70% of total sales . . . while the high-priced and low-priced segments accounted for 20% and 10% respectively.

Information presented to the Commission indicated that the high-priced products are usually restricted to major department stores and selectively franchised pharmacies; that the medium-priced products generally have high-quality packaging and are sold through pharmacies, major department stores and by direct selling methods; and that the low-priced products are less expensively packaged, are sold through chain stores, department stores and pharmacies, and include the house brands supplied by the local contract manufacturers in competition with low-priced cosmetics imported mainly from Britain and the United States.

It is understood that most of the major cosmetics suppliers trade on a franchise system where one or two pharmacies in a district are given the exclusive agency for a particular brand. Avon, however, distributes solely by direct selling methods using mainly part-time representatives.

The report quotes figures from the Australian Bureau of Statistics for 1968–69 which show that pharmacies made 51% of all retail sales (excluding direct selling), and department stores made 18% and continues:

The limited information available revealed retail mark-ups of up to 150% on estimated total cost . . . Mark-ups on imports appear to be significantly higher and in some instances retail prices were found to be more than 2000% of declared [import] value . . . the Commission noted that as the mark-up increased, the significance of the duty in the final price of the product decreased . . .

Avon stated that its experience of the Australian cosmetic market had shown that it was not particularly sensitive to price changes . . .

Consumer behaviour in Australia appears to conform with observed overseas patterns in that consumers associate high quality and prestige with higher prices. *Avon stated that there was little difference in the functional quality of the high and low priced products* [italics added]. The Commission believes that the industry uses packaging, advertising and promotion, and pricing to create, in the mind of the consumer, distinctions between basically similar goods.

It is only through inquiries of this type that the public is able to gain easy access to detailed information on an industry.

The final recommendation of the Industries Assistance Commission in its report was that most cosmetics be dutiable at a rate of 15%. However, industry lobbyists tended to adopt strategies to delay the introduction of a lower tariff, so that it could then be argued that the data on which the recommendations were based were out of date. As a consequence of the government's endorsement of IAC recommendations, the import duty on finished cosmetics was set at 15% from the introduction of the harmonised tariff on 1 January, 1988.

In 1986, sales totalled A$854 million at ex-factory or wholesale prices, excluding sales tax. The A$254 million (42%) increase in the two years since the 1984 edition of this book reflects the poor quality of the earlier data, rather than exceptional market growth. Similarly the breakdown variations are a result of more accurate data now available.

THE CHEMISTRY OF COSMETICS

Having discussed the marketing of cosmetics, let us now have a closer look at the cosmetics themselves. The range of goods covered by the term *cosmetics* is very large. It includes skin-care products, eye and face make-up, fragrances, hair-care products, hand creams and lotions, nail-care preparations, bath products, deodorants, depilatories, toothpastes and mouthwashes, shaving lotions and soaps, and sunscreens (see Table 4.1).

TABLE 4.1
Break-down of cosmetic sales in Australia, 1986

Category	Percentage
Women's perfumes and colognes	9.0
Women's colour cosmetics	16.0
Women's skin-care products	16.0
Hair preparations and shampoos	25.5
Men's toiletries	5.5
Soaps, toothpastes etc.	16.5
Baby-care products	3.5
Sun-protection products	2.0
Women's and family deodorants	6.0

Source: Cosmetic, Toiletry and Fragrance Association of Australia.

By way of introduction, before discussing several of these products, here is some background information on the skin. Because cosmetics are applied to the skin, it is helpful to understand its functions.

The skin

The skin is an important organ with many tasks to perform. It encloses the body, preventing some internal materials from escaping while allowing others to pass through, and at the same time, it keeps most external materials out. It regulates body temperature by controlling the escape of water (as sweat); it regulates the penetration of sunlight by allowing sufficient for the production of vitamin D, but not so much that the underlying tissues are damaged; and it senses and transmits information on temperature and pressure.

As Figure 4.2 shows, the skin comprises a large number of components. For convenience, it can be considered as four layers—subcutaneous tissue (underlying fatty material), on top of which is the dermis or true skin, which in turn is covered by an outer layer called the epidermis, on top of which is a horny layer, the stratum corneum.

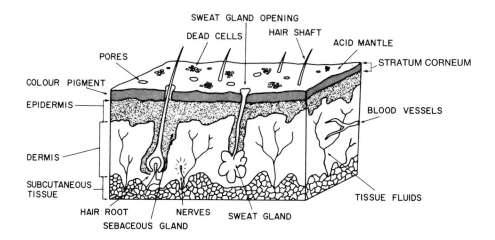

Fig. 4.2   Cross-section through the skin

The *stratum corneum* or the horny 10 μm outer layer of the skin consists of about 20 layers of dead cells containing mainly a protein called keratin. These cells are overlapped like roof shingles. Water-soluble substances cannot penetrate this layer, but oil-soluble substances (organic solvents but not ethanol) can dissolve the sebum (oil) that surrounds each hair follicle and then move along the follicle. These outer cells are being continuously sloughed off. This is obvious on the scalp, where the cells stay caught in the hair and we see the excess as dandruff. Underneath the stratum corneum lies the living epidermis, whose outer cells replenish the stratum corneum as it wears away. There are some chemicals that can destroy the stratum corneum, such as sodium hydroxide, formic acid and hydrogen fluoride. Keratinised growths (warts etc.) are removed with agents such as silver nitrate, acetic acid, lactic acid and resorcinol. Stronger agents are also used, such as trichloroacetic acid and phenol. Soap, detergents and organic solvents remove the sebum, after which the skin tends to crack and then dehydrate. This can lead to chemical dermatitis and infection.

The *epidermis* contains the pigment-producing cells that determine the degree of darkness of the skin and has a deep layer of growing cells.

The *dermis* contains the working elements of the skin—sensory nerves, hair follicles, blood vessels and sweat glands. Associated with each hair follicle are several sebaceous glands, which excrete, through the pores, an oily material called sebum. This is how the oiliness of the skin is controlled. Sebum is approximately 50% fats, 20% waxes and about 5% free fatty acids, which give the fluid a slightly acidic pH, helping a little to combat bacteria. This acidity would normally be neutralised by excess soap. Blackheads are formed when sebum dries in the skin ducts and blackens on reaction with the air. Pimples are inflammations caused by the irritation of sebum escaping into surrounding tissue. They are not initially caused by infection, although secondary infection can occur and make them worse.

Under normal conditions the water content of the skin is higher than that of the surrounding air, so that evaporation occurs. If this water is not replaced, the skin becomes dry. However, direct application of water to the skin does not replace the water in the skin. An oily vehicle such as used in a moisturiser is required to hold the water in contact with the skin and to hinder evaporation of the water.

Skin penetration by drugs

In treating the skin with drugs, the great problem is to arrange for the substance to enter into the skin without going through the skin. Salicylic acid ointments (10%), to suppress chronic skin conditions (e.g. dandruff, psoriasis), are often applied daily for years. Symptoms of poisoning by salicylic acid are chronic gastritis, singing in the ears, irregular heartbeats and even impaired hearing. Hexachlorophene (pHisoHex), while safe for adults, has killed new-born babies. Treatment of burns in children with a 3% solution has also caused death.

Lindane (see 'Pesticides and alternatives—Organochlorine compounds' in Chapter 5), which is used extensively to rid children of lice and scabies, can also penetrate the skin and has caused convulsions in patients who have been too liberally treated. Boric acid (boracic acid) has also been a health hazard as an additive in talcum powder used on babies' nappies. The symptoms produced ranged from vomiting to circulatory collapse and death.

The skin can carry out metabolic reactions similar to but much weaker than those of the liver. Drugs can be administered to the skin to avoid their inactivation through other modes of administration or to achieve a prolonged effect. Nitroglycerine ointment (2%) is applied to a patient's chest to treat an attack of angina pectoris, and this can be effective for about three hours. Seasickness can be kept at bay by applying hyoscine in a plaster affixed behind the ear.

Skin types

Cosmeticians and dermatologists disagree about the validity of classifying skins into distinct types. Dermatologists think in terms of a normal distribution about an average skin type, with most people showing minor variations; that is, some have more oil and others less oil than the average. For the cosmetics industry, however, the greater the degree of classification, the greater the variety of products that can be offered. Therefore, cosmeticians divide skins into the following types:

- *Normal*—although 90% of children have this ideal skin, very few adults can boast of a skin that is smooth, soft, moist and of healthy appearance.
- *Oily*—shiny, with enlarged pores, a tendency to blemishes and coarse texture, and sometimes a flakiness caused by the accumulation of dried out oils.
- *Dry*—fine-textured, flaky and taut; expression lines around eyes, mouth and throat; with ageing, dry skin loses elasticity and develops pronounced wrinkles.
- *Combination*—areas of both dryness and oiliness.
- *Sensitive*—florid, with tiny broken capillaries; usually fine-textured and with many of the characteristics of dry skin (a rather different definition from that used by dermatologists).
- *Blemished*—excessively oily, with blemishes such as pimples and blackheads.

Ageing

With age the skin loses its elasticity and becomes thinner and drier. Folds appear and wrinkles fail to vanish. Sometimes changes in colour occur. The process is universal and irreversible, yet there are some people who look younger at 50 than others do at 40. Although ageing is partly determined by factors such as heredity, dermatological research has shown that exposure to sunlight (i.e. to ultraviolet light) is an important contributor. Evidence of the effect of sunlight can be seen in the contrast between the leathery skin that is normally exposed and the soft, smooth skin that is usually covered, especially with outdoor workers and elderly people. Persistent and repeated exposure to sunlight can also cause skin cancer. The incidence of skin cancer is much higher in Australia than in Britain for people of comparable heredity.

The most obvious signs of ageing are wrinkles. Their cause is not known, but it is probable that they are the result of changes in the dermis or in the subcutaneous tissue. The dermis becomes thinner and less elastic, and the sebaceous glands are less active, leading to dryness.

Hair[2]

The reason we are 'naked apes' is not that we have fewer hair follicles than our chimpanzee cousins, but that we produce fine, short hair rather than fur. (Some simple genetic engineering is called for here!) Our heads, however, have about 150 000 individual hairs, each about 70 μm in diameter.

Hairs develop in long pits in the epidermis, extending deeply into the lower active dermis. At the base of the hair follicle is a region supplied with nutrients by blood vessels called the papilla. The rapidly multiplying cells are surrounded by an inner root sheath and as they move up they gradually fill with keratin, dry, harden and eventually die. As the hair emerges through the skin, it is compressed in a roller action which seals the cuticle, with cells overlapping, shingle style as in skin, or more accurately like a set of plant tubs or barrels stacked inside one another.

The growth of scalp hair is complex and fascinating. The growing phase lasts about six years in women and three to four years in men. The length attained, if uncut, is 70 to 80 cm and 40 to 50 cm, respectively. The growth then ceases and roots develop attaching the hair to the follicle. The follicle shrivels and rests for three to six months, then starts to produce a new hair, which pushes the old one out. So every day, about 100 mature hairs are pushed out (150 000 follicles in a hair production time of 365×4 days for a man), as normal turnover. At any one time, about 90% of the hairs are actively growing. The cuticle (of shingled cells) protects the hair shaft by forming a sheath around it. If a hair is cut, the cement that holds the overlapping cells together oozes out and seals the end. The structure of the hair shaft is a complex yarn (literally). There is a tangle of keratin protein chains, about 1 nm (10^{-9} m) long, imbedded in a protein matrix. Five of these threads then twist together to form a yarn. The yarn in turn is bundled into cables. A single hair will support about 80 g. So if hairs did not pull out, a mere thousand (less than 1% of the scalp) could support a person weighing 80 kg!

The pigment melanin, which gives skin its colour, also colours hair. Different amounts of the same pigment and in different places account for light blonde to blue-black. Only redheads are different, in that their hair contains an additional unique iron-based pigment, which indeed makes their hair 'rusty'. The melanin-containing granules are present mainly in the cortex of even the darkest European and Japanese hair, but in Bantu hair they are present also in the surrounding cuticle. Dark hair has both more granules and more melanin per granule.

The clue to shape of a mop of hair came from a study of the adsorption of dyes on wool. Basic dyes and metal-based stains adsorbed at two different regions of the fibre cortex, now called orthocortex and paracortex. The orthocortex is mainly in the form of fibrils and takes up basic dyes, while the paracortex has a matrix of protein with many exposed disulfide bridges between protein molecules which can take up the metal dye. In wool, the ortho and para forms tend to line up on each side of the fibre, which leads to the characteristic crimp of wool, because the para form crosslinks to form the inside of a bend. The most crimped human hair of some Africans has a similar structure, while the straight hair of the Japanese has the ortho and para forms arranged in concentric circles to give symmetric disulfide bridges. Other hair lies somewhere in between in structure.

A HAIRY TALE

The strong bonding of metals by the sulfur group on the protein in hair creates an interesting forensic possibility. The metal content of a particular section of hair reflects the metal intake of the person at the time that section of hair was actively growing. A classic historical case relates to the death of Napoleon in 1821 in (his second) exile on the island of St Helena. Examination of his hair at a later date showed significant levels of arsenic, and the position on the hair gave an approximate date of exposure. Arsenic was a common component in dyes used for the popular green wallpapers of the time and microbes digesting the paper could disperse the arsenic into the surroundings. There is no need for a deliberate poisoning hypothesis.

The condition or gloss of hair depends on the outer cuticle, whose stacked transparent plates reflect the light. Harsh chemicals, too much heat and excessive brushing dislodge the protective tiles, reduce reflection and give that dull appearance seen in the 'before' part of advertisements for hair products. The raised scales can catch one another and cause tangles, while the escape of moisture from the cortex causes the hair to appear dry and brittle.

The basic aim of all hair conditioners is to smooth the surface of the hair, and thereby avoid all the other negative effects. The newest silicone-based conditioners, containing polydimethylcyclosiloxanes are extremely good lubricants for hair. The reduction in friction also reduces static, and thus 'flyaway hair'. The displacement of moisture on the surface (similar compounds are used for waterproofing camping and wet-weather wear) allows hair to dry more quickly.

Making your hair darker is relatively easy, particularly if you don't mind doing it fairly often. Semipermanent rinses and washes all share the property of attaching the colour to the outside of the hair. The bright fashion colours use large granules, which vanish with one washing. The finer particles of more traditional rinses bond to the scales of the cuticle and survive many shampooings. With permanent dyeing, the idea is to move through the cuticle and enter the cortex cells. Hydrogen peroxide is used to soften the cortex and thus allow very small granules to pass through. It also bleaches the melanin and so allows lighter shades. Thirdly, its oxidation action clumps the granules in the cortex and thus hinders their migration out again. Hydrogen peroxide probably weakens the hair in the longer term. A popular preparation for camouflaging stray grey hairs works by depositing a layer of metal salts on the hair. The metal tarnishes in air to give a dull green-black result of dubious merit.

The chemistry of adjusting the curl again falls into groups depending on the degree of permanency required. The simplest procedure is to wet the hair, which protonates and breaks some of the disulfide bonds holding the proteins in shape. Drying the hair while it is held in either a straighter or curlier shape than before sets it temporarily. Indeed even heat alone on dry hair causes some changes. The slightest increase in humidity causes the hair to revert. This sensitivity of hair curl to heat and humidity is the basis of the picturesque 'weather houses' where two figures balanced on a horizontal axle are suspended on a (human) hair. The changing weather conditions twist the hair, which causes the figures to move in and out of a 'doorway' (Fig. 4.3).

A longer-lasting wave is obtained by spraying the hair with a solution of polymer. On evaporation of the solvent, the hair remains coated with a resin. (The most common polymers are poly(vinyl pyrrolidine)polymers, or co-polymers with vinyl acetate; co-polymers of maleic anhydride and vinyl acetate etc.) However, even with heavy, thick coatings, the effect does not last long.

Fig. 4.3   Weather house

Chemical attack on the internal proteins is the only way to obtain a 'permanent' wave. In about 1930, scientists at the Rockefeller Institute demonstrated that the disulfide bonds that give proteins their spatial 3D macroscopic structure can be split at ambient temperature and slightly alkaline pH by the action of sulfides or mercaptans. Thioglycollic acid is used in a suitable pH buffer together with cuticle softener. The smell of rotten eggs accompanies this process to a greater or lesser extent. Having rendered the internal structure of the hair floppy, the hair is set, the effect of the thioglycollic acid neutralised with peroxide, and the disulfide bonds formed to hold the hair in its new shape. (See Experiment 13.11 on permanent setting of wool.)

Incidentally, depilatory creams and lotions work on the same principle and with the same ingredients. They soften and loosen the unwanted hair, which is then easily removed.

In this fourth edition of *Chemistry in the marketplace*, the author's interest has shifted from topics such as nappy rinses to baldness. The orang-utan and a particular species of macaque monkey go bald in middle age like the 40% of men who show 'male-pattern baldness'. Both the tendency to baldness and its pattern tend to be inherited. More reassuring is that in male gorillas, social dominance is strongly associated with age, and the leader of a troupe often has grey hairs on his back and flanks.

The reason for this phenomenon lies at the other end of life. In the developing fetus, both forehead and scalp contain hair follicles and in fact form one tissue structure. At about the fifth month, the follicles on the scalp continue to grow while those on the forehead do not. After birth the hair follicles regress over most of the baby's body, including the forehead, and the hairs there become finer and almost invisible. In older age, under the influence of the male sex hormone testosterone, some follicles on the scalp become infantile again and the fine downy hair they produce is called lanugo. And a cure for baldness? An unexpected side-effect from a drug used to treat high blood pressure, minoxidil, sold under the brand name Loniten (Upjohn Co.), caused reasonable hair regrowth in about a third of patients, downy growth in another third and nothing in the remaining third. An ointment with the active ingredient is about to go on sale, after, presumably, the usual tests on the flanks of orang-utans.

Emulsions

The basis of most cosmetic products is an *emulsion*, which is the combination of two materials that do not normally dissolve in each other. Because they do not mix, the components are called phases. The emulsion system provides a convenient means of applying both an oily material and water simultaneously to the skin and hair.

In order for oil and water to form an emulsion an *emulsifier* must be added (Fig. 4.4). The purpose of the emulsifier is to reduce the difference in surface tension (i.e. the mutual repulsion) between the two phases. For example, in making mayonnaise from oil and vinegar, egg yolk acts as an emulsifier. Emulsifiers reduce the interfacial tension between the two immiscible phases; they allow the oil and water to 'wet' each other.

The selection of the correct emulsifier is critical. Emulsifiers are characterised

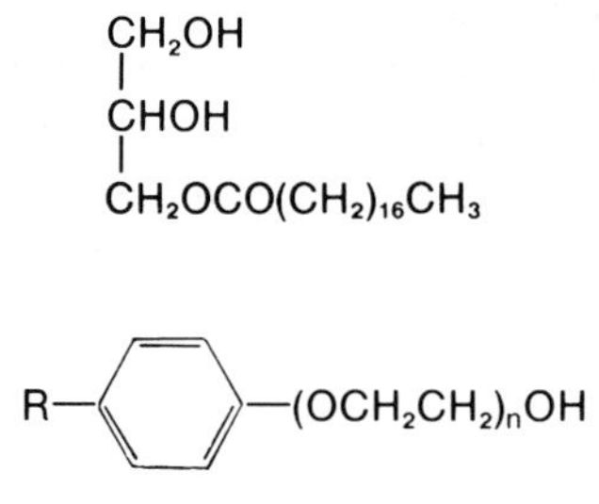

CH₂OH
|
CHOH
|
CH₂OCO(CH₂)₁₆CH₃

glycerol monostearate

a 'fat' with two of the fatty
acids missing; a *mono*glyceride.

R—⟨ ⟩—(OCH₂CH₂)ₙOH

polyoxyethylene alkyl phenol

a non-ionic detergent
e.g. triton X-100, teric X10

Fig. 4.4   Typical emulsifiers used in cosmetics

on a scale called HLB, the Hydrophilic-Lipophilic Balance (hydrophilic = water loving; lipophilic = oil loving). In this system the relative affinity of an emulsifier for the oil phase is expressed as a number ranging from 1 to 20. Propylene glycol monostearate has a low HLB number—it is more at home in the oil phase. Polyoxyethylene monostearate $CH_3(CH_2)_{16}COO(CH_2CH_2O)_n$ H, which has a long polyoxyethylene chain with lots of polar oxygen atoms, has a high HLB value and is quite at home in water. In general, emulsifiers with HLB values of 3 to 6 will produce emulsions of water dispersed in oil, whereas those with HLB values of 7 to 17 give emulsions of oil in water. Cosmetic emulsions can be either dispersions of oil in water—o/w—or water in oil—w/o (see Fig. 4.5).

The effect of the two different types of emulsions is quite different. Water evaporates from an o/w emulsion and this causes cooling and leaves a film of the oily ingredients (e.g. oils, waxes, emulsifiers, humectants). On the other hand, a w/o emulsion permits direct immediate contact of the oil phase with the skin. No cooling effects occur because the evaporation of the emulsified water is much more gradual. These are 'warm' emulsions in the sense of their apparent effect on skin temperature. As most cosmetic ingredients (such as perfume) are in the oil phases, there are greater formulating difficulties for w/o systems than for o/w types. The minimum concentration of the continuous phase must be at least 26% of the total.

In an emulsion, the finer the particle size is, the more stable the emulsion and the higher the *viscosity* (resistance to flow). (The lower the viscosity of the emulsion, the more runny it is.) Large particle size increases the tendency for the particles to coalesce and hence, finally, for the emulsion to separate into the two phases.

The easiest way to test which type of emulsion is present is to measure electrical resistance. Most oils have a much higher electrical resistance than aqueous solutions. The w/o type are relatively non-conducting compared to the o/w type. A simple resistance meter is often all that is required for this test, or a

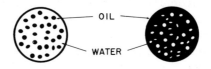

Fig. 4.5   Oil in water and water in oil emulsions

neon glow lamp and two electrodes can be used with a mains supply. Another test is to use an oil-soluble dye (e.g. fuchsin), which will spread and colour the surface only if the oil is in the continuous phase (i.e. w/o). Conversely, water-soluble dyes (e.g. food colours or methyl orange) will colour only o/w emulsions. Oil/water emulsions are used in 'vanishing creams' to assist penetration into the skin and to conceal the oily nature of the preparation.

Skin-care cosmetics

Moisturisers
Moisturisers are preparations that replace the water lost from the skin, and both o/w and w/o emulsions are used. The dryness and reduced flexibility of the skin cannot be corrected by adding oily materials, but the skin will become more flexible when water is replaced. The skin can be protected and skin dryness prevented or relieved by emollient creams and lotions, which slow the evaporation of water from the outer layer of the skin. Detergents cause dry and chapped skin because they dissolve some of the water-attracting components of the skin. As we get older these components are reduced in concentration anyway.

Cleansing creams and lotions
Even though adequate washing with soap and soft water will achieve the same result, there are possible advantages in using a cleansing cream to remove facial make-up, surface grime and oil from the face and throat. The specific chemical design of a cleansing cream allows it to dissolve or lift away more easily the greasy binding materials that hold pigments and grime on the skin. Most emulsified cleansing creams can be considered as cold-creams that have been modified to enhance their ability to remove make-up and grime.

The increased use of eye cosmetics has created a need for a means of removal. Mineral oil, alone, is a safe, effective agent.

Cleansers for oily skin
For the usual type of oily skin where there is no associated acne, the use of ethyl alcohol or isopropyl alcohol can provide temporary relief from excessive oil flow and the resulting shiny skin. The concentration of alcohol should not exceed 60% and preferably should be below 50% or it could be too drying or irritating. Other modifying ingredients are included in the formula to balance the harshness of the alcohol.

Deodorants and antiperspirants

The main ingredients used in antiperspirants are compounds of aluminium and zinc, and occasionally zirconium. The most common ingredient is aluminium chlorohydrate. The mode of action is unclear, but aluminium salts coagulate proteins and (in the form of alum) are used to stop bleeding from small cuts during shaving. The high charge of the ion (trivalent for Al) is important. See also Experiment 2.1 in Chapter 2 and p. 133. Something similar must be occurring in the sweat glands. Zirconium is considered hazardous in aerosol form because there is evidence that it causes lung tumours in animals. In fact aerosols themselves are suspect, irrespective of the compound. Although these antiperspirants have slight antiseptic action and thus act as deodorants, specific materials, such as chlorhexidine, triclosan (Irgasan) and zinc phenosulfonate, are generally added for this purpose. Except for silk and rayon, staining of fabrics is no longer a problem.

Lipsticks

The requirements of a good lipstick are: uniform, intense colour with good coverage; shiny but not too greasy; retention of form and consistency in reasonable temperatures; usable in cold temperatures without crumbling or breaking; stable to light, moisture and air; nontoxic and nonirritant; and neutral in taste.

Your tube of oil-wax base with antioxidant, preservative, perfume and colour has to perform all the above functions! Some of the properties are easier to test than others. For example it is easy to establish the 'droop point', the temperature at which lipstick lying flat in its case will droop against the case and ooze oil or flatten out. This should be over 45°C and preferably over 50°C. The colours and dyes in lipsticks are not regulated in Australia and the materials used include some that can cause skin allergy, especially on exposure to light (see also Sunscreens, below). Colours used include brilliant blue, erythrosine, amaranth, rhodamine, tartrazine and eosin (tetrabromofluorescein).

Lipsticks in their modern form were introduced after the First World War. They were coloured with carmine, a dye made from cochineal, a small red insect, by powdering the dried insect and treating it with ammonia. Indelible lipsticks were introduced in the 1920s. The dyes in these lipsticks had little colour in the tube but became coloured on reacting with the lips and stayed on for many hours. Tangee natural lipstick remained popular until the 1950s. In the 1960s the pale, lipless look was popular and there was no need for the long-lasting qualities. However, today we are back to the twenties, but with the additional twist of colour-changing lipsticks.

The body of a lipstick is a mixture of castor oil and wax, generally beeswax or carnuba wax (popular as a car wax because of its high melting point, 85°C). The aim is to have a mixture that is thixotropic; i.e. it should remain stiff in the tube but flow easily when under the pressure of applying it to the lips. Esters such as 2-propyl myristate (14 carbon carboxylic acid) are added to reduce 'stickiness'.

The colours used must be insoluble in water, otherwise you would lick them off in no time. Thus the dyes used are either oil-soluble or water-soluble dyes that have been combined with aluminium oxide (Al_2O_3) to precipitate as an insoluble 'lake' a pigment, usually solid. This precipitate can be suspended in castor oil, but does not actually dissolve in it. Such lake dyes are often brighter and more vigorous than the original water-soluble dyes from which they were made.

The colour-change lipsticks go back to the indelible technology of the 1920s and use a dye such as eosin (tetrabromofluorescein). This dye is lightly coloured, but becomes red when combined with the free amine ($-NH_2$) groups on the protein present in the skin. In the tube, this dye is masked by a lake dye (green or blue are popular), which is the colour you see when you apply the lipstick. This colour is overtaken by the eosin as it turns red and indeed is removed when the lipstick cream is wiped off, while the reacted dye remains. As the colour-change dyes do not work well at a basic pH, the lipsticks often contain citric or lactic acid.

Use your mood-change lipstick to test the pH of apples and potato slices or chips; it also reacts with some paper plates. Why?

Toothpaste

The human mouth contains a large number of bacteria. Within minutes of the tooth being cleaned, it is coated by a thin film (called pellicle) that is derived mainly from saliva. The pellicle is colonised by bacteria, which produce a sticky, gel-like substance called plaque in which to live. The bacteria in the plaque ferment sugars to produce acids, which attack the tooth enamel and eventually form a hole. Plaque builds up on the gum and causes gum inflammation (gingivitis) and later attacks the deeper tissue and bone (periodontal disease).

Toothpaste is meant to remove mouth odours, but 'bad breath' can originate in the mouth or come from the lungs. When you eat garlic, your social unease comes not from bits of garlic left in your mouth but from garlic entering the blood stream and releasing volatile sulfur vapours through your lungs. Why does an intact garlic bulb have virtually no odour? Enzymes are released when the plant is damaged, and they attack stored precursor molecules and release the gases.

Now, back to the bathroom. Toothpaste has a solid phase (a polishing agent or blend of agents) suspended in aqueous glycerol, sorbitol or propylene glycol (all polyalcohols) by means of a suspending agent (e.g. sodium carboxymethylcellu-lose: see also 'Domestic laundry detergents—The organic builder' in Chapter 2; 'Paints—Plastic or latex paints' in Chapter 8; Table 11.1). (A similar chemistry is used in the 'froth flotation' method; see 'Chemistry of surfaces', in this chapter.) The mixing is done in a vacuum to avoid air bubbles and to give a consistency to the paste. Air bubbles can also lead to deterioration in flavour. If the suspension is destroyed, the product becomes watery. A small percentage of anionic detergent (sodium lauryl sulfate) is added as a foaming agent.

An Australian draft standard for toothpaste stipulates a number of important requirements. It prohibits the use of sugar or other readily fermentable carbo-hydrate. The pH range allowed is quite wide (4.2 to 10.5). When heated to 45°C and kept there for 28 days, toothpaste must not form gas, separate or ferment. The container must withstand this temperature also, because this can be an important requirement in some parts of the country. Toothpaste should not run out of the tube when the cap is left off and the tube is left lying on its side, nor should it sink into the bristles of a toothbrush as soon as it is applied. However, it should not be so firm as to roll off the brush under normal use.

Because of consumer concern about the abrasive action of toothpaste, an optical illusion is used to disguise the abrasive in translucent toothpastes. The abrasive and the surrounding medium have approximately the same *refractive index* and so the abrasive particles cannot be seen. (The refractive index is a measure of a substance's ability to bend light.) If the abrasive and the surrounding medium had different refractive indices, you would be able to see the solid particles. (See also 'Paints—Hiding power' and box 'Disappearing tricks' in Chapter 8.)

However the abrasive function of a toothpaste is most important. In ascending

TABLE 4.2
Mohs' Hardness Scale (modified)

Hardness number	Material	Hardness number	Material
1	Talc	9	Topaz
2	Gypsum	10	Garnet
3	Calcite	11	Fused zirconia
4	Fluorite	12	Fused alumina
5	Apatite	13	Silicon carbide
6	Orthoclase	14	Boron carbide
7	Vitreous silica	15	Diamond
8	Quartz or shellite		

order of difficulty, the paste has to remove food residue, plaque, pellicle and calculus (or tartar). Calculus is plaque that has calcified or hardened. It can come to look and feel like an extra bit of tooth. Calcification can start within two to 14 days of plaque formation. The correct hardness of a toothpaste is critical. Hardness is measured according to a geological scale called Mohs' Scale (Table 4.2). Tooth enamel is quite hard (5.5–7), but the roots are soft (3.5–5) and can be worn away if gum disease has exposed them. You can check toothpastes to see if they will scratch against glass (which rates 5.5). Toothpaste should be below 5.5, and thus not scratch glass.

Baby-care products

The skin of a baby is a much less effective barrier than that of older children (from age 3) and of adults. Because it is thinner and softer and contains more water, material can pass through it both ways more easily. Because of the concentration of moisture and soil in the nappy (diaper) area, bacteria breed quickly and irritants remain in contact with the skin. This is the major cause of nappy rash. In 1921 it was proved conclusively that ammonia, liberated from urinary urea by the action of an enzyme present in a bacterium (*Bacillus ammoniagenes*) inhabiting the colon, was a cause of nappy rash:

$$CO(NH_2)_2 + 2H_2O \rightarrow CO_3(NH_4)_2$$
urea

$$CO_3(NH_4)_2 \rightarrow 2NH_3 + H_2O + CO_2$$
ammonia

Other bacterial organisms act similarly on urea. These bacteria are present in the faeces under normal conditions. When the intestinal contents are of a low acidity, the number of such organisms is greatly increased, and consequently the nappy region becomes infected with the bacteria. They grow most rapidly under neutral to alkaline conditions (pH 7–9), whereas there is practically no growth at pH 6.

Nappy rash can be a frightening condition to a young mother. It happens only rarely during the baby's first months, but later it can be very severe. Fortunately, in most cases, nappy rash responds readily to treatment, but it can persist or occur again. To avoid the problem the following suggestions should be followed:

1. Plastic pants are great from mum's point of view but they also suit the bacteria which form the ammonia and the nappy rash—so they generally make things worse.
2. Obviously, the more often nappies are changed, the less likely the problem will be. The skin should be bathed with water each time the nappy is changed. Do not wipe the baby's bottom with the unsoiled part of the nappy. Pat dry with a fresh towel or nappy.
3. After scraping soil off the nappies, launder in *hot* water (above 60°C), because then most of the relevant bacteria will not survive.
4. Rinse nappies *very* thoroughly to remove all chemicals—the (cationic) nappy washes do give softer nappies but their bacteria killing ability *in practice* has been disputed in *Choice*.
5. Sun drying is far preferable to other methods because of the sanitising effect of sunlight.

Manufacturers make nappy conditioners, nappy sterilisers and nappy washes to assist in combatting nappy rash. Some claim to sterilise, some to sanitise and some to soften. The compositions of two typical nappy sanitisers are given in Table 4.3. These are approximate ingredients only, true at the time of analysis.

After many complaints by consumers about the effectiveness of nappy treatments, an Australian Standard (AS2351, 1980) was published. The standard uses the organisms *Escherichia coli* and *Staphylococcus aureus* as its testing species. Other testing organisations have tended to use a wider selection of microbes and have therefore found some preparations less than satisfactory. On the other hand, nappy softeners (cationic detergents added in the final rinse), while not being very effective destroyers of bacteria, do remain absorbed on the nappy, and when the nappy is used next time, they may inhibit the bacteria from forming ammonia in the wet nappy.

The standard commercial disposable nappy consists of two main components: highly absorbent pulp on the inside and an outer cover of non-woven fabric. Washable nappy liners generally consist of a fabric measuring 15 × 40 cm,

TABLE 4.3
Compositions of nappy sanitisers

Napisan	Nursil
Sodium tripolyphosphate (see Fig. 2.7)	Sodium perborate
Sodium chloride (common salt)	Sodium tripolyphosphate
Potassium persulfate	Sodium bicarbonate
Surfactant	Sodium chloride
Fluorescers	Surfactant
	Optical brightener, perfume and paraffin
This product operates on the principle that potassium persulfate and salt are stable as a mixed powder but on dissolving in water react to form chlorine bleach, which is constantly generated.	This is similar to a built laundry detergent (surfactant plus additive) but with extra perborate bleach and less surfactant.

specifically knitted, not woven, from a specially prepared type of polypropylene yarn, which allows the urine to drain through the interstices of the knitted plastic.

Baby oils are generally based on a mineral oil. They may also contain vegetable oil, lanolin, antioxidants and germicides. Products such as pHisoDerm are emulsions and lather like soap. Their pH is slightly acidic so that it matches that of the skin. The product pHisoHex contains hexachlorophene, which is no longer recommended for continuous use on infants since a large number of babies died when a sample talc used in France had a many-fold excess of this material included by mistake. It is difficult to ascertain from the technical literature just how baby cosmetics differ from those made for adults.

The US Food and Drug Administration (FDA) announced that all labels affixed to cosmetic and toiletry products after 15 April 1977 must list the ingredients in descending order of predominance. An ingredient list allows shoppers to make comparisons and helps consumers avoid ingredients to which they are allergic or sensitive.

Sunlight on skin

The sun irradiates the surface of the earth in the wavelength range from 290 nanometres ($1nm = 10^{-9}$ m) in the ultraviolet through the visible into the infra-red or heat end of the spectrum. The amount of ultraviolet radiation reaching us is limited by its absorption by ozone in the upper atmosphere—hence the concern over the possible destruction of the ozone layer by the emissions from supersonic, high-flying commercial and military aircraft and fluorocarbon propellants in aerosols.

Ultraviolet light damages the hereditary material in the cell (DNA) by causing two pyrimidine bases, usually, but not always, on the same strand, to become covalently linked as a dimer. In ultraviolet-sensitive bacteria this linkage inhibits DNA replication and thereby stops further growth. It appears that normal human skin cells have an enzyme system for the repair of ultraviolet damage by excising the dimers and repairing the gap. A rare genetic disease where this enzyme is lacking makes its sufferers very liable to skin cancer.

The natural skin colour of humans evolved to match the intensity of sunlight according to the region of the earth that they inhabited. However, with migration and colonisation, this neat balance has been upset: Caucasians in hot climates suffer from sunburn and increased skin cancer, and Negroes in cold climates have problems caused by insufficient vitamin D synthesis in the skin. So do Bedouin women, who wear veils as well as the traditional black robes, which act as radiation shields while allowing convective cooling. Tanning is the negative-feedback system developed by nature to control the level of sunlight activity on the skin. On exposure to sunlight the ultraviolet rays cause melanin, a pigment in the skin, to darken at the surface. More melanin is produced in the lower layer, and it darkens and moves to the surface. The skin also tends to thicken. The processes of tanning, sunburn and vitamin D production are closely linked, as Figure 4.6 shows.

The ultraviolet range of the spectrum is classified into three regions, the approximate range of wavelengths (nm) being as follows: near UV, 'A', 320–400;

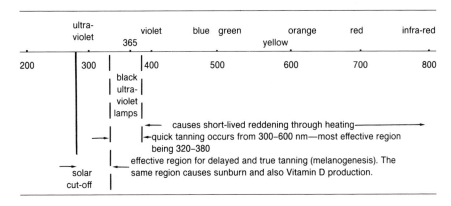

Fig. 4.6　The effect of different regions of the sun's radiation on skin

mid UV, 'B', 280–320; far UV, 'C', 200–280; actinic UV, 200–320. The intensity of sunlight at various wavelengths is shown in Figure 4.7. A great deal of the mid UV is removed by the earth's atmosphere. (The wavelength sensitivity of the human eye is shown for comparison.) The peak sensitivity of the skin occurs at about 297 nm. For wavelengths greater than 320 nm, the sensitivity drops to less than one thousandth of the value at the peak at 297 nm. When the dependence of the solar radiation on wavelength is combined with the dependence of skin sensitivity on wavelength, the product curve shows a maximum at about 305 nm (Fig. 4.8). Although much is said about 'broad spectrum' effects, the increase in the intensity of solar radiation in the near UV compared to the mid UV is much less than the factor of one thousand drop in skin sensitivity.

A historical view of sunbathing amongst Caucasians is revealing. For most of recorded history, white skin implied a lofty position in society. While workers, serfs and slaves spent most of their time in the sun, aristocrats sought shade by carrying parasols, wearing hats and sun bonnets and staying indoors. However, for most, the Industrial Revolution did away with the pursuit of pallor. Workers, herded in factories, spent long hours indoors. Industrialisation made shade cheap and sunlight expensive. A suntan showed that its wearers had the wealth and leisure time to travel to places where they could get a lot of sun.

The tanning of our skin involves a complicated polymer called melanin. This polymer is the pigment of our skin, and we are all born with different amounts of it. Fair-skinned people have a little, swarthy-complexioned people have more and black people have a lot.

Melanin reacts to the sun in two stages. In the first stage, pale (unoxidised) melanin granules near our skin's surface are changed by ultraviolet light to their dark-brown (oxidised) form. This gives an immediate tan, usually within an hour. It fades within a day. A more lasting tan results from the second stage. In this process, new quantities of melanin are produced from tyrosine, an abundant amino acid in our skin's protein. This second-stage tan endures for several days without further exposure. Additional sunbathing not only produces more melanin but also lengthens the polymer chains and deepens the colour.

However, the final effect of ultraviolet radiation is the damage to the proteins that make up the skin's connective and elastic tissue. This leads to an irreversible wrinkled, leathery and sagging skin. The word 'tanning' for the effect of sun on skin is well chosen (see 'Leather' in Chapter 7).

There are formulations that induce a suntan-like darker skin without sunlight. The tan is artificial because no melanin is involved in the process. This browning offers no protection against sunburn. A brown complexion is formed with the skin protein by the active ingredient (usually dihydroxyacetone). This material is produced commercially from glycerine using the bacterium that converts alcohol into vinegar.

Of the three types of skin cancer, the least common but most dangerous is melanoma. Death from melanoma has increased since the 1920s, and its victims

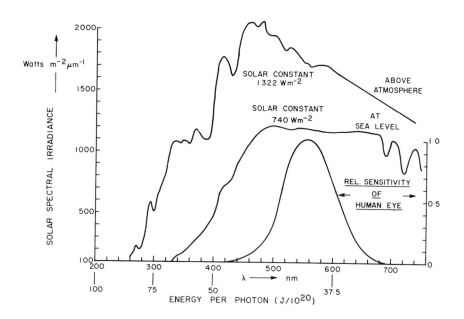

Fig. 4.7 Sunlight above the atmosphere and at sea level. Also shown is the colour sensitivity of the human eye. Note the solar irradiance P expressed in units of wavelength (λ) μm (micrometres) is watts m$^{-2}\mu m^{-1}$ and is related to that expressed in units of frequency (ν) watts m^{-2}Hz^{-1} through the relationship $P_\nu(\nu) = \dfrac{d\lambda}{d\nu} P_\lambda(\lambda) = \dfrac{c}{\nu^2} P_\lambda(\lambda)$ where c is the speed of light ($2.998 \times 10^{14}\ \mu ms^{-1}$). The conversion of either of these to the spectral energy density is given by $\rho(\nu)$ [joule m^{-3}Hz^{-1}] $= \dfrac{1}{c} P_\nu(\nu)$ and $\rho(\lambda)$ [joule m$^{-3}\ \mu m^{-1}$] $= \dfrac{1}{c}P_\lambda(\lambda)$ where c is now $2.998 \times 10^8 ms^{-1}$.

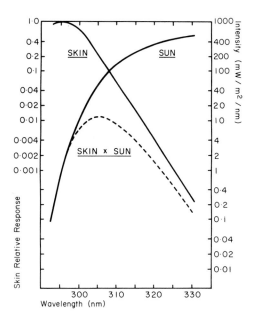

Fig. 4.8 The skin sensitivity and sun spectrum combination showing at each wavelength the equivalent intensity at 295 nm for the same sunburn result (After D.F. Robertson, University of Queensland)

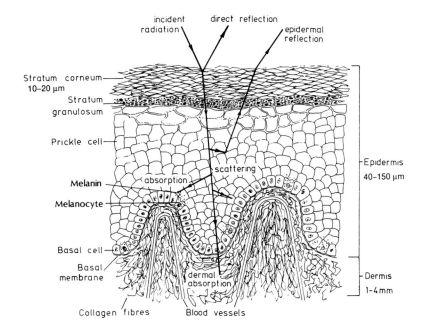

Fig. 4.9 Possible pathways for incident radiation through the human skin: release of melanin from melanocytes (Courtesy Dr Keith Lokan, Director, Australian Radiation Laboratory, Melbourne)

are often professional or managerial workers, not workers who spend their days in the sun. Epidemiology work by Dr B. K. Armstrong, Department of Medicine, University of Western Australia, has borne this out. In Queensland, one new case of melanoma occurs each year for every 6000 inhabitants. In cloudier Britain the incidence rate is one new case per year per 37 000. The Queensland rate is 50% higher than in the US, where nearly 20 000 will be diagnosed with the disease this year and more than half of these will die from it (mainly men).

While melanoma is increased by exposure to sunlight it does not necessarily occur on exposed parts of the body. Women in Australia develop melanomas mainly on the legs, while for men they occur anywhere. The rare instances of melanoma in dark-skinned people tend to be on the palms of the hand or soles of the feet, another clear example of the occurrence of melanomas in parts of the body not exposed to the source of radiation.

The 290–320 wavelength range is the active one that causes sunburn, triggers skin cancer, and produces vitamin D in the skin (which takes place through a photochemical reaction).

Sunscreens

From the above argument, it follows that, except for the immediate short-lived tanning response, it is not possible to screen selectively for true tanning while protecting against sunburn and skin cancer. Pigmentation requires preceding sunburn (which can be kept at a low level) to trigger the process. Sunscreens are used to lower the dose of light received by the skin to the point where tanning has a chance to catch up. The amount by which the dose must be lowered will depend on the intensity of the light and susceptibility of the skin. Sunscreens that are formulated to give a quick tan but to suppress sunburn (and true tanning) will absorb light more strongly in the 290–320 nm range than in the longer wavelength range. This can be checked by measuring an absorption spectrum. For the suppression of sunburn it is only the absorption in the 290–320 range which is of interest. Table 4.4 gives some results, and reveals that the p-aminobenzoates show an optimum ratio of quick tanning transmission to sunburn suppression. The safety and effectiveness of sunscreens have been reviewed by the US FDA Over the Counter Review Panel. Some of the accepted ingredients are listed in Table 4.4.

It is interesting that glass does not transmit much light with wavelength below

TABLE 4.4
Tanning range transmission of commercial sunscreening compounds

Sunscreen ingredient	Percentage of ingredient required to reduce sunburn transmission to 7%	Percentage transmission in the quick-tanning range of such a screen
A. p-dimethyaminobenzoate (PABA) derivatives	2.0	87.0
B. isoamyl and amyl N,N-dimethyl p-aminobenzoate (Escalol 506)	1.3	71.0
C. phenyl salicylates	11.0	68.0
D. 2-ethoxyethyl p-methoxycinnamate (Giv-tan F)	1.4	5.10

350 nm. If you sunbake behind a window, the main effect is reddening of the skin by heating. The same is not true of perspex, which does transmit light of shorter wavelengths.

For sunburn, it is also important to take reflected light into account. Typical reflection values for 300 nm radiation from specific surfaces are: snow, 0.85; dry sand, 0.17; water, 0.05; grass, 0.025 (the higher the index, the greater the reflection).

An Australian standard for sunscreens was issued in 1983. It was produced after an extensive six-year survey of the literature and critical discussion. At the beginning of this exercise the complexities of devising a reliable, yet not too expensive test for a sunscreen were hard to imagine.

What is important, however, is that an adequate amount of sunscreen be used. Because of the importance of adequate application of sunscreens, their sale in small packs and at high prices, encouraging sparing use, is counterproductive.

People's skins vary a little in the range of ultraviolet to which they are sensitive and so a quoted 'protection factor' can vary for different people for some sunscreens but not for others. The most effective sunscreens are zinc and titanium oxide creams. Although they may not look very attractive, at least you can tell they are still there, which is not the case with most screens. Some sunscreens are supposed to be swim-resistant, but they will wash off eventually, and it should be noted that fresh water dissolves sunscreens more effectively than salt water.

There are many parameters that are important in a sunscreen. The ability to absorb or reflect radiation is perhaps the easiest parameter to measure and it used to be quoted for commercial products. A sunscreen must be chemically and photochemically stable, otherwise its absorption ability changes with time. It must be soluble in the cosmetic base but insoluble in water or perspiration. The compounds given in Table 4.4 are modified versions of older compounds that reacted with and stained fabrics and towels. The self-plasticising properties of commercial compound A allow the formation of a continuous plastic layer on the skin. The benzophenones are optically more efficient than the aminobenzoates (i.e. they screen against sunburn more efficiently) but they crystallise easily and hence do not form a film and adhere as well to the skin.

ALEXANDER'S RAG TIME BAND

The story is told that Alexander the Great made use of the fact that some colours are photochemically unstable and bleach rapidly in sunlight. Because his commanders didn't have watches to synchronise their attacks, he gave them bleachable coloured rags to put around their arms so that they could measure time during the day. Thus came into being Alexander's rag time band.

Another problem is absorption of chemicals through the skin. When sunscreens containing benzoate or salicylates (~5%) are applied to the skin, these chemicals can be detected within 30 minutes in the urine of the user. Those people who need to avoid salicylates in food should also avoid absorption of them from sunscreens.

Evaluation of screening

The amount of light a material will let through (at a given wavelength) depends on the thickness of the material, d cm, the concentration of the material, c g/L and a property of the material called the absorption coefficient, a. The relationship between the light transmitted I, and the light falling on the surface I_0, is given by Beer's Law:

$$\log \frac{I}{I_0} = -a.c.d$$

I/I_0 is called the transmittance. It can be expressed as a percentage, $100\ I/I_0$, called the per cent transmittance:

$$\% \ T = 100 \ \frac{I}{I_0}$$

Thus a 1% transmission requires I/I_0 to be 0.01 and then $\log I/I_0 = -2$ and $a.c.d. = +2$. For a thickness of 0.01 mm, $a.c. = 2000$. Thus knowing the characteristic value of a for a material allows c, the necessary concentration, to be determined. Because the relationship is logarithmic, a 0.1% transmission requires 1½ times (not 10 times) the concentration required by a 1% transmission. However halving the concentration increases the transmission from 1.0% to 10%. What we have said about concentration is equally applicable to path length (thickness), d. These two parameters occupy equivalent positions in the equation. The percentage of light transmitted increases and decreases exponentially with both variables.

Other examples where this relationship applies are the thickness (normally given as weight) of window awnings used to provide shading from sunlight and the thickness of thermal insulation for 'shading' of heat radiation.

After about 20 minutes exposure to the midday sun, an average untanned white skin will be affected by sunburn although the actual reddening will not appear until after about six hours. This reddening will still be visible 24 hours later. The exposure needed to give this effect is called the minimum erythemal dose (MED) and it depends on the intensity of the radiation and the time of exposure. By comparing the time necessary to produce this MED on unprotected skin to that needed to produce it on skin protected with a standard amount of sunscreen, it is possible to give a protection factor (PF) for the sunscreen (independent of the absolute intensity of the radiation):

$$PF = \frac{\text{exposure duration for minimum erythema in } protected \text{ skin}}{\text{exposure duration for minimum erythema in } unprotected \text{ skin}}$$

A protection factor of 10 means that, if a sunscreen is used, a person can stay out in the sun about 10 times longer than without a sunscreen and achieve the same effect. The protection factor should be proportional to the quantity of UV light transmitted through the layer of sunscreen on the skin. That is, if the sunscreen has a transmittance (T) of 50%, it should provide a PF of 2. Conversely, a PF of 10 should correspond to a transmittance of 10%. When such a screen is tested *in vitro* (on an absorption instrument), the actual transmittance is closer to

$10^{-8}\%$! The most probable reason for this enormous difference is that the roughness of the skin can reduce the film thickness from the value calculated on the basis of the amount of material used divided by the surface area (normally about 20 μm) to about one tenth of that (average of 2 μm). With an assumption of this much thinner film, the PF and transmittance are in better agreement. It is problems such as this that have led to the rejection of laboratory measurements on transmittance of sunscreens to actual testing being carried out on human backs (*in vivo*) to determine the protection factor (Fig. 4.10). For a variety of technical reasons, the preferred sources of radiation are artificial (xenon and mercury UV sources) rather than sunlight. The largest source of error is how accurate and uniform the application of the sunscreen is.

Few cosmetic preparations are used as extensively as sunscreens or cover such a large area of skin. If one estimates the total skin area of an adult (which is about the same area as the area to be covered) to be about a square metre and assumes that a sunscreen lotion with an active concentration of 1.5% is used, then about 0.3 g of lotion will be deposited on the skin. This increases to 2.2 g for a lotion with 11% active concentration.

There are some very subtle effects of sunlight that are now being explored. Levels of the natural opiate, β-endorphin, rise after exposure to UV, which might explain the sensation of wellbeing that sunbathing can promote. On the other hand UV appears to weaken the body's immune system by deactivating certain cells in the skin called Langerhans cells. These cells act as 'presenters' that stimulate the proliferation of a particular kind of white cell in the blood called T

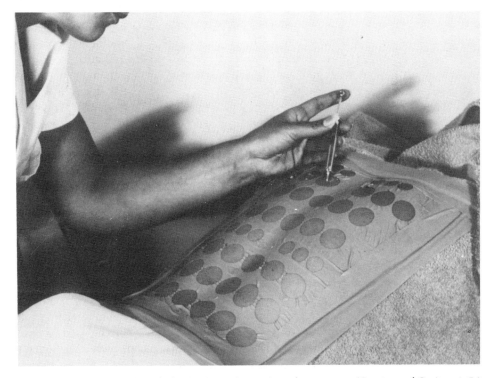

Fig. 4.10 *In vivo* testing of sunscreens (Courtesy of Greiter, A.G.)

cells. These T cells recognise foreign (protein) molecules (called antigens) and proliferate to overcome and destroy the invaders. If the T cells fail to proliferate, then our immunity is lowered. The precise mechanism of this photoprocess is still in dispute.

Sunglasses

In the *Canberra Consumer* (no. 63, September 1978), there was a report giving the results of optical transmission measurements on a wide range of sunglasses supplied by a local pharmacist. The work was carried out by the author and the pharmacist. The most surprising result was that a number of sunglasses had widely differing optical properties in the right- and left-hand lenses in the ultraviolet region. Some sunglasses showed evidence that the colouring had faded with time or exposure while on display and one pair showed a greater reduction in the visible light transmission than in the ultraviolet region.

The problems with sunglasses arose when they dropped the 'glass' and produced 'sunplastics'. Glass absorbs ultraviolet light (as we saw in the discussion of sunscreens) and infra-red (and hence is useful in greenhouses). Plastics on the other hand have variable optical properties in these regions.

The pupil of the eye opens up in response to a reduction of visible light intensity caused by absorption of light by a plastic sunglass lens. However if the plastic cuts back disproportionately less ultraviolet or infra-red radiation than it does visible light, the overall exposure of the eye lens (and retina) to these radiations will be increased. Excessive ultraviolet and infra-red radiation can cause eye damage. However the eye protects itself quite adequately in normal sunlight if *no* sunglasses are worn, and for comfort in the sun, hats are far better anyway, and give protection for the face as well (see Plate III).

Sunglasses of any type are dangerous for night driving.

The transmission of ultraviolet light at different wavelengths in the eye is shown in Figure 4.11.

The predominant emission of the ultraviolet (mercury fluorescent) lamps used in discos occurs at 365 nm. This wavelength causes the lens of the eye to fluoresce, and that is why these lamps appear fuzzy when you look at them. No damage is caused to the eyes at this wavelength. The same wavelength stimulates blue fluorescence from the whiteners used to launder shirts (see 'Domestic laundry detergents' in Chapter 2).

A report on the sunglasses results was prepared at the request of the Commonwealth Department of Business and Consumer Affairs in November 1978, which recommended a revision of the Australian Standard AS 1067–1971, and its enforcement under Section 62 of the Trade Practices Act. In early 1979, Standards Australia reconvened the committee and proceeded to develop a new and enforceable standard. On 1 October, 1985, a mandatory safety standard passed by the Commonwealth parliament came into effect, which required all sunglasses (local and imported) to meet the safety aspects of the new Australian Standard. Three categories are defined:

- general purpose sunglasses, to reduce glare;
- specific purpose sunglasses (ski etc.), to meet more stringent standards;
- fashion spectacles, worn for their appearance, rather than to protect the eyes.

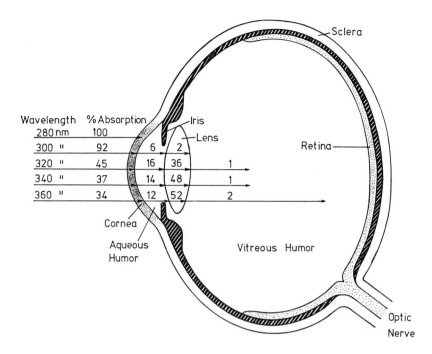

Wavelength 280 nm	% Absorption 100
300 "	92
320 "	45
340 "	37
360 "	34

Fig. 4.11 Ultraviolet radiation absorption in the eye. The values given are the percentages of the incident radiation absorbed in each layer of the eye. (Courtesy of Dr Keith Lokan, Director, Australian Radiation Laboratory, Melbourne)

Suppliers had three years to dispose of an estimated three million pairs of non-standard sunglasses, that is until 1 October 1988. This means that exactly 10 years will have passed between the publication of the experimental results and complete consumer protection—in the author's experience 'par for the course' (see 'Food additives—Labelling' in Chapter 11).

CHEMISTRY OF SURFACES

'Superficial' means pertaining to the surface. It has a derogative connotation because it implies the ignoring of important aspects below the surface. However surfaces are very important in themselves and often determine many aspects of chemical and physical processes. Our own surfaces are very important, as our use of cosmetics would indicate. To the comment 'Beauty is *only* skin deep', the most appropriate answer might well be 'so what!'

Consider a cube with 10 cm sides. It has a surface area of 0.6 m² (600 cm²). If you cut this cube up into cubes with 1 cm sides (1000 cubes), the surface area increases to 6 m². Further subdivision to cubes with 1 mm sides increases the surface area to 6000 m². Thus, the surface area of the porous particles of activated charcoal used in cleaning up chemicals and stomach upsets is quite enormous, maximally 800m²/g.

Such a large surface area can completely change the chemical reactivity of a material. A lump of coke is quite stable in air, but if you suck up the fine carbon powder used in some photocopiers into a vacuum cleaner, you can cause quite an explosion. Flour in flour mills can ignite spontaneously. In fact flour can be used as a fuel in a diesel engine—it will diesel (spontaneously combust) on compression (and hence heating) of air (see 'Chemistry of the car—Diesel engine' in Chapter 10).

Dragon's Breath, available from magician's supply houses, is a powder which when sprayed into a fire burns with a large billow of flame. It is composed of lyco-podium, a fine yellowish powder of club-moss spores (also available from laboratory supply houses). Lycopodium does not burn in bulk.

Small is better

If you look at a glass of soda or beer, you see bubbles of gas rising in the liquid because the liquid can flow. At the top of the beer there is foam in which gas is dis-persed as very small bubbles in a liquid. The gas does not move up and, in fact, you can't really distinguish the liquid and gas states. You can cut the foam with a knife and examine a slice of the colloidal state. Incidentally, the stabilising agent in *real* beer is the protein derived from the grain. However, when large amounts of adjunct are used (such as sugar, starch, potatoes), a protein extract of seaweed is needed to maintain a reasonable head. The protein 'protects' the air particles by denaturing at the interface. Next time you are at the beach you can lament over all that seaweed going to waste forming foam that could be on your beer (see also 'Alcoholic products—Beer' in Chapter 11).

If you are selecting a dry battery (Lechanché) in the supermarket, you may well be bewildered by having to choose between, say, Eveready silver and red. The dif-ference is only in the nature and area of the surface of one of the components, manganese dioxide. The larger the surface area the more extensive the reaction and hence the greater the current that can be drawn. The alkaline cell (gold colour) has powdered zinc as well. The alkaline reaction is also more efficient in the chemical use of the manganese dioxide (see also Fig. 10.20).

While on the topic of manganese, a nice (but dangerous) experiment demon-strates the rate of reaction as a function of particle size which involves the reaction of potassium permanganate (Condy's crystals) with oxidisable organic material such as glycerol or glycol (brake fluids, etc.). The delay before the reaction starts is a function of particle size. This reaction is the basis of the air-borne dropped capsules used in controlled burning in forests and is the so-called 'insurance' reaction, where people burn down unprofitable businesses using the delayed reaction in an attempt to establish an alibi. (Actually, the reaction of swimming pool 'chlorine', $Ca(OCl)_2$, with brake fluid is more spectacular—and more dangerous. See also 'Swimming pools' in Chapter 5.)

Let's look at some other examples. Why is cream more palatable than butter, which in turn is more palatable than lard? Lard is solid fat, whereas the butter is an emulsion of *water in fat*, which provides the fat with a large surface area with which to react with our digestive juices. Cream is an emulsion of *fat in water*, which provides an even greater surface area for the fat. Making butter from cream involves inverting a fat-in-water emulsion to a water-in-fat emulsion by denatur-ing the protein that 'protects' the fat globule and prevents it from coalescing. Our digestive detergents—the bile acids—form emulsions with fat and so increase the rate of reaction. (See 'Emulsions' earlier in this chapter.)

A most important emulsion in the Third World is that produced by the rubber tree—an emulsion of rubber latex in water. This emulsion is collected from the trees in the early morning when the flow is greatest. The flow is stimulated by a synthetic plant hormone, chlorethylpropionic acid. The latex is coagulated (curdled) with acetic acid (vinegar) and the excess water removed in several stages to produce mats, which are sun dried. These are bought by the traders, who take them to a factory where the rubber is macerated and pushed into blocks. The whole process is highly labour intensive and uses a potentially renewable source. However, the rubber collectors do not own the trees and so have no great incentive to conserve the trees as a resource.

Apropos emulsions (but otherwise not really relevant) is a rather attractive artificial opal that is made by 'setting' an emulsion of microscopic uniform polystyrene spheres in plastic. The small spheres pack themselves in a regular array to form a diffraction grating which obeys the Bragg equations, $n\lambda = d \sin \theta$ where $n = 0, 1, 2$, etc., λ is the wavelength of the light, d is the spacing of the spheres and θ is the angle between the light shining on the opal and the direction from which you look at the opal. This plastic replica mimics a real precious opal stone in which spheres of *silica* pack in a regular array. Smaller spheres give a smaller pattern and will diffract shorter wavelengths (blue, green), while larger spheres (or larger angles) will diffract longer wavelengths (orange, red). So small is beautiful!

In a polymer bead paint, hollow micro-beads of hard (cross-linked) latex replace pigment and so save money. The beads give whiteness to the paint by increasing the amount of light scattered (which is after all whiteness!) by providing an increase in film-air surfaces (see 'Paints—Hiding power' in Chapter 8).

Australia's most famous foam is probably not from a 'tinny' but is the pavlova, named after the ballerina. It is vital to add a *pinch* of salt and a *squirt* of vinegar. However, too much of either spells disaster in the form of a collapsing crust and a panic journey to a cake shop. Too much vinegar lowers the acidity (pH), which removes the charge of the protecting proteins (pK_a of COO^- is 4.8); too much salt collapses the atmosphere of counter-ions (Debye-Hückel layer) around the charged proteins. Both of these excesses remove the mechanisms that keep the bubbles of air apart, and hence keep the foam stable.

An old cook's tale recommended copper bowls for beating eggs into a better, creamier foam. In fact, experiments show that under identical conditions, whipping egg whites in a glass bowl produces a grainy and dry-looking foam after 10 minutes, whereas beating in a copper bowl produces a stiff and smooth foam after 20 minutes. Apparently a certain amount of copper is released and this reacts with the egg protein and stabilises the partially denatured protein film that constitutes the foam. So if your pavlova has collapsed, here is the chemistry to explain it. I am sure your guests will be impressed.

As they say in the trade, *pulchritudo per scientam* (the motto of the Australian Society of Cosmetic Chemists): 'beauty through science'. In cosmetics we often use emulsions. Oil in water emulsion gives 'cold' (aqueous) creams, where the evaporation of the outer water causes evaporative cooling, and water in oil emulsions give 'warm' creams.

Toothpastes are sold in liquid (glycerol) suspensions and in the 'transparent' toothpastes the refractive indices of the solid and liquid are matched so that the solid particles 'disappear' from sight.

As we have seen, sunscreens represent a very important cosmetic because of the danger of skin cancer and also because of the volume of material used on our bodies. The total skin area of an adult is about 1 m^2. The skin can absorb some chemicals very efficiently. In fact drugs are occasionally given by absorption behind the ear or in the nose.

While the ultraviolet screen in a sunscreen is the agent that absorbs the light, the quality of the spreading agent is vital. If the emulsion or solution does not spread to form a continuous, coherent, stable film on the skin but breaks up into globules, you could come out of the sun looking like a micro-dotted dalmatian!

The science of being small

What is a surface anyway? It is what separates two *phases*—generally gas (g), liquid (l) and solid (s). This gives five types of surfaces, s/s, s/g, s/l, l/l, l/g, noting that all gases are miscible (i.e. mix in all proportions). One of the characteristics of surfaces is that their production requires energy. You have to do work to create a surface (e.g. to blow a soap bubble). It is not so obvious, but equally true, that you need work to create a solid surface. That is why surfaces spontaneously contract— that is, reduce the amount of surface. Soap bubbles contract, and lots of small crystals in a crystallising dish dissolve and allow big crystals to become bigger. A small soap bubble contracts to allow a large bubble to grow (see 'Discussion topic 11', and Figure 2.16 in Chapter 2).

The *smaller* the bubble, the *higher* is the internal pressure. Actually the French mathematician, Laplace, is responsible for the precise relationship. The excess pressure inside a bubble, ΔP, is equal to twice the surface tension, γ, divided by the radius r of the bubble:

$$\Delta P = \frac{2\gamma}{r}$$

That raises an interesting question. When you form a bubble, say in a beaker of boiling water, the initial pressure required when $r = 0$ is infinite! In fact it is very difficult to boil very pure water in a very clean, blemish-free beaker. If you observe boiling water in a beaker closely, you will see that tiny air bubbles are released from flaws in the beaker. Into these air bubbles water vapour rapidly moves to form an expanding bubble of steam. The same flaw will continue to supply nuclei of air for some time. In a perfect beaker, with no nuclei, the water heats up and leaves the beaker in one go in a process called 'bumping'. A boiling chip (porous pot) is added deliberately to provide nuclei and prevent bumping.

The lability of small compared to large is true of 'solid bubbles' (i.e. drops). Small drops coalesce to form larger drops. Small drops have a higher vapour pressure than large drops. Special efforts are needed to keep small stable with respect to large; for example water drops in clouds, in emulsions and in precipitated crystals.

The increased pressure behind a curved surface has many consequences. A cylindrical column of water from a tap breaks up into droplets when the slightest disturbance causes a curvation in the surface along the cylinder length. As you lower the water pressure the smooth column of water becomes shorter. Columns of water formed along the threads of a spider's web as dew break up into consecutive microscopic small and large drops.

A somewhat dubious mechanical analogy is illustrated by the experiment in which you carefully stand with one foot on an empty aluminium drink can. If the can is undented, an average adult should be able to stand on the can without crushing it. A hollow column has considerable compressive strength.

However a sharp tap with a stick on the side of the can producing a minor kink will cause instant collapse, as the inhomogeneity propagates rapidly.

Liquid rises up a capillary tube if the liquid wets the surface of the tube (e.g. concave water/air surface) but is depressed down the capillary when the liquid does not wet the surface (e.g. convex mercury/air surface).

The liquid drop was the model used in explaining nuclear fission. As a nucleus becomes larger, the charged repulsion of the positive protons increases and *'lowers the surface tension'*. The nuclear drop then reverses the coalescing process and breaks up into two drops (fission products).

Emulsions such as mayonnaise are *thermodynamically* unstable. This means there is a natural tendency for the suspended droplets to coalesce and reduce the overall surface area and hence surface energy. The two components then separate into two layers.

We know that adding surfactants to water reduces the surface tension. There are also surfactants that, when added to oil, lower its surface energy. By choosing very effective agents we can reduce the surface energy between oil and water to zero—or even negative! Then the whole situation will reverse and the surface area will tend to *increase* and we find spontaneous emulsification to a *thermodynamically stable*, so-called *micro-emulsion* (in which the droplet size tries to become as small as possible, limited by the amount of surfactant added, which limits the total surface area possible). Micro-emulsions of oil in water are used experimentally in tertiary oil recovery from a Texas oil well, where about 70% of the oil remains in the well after it has been pumped 'dry'. So the use of micro-emulsions could be a very important new technology.

Suspending small particles

The 'behaviour' of soils, particularly clay soils, depends on the *surfaces* of the soil particles. The physical properties and behaviour of a soil are strongly influenced by the relative proportions of calcium, magnesium, sodium and potassium adsorbed onto the surfaces of its clay minerals as well as the amounts of organic matter, oxides and carbonate present. (See 'Soils ain't just soils' in Chapter 5.)

You can devise an interesting experiment using a swelling clay from cracked, dry clay pans (montmorillonite or fine-grained illite). Pack the clay in tubes with a fitted filter base. Add sodium chloride to one and calcium chloride to the other; then pass water through. Note the different rates at which the water passes through. This explains the effectiveness of the practice of throwing gypsum ($CaSO_4 \cdot 2H_2O$) into pig pens to reduce the sloppiness.

The best agricultural soils have a low proportion of 'exchangeable' sodium. As the proportion of exchangeable sodium increases (particularly if the exchangeable magnesium is in greater proportion than calcium), soils become difficult to work. In surface soils with a high exchangeable sodium, rain breaks down aggregates to a dispersed layer of soil in which water-conducting cracks are blocked and this leads to waterlogging. On drying, the clay sets hard, shrinks and cracks.

In washing clothes we have exactly the opposite problem. We want to *disperse* clay particles and hold them in suspension. So the opposite tactic is used. Washing soda ($Na_2CO_3 \cdot 10H_2O$) is added so that *sodium* ion can displace the *calcium* ion. Modern synthetic detergents use polyphosphates, which 'tie-up' calcium ion and remove its influence from the system (see 'Domestic laundry detergents— Inorganic builders' in Chapter 2).

When a swimming pool is full of suspended clay washed in by rain, we can co-agulate the fine particles by going one better than calcium. We use aluminium (which has a charge of three compared to calcium with two and sodium with one). The addition of alum (a salt of aluminium) 'flocculates' the clay into large particles and allows it to be caught in the filter and so the pool can be clarified again. (See also Experiment 2.1 in Chapter 2.) Some of the cationic detergent algaecides sold for swimming pool maintenance (which, incidentally, were originally developed for clearing yeast from beer-lines) can cause problems by the positively charged surfactant chain attaching itself to the negative clay particles and keeping them in suspension.

Actually, Australia's most important foam is neither beer head nor pavlova but the process used for separating crushed mineral ores using *froth flotation*. This process was developed by the CSIRO in the 1940s. Most ores consist of a number of mineral species, some of which are valuable and some worthless (gangue). Froth flotation is a technique for separating ground particles based on the different properties of their surfaces. If a material that is not wetted by water (such as finely ground candle wax) is placed in water and air is bubbled through, the bubbles will attach themselves to the wax particles and lift them to the surface. Air 'wets' the surface of wax better than water does.

Some minerals have 'waxy' surfaces and are naturally floatable. These are graphite, talc, sulfur, molybdenite (MoS_2), orpiment (As_2S_3). Most minerals are fairly wettable by water and won't float. If we can *selectively* render particular minerals 'waxy', then we can float those minerals. Originally pine oil was used as both a waxing agent and as a frothing agent. Today more selective materials are used.

Hair conditioners and laundry wash final rinses for wool both work by coating fibres with a waxy outer layer one molecule thick. As we saw in Chapter 2 the active ingredient is a cationic detergent with a positive charge (see also 'Swimming pools' in Chapter 5) that is attracted to the negative charge on the fibre. Glass also has a negative charge and is made waxy by these detergents. It follows that ground silica and silicate minerals can be floated using these detergents. On the other hand, minerals with *positive*-charged surfaces will be made waxy with *negative*-charged soaps and detergents, just as plastic plates are affected in the same way.

A surface can be modified more effectively by changing the surface with a chemical reaction rather than just relying on attraction of opposite charges. For floating sulfide ores of copper, lead, nickel and zinc, sulfur-containing compounds such as thiocarbonate (xanthate) or thiosulfate (photographic 'hypo') are used to change the surface chemically.

The children's toy material called Magic Sand (Wham-O Manufacturing Co., California: US Patent 3,562,153) is just ordinary sand that has been treated with a dye and coated with a hydrophobic silicon coating. When Magic Sand is placed in water, it can form underwater columns and other structures but is perfectly dry on removal. The 'cohesion' of Magic Sand under water is a form of phase separation that reduces the surface area of the water-sand interface—exactly the opposite effect we try to achieve by using detergents so that soil is wetted by water and deaggregates. The material was originally developed for large-scale clean-up of oil on water pollution. Why would it be effective for this purpose? What advantage would it have over the use of detergents?

Isn't it annoying when a baked cake adheres to the cake tin. This tends to happen when the water in the dough wets the metal surface and then dries during baking, depositing dissolved materials that then bond the product to the metal. Oiling the surface is intended to prevent wetting and hence adhesion. You can measure the degree of wetting by a liquid on a solid by studying the contact angle that the liquid and solid form (see Fig. 4.12). For further details on adhesion, see 'Adhesives' in Chapter 8.

Metals are characterised by high surface energies (replaces the older term of surface tension) ranging from 0.5 to 5 J/m^2, while water has a value of 0.07 J/m^2 and oil about 0.02 J/m^2. A material with a low surface energy will wet a material with higher surface energy. Both water and oil will wet metal. Oil will spread on water to a greater or lesser degree, but water won't spread on oil. Commercially a lecithin/white oil mixture is used to coat aluminium baking trays to prevent 'stickers'. Trays contaminated with polymerised processing oil when they are made may stick and need to be scrubbed with a solvent such as trichloroethylene. The build-up of polymerised oil on bread-baking pans is a major problem as it increases the incidence of sticking loaves and stained bread.

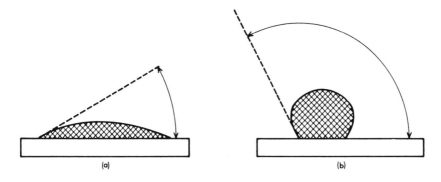

Fig. 4.12 Contact angles of a liquid on a solid surface: (a) small contact angle associated with a high degree of wetting; (b) large contact angle associated with a low degree of wetting. (From *Food Research Quarterly* 39, 1979, 30–33)

REFERENCES
1. This section is an edited version of a talk given by Ms M. Elfverson to the Australian Cosmetics Association.
2. From J. Cherfas, 'A hair piece for Christmas', *New Scientist* **19** (26) December 1985.

BIBLIOGRAPHY

General
Alexander, A. E. and Johnson, P. *Colloid science*, Oxford University Press, Oxford, 1950, ch. 23. In spite of its age, this is still a classic.

Balsam, M. S. *et al.* (eds). *Cosmetics: science and technology*, 2nd edn, vols 1,2,3. Wiley-Interscience, New York, 1972–74. Excellent reference material, somewhat repetitive in places, but will answer most questions on cosmetics.

'The consequences of exposure to ultraviolet radiation', Seminar, Cosmetic and Toiletry Manufacturers Association of Australia and Proprietary Association of Australia, Sydney, June 1982.

Consumer protection against toxicity of cosmetics and household products. OECD, Paris, 1974.

CTFA cosmetic ingredient dictionary. Cosmetic, Toiletry and Fragrance Association Inc., Washington DC, 1973.

Curry, K. V. 'Manufacture of cosmetics'. *Education in Chemistry*, November 1972, 233, (Chemical Society, London).

Groves, G. A. 'Cutaneous photobiology, Parts I, II and III'. *Australian Journal of Pharmacy*, October 1975, 547.

Gostelow, J. *Cosmetics in the school laboratory*. Shell, London. Contains recipes for a wide variety of cosmetic products. They are very simple. More sophisticated (but not necessarily so much better) recipes can be found in specialist books. Preservatives are not included, so the products should not be stored for lengthy periods before use.

Tooley, P. *Experiments in applied chemistry*. John Murray, London, 1978.

Face cosmetics
'Facial cleansers'. *Choice*, ACA, October 1985.
'Lipstick'. *Choice*, ACA, April 1983.
'Moisturisers'. *Choice*, ACA, **14** (8), 1973; February 1982.

Hair
Cherfas, J. 'A hair piece for Christmas'. *New Scientist* **12** (26), December 1985.
'Hair conditioners'. *Choice*, ACA, January 1985.
'Hair removal' *Choice*, ACA, November 1984.

Skin
'Cosmetic care for your skin'. *Consumers*, Consumers' Institute of New Zealand, February 1977.
'Eye make-up'. *Choice*, ACA, April, 1985.
'Headlice treatments'. *Choice*, ACA, February 1985.
'Nappy wash'. *Choice*, ACA, **16**(2), 1975; March 1979.

Sun and skin
Hughes, J. and Beggs, C. 'The dark side of sunlight'. *New Scientist*, 21 August 1986.
'Screening out the sun'. *Choice*, ACA, October 1984.

Selinger, B. 'Sunscreens help avoid a real tanning'. *Canberra Times*, 29 December 1980. Reprinted as 'Acquiring a suntan: healthy or hazardous', in *Australian Standard*, July 1981, SAA; reprinted in *ACT Cancer News*, February 1981.

Smith, W. D., 'Suntans: good news and bad'. *Sciquest* **52**(5), May/June 1979, US Department of Health, Education and Welfare; *News*, August 24 1978/ *Federal Register* August 25, 1978. Note that the sun protection factor was introduced in the US in 1978, well before its introduction in Australia in 1984.

'Sun on skin: investigating the costs'. *Ecos* **48**, Winter 1986, CSIRO.

Sunscreen products: evaluation and classification, Australian Standard AS260—1983. Being revised—draft revision DR84280.

Toothpastes

'What's in toothpaste'. *Choice*, ACA, May 1982.

Chemistry in the garden

5

SOILS AIN'T JUST SOILS

Soils

If you stick your hand in a bit of dirt in the backyard, you will come up with a paw-ful of something that will differ widely in physical and chemical composition, depending on where you live and how keen a gardener you are. Let us think of the soil as we might consider a wine.

Colour

Organic matter darkens any soil, so that topsoils are usually darker than subsoils. Light or grey soils are often leached of their hydrated iron oxides, which give the yellow through orange, brown and red colour to soil. The greenish-grey colour of some waterlogged soils is due to the reduced ferrous form of iron.

Texture

Soil particles range in size and these are defined in Table 5.1.

The proportions of these, and especially the clay content, give a soil its typical feel or texture. Field texture is determined by working the moistened soil in the hand and comparing its feel with that of a series of standard soils. A triangular graph (Fig. 5.1) can show the definition of soils based on the proportions of three components.

Water percolates readily through sands with little silt or clay, so saltier water can be tolerated in irrigation; but fertilisers leach out quickly too (slow release formulations should be used).

TABLE 5.1

Particle	Equivalent diameter mm	Surface area cm^2/g
Gravel	>2	
Coarse sand	2–0.2	23
Fine sand	0.2–0.02	90 to 230
Silt	0.02–0.002	450
Clay	<0.002	8 000 000

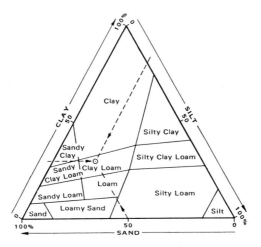

Fig. 5.1 Some grades of soil textures: The clay loam soil represented by ⊙ has 50% sand, 19% silt and 31% clay. (From *Soils: an outline of their properties and management*, CSIRO, Melbourne, 1977, p. 12)

Generally clay soils are difficult to cultivate, being sticky when wet and setting hard on drying, because of lack of sand-size particles, too little humus, and too much sodium. They are great for sealing earth-fill dams, but a pain for draining septic tank effluent. The type of mineral present is also very important. Adding sand to clay can open up the structure but, as Figure 5.1 indicates, amounts to the order of 60% are needed to make loams from clays. This is impractical except for small areas. The problem of excess sodium can be solved by adding calcium ion in the form of gypsum or lime. The calcium ion pushes the sodium ion off the clay particles and allows them to aggregate into larger clumps which do not pack as tightly, allowing easier circulation of air and water. A typical application rate is 0.5 kg/m^2, but it must cover the whole surface.

To test your soil for excess sodium, drop an approximately 6 mm (~ ¼ inch) diameter fragment of soil into a glass of distilled water (rainwater or defrost refrigerator water). Leave to stand for 24 hours. If the fragment is surrounded by a milky-brown halo, the soil will certainly respond to gypsum.

Mineralogy

Sand and silt contain a variety of minerals that are similar in physical properties although they differ chemically. This is not true of clay minerals, whose two major types, kaolinite and montmorillonite, have very different physical properties. Because clay minerals have crystal shapes with a very large surface-to-volume ratio, the shape and chemical bonding between particles determines the physical behaviour. Montmorillonite clays are called reactive clays and swell by as much as three-fold as water molecules enter between the sheets of atoms in their crystal lattice. They cause engineering problems for buildings, as they shrink at different rates on drying out. Trees can cause cracking of buildings by drying out the soil under the foundations. The physical condition of the soil in relation to plant growth is called tilth.

Chemistry

Soils provide an anchor for plant roots and also provide the chemical elements necessary for growth. The elements nitrogen, phosphorus, potassium, calcium, magnesium and sulfur are called the major elements, while manganese, boron, iron, copper, zinc, molybdenum, sodium, chlorine and cobalt are called trace elements. Other elements are not essential for the plants but are taken up and are essential for animals feeding on the plants. The availability of elements depends not only on the amount present, but also on the form in which it is present, the rate it is released from the minerals and the pH of the soil. Positive ions are held on negatively charged clay and humus particles. Excess potassium can prevent the uptake of magnesium, and vice versa. Even with adequate zinc in the soil, excess superphosphate can induce zinc deficiency. Excess manganese can inhibit cobalt uptake, essential for the bacteria that fix nitrogen in the root nodules of legumes.

The pH of the soil has a critical effect on element availability, as seen in Figure 5.2. Thus heavy liming will reduce the availability of iron and manganese, while ammonium sulfate causes the pH to drop and can cause molybdenum deficiency and possibly manganese toxicity.

Food for plants

What *is* a cabbage? We see the top as a symmetrically packed set of leaves, while a worm would see a root system exploring the soil. Both leaves and roots 'catch' food

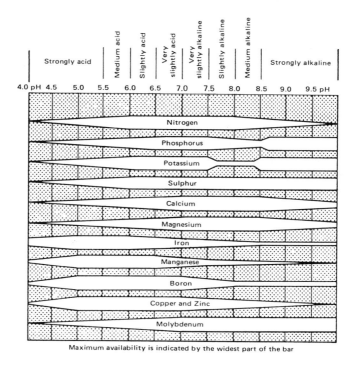

Maximum availability is indicated by the widest part of the bar

Fig. 5.2   The availability to plants of nutrient elements varies with soil pH (From *Soils: an outline of their properties and management,* Discovering Soils 1, CSIRO, Melbourne, 1977, p. 20)

for the whole plant. If you dry a 2 kg cabbage in an oven, driving off all the water, you will be left with about 160 g dry weight. Burning will reduce this to a grey-white ash of about 12 g, and mainly carbon dioxide gas is given off. The ash is only 0.6% of the original mass of the cabbage and contains a wide variety of elements.

Carbon dioxide in the air is present at a level of only 0.033% and yet this is the sole source of organic carbon in plant and animal. One form of fertilising in greenhouses is to artificially increase the carbon dioxide level to around 0.1 to 0.15%. In contrast 78% of the air is nitrogen, but plants cannot use this until it is 'fixed' as ammonium cation or nitrate anion. Micro-organisms such as *Rhizobium* that live in the nodules on the roots of legumes fix the equivalent of at least 10 million tonnes of ammonium sulfate every year.

Phosphorus is vital to virtually all processes of life, particularly in the processes of energy usage. Australian soils are nearly all deficient in this element. Agricultural products remove it on cropping: e.g. 3 to 6 kg/ha for cereal grains, 5 to 15 kg/ha for hay and lawn clippings. As we transfer nutrients from soil to sea via sewage, we need to provide a continuous input to the soil. Superphosphate is made from imported rock phosphate and sulfuric acid. When added to the soil some is 'fixed' by soil minerals and some is available to plants. The fraction fixed decreases with each addition. Blood and bone and other organic fertilisers provide slow release formulations of nutrients.

Potassium is important in the balance of sap minerals and is involved in extension of stems and thickening of the outer walls in the outer cells of the plant. Its absence leads to accumulation of sugars and nitrate in plant tissues. Sugars provide food for attacking organisms and so make the plant more susceptible to disease. Potassium occurs in micas and on the surface of clay particles. Deficiencies occur mainly in the wet, coastal regions. Good sources are wood ash (hence the name 'potash'), flue dust from cement works, seaweed and urine.

Calcium is needed for cell division and cell walls. Only sandy, acid soils in high rainfall areas are likely to be deficient.

Magnesium is needed to produce chlorophyll and hence is vital for photosynthesis. Deficiencies are the same as for calcium, but high doses of potassium can cause magnesium deficiencies.

Sulfur is essential for the amino acids cysteine and methionine, and so is essential for protein production. Sulfur is also a constituent of many flavours and odour components in plants, e.g. brussels sprouts, cabbages and onions. The smell of urine after eating asparagus is due to sulfur compounds. Sulfur loss is a problem in sandy soils. It is readily available from sulfate and often the response to superphosphate is as much due to the sulfur as the phosphorus. Sulfate from sea-spray can be a major source near the sea, while burning fossil fuels add sulfuric acid to rain. In highly industrialised countries amounts of 30 kg/ha can reach the soil each year from this source and acidify the rainwater to such an extent that trees and other plants are damaged.

Trace elements

Iron is essential to the function of the chloroplasts in photosynthesis. Iron is present in abundance in all soils as oxides and silicates. However other soil components can interfere with its availability, particularly lime, either added or naturally occurring as limestone. 'Lime-induced chlorosis' is the name given to the disease this causes in such plants as citrus, soft fruits, beet, spinach, peas, and in many native plants transferred from their natural acid-soil environment to an alkaline soil.

Manganese is important for reasons that are not well understood. It is present in soils as insoluble oxides formed from soluble compounds by bacteria (particularly in alkaline soils).

Copper is essential to the production of enzymes, critical to plant cell function. Deficiency occurs in acid sands, loams and gravelly soils. Copper toxicity can occur through the use of copper-containing sprays, such as Bordeaux mixture.

Zinc is involved in producing the plant hormone auxin responsible for stem elongation and leaf expansion.

Boron is important for the uptake of calcium. Boron levels are high in sea water and so also in soils with a marine sediment origin, sometimes to a toxic level. Liming can reduce boron toxicity.

NUTRIENT DEFICIENCIES IN PLANTS—A QUICK GUIDE

Symptoms appear first in the OLDEST leaves

Nitrogen	general yellowing; stunting; premature maturity
Magnesium	patchy yellowing; brilliant colours especially around edge
Potassium	scorched margins; spots surrounded by pale zones
Phosphorus	yellowing; erect habit; lack-lustre look; blue-green, purple colours
Molybdenum	mottling over whole leaf but little pigmentation; cupping of leaves and distortion of stems
Cobalt	legumes only, as for nitrogen
(Excess salt)	(marginal scorching, generally no spotting)

Symptoms appear first in either the OLDEST or YOUNGEST leaves

Manganese	interveinal yellowing; veins pale green, diffuse; water-soaked spots; worst in dull weather

Symptoms appear first in the YOUNGEST leaves

Calcium	tiphooking; blackening and death
Sulfur	yellowing; smallness; rolled down; some pigmentation
Iron	interveinal yellowing; veins sharply green, youngest leaves almost white if severe
Copper	Dark blue-green; curling; twisting; death of tips
Zinc	smallness; bunching; yellow-white mottling
Boron	yellowing margins; crumpling; blackening; distortion

Source: 'What's wrong with my soil?' *Discovering soils* no. 4, CSIRO, Melbourne, 1978, p. 4.

Molybdenum is needed by the bacteria that fix atmospheric nitrogen and so is particularly important for legumes. In addition, it is needed to form proteins from soluble nitrogen compounds. Molybdenum is associated with iron minerals. It dissolves more rapidly in alkaline soils.

Cobalt is also needed by the nitrogen-fixing bacteria (as well as directly by some plants). Thus cobalt deficiency, like molybdenum, will be seen in legumes as a nitrogen deficiency. Cobalt in soils may be deficient, or just mopped up by manganese oxides.

Fertilisers

Labels on fertilisers can be confusing. A shorthand is used whereby only the main nutrient elements are listed, leaving out oxygen, chlorine and water, so the percentages never add up to 100%. Here are some of the things you will read:

N or N(t) is the percentage total nitrogen in the fertiliser, while N(a) is the percentage nitrogen as ammonium (there as ammonium sulfate, nitrate or phosphate). N(n) is the percentage nitrogen as nitrate (usually as calcium or ammonium nitrate). Nitrogen as urea is indicated by N(u). This rapidly breaks down to ammonium salts. N(o) is the percentage nitrogen in organic form, e.g. in blood and bone, animal manure. This is more slowly available to plants during the growing season.

P(t) is the total percentage of phosphorus in all forms in the fertiliser. P(w) is the percentage of phosphorus available in a water soluble form, while P(cs) is the percentage soluble in ammonium citrate. Both of these are considered to be available to the plants. Citrate insoluble phosphorus is considered to be unavailable.

Other elements are generally given as their total percentage.

An older labelling method still used overseas gives the percentages of phosphorus and potassium as P_2O_5 and K_2O (potash) respectively. You need to know the atomic masses of the elements to convert these values.

Exercise: What percentages (by mass) of potassium and phosphorus does a fertiliser labelled 19% potash and 19% P_2O_5 contain? (Answer: 15.8% (by mass) potassium and 8.4% phosphorus.)

Fertiliser is used to supply plants with nutrient elements that are not present in adequate amounts in the soil. What is required depends on the soil and on the growing cycle. For all vegetable crops grown in southern Australian soils not receiving organic matter, CSIRO recommends an NPK mixture of composition $5:8:4$ applied at about 100 g/m^2, preferably placed in a band below and to the side of the seedlings. For further details consult CSIRO booklets (see Bibliography).

Composting

Composting is really just a method of speeding up the natural process of rotting under controlled conditions. The essential ingredients are organic materials, micro-organisms, moisture and oxygen (and a little soil). The organic matter is the food for the bacteria and it must meet their nutritional requirements. The most

TABLE 5.2
Approximate composition of composting materials

Material	C/N ratio	% moisture	gC/100 g (moist)	gN/100 g (moist)
Lawn clippings	20	85	6	0.3
Weeds	19	85	6	0.3
Leaves	60	40	24	0.4
Paper	170	10	36	0.2
Fruit wastes	35	80	8	0.2
Food wastes	15	80	8	0.5
Sawdust	450	15	34	0.08
Chicken litter (typical)	10	30	25	2.5
Straw	100	10	36	0.4
Cow dung	12	50	20	1.7

important of these is the ratio of carbon to nitrogen (C/N) which needs to be about 25:30. With a lower ratio, nitrogen will be lost from the soil; with a higher ratio, the bacteria are starved and activity is reduced. The compost mix should be selected to provide a balanced diet. Table 5.1 shows average C/N ratios (which can vary quite a bit) for various materials.

Thus a mixture of leaves : sawdust : cow-dung in the ratio of 2:1:2.5 will give a C/N ratio of

$$\frac{C}{N} = \frac{(2 \times 24) + (1 \times 34) + (2.5 \times 20)}{(2 \times 0.4) + (1 \times 0.08) + (2.5 \times 1.7)} = 26$$

Phosphorus can be important and a C/P ratio of 75 to 150 is optimum. Leaves (especially gum leaves), woody plant residues and occasionally lawn clippings can be low. A sprinkling of superphosphate can be helpful but if more than 2% is used, it can inhibit the composting process.

Moisture
The moisture content of the heap is very important. Below 40% and decomposition does not occur. Above 60%, airflow is reduced and the heap becomes anaerobic. The microbes in a reasonable-sized heap go through many cubic metres of air a day. The right moisture content of 55% feels damp but not soggy, like a squeezed sponge.

Temperature
If you place a thermometer in the compost heap you will notice some interesting changes. The working microbes generate heat and in two or three days the temperature will rise to 55°–60°C (bigger heaps get hotter and are better in winter). At about 40°C there is a change in shift as the starting microbes, who like the same temperatures as we do, flake out and the work is taken over by other microbes that like higher temperatures. Compost heaps are turned to aerate them and this causes a temporary drop in temperature of 5°–10°C. Temperatures higher than 60°C start to kill off bacteria and the process slows down until the heap cools again.

pH

The pH of the heap starts off slightly acidic from the cell sap. As fermentation starts, the acidity increases and the pH drops further. In the hot-heap stage the production of ammonia causes an increase of pH and the heap becomes alkaline. Finally the pH drops to almost neutral as the ammonia is converted back to protein and the natural buffering capacity of humus takes control. Lime causes loss of ammonia and is best added to soil and not to compost heaps.

Microbiology

In the early stages of decomposition acid-producing bacteria and fungi dominate, consuming the sugars, starches and amino acids. The high-temperature bacteria decompose protein, fats and hemicelluloses (similar to cellulose but composed of mannose and galactose as well as glucose—see Fig. 11.13). The high temperature fungus, *Actinomyces*, decomposes the cellulose. An important function of the high-temperature stage is to kill parasites and dangerous organisms as well as weed seeds. Much of the carbon of the original heap is 'burnt' by the microbes in their life processes and escapes as carbon dioxide. The dry weight of the heap is reduced by 30–60% and the volume drops by about two-thirds.

If you look inside a compost heap that has only partially completed its task, you will often notice that the materials have turned white or grey-white. This is because *Actinomyces* has been hard on the job. If the compost heap dies out, these white spores can disperse and this can be irritating, so keep the heap moist.

If green grass is piled in a pit and air sealed out with a layer of plastic sheeting or soil, then *anaerobic* bacteria dominate and the main acids produced are lactic and acetic acids. No other processes occur. The pH is around 4 to 5. This product is called *silage*. The grass has been pickled or preserved by fermentation in the same way as peat and peat moss.

PESTICIDES AND ALTERNATIVES

Introduction

Insects may be small in size but that's all. Insects represent about 76% of the total animal mass in the world today and probably have always been the major form of animal life. Egyptian hieroglyphics mention the frightening effect of locust swarms, which can have a mass of 15 000 tonnes. Locust swarms are highly visible, but most insects are much less conspicuous. If you examine a few square metres of typical sheep-grazing soil and count all the grass grubs, the chances are that the weight of the grubs eating the grass from below will be greater than the weight of the sheep eating it from above. The development of agriculture and the concentration of growing plants was a marvellous step forward—for insects. It concentrated their feeding and breeding. The storing of produce and the keeping of reservoirs of water for domestic use also helped insects.

Apart from these primary effects of agriculture, there were also secondary effects of human settlement. For example, in contrast to other primates, humans have fleas because they have a permanent residence, one of the requirements in the life-cycle of the flea. It is no wonder that the war against insects is waged with vigour. Pesticides were the first weapons to be used but they are failing rapidly. It is no longer practical or economical to eliminate insects completely. The concept of pest management has been adopted instead. Management entails the acceptance of a certain level of damage.

The consumer is still presented with a bewildering array of dangerous chemicals in the gardening section of the retail store, often in little bottles with even littler instructions. There are also reassuring warnings, such as *Keep out of reach of children* and *If swallowed seek medical advice*, but they appear on so many products that it is difficult to assess the degree of dangerousness. Scheduling of chemicals is discussed in Chapter 9.

If you are intent on following the advice of your gardening books and the helpful hints in the gardening section of your newspaper and wish to continue the chemical warfare in the garden, perhaps this chapter will help you understand what you are using and help to ensure that it is the insect rather than you that ends up knocked out.

A *pesticide* is a material that is selective to a degree in killing a pest. Pesticides are classified by their intended use; for example, insecticides, fungicides, herbicides (weed killers), rodenticides and acaricides (mites, ticks, spiders, etc.). This list is by no means exhaustive and some classifications overlap. These major divisions are subdivided according to chemical type (generic group), mode of action or specificity, and these subdivisions are again not mutually exclusive.

Historical perspective[1]

The control of insects by chemicals goes back to antiquity. The fumigant value of burning sulfur is mentioned by Homer, and Pliny the Elder was aware of the use of soda and olive oil for seed treatment of legumes and the use of arsenic to kill insects. 'Sulfur and the gross air' was how Pliny the Younger described the death of his uncle, Pliny the Elder, in Pompeii in the year 79. The Chinese were known to be using arsenicals and tobacco extracts (nicotine) in the sixteenth century. By the nineteenth century both pyrethrum and soap had been used for insect control.

Chemical pesticides today have their detractors, but this was also the case in earlier times. A method attributed to Pliny involved enticing caterpillars off infected trees by persuading an unadorned maiden of appropriate charm to dance around the orchards. The High Vicar of Valencia preferred a sterner legal approach when in 1485 he indicted caterpillars to appear before him (he provided a defence counsel) and then proceeded to condemn them to exile from the area!

The first scientific studies in the middle of the nineteenth century included the introduction of Paris green in 1867, an impure copper arsenite. Its use in the USA to check the spread of the Colorado beetle led to the introduction of probably the world's first pesticide legislation in 1900. Lime sulfur (Bordeaux mixture) was introduced in Europe in 1885 to control downy mildew on vines. In 1896 a French grape grower noticed that the leaves of yellow charlock growing nearby turned black and realised that it was the Bordeaux mixture he was using. Thus was born the first chemical herbicide. Iron sulfate was soon shown to kill dicotyledonous weeds but not cereal crops when both grew together. The organomercury compounds that had been used to combat syphilis were found in 1913 to be effective in protecting seeds from insect attack. Tar oil used to control aphids by killing their eggs on dormant trees was introduced between the wars. The control of weeds in cereals using dinitro-orthocresol was patented in France in 1932 and thiram was the first dithiocarbamate fungicide. During the Second World War both DDT and organophosphorus compounds were developed in Germany, while the phenoxyacetic acid (2,4-D) types of weedicide were developed in the UK. After the war came the soil-acting carbamate herbicides in the UK and the organo-chlorine insecticide chlordane in the USA and Germany. Switzerland introduced carbamates as insecticides soon afterwards. The period 1950 to 1955 saw the introduction of urea as a herbicide, captan and glyodin as fungicides, and the insecticide malathion. Then from 1955 to 1960 came the triazine herbicides from Switzerland.

The important systematic fungicide benomyl and soil-acting herbicide glyphosphate are US products of the late 1960s, but relatively few actual new *groups* of pesticides have appeared in the past few decades.

TOXICITY

An *acute effect* is an effect that occurs shortly after contact with a *single dose* of poison. The effect depends on how poisonous the substance is and on where it is applied. A drop of sulfuric acid is more dangerous on the eye than on the skin; and arsenic compounds are more toxic than those of sodium. Because the way in which an acute poison acts is generally understood, acute toxic responses can usually be given in a quantified form and it can be generally argued that below a certain dose a substance is not toxic.

A *chronic effect* is the result of exposure to *repeated, small, non-lethal doses* of a potentially harmful substance that causes cumulative damage over a long period of time. Asbestosis is a classic example.

A measure of acute toxicity commonly quoted is the LD_{50} value. In a randomly chosen batch of test animals, the dose that on the average kills half the sample when applied in a particular way under stated experimental conditions is the LD_{50} and is expressed in terms of mg of poison per kg of body mass of the species. Such a method of representation eliminates the variation of body mass between species, so that when rats and people are compared, at least the mass ratio is taken into account, even though lots of other characteristics may differ. When, for example, adult male rats are used, it should be remembered that female rats are on average more sensitive than males, as are also young and sick rats. Not all people exposed to chemicals are adult male and healthy!

An LC_{50} is a *concentration* dose (usually as a vapour in air for mammals or solution in water for fish). ED_{50} (or EC_{50}) refers to *effective dose* when some criterion other than death is used.

It should be noted that the *smaller* the LD_{50} the *more toxic* the substance. The units of mg/kg are equivalent to μg/g or ppm body mass. An approximate description of toxicity is:

LD_{50} < 1 mg/kg is *extremely* toxic;

1 < LD_{50} < 50 mg/kg is *highly* toxic;

$50 < LD_{50} < 500$ mg/kg is *moderately* toxic;
500 mg/kg $< LD_{50} < 5000$ mg/kg is *slightly* toxic.

In common with many other variable factors of a biological nature, the susceptibility of individual members of a group within a species varies. Thus variation follows an approximately normal (bell-shaped) distribution *if* the variable, such as dose of poison, is plotted on a logarithmic scale (the so-called log-normal distribution; see Fig. 5.3).

This distribution arises when the variability (variance) of the population is not constant but increases linearly with the dose. A plot of the number of individuals responding (on average) as the logarithm of the concentration increases is shown in Figure 5.4 and follows a sigmoidal curve. The greater the dose, the more animals responding, and the greatest change occurs around 50% response and therefore LD_{50} rather than say LD_5 is preferred (although where cost or humanitarian considerations apply, a lower LD may be preferable). Note that the sigmoid curve is just the plot of the cumulative area under the normal curve as you go from left to right. Remember that the concentration scale is logarithmic and not linear, so high concentrations are compressed.

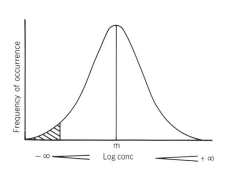

Fig. 5.3   The log-normal distribution

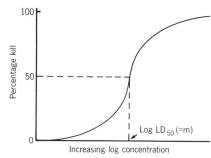

Fig. 5.4   The cumulative log-normal distribution

Insecticides

Insecticides are sometimes divided *functionally* as:
- stomach poisons (require ingestion);
- contact poisons (absorbed through the cuticle);
- fumigants (gases).

Chemically the more important groups of insecticides are:
- inorganic group;
- organochlorine compounds (chlorinated hydrocarbons), e.g. DDT, aldrin;
- organophosphorus compounds (e.g. parathion) and carbamates, which are similar acting (e.g. sevin);
- plant extracts (botanical), e.g. pyrethrins and the pyrethroids, rotenoids.

The inorganic group are almost exclusively stomach poisons, which are active only after ingestion and are thus restricted mainly to chewing insects. They are not very effective against sucking insects such as aphids and mosquitoes. Inorganic pesticides are the most persistent.

Other classes of chemicals used either alone or in conjunction with insecticides are attractants and repellents and synergists.

Inorganic insecticides

Typical inorganic insecticides are usually heavy metal compounds, particularly of lead, mercury, arsenic and antimony, although some others, such as fluoride salts (e.g. NaF), sulfur and polysulfides, and borax, have limited use.

Lead arsenate ($PbHAsO_4$). A typical heavy metal compound. Being water-insoluble it is not readily absorbed by plants on contact and is effective only by ingestion. The lead ties up essential sites on enzymes and is thus non-specific. It is toxic to all living systems as well as being extremely persistent. Sodium arsenite was once used as a cattle dip against ticks, but it is no longer used.

Sodium fluoride (NaF) and *cryolite* (Na_3AlF_6). These compounds liberate fluoride ion, which precipitates Mg^{++} as fluorophosphate and upsets magnesium-dependent enzymes. It is non-specific and toxic to animals.

Borax ($Na_2B_4O_7$). Used as a cockroach and ant poison but is not very toxic to mammals.

Elemental sulfur (S) and *lime sulfur* (CaS_n, n ~ 5). Lime sulfur is a solubilised form of sulfur (a polysulfide). Its mode of action is through aerial oxidation to SO_2, which is an effective fungicide and acaricide, but is of limited use as an insecticide. It is one of the safer fungicides.

Copper compounds. Two copper compounds are widely used by home gardeners: Bordeaux mixture, a combination of copper sulfate and lime; and Paris green, copper aceto-arsenite, which is very poisonous to humans. Copper sulfate causes vomiting and promotes its own elimination.

Timber for garden use is impregnated by the pressure pot method with copper chrome arsenate. Timbers treated with various brands of copper-chromium arsenic (CCA) appear greenish in colour. Pine logs treated this way are used as fence posts and in children's playground equipment. The arsenic in CCA-treated timbers is relatively non-toxic and readily excretable, and is strongly locked into the wood. Toxic doses would require the ingestion of about 10–20 cm^3 of treated wood. The only place in which a problem might arise is where such treatment is taking place or the timber is being sawn and livestock have access to the sawdust. Exhaust ventilation should be provided when sanding or finishing treated timber, and the urine of regular workers should be analysed for arsenic content. Dermatitis can result from chronic skin contamination.

The whitish crystals or powder-like substances that appear on newly purchased timber are mostly sodium sulfate, which is harmless. Timber that has been aged properly after treatment and washed should not show this effect.

Burning treated timber or sawdust is dangerous because arsenic is released as a vapour.

Natural insecticides

There are several well known insecticides of plant origin. Examples are the rotenoids (sold as Derris Dust) and nicotine. (Derris Dust is not persistent and must be sprayed every three days. It is primarily rotenone, which is also used to poison fish.) Even garlic oil has been shown to be effective against the larvae of mosquitoes, houseflies and other pests. Some of the oldest and best known of the natural insecticides are those that make up the class of compounds known as pyrethrins, which are extracted from the pyrethrum flower, a daisy-like member of the *Chrysanthemum* genus, which originated in Persia (Iran) but was being cultivated in Yugoslavia by the mid-nineteenth century. Pyrethrins have a remarkable 'knock-down' effect on flying insects, but a second poison is needed to kill the insect. They have extremely low toxicity to warm-blooded animals. Currently, the major production is in East Africa. They are expensive and photosensitive and thus commercially unreliable. Synthetic pyrethroids are now produced and used widely on fields because they are not photosensitive.

Organochlorine compounds

Bubonic plague is estimated to have killed 100 million people in the sixth century, topped up with another 25 million in the fourteenth century. Typhus killed 2.5 million Russians during the First World War and was a critical factor in the collapse of the Balkans and in the defeat of the Russian armies on the Eastern front. Malaria has long been *the* most fatal and debilitating human disease. As late as 1985 three million people died in an Ethiopian epidemic, and in 1968 Sri Lanka (then Ceylon) suffered more than one million cases.

What have these diseases in common? They are all spread by insects. The flea (plague), the louse (typhus) and the *anopheles* mosquito (malaria) have all been controlled since 1939 with DDT (1,1-bis(4-chlorophenyl)-2,2,2-trichloroethane). Only in the developed world has DDT been replaced by other chemical, medical and physical means.

The organochlorine pesticides are so called because they have one or more chlorine atoms attached to the carbon atoms, replacing hydrogen. DDT was the first, the most dramatic in terms of human life saved and the safest in terms of immediate short-term effects. The organochlorines are cheap to make. Often only a couple of chemical steps are needed, including the famous Diels-Alder reaction (hence dieldrin and aldrin).

As well as meeting an important human need, the organochlorines solved an industrial problem. At the very base of an emerging chemical industry is the process of passing electricity through a solution of salt to produce caustic soda and chlorine, just as occurs in salt-water swimming pool electrolytic chlorinators, except that the concentration of salt is very high (see later in this chapter). The demand for caustic was very high for making soap and paper, and more recently producing bauxite for the manufacture of aluminium. The demand for chlorine was low. Organochlorines and later chlorinated plastics such as PVC restored a profitable balance.

The discovery of the insecticidal properties of DDT in the Swiss laboratories of Geigy Pharmaceuticals in 1939 was an event of major importance. Before this discovery the main insecticides available were natural products. DDT is a highly effective insecticide, both by contact and by ingestion, and is of very low toxicity to mammals. Oral LD_{50} is 300–500 mg/kg; dermal LD_{50} is 2500 mg/kg. It is practically odourless and tasteless, and it is chemically stable, which is now seen as a major disadvantage. It is made in a one-step reaction (Fig. 5.5) from low-cost raw materials and therefore is cheap.

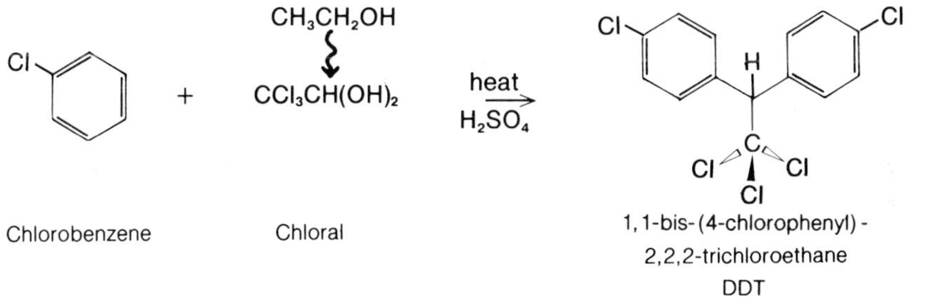

Fig. 5.5 Preparation of DDT

Increasingly, many insects are developing resistance to DDT by the natural selection of surviving members whose enzymes can detoxify the DDT to a substance known as DDE (Fig. 5.6). DDE is flatter than DDT and the change in shape alters the toxicity. The chemically stable DDT and its metabolic breakdown product DDE tend to accumulate in the fat of birds and fish because these creatures are at the end of the biological chains. The mosquito eradication campaign on Long Island, New York, showed levels of DDT in the sea water of 3×10^{-6} mg/kg (non-toxic); in the fat of plankton, 0.04 mg/kg; in the fat of minnows, 0.5 mg/kg; in the fat of needlefish, 2 mg/kg; in the fat of cormorants and osprey, 25 mg/kg. This progression is called biological magnification. The level in the birds is biologically active and is believed to upset the metabolism of the female hormone oestrogen. It has been alleged that this has resulted in eggs being produced with very thin shells that break easily.

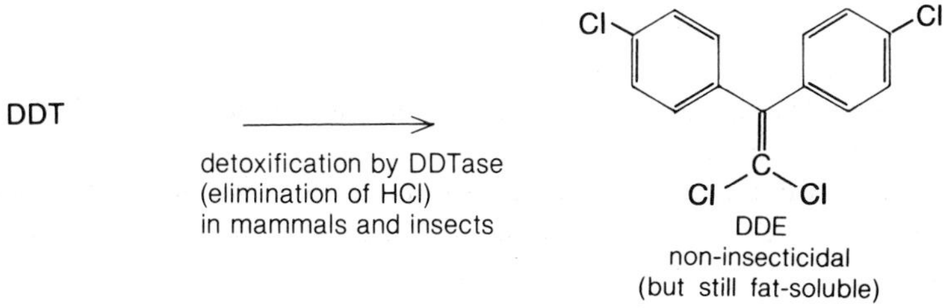

Fig. 5.6 Detoxification of DDT

BEES JUST ADORE DDT!

Scientists in Brazil have discovered a species of bees (*Eufriesia purpurata*) in the Amazon region that just loves DDT.[2] The bees were observed collecting DDT from houses that have been sprayed against malaria-carrying mosquitoes. Some bees were found to have levels in their bodies of 42 000 mg/kg (4.2%), which is amazing, as for normal bees the LD_{50} dose is only 6 mg/kg. There may be a connection with sex-attractants, as the attractants of male bees of this species are characteristically aromatics and monoterpenes.

Fig. 5.7

Development of biodegradable analogues of DDT is well advanced. This involves including groups on the molecule that can be metabolised to polar groups, thus giving water solubility and allowing their excretion. An example is the insecticide *methoxychlor*, a biodegradable analogue of DDT (illustrated in Fig. 5.8). This method is the basis of *building in* biodegradability. Non-polar compounds stay dissolved in the fat of the body, whereas polar substances are water soluble and can be excreted and also can be further attacked by other organisms.

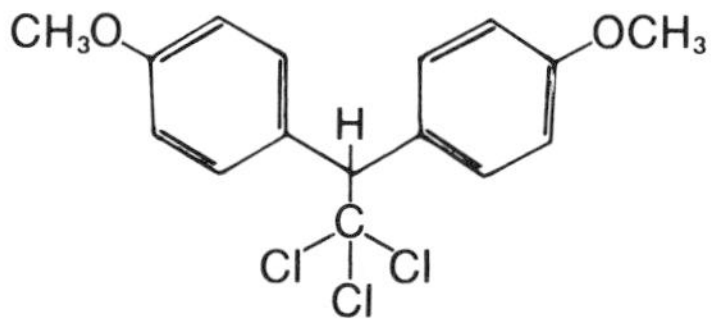

Fig. 5.8   Methoxychlor. Oral LD_{50} is 5000–7000 mg/kg

The mode of action of DDT is that its three-dimensional shape props open the sodium channel of the nerve cell and allows sodium ions to leak in (see Plate IV, which shows the shape of the molecule in a space-filling model). This causes continuous transmission of nerve impulses and the insect dies of exhaustion. In warm-blooded animals the binding between the pesticide molecule and the ion channel of the nerve cell does not occur. DDT-like action is clearly related to a specific size and shape of the molecule because the same effects can be elicited by an analogue of DDT in which methyl groups replace the chlorines (Fig. 5.9). This methyl analogue is even more persistent than DDT because it cannot be detoxified by the elimination of HCl, but it can be destroyed by oxidation. The pyrethrins owe their activity to exactly the same shape property.

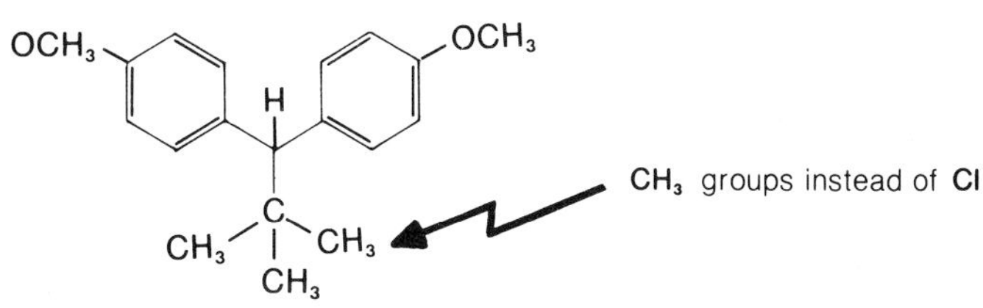

Fig. 5.9   The methyl analogue of methoxychlor

SYNERGISM

DDT, methoxychlor, and particularly the pyrethrins are much more efficient in the presence of piperonyl butoxide (Fig. 5.10). This is called a *synergistic* effect. Synergism differs from a similar effect—namely potentiation (see Fig. 2.9 in Chapter 2) because in this case the additive is not itself an insecticide. The piperonyl butoxide helps to *deactivate* certain enzymes (mixed function oxidases) which are important for detoxification of the pesticide in the insect.

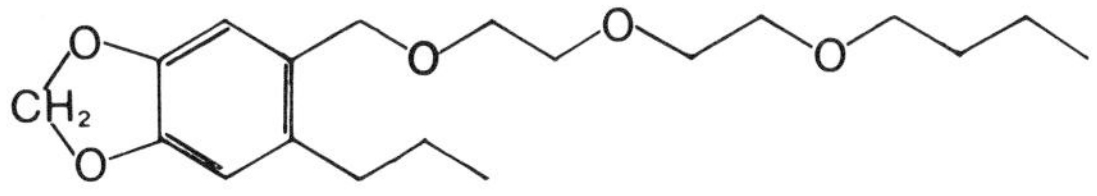

Fig. 5.10   Piperonyl butoxide (common name)

Lindane/gammexane/BHC (benzene hexachloride)

This substance, known by many names, was first discovered in 1825 by Faraday. It is made simply by the addition of chlorine to benzene in the presence of light (Fig. 5.11). A mixture of nine isomers is formed, but the only active one is the γ-isomer, which forms only 13–18% of the mixture. The insecticidal properties of the γ-isomer were discovered in 1943. Detoxification is again accomplished by elimination of HCl.

Lindane is accepted for medical use to treat humans; for example to get rid of lice in children. The following related compounds, however, are considered dangerous to humans.

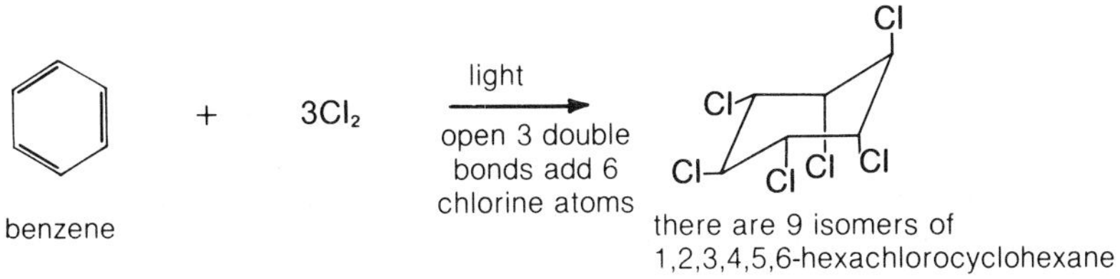

Fig. 5.11   Chlorination of benzene to form lindane. Oral LD_{50} is 230 mg/kg; dermal LD_{50} is 500–1000 mg/kg

Aldrin, chlordane, dieldrin, heptachlor, endrin

These compounds form a closely related group of insecticides and are made by Diels-Alder reaction with certain alkenes from a common material, hexachloro-cyclopentadiene (Fig. 5.12). Chlordane is very easy to make and is toxic to insects and people, though less toxic than dieldrin or aldrin. When the CSIRO was given the task of protecting telephone cables from termites that 'ate' through lead sheath, dieldrin spread around the cable proved satisfactory.

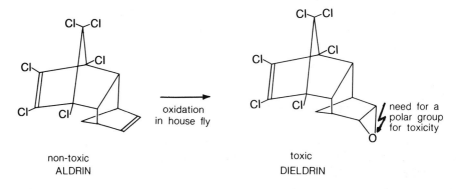

Fig. 5.12   Diels-Alder reaction to form chlordane. Oral LD_{50} is 280 mg/kg; dermal $LD_{50} > 1600$ mg/kg

These compounds are all broad-spectrum insecticides, are highly toxic to both insects and mammals, and have a high persistence. Like DDT they accumulate in body fats and act on the central nervous system. In some cases the actual compound itself is non-toxic but is converted to a toxic material in the body (Fig. 5.13).

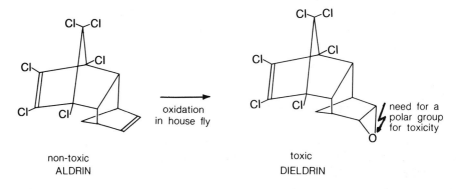

Fig. 5.13   Oxidation of aldrin to dieldrin, which has an oral LD_{50} of 40 mg/kg and a dermal $LD_{50} > 100$ mg/kg

Tests for chlorinated hydrocarbons

There is no simple and generally applicable method of testing whether excessive chlorinated hydrocarbon absorption has occurred. When absorbed into the body, such pesticides enter the bloodstream and remain there for varying periods (measured as a half residence time). This varies from only a few hours in the case of endrin (which is one of the most toxic chlorinated hydrocarbons) to approximately 80 days for dieldrin and up to 100 days in the case of DDT. This group of pesticides is detoxified in the liver and excreted by the kidneys. However a certain amount is retained and stored for long periods in the body fat and is re-leased in times of illness and stress. It is also released into maternal milk. There was a time when it was claimed that mother's milk in the USA would not be al-lowed for sale (if it were sold) because of the high levels of DDT it contained.

Surveys in Western Australia have found high levels of dieldrin in samples of human milk, where aldrin had been used to protect the home against attack by white ants.[3] The levels continued to rise until the seventh or eighth month after treatment at which time they levelled off and start to decline. Urban use rather than contamination of foodstuffs appears to be the major source.

The high persistence of the chlorinated pesticides has led to increased restriction of their use, and they are being increasingly replaced by the organophosphorus group, which is now the fastest growing group of insecticides. As each new class of broad-spectrum insecticide is introduced, the safety margin between insect toxicity and human toxicity diminishes.

SUPERTERMITE

Of the 200 or so termite species in Australia, one stands out in size and diet—*Mastotermes darwiniensis* Froggat. It can grow to about 15 mm— three times the size of other termites. It is found north of the Tropic of Capri-corn and, although it prefers a diet of timber, it has been known to attack plastic cables, cow-dung pads, paper, wool, corn, bagged salt, ivory, bitumen, pebbles, ebonite, lead and even billiard balls. What a super recycler! The actual digestion is done by a bacterium that lives symbiotically in the termite's gut.

Organophosphorus insecticides

The organophosphorus insecticides are a wide variety of compounds with a tremendous range of activity, persistence, specificity and function. All have the general formula $(RR'X)P = O$ (Fig. 5.14), where R and R' are short-chain groups and X is a leaving group especially selected so that it is easily removed from the molecule either directly or after a reaction in the body. This group is built in so that the persistence of the substance is reduced.

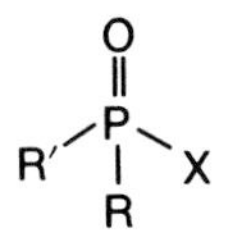

$$\underset{\underset{R}{\overset{\displaystyle |}{R'}}{\overset{\displaystyle O}{\overset{\displaystyle \|}{\overset{\displaystyle P}{\diagdown}}}}X$$

Fig. 5.14 Organophosphates. P has valencies 3 or 5. It is the 5-valent state that is of importance for insecticides

The five compounds depicted in Figure 5.15 are among the best known organophosphorus insecticides. Some are very easily and cheaply made. An

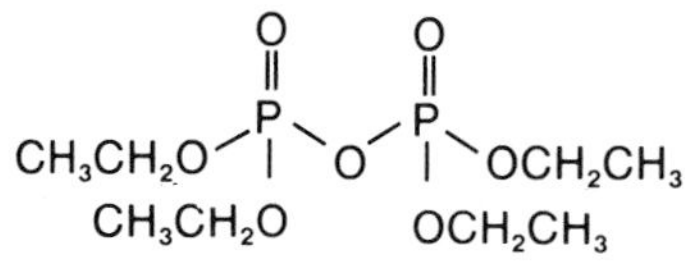

TEPP
(tetraethylpyrophosphate)
oral LD_{50} rat 1 mg/kg
no longer in use

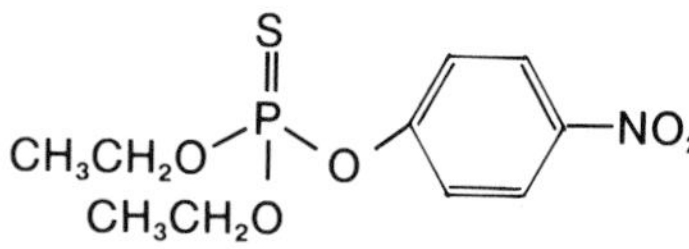

parathion (code number E 605)
oral LD_{50} 3-6 mg/kg
dermal LD_{50} 4-35 mg/kg

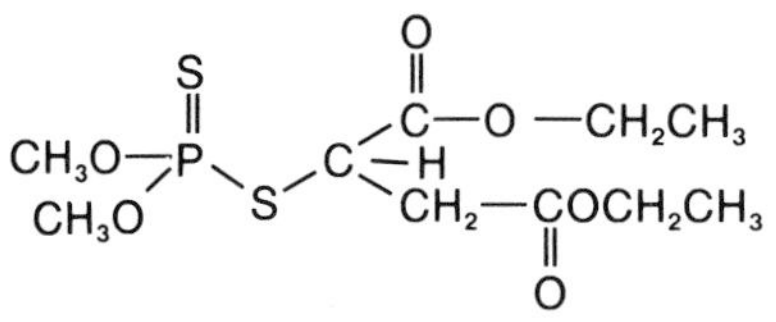

maldison
(Malathion)
oral LD_{50} 1400-1900 mg/kg
dermal LD_{50} > 4000 mg/kg
toxicity ratio for
insect: human is about 100, i.e. the
LD_{50} value for humans
is 100 × that for insects.
insects are 100 × more susceptible.

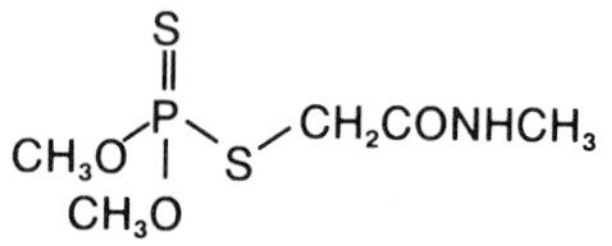

dimethoate (Rogor)
oral LD_{50} 200-300 mg/kg
dermal LD_{50} 700-1150 mg/kg

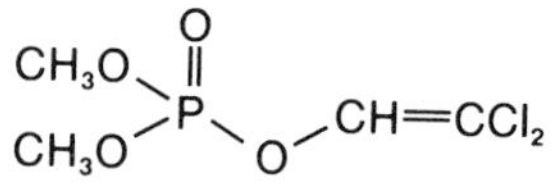

dichlorvos (Shelltox strips)
oral LD_{50} 25-30 mg/kg
dermal LD_{50} 75-900 mg/kg

Fig. 5.15 Some organophosphorus insecticides

example is the production of parathion (Fig. 5.16), which is still used in developing countries because it is cheap—but it is also dangerous. Small modifications to the molecule make it safe but also more expensive.

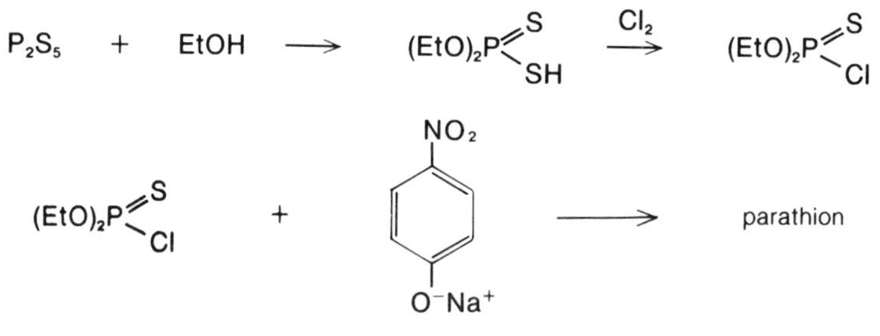

Fig. 5.16 Synthesis of parathion

The other important characteristic, apart from ease and cheapness of production, is persistence—how long the material stays around. For crop-dusting insecticides this is specified by a withholding period—the number of days between application and harvesting. Although the exact period varies with the climate and the time of the year, the average value generally offers a good guide to the persistence of a substance. Sometimes a time is specified for a certain fraction of the material to have disappeared:

$t_{1/2}$ = time for half the material to disappear
$t_{99}\%$ = time for 99% of the material to disappear

Thus, for *TEPP* the LD_{50} oral/rat value is 1 mg/kg, which is highly toxic. The half-lives are:

$t_{1/2}$ (25°) = 6.8 hours
$t_{99}\%$ (25°) = 45 hours—quickly detoxified

For *parathion*, the LD_{50} oral/rat is 3–6 mg/kg—highly toxic.

$t_{1/2}$ (28°) in water = 120 days—very persistent

Parathion is used as both methyl and ethyl ester. It has probably been responsible for more deaths than any other pesticide. The methyl ester (MSO) has not caused many deaths but this is probably because of lower usage.

Maldison, like parathion, is persistent but has low *mammalian* toxicity. The LD_{50} oral/rat is 1300 mg/kg, because mammals and insects detoxify it differently and with very different efficiency (see Fig. 5.17). (For trade mark reasons, Malathion cannot be used as a generic name in Australia.)

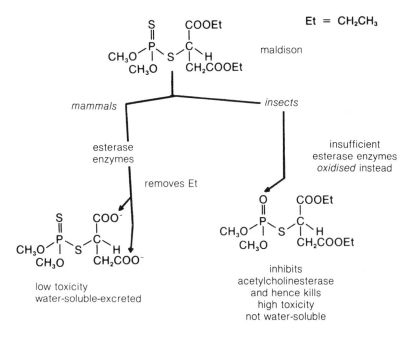

Fig. 5.17 Detoxification of maldison

Dichlorvos (dimethyl dichlorovinyl phosphate—DDVP) is relatively volatile (low molecular mass), broad-spectrum, but is rather toxic to mammals: LD_{50} oral/rat is 30 mg/kg. It is not very persistent: $t_{1/2}$ (25° in water, pH 7) is eight hours. This compound is used in the Shelltox pest strip in a slow-release formulation. A

Shelltox strip weighing 21.3 g was stated to contain 18.6% by weight of dichlorvos. This means that the strip contained 4 g of insecticide. Using the LD_{50} value of 30 mg/kg, the dose for which, on the average, half of a group of 10 kg children could be expected to die would be $10 \times 30/1000$ or 0.3 g, equivalent to the contents of less than one-tenth of a strip per child. Note: the material would have to be swallowed.

Unlike aerosol insecticides, which kill flies with a quickly dispersed spray, the strips continuously saturate the domestic atmosphere with DDVP vapour. The concentration of this vapour is lethal to insects; whether breathing the same concentration for hours on end is also harmful to humans has been a matter of dispute since the strips were first marketed in the mid-1960s. The strips were banned in Holland in January 1974 and were under study in the USA, where the EPA (the Environmental Protection Agency) has now cleared them.

There were a number of reasons a ban was sought. The manufacturer had failed to list all the insecticidal components in the strips, which were alleged to include two known human mutagens. (The specification has improved substantially over the years, and the typical active component is said to be now 99% pure.) Many consumers were in the habit of hanging the strips from light fittings (just as people did with fly-paper years ago); but heat can cause an increase in vapour release. Surveys had also showed that the strips were being used where they shouldn't be— in spite of printed warnings on the labels. For example, they were being used in domestic kitchens, dining rooms and sick rooms. They should not be used in restaurants or in other areas where food is prepared or served. But this is where they are very commonly used.

The Shell company's own results showed no evidence of carcinogenic properties. However, when the strips were used according to instructions in the houses of their own employees, they also found that the average concentrations were higher than necessary to be effective. Note that these instructions state that only one strip should be used per 30 cubic metres of room (say $4 \text{ m} \times 3 \text{ m} \times 2.5 \text{ m}$). The concentrations are very sensitive to humidity because water destroys the DDVP and lowers the concentration. In dry areas the level can be very high. Certain groups in the community—such as asthmatics—could be particularly sensitive because DDVP vapour increases the resistance of air passing into the lungs.

In most industrialised countries flies are an aesthetic rather than a medical problem in most areas and so 'blanketing our homes with a 24-hour nerve gas which may well be causing mutations, altering our chemistry, and reducing the breathing ability of people with severe lung disorders'[4] should be considered carefully. (Note that the quoted statement shows the *New Scientist* in one of its more vindictive moods. It has had a long history of non-cooperation from Shell.)

All these organophosphorus insecticides are contact poisons that are rapidly absorbed through the insect cuticle. However, they can also be absorbed through human skin during spraying operations. One way of solving this problem is to use *systemics*. Systemic insecticides are absorbed into the plant (e.g. through the leaves) and are ingested by sap-sucking insects along with the plant juices (but they do not harm the plants because they do not have a nervous system). They are equally effective against both chewing and sucking insects and are administered as a spray, or to the plant roots as granules, or injected into the plant trunk. Two typical organophosphorus systemics are demeton and dimethoate.

Demeton (Systox) is a mixture of two components:

$(EtO)_2P(=S)—OCH_2CH_2SEt$ (Demeton O) and
$(EtO)_2P(=O)—SCH_2CH_2SEt$ (Demeton S)

It is a broad-spectrum, persistent systemic, very toxic to mammals: LD_{50} oral/rat is 9 mg/kg. It is commonly used by home gardeners, and cases of poisoning have resulted from absorption through the skin.

Dimethoate (Rogor) is also broad-spectrum, but is much less persistent than demeton because it is detoxified by the plant fairly rapidly. It is formed by the reaction

$$(MeO)_2P{\overset{S}{\underset{S^-Na^+}{}}} + ClCH_2CONHMe \longrightarrow (MeO)_2P{\overset{S}{\underset{S—CH_2CONHMe}{}}} + Na^+Cl^-$$

dimethoate

A useful alternative to spraying large (or small) trees and shrubs is injection into the trunk. Fine holes bored into the sapwood (i.e. through the bark but not into the heartwood) in a downward direction near the base of the tree can be filled with neat insecticide using a syringe and sealed with plasticine or putty. This technique requires less chemical and localises it on the plant needing protection. The chemical is protected from the weather and remains effective for much longer (2 months) than sprayed chemicals. Birds do not come in contact with the insecticide and, even if they eat the insects affected by the insecticide, the chemical has by then changed in the gut of the insect to a non-toxic form. *Note*: do *not* use if koalas, possums or stock eat the foliage. Fruit trees should *not* be treated this way. Large trees require several holes around the trunk and for multi-trunked trees, each trunk may need to be treated separately.

Dangers

Poisoning by organophosphates may occur if they are inhaled, swallowed or come in contact with the skin. The first symptoms of poisoning come on within minutes or at the most hours after exposure. They are usually headache, fatigue, giddiness, saliva formation, sweating, blurred vision, pin-point pupils, chest tightness, nausea, abdominal cramps and diarrhoea. If someone who has been using one of these insecticides develops these symptoms, *they should seek medical help immediately*. When absorbed these pesticides inactivate an enzyme in the body called *acetylcholinesterase*. This enzyme is normally responsible for returning to normal a compound (acetylcholine) used to activate muscles. When the enzyme is inactivated, muscular spasms in involuntary muscles occur first, and overactivity of certain glands can take place. A second enzyme, called 'non-specific esterase', occurs in the bloodstream. Monitoring of blood cholinesterase of exposed persons has been used as a safety measure. After exposure, several days can be needed to restore normality in one enzyme, while the other requires regeneration of red blood cells and this occurs at 1% of the normal value per day. Treatment involves the inducement of vomiting (for swallowed poison) and the removal of clothing and thorough washing of skin.

In 1937, while carrying out research on organophosphorus insecticides, Schrader, in Germany, discovered the nerve gases. The first of these nerve gases was called tabun (Fig. 5.18). These compounds will kill humans as well as insects. A 0.2 mg splash on the skin is fatal; in fact, every accident the workers had was fatal. Although outlawed by definition by the Geneva Convention of 1925, tabun appears to have been used in the Gulf War in 1984.

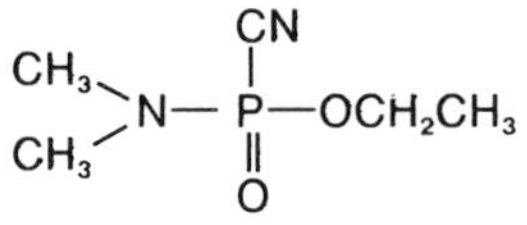

$$\underset{CH_3}{\overset{CH_3}{>}}N-\underset{\underset{O}{\parallel}}{\overset{\overset{CN}{|}}{P}}-OCH_2CH_3$$

Fig. 5.18   Tabun

The persistent organophosphorus insecticides are mostly thiophosphates (Fig. 5.19) (thio = sulfur). The presence of sulfur makes them more resistant to decomposition by water and lengthens their shelf-life as well as their persistence. But these thiophosphates are not reactive enough to be toxic. Except for insects, living organisms lack enzymes that can convert the thiophosphates to their toxic reactive phosphate analogues, which are the active insecticides. This is an example of *biological priming* or activation of an inactive precursor.

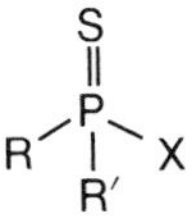

$$\underset{R'}{\overset{S}{\underset{|}{\overset{\parallel}{\underset{R}{P}}}}}X$$

Fig. 5.19   Organothiophosphates

Tabun and the related sarin completely eliminate phylloxera in grapevines in one week when a 0.1% solution is spread in the ground near the vine root. However, they are *far too dangerous* to use in practice.

Water solubility for most of the contact organophosphorus insecticides is very low (typically 0.000 01%), but the water solubility is improved by selecting a suitable group, X, which changes with time and allows the compound to become water-soluble.

Systemics are necessarily fairly persistent to allow time for the slow plant transportation process.

Carbamates

Carbamates act in a similar way to the organophosphates. They are based on carbamic acid (Fig. 5.20). A summary of pesticide use for homes and alternatives prepared by the Total Environment Centre, Sydney, is given in Table 5.3.

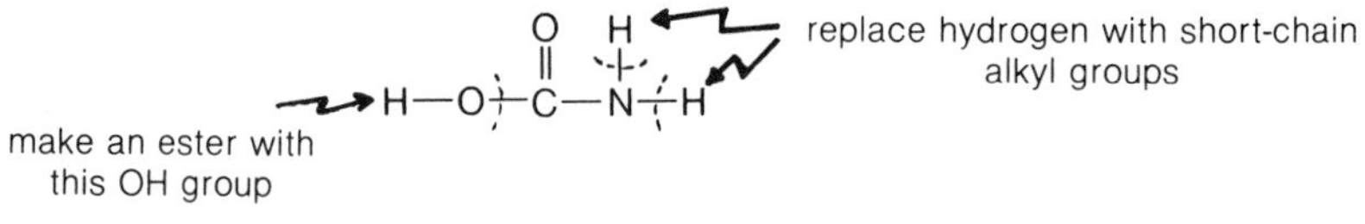

$$\underset{\text{make an ester with}\atop\text{this OH group}}{H-O}\overset{\overset{O}{\parallel}}{-C}-\underset{}{N}\overset{H}{\underset{H}{\big<}}\quad\text{replace hydrogen with short-chain}\atop\text{alkyl groups}$$

Fig. 5.20   Carbamic acid and derivatives

TABLE 5.3
Pesticide use in the home

Pest	Commonly used pesticides—active constituents	Popular consumer products	Common method of application	Alternative controls
INTERNAL Cockroaches	chlordane, dichlorvos, diazinon, chlorpyrifos, bendiocarb, propoxur, pyrethrins, synthetic pyrethroids	Lane Chlordane Baygon Surface Spray Mortein Surface Spray Embassy Surface Spray Pea Beu Surface Spray Johnson's Protector Probe Johnson's Cockroach Baits	Aerosols, dusts, low-volume liquid spray (interior), high-volume liquid (subfloor, exterior) baits, lacquers	• Hygiene • Baits: red wine in dish/cup (for large varieties) • Hortico liquid permethrin in atomiser • 'Zoro-Zoro' pheromone traps • Synthetic pyrethroid pressure packs containing permethrin (with extension nozzle, e.g. Protector Probe) • Reduce potential harbourage areas • Treat only harbourages, cracks/crevices
Fleas	dichlorvos, carbaryl, bendiocarb, propoxur, pyrethrins, synthetic pyrethroids, diazinon, maldison	Baygon Surface Spray & Dust Bayer Blue Cross } Products for pets Excel Go-pet	As above Flea collars, medallions, shampoos, aerosols (for pets) 'bombs'	• Regular vacuuming (empty collection bag outside, away from house) • Herbal shampoo (pennyroyal) on pets • Use flea comb on pets regularly • Plant mint and wormwood in backyard • Use of permethrin liquid on carpets etc.
Silverfish Clothes moths	chlordane, dichlorvos, bendiocarb, permethrin, naphthalene	Embassy Naphthalene Flakes & Moth Balls Shelltox Ministrips	Low-volume liquid spray (interior), dusts, pest strips, mothballs, flakes	• Lavender, eucalyptus • Regular turnover of clothes, reduce dampness, improve hygiene • Low toxicity synthetic pyrethroids

Continued

TABLE 5.3 — *Continued*
Pesticide use in the home

Pest	Commonly used pesticides—active constituents	Popular consumer products	Common method of application	Alternative controls
INTERNAL				
Ants	dichlorvos, chlorpyrifos, diazinon, bendiocarb, propoxur, borax, chlordane, heptachlor, deltamethrin	Ant-Rid Baygon Surface Spray & Dust Lane Chlordane Pea Beu Ant and Roach Dusting Powder Johnsons Protector Probe	Aerosols, low-vol. liquid spray (interior), high-vol. liquid spray (exterior and subfloor), dusts, baits	• Baits: borax with honey, jam or liver • Scatter equal parts borax and icing sugar or pepper on trail • Tansy, pennyroyal planted near doors, scatter on shelves • Eucalyptus oil • Permethrin liquid • Synthetic pyrethroid pressure pack with extension nozzle for cracks/crevices.
Flies, mosquitoes	pyrethrins, synthetic pyrethroids	Protector Fast Knockdown Mortein Low Irritant Scotts Insect Killer Pea Beu Low Irritant	Aerosols, mists	• Eliminate breeding areas • Flyscreens • Swats
Rats, mice	warfarin bromaldiolone diphacinone thallium sulfate	Ratsak Diphacin Baits Bromakil Baits Rat and Mouse Traps	Baits, gels, glues	• Good hygiene • Proof house • Traps
STRUCTURAL				
Termites (white ants)	chlordane, dieldrin, heptachlor, aldrin, chlorpyrifos, arsenic trioxide (dust)	Hortico Lane Chemspray Rentokil Nufarm } chlordane and dieldrin products	High-volume sprays and injection, dusts	Preventative measures are: • Regular inspection of property • Improve subfloor ventilation • Good building practices
Borers	chlordane, heptachlor, aldrin, dieldrin, chlorpyrifos, methyl bromide (fumigant)	Hortico Lane Chemspray Rentokil Nufarm } chlordane and dieldrin products	High-volume sprays, fumigation	• Selection of timbers (e.g. cypress pine ideal for flooring as it is resistant to furniture beetle) • Small items: heat in microwave oven to destroy larvae

Pest	Chemical	Product	Formulation	Control measures
Fungi (wood decay)	copper (in solution) pentachlorophenol chromium creosote	Crop King Bluestone Hortico Bluestone Celcure (Rentokil) Mobil Cellavit Caltex Koppers } Wood preservatives	Paints and pastes	• Furniture, doors, etc: seal in black plastic and leave outdoors in strong sunlight for full day to destroy larvae • Eliminate plumbing leakages under house • Maintain good guttering • Improve subfloor ventilation
GARDEN Funnel-web spiders	heptachlor, chlordane, diazinon, chlorpyrifos	Lane Garden Insecticide Rentokil Chekpest Lane Spider Killer	Low to high volume sprays	• Make barriers to prevent entry into house (e.g. slippery gloss painted vertical section of steps) • Clear backyard, eliminate harbourage areas (dark, moist) • Spot treat harbourage areas with permethrin • Pour boiling water down burrows (where found) • Not to be confused with trap-door spiders
Red-back spiders	dichlorvos, diazinon, chlordane, heptachlor, chlorpyrifos	Lane Garden Insecticide Rentokil Chekpest Lane Spider Killer	Low to high volume sprays	• Eliminate harbourage areas (e.g. tidy, clean garden sheds, toilets) • Spot treat harbourage areas with permethrin
Snails, slugs	methiocarb metaldehyde sodium fluosilicate	Defender Pellets Baysol Mesurol } Spray/Bait	Pellets, sprays	• Sawdust or grit barriers around garden • Baits: milk or beer traps • Catch under upturned grapefruit, potato shells, or in terracotta pipes • Ducks, frogs, lizards • Diatomaceous earth (found in pool shops)

Continued

TABLE 5.3 — *Continued*
Pesticide use in the home

Pest	Commonly used pesticides—active constituents	Popular consumer products		Common method of application	Alternative controls
GARDEN Fruit and vegetable pests	maldison, carbaryl, pyrethrins, captan, permethrin	Garden King Hortico Rentokil Cliftons Lane Nufarm Chemspray	Products containing organo-phosphates (e.g. maldison) and carbamates (e.g. carbaryl, captan)	Low to high volume sprays, dusts	• Herbal sprays, e.g. garlic, eucalyptus • Attractants, e.g. hydrolysed protein for fruit flies • Physical means, e.g. hand picking, vacuuming, high pressure hose, dust with flour or clay • Soapy water, e.g. 'Sunlight' soap • Encourage diversity for natural enemies

Source: An alternative guide to pest control in the home. Toxic and Hazardous Chemicals Committee of the Total Environment Centre, 18 Argyle St, Sydney 2000, Australia.

The chemical compounds in commercial use as pesticides number about 350. At first everybody was looking for broad-spectrum chemicals that could be used extensively and marketed economically (e.g. lindane, DDT, aldrin). This increased the hazard of undesirable side-effects. Then, since rapid penetration of the living cell was desirable, the chemist turned more and more to designing compounds of high fat/oil solubility, and these tended to accumulate in fatty tissues and hence become concentrated in the last links of a food chain.

The pesticide industry is moving towards the pharmaceutical industry. There is a steadily increasing number of products, each catering for a smaller and more select market, and requiring lower dosages, stricter quality control and higher purity. An example of such developments is Pirimicarb (a carbamate, Fig. 5.21), a selective and systemic insecticide for aphids. It has a very low toxicity to predators that do not suck the sap; for example, it has no effect on ladybirds, lacewings or bees. It is rapidly metabolised and leaves no lasting residues in plant materials. It is an expensive, highly specialised chemical. Its synthesis, which is complex and involves five or six steps, requires sophisticated chemical plant and handling procedures.

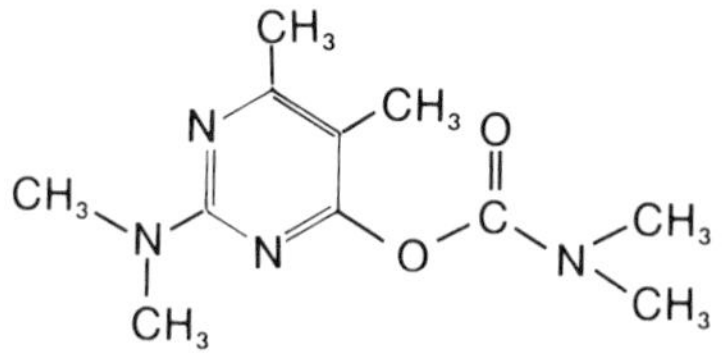

Fig. 5.21 Pirimicarb (ICI)

Fungicides

Interestingly, it is the fungi (rather than bacteria, viruses or insects) that have had the most devastating effect on our crops. Potato blight, wheat rust and smut, and wine mildew are a few examples whose historical consequences will be known to the reader.

Contact fungicides are designed to attack the fungus directly before it penetrates the plant, and inorganic chemicals were the earliest ones used. The use of sulfur goes back 2500 years. Bordeaux mixture (lime and copper sulfate) was introduced in 1885. In 1934 the iron salt (ferbam) and zinc salt (ziram) of dimethyldithiocarbamic acid were introduced and found to be non-toxic to humans, cattle and higher plants. A related compound, captan, is widely used on fruit trees and works by liberating thiophosgene ($CSCl_2$) inside the fungus. Naban is a related compound. 8-hydroxyquinoline (Quinolinal, Oxine), which needs iron to activate it, is used for mould-proofing outdoor equipment.

Treating plants with systemic fungicides is more difficult than treating animals with oral drugs because plants lack true circulation or excretion mechanisms. The compound benomyl (Benlate) was introduced in 1968 with great success. It has been particularly useful for protecting stored seeds (used for planting), for which purpose it has replaced the very hazardous mercurials and hexachlorophene, which have killed people who have eaten the seeds by mistake.

A sulfur-containing antibiotic in garlic is known to be active against a variety of bacteria and fungi, and garlic juice has been shown to inhibit the growth of fungus spores on nutrient plates in the laboratory. Garlic extract dramatically reduces disease in germinating bean seeds inoculated with fungus spores and also increases growth of the root system. So, eat salami and breathe on your roses!

Herbicides

The use of chemicals that selectively destroy plants had a very slow beginning. During 1895–97, copper sulfate solution was found to kill a weed (charlock) without causing injury to crops, and in 1911 dilute sulfuric acid was used in a similar situation. Dinitro-*o*-cresol was soon found to be useful as well. The 1930s saw the introduction of chemicals that mimic the natural auxin (plant hormone) indoleacetic acid. These were the phenoxyacetic acids, which, unlike the natural auxins, were not destroyed by the plant. They were also very effective in setting unfertilised fruits and promoting root growth.

Their use in excess showed that they were very effective herbicides. The most used of these is 4-chloro-2-methylphenoxyacetic acid (MCPA) (although the 2-methyl-4-chloro- is preferred in the UK); it is used to kill weeds in cereal crops (Fig. 5.22). The selectivity for killing dicotyledons but not monocotyledons (when equal amounts are *absorbed* by weeds and cereals, the cereals remain unharmed) is explained by the differences in their growing shoot structure.

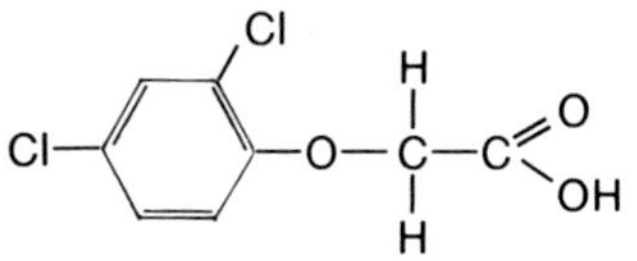

2,4-D
oral LD$_{50}$ 400–500 mg/kg
dermal LD$_{50}$ 1500 mg/kg

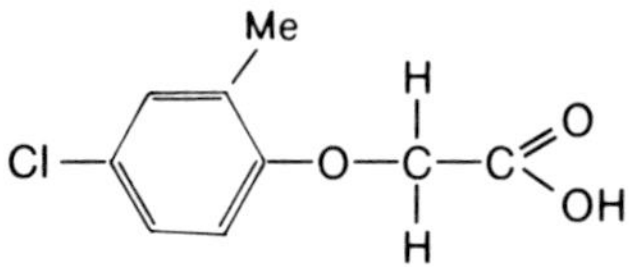

MCPA
oral LD$_{50}$ 800 mg/kg
dermal LD$_{50}$ > 1000 mg/kg

Fig. 5.22   2,4-dichlorophenoxyacetic acid (2,4-D) and 4-chloro-2-methylphenoxyacetic acid (MCPA), Methox, Methoxone and Tuloxone

It was found later that orthocarboxylic acids often behave in a similar manner. Picloram (Fig. 5.23) was introduced in 1963 and found to be more potent than 2,4-D.

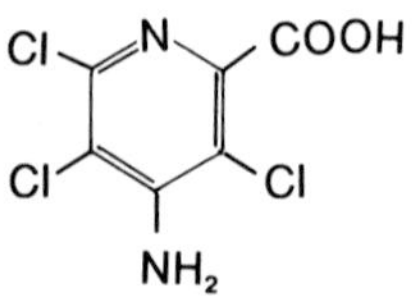

Fig. 5.23   Picloram

The compound 2,4,5-trichlorophenoxyacetic acid was found to be able to deal with such noxious weeds as privet and blackberry and thus became an urban and domestic herbicide (Fig. 5.24).

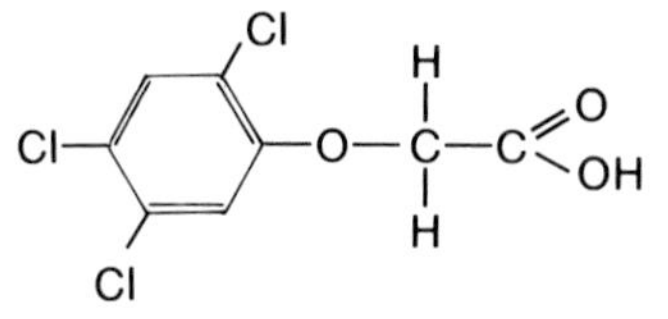

2,4,5-T

Fig. 5.24   2,4,5-trichlorophenoxyacetic acid (2,4,5-T)

2,4,5-T was implicated in causing birth defects in animals. It turned out that the cause of the problem was a small quantity of impurity—a dioxin formed during manufacture (Fig. 5.25)—for which LD_{50} guinea pig is 0.0006 mg/kg. It is fat soluble and may be concentrated like DDT. Adequate manufacturing control can keep the level of this impurity below 0.1 mg/kg of 2,4,5-T, which is considered acceptable. (In 1965 it was 5–32 mg/kg.)

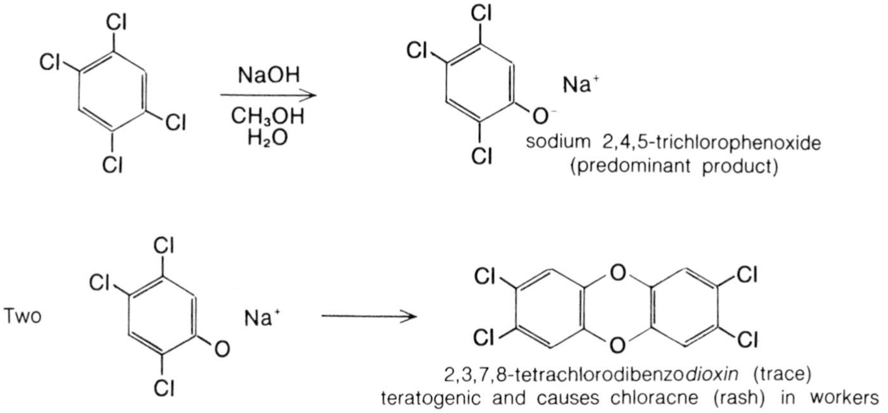

sodium 2,4,5-trichlorophenoxide
(predominant product)

2,3,7,8-tetrachlorodibenzo*dioxin* (trace)
teratogenic and causes chloracne (rash) in workers

Fig. 5.25   2,3,7,8-tetrachlorodibenzo*dioxin* (trace)

Herbicides with quite opposite modes of action include 2,2-dichloropropionic acid (dalapon, Dowpon). Chlorpropham is a carbamate and acts by preventing mitosis through inhibiting synthesis of RNA and protein. The redox indicator methyl viologen has been used by chemists since 1933 but its herbicidal properties were not discovered until 1955 (by ICI), when it was renamed Paraquat. Both uses depend on its ability to form a stable, free radical. (A molecule with a single un-paired electron is called a radical. Such a species is very reactive and chemically ferocious.) Paraquat, Diquat and Cyperquat are quaternary ammonium herbi-cides (related to cationic surfactants) and are unique in that they kill by contact with the plant foliage and cannot enter the plant via its roots because they are very strongly adsorbed onto clay and soil particles (for water growing plants this no longer holds). The quats are used as chemical ploughs to kill weeds between crops or just before a crop emerges.

The quats are moderately toxic. Small amounts can irritate the skin or, if inhaled can cause nose bleeding, but such amounts are rapidly and completely eliminated from the body. However larger amounts cause a terrible death. The symptoms are minimal for several days but later a non-cancerous proliferation of lung cells occurs, which continues long after all traces of herbicide have disappeared, leading to respiratory failure. There is no known antidote.

In 1977 I noticed that a bottle of 2,4,5-T that I purchased to control some over-active raspberry was labelled:

XXXX's Blackberry and Brush Killer. Buyer assumes all risk of use and handling *whether or not product is used in accordance with directions given.*

Perhaps it would be wise to let the vines take over after all. I made an application to the Trade Practices Commission to have such exclusion clauses deleted, and it has since rendered them illegal.

On 10 July 1976 a cloud of gas escaped after an explosion at the Swiss-owned ICMESA chemical plant at Seveso in northern Italy. The cloud contained dioxin, which was reported to have been formed because of a sudden rise in temperature within a reactor producing trichlorphenol.[5] Professor Fritz Mori, who designed the plant, stated that the poison could not have been created in his plant without the highly improbable chain of events leading to the disaster.

A chemical detective story

It all started with a search for a suitable example for a student exercise. I was look-ing for a piece of chemistry involving statistics with a social aspect. A government inquiry and report on apparent high birth abnormalities in Yarram (in Victoria) and their possible link with herbicide use seemed a suitable candidate.

However, this report soon showed some very shaky statistics, and I called in a statistician colleague, Dr Peter Hall, for a second opinion. This was the beginning of a long and productive partnership. We soon produced an article in which we showed up the errors in the report and pointed out that there was, for Australia as a whole, a statistically significant tendency for the rate of neonatal mortality (that is, infant deaths) from congenital abnormalities of the central nervous system to increase over the period 1968–1977. We could not link this to herbicides because in fact 'proof' of a cause is impossible to obtain without direct human experimen-tation on a very large scale.

But in searching for herbicide usage data in Australia for the early 1970s, we stumbled on a fascinating story, which we hinted at in the first article and expanded in the second one, published in *Chemistry in Australia*, and later in a letter to *Nature*. (We asked Professor Charles Kerr and Dr Barbara Field to join us in writing this letter because of their concern about this problem.) The second article looked at two birth abnormalities, spina bifida and renal agenesis, which were of concern at Yarram and which had been linked to the herbicide. Only mor-tality data were available on a sufficiently wide basis. Incidence data are much more sparse. This introduces some problems, which we considered in detail. These abnormalities show a highly significant tendency to increase between two five-year periods, 1968–1972 and 1973–1977. In addition, we tracked down some large

TABLE 5.4
Weight in kg × 1000 and value in thousands of US$ of imports into Australia with 1972–73
SITC code 512.28.09, by stated country of production and financial year

| | Country of origin | | | | | | | |
| | Singapore | | UK and USA | | Others | | Total | |
Year	'000 kg	'000 $	'000 kg	'000 $	'000 kg	'000 $	'000 kg	'000 $
1967–68	0	0	na	152	na	71	na	223
1968–69	na	51	na	182	na	100	na	333
1969–70	168	161	72	141	37	44	277	346
1970–71	142	140	116	177	75	79	333	396
1971–72	0	0	200	273	48	60	249	333
1972–73	0	0	171	252	31	57	202	309

na not available

imports of related material (as defined by the Standard International Trade Classification (SITC) code) into Australia from Singapore in the earlier period. These were listed in Australia's import statistics but *not* in Singapore's export statistics (see Table 5.4).

There had been considerable dumping on the Australian market of phenoxy-acetic acid herbicides (2,4-D and 2,4,5-T) and the precursor of 2,4,5-T, which is 2,4,5-trichlorophenol (TCP), when the use of Agent Orange and other sprays ceased in Vietnam (1969–1971). This dumping was stopped by a number of anti-dumping gazettals in the Commonwealth Gazette but no Customs Department investigation was ever carried out in Singapore itself. We speculated that Agent Orange may have been 'factored' through Singapore, and we carried out a sustained investigation of this and published what we had established. We believed that was as much as we could do.

However, amongst the voluminous correspondence we had with the bureaucracy was a suggestion that we should look at the relevant tariff report on the dumping and the related transcript of public evidence. This gave us a critical clue.

There was a very large import of potassium 2,4,5-trichlorophenolate (KTCP). This compound was produced in Singapore from TCP in order to avoid the tariff duties on TCP and its sodium salt. The lawyers had failed to include the potassium salt in the tariff regulations and it came in duty-free for a couple of years. When this was discovered by a large multinational competitor, that company agitated for an inquiry and the whole matter was considered by a tariff inquiry.

This revealed, as a side issue, a large shipment of *fire-damaged* KTCP. To the tariff officers, the significance of this was merely that the declared value of the import was less and hence the duty should be lower. A sample of material had been submitted to support the case for lower duty. This sample was labelled as 2,4,5,-T that had been made from the fire-damaged KTCP, and it was alleged that it blocked the nozzles of sprays and hence was difficult to use.

For the author, as a chemist, the heating of a sample of alkaline 2,4,5-trichlorophenol recalls the industrial accidents with it, the latest of which occurred at Seveso. (The conversion to TCDD (dioxin) occurs at normal pressure and elevated temperature.)

We now had an authenticated sample of the material. And what is more, we were looking after it until the government decided what to do. Our suggestion was that it should be analysed. The Industries Assistance Commission (IAC) informed various high levels of government of the situation and asked the Australian Government Analytical Laboratories (AGAL) to pick up the sample and analyse it for the IAC. We were given a formal receipt and, as it left our hands, we wondered whether in fact it would ever be analysed.

After some political shenanigans it was finally analysed in Perth and the results were tabled in the Commonwealth Parliament as follows:

MINISTER FOR BUSINESS AND CONSUMER AFFAIRS
HOUSE OF REPRESENTATIVES QUESTION
(Question No. 2059) 19 June 1981

MR WEST asked the Minister for Business and Consumer Affairs, upon notice, on 6 Ma 1981:

(1) Were samples of suspected fire-damaged KTCP (potassiumtrichlorophenolat a potassium salt of 2,4,5,-T [sic]) sent by the Industries Assistance Commissio to the Australian Government Analytical Laboratories for analysis during about March 1981.
(2) If so, has it been analysed and what are the results.
(3) From where did the IAC obtain the sample.
(4) Is he able to say if more of the sample is present in Australia; if so, where is i
(5) What does the IAC propose to do with the sample.

MR MOORE—The answer to the Honourable Member's question is as follows:
The Chairman of the IAC has advised me that:
(1) A sample which the Commission then believed to be fire-damaged KTCP wa forwarded to AGAL during March 1981 for analysis.
(2) Yes. The analysis showed that the sample was mainly isobutyl ester 2,4,5,- with a lesser amount of the n-butyl ester. The sample does not contain KTCP c 2,4-D. Dioxin (TCDD) content of the sample is 19 mg/kg, which is equivalent t 26 mg/kg on a 2,4,5-T acid base.
(3) The sample of about 130 mL was received about January 1973 from Chemica Industries (Kwinana) Pty Ltd as part of its evidence in relation to the Tarin Board inquiry into 2,4,5-T products and 2,4,5-trichlorophenol and its salt (Dumping and Subsidies Act).
(4) The sample was forwarded to AGAL.
(5) See [the answer to] (4).

The levels provided in the above answers should be compared to the levels used in Vietnam (Table 5.5).

TABLE 5.5
TCDD levels in 200 random samples of Agent Orange returned from Vietnam to Johnston Island.

TCDD range (mg/kg)	<0.05	0.05 to 0.10	0.11 to 0.50	0.51 to 1.00	1.1 to 2.0	2.1 to 3.0	3.1 to 5.0	5.1 to 7.0	7.1 to 10.0	10.1 to 20.0	>20.0*
% of samples	12.5	21.0	35.0	8.5	4.0	3.0	5.0	2.5	6.0	1.0	1.5

*3 samples of 22, 33 and 47 mg/kg.

Correspondence with the Singapore High Commission (18 June 1981) revealed the following answers to some questions:

(a) The Government was aware of the exact nature of the operations of Xxxxxx Pty Ltd *after* the fire. We have no information as to whether the Government was aware *prior* to the fire.

(b) Inspections [of the factory] were carried out but no sampling was done.

(c) Yes, there were reports of chloracne in the factory workers.

(d) The neighbouring factory complained of symptoms subsequent to the fire. No neighbourhood survey [of chloracne, birth abnormalities, soft tissue cancer etc.] was done.

Bhophal

Until a cloud of poisonous gas enveloped Bhophal on 2–3 December 1984, few people would have heard of this Indian city or of methyl isocyanate. Union Carbide was manufacturing carbaryl (Sevin), a carbamate pesticide (see 'Carbamates', p. 160). The process proceeds in several stages (Fig. 5.26). Carbon monoxide is produced by the reaction of petroleum coke with oxygen. Carbon monoxide is reacted with chlorine to produce phosgene. Phosgene is reacted with (mono) methylamine to produce hydrogen chloride and methyl carbamoyl chloride. Methyl carbamoyl chloride is then pyrolysed to produce methyl isocyanate. The final step is to react methyl isocyanate with 1-naphthol to produce carbaryl as shown in Figure 5.26.

1. $2\,C + O_2 \rightarrow 2\,CO$
2. $CO + Cl_2 \rightarrow COCl_2$
3. $COCl_2 + CH_3NH_2 \rightarrow HCl + CH_3NHCOCl$
4. $CH_3NHCOCl \rightarrow CH_3NCO + HCl$
5. $CH_3NCO + OH \rightarrow$ $OCONHCH_3$

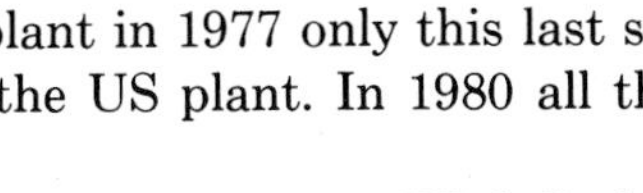

Fig. 5.26

At the opening of the plant in 1977 only this last step was carried out, using materials imported from the US plant. In 1980 all the reactions were done at Bhopal.

All the gases involved are very poisonous. While little was known about the toxicity of methyl isocyanate itself, under the action of heat it forms hydrogen cyanide, a most toxic gas used in US execution chambers. Phosgene was used in war and is more potent than chlorine. Carbon monoxide is toxic. Altogether (literally) not a very pleasant soup to spread over 40 km^2 and 200 000 people. Thus there was considerable confusion as to what the actual poison was and therefore how to treat its victims. At first doctors in Bhophal thought the gas was from a leak and therefore phosgene, and based their treatment accordingly. On a cold December night phosgene (b.p. 8°C) was more likely to escape than methyl isocyanate (b.p. 39°C). Also there was damage to crops for kilometres around the factory, which was to be expected from phosgene but not from methyl isocyanate. On the other hand phosgene, like chlorine, slowly fills the lungs with fluid and would not kill people immediately, as happened in Bhophal. Then again methyl isocyanate is stabilised with 200 to 300 mg/kg of phosgene to slow the reverse breakdown reaction during storage. Autopsies revealed that the blood of victims was a dark, cherry red and that lungs and other organs were also red. If fluid in the lungs had caused death by suffocation, then the blood would be bluish. Hydrogen cyanide was suspected, for which the antidote is sodium thiosulfate (photographic hypo). It reacts with cyanide to produce thiocyanate, which is excreted in the urine. This is a safe compound to administer even if the diagnosis is wrong, but its use was forbidden because it was not proven that cyanide was involved. However by giving this treatment and monitoring for thiocyanate you can prove that cyanide (above the tiny amount that is present in the body naturally) has been absorbed. Cyanide does not accumulate in the body and so treatment must be fairly rapid. On the other hand when sodium thiosulfate was finally administered after the Catch-22 argument was resolved months later, marked improvements were found, probably because of damage still present in the body from the methyl isocyanate gas itself. When Union Carbide published their report in March 1985, no mention was made of the nature of the gases that escaped from the tank.

In this edition, I have dropped the detailed discussion of the Seveso accident and included Bhopal. Will such a change become a constant feature of further editions?

Biological control

Instead of using chemicals against plant or insect pests, an alternative is to use their natural (or imported) enemies. Probably the best known and most successful application of this approach was against a plant—the prickly pear—in southern Queensland. During the late 1920s caterpillars of moths brought in from Argentina literally ate their way through some 25 million hectares of prickly-pear infested country. The Cactoblastis Memorial Hall near Chinchilla in central Queensland, on the Warrego Highway, commemorates the little grub's feat.

From the time they were introduced by the First Fleet, cattle have been upsetting the ecological balance in Australia with their dung, which gives bushflies, and the blood sucking buffalo flies, copious breeding places. Australia has native beetles that bury dung, but they evolved to cope with the pellet-like droppings of the native marsupials, not the massive and sloppy offerings of cattle. Dung beetles from Africa have lived on the droppings of large plant-eating animals for millennia and appear to be finding the local drop quite digestible. They were introduced into Australia in the early 1970s.

Insects have been used to control lantana, a fungus has been found effective against skeleton-weed, and a bacterium is being used commercially to control caterpillars.

Sex attractants

One of the most fascinating approaches to insect control involves *pheromones*. These are chemicals excreted externally and may serve to mark a trail, to send an alarm or to attract a mate. The last, the sex attractants, are usually excreted by the female to attract males. These compounds are detectable in extremely low concentration by males and can be used to lure males into traps or to disorient them. Field tests have shown that sex attractant of the gypsy moth is effective at amounts of 0.000 000 000 000 1 g (10^{-13}) in the field. It is interesting that the first claim, in 1961, for having discovered the structure of the attractant was wrong. Research in this area is difficult. In 1967 researchers used the abdominal tips—which contain the glands that produce the sex attractant—of hundreds of thousands of female gypsy moths to isolate a minute amount of attractant, and it was synthesised three years later (see Fig 5.27).

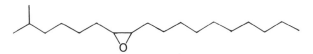

Fig. 5.27   Sex attractant of the gypsy moth

Some sex attractants are simple; others are complex. In some complex mixtures, the *ratio* of chemicals is important. The sex attractant for the common house fly is now known to be (at least) a two-compound system. One component is $CH_3(CH_2)_7CH=CH(CH_2)_{12}CH_3$, which is fairly easy to make. The use of sex attractants now looks promising for controlling insect pests in the mobile stage of their life-cycles.

Juvenile hormones

Juvenile hormones control the rate of development of the larval stage of insects and are switched off to allow development of the adult. Application of a mosquito juvenile hormone will keep mosquitoes in the harmless larval stage. Synthetic hormones that are much more potent that the natural compounds have been produced. This technique is effective for insects which are pests in the adult stage. It is not effective against insects that are pests in the larval stage (e.g. the caterpillars that are a pest in farmers' fields).

Sterilised male

In some cases, male insects are sterilised by irradiation from a radioactive source and then let loose in the insect populations. The females of many insect species mate only once, so that it is statistically possible to wipe out a species in an area quickly. This technique has been used successfully against the fruit fly in an *isolated* area. The biggest control program is the one in Texas against the screw-worm (a fly), but there the aim, at the moment, is to erect a 'barrier' to prevent the screw-worm moving north out of Mexico.

Resistance

The main reason for the development of new insecticides and the search for biological control measures is the development of resistance by insects to present methods. Some insects are resistant to organochlorine, organophosphorus, and carbamate compounds—at least at levels that do not give problems of residues in crops. One method of developing resistance depends on the Darwinian concept of the survival of the fittest. If an insecticide wipes out 99% of a population of insects, then the 1% that survive contain those best suited to deal with the poison, and it is these that breed the next generation. So insecticides carefully breed resistant insects!

Insect repellents

On a less drastic level there are several products that will keep insects away, at least from you. In the past, insect repellents contained strong-smelling oils such as citronella. These products kept away friends as well as the insects. It was later found that *contact* with the product was needed, not smell. In a study of more than 7000 chemicals, the US Department of Agriculture found only a few were really effective repellents. *N,N*-diethyl-*m*-toluamide, deet (Fig. 5.28), was found to last twice as long as any other. Dimethylphthalate (DMP, Fig. 5.29) is also very effective, but it is a rather good solvent for plastic watch 'glass' and spectacles as well (see also 'Chemistry of plastics—Vinyl polymers', Chapter 6). Other compounds found by the study to be effective were ethyl hexanediol (E-Hex) and Indalone (registered trade name for butyl 3,4-dihydro-2,2-dimethyl-4-oxo-2H-pyran-6-carboxylate).

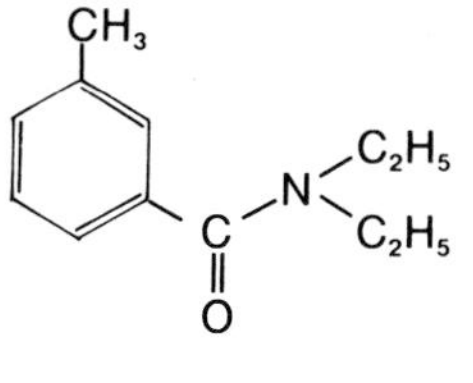

Fig. 5.28   Deet

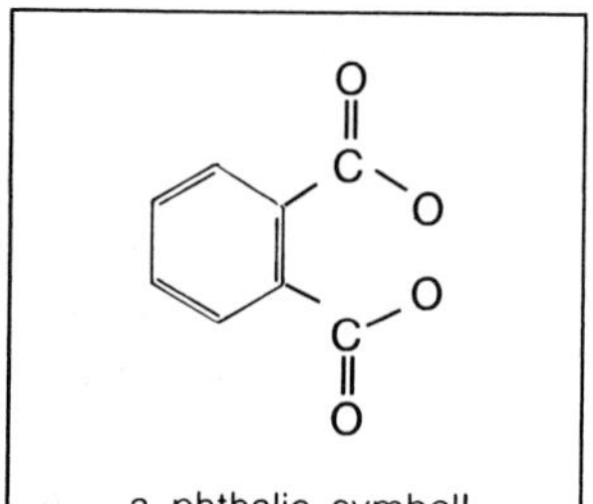

a phthalic symbol!

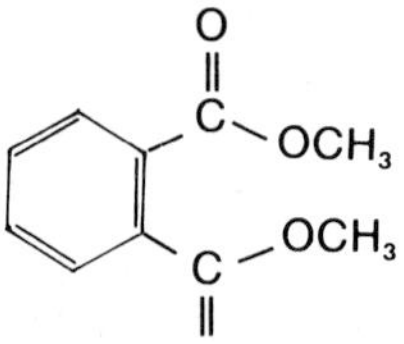

Fig. 5.29   Dimethylphthalate (DMP)

Choice (the journal of the Australian Consumers Association) found that the concentration of deet and/or ethyl hexanediol determined the *duration* of repellent action. Deet has somewhat better stay-put ability than the other preferred chemicals. On fabric, repellents are not rubbed off as quickly as they are from skin and will last for days instead of hours, but they can stain permanently and damage some synthetics and plastics.

One point that is not adequately emphasised is that you should know which insect you are intent on repelling. The Australian bush fly, *Musca vetustissima*, is a notorious nuisance to people throughout Australia (a rumour that the CSIRO breeds them in the wilds around Canberra is hotly denied!) It has been known since 1947 that effective mosquito repellents such as DMP are of no use and that deet, which is one of the best mosquito repellents, is of little value. An aerosol with an effective repellent was introduced on the Australian market in 1961 (Scram— David Gray Co.). It contains 5% di-*n*-propyl isocinchomeronate, the persistency of which is extended by the addition of the pyrethrum synergist *N*-octyl bicycloheptenedicarboximide. The use of such aerosols a few times a day prevents the flies from *settling* but not from momentary contact to test the site. Other Australian manufacturers market aerosols containing this repellent, along with others to repel mosquitoes, sand flies, etc. If it's bush flies you're after (or not after!) then 'isocinch' plus synergist is what to look for on the can.

SWIMMING POOLS

The view of any Australian city from the air presents a myriad of blue spots, mostly round but many oblong. These are backyard swimming pools, which have increased in number dramatically in recent years because of cheap systems of construction (e.g. fibreglass or PVC liners with iron retaining wall).

Along with every pool there comes a test kit for measuring pH and, above all, the level of chlorine. The test for chlorine has been taken over commercially from the accepted method of analytical chemists without any consideration of whether the two situations are comparable. The accepted analytical method involves the use of *o*-tolidine (4,4'-diamino-3,3'-dimethylbiphenyl) reagent. In the UK, *o*-tolidine is controlled by the Carcinogenic Substances Regulations 1967 (N:897). Even though alternative dyes are available (diethylparaphenylenediamine), they are not always used.

An analytical test must be sensitive, accurate, reproducible and immune, as far as possible, from interference. The *o*-tolidine test was selected by analytical chemists on this basis. In home swimming pools, however, interference from cyanide and thiocyanate could well be considered unlikely. In addition to a much lower problem of interference, a home test must involve safe chemicals, be easy to use and be unambiguous. The ability to distinguish between free chlorine and chloroamines (formed in the pool from reaction of chlorine with nitrogenous waste materials) is critical. With the *o*-tolidine test the skill of a good analyst is required and there is no evidence that any meaningful results are obtained by home users with this test. It is feasible that, before the solution is used by the consumer, it may have deteriorated, and it certainly will between seasons. If a test involves a hazard and also gives the wrong answer, there is really very little left to commend it.

Three years after I submitted a paper to the NH & MRC (followed by a little subtle external pressure), o-tolidine was placed in Schedule 7 of the Uniform Poisons Standard Act, 1975 (see 'Schedules', Chapter 9). A few years later the material was quietly removed from the schedule and allowed back on the market because there was no commercial alternative. Included in the submission was an alternative test method. The method, based on a recognised (but not standard) method for testing for chlorine, was found over a period of three years to work very well in unskilled hands (see question 15, 'Swimming pool chemistry').

Swimming pool chemistry

A remarkable variety of chemical concepts can be taught using the swimming pool as the medium. These concepts include pH, acids and bases, buffers, equilibrium, oxidation-reduction, properties of chlorine, electrolysis, photochemistry, mole concept and Faraday's law. On a more practical level a number of preliminary questions can be asked.

Question 1
a. What chemicals used in the school laboratory or discussed in class are available directly to the public through supermarkets, pharmacies, art supply shops or hardware stores (a suggested list is given in Table 13.9).
b. What chemicals do people buy for their home pool and what is the purpose of each of these chemicals?
c. Are the chemicals cheap or expensive? Who manufactures them? Why are these particular chemicals used rather than other chemicals with similar properties?
d. Do you think that any of the chemicals are *by-products* in the manufacture of some other product and hence a waste product being sold and producing unexpected profit?
e. Are you aware of any problem associated with pool chemicals? Any newspaper reports of fires? Can you remember if there are warnings on the containers? What do they say?

Then we come to the pool

Question 2
Why do we need to chlorinate our swimming pools?

Answer: Even water that is not meant to be drunk must be kept free of micro-organisms. When the same water is used over and over again, it quickly becomes contaminated with bodily excretions and garden debris. We use a strong oxidising agent such as chlorine to sterilise the water.

Question 3
What happens when chlorine is bubbled through water?

Answer:

$$Cl_2(aq) + 2H_2O \rightleftharpoons HOCl + Cl^- + H_3O^+$$

When chlorine is added to chemically pure water, a mixture of hypochlorous (HOCl) and hydrochloric (HCl) acids is formed. At ordinary water temperatures this reaction is essentially complete within a few seconds. In dilute solution and at pH levels above about 4, the equilibrium (in the equation) is displaced to the right, and very little Cl_2 exists in solution. In normal practice (except for the concentrate from solution feed chlorinators), the amount of chlorine supplied to water does not produce a concentrated solution of such strength as to yield such a low pH. The oxidising property of the chlorine is, however, retained in the HOCl formed, and this is the form that has the principal disinfecting action of chlorine solutions.

Hypochlorous acid ionises or dissociates, in a practically instantaneous reaction, into hydrogen and hypochlorite ions (note that the reaction is reversible), the degree of dissociation depending on pH and temperature:

$$H_2O + HOCl \rightleftharpoons H_3O^+ + OCl^-$$

Question 4
Why don't people chlorinate their pools with chlorine gas?

Answer: Chlorine is a yellowish green, dense gas. Chlorine gas bubbled through the pool water is an obvious way of treatment. However, it is inconvenient to use as a gas because it causes rapid corrosion of metals and destruction of plastics. It is also a dangerous gas, which attacks the mucous membrane linings of the eyes, nose, throat and lungs. It causes the lungs to fill with fluid and the victim drowns. During the First World War, chlorine gas was used as a chemical weapon.

Question 5
Is there a more convenient way to chlorinate a pool?

Answer: All forms of chlorine, such as pool chlorine (70% available chlorine), calcium hypochlorite, $Ca(OCl)_2$, and sodium hypochlorite, NaOCl, chlorine gas etc. ionise in water and yield hypochlorite ions:

$$Ca(OCl)_2 + H_2O \rightleftharpoons Ca^{++} + 2 OCl^- + H_2O$$
$$NaOCl + H_2O \rightleftharpoons Na^+ + OCl^- + H_2O$$

The hypochlorite ions also establish equilibrium with hydrogen ions, depending on the pH. Thus, the *same* equilibria are established in water regardless of whether elemental chlorine or hypochlorites are employed. The important distinction is the resultant pH and, hence, the relative amounts of HOCl and OCl^- existing at equilibrium. Chlorine tends to decrease the initial pH, and hypochlorites tend to increase it.

Question 6
What is this effect of pH?

Answer: Hypochlorous acid is weak and dissociates poorly at pH levels below about 6; thus, chlorine exists predominantly as HOCl at low pH levels. Between a pH of 6.0 and 8.5, there occurs a very sharp change from undissolved HOCl to almost complete dissociation. At 20°C above a pH of about 7.5, hypochlorite ions (OCl^-) predominate and they exist almost exclusively at levels of pH around 9.5 and above (Fig. 5.30).

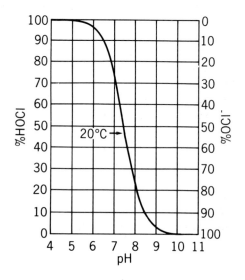

Fig. 5.30   Distribution of HOCl and OCl^- in water at indicated pH levels

The HOCl is much more effective in killing bacteria than OCl^- because the negative charge on the ion hinders it entering into the bacteria and oxidising the contents.

Question 7
How do you measure the amount of chlorine in water?

Answer:

$$HOC\ell + 2I^- + H_3O^+ \rightarrow C\ell^- + I_2 + 2H_2O$$
$$I_2 + 2S_2O_3^{2-} \rightarrow S_4O_6^{2-} + 2I^-$$

The amount of oxidant is measured by its ability to liberate iodine from acidified iodide solution. A sample of water can be titrated with standard iodide solution.
    The iodine released is detected by a blue colour formed with a fresh starch indicator (see also 'Quantum mechanics in the kitchen', p. 82, and Experiment 3.2). The amount of iodine released is determined by back titration with sodium

thiosulfate (hypo). Note that the reagent is only oxidised to tetra-thionate by the iodine whereas with chlorine it is oxidised to sulfate. Calculate the oxidation number of sulfur in these various compounds.

This titration is still the standard method for determining free chlorine, although other tests are available. (Potassium iodide releases iodine with chloramines as well, so these are included in the titre if present. See question 15.)

Question 8
What does 'available chlorine' mean and why is chlorine gas 100% available when it isn't?

Answer: When chlorine gas dissolves in water, it forms one molecule of hydrochloric acid and one molecule of hypochlorous acid. Only the latter is 'active'; so only half the chlorine is usable. Nevertheless, chlorine gas is set at 100%. So compounds for which all the chlorine in solution is active will have percentages twice the value based on composition. Table 5.6 gives some typical values of available chlorine. Note the entry for $Ca(OCl)_2$, which is 99.2% for the pure material. This is generally quoted as 100% and raises some confusion. $Ca(OCl)_2$ produces two moles of active chlorine compared to only one from Cl_2. However it has just over twice the molecular mass (a ratio of 143 : 71). So, on a mass basis, both materials are equally effective. Materials releasing other oxidising agents when dissolved in water are measured on the basis of the same redox reaction. Chlorine dioxide thus gives a value in excess of 200%.

TABLE 5.6
Per cent available chlorine of various products

Material	Available chlorine %
Cl_2 chlorine	100[a]
Bleaching powder (chloride of lime, etc.)	35–37
$Ca(OCl)_2$ calcium hypochlorite	99.2
Commercial preparations	70–74
NaOCl, sodium hypochlorite (solution 100%)	95.2
Commercial bleach (industrial)	12–15
Commercial bleach (household)	3–5
ClO_2, chlorine dioxide	263.0
NH_2Cl, monochloramine	137.9
$NHCl_2$, dichloramine	165.0
NCl_3, nitrogen trichloride	176.7
$HOOCC_6H_4SO_2NCl_2$ (Halazone)	52.4
CONClCONClCONCl, trichloroisocyanuric acid	91.5
CONClCONClCONH, dichloroisocyanuric acid	71.7
$CONClCONClCON^-$ Na^+, sodium dichloroisocyanurate	64.5

[a]by definition.

Source: W.H. Sheltmire, 'Chlorinated bleaches and sanitizing agents', Ch. 17 of ACS Monograph *Chlorine*, Reinhold Publishing Co., New York, 1962.

Question 9

Are those the only ways of obtaining chlorine in a pool? What about electrolysis?

Answer:

cathode $2C\ell^- \rightarrow C\ell_2(g) + 2\varepsilon^-$

anode $2H^+ + 2\varepsilon^- \rightarrow H_2(g)$

followed by: $C\ell_2(g) + H_2O \rightleftharpoons HOC\ell + H^+ + C\ell^-$

Note: ε^- in a half-equation represents an electron that is transferred. When the two half-equations are added, the electron cancels out. (g) means that the substance is a gas.

If salt is added, it is possible to generate the hypochlorous acid continuously in the pool by using an electrolysis cell. With this method of chlorination, the pool water will also gradually become acidic and may need some alkali added to adjust the pH. Commonly used alkalis are sodium carbonate and sodium bicarbonate. Sometimes the electrolysis of salt is carried out beforehand and a solution of sodium hypochlorite, NaOCl (liquid bleach), is sold for adding to the pool. Sodium hypochlorite is not stable as a solid.

Question 10

Why not drop the pH of the pool well below 7 if the oxidising strength of HOCl increases by doing so?

Answer: For a start, more acidic solutions will corrode the pool components. In marblesheen and tiled pools, the corrosion is even greater and the recommended pH range for these is 7.4–8.0. In addition, calcium chloride is added (100–200 mg/L) to counter the removal of calcium salts from the grouting.

More interesting, though, are the reactions of chlorine with ammonia and ammonia-like compounds that are formed from organic waste to form chloramines.

$$NH_3 \;\; + HOCl \rightleftharpoons NH_2Cl + H_2O \qquad \sim 1 \text{ minute}$$

$$NH_2Cl + HOCl \rightleftharpoons NHCl_2 + H_2O \qquad \text{slower}$$

The chloramines and dichloramines also react with each other:

$$NH_2Cl + NHCl_2 \rightarrow N_2(g) + 3HCl$$

The oxidation of chloramines in total is therefore given by:

$$2NH_3 + 3HOCl \rightarrow N_2(g) + 3HCl + 3H_2O$$

Further additions of chlorine beyond the breakpoint can result in the formation of nitrogen trichloride:

$$NHCl_2 + HOCl \rightarrow NCl_3 + H_2O$$

This gives rise to the so-called 'smell of chlorine', because it escapes easily from the water when the water is stirred up. It causes severe eye irritation.

A point of major confusion is whether a test for chlorine in water measures the free residual chlorine or the free chlorine *plus* the chloramines (combined chlorine).

The lower the pH, the more readily chloramines form. Above a pH of 7 their formation is minimal. This is another reason for keeping the pH above 7.

Question 11
What is superchlorination?

Answer: When chlorine in its various forms is added to water, it is used up in oxidising any material for which it is a sufficiently strong oxidising agent (iron II, sulfide, nitrite etc.) After this demand is satisfied, chloramines (combined chlorine) are formed from reaction of chlorine with organic nitrogen compounds. Additional chlorine (at pH greater than 7) will be used up in oxidising these chloramines to nitrogen gas and nitrate ion. The breakpoint occurs when all this material has been oxidised (Fig. 5.31). Further additions of chlorine are not used up and remain as residual chlorine ready to react with any material *now* added to the pool. This process of superchlorination carried out at regular intervals involves precisely this piece of chemistry. Strong oxidation can also be carried out using potassium monopersulfate. This will also oxidise any chloride back to chlorine (see also nappy sanitisers, which use the same reaction, in 'The chemistry of cosmetics—Baby-care products', Chapter 4).

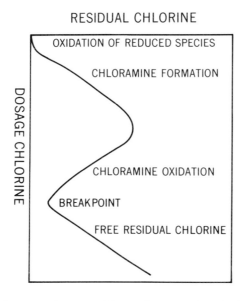

Fig. 5.31 Relationship between chlorine dosage and residual chlorine for breakpoint chlorination

Question 12
How do we adjust the pH of a pool?

Answer: We now see the need to control the pH of the water in a swimming pool. We can measure the pH with a suitable indicator that changes colour at the pH in which we are interested. The indicator is itself a weak acid that shifts from one colour to the other just as hypochlorous acid shifts from HOCl to OCl$^-$. That is, we choose an indicator with the same pK$_a$ as hypochlorous acid. The pH of natural water is generally about 6 because dissolved carbon dioxide from the air forms carbonic acid and lowers the pH below the neutral value of 7. The continuous addition of hypochlorite powder will gradually raise the pH, so that the addition of acid may be necessary after some time.

Solids are easier to store and use than liquids, so the 'acid' generally used is sodium hydrogen sulphate, NaHSO$_4$, although hydrochloric acid is supplied for salt-water pools with electrolysers. Why?

'Alkalinity' is a term used as a measure of the buffer capacity of water, or the degree of resistance to change in the pH of water on the addition of strong acids or bases. If the alkalinity is too low, pH control is difficult, because the pH is sensitive to small amounts of acid and base. From the volume of water in the pool, calculate how much 0.1 M hydrochloric acid needs to be added to change the pH by one unit if there is no buffering capacity.

Sodium bicarbonate in amounts of 80 to 120 mg/kg is added as a buffering agent. The pH of sodium bicarbonate in water (8.4) lies between the pK$_a$ values of the first and second dissociation constants of carbonic acid (6.35 and 10.33).

The term 'alkalinity' comes from the common use of sodium bicarbonate as a buffer and its alkaline pH of 8.4. However, acid buffers can also be used and 'buffer capacity' is a better term than 'alkalinity'.

Very high buffer capacity makes it difficult to change the pH when you want to, because very large amounts of acid or base are then required. Sodium bicarbonate

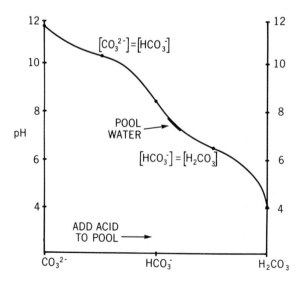

Fig. 5.32 Titration of sodium carbonate with hydrochloric acid

is also used as a base to raise the pH, but sodium carbonate is more effective and more common. The addition of sodium bicarbonate to water gives a pH of 8.4, whereas the addition of sodium carbonate to water gives a pH of 11.6. The addition of acid changes the pH according to the titration curve (Fig. 5.32).

Question 13
What do we mean by more powerful oxidising agent?

Answer: The ability of a material to oxidise is measured by the standard half-cell reduction potential (given in volts). This is an equilibrium value given for very specific conditions and so gives only a general indication for a practical situation. The larger the potential is, the stronger the oxidising agent. The standard electrode potentials, $E°$, are just two points on a potential versus pH curve ($E° = 1.49$ at $pH = 0$, $E° = 0.94$ at $pH = 14$). The equations and Figure 5.33 show that HOCl is a stronger oxidising agent than OCl^-.

$$\begin{array}{lr} & E°/V \\ HOCl + H_3O^+ + 2\varepsilon^- \rightleftharpoons Cl^- + 2H_2O & 1.49 \\ ClO^- + 2H_2O + 2\varepsilon^- \rightleftharpoons Cl^- + 2OH^- & 0.94 \end{array}$$

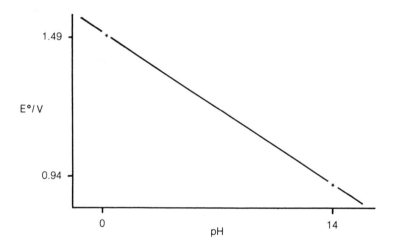

Fig. 5.33 Reduction potential of chlorine in water as a function of pH

Question 14
Is it dangerous to store pool chlorine near materials that can burn because it is a strong oxidising agent?

Answer: Yes! Hypochlorite is chosen for sterilising water, including swimming pools, because HOCl is a strong *oxidising* agent. If the material comes in contact with something that is able to be oxidised (i.e. burnt, but with chlorine, not oxygen, as the agent) then a fire can occur.

Hypochlorite powder + brake fluid → flame.

(WARNING: The experiment described here is *very dangerous*.)

Note that chlorine gas is released when solid hypochlorite powder comes in contact with moisture (at any pH). The constant escape of chlorine as a gas means the reaction continues to maintain equilibrium and thus produces more chlorine gas.

Question 15
How do we measure chlorine levels?

Answer: It is important to ensure that excess 'pool chlorine' is present in the pool after all the organic material has been oxidised. Some method of measuring free chlorine is needed.

The level of chlorine in water is a difficult analysis to carry out satisfactorily. A number of commercial test-kits are available. One of the most common is based on ortho-tolidine (4,4'-diamino-3,3'-dimethylbiphenyl—not to be confused with ortho-toluidine, which is only half the molecule; see Fig. 5.34). Unfortunately this compound is a close relative of benzidine (without the methyl groups), which is a known bladder carcinogen.

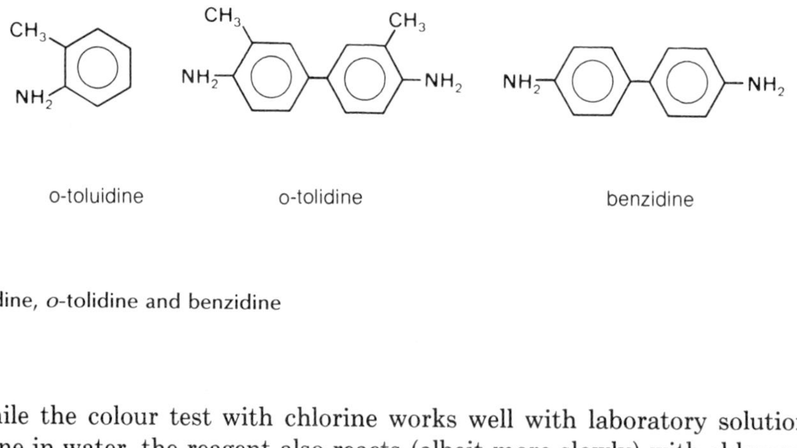

o-toluidine　　　　　　o-tolidine　　　　　　benzidine

Fig. 5.34　*o*-toluidine, *o*-tolidine and benzidine

While the colour test with chlorine works well with laboratory solutions of chlorine in water, the reagent also reacts (albeit more slowly) with chloramines, the compounds which chlorine forms when reacting with nitrogenous body waste products in pools. It is critical to measure the 'free' chlorine, over and above what has reacted with waste, otherwise underdosing will occur and the consequent danger of ear, eye and throat infections increases. Excess chlorine (breakpoint or superchlorination) oxidises the chloramines away. This process is essential in stabilised pools (see questions 18 and 19).

The methyl orange test measures *free* residual chlorine (HOCl + OCl⁻) whereas the OTO commercial kit measures the chloramines as well (at the rate of 3% per second, so you have to be fast to obtain the free chlorine value—this test is really quite unsuitable). You can demonstrate this elegantly by adding chlorine solution to ammonia to form chloramines as in question 10. Make up a solution of 0.2 g of $Ca(OCl)_2$ per litre (1.78×10^{-3} M) as stock solution. Dilute by a factor of five for use (3.5×10^{-4} M or 25 mg/kg Cl_2 equivalent). (This high concentration is

DO-IT-YOURSELF TEST-KIT FOR CHLORINE LEVELS

VERSEL methyl orange test kit (based on a titration method of methyl orange).

A test-kit you can make up very easily works on the principle that HOCl will decolourise methyl orange quantitatively and so can be titrated against it.

Theory

Free chlorine bleaches methyl orange solution quantitatively. At pH 2 the rate of reaction with chloramines is very slow and so only the free chlorine is measured. (The reaction was tested by preparing standard solutions and calibrating them iodometrically immediately before use (question 7). A spectrophotometric analysis of the change in absorbance of methyl orange at 510 nm was checked against a calibration curve. This is, of course, not convenient for home use.)

Method

A stock solution is prepared as follows: 0.05 g methyl orange is dissolved in 100 mL of water. A standard solution is made by diluting 10 mL of stock to 100 mL after adding 0.2 g NaCl. The reagent appears to be stable in the dark for years. A test solution consists of 24 mL of standard methyl orange solution to which is added 3 mL of 6 M hydrochloric acid (to lower the pH of the solution to 2 after the addition of pool water).

Procedure

It is essential that all samples from swimming pools be tested immediately. Place 0.25 mL of test solution in a test tube. Add 10 mL of pool water.

A. It decolourises instantly. There is at least 1 mg/kg chlorine present, which is sufficient.

B. It does not decolourise instantly. There is insufficient chlorine present; add more chlorine to pool and repeat test.

Repeat the test but use 0.5 mL of test solution.

A. It decolourises instantly. There is at least 1.5 mg/kg chlorine present.

B. It does not decolourise instantly. You have between 1.0 and 1.5 mg/kg of chlorine. This is ideal.

Repeat the test but use 0.75 mL of test solution.

A. It decolourises instantly. There is at least 1.75 mg/kg of chlorine present. This is too high.

B. It does not decolourise instantly. You have between 1.5 and 1.75 mg/kg of chlorine. This is high.

used to reduce losses and errors—stopper all solutions.) Add this in 20 mL aliquots to 1 mL of 0.015 M NH_3 solution (255 mg/kg)—a concentrated ammonia solution keeps the volumes almost constant. Test with indicators. Methyl orange will not bleach until about 2 mole equivalents of chlorine have been added (about 80 mL) and free HOCl is present. OTO, on the other hand, will show off scale from the beginning, indicating reaction with chloramines as well. Due allowance must be made for the time taken by the oxidation reactions (see question 10) and a chlorine solution should always be measured as a control for the one added to ammonia to monitor any loss of chlorine. The pH is also a factor.

Question 16
What about chlorine meters?

Answer: The 'chlorine' meter shown in Plate V is said to indicate the HOCl level. The unit does not contain a battery, so the current activating the meter must come from some electrochemical reaction. What do you think it is? If you look closely at the two electrodes of the meter you will notice that they are of dissimilar metals, probably aluminium and copper. Can you work out what the electrochemical processes are that could give rise to the electric current? Why would such a device be unreliable?

Question 17
Why is chlorine lost from pools in sunlight?

Answer:

$$OCl^- \xrightarrow[\text{light}]{\text{UV}} Cl^- + \tfrac{1}{2}O_2(g)$$

Chlorine is rapidly lost from pool water in the presence of sunlight. It is estimated that about 90% of the chlorine consumed in pools is the result of this reaction, which is an example of *photolysis* (chemical reaction brought about by the energy of light). Shallow pools require frequent addition of chlorine.

The ultraviolet range in sunlight is from 290 nm to 350 nm and all the chlorine species absorb in this region, with the OCl^- (pH 7–8) showing the strongest absorption (see Fig. 5.35).

Question 18
What do we do to stop loss of chlorine in sunlight?

Answer: There are compounds such as cyanuric acid (Fig. 5.36) which can be used to stabilise chlorine in swimming pools.

Cyanuric acid is made by heating urea or some of its derivatives. (It is a selective herbicide, very toxic to barley and radishes.) It exists as two tautomeric forms in equilibrium—an alcohol and a ketone.

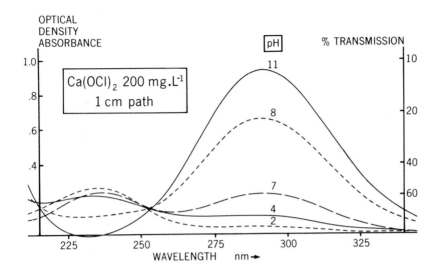

Fig. 5.35 Ultraviolet absorption spectrum of pool chlorine for the equivalent of 1 m depth at recommended concentrations

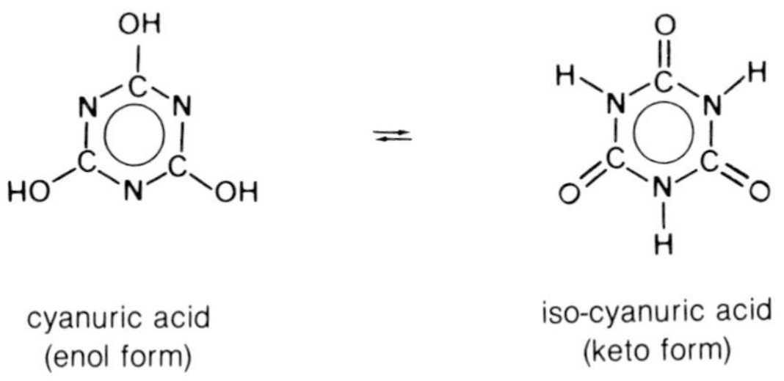

Fig. 5.36 Cyanuric acid exists as two tautomeric forms in equilibrium

Question 19
How does cyanuric acid work?

Answer: Cyanuric acid reacts with chlorine to give dichloro(iso)cyanuric acid in a chemical equilibrium (see Fig. 5.37).

As chlorine (as OCl^- or $HOCl$) is used up in the pool more OCl^- is released from the dichloro(iso)cyanuric acid to re-establish the equilibrium. Thus a constant amount of chlorine is maintained in the pool until all the dichloro(iso)cyanuric acid has lost its chlorine.

The general commercial chemicals are trichlorocyanuric acid and sodium dichlorocyanuric acid. Which would you expect to be more water soluble and why? These compounds are not photosensitive. The same effect can be obtained by adding just cyanuric acid and the chlorine from another (cheaper) source. This

(iso)cyanuric acid   dichloro(iso)cyanuric acid

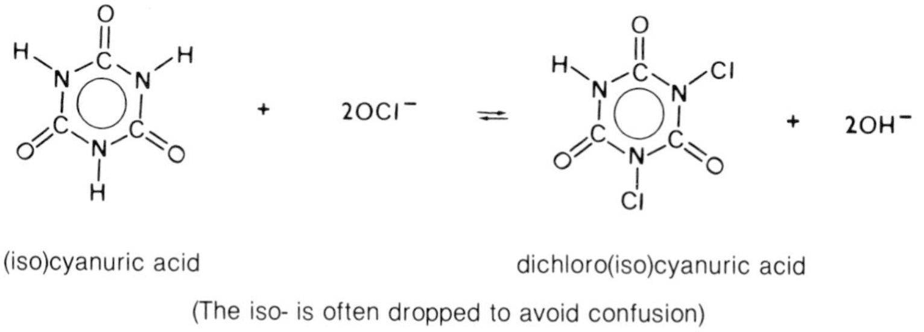

(The iso- is often dropped to avoid confusion)

Fig. 5.37 Equilibrium of cyanuric acid with chlorine and dichlorocyanuric acid

produces the chlorocyanuric acid as shown in the reaction. The amount of cyanuric acid is kept constant in the pool between 30 and 80 mg/L by the initial addition of any of the three compounds. From then on, only hypochlorite is added (to keep the level of free chlorine at 2 mg/L). The *equilibrium* in the reaction shown *stabilises* the concentration of OCl^-. What happens if excess cyanuric acid is used? (Consider the equilibrium and the effect on OCl^-.) The chlorocyanuric acids do not absorb the sun's ultraviolet light and thus their use retards the loss of chlorine through sunlight.

Question 20
Can we measure the level of cyanuric acid in a pool?

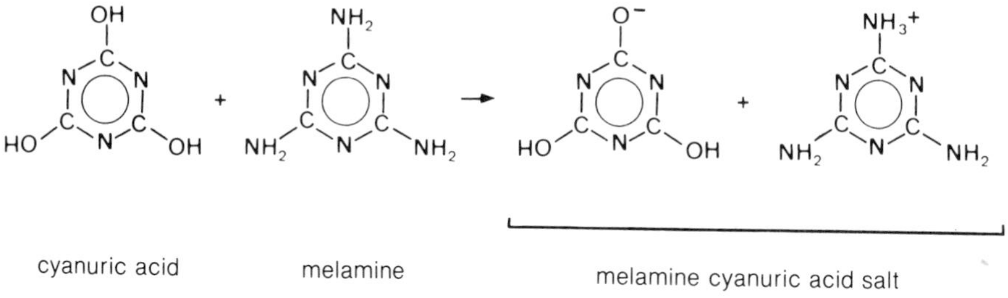

cyanuric acid   melamine   melamine cyanuric acid salt

Fig. 5.38 Salt formation by cyanuric acid and melamine

Answer: To determine the level of cyanuric acid in the pool you can carry out a reaction with melamine (see Fig. 6.2) to form a salt that precipitates and scatters light. The amount of turbidity is proportional to the amount of cyanuric acid. The turbidity is measured by the depth of solution required in a standard (Nessler) tube to just obliterate an object at the bottom of the tube.

Interestingly, anionic surfactants are measured by a similar salt formation using methylene blue. The salts formed are sufficiently lyophilic (non-polar portion overwhelms ionic contribution) to transfer into organic solvents and to be estimated by the depth of colour in the organic solvent.

Incidentally, if chlorine gas is passed through an ice-cold water solution of calcium chloride, feathery crystals of an unusual complex, $Cl_2 \cdot 7 \cdot 3H_2O$, separate. Such complexes in which a molecule is trapped in the crystal structure of a host ('ice' here) are called *clathrates* (from Latin *clathratus*, meaning 'enclosed or protected by cross bars or grating'). (See also the discussion on the structure of iron and steel in Chapter 6.)

Question 21
What about algae?

Answer: In addition to the need to sanitise the water against body waste and bacteria, algae can also be a problem. Small concentrations of copper or silver ions produced by electrolysis can be effective. Materials similar to cationic surfactants are also used as algaecides (see 'Soaps—How do surfactants work?' in Chapter 2). Clarification of muddy pool water is carried out with a flocculant, commonly a highly charged ion such as Al^{3+} from alum. (As well as coagulating clay suspensions, aluminium ion is used in sticks to coagulate blood from shaving cuts and in antiperspirants to coagulate sweat from pores under the arm. See also 'Chemistry of cosmetics—Deodorants and antiperspirants'; 'Chemistry of surfaces', Chapter 4.)

Question 22
How much does it cost to chlorinate a 60 000 L pool to a level of 1.5 mg/kg chlorine every day with pool chlorine (70% available chlorine) costing $6/kg for 1 year?

Answer: It is said that pools should contain about 1.5 ppm of chlorine (i.e. 1.5 mg/kg of Cl_2), each mole of which produces one mole of HOCl in solution. If $Ca(OCl)_2$ is used instead, then, because it has twice the molecular weight but also produces twice the amount of active chlorine, the same weight applies; i.e. 1.5 mg/kg. In a pool of 60 000 L needing an amount of chlorine to bring the pool from zero to this level each day, the quantity needed is $1.5 \times 10^{-3} \times 6 \times 10^4$ = 90 g per day. For pool chlorine of 70% available chlorine, the amount increases to 130 g per day. For 365 days and $6/kg the cost is

$$\frac{130}{1000} \times 365 \times 6 = \$280.$$

Question 23
How much does it cost to chlorinate a 60 000 L pool to a level of 1.5 mg/kg chlorine every day with an electrolysis system with an input power requirement of 160 W.

Answer: If you produce chlorine by electrolysis you have to calculate how much chlorine is formed by the current. One mole of chlorine gas (71 g) requires two moles of electrons or $2 \times 96\,500$ coulombs. Thus, 71 g of Cl_2 is produced by 193 000 coulombs. One ampere flowing for one second provides one coulomb of charge. To provide 193 000 coulombs from, say, a 15 A current will require

$$\frac{193\,000}{15 \times 60 \times 60} = 3.6 \text{ hours for 71 g of } Cl_2.$$

For the 90 g of chlorine shown to be needed in question 22, it will take 4.6 hours. This assumes the electrolysis is 100% efficient in producing chlorine rather than, say, producing some oxygen. Assume 80% efficiency, which increases the time to 5.7 hours—say 6 hours. The cost of electrolysis is determined by the input power of the unit, 160, for 6 hours every day of the year to give

$$\frac{160 \times 6 \times 365}{1000} = 350 \text{ kWh}.$$

At 7.5 c/kWh for electricity, the cost per annum is $26.25. The capital depreciation on the electrolysis unit must be added.

In both cases, the cost of running a 750 W filter/pump for 6 hours a day must be included. As well, the depreciation of the unit, cost of replacement electrolysis cells and the foregone return on investment in capital cost of the unit must be added to the running cost of the electrolysis system.

Water treatment products are discussed in Appendix I(b).

REFERENCES

1. Based on K. A. Hassal, *The chemistry of pesticides: their metabolism, mode of action and uses in crop production*, Verlag Chemie, Weinheim, 1982.
2. See *Nature* **297**, 1982, 62.
3. C. L. Stacey, *et al.*, *Arch. Environ. Health* **40**(2), 1985, 102; *Bull. Environ. Contam. Toxicol.* **35**, 1985, 202.
4. J. Tinker, 'Vapona: overkill for household flu', *New Scientist* **66**, 1975, 196.
5. *Times* (London), 3 August 1976.

BIBLIOGRAPHY

Household pesticides

Chemicals in the house, ACA and Total Environment Centre, Sydney, 1984.
'Lice are lousy. . . but so are some treatments'. *Choice*, February 1985, 16.

Pesticide poisoning

'Case-control study of congenital anomalies and Vietnam service' (Birth defects—report to the Minister of Veterans Affairs). AGPS, Canberra, 1983.
Cawcutt, Linda, and Watson, Catherine. *Pesticides, the new plague: the victims, the politics, the alternatives.* Friends of the Earth, Victoria, 1984.
Hazardous chemicals in the Australian environment, Proceedings of the conference, August 1983, University of Sydney. ACA and Total Environment Centre, 1984.
Inquiry into hazardous chemicals, 2nd report. House of Representatives Standing Committee on Environment and Conservation. AGPS, Canberra, 1982.
The pesticide handbook: profiles for action. IOCU, 1984.
Pesticide residues survey in the total diet. NH & MRC reports, Department of Health, Canberra, 1971, 1974, 1975.
Poisoning by organo-phosphorus compounds. NH & MRC booklet, Department of Health, Canberra, 1969.
Poisoning by pesticides. NSW Health Commission, 1976.

Senate Standing Committee on Science and Environment. *Pesticides and the health of Australian Vietnam veterans*, First Report. AGPS, Canberra, 1982.

Vettorazzi, G. 'State of the art of toxicological evaluation carried out by the joint FAO/WHO meeting on pesticide residues II. Carbamate and organophosphate pesticides used in agriculture and public health', *Residue Reviews* **63**, 1976, 1.

Pesticides

Caley, V. T. *Pestkem manual.* Queensland Agricultural College, Gatton. Lists the chemicals used extensively in NSW and Queensland for each fruit and vegetable group.

Caufield, C. 'Pesticides: exporting death'. *New Scientist*, 16 August, 1984, 15. Britain is now tightening its control over the use of pesticides, but the new legislation will do nothing to protect vulnerable farmers in the Third World.

Chemistry in the economy American Chemical Society, Washington DC, 1973. Chapter 11, 'Pesticides'.

'Guidelines to the use of the WHO recommended classification of pesticide by hazard', World Health Organization. This was issued in 1975 and has been revised on several occasions since. The final classification of any product is intended to be by formulation.

Hassal, K. A. *The chemistry of pesticides: their metabolism, mode of action and uses in crop protection.* Verlag Chemie, Weinheim, 1982. A far-ranging and well-written scientific treatise on the subject.

Martin, H., and Worthing, C. R. *Insecticide and fungicide handbook*, 6th edn. British Crop Protection Society/Blackwell Scientific Publications, London, 1980. This very useful reference lists the structure, syntheses and use patterns for a very large selection of pesticides.

The pesticide handbook: profiles for action, IOCU, 1984. This is a very well organised reference book giving information on Third World regulations as well as the usual Western sources.

Pesticides: synonyms and chemical names, 6th edn. Department of Health, Canberra, 1983. You need this to work your way through the morass of trade names, generic names, chemical names and the approved names of pesticides.

Pesticides: the chemical control of crop pests and diseases. ICI Educational Publications, London, 1975. Beautifully produced.

Rae, G. H. 'Towards safer methods of pesticides application'. *Chemistry and Industry*, 19 May, 1986, 354.

Report on the working party on pesticide residues (1977–1981). Food Surveillance paper no. 9, 1982, HMSO, London.

Short review of pesticides. Hodogaya Chemical Co., Tokyo, 1980.

Termite treatment of existing buildings, Australian Standard AS2057, 1981. Standards Association of Australia, Sydney, 1981.

Soil

The following publications from the CSIRO were used in the section on soils.
1. *Soils: an outline of their properties and management*, 1977.
2. *Soil—Australia's greatest resource*, 1978.
3. *Composting: Making soil improver from rubbish*, 1978.
4. *What's wrong with my soil?* 1978.
5. *Earthworms for gardeners and fishermen*, 1978.
6. *Food for plants*, 1978.
7. *Organic matter and soils*, 1979.
8. *When should I water?* 1979.

2,4,5-T and dioxin

'Dioxin report'. *C & EN* 6 June 1983, 20.

Esposito, M. P. *et al. Dioxins*. US Environmental Protection Agency, 1980. A very comprehensive and clear review of the topic.

Hall, P. and Selinger, B. 'Australian infant mortality from congenital abnormalities of the central nervous system. A significant increase in time: is there a chemical cause?' *Chem. Aust.* **47**, 420. Reprinted *AFCO Quarterly* **20**, 13.

Hall, P., and Selinger, B. 'Antipodean 2,4,5-T'. *Nature* **290**, 1981, 8.

Hall, P., and Selinger, B. 'Australian 2,4,5-T'. *Nature* **292**, 1981, 286.

Hall, P., and Selinger, B. Submission and Evidence, Senate Standing Committee on Science and the Environment, 'The use of pesticides', Federal Hansard, December 9, 1981. Second Submission, March 1982.

'Herbicide statistics: a critical examination of the role of statistics in some topical decisions'. *Search* **13**(11/12), 1983, 300.

Bhophal

Agarwal, A. 'The cloud over Bhophal'. *New Scientist*, 28 November 1985.

'All the world gasped'; 'India's night of death: more than 2 500 people are killed in the worst industrial disaster ever'; 'Hazards of a toxic wasteland'; 'An unending search for safety; 'A calamity for Union Carbide'. *Time Magazine*, December 17, 1984, 6–23.

'Report on scientific studies on the factors related to Bhopal toxic gas leakage', December 1985. High Commission of India, Canberra.

Weir, D. *The Bhophal syndrome: pesticide manufacture and the Third World*. IOCU, Penang, 1986.

User-friendly pesticides

Francis, D. F. and Southcott, R. V. *Plants harmful to man in Australia*. Miscellaneous Bulletin no. 1, Botanic Gardens, Adelaide, 1967. (Out of print; available from National Library, Canberra.) Excellent reference. The contrary view—how plants try to wipe out man!

'Getting away from pesticides'. *Ecos*, Canberra (CSIRO), **2**, 1974, 19.

'Insect repellents'. *Choice*, ACA, **10**(1), 1969; **18**(1), January 1977.

Pumpkins, poisons and people. Conservation Council of Victoria (publ. no. 3), Melbourne, 1978.

Waterhouse, D. F. and Norris, K. R. 'Bushfly repellents'. *Aust. J. Science* **28**, 1966, 351.

Wood, W. F. 'Chemical ecology: chemical communication in nature'. *J. Chem. Ed.* **60**(7), 1983, 531.

Chemistry of hardware and software (I)

6

CHEMISTRY OF PLASTICS

When molecules are strung together like a string of paper-clips, in one or three dimensions, the resulting compound is called a *polymer* or *macromolecule*. The synthetic members of this group include the plastics, fibres and rubbers. The small, basic building-blocks are called *monomers*. Differences in the chemical constitution of the monomers, in the structure of the polymer chains, and in the interrelation of the chains determine the different properties of the various polymeric materials.

There are many ways of classifying plastics. One method is by division into two groups: those polymers that extend in only one dimension (that is, those that consist of linear chains); and those polymers that have cross-links between the chains, so that the material is really one three-dimensional giant molecule (an example is given in Fig. 6.1). The first group of plastics are called *linear polymers*, and are *thermoplastic*; that is, they gradually soften with increasing temperature and finally melt because the molecular chains can move independently. An example is polythene, which softens at about 85°C. The second group are the *cross-linked polymers*, which are *thermosetting*. They do not melt on heating, but finally blister, the result of gases being released, and char. An example is bakelite.

One readily available linear polymer is made from milk. The milk protein casein is separated out (using rennet from calves' stomachs, or acid) and moulded and then cross-linked (hardened) with formalin. Casein plastic is used to make buttons and knitting needles.

HISTORY OF THE DEVELOPMENT OF SYNTHETIC POLYMERS

1839   Styrene polymerisation observed.

1843   Caoutchouc (native rubber) vulcanised.

1844   Linoleum produced.

1864   Christian Schonbein prepares cellulose nitrate by treating paper with nitric acid.

1865   Alexander Parkes prepares the first plastic—parkesine—composed of cellulose nitrate, vegetable oils and camphor.

1869   John Hyatt patents celluloid, an improvement on parkesine.

1880   Acrylates first prepared in the lab. Related to perspex.

1897   Casein-formaldehyde polymer.

1897	Urea-formaldehyde first prepared in the lab.
1898	Einhorn prepares polycarbonate in the lab.
1905	Schnitzenberger prepares cellulose acetate.
1907	Leo Baekeland patents bakelite. Produced from phenol and formaldehyde, this was the first wholly synthetic plastic.
1911	Cellophane developed by Brandenburger.
1912	Ostromislensky produces PVC.
1915	Dimethylbutadiene rubber produced.
1918	Urea-formaldehyde patented.
1921	First plastic injection moulder developed.
1922	Staudinger shows that rubber is a chain of isoprene units.
1924	Poly(vinyl alcohol) first prepared.
1927	Perspex first prepared in the lab.
1928	Du Pont develops 'superpolyamide', now known as nylon.
1930	Poly(vinyl acetate) developed. Epoxy polymers produced from phenol-polyalcohols.
1931	Neoprene (polychloroprene), poly(vinyl acetal) and polyisobutylene first produced.
1932	ICI develops polyethylene.
1933	Roy Plunket discovers teflon, polytetrafluoroethylene.
1935	First extruder for thermoplastic polymers developed.
1936	ICI produces perspex commercially.
1937	Polyurethane developed.
1939	Polyethylene produced commercially.
1939	Melamine formaldehyde begins to replace urea-formaldehyde because of superior properties.
1940	Polyethers (acetyls) first prepared.
1942	Saran fibres (poly(vinylidene chloride)) are produced.
1943	Silicones become commercially available.
1943	Epoxy resins are developed.
1950	Orlon (poly(acrylonitrile)) developed.
1956	Polycarbonates commercially produced.
1960	Polyacetal first prepared.
1965	Poly(phenylene oxide) produced.
1965	Polysulfone developed.

Source: CSIRO Science Education Centre

In 1907, when Leo Baekeland, a Belgian working in the United States, was looking for a shellac substitute, he discovered the first artificial plastic—bakelite—by mixing phenol and formaldehyde. Bakelite is a good electrical insulator and is still used today for power plugs, points and switches, and for electric jug lids etc. It is still the most important cross-linked plastic because it is cheap and rigid (but it is available only in dark colours—white double-adaptors were originally more expensive than brown).

Plastics related to bakelite are made from a combination of urea and formaldehyde (see Fig. 6.1). This mixture is also widely used for adhesives that are not water-soluble.

Urea-formaldehyde is usually formulated as a moulding powder with α-cellulose (purified wood pulp) filler. Because the refractive indices of the resin and cellulose are similar, there is little light scattering at the phase boundaries and so it is easy to produce pastel-coloured products (cf. 'Chemistry of cosmetics—Toothpaste', in Chapter 4; 'Paints—Solid paints: Hiding power', in Chapter 8). They are less water and heat resistant than the phenolics.

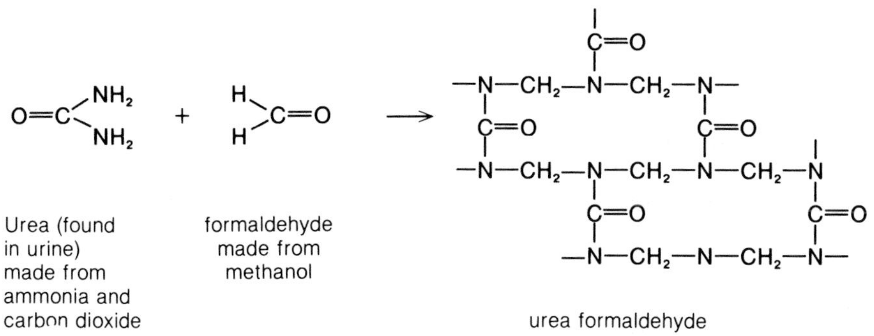

Urea (found in urine) made from ammonia and carbon dioxide

formaldehyde made from methanol

urea formaldehyde

Fig. 6.1 Three-dimensional giant molecule of urea-formaldehyde

In melamineware (used for dinnerware) the compound melamine (Fig. 6.2) is used instead of urea. This gives a product that combines the good properties of the other two resins. Melamine formaldehyde is water resistant, can come in pastel colours and can be kept continuously at above 100°C (dishwasher proof).

Some interesting economics emerge. Urea-formaldehyde resins dominate the adhesive market because they are cheap. However, when a filler is needed, the phenol-formaldehyde can tolerate cheaper fillers, so the moulding powder market is dominated by the phenolics. When a water-resistant chipboard is required or a moulding in a pastel colour, you have to pay extra for the urea resin. If you want a plastic that is both water resistant and pastel coloured, you will have to buy melamine, which is more expensive.

Laminated plastics and veneers (e.g. Formica and Laminex) are made by impregnating several sheets of materials (usually paper or cloth) with plastic, then pressing the sheets together and hardening them in the oven.

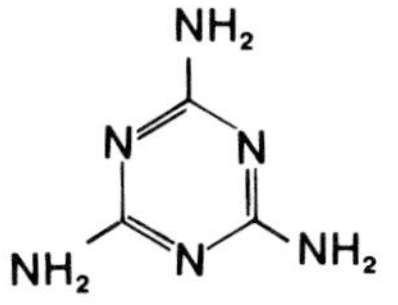

Fig. 6.2 Melamine

Addition polymerisation

The *linear* polymers can be made by joining a sequence of monomers, but here the possibility for cross-linking in two or three dimensions does not exist. The simplest and most widely used is polyethylene ($\ldots$ $-CH_2-CH_2-CH_2-$ $\ldots$). When the chain length exceeds about a thousand units, the material becomes relatively rigid and can be used in a variety of ways. Apparently the linear molecular chains are partly tangled with each other to form amorphous (without structure) regions that impart strength and a higher melting point to the material. Polyethylene (polythene) is made by joining together monomers of ethylene in a process called *addition polymerisation*, according to the reaction:

$$R-O-O^{\bullet} + CH_2{=}CH_2 \longrightarrow R-OOCH_2-CH_2^{\bullet}$$

$$+ \text{ next } CH_2{=}CH_2 \longrightarrow R-OOCH_2-CH_2-CH_2-CH_2^{\bullet} \quad \text{etc}$$

where $R-O-O^{\bullet}$ is a free *radical* and acts as an *initiator*. (When a molecule has an odd electron it is called a radical, and this is indicated by a dot at the point where the electron is expected to be.) The final product of the reaction is polyethylene, $(CH_2)_n-CH{=}CH_2$, termination of the chain often occurring by a reverse of the initiation step. About every tenth unit has a CH_3 branch. Polythene is fairly inert but biodegradability can be built in by increasing the number of double bonds, which gives microbes somewhere to chew (see 'Domestic laundry detergents— Biodegradability', Chapter 2).

Under normal conditions, using the usual catalysts, the spatial arrangements of the branches in polymer products are random. Such polymers are called *atactic* (an example is given in Fig. 6.3).

Ionic polymerisation

When a suitable acid (A) is used to initiate the reaction, the process becomes a particular type of addition polymerisation called *ionic polymerisation*. The reaction proceeds as follows:

$$A + CH_2{=}CH_2 \longrightarrow \bar{A}-CH_2-\overset{+}{C}H_2$$

$$\bar{A}-CH_2-\overset{+}{C}H_2 + CH_2{=}CH_2 \longrightarrow \bar{A}CH_2-CH_2-CH_2\overset{+}{C}H_2$$

This process allows the use of suitable catalysts to give products in which the branches are arranged in an orderly manner. These plastics are called *isotactic* polymers. An example, polypropylene is illustrated in Figure 6.3. For polyethylene, this form of polymerisation has the advantage of reducing the number of branches that are formed. The process uses lower temperatures and pressures than radical polymerisation and consequently little cross-linking or branching occurs. The truly linear chains pack together to give a *high-density* polymer that has a high melting point. By using anionic catalysts mounted on a crystalline solid, the geometry (stereochemistry) and shape of the product can thus be determined.

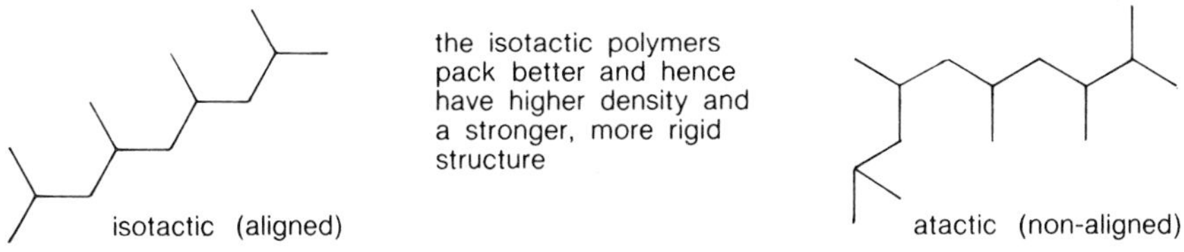

the isotactic polymers
pack better and hence
have higher density and
a stronger, more rigid
structure

isotactic (aligned)   atactic (non-aligned)

Fig. 6.3 Polypropylene

The molecules in linear isotactic polyethylene can line up with one another very easily, to yield a tough, high-density compound that is useful for making toys, bottles etc. The polyethylene with irregular branches is less dense, more flexible, and not nearly as tough as the linear polymer because the molecules are further apart. Polyethylene is the most widely used plastic. It was discovered in the early 1930s and gained prominence during the Second World War as an insulator for high-frequency cables used in radar installations.

The factors that affect the rigidity of the polymer are the packing of the molecules, which is seen as variations in *density* and *molecular mass*—the number of monomer units combined to give a polymer. With increasing molecular mass, there is a corresponding increase in strength, toughness and chemical resistance.

A third basic parameter is molecular mass *distribution*. Because of the statistical nature of polymerisation, polymer molecules show a variation in mass. Although the chemical properties of a polymer remain much the same regardless of mass, its mechanical properties are affected by variations in the mass of its molecules. If most of the molecules of a resin fall within a very narrow mass range, products made from it will have better mechanical properties than those made from resins whose molecules vary over a large range of masses. It is the balance in these three characteristics that provides the variation in properties of the different plastics formed from the same basic monomer.

Certain properties can also be controlled by adding small amounts of other monomers to the ethylene monomer. Vinyl acetate or ethyl acrylate in low-density resins and hexene or butene in high-density resins will increase branching—thus increasing flexibility and elasticity. Finally, polyethylene can be changed from a linear polymer to a cross-linked polymer by incorporating particular reagents. Additives are also used in polyethylene. These include slip agents (to decrease frictional properties), anti-block agents (to prevent sheets of molecules from sticking together), and antioxidants.

Most polymers undergo oxidation and photo-initiated degradation, and this can be retarded by antioxidants. Low-density polythene requires only a very small amount of antioxidant. High-density polyethylene, polystyrene and, in particular, polypropylene are much more sensitive, both during processing and on exposure to the environment. Oxidation causes the links in the polymer chain to break, which lowers the molecular mass of the polymer and makes it less tough. It also causes cross-linking of the molecules, which, although resulting in a higher molecular mass, in this case decreases toughness and makes the plastic more brittle. To combat oxidation, various materials are used: primary antioxidants, BHT and BHA (see Fig. 3.4), and other materials. Polypropylene used in, for example, washing-machine agitators, requires antioxidants capable of withstanding hot detergent. High molecular mass (> 800) derivatives of BHT and BHA are used.

Both high-density and low-density polyethylenes have the same density in the molten state, so that in a mould the high-density polymer shrinks much more than the low-density one. The consequent warping of the high-density material has been a problem. Shrink-wrapping of products is achieved by allowing polyethylene to shrink on heat treatment. A paper-like film resembling parchment is 'blown' from high-density polyethylene. Corrugated agricultural drainage pipe with small slits is made from polyethylene by using a continuous extrusion blow-moulding process. Polyethylene has replaced wax on the inside of cardboard milk cartons.

Vinyl polymers

If one of the hydrogen atoms in the ethylene molecule is replaced by chlorine, *vinyl chloride* is formed. If two hydrogens on the same carbon atom are replaced, *vinylidene chloride* is formed. If we replace a hydrogen in ethylene with acetate (from acetic acid) we form *vinyl acetate*. These monomers can all be readily polymerised. Their molecular structures are given in Table 6.1.

Vinyl polymers and copolymers make up one of the most important and diversified groups of linear polymers. This is because PVC (poly(vinyl chloride)) can be *compounded* to produce plastics with a wide range of physical properties. PVC is used to make products ranging from guttering and water pipes to the very thin, flexible surgeon's gloves.

Polymerisation of vinyl chloride to PVC can be carried out in three different ways. In the *suspension* process, droplets of monomer are dispersed in water and polymerised. In the *mass* process special agitation is used to polymerise liquid vinyl chloride monomer without water present. The commonest method of making PVC is to disperse vinyl chloride in water as an *emulsion* (using surfactants) using catalysts and heat. The monomer is polymerised to solid particles of polymer that emerge as a suspension in water. This is centrifuged and dried. The important parameters here are the average molecular mass (i.e. length of polymer molecule) and chemical purity. A large molecular mass means a stronger and more rigid polymer, but this type of polymer is more difficult to work. PVC that is to be used as an electrical insulator is generally not made by this method, which is known as *emulsion polymerisation*, because the soap that remains in the polymer as an impurity decreases the insulating properties. However vinyl acetate is generally polymerised by emulsion polymerisation because most PVA (poly(vinyl acetate)) is used in making emulsion (latex or water-based) paints. Plasticiser and pigment are added. When the paint is applied to a surface, the water evaporates and leaves a polymer film containing pigment and plasticiser.

Depending on the processing required and the end-use, a wide variety of additives are added to PVC resin. For example, rigid plastics do not contain a plasticiser because plasticisers increase malleability; flexible plastics do not require impact modifiers to improve resistance to impact; and transparent plastics do not contain fillers, which are used only if a cheap, rigid plastic is required. The plasticisers used (and about 80% of all plasticisers used are used in PVC) are organic liquids of low volatility. They are used to facilitate internal movement of the molecular chains; that is, to make the PVC more flexible. The esters of phthalic acid are most commonly used, particularly di(2-ethylhexyl)phthalate

TABLE 6.1
Vinyl polymers

Monomer	Monomer structure	Polymer	Main uses
A. Common polymers			
Ethylene	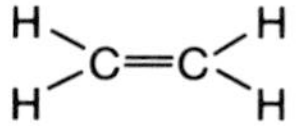 $CH_2=CH_2$	Polythene: low density, high density	Bottles, tubing, sheets, and other moulded objects
Vinyl chloride	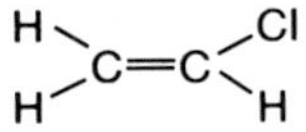 $CH_2=CHCl$	Poly(vinyl chloride) (PVC)	Raincoats, shower curtains, gramophone records, garden hose, rigid clear bottles, swimming-pool liners
Propylene	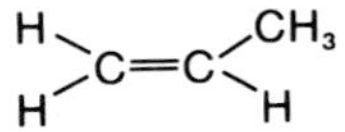 $CH_2=CHCH_3$	Polypropylene	As for polythene, and carpets (isotactic)
B. Specialised polymers			
Vinyl acetate	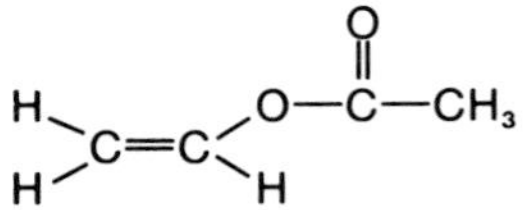 $CH_2=CH{-}O{-}\overset{\displaystyle O}{C}{-}CH_3$	Poly(vinyl acetate) (PVA)	Adhesives, latex paints
Vinylidene chloride	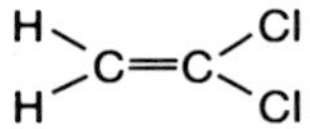 $CH_2=CCl_2$	Poly(vinylidene chloride), co-polymer with PVC in Saran	Some clinging wraps, some freezer bags
Tetrafluoroethylene	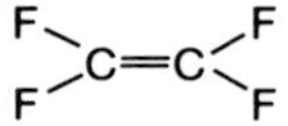 $CF_2=CF_2$	Poly(tetrafluoroethylene) (PTFE), Teflon	Bearings, gaskets, non-stick pan lining, chemical resistant films
Styrene	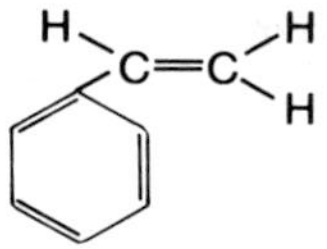 $CH_2=CH{-}C_6H_5$	Polystyrene	'Rigid foams', moulded objects, electrical insulation

Continued

TABLE 6.1 — Continued
Vinyl polymers

Monomer	Monomer structure	Polymer	Main uses
Methyl methacrylate	$H_2C=C(CH_3)-C(=O)-O-CH_3$	Poly(methyl methacrylate), perspex, lucite, plexiglas	'Safety glass' but PV Butyral is the adhesive in Triplex and Pilkington's windscreen glass
Acrylonitrile	$H_2C=CH-C\equiv N$	Poly(acrylonitrile)	Orlon, acrilan textile fibres
C. Mixed polymers			
Styrene and acrylonitrile		SAN	Latex paints, plastic plates, etc.
Acrylonitrile, butadiene and styrene		ABS	Rigid: telephone sets, shoe soles, automobile parts

(DEHP), also called dioctylphthalate (DOP). Combinations of compounds are used as plasticisers and the amount can form up to 50% of the finished product. Because plasticisers are not bound chemically, they tend to migrate to the surface and be lost by abrasion, solution or slow evaporation, leaving a more brittle, stiff product. The use of non-compatible plasticisers results in the appearance of a 'spew', an oily exudate, on a vinyl product surface.

PVC is used industrially for piping, and for building products (panels, window sashes, gutters, downpipes, conduit). The automobile industry uses large quantities of flexible PVC, for wire insulation, injection moulded knobs and upholstery. (During hot weather you end up with a partly opaque film of plasticiser on the inside of your car windows, particularly the windscreen and rear window, because the plasticiser distils out of the upholstery in hot weather and condenses on the windows forming a film that is hard to remove). Bottles produced from rigid PVC are particularly useful for holding alcohol or oily products. (Polythene is unsuitable because it is permeable).

Polystyrene is an amorphous, transparent polymer (the bulky phenyl groups inhibit crystallisation) that softens at 94°C and so can't be sterilised. When it is hit, it gives a metallic ring. Products made from it sparkle and are quite attractive because it has a high refractive index (1.6). (See also 'Paints—Solid paints: Hiding power', in Chapter 8.) The brittle nature of the polymer can be overcome by adding 5 to 10% of butadiene monomer (see Fig. 1.6) to give 'impact polystyrene', but then light scattering occurs at the phase boundaries and the product is opaque. However polystyrene is flammable, softened by many solvents and light sensitive. Polystyrenes that contain flame-retardant are used as ceiling tiles, and those containing UV absorbers are used as fluorescent light diffusers. Because of the wide gap betwen the glass temperature ($Tg = 94°C$) and the melting point ($Tm = 227°C$), polystyrene is a pleasure to process. Injection moulding is used for making bottles and jars. Extrusion moulding is used to produce sheeting, which can then be thermally shaped to make refrigerator linings and three-dimensional contour maps. Rigid polystyrene foam accounts for about 15% of this plastic's use and a considerable amount of this is used in packaging (see later in this chapter). Co-polymerisation with acrylonitrile to give styrene acrylonitrile (SAN) provides a slightly better product.

Polystyrene burns with a hot smoky flame and melts while burning. If the molten, burning polymer falls on the skin, it tends to stick and cause severe burns. Because polystyrene dissolved at high concentrations in petrol forms a gel, this mixture is used in the modern formulation of that nasty weapon called napalm. The name napalm comes from the original gelling agent for petrol—the aluminium salt of naphthenic and palmitic acids (the aluminium salt of household soap also works well). However increasing demand for margarine diverted palm oil (the source of palmitic acid) from both napalm and soap.

Condensation polymerisation

Another form of polymerisation involves the joining together of monomers by removing a small molecule in the joining process. This is called *condensation polymerisation*. An example of such a process is the formation of a *polyamide*. The amide bond is illustrated in Figure 6.4.

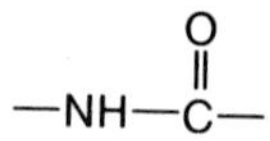

$$-NH-\overset{\overset{\displaystyle O}{\|}}{C}-$$

Fig. 6.4 The amide bond

The original polyamide was 6,6-nylon, which was developed as a replacement for silk in parachutes. The reaction is:

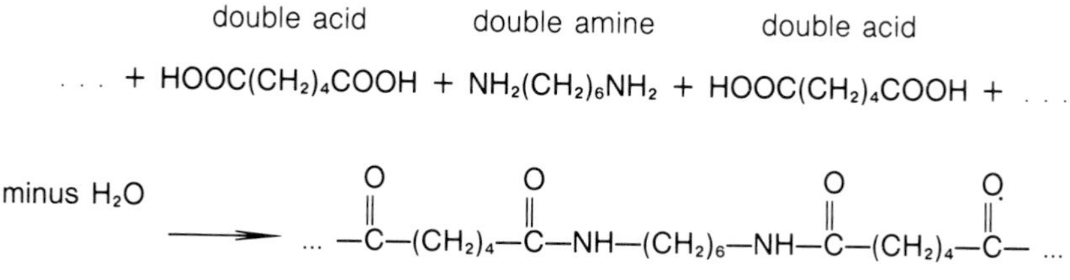

   double acid   double amine   double acid

... + $HOOC(CH_2)_4COOH$ + $NH_2(CH_2)_6NH_2$ + $HOOC(CH_2)_4COOH$ + ...

minus H_2O

$$\longrightarrow \quad ... -\overset{\overset{\displaystyle O}{\|}}{C}-(CH_2)_4-\overset{\overset{\displaystyle O}{\|}}{C}-NH-(CH_2)_6-NH-\overset{\overset{\displaystyle O}{\|}}{C}-(CH_2)_4-\overset{\overset{\displaystyle O}{\|}}{C}- ...$$

Stretching aligns the chains, and additional weak (hydrogen) bonds between the chains strengthen the fibre. The two different monomers each have six carbon atoms—hence the polymer is called 6,6-nylon.

Why not have *one* monomer with *two different* functional (or end) groups? Why not! Using $NH_2-(CH_2)_5COOH$ (illustrated in Fig. 6.5), the polymerisation product is 6-nylon.

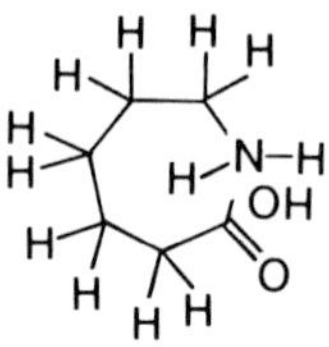

Fig. 6.5 6-nylon monomer

The amide bond is similar to that formed when proteins (which are polymers) are formed from individual (but not identical) amino acids. (However, in protein, instead of there being only one type of monomer, there is a choice of about 20 amino acids to string together; that is, the paper-clips can be different.) The shorter the $-CH_2-$ chain the more hygroscopic the nylon is. For industrial applications (nylon bearings etc.) long $-CH_2-$ chains are used to reduce water absorption. Nylons are amongst the toughest plastics: they can withstand repeated blows, and they also have the advantage of low frictional properties. Nylons also resist many solvents, but they are soluble in formic acid and phenols. Oven bags for cooking poultry, roasts etc. are made from cross-linked nylon as well as from polyester. (I once used a polyethylene bag by mistake—and provided a dinner of shrink-wrapped chicken with little gastronomic appeal). Although the two different bags look the same to a hungry cook, the cross-linked plastic has a higher melting point. It is an interesting exercise to do a dummy run without any food in the cooking bag and to smell what comes out of the bag when it's used by itself.

A *polyester* can be obtained by forming an ester bond instead of an amide bond.

(Compare the ester bond in Fig. 6.6 with the amide bond in Fig. 6.4.) The reaction is shown in Figure 6.7. The polyester product of this reaction has numerous trade names. Examples are Terylene and Crimplene in the United Kingdom, Dacron in the United States, and Trevira in West Germany. Polyester film material has unusual strength and electrical resistance (mylar film, magnetic recording tape, frozen food packaging).

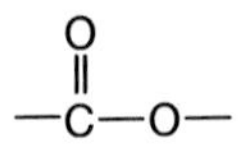

Fig. 6.6 The ester bond

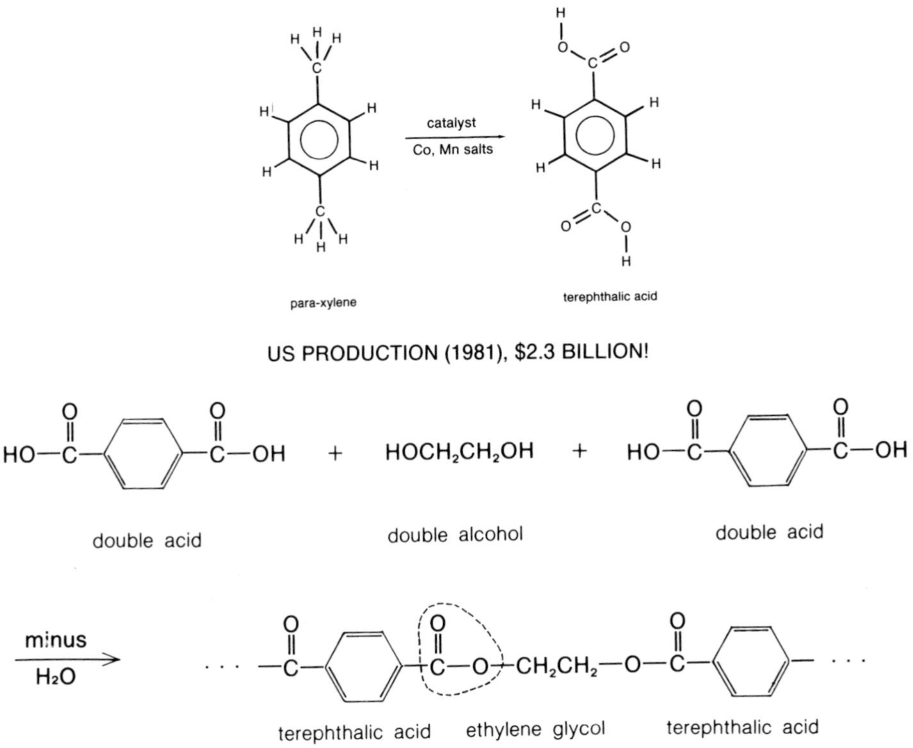

para-xylene

terephthalic acid

US PRODUCTION (1981), $2.3 BILLION!

double acid   double alcohol   double acid

terephthalic acid   ethylene glycol   terephthalic acid

Fig. 6.7 Polymerisation to a polyester

An unsaturated polyester resin consists of a linear polyester whose chain contains some double bonds together with a monomer, such as styrene, that co-polymerises with the polyester to provide a cross-linked product. It is, in effect, a *cold*-curing, *thermo*setting plastic (if you will excuse the contradiction!). The formulation has inhibitors (to prevent reaction during storage), UV absorbers, extenders, thixotropic agents etc. Curing is brought about with a free radical initiator. Small heat-sensitive objects (e.g. electronics components) can be 'potted', and large objects such as boats and swimming pools can be made without having to use huge ovens. As well as the fibreglass layering technique, which is a slow process, injection moulding, which is much faster, can be used by mixing the polyester and short glass fibres into a 'dough'. This overcomes the problem of delamination in poorly fabricated fibreglass products. For short-run production of custom car bodies fibreglass is more economical than metal stamping. A rapidly increasing use of fibreglass is in the manufacture of simulated wood for doors and imitation marble. Products as diverse as pistol grips, explosion barriers for TV tubes and bowling balls are made from polyester. A cunning way of providing a smooth skin is to include an insoluble thermoplastic additive (such as perspex) that migrates to the surface of the mould and layers out as a smooth 'skin' that may be readily decorated.

Epoxy resins are dealt with in Chapter 8 (under 'Adhesives'). They adhere better to metals than polyesters do.

If each monomer has only two functional groups, the resulting polymer is a linear chain. If there are more than two functional groups on the monomers, cross-linkage can occur and this will result in a more rigid lattice. An example is the reaction shown in Figure 6.8. Cross-linked polymers of this type are known as *alkyd* resins. They are used in paint enamels and in the manufacture of false teeth.

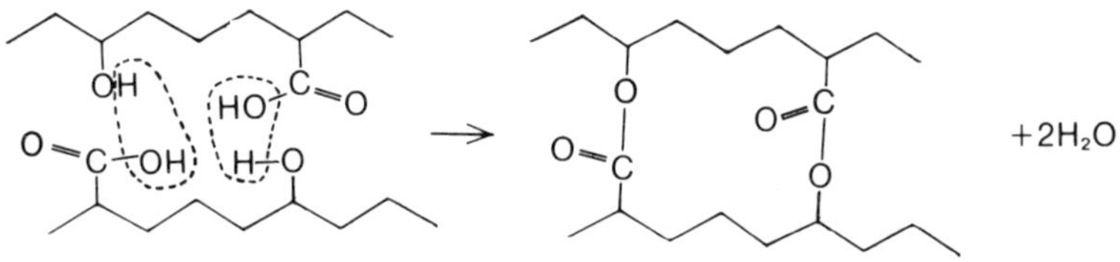

Fig. 6.8   Cross-linking in a polyester polymerisation

Molecules containing the isocyanate group, —NCO, will react with other molecules containing, say, an —OH group to give a *urethane* linkage (Fig. 6.9), which is similar to the amide bond in nylons. Polyurethanes can be formed by using bifunctional molecules (Fig. 6.10). Because the isocyanate monomer decomposes with water to form gaseous CO_2, judicious amounts of water can be added during

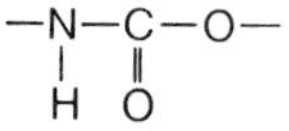

Fig. 6.9   The urethane linkage (compare Fig. 5.20).

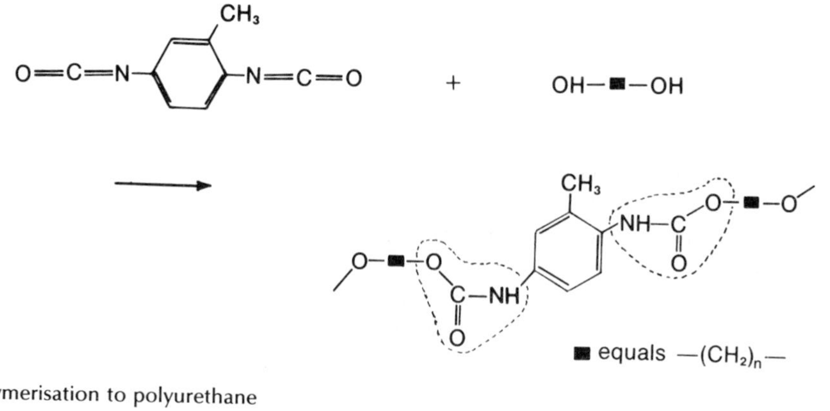

■ equals $-(CH_2)_n-$

Fig. 6.10   Polymerisation to polyurethane

polymerisation to form polyurethane foam rubber. Heating polyurethanes produces unpleasant vapours containing nitric acid (HNO_3), nitrogen dioxide (NO_2) and even hydrogen cyanide (HCN). A burning pillow will quickly fill a room with thick, dense, toxic fumes.

Elastomers

Natural rubber (Fig. 6.11) contains a linear polymer of isoprene, $CH_2=C(CH_3)$ $CH=CH_2$, in which all the $-CH=CH-$ groups are *cis* (Fig. 3.2).

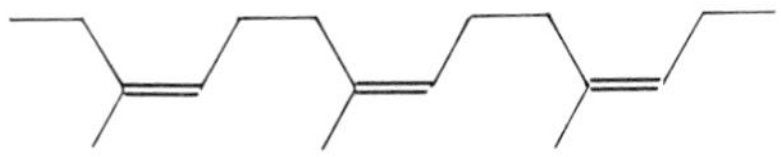

50 000 to 3 000 000
units of isoprene

Fig. 6.11 Natural rubber

Strictly speaking, the property of being 'elastic' means the degree to which a material returns to its original shape after stretching. However, in general usage, it also means the degree to which a material can be stretched. The polymer chains in elastomers are elastic in the second sense, because the chains can be unravelled without coming apart.

Elasticity, in the first sense, was improved by cross-linking with sulfur (as discovered by Goodyear), which produces *vulcanised* rubber.

Synthetic rubbers are produced from related monomers. Polymerisation of *butadiene* ($CH_2=CH-CH=CH_2$) using a sodium (*Na*) catalyst produces *Buna* rubber. The monomer chloroprene, $CH_2=C(Cl)-CH=CH_2$, on polymerisation, gives neoprene, an oil-resistant rubber. Co-polymers such as SBR (styrene-butadiene rubber) are produced in even greater quantities for vehicle types (90% of Australian-made tyres). To produce all *cis* polymers (which have much higher elasticity than those which are not entirely *cis*) a special stereo-regulating catalyst must be used. (The *trans*-isoprene polymer also occurs in nature, in various Malayan trees, and is called gutta-percha).

Silicone rubbers are thermally stable, resist oxidation, are chemically inert, are flexible over a wide range of temperatures ($-90°$ to 250°C) and are virtually non-stick. They are usually defined in three categories:

1. 1-component RTV (room temperature vulcanisation). These are mainly listed as sealants.
2. 2-component RTV. These are used as mould-making/impressed materials, for potting of electronic components etc.
3. Heat-curing silicones.

Each division is further subclassified. Some other uses are for oven-door seals, heart valves (and for other medical applications), moulding, and special tyres used in high-flying aircraft, where normal rubber would be too brittle if not thawed in time for landing. An RTV resin can be obtained for experiments from Dow-Corning or Rhone Poulenic Silicones. The material is supplied as a very viscous liquid silicone polymer of 1000–9000 monomer units cross-linked at only every 1000 units on average. This low level of cross-linking maintains the rubber characteristic (see Fig. 6.12).

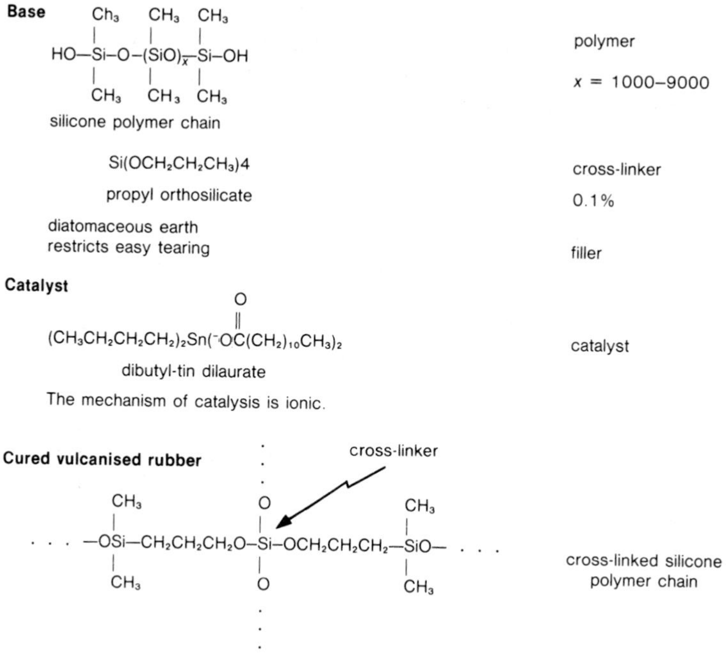

Base

HO—Si—O–(SiO)$_{\overline{x}}$–Si—OH

with CH$_3$ groups on each Si

silicone polymer chain

polymer

x = 1000–9000

Si(OCH$_2$CH$_2$CH$_3$)4

propyl orthosilicate

cross-linker

0.1%

diatomaceous earth
restricts easy tearing

filler

Catalyst

$$(CH_3CH_2CH_2CH_2)_2Sn(^-OC(CH_2)_{10}CH_3)_2$$

dibutyl-tin dilaurate

catalyst

The mechanism of catalysis is ionic.

Cured vulcanised rubber

cross-linker

. . . —OSi—CH$_2$CH$_2$CH$_2$O–Si–OCH$_2$CH$_2$CH$_2$—SiO— . . .

cross-linked silicone
polymer chain

Fig. 6.12   Silicone rubbers (Courtesy of D. Hyatt, CSIRO Science Education Centre)

Polymer foams

Polyurethane and polystyrene foams and rubber foam latex dominate the market. The glass transition temperature of rubber is so low that only flexible foams are possible; it is so high for polystyrene that only rigid foams are possible. However polyurethanes can be formulated to provide either type. These foams can be cured in situ, which gives them great versatility in the packaging industry and in producing items such as shoe soles.

The foaming agent is variable. In foam rubber, the latex is mixed at very high speeds and the foam contains air. PVC foam is produced when a 'blowing agent' that decomposes to a gas on heating is included. Polystyrene beads are impregnated with pentane (C_5H_{12}), which boils at 36°C. In practice, the beads are heated to 100°–120°C by high-pressure steam, but immersion in boiling water is usually adequate. The polystyrene softens above 100°C and the dissolved pentane boils and expands the beads. Cooling below 100°C causes the polystyrene to harden before the pentane condenses and so the beads remain expanded. Polyurethane is foamed with freon (which boils at 24°C) or with carbon dioxide, which is formed during the reaction. A catalyst, tin (II) octoate, speeds up the reaction.

Urea-formaldehyde foam

Under certain conditions, some polymers can break apart into their component monomers. Examples of these are poly (methyl methacrylate) which reverts to the monomer on distillation, and polyurea-formaldehyde, which reverts to its monomers, to a small extent, in the presence of acid. Other polymers do not decompose into their monomers but into other products. An example of this is poly(vinyl chloride). The only monomer released by this polymer is the one that has been left behind from the initial reaction forming the polymer. On decomposing, the polymer forms hydrochloric acid.

Because formaldehyde is a suspected carcinogen and provokes allergic reactions in some people, the use of polyurea-formaldehyde foam for heat insulation of buildings has raised worldwide concern. It is interesting to compare the response of different countries.

By December 1980, the Canadian Government had imposed a temporary ban on the use of the product pending further study. In Australia the matter was 'raised' at the Consumer Product Safety (Standing) Committee. After a later meeting with industry representatives, certain recommendations covering its use were made.

In February 1981, the US Product Safety Committee announced its intention to ban the sale of the product until the results of a public inquiry were known.

In April 1981, the Canadian Government announced its findings and continued its temporary ban until such time as the industry could demonstrate that it could produce a stable and safe product, while in Australia the Public Health Advisory Committee noted that the product, if properly installed, presented no danger to health.

In June 1981, in Australia, the Industry Liaison Committee recommended that a working party be established to 'further examine the potential health hazards associated with the product'. This recommendation was acted upon in December 1981 and the working party met for one day in February 1982. In the USA, at the same time, the US Product Safety Committee confirmed its intention to ban the product, which it did in April 1982 (to take effect in July). The working party in Australia passed its recommendations on to the Consumer Product Safety Committee, which passed its recommendations on to the Public Health Advisory Committee, which passed its recommendations on to the National Health and Medical Research Council, which, in June 1982, *finally* brought down the very first formal recommendation on urea-formaldehyde foam. The various States and Territories then began (and are continuing) their fumbling path in dealing with this recommendation.

In the USA there were various court actions. The US Federal Court ruled the banning order as invalid and the Consumer Product Safety Committee appealed to the Supreme Court. The Supreme Court rejected the appeal.

The Canadian temporary ban of 1980 *is still in effect*, awaiting a satisfactory product. In Canada, the Environmental Secretariat of the National Research Council is the operative advisory body and I had the pleasure of working with it during the (northern) Autumn of 1981. It manages to bring the top research potential of the NRC to bear on issues of public health without the academic status distinction evident in the USA and overwhelming in Australia (see also the conclusion to Chapter 11).

In November 1983, the US Environmental Protection Agency (EPA) took a first step that could lead to tighter restrictions on all kinds of formaldehyde emissions. The agency decided to rescind a February 1982 determination that formaldehyde does not warrant priority review for health hazards.

PLANTS EAT FORMALDEHYDE

Scientists working for NASA have found that spider plants are best for removing from the air formaldehyde, a possible cancer agent found in many homes. They say that an average home could be kept completely free of formaldehyde gas by the installation of 70 spider plants.

The researchers, working on NASA's enclosed ecological life-support system project, looked at a number of house plants, but the spider plant was five times better at absorbing the gas than any rival. Nobody knows quite how it does it. The scientists recommend one plant for every 2.5 square metres in homes or offices.

Source: New Scientist, 8 November 1984.

The consumer movement has been criticised for advocating a ban only on the use of urea-formaldehyde foam for home insulation, without also urging that formaldehyde be taken out of other products. While these products may not emit formaldehyde gas at the same rate as UF foam, they are significant contributors to the artificially high levels of formaldehyde in modern homes.

The Swedish Government has already introduced restrictions on the use of formaldehyde in particle board. The latest EPA move suggests that the US Government might take this much further. The February 1982 decision was based on the Reagan Administration's then philosophy that the results of animal tests indicating cancer should not be so heavily relied upon in human health decisions. It seems the administration is now more willing to accept that animal studies have implications for humans and consequently the decision has been overturned.

Recent research on formaldehyde suggests that, when used at customary levels, formaldehyde is not a human carcinogen.[1]

An Australian Standards Committee (BD/58/9) is still considering proposals for an acceptable standard for urea-formaldehyde insulation, but in the meantime the product has virtually disappeared from the market. Even if the product can be made to perform successfully, it may well never reappear—the price of poor pre-marketing research.

Polyester embedding resin

Apart from being used to make fibres such as Dacron and Terylene, polyester is used as a clear, crystalline, embedding resin. This resin is supplied as a viscous liquid of the already linear polymerised unsaturated polyester dissolved in styrene monomer. The catalyst generally used is methyl ethyl ketone peroxide, which cross-links the styrene to the double bonds in the polyester to give a three-dimensional structure (see Fig. 6.13). The amount of cross-linking can be controlled by using different proportions of saturated and unsaturated polyester.

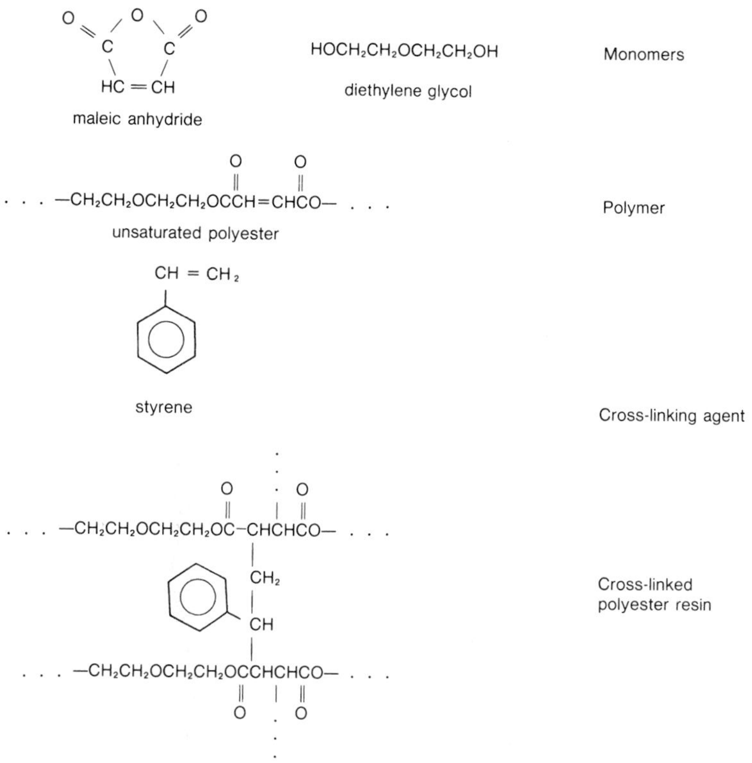

Fig. 6.13 Cross-linked polyester

PHYSICS OF PLASTICS

Plastics[2] can be subdivided roughly into groups on the basis of their mechanical properties. One way of grouping plastics is by measuring their rigidity as a function of temperature (Fig. 6.14). In another method the *stress* versus *strain* curves for plastics place them approximately into different categories. Some of the plastics not mentioned in the previous section, fall into one of these categories—*hard and tough.*

Plastics that are hard and tough have a high tensile (elongation) strength and stretch considerably before finally breaking. The top performers in this grade are called *engineering plastics.* These are relatively new and expensive and have specialised applications. *Polyacetals* are an example. They are very abrasive resistant and resist organic solvents and water. These plastics are used in plumbing to replace brass or zinc in shower heads, valves, etc. Furniture castors, cigarette lighters, shavers and pens are also often made from polyacetals as they give a non-stain, satin finish. That great Australian invention, the totalisator wheel, is moulded by Du Pont from an acetal resin and replaces a 23-part metal assembly.

Polycarbonates, another group of hard, tough plastics are often used instead of glass because they are transparent and retain their dimensions and their resistance to impact even when subjected to a wide range of temperatures. Babies' bottles, bus-shelter windows, plastic sheeting for display signs and telephone dials are just a few examples of where you will find this type of plastic. Because of their fire-resistance, they are used in firemen's masks, interior mouldings in aircraft and in electronic equipment. In sporting equipment you will find them in helmets used by cricketers and motor cyclists, as well as in baseball helmets and snow-mobiles. The most common plastic optical lens material is a poly-diallylcarbonate (Columbia resin CR39) and related materials (CR64, EX80), which are often used for embedding (of flowers, etc.) and casting. Polycarbonate binoculars not only resist breaking when dropped but they also float in water.

Polycarbonates are produced by an interesting process called phase transfer

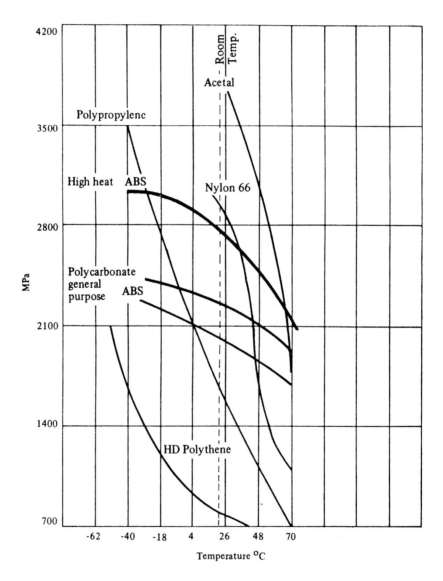

Fig. 6.14 Comparison of flexural modulus (rigidity) versus temperature for various plastics (From *Know your plastics*, Plastics Institute of Australia, 1980)

catalysis. The two components of the polymer are each in separate non-mixing phases—one in water and the other in an organic solvent such as methylene chloride—so that they do not react directly. A quaternary ammonium positive ion (a short, cationic surfactant—see Chapter 2) forms a neutral salt with the negative ion of the water-soluble component and takes it across into the organic layer, where it can react with the component there, releasing the quaternary ammonium ion, which can then go back into the water. (Such a salt was mentioned in Chapter 5 in regard to measuring cyanuric acid levels in swimming pools.) An example of this procedure is given in Figure 6.15.

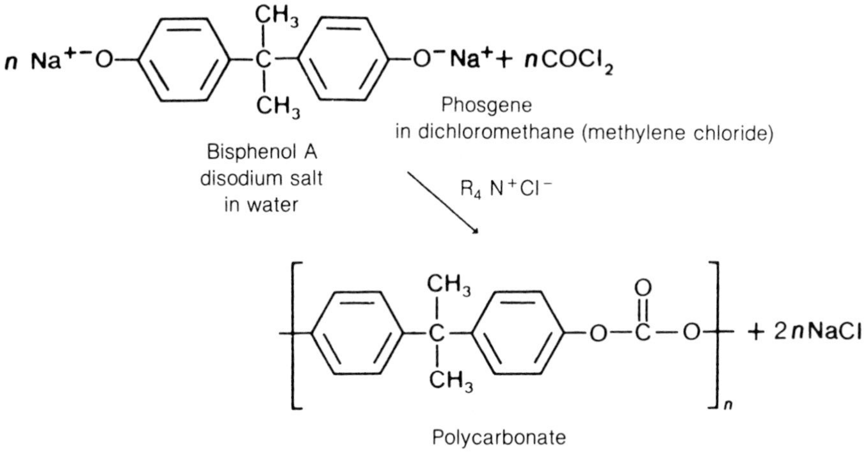

Fig. 6.15   Formation of polycarbonate

One plastic that is not considered an engineering plastic but is still hard and tough is *high-density polyethylene* (HDPE). However, during injection moulding, stresses are set up inside the material which cause bonds to rupture. These cracks propagate (see below) until visible cracks appear and the surface of the polymer roughens. This is called *environmental stress cracking* (not to be confused with personal problems at work!). If small amounts of propylene as co-monomer are added, this problem is reduced. Polypropylene itself has superior properties but is expensive and light-sensitive.

Moulding of plastics allows new forms of design, such as the one declared in US patent 2 487 400 (1949), lodged by one Earl S. Tupper (Fig. 6.16):

> ... the invention herein provides a sealing enclosure for containers in the form of a hollow finger-engageable stopper having elasticity and flexibility with a slow rate of recovery to provide a non-snapping and noiseless type of cover which is applicable to the lip of a container by hand conformation and removable therefrom by a peeling-off type of procedure.

The patent does *not* specify that you can buy Tupperware only at special parties!

An interesting plastic system is *acrylonitrile-butadiene-styrene* (ABS). This is in fact a two-phase system in which a styrene-butadiene rubber is dispersed in a glassy stryene-acrylonitrile (SAN) matrix. Without fire-retardants, ABS burns. It has a high gloss and keeps its shape and finds uses in children's toy block building

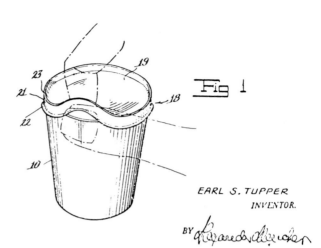

EARL S. TUPPER
INVENTOR.

BY

ATTORNEY

Fig. 6.16   Tupperware

kits, housings for radios, calculators, telephones, computers and the better quality refrigerators. It is also used in lawn-mower housings, high-quality pipes and fittings, luggage and the top layers of skis. ABS tends to be used as a replacement for metals (e.g. telephone dials).

Nylon has excellent mechanical properties, resists solvents and is an ideal material for gears and bearings that cannot be lubricated. About 50% of all mould-ed nylon fittings go into cars in the form of small gears (wipers), timing sprockets and all sorts of clips and brackets. A special nylon was developed for the important use of sealing the side of steel cans to replace solder. The nylon seal is mechanically stronger and does not present problems of lead release. The nylon used has a very long chain (C_{36} dibasic acid), which increases its adhesive power. It bonds directly to steel and does not require a tin surface, unlike solder, and the seam itself is a simple overlap of the two ends of the steel rather than a soldered join of the two edges.

Fluoroplastic (e.g. PTFE or Teflon) is very hard and tough and is also non-flammable and has unique resistance to chemical attack. However it does lack dimensional stability and is subject to creep. The non-stick frypan is the classic example of this plastic's chemical resistance.

Plastics for packaging

Of the wide variety of commercial organic polymers, several major types occupy a commanding position in packaging. Low-density polyethylene is fairly flexible and is outstanding for its extensibility, toughness, chemical resistance and low cost. Its limitations are low strength, inability to recover after stretching, low melting point, translucency and sensitivity to oxidation—it requires the addition of hindered phenolic antioxidants (see Chapter 3) for stability during processing. It is particularly useful as film packaging, for tubing and for squeeze bottles.

In contrast, high-density polyethylene is more rigid, stronger, less extensible and less tough, higher melting and more opaque. It finds its major use in blown bottles. Polypropylene is stronger still, and even more rigid, less tough, higher melting, and is much more sensitive to oxidation, requiring multicomponent anti-oxidant systems to stabilise it during processing and use. These multicomponent antioxidant systems generally include hindered phenols and organic sulfides such as dilauryl thiodipropionate, along with a variety of other synergists. Polypropylene is finding growing use in packaging films, fibres and moulded forms such as caps for bottles and jars. Shrink films are formed by stretching the film, which orientates the molecules. Upon heating, the film shrinks back to its original size. In this way, close-fitting wraps can be achieved.

Clear, unmodified 'crystal' polystyrene has found wide use as vials for pharmaceutical tablets. Oriented polystyrene foam is used in packaging of fresh produce. Rubber-modified polystyrene, generall called 'impact styrene' or even just 'polystyrene', is the most widely used plastic for moulded and heat-pressed food packaging, because it is reasonably rigid, strong, tough, chemically stable and relatively cheap. Windows in window envelopes are generally made from polystyrene.

Poly(vinyl chloride) is noteworthy for its compatibility with high-boiling point liquid plasticisers, permitting the plastics manufacturer to use it in a wide range of formulations from rigid, strong products such as clear bottles to soft, flexible tubing and film. The rigid unplasticised product, generally blended with rubbery polymers for toughness, requires careful stabilisation to prevent liberation of hydrochloric acid and discolouration during melt processing. The most effective stabilisers are lead compounds, which present serious toxicological problems, and organo-tin compounds, of which the dioctyl tins have received hesitant toxicological approval, while the more common dibutyl tins and the newer dimethyl tins raise more serious questions. The safest stabilisers, based on calcium and zinc, are only moderately effective, and are relatively unreliable in production. Rigid poly(vinyl chloride) is characterised by its high density, low softening temperature and good chemical resistance.

Soft, flexible poly(vinyl chloride) is produced by the addition of liquid plasticisers in the proportion of 30 to 120 parts of plasticiser per 100 of resin (the polymer before formulation with additives). The most common plasticisers are esters of phthalic, phosphoric and adipic acids, all of which tend to be extracted from the plastic and migrate into adjacent materials. Some have found fairly wide acceptance in food and drug packaging, but questions concerning their safety still remain.

The ester-type plasticisers can be replaced by polymeric plasticisers—polycaprolactone, nitrile rubber—which entirely eliminates the problem of migration. The polymeric plasticisers are used in plastic pilchers and in other products that have to withstand repeated extraction by laundry detergent. Even so, the pilchers become brittle and tear after a number of washes. The problem of thermal stability is reduced by the lower processing temperature of the plasticised polymer, permitting use of less harmful stabilisers such as calcium and zinc soaps, epoxidised soyabean oil, and organic phosphite esters.

Plasticisers also markedly increase permeability, which is undesirable in certain types of packaging. Unlike polyethylene, plasticised poly(vinyl chloride) combines high extensibility with fairly complete retractability, giving better form stability in film packaging. However when plasticised PVC is used in, say, car upholstery, the plasticiser distils out in hot weather and condenses on the windows in a film that is difficult to remove.

Cellophane—regenerated cellulose in plasticised film form—is, of course, the original manufactured packaging film, and it still enjoys wide usage because of its strength, clarity, non-toxicity and moderate cost. Cellulose ester plastics, such as the acetate and butyrate, generally plasticised by phthalic esters, are outstanding for their toughness, clarity and gloss, but cost considerably more. Their use in packaging is restricted to those applications in which a glistening appearance is important for marketing. They find considerable application in blister packaging and as see-through windows for boxes. (Note from Table 6.2 how permeable to water vapour they are.)

While the plastics that have been discussed are the most common that are currently used in packaging, other speciality polymers, such as poly(vinylidene chloride), polyester, polyurethane and polyamide films, are already in limited use and may find growing application as time goes on. Many newer plastics are also bidding for their share of the growing packaging and medical markets. It should be remembered that each new polymeric material, along with its positive qualities, may introduce unexpected problems. These problems will require unforeseen tests and specifications before the material can be safely used in food and pharmaceutical packaging. Those developing new packaging should never be content to rest secure in meeting the present specifications drawn simply for the present materials and their present problems.

Potential problems of plastic containers

Desorption

Desorption, or the leaching of plastic components into the contents of the container, has received a great deal of attention. Naturally enough it is the potential toxicity of extractives that has been of concern and has led to the adoption of various tests and specifications by regulatory authorities. The design of extraction procedures that yield data suitable for extrapolation to the variety of compositions existing in foods or pharmaceutical preparations is an exercise to be undertaken with caution.

Nevertheless, a great deal of work has been undertaken in connection with the biological effects of extracted materials, and several pharmacopoeias describe specific tests and specifications. The important role of the extractant in such methodology has often been pointed out. For example, one research worker quotes that after six hours of recirculation of physiological saline solution in poly(vinyl chloride) tubing no di(2-ethylhexyl)phthalate (DEHP) was extracted. On the other hand, whole human blood stored at 4°C for 21 days in DEHP-plasticised blood bags *contained 5–7 mg of plasticiser per 100mL of blood*, almost all of it associated with the lipo-protein portion of plasma. A 4% bovine serum albumin was able to extract DEHP from PVC tubing. Significant quantities of DEHP were found in the spleen and abdominal fat of two patients who had received blood transfusions from blood bags of this type.

In the case of food packaging, the NH & MRC standard for leaching of monomer sets a maximum of 0.05 mg/kg for vinyl chloride and 0.02 mg/kg for acrylonitrile monomer. (See also 'Plastics for food contact', below, and 'Food additives' in Chapter 11.)

Photodegradation

Most plastics exhibit varying degrees of degradation upon prolonged exposure to sunlight. Poly(methyl methacrylate) (perspex) is one of the few exceptions. The ultraviolet and blue parts of sunlight are sufficiently energetic to cause polymer bonds to break. One way of retarding this effect is to add a compound that will absorb the radiation and convert it efficiently to heat. Derivatives of benzophenone are used (particularly in polypropylene) because of their high absorbance of light in the 290 to 400 nm range (see 'Chemistry of cosmetics—Sunscreens', Chapter 4). Substituted benzotriazoles are also used.

Stress cracking

Polyethylene and polystyrene are examples of plastics that are subject to environmental stress cracking (see page 211). Crack-resistance tests have shown that alcohols, organic acids, vegetable and mineral oils, ethers and surface active agents provide an active environment for stress cracking polyethylene.

Permeability of plastic films

One of the important properties of plastics when used as flexible films is their permeability to various gases. This is particularly true when they are used for wrapping food, because if they are permeable to oxygen from the air the food will spoil. Table 6.2 gives gas permeabilities for a number of compounds. Figure 6.17 illustrates the meaning of permeability using SI units. Conversion of units is given in Table 6.4.

TABLE 6.2
Gas permeability of plastics[a]

| Name | Permeability to | | | |
	Oxygen	Carbon dioxide	Water[b]	Nitrogen
Cellulose (PVDC-coated)	0.06	0.4	20	0.09
Cellulose acetate	7.8	68	75 000	2.8
Rubber hydrochloride	0.3	1.7	240	0.8
Polyethylene (low density)	55	352	800	19
Polyethylene (high density)	10.6	35	130	2.7
Polypropylene	23.0	92	680	—
Poly (vinyl alcohol)	0.3	low	soluble	—
Poly (vinyl chloride) (PVC)	1.2	10	1560	0.40
Poly (vinylidene chloride) (PVDC)	0.053	0.29	14	0.0094
Polystyrene	11.0	88	12 000	2.9
Polyamide (nylon)	0.38	1.6	7000	0.10
Polyester	0.3	1.53	1300	0.5
Chlorotrifluoroethylene	0.10	0.72	2.9	0.03

[a] The permeability P is the volume of gas in cc which passes through a square centimetre of plastic surface 1 mm thick in 1 sec for a pressure difference of 1 cm height of mercury—i.e. the lower the number the slower the gas moves through the plastic film. Figures given are $P \times 10^{10}$ cc cm^{-2} mm^{-1} sec^{-1} (cm Hg)$^{-1}$
[b] Water at 90% relative humidity and 25°C.

Source: After Sacharow, *Drug Cosm. Ind.* **97** (3), 1965, 359.

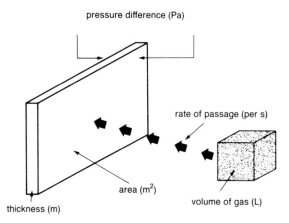

Fig. 6.17   Schematic representation of gas permeability of plastics (SI units)

Oxygen diffuses through most plastics about four times as fast as nitrogen; carbon dioxide diffuses through most plastics about 25 times as fast as nitrogen. The best oxygen barrier is provided by poly(vinylidene chloride) (PVDC), while for water vapour, chlorotrifluoroethylene is best. Another set of gas permeability data is given in Table 6.3 (but take care—the units are not the same as those in Table 6.2).

PERMEABILITY OF PACKAGING FILMS

The mathematical theory of the permeability of homogeneous polymer films is based on Fick's law. In the steady state this has the form

$$\frac{q}{At} = \frac{-DdC}{dx} \qquad (1)$$

where q is the quantity of permeant passing through the film, A is the area of the film, t is the time, D is the diffusion coefficient, C is the concentration of penetrant in the film, and x is the thickness.

Provided that (1) dC/dx is linear, (2) diffusion occurs in one direction only, and (3) D is independent of concentration, equation 1 can be integrated with respect to thickness to give:

$$\frac{qL}{At} = D\Delta C \qquad (2)$$

where L is the total film thickness, and ΔC is the difference in concentration of penetrant on the inner and outer surface of the film. Henry's law is used to convert to partial pressures of penetrant and a concentration gradient becomes a partial pressure difference Δp.

$$\frac{qL}{At}\Delta p = DS \qquad\qquad (3)$$

where S is the solubility coefficient of the penetrant; $C = Sp$.

TABLE 6.4
Factors for converting a variety of units used in permeability
data to SI units (Change one unit at a time)

Symbol	Given unit	SI unit	Conversion factor
q	g	L (STP)	22.4/mol mass (g)
L	cm	m	10^{-2}
	0.001 in.	m	2.54×10^{-5}
A	cm^{-2}	m^{-2}	10^4
	100 in.$^{-2}$	m^{-2}	15.5
t	min.$^{-1}$	s^{-1}	1.67×10^{-2}
	h^{-1}	s^{-1}	2.78×10^{-3}
Δp	$mm^{-1}Hg$ (Torr)	Pa^{-1}	7.5×10^{-3}
	$cm^{-1}Hg$	Pa^{-1}	7.5×10^{-4}
	atm^{-1} (760 Torr)	Pa^{-1}	9.87×10^{-6}

Example
(Convert into SI units a water permeability value expressed as g $\times$ 0.001 in. $\times$ 100 in^{-2} $\times$ 24 h^{-1} at a vapour pressure gradient of 90% relative humidity at 100°F. The molecular mass of water is 18.02 and the vapour pressure at 100°F ($=37.8°C$) is 4.915 Torr. At 90% RH this gives a gradient across the membrane of 4.424 Torr. The conversion factor is:

$$\frac{22.4}{18.02} \times 2.54 \times 10^{-5} \times 15.5 \times 1.157 \times 10^{-5} \times \frac{7.5 \times 10^{-4}}{4.424}$$

Source: E. G. Davis and M. L. Rooney, *Instrumental techniques: specialist techniques for the food industry,* no. 1, CSIRO, Sydney, 1971, p. 30

It is no wonder that the CSIRO, in its guide to food storage, *Handling food in the home,* recommends flexible film coated with poly(vinylidine chloride) rather than the common cling wraps or bags made from polyethylene (see also Experiment 13.15). Note that, for the nylons, the permeability for water decreases while that for O_2 and CO_2 increases from nylon 6 to nylon 12. The reason is that there is a change from one polar group for every six non-polar CH_2 groups to one for every twelve. So, overall, nylon 6 is more polar than nylon 12, and hence for polar water the permeability is greater in the more polar nylon 6, but the opposite is true for the non-polar O_2 and CO_2.

The gas permeability problem is not restricted to film packaging but applies to thicker bottles as well. For a research chemist the storage of standard reagents in polythene bottles results in a loss of water.

TABLE 6.3
Comparative properties of flexible plastic films

Product	Method manufacture	Density g/cm^3	Yield m^2/kg per mm	Thickness range—μm	Effective temperature range °C	Heat sealing range °C	Water permeability $g/24\ h/m^2$ per mm at 35°C 90% RH	Resistance to grease and oils	Tensile strength MPa
1. Cellulose acetate	Casting extrusion	1.43-	760–790	12–760	Sub-zero to 104	Non-sealing	8.8 550–3500	Good	37–96
2. Polyester	Biaxially oriented extrusion	1.38–1.395	720	6–25	−106–150	160–180	20–60	Excellent	117–172
3. Polyethylene (low density)	Extrusion casting	.910–.929	1083	100–6250	−45–93	110–200	470–710	Fair to good	9–17
4. Polyethylene (medium density)	Casting extrusion	.926–.940	1065	100–6250	−50–93	120–190	200–275	Good	14–24
5. Polyethylene (high density)	Casting extrusion	.941–.965	1047	100–6250	−57–120	125–190	120–160	Good	17–42
6. Polypropylene	Casting extrusion orientation	.9	1107	12 and up	−50–168	95–230	24–275	Good	22–70
7. Polystyrene oriented	Biaxially-oriented extrusion	1.05–1.06	942	8–500	−62–80	120–150	1732	Good	62–83
8. Poly(vinyl alcohol)	Casting extrusion	1.21–1.31	780	12–300	−40.....	115–190	3937 estimated	Excellent	48–70

TABLE 6.3 — *Continued*
Comparative properties of flexible plastic films

Product	Method manufacture	Density g/cm³	Yield m²/kg per mm	Thickness range—μm	Effective temperature range °C	Heat sealing range °C	Water permeability g/24 h/m² per mm at 35°C 90% RH	Resistance to grease and oils	Tensile strength MPa
VINYLS									
9. Poly(vinyl chloride) (rigid PVC)	Calendering extrusion casting	1.35–1.45	720–830	25–760	−40–93	170–205	1575	Good	48–70
10. Poly(vinyl chloride) (flexible PVC)	Calendering extrusion casting	1.24–1.45	720–830	25–760	−40–93	135–180	1300–7100	Poor to excellent	10–44
11. Vinylidene vinyl chloride co-polymer	Calendering	1.20–1.68	580–830	12–250	−46–93 −46–150(wet)	115–160	4–12	Good to excellent	55–140
12. Regenerated cellulose film	Casting	1.40–1.55	415–900	20–45	−46–190	90–180	Non-moisture-proof 550–1060 moistureproof 3–12	Excellent	48–207
13. Rubber hydrochloride	Casting	1.11	903	10–63	−30–96	—	2.75–90	Excellent	24–35

Source: Plastics Institute of Australia

From the properties of polymers studied so far you may be able to make some educated guesses about the reasons for the distribution in Table 6.5, which gives the use of plastic films in Australia for packaging a variety of foods (cost is also a factor).

Plastics for food contact

Australian standards for plastics intended for food contact have been prepared: AS2070 Part 1, 1980, polyethylene(PE); Part 2, 1977, poly(vinyl chloride) (PVC) compound; Part 3, 1979, styrene plastic materials; Part 4, 1979, acrylonitrile plastic materials; Part 5, 1981, polypropylene. The introduction to this standard outlines the scope of the problem. Table 6.5 lists a number of plastic films used for packaging food.

TABLE 6.5
A variety of plastic films used for packaging food

Food	Cross-linked PE	HDPE	LDPE	Polypropylene	EVA	PVC 10% plasticiser	PVC 10%-20% plasticiser	PVC 20% plasticiser	PVDC/PVC copolymer	NC coated cellulose	PVC coated cellulose	PVDC coated cellulose	Polyester	Nylon 6	Nylon 11	Nylon 12
Bread	×		×	×		×	×	×		×	×					
Cakes	×			×		×	×	×	×	×	×					
Crumpets						×	×	×			×					
Biscuits												×				
Milk		×	×		×											
Powdered milk			×									×				
Butter		×	×													
Cheese						×	×	×								
Fresh meat	×	×	×			×	×	×		×						
Frozen meat	×	×	×			×	×	×								
Chilled meat						×	×	×	×							
Processed meat	×	×	×			×	×	×							×	×
Fresh poultry	×	×	×		×	×	×	×								
Frozen poultry	×	×	×		×	×	×	×	×							
Cooked poultry														×	×	
Cereals			×	×												

TABLE 6.5 — *Continued*
A variety of plastic films used for packaging food

Food	Cross-linked PE	HDPE	LDPE	Polypropylene	EVA	PVC 10% plasticiser	PVC 10%-20% plasticiser	PVC 20% plasticiser	PVDC/PVC copolymer	NC coated cellulose	PVC coated cellulose	PVDC coated cellulose	Polyester	Nylon 6	Nylon 11	Nylon 12
Chocolate	×	×	×	×		×	×	×		×	×	×				
Non-chocolate confectionary	×	×	×	×		×	×	×		×	×	×				
Crisps		×	×							×	×	×				
Nuts		×	×							×	×	×				
Crackers		×	×							×	×	×				
Soft drinks				×												
Squash				×												
Jam						×										
Marmalade						×										
Honey						×										
Prepared meals		×														
Fresh fish	×	×	×			×	×	×								
Frozen fish	×	×	×			×	×	×	×							
Fresh fruit	×	×	×			×	×	×								
Frozen fruit	×	×	×													
Fresh vegetables	×	×	×			×	×	×								
Frozen vegetables		×	×			×	×	×								
Flour				×												
Sugar				×												
Rice				×							×					
Coconut				×							×					
Noodles etc.				×												
Salt				×												

HDPE   High-density polyethylene
LDPE   Low-density polyethylene
NC   Nitrocellulose
EVA   Ethylene-vinyl acetate (copolymer)
PVDC   Poly (vinylidene chloride)
PVC   Poly(vinyl chloride)

Source: Plastics Institute of Australia.

The packaging and processing of food introduces the possibility of migration or transfer of substances from the packaging or wrapping materials to the food. It is essential that the formulation of the packaging material is selected so that the migration of substances from the package is minimised and if migration occurs no toxic hazard exists to the consumer of the food.

The occurrence of 'acute toxicity' resulting from plastics being in contact with food is most unlikely since only trace quantities of the potentially toxic materials are likely to migrate. 'Chronic' effects are possible, however, where small quantities of a biologically active substance transfer from packaging materials and are ingested in small amounts over a long period of time. It is necessary to consider the intrinsic toxicity of each ingredient in the plastic, its ability to migrate under extreme conditions in an original or altered form and the amounts that may be safely ingested.

The high molecular mass polymer does not itself pose a toxic hazard because it is inert and essentially insoluble in food. Furthermore, if accidentally ingested it does not have any toxic effects in its passage through the digestive system.

In the preparation of the polymer, numerous additives are added and the nature of these depends on the polymer being produced (e.g. poly(vinyl chloride) or polyethylene). Examples of the additives that may be used are catalysts, suspension and emulsifying agents, stabilisers and polymerisation inhibitors. These additives are bound either chemically or physically into the polymer and may be present in their original or an altered form. In addition, the polymerisation may leave trace quantities of residual monomer or low molecular mass polymer in the product. It is thus necessary to specify the purity of the polymer to be used as a raw material in a plastic used to package food.

The discussion on plastics could continue indefinitely. A suitable topic on which to conclude could be plastic garbage bins. These have traditionally been made from left-over plastic (off-cuts) and hence their composition can vary greatly. One result of this is often a relatively high *glass transition temperature* (see 'Paints—Solid paints: Flexibility', Chapter 8). During cold weather, if the plastic is below its glass transition temperature, it becomes rigid and tends to split under stress. Australian standard AS1535, 1975, defines two types of plastic bins. One is a cold-climate version for which the stress testing is done at $-10°C$ to $-14°C$. Dibutyl sebacates are particularly effective plasticisers for producing a low glass transition temperature; however they tend to be too volatile for warm climates and hence there is a second version in the standard for a regular plastic garbage bin.

GLASS[3]

When it is cooled a hot solution of sugar in water can form crystals of sugar or just stay clear and viscous like molasses or honey. Molasses is a super-cooled liquid because it doesn't form crystals at the temperature one would expect. In the same way, a mixture of silicon dioxide and metal oxides forms glass, which is also a super-cooled liquid that has not crystallised. Glass may crystallise over a period of many years and then become brittle. On the other hand, pieces of ancient Egyptian glass have remained uncrystallised for 4000 years.

Egypt is one of the few places on earth where sodium sesquicarbonate (trona) occurs as a natural mineral. It is formed by the action of concentrated salt solutions on limestone.

Le Chatelier discovered the idea of reversible chemical reactions and equilibrium when contemplating these Egyptian mineral deposits. He decided that they must have been formed by the reaction

$$2NaCl(aq) + CaCO_3(s) \rightarrow Na_2CO_3 \bullet 10H_2O(s) + CaCl_2(aq)$$

saturated   limestone
brine

which is the opposite of the known precipitation reaction

$$CO_3^{2-}(aq) + Ca^{2+}\,aq \rightarrow CaCO_3(s)$$

The direction of the reaction is determined by the concentration of reactants and products.

In South Australia the reaction is used today to produce sodium carbonate. The waste calcium chloride liquor finds use as an energy storage fluid for solar heat.

With temperatures up to 1000°C in charcoal furnaces, the ancients discovered that mixtures of sand, limestone and sodium carbonate fuse to form glass, in which all the crystalline order of the added minerals has been lost. With only sodium carbonate and sand, a glass is also formed, but this is soluble in water:

$$Na_2CO_3(s) + SiO_2(s) \xrightarrow{\text{heat}} Na_2SiO_3 + CO_2(g)$$

heat   glass

This material is waterglass, which is used for chemical gardens and preserving eggs. When the glass is made with calcium as well, it becomes insoluble in water and much harder. A typical glass contains (by weight) 70% silicon dioxide (SiO_2), 15% sodium oxide (Na_2O) and 10% calcium oxide (CaO), with 5% other oxides. As we saw, the sodium and calcium are added as carbonates and lose carbon dioxide to form the oxides in the glass. The glass formed in this process softens at 650°C.

There are two distinct constituents of glass. The non-metal oxide is usually silicon, but it can be boron, aluminium or phosphorus. This is the network former. The network modifiers are usually sodium, potassium, calcium and magnesium. Without the metal oxides, silicon oxide or silica is a crystalline material called quartz (although fused silica can be made at 1700°C). The metal oxides help destroy the three-dimensional lattice and lower the melting point by about 1000°C. Particular metal oxides are used for particular effects. Lead gives the glass a high refractive index and more brilliance because of the greater internal reflection of light. Cobalt gives a blue glass, manganese gives purple, chromium gives green and copper gives either red or blue-green.

The natural colour of 'glass' (best seen edge on) is green to yellow because it contains iron impurities. By adding manganese dioxide, MnO_2, the coloured iron (II) is oxidised to the paler iron (III).

BORAX BEADS

Platinum wire has traditionally been used for borax bead tests. The free end of the platinum wire is coiled in a small loop through which a match will just pass. The loop is heated in a Bunsen flame until it is red hot and then quickly dipped into powdered borax, $Na_2B_4O_7 \cdot 10H_2O$. The adhering solid is held in the hottest part of the flame; the salt swells up as it loses water of crystallisation and shrinks upon the loop forming a colourless, transparent, glass bead. The bead is moistened and dipped into the finely powdered sample, allowing only a very *small* amount of material to adhere (otherwise the colour to be formed will be too dark). The bead is then reheated in the oxidising (outer purple) or reducing (inner blue) part of the Bunsen flame.

After each test, the bead is removed from the wire by heating it again to fusion, and then flicking it off the wire into a vessel of water. The platinum can be cleaned by melting a pure borax bead and running it up and down the wire. A white ceramic tile (saucer) can be used instead of a platinum wire. Borax beads illustrate glass formation at a lower temperature.

TABLE 6.6
Colour test using borax beads

Metal salt[a] (solid)	Oxidising flame Hot	Cold	Reducing flame Hot	Cold
Copper	green	blue	—	opaque red-brown
Iron	yellow-brown	yellow	green	green
Chromium	yellow	green	green	green
Manganese	violet (amethyst)	amethyst	—	—
Cobalt	blue	blue	blue	blue
Nickel	violet	red-brown	grey	grey
Gold	rose violet	rose violet	red	violet
Tungsten	pale yellow	—	green	blue (but blood red if iron is present)

[a] Minerals work quite well.

Note: More pronounced colours can be obtained by using 'microcosmic' salt, $Na(NH_4)HPO_4 \cdot 4H_2O$, but the glass is less viscous and more difficult to handle. In any event, some practice is needed.

In 1912, the Corning Glass Company in the USA found that adding 10% to 15% of boron oxide, B_2O_3, gave a glass that was more shock-resistant. This borosilicate glass (Pyrex) was found to be resistant to chemical attack from virtually everything except hydrofluoric acid (HF). A glass composed of 96% SiO_2, 3% B_2O_3 and 1% of another oxide produced a glass called Vycor, which could withstand very sudden changes in temperature (e.g. it did not break when plunged red hot into cold water). Vycor found use in the glass panels of oil heaters and for scientific work, as it was a cheaper alternative to fused quartz. Aluminium oxide gives aluminosilicate glass, which is used in glass reinforcing fibres. Glass can be 'chemically' toughened by rapid cooling and dipping into potassium chloride (KCl) solution. This replaces some of the sodium atoms near the surface with the larger potassium ions and puts the surface under compression stress (see 'Chemistry and design'). Titanium compounds are also used. They allow the production of lighter glass bottles.

An interesting development was the incorporation into glass of silver chloride (AgCl), which darkens on exposure to the ultraviolet component in sunlight (just as it does in a photographic film). The darkening is the result of the silver ions (Ag^+) converting to metallic silver (Ag) by picking up an electron. This colour is lost again in the dark and so we can produce photochromic sunglasses.

An accidental overheating of this photosensitive glass at Corning led to a radical new discovery. The glass became permanently opaque and apparently useless. However, it was quickly found that the glass was now virtually unbreakable. Glass will crystallise on its own (devitrify) over centuries or at 600°C in a matter of months. With silver (titanium, phosphorus etc.) the process is speeded up and occurs in hours and yields a very strong crystal (with the dislocations locked—see 'Chemistry and design'). A formulation of 74% silicon dioxide, 16% aluminium oxide, 6% titanium dioxide and 4% lithium oxide doesn't even expand on heating. These are *glass ceramics*. Small changes in composition can cause large changes in properties. Some ceramics do not expand on heating; some have high resistance to fractures. Glass ceramics out-perform glass in almost every respect except one, transparency.

New formulations for glass bring new surprises. A glass that has phosphorus oxide instead of sodium and calcium dissolves very slowly. It is being tested as a medium for the controlled release of other substances such as copper into ponds for controlling the disease schistosomiasis (bilharzia). Then there is the continuing debate about the type of glass or synthetic material that should be used to store nuclear waste material.

CHEMISTRY AND DESIGN[4]

A light breeze, an expanse of water and a few hours to spare—why not go sailing? The physics of sailing has remained unchanged since the first human being ventured onto water. However, the materials from which boats and sails are made have changed dramatically. The chemistry of new materials has encouraged new designs that have drastically reduced costs and enormously increased efficiency.

The great division in technology has always been between metals and non-metals. Their properties are clearly very different (and chemistry explains why), and the mental processes involved in using them are also very different. The history of Western technology has very much been the history of steel.

The price of steel *dropped* by a factor of more than 10 during the reign of Queen Victoria. This was the most important event of its kind in history until perhaps today, when something similar is occurring with micro-electronics.

For generations, engineers had no idea at all why steel and concrete behaved in the different ways they do. However they described their behaviour in minute detail and filled many unreadable books with their results. These 'properties' of materials were essential to design but useless in logically devising new materials.

As an exercise you might try to describe, objectively, the *mechanical* differences between chalk and cheese. Only chemistry can be used to explain the differences.

In the absence of fundamental understanding, superstition and craftsmanship became gruesomely intertwined. In ancient Babylon, the making of glass required the use of human embryos; Japanese samurai swords were said to have been quenched by plunging them, red-hot, into the bodies of living prisoners. Cases of burying victims in the foundations of buildings and bridges were common throughout history. In Roman times a doll was substituted!

It is a mistake to exaggerate the virtues of traditional design even though the workmanship may have been excellent. The wheels really did keep coming off coaches and wooden ships always leaked, quite unnecessarily, because shipwrights did not understand shearing stress.

The understanding began with the work of Robert Hooke and with what is now known as Hooke's Law. Hooke did not suffer unduly from modesty and staked his claim to priority in a number of fields by publishing in 1676 'A tenth of the hundredth of the inventions I intend to publish', among which was an anagram which revealed *Ut tensio sic vis*—'as the extension, so the force'. (The Latin *tensio* means 'extension' not 'tension'. The Romans confused these ideas and literary writers probably never thought much about them.)

It took over a century to move from Hooke's Law to Young's modulus E, which described the elastic or stiffness properties of a *material* rather than the properties of an *object* (of undefined dimensions). Thomas Young published the idea of his modulus in a rather incomprehensible paper in 1807 after he had been dismissed from his lectureship at the Royal Institution, London, for not being sufficiently practical. Thus was born the most famous and most useful of all concepts in engineering: the modulus E determines the deflection of structures when loaded. The use of the modulus predicted that, although a plastic railway carriage designed for British Rail would have properly fitted doors when the carriage was empty, the doors would neither open nor close when the carriage was full of passengers!

Attempts have been made to replace steel with aluminium in tanks, such as the hull of the US M551 Sheridan. However, to achieve the same degree of ballistic protection as steel armour, the aluminium-magnesium alloy must be about three times as thick so that the weight saving is small (about 6%). What real weight savings there are come from the increased rigidity or stiffness of aluminium hulls with their thicker walls and so the purely structural components can be lighter than their steel counterparts.

The stiffness of a material is not a measure of its strength. A biscuit is stiff but weak; steel is stiff and strong; nylon is flexible and strong; raspberry jelly is flexible and weak. The two properties together describe a solid about as well as you can reasonably expect two parameters to do.

Strength is usually thought of as the stress needed to pull materials completely apart, although materials are most often used in compression. When a squat column of material is compressed it can squash out sideways like plasticine. Copper behaves this way. Brittle materials like stone and glass will explode sideways into dust and splinters. A long, slender column may buckle like a walking stick or tin can.

During the 1920s an engineer named Griffith was working at the Royal Aircraft

Establishment at Farnborough. He asked, 'Why are there large variations between the strengths of different solids? Why don't all solids have the same strength? Why aren't they much stronger? How strong 'ought' they to be anyway? These were regarded as quite silly questions.

Liquids like water have surface tension—the skin-like property of the surface that allows drops to form and insects to walk on water. Solids also have surface tension, or surface energy as it is usually called (see 'Chemistry of surfaces', Chapter 4). Just as it is possible to calculate the weight of the largest insect than can walk on water, it turned out to be quite simple to calculate the stress needed to separate atom layers inside a material. The problem was that most materials came nowhere near these theoretical results. Griffith did some experiments on glass fibres because glass was easier to handle than steel. However, when his superiors became aware of his work (after a fire wrecked his laboratory), he was immediately transferred to other duties. 'Glass fibre was not a structural material, sir!'

Not only did this attitude turn out to be incredibly wrong, but glass was an ideal material for theoretical study. Because glass is a solidified liquid, the idea that strain must overcome surface tension to cause a break followed very naturally.

Another fascinating result was that the strength of glass fibres increased enormously with *thinness*—by a factor of about 100! In fact, thin glass fibres are stronger than the strongest steel and close to the theoretical limit of strength. Bulk materials are actually much weaker than expected is because they have local high-stress points. It was a small rivet hole, perhaps 4 mm diameter, that caused the Comet aircraft disasters in the 1950s. Chemical corrosion also occurs most readily at points of high stress. Glass is cut by using a shallow but sharp scratch on the surface and then breaking it along the stress line. The sharper the angle of the cut the greater the stress.

The chemical molecular picture of solids now provides the theoretical framework for explaining the real strength of materials and the methods by which weakness may be overcome or compensated for. By making glass fibres thin enough, the room for surface cracks to form was reduced and the material became stronger. From this grew the research that led to the development of ultra-strong whiskers of metals and non-metals. Purity of material was of great importance because impurities would sit at so-called grain boundaries (the surfaces between pure microcrystals packed together) in the crystal structure and cause weakness. When sea water freezes, the ice formed is substantially fresh because the crystallising forces out the impurities. The wrong impurity in an alloy can completely ruin its strength by being forced into grain boundaries.

An interesting application of this principle is in the use of glycol antifreeze in car radiators. While it is true that glycol lowers the freezing point and postpones freezing, the more important aspect is that, when the coolant does eventually freeze, it forms mushy ice without much mechanical strength and is unlikely to burst the radiator or head.

So the worst sin in an engineering material turns out to be not lack of *strength* or lack of *stiffness*, desirable as these properties are, but lack of *toughness*; that is, lack of resistance to the propagation of cracks. The lack of strength or stiffness can be overcome by adequate design considerations.

Strain induced in a material by a sudden blow can propagate through the material and cause very high local stresses far removed from the impact point.

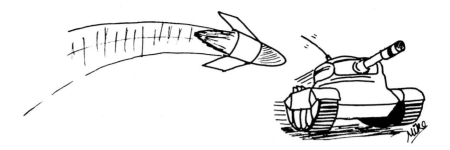

Fig. 6.18

The material has to resist the effect of these stresses. A shell fired at armour plate, but not penetrating, can throw off a 'scab' on the inside of the plate that can bounce around causing great damage. The reason for the internal head-band in crash helmets is not for ventilation but to cushion shock waves and prevent damage at the back of the skull, opposite the usual point of impact.

Now for some examples. Wooden sailing ships were used to explore the world and, later, to survey it. The last major wooden *passenger* sailing ship seems to have been the *Torrens,* which carried passengers to Adelaide in 1903—just within living memory. Large ships were built up of naturally curving wood, chosen to have the right shape. The watertight skin and deck were laid lengthwise, at right-angles to a closely spaced framework of ribs. The grooves between the timbers were filled with *oakum* (made by picking old rope to pieces in the prisons and work-houses), driven in by mallet and chisel. Decks were paved with hot pitch and the surplus chipped off when cold. (The Devil, incidentally, was a particular seam and hence 'the Devil to pay and only half a bucket of pitch!') With this design the shear stresses were not taken up and the strains were left to the caulking, which always leaked and occasionally fell out. Generally the crew became exhausted from having to continually pump water. Up till 1914, Norwegian ship-owners were still making a living by buying up British sailing ships that had become too leaky with age and converting them to run with windmill pumps. The quality of ropes and spars was awful, and this created problems particularly when ships needed to go into the wind. Bligh's crew on the *Bounty,* for example, mutinied after one appalling attempt to beat around Cape Horn in which the ship almost fell apart, forcing Bligh to turn around and go right around the world the other way in order to reach the Pacific. With the discovery of gold in California, the trip around the Horn was in great demand. It is clear that the West was largely won with better rope.

Slow ships and lack of facilities ashore meant that the fouling of ships' bottoms by weed and worm was a very serious problem. Copper sheathing was introduced in 1770 and the slow release of small amounts of copper ions proved highly toxic to marine organisms. So successful was this piece of chemistry that it delayed the replacement of wooden hulls by iron because the copper sheathing on iron set up an electrochemical cell in sea water (a battery), which rapidly corroded the iron. In order to save copper, Sir Humphrey Davy applied blocks of zinc to the copper-sheathed hull of the frigate *Samarang* in the 1820s. The idea was to give cathodic protection to the copper in the same way that zinc does for iron in galvanised steel. The experiment was a great success in preventing the copper from dissolving, but then the copper failed completely to kill the growth of fouling. Some iron hulls were even sheathed with wood and then coppered. The *Cutty Sark* (launched 1869) was planked with teak and greenheart bolted to wrought-iron frames, with adequate shear bracing, and the bottom sheathed with a copper alloy called Muntz metal. With the drop in the cost of iron and steel in the 1870s, steel hulls replaced composite construction and, when steam replaced sail, fouling became less of a problem as it occurs most rapidly in becalmed ships.

In pure strength (apart from flexibility) the bindings, lashings and sewings of primitive people and older seamen were more efficient than metal fastenings. In fact the wood screw forms the least efficient of all joints. In 1920, flying-boat hulls were sewn together with copper wire. Modern glues have made most of this argument academic anyway. Any two solids can be glued together if we can find a liquid that will wet them both and then harden. Wood glued with water which is then frozen passes most of the tests in the specifications for wood adhesives! Carpenter's glue is a variant of ice in that the melting point is raised to a more practical temperature (70°–80°C).

Whether a liquid wets a solid, that is, spreads thinly over it, depends only on the surface energy of the liquid and the solid. This is the same surface energy we discussed earlier in relation to the ease of formation of surface cracks in a solid. When a liquid wets a solid some of the solid–air surface and liquid–air surface are replaced by solid–liquid surface. The arithmetic is quite straightforward. Incidentally, the strength of a glued joint depends mainly on its width not its area, because of the way stress concentrates. Casein glues (from milk protein) used to be very popular but, as casein is more or less a mixture of lime and cheese, on becoming moist it deteriorates and its last hours are like those of a Camembert. Great efforts were made to find synthetic substitutes. In 1907, Dr Baekeland's discovery of phenol-formaldehyde resin (Bakelite) proved excellent for the manufacture of plywood, and urea-formaldehyde (which can be set cold) was used for joints. Other variants include resorcinol-formaldehyde and epoxy resins.

Baekeland had already made a fortune from his 'Velox' photographic printing paper and he started his great Bakelite Company making synthetic varnish in Edwardian times with the registered name 'Dammard Lacquer Company'. It is alleged to have marketed three grades—'Dammard', 'Dammarder' and 'Dammardest'.

It was as a moulding powder, in which the resin was mixed with short cellulose fibres (wood flour), that the new material made its immediate impact. The gear-lever knob for the 1916 Rolls was allegedly its first commercial success, and resulted finally in the extinction of the Birmingham brass trade. After the Second World War, melamine formaldehyde (which is colourless and can be coloured to taste) ushered in the great 'Kitchen Revolution' and the seeds of feminine emancipation. Previously there had been no suitable surface for table tops, and wood, being porous, strongly absorbed dirt. The many hours spent scrubbing kept the fair sex from entertaining loftier thoughts.

Glass fibre technology also dates from the end of the Second World War, when it was developed for housing radar scanners. A material was needed that was transparent to radar microwaves and which therefore had to be electrically non-conducting and non-polar. It was also wartime radar needs that gave impetus to the development of polyethylene, discovered at ICI in the UK in 1932. It provided the ideal low-cost dielectric insulator for microwave cables.

Nylon, discovered at Du Pont in the USA a year later, served initially as a plastic. It replaced hog bristles in toothbrushes in 1937, but in 1939 women's nylon woven stockings became the rage. With the outbreak of war, nylon replaced silk in parachutes. It is less popular today as a textile fabric and has found new important uses as a very tough plastic in machine parts such as gears and bearings.

Today the industrial designer is faced with a wide selection of plastic materials and a bewildering array of performance parameters to evaluate against each other.

The design of sailboards[5]

To most people names like Mistral, Cobra, Ten Cate, Dufour, Hi-Fly and Sodim have no common link, but add Windsurfer and they are at once identifiable as brand names of sailboards. Board sailing is listed as the world's fastest growing sport and few who have mastered it or have seen accomplished sailors in action in high wind will question the reasons why. Annual sales in Europe alone in the mid-1980s topped the 200 000 mark, and manufacturers of sailboards count their total sales in the hundreds of thousands—remarkable when one considers that Hoyle Schweitzer developed the first windsurfer board as recently as the late 1960s and most manufacturers only entered the market from the mid-1970s onwards. The sport achieved Olympic status in 1984 when the Los Angeles sailing events featured a windglider sailboard class.

The concept is simple. A floating hull is propelled by a sail fixed though its mast by some universal joint such that the sail can be pivoted to fore and aft, to windward and to leeward. The wishbone-shaped boom, fixed to the mast at one end and the point of the sail at the other, serves to keep the sail at full stretch and provides a hand hold for the sailor, who uses body weight to counteract the force of the wind in the sail. The net forward propulsion developed by the shape of the sail is transmitted to the board through the universal. Fore and aft tilting of the 'rig' (as the mast, sail and boom are called) moves the centre of effort of the sail in front of and behind the pivotal centre of the hull and results in steering the board to leeward or windward respectively. Stability is enhanced by a central dagger board which also prevents sideways slip on reaches or when close-hauled. With two weeks practice under ideal conditions—5 to 8 knot (9 to 15 km/h) winds on calm water—most people should learn to sail.

The rapid growth of the sport has meant a sizeable new outlet for chemical companies such as Ciga-Geigy, BASF and Du Pont. With the exception of booms and some masts which are made from aluminium alloys, almost every component is derived from organic chemicals. The function of each has dictated the type of plastic used in its construction and often the manufacturing process itself. New polymers have been developed to meet certain requirements and, more recently, the expansion of the industry to 'third generation' jumpboards has opened up further avenues for research and development.

Three basic processes are used in board manufacture. The windsurfer is formed from cross-linked polyethylene using a seamless process called 'Roto-moulding', in which the polymer powder is melted inside a hot mould which, when rotated, distributes the polymer evenly on the inner surface. The hollow hull is then filled with a polyurethane foam.

The Hi-Fly manufacturers, Akutec, in conjunction with West German giant BASF, have installed a plant to produce polyethylene blow-moulded hulls and in 1979 produced the world's largest single-piece blow-moulded object in the form of the Hi-Fly III (a step up from mineral water bottles!). The seamless hull is likewise filled with polyurethane foam. Such fillings impart rigidity to the boards, which are at the same time highly durable and impact resistant (particularly important at rocky beaches).

The lower the flexibility in the hull, however, the faster it sails, and more advanced racing hulls are made of fibreglass. In general, a shaped polystyrene foam core is encased in resin impregnated fibreglass sheets which, when cured, produces a seamless hull of great rigidity. The up-market Mistral boards employ a pressure bonding process, the epoxy resins for which were designed specifically by Ciba-Geigy.

However, lower-cost boards are made from moulded ABS (see Table 6.1) plastic sheets, which are welded together and filled with polyurethane foam. Hulls of this type are of consistent thickness and are light and very durable.

Lightness is indeed a key to speed, and recent manufacturing techniques such as variable-density foam filling (low in the core and high closer to the skin) as well as the use of specifically designed polymers has enabled some manufacturers to make boards of 14 to 18 kg in weight. For instance, Du Pont has developed the lightweight fibre matrix Kevlar, which has been utilised in the manufacture of the ultra-light Cobra sailboards.

The worldwide distribution of specifically designed boards sometimes creates problems. One manufacturer of a polypropylene hull found the product to be unsuitable for Australian conditions owing to its poor impact resistance, although the same problems were not encountered elsewhere. An acrylonitrile-styrene-butyl acrylate co-polymer (ASA) laminated board introduced to Australian three years ago was likewise withdrawn because of unexpected delamination problems, although other ASA boards are marketed with success.

The tortuous exposure to UV radiation requires that all plastic sailboard components contain heavy doses of UV stabilisers. Local Australian levels of radiation may require better stabilisation of polymers than is needed in higher latitudes.

Sails are most often made from polyesters, but recently Du Pont have developed a clear Mylar (terephthalic polyester) coating for sails which provides minimal elasticity. Used in a laminate with polyesters such as Dacron or by itself with a fibre reinforcing, it reduces stretching (bagging) of sails, thus maintaining the optimum shape. It also provides a water-resistant finish, which has an added advantage for sailboard sails because they are lighter to haul out of the water. Masts are generally made of glass-fibre construction employing polyester or epoxy resins. Carbon fibre and carbon-fibre-reinforced glass fibre are new innovations. Design of masts must allow some forward flexibility without sideways bending. Even universal joints have moved away from the traditional mechanical type used in windsurfers to a rubber 'hour-glass', or polyurethane connecting cords.

Neoprenes and foams of varying weights are used as boom covers to increase traction and comfort when pulling against strong winds, and the application of specially formulated coatings to the upper surfaces of the hulls is often employed to prevent skidding.

Dagger boards, which were originally made from wood, are now made from light-weight plastics such as polypropylene or reinforced polyurethane. Skegs attached to the rear of the hull must be rigid and here polycarbonates are typical plastics of choice. Mouldings which house the universal joints and boom ends are injection moulded with nylon or fibre-reinforced nylon.

The booming sailboard industry is providing new challenges to the polymer chemist and chemical technologists. The dictum 'rigid and light equals fast' holds, and board manufacturers are likely to vie to outdo one another in attaining these objectives, particularly as the market becomes saturated and purchasers become more selective. They could of course resort to less orthodox sales techniques. One manufacturer is already marketing a 'pineapple-scented hull'!

METALS

Technology is what is around us in everyday life as a result of the application of acquired knowledge and experience, occasionally and more increasingly obtained as the result of scientific understanding. We find technology in the supermarket and the home.

In a supermarket we can buy aluminium foil. In the hardware store we can buy lead and copper foil, but we cannot buy zinc foil even though zinc is cheaper than copper! Nickel, chromium and iron can come as wire, but never their close neighbour cobalt. One type of brass you can machine, while the other must be cast like bronze.

TABLE 6.7
Metal prices (October 1988)

	US$/lb(New York)	A$/kg
Platinum	8190	24 220
Gold	6370	18 230
Palladium	1920	5153
Silver	100	262
Tin	3.36	11.40
Copper	1.18	3.2
Aluminium	1.06	3.4
Zinc	0.70	1.8
Lead	0.40	0.8

Source: Financial Review 5 October 1988

Play with some pieces of different metals such as copper, soft iron, zinc, aluminium and magnesium. They look different: magnesium, aluminium and zinc are silvery; iron is dark, while copper is salmon-brown. The really striking difference is in the way they behave physically, when we hammer, stretch or twist them. Why?

The answer can be obtained using a heap of ping-pong balls, polystyrene balls, or marbles. Rack some together tightly in one layer, as at the start of a snooker game. Then place another layer on top of the first, in the indentations created by the first layer. The two layers are of course not directly on top of one another. Then place a third layer on top of the second. *Wait!* There are two ways we can do this. They can be placed so that the third layer is directly over the first layer, or they can be placed so that all three layers are each in a different position. In the first case, we can continue to pack layers alternately, while in the second case we repeat the pattern every three layers. We can call these different packing arrangements *ab* and *abc*. The packing efficiency for the *ab* and *abc* arrangements is the same, and they are the most efficient possible. In both cases, each sphere is surrounded by 12 nearest neighbours. For equal-sized spheres, geometry shows that in both cases the spheres occupy 74% of the total volume. If your geometry is not up to proving this, you can 'use your marbles' and pack them into an empty ice-cream tub (or other large container) and fill it with water to a level near the top of the marbles. Then pour the water out, measure the volume, and compare it to the volume of the empty tub at the same height. (You will have an error due to 'edge effects' at the bottom, sides and top, which will be small if the marbles are small compared to the size of the container.)

The best way to obtain a feel for the structures is to prepare triangular planes of balls (just as in snooker), and stick these together. (Polystyrene balls can be stuck with dichloromethane solvent.) Make the triangles of decreasing size. Then place one triangular layer on top of another. If you pile up the layers in an *abc* arrangement, you will form a pyramid on the triangular base. In fact it is a tetrahedron where three different faces have the atoms in layers just like the base layer (See Plate VI(a).) There are thus four directions in which the balls are layered. On the other hand if you pile the layers up *ab* as in Plate VI(b), the layers will jut in and out alternately as you go up. If you look down from the top to the base you will see channels going through the whole structure in position *c*, which has been kept clear. There will be no new planes of atoms formed, only the horizontal ones you have been placing.

Metals (or alloys) that pack *abc* are much easier to shape than those that pack *ab*, because it is the planes of atoms that move and make metals ductile and malleable. Only metals that pack as *abc* are found in foils and wire.

The *abc* structure is called face-centred cubic close packing (fcc), while the *ab* structure is called hexagonal close packing, (hcp). The reasons for these names can be seen by studying the models more closely. The way different metals pack is shown in Table 6.8.

Another but less efficient way of packing spheres, involves surrounding a sphere with only eight nearest neighbours. Geometry shows that, for equal-sized spheres, only 68% of the volume is occupied. This packing arrangement is called body centred cubic or bcc. This does not have quite as many slip planes as the fcc but many more than the hcp. Amongst metals, fcc and hcp are about equally as common, while bcc is about half as common. Other structures constitute about 10% of the total.

TABLE 6.8
Interatomic distances of common metals (298 K)

Metal	Structure	Interatomic distance (nm)
Ag	fcc	0.2888
Al	fcc	0.2862
Au	fcc	0.2882
Cd	hcp	0.296^a
Co	hcp	0.250^a
Co	fcc {> 417°C}	
Cr	bcc	0.2498
Cu	fcc	0.2556
Fe (α)	bcc {< 912°C}	0.24824
Fe (γ)	912°C < fcc < 1400°C	0.2540
Mg	hcp	0.322^a
Mo	bcc	0.2725
Pb	fcc	0.3499
Pt	fcc	0.2775
Sn^b		
Ti (α)	hcp {< 880°C}	0.293^a
Ti (β)	bcc {> 880°C}	0.285
W	bcc	0.2734
Zn	hcp	0.278^a

^{a} In hcp the hard-ball atom is not quite spherical, but ellipsoidal. (Why?) Thus the radius varies somewhat with direction and the average value is given.
^{b} Tin (Sn) has three crystalline structures:

	13.2°C		161°C		232°C	
α-Sn	$\leftrightarrow$	β-Sn	$\leftrightarrow$	γ-Sn	$\leftrightarrow$	Sn (liq)
'grey' (diamond structure)		'white'(metallic-tetrahedral)		also metallic		
density is 5.75		density is 7.31				

Tin

Tin is the Oscar Wilde of metals. It is a little uncertain as to its status as a 'real' metal and shows some outrageous behaviour. Because tin is available as a foil you would predict that it packs abc, and it does, almost. Tin is soft and weak; it melts at 232°C. Lead added to tin lowers the melting point even further and gives us pewter, although for health reasons modern 'pewter' tends to tin with no lead but with just a little antimony and copper to keep it healthy. On the other hand a small amount of tin added to copper strengthens it enormously and this piece of technology which science now makes understandable, gave us the Bronze Age which lasted a millennium and allowed the development of civilisation as we know it.

When a piece of pure tin is bent backwards and forwards near your ear, it can be heard to produce a plaintive 'cry'. This is due to the presence of 'twinned' crystals in the metal, which you are cruelly separating. Tin also suffers from a thermo-dynamic 'disease'. Below 13°C, the atoms of tin very slowly change their packing arrangements. In practice, you need a much lower temperature to see this happen in a reasonable time. The new structure is *not a metal* structure at all (tin is as closely related to the non-metal semiconductors such as germanium as it is to the metals), and tin in this form is grey and crumbly. For this reason pure tin is not much used today, but historically tin 'pest' has had its consequences. Napoleon's army lost their trousers in the freezing winter of 1812, when their tin buttons crumbled away. Another cold Russian winter was said to be responsible for the rumour that the Russian hoard of silver had turned to powder in 1867–68. It was actually tin.

We see tin mainly in tin-plated cans. Unless these are lacquered internally, a little tin dissolves in the natural vegetable acids and gives, say, tomato soup its special colour and flavour. One cause of a can 'blowing' is the production of hydrogen from the acid reaction with the tin. Elementary science tells us that tin is just above hydrogen in the activity series and therefore does not normally react with dilute acid. Do the food acids have special properties? Well yes, they do. They are strong complexing agents for metals and so keep the concentration of free metal ion in solution low. By Le Chatelier's principle, more tin will dissolve and be complexed and so on. A maximum level of 250 mg per kilogram has been set.

Mixing metals

We shall consider below what happens when we make a mixture of two metals.

Discover why we are losing our copper coins. Look up the price of metals quoted in the newspapers (see Table 6.7). Converting them to common units (A\$/kg), compare the copper price with the face value of the 'copper' coins (the 1 cent coin weighs 2.60 g, the 2 cent coin 5.20 g). The first sign of a banana republic is when the currency is worth more when melted down!

The first chemistry experiment that most students carry out is the burning of magnesium in air. The aim of the experiment is to see the mass increase that occurs when oxygen is added to magnesium. However how does the volume change and why?

Magnesium, with a density of 1.74 g/cm^3, oxidises to MgO with a density of 3.58 g/cm^3.

$$Mg + \frac{1}{2}O_2 \rightarrow MgO$$

The per cent volume change occurring in the reaction is negative, i.e. -19%.

Less volume is required for MgO than Mg, even though oxygen has been *added*! The ionic Mg^{2+}—O^{2-} bond is so strong it pulls the ions close together (0.21 nm, centre to centre). In magnesium the metallic Mg—Mg bond is weaker (with a centre to centre distance of ~0.34 nm). The stronger bonds in the oxide lead to a higher melting point, 2800°C compared to that of metallic magnesium, 650°C.

Phase diagrams—or have a dialectic with a eutectic

To understand the behaviour of mixtures of solids (and liquids) we must understand equilibrium phase diagrams. A phase, in a chemical sense, is a uniform, physically distinct form of matter that can be present in a system. Ice, water and steam are different phases of a single component, water.

We will examine the equilibrium phase diagram for mixtures of the metals tin and lead as a function of temperature (Fig. 6.19). Such a diagram is obtained by taking a series of mixtures (0–100% tin/100–0% lead), heating them up so that they completely melt, and then cooling them very slowly and plotting what different solid and liquid mixtures appear as the temperature drops.

In Figure 6.19a, the bottom axis shows increasing per cent of tin by mass. (Along the top axis it is given in terms of per cent of atoms of tin.) The vertical axis is temperature and you will note that pure lead melts at 327°C (left-hand side) and pure tin melts at 232°C (right-hand side).

In between, you will notice that adding tin to lead lowers the melting point of the lead, and adding lead to tin lowers the melting point of tin. These two lines meet at a minimum of 183°C for a composition of 61.9% tin. This temperature is called a *eutectic* temperature, and this tin–lead mixture is called a *eutectic* mixture (Greek *eu* good; *tektos* melted). To the immediate left and right of this point are regions where solid and liquid are both present. To the left, there is a solid called phase α, plus liquid. To the right, there is a solid called phase β, plus liquid.

Further to the left is an all solid region with phase α and further to the right is an all solid region with phase β. Below this point is an all solid region with two solid phases α + β. The solid phases α and β are each solid *solutions*. Phase α is lead with some tin dissolved in it and phase β is tin with some lead dissolved in it. In the region α + β these two phases are *mixed* together. Note the distinction between a *solution*, where atoms are mixed, and a *mixture*, where crystals (both containing both tin and lead atoms) are mixed. (Figure 6.19b) shows the nature of the phases formed. So far so good.

In phase α, how much tin is dissolved in the lead? Well this depends on the temperature at which solid phase α forms. The lines on the left show that starting at low temperature this increases from a couple of per cent to a maximum of 19.2% at the eutectic temperature of 183°C and then drops again at higher temperatures to be zero at 327°C. (A similar argument applies to phase β on the far right.) Tin is overall more soluble in solid lead than lead is in solid tin.

Follow what happens when a 60:40 mixture of tin and lead is cooled from the liquid. Not only does it freeze at 183°C but it freezes *sharply*. That is, below 183°C it is a solid and above 183°C it is a liquid. A mixture that has a sharp melting point (like a pure substance) is called a *eutectic* mixture. The eutectic mixture of tin and lead is in the ratio 60:40. It is this composition of tin and lead that constitutes electricians' solder. Electricians need a low melting metal that conducts electricity, 'wets' copper wire and sets quickly, so that the electronic components do not overheat. Note, however, that although the eutectic mixture behaves like a pure substance (in that it has a sharp melting point), it is in fact a solid mixture of two phases (α + β), the different crystals of which can be seen in a cut and polished sample under a microscope.

Now follow what happens when a 30:70 mixture of tin and lead is cooled from the liquid. It begins to solidify at 250°C, but now the freezing point is not 'sharp', because both liquid and solid (phase α) are present and stay present over a temperature range of 70°C. It is not until it reaches 183°C that the mixture freezes completely to form a total solid. This solid is a mixture of phase α (which has been crystallising out during cooling) and feathery crystals of the eutectic (which solidifies suddenly when the temperature drops to 183°C), which can be seen

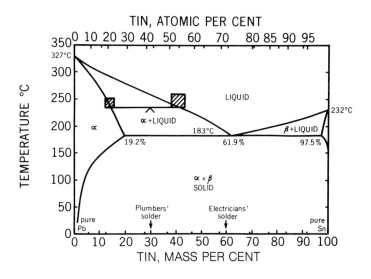

Fig. 6.19a Phase diagram for the lead–tin system.

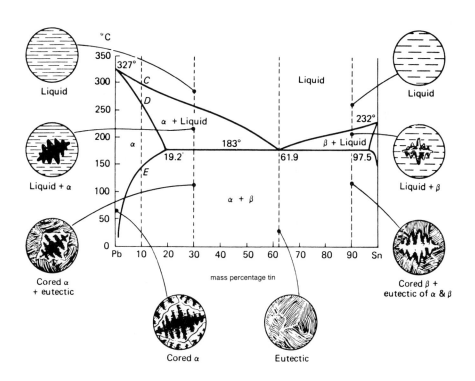

Fig. 6.19b The tin–lead equilibrium diagram. The cored alpha and beta grains result from industrial cooling. (Courtesy of R. A. Higgins, from *Engineering Metallurgy*, vol. 1, Hodder and Stoughton, London, 1972)

packed in between the phase α crystals when viewed under a microscope. The feathery eutectic crystals are a solid mixture of phase α and phase β.

This solid is *plumbers' solder* (Latin *plumbum* lead). When the plumber solders a joint with molten solder at 250°C, the solder takes a long time to cool to 183°C, at which point the mush of liquid and solid finally sets completely to solid. This gives the plumber time to set the joint. The mechanical properties of this sludge are such as to permit it to be handled by the plumber in an effective way. The cooling sludge is interesting because it is not a liquid from which more and more of the *same* crystals precipitate. If you look at the line enclosing the α phase at the left from the point 19.2% tin, 183°C to 0% tin (pure lead), 327°C, you will see how the composition of the solid phase α changes with temperature. You can see that the higher the temperature the less tin the phase α contains. As our liquid plumbers' solder begins to form crystals below 250°C, the composition of the crystals changes as the solder cools. Each new batch of crystals forming has more tin in it, and liquid remaining behind is also richer in tin. In fact the changing composition of the *liquid* phase can be obtained from looking at the line from the eutectic (61.9% tin, 183°C) to melting lead (0% tin, 327°C). Thus, molten plumbers' solder at 250°C has 30% tin but, as it cools, the composition of the crystals forming is obtained by taking a horizontal line across to the left to where it hits the phase α curve, and the composition of the liquid remaining is obtained by taking a horizontal line across to the right to where it hits the liquid curve. The complete horizontal line is called a *tie line* and it ties together the composition of the solid formed and the liquid remaining. As the mixture cools the tie line drops lower, and the composition of the crystals (phase α) and the liquid are *both* richer in tin! (Good grief, the alchemist's dream come true! We should start working on a *gold* lead alloy!) Alas, we have considered only the *composition* of each phase, not the *amount* that is there. We start off with liquid and on cooling we have more and more solid and less and less liquid. So, even though both phases become richer in tin, the phase with less tin overall (solid α) is increasing in amount, while the phase with more tin overall (liquid) is decreasing. The change in amount of each phase is described by placing a hypothetical fulcrum at the point you are at (temperature, overall composition) and balancing the amount of liquid and solid on the tie line where it touches the respective curves (See Fig. 6.20). The lever shifts horizontally as the temperature changes.

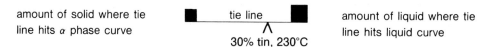

amount of solid where tie line hits α phase curve   tie line   amount of liquid where tie line hits liquid curve

30% tin, 230°C

Fig. 6.20

This is called (surprise!) the *lever rule*. The changing amount of each phase and the changing composition of each phase combine to keep the *overall* composition the same (30% tin in this example), and so we have not yet found the philosopher's stone.

We could go on and look at the right-hand side of the diagram, where we have lead dissolved in tin. This material is called *pewter*. However the lowering in melting point that alloying with lead achieves is no more than 40°C. While this lowering was important in older times, today the increased awareness of the toxicity of lead has meant that pewter (at least for food purposes) tends to be tin with copper, to give a tarnished appearance but with no lead content. Well if you are not yet 'phased', there are plenty of even more interesting diagrams to study— such as, for example, the phase diagrams of iron and carbon to form steel and the effect on steel of adding additional metals. Phase diagrams give the results of experiments carried out at *equilibrium*. What happens when materials are heated and then cooled quickly is even more complicated, especially if the materials are metals that can take on a variety of crystal structures.

Iron and steel

Metals were brought into use throughout history in a sequence determined by the increasing difficulty of winning them from their ores, or in a sequence that parallels their thermochemical properties. Whereas bronze melts between 900° and 1000°C (within the reach of an ordinary wood fire), pure iron melts at 1535°C, which, for a large part of our history, was out of reach of human technology.

When an iron ore is heated in a charcoal fire, two reactions occur—reactions that together ushered in the iron age.

$$3Fe_2O_3 + 11C \rightarrow 2Fe_3C + 9CO(g)$$

The oxygen in the ore is removed as carbon monoxide gas and the iron reacts to form iron carbide (cementite), a compound with 6.7% carbon. Some pure iron is formed at the same time.

$$Fe_2O_3 + 3C \rightarrow 2Fe + 3CO(g)$$

Now the crux of the matter is that iron and iron carbide are mutually soluble and form a *eutectic*. As we saw, the mixture of the two compounds has a *lower* melting point than each of the components. Whereas pure iron melts at 1535°C, the mixture with, say, 4% overall carbon, melts at only 1150°C. The fact that the above two reactions occur together producing a mixed product as a melt is the reason that the reactions occur at all.

'Cast iron' contains about 4% carbon, which is as much carbon as iron will hold. As well as carbon, other impurities are present because the original ore is not pure. These are removed by adding a 'flux' in the form of limestone, which combines with the other metals to form a low melting glassy slag. Today this material is sometimes made into fibres (rock wool) for home insulation.

Crude cast iron is thus a mixture of metal and non-metal crystals, which makes it brittle when cold but malleable above 250°C because at that temperature the crystal dislocations can move.

Cast iron is 'wrought' at 800–900°C by hammering. The hammering mechanically squeezes out most of the slag and solid impurities and also burns out most of the carbon. The heated iron forms an oxide layer (FeO) on the surface. When the iron is beaten and folded over like pastry, the surface layer can react with the bulk material.

$$Fe_3C + FeO \rightarrow 4Fe + CO$$

This long and laborious process was repeated perhaps one thousand times to produce almost pure iron with a few streaks of inclusions such as slag, which provide the traditional wavy patterns of wrought iron. To produce swords, the wrought iron had to be 'case hardened' by inducing a certain amount of carbon back in again, but only on the surface.

For best results the 'steel' was quenched by cooling it suddenly in a liquid. The mechanism of this process is now fully understood and is related to the packing of the spheres.

Below 910°C iron metal has the bcc structure. This form of iron is called *ferritic* and it is magnetic. Above 1400°C iron again has a similar ferritic structure. But between 910°C and 1400°C iron has the fcc non-magnetic *austenite* structure.

Fig. 6.21 Left: body-centred cubic structure with some interstices filled, e.g. low-temperature (ferritic) form of iron with space for carbon. Right: face-centred structure with interstices filled, e.g. high-temperature (austenite) form of iron with space for carbon

As we noted with close-packed atomic spheres, there is 26% vacant space in between. In 1831, Gauss calculated the percentage occupancy as being $\pi/3\sqrt{2} = 74.05\%$. Lagrange had already solved the two-dimensional case in 1773. (The answer is $\pi/2\sqrt{3} = 90.7\%$.) Simple geometry shows that if the spheres have a radius of, say, 1.0, then for each sphere in the crystal there will be eight holes of radius 0.155, two holes of radius 0.225 and one hole of radius 0.414. A large number of structures can be rationalised on the basis of little atoms fitting in the spaces provided by packed big atoms.

Carbon can dissolve in the holes between the iron atoms as well as combine with iron chemically to form the compound cementite (Fe_3C). More carbon can dissolve in the high-temperature form of iron (austenite) than in the low-temperature form (ferritic).

If austenite steel is cooled slowly, it results in a banded structure consisting of layers of pure iron (ferrite) and iron carbide (cementite). This steel is tough and strong but not particularly hard. However, if the austenite steel is cooled very quickly, the carbon atoms cannot move quickly enough from the holes between the iron atoms to react with the iron atoms and form cementite. They therefore remain trapped in a ferrite structure, but now without sufficient room. This means we have produced a structure in which dislocations cannot move. This quenching process is called *martensite* hardening and provides a very hard ferrite steel.

Biological fluids used to be preferred to water as the quenching agents. For example, prisoners were used in the production of Japanese samurai swords and urine was used in Europe—a somewhat more renewable resource. Technological innovation of this type often waits a long time for chemical explanation.

It is important that the steel be cooled as quickly as possible to prevent the carbon atoms from moving. When water is poured on hot metal, steam is formed, which prevents the liquid from touching the metal, and the transfer of heat from the metal to the water is poor. You can demonstrate this by dropping water on an electric hot-plate and watching the water droplets bounce around. With urine, however, crystals form on the steel and break up the steam barrier. (Don't try it, the smell is awful!) The nitrogen in the urine (as urea) also reacts with the iron to form hard needle crystals of iron nitride (Fe_2N) and also nitrogen molecules enter the vacant holes in the same way as carbon atoms and help to pin the dislocations. The modern method uses several days of urea or ammonia treatment on the hot metal.

Tempering of steel at between 220° and 450°C oxidises the carbon in the steel, which softens the steel to make it more ductile. Quenching and tempering are alternated and the art is to provide just the right balance of hardness and ductility for the purpose. When a process is scientifically understood, art is replaced by science and skilled artisans are replaced by automated processes.

The military history of Europe has in many ways been a history of improvements in iron and steel production. Cast iron was a poor material for any purpose involving tension, such as cannons. The main armaments of HMS *Victory* were 32-pounders weighing 3 to 4 tonnes. (These cannons have been replaced in the ship's museum by wooden replicas because otherwise the ship would collapse under their weight.) The *Victory's* opponent at Trafalgar, the *Redoubtable*, surrendered after two of her guns burst. Even by the time of the Crimean War, the guns had still not improved much and the Light Brigade were probably more at risk from their own guns.

Henry Bessemer had the brilliant idea of blowing air through liquid pig-iron to remove carbon and other impurities as their oxides and so 'convert' the iron to steel. The addition of manganese, the subject of another man's patent (Mushet), ensured the success of the Bessemer process. The original process would have failed because sulfur impurities react with iron to form iron sulfide (FeS), which can dissolve in *molten* iron but will crystallise out of the *solid* iron (like salt from freezing sea water) to settle at the boundaries between the iron crystals and weaken the steel considerably. The addition of manganese changes FeS to MnS, which is insoluble in liquid iron and so passes into the slag. There was considerable legal argument over whether this represented a patentable new process.

Stainless steel

The rusting of iron and steel occurs because an incoherent oxide layer is formed. This is a major drawback. In contrast, similar but far more expensive metals such as chromium form coherent oxide layers that do not allow further attack.

Chromium atoms are about the same size as iron atoms and form the same crystal structure (bcc). In alloys, the chromium atoms just replace some of the iron atoms. At least 12% chromium is needed to give corrosion resistance to iron, and 18% is usual. However, the effect of chromium on iron is to stabilise the low-temperature bcc (ferritic) structure even at high temperatures. Add much above 12% chromium and no fcc (austenite) structure is formed at *any* temperature, so that martensite hardening, which relies on a change in structure on quenching, cannot be used. The addition of nickel stabilises the fcc (austenite) structure. Stainless steel type 302 is the popular 18:8 (18% Cr, 8% Ni) product. This steel cannot be hardened and attempts to harden it produce ferritic (magnetic) steel.

Questions

Test your stainless steel 18:8 cutlery with a magnet and compare the knife handle and blade. The 'chrome' used on a car trim is steel with about 17% chromium. Is it magnetic? What is its structure?

A paper-clip on a piece of paper is left exposed to the elements. It corrodes underneath the clip, where the clip is in contact with the paper. This effect has nothing to do with the chemistry of the paper but with keeping oxygen *away* from the metal. Why does iron rust where the oxygen level is reduced? What does that tell you about the way you should store your stainless steel cutlery?

REFERENCES

1. V. Turoski (ed.), *Formaldehyde: analytical chemistry and toxicology*, American Chemical Society, Washington DC, 1985.
2. H. A. Wittcoff and B. G. Reuben, *Industrial organic chemicals in perspective. Part 2: technology, formulation and use*, Wiley, New York, 1980.
3. This section is based on J. Emsley, 'Glass: past elegant, future thin', *New Scientist*, 8 December 1983.
4. The major source of this section was J. E. Gordon, *The new science of strong materials,* 2nd edn, Penguin, Harmondsworth, Middlesex, 1982.
5. From Dr. S. Glover, *RSC News* **15** (9), 1984, 3, (with thanks to Brian Raison of Aurora Sailboards for technical information).

BIBLIOGRAPHY

General

Carraher, C. E., Hess, G., and Sperling, L. H. 'Polymer nomenclature: or what's in a name'. *J. Chem. Ed.* **64**(1), 1987, 36.

Carraher, C. E. and Seymour, R. B. 'Polymer properties and testing: definitions'. *J. Chem. Ed.,* **64**(10), 1987, 866.

Encyclopedia of polymer technology. Wiley, New York, 16 vols. 1962–72.

Gordon, J. E. *The new science of strong materials or why you don't fall through the floor.* 2nd edn, Penguin, Harmondsworth, Middlesex, 1982. This is a beautiful book and much use was made of it for this edition.

Schlenker, B. R. *Introduction to materials science, SI edition.* The Jacaranda Press, Brisbane, 1986.

Packaging

Code of practice for the manufacture of plastics items for food contact applications, AS2171. Standards Association of Australia, Sydney, 1978.

Plastics materials for food contact uses: polyethylene, polyvinylchloride (PVC) compound etc. AS2070. Standards Association of Australia, Sydney.

Plastics

Braun, D. *Simple methods for identification of plastics.* Hanser Publishers, 1982.

Handbook of chemistry and physics, 49th edn. 'Properties of commercial plastics'. p. C726. Chemical Rubber Co., Ohio, 1968–69.

Haslam, G., Willis, H. A. and Squirrell, D. C. *Identification and analysis of plastics,* 2nd edn. Butterworths, London, 1972.

Know your plastics. Plastics Institute of Australia, 1980.

Plastics materials identification kit and manual. The National Plastics Industry Training Committee, 121 Bluff Road, Black Rock, Victoria 3193.

Principles of chemical processes. Polyvinylchloride: a case study. The Open University ST294 CS1. Open University Press, Milton Keynes, UK, 1975.

Sikdar, S. K. 'The world of polycarbonates'. *Chem. Tech.* 1987, 112.

Sothman, R. D. 'Optical plastics'. *PRACI,* **44**, 1977, 47.

Stern, E. S. (ed.). *The chemistry in industry I, fine chemicals for polymers.* Oxford Chemistry Series, Oxford University Press, Oxford, 1973.

Trade designations of plastics and related materials (Plastic Note N9B). Plastics Technical Evaluation Center (New Jersey), October 1974.

Waddams, A. L. *Chemicals from petroleum,* 4th edn. John Murray, London, 1978.

Wittcoff, H. A. and Reuben, B. G. *Industrial organic chemicals in perspective. Part 2. Technology, formulation and use.* Wiley, NY, 1980.

Popular

Chemistry in the economy, Chapters 3, 4. Am. Chem. Society Study, Washington DC, 1973.

ICI Plastics Division, 1974, Schools publications.
   Identification of plastics for schools, 7 pp.
   Polymer and plastics in school science laboratories, 20 pp.
   Thermoplastics in school craftwork, 15 pp.
   The uses of plastics in school laboratories and craftrooms, 8 pp.

Magat, E. E. and Morrison, R. E. 'The evolution of man-made fibres'. *Chem. Tech.,* November 1976, 702.

Mason, P. *Cauchu—the weeping wood: a history of natural rubber.* Australian Broadcasting Commission, 1979.

National Plastics Industry Training Committee. *New dictionary of plastics,* 2nd edn (free of chemical formulae, etc.). *Plastics laboratory procedures* (for industrial plastics course). NPITC, Black Rock, Victoria.

Penfold, R. C. *A journalist's guide to plastics.* The British Plastics Federation, London, 1969.

Problems

Fanger, P. O. and Valbjorn, O. *Indoor climate—effects on human comfort, performance and health ... in buildings.* Danish Building Research Institute with WHO, Copenhagen, 1979.

Hardwick, R., and Cole, R. 'Plastics that kill plants'. *New Scientist,* 30 January 1986, 88.

Indoor pollutants. National Academy Press, Washington DC. 1981.

'Should UFE foam be banned?' *Choice,* ACA, March 1982.

The use of urea formaldehyde foam for insulation. Experimental Building Station, Department of Housing and Construction, Sydney, 1981.

Chemistry of hardware and software (II)

FIBRES, YARNS AND FABRICS

When the material in a fabric is given a name, it can refer to a number of different things. It may be the name of the *type* of fibre—nylon or acrylic, for instance. It may be the *brand* name of that particular fibre; for example, Crylon is Courtauld's nylon and Orlon is Du Pont's acrylic. It may be the brand name of the *yarn* into which the fibres are usually formed before the fabric itself is made; for example, Agilon is the brand name used for a particular stretch nylon yarn. It may be the brand name of a special *treatment*; for example, Koratron is one for a durable press finish. It may be the brand name of the *fabric*; for example, the Moygashel range can be woven from natural or synthetic fibres. Or—a more recent development—it may be the brand name used for the fabric in a particular type of *garment*. Thus Tricopress is the brand name used only on approved shirts and pyjamas made from Bri-Nylon, ICI's brand name for its nylon.

In France and the USA, fibre brand names can be used only alongside the name of the type of fibre. Different brand versions of the same fibre type vary somewhat, but their similarities are very much greater than their differences. Some characteristics—notably shrinkage—depend very much on the way the fabric is made up. The type of yarn is also important. Yarn can be bulked to make it feel softer and warmer. Crimplene is the yarn made by bulking Terylene (polyester). It is crimped (coiled or crumpled) to give bulk or stretch, or is given stretch characteristics in other ways, including twisting. Helanca covers a range of crimped or twisted yarns, mainly of polyester or nylon.

Fibres

Cotton is a natural cellulose fibre obtained from the boll of the cotton plant. Its quality depends on fibre length, fineness, colour and lustre. Cotton is still the world's major textile fibre and is used alone and in blends.

Linen is produced from the fibrous materials in the stem of the flax plant. These fibres, in a great variety of thicknesses, tend to cling together, giving linen its characteristic 'thick-thin' quality. Linen does not lint. It has poor resistance to flex abrasion, and may crack or show wear along seams and edges, where fibres are bent.

Silk is the only natural continuous filament fibre and is obtained by unreeling the cocoon of the silkworm. *Raw* silk contains the gum that binds the fibres to the

245

cocoon. *Wild* or *tussah* silk comes from uncultivated silkworms. *Spun* silk is made from pierced cocoons or waste silk, the tangled fibres on the outside of the cocoon. *Weighted* silk contains metallic salts, which make the silk cheaper and more drapable but less serviceable. Silk is weakened by sunlight and perspiration, and is yellowed by strong soap, age and sunlight.

Acetate is made from cellulose but is considered a synthetic fibre because of the significant alteration in properties. It is closely related to rayon. The fabric dissolves in nail polish, paint remover and some perfumes (e.g. Acete from Du Pont).

Triacetate is similar to acetate but has lower strength when wet and has low resistance to abrasion. It is much more heat resistant and hence shows durable crease and pleat retention, dimensional stability and resistance to glazing during ironing. It can be machine washed, tumble dried and ironed at temperatures up to 230°C (e.g. Arnel and Tricel from Celanese).

Acrylic is polyacrylonitrile. Brand names include Acrilan, Creslan, Orlon, Zefran (all of which have 10% to 14% of other monomers added to improve dyeing), Verel (which is 40% to 50% vinyl chloride), Dynel (more than 50% vinyl chloride), and Belson, Cashmilon, Courtelle, Crylor, Darvan, Dralon, Teklan and Vonnel. The major contributions of acrylic fabrics are their wool-like qualities and their easy care properties. In comparison with wool, acrylic fabrics are stronger, easier to care for, softer, do not felt and provide more warmth for less weight. However, they pill worse than wool (i.e. gentle rubbing forms small but unsightly nodules or pills as the surface fibres are raised) and the pills do not break off. Acrylic does not bounce back in the same way as wool, and garments do not retain the snug fit of wool. They are also more flammable (see below). Acrylic is not attacked by common solvents, bleaches, dilute acids or alkalis, and is resistant to weathering.

Spandex (polyurethane) usually occurs as a 5% to 10% contribution by weight in blends with other fabrics such as wool, cotton, linen, nylon or silk, without changing the look and feel of the basic fibre, but it contributes properties of stretch and recovery. For example, in upholstery, the fabric can stretch to conform to the contours of furniture. The fibres are used in lightweight garments and swimsuits. Note that they are vulnerable to bleaching agents such as chlorinated swimming-pool water.

Modacrylic is a fibre composed of less than 85% and more than 35% by weight of acrylonitrile units. Dense, fur-like fabrics are produced by combining fibres with different heat shrinkage capacities, so that a surface pile that resembles the hair and undercoat fibres of natural fur can be formed. The use of 15% to 65% vinyl chloride or vinylidene chloride co-monomer reduces the flammability of the fabric.

Polyamide occurs in various types: Nylon 66, Nylon 6, Nylon 11 (Rilsan), Antron, Bri-Nylon, Caprolan, Enkalon, Nylon 4.

Nylon is one of the strongest of all synthetic fibres in common use. It neither shrinks nor stretches on washing and is abrasion resistant. It is however degraded by ultraviolet light and is useless for curtains. White nylon garments yellow with age. Nylon pills easily. Its use in garments has declined and it is now used where its strength is important (e.g. in tyres).

Polyolefins occur as polyethylene with brand names Courlene, Nymplex, Pylen; and polypropylene with brand names Drylene, Herculon, Marvess, Merkalon,

Moplen, Polycreast, Pylene, Reevon, Ulstron. Polyethylene and polypropylene are the lightest weight of all fibres and are difficult to dye. They are generally used in blankets, carpets, upholstery, and also apparel. When mixed with wood they provide better thermal insulation than wool alone, but when used as a filler in quilted pads (and not treated with wash-resistant antioxidant) they can catch fire in a tumble drier.

Polypropylene has virtually replaced sisal in cheap ropes (nylon and polyester for high quality). It is also the fibre used in synthetic grass, webbing and carpet backing.

Polyester is made under the brand names Dacron, Diolen, Fortel, Kodel, Tergal, Terlenka, Terylene, Tetoron, Trevira, Vycron. It does not shrink or stretch appreciably in normal use; heat-set pleats and creases last well and water-borne stains may be quickly and simply removed.

Like nylon and acrylics, polyester tends to pick up static charge, which attracts dust particles and it has an affinity for oils, fats and greases. It has a high density (1.38 compared with 1.14 for nylon and 1.18 for acrylics) and so the cost per unit area of cloth is high. The close-packed rigid structure without highly polar groups makes it difficult to dye. However, the stress-strain curve of the polyester fibre can be varied to match that of other fibres with which it is to form a blend, and it was the development of this technology that made possible the polyester-cotton blends now so widely used.

Rayon occurs as two types, both of which are made from cellulose: cuprammonium and viscose. In recent years several 'new' rayons, called polynosic rayon, have been developed with greater wet strength. Rayon is one of the cheapest 'synthetics' and is easily blended.

Chloro-fibres are poly(vinyl chloride) and co-polymers. Brand names are Vinyon, Geon, Krekalon, Movel, Pe-Ce, Rhovyl, Saran, Teviron, Tygan, Velon.

Vinal: poly(vinyl alcohol), poly(vinyl acetate) and co-polymers are made under the brand names Kuralon, Mewlon, Vinylon.

FROM MESS TO MILLIONS

In January–February 1939, a consumer product hit the US market. It is without equal in its impact before or since. Nylon stockings were exhibited at the Golden Gate International Exposition in San Francisco and were sold first to employees of the inventor company, Du Pont de Nemours. Some months earlier nylon had made its first debut in a less published manner as 'Exton' bristles for Dr West's toothbrushes.

On May 15, 1940, nylon stockings went on sale throughout the US, and in New York City alone four million pairs were sold in a matter of hours.

Two years earlier, just three weeks before the basic patent on nylon had been filed, the discoverer of nylon, Wallace Hume Carothers, suffering from one of his increasingly frequent attacks of depression, caused by his conviction that he was a scientific failure, drank lemon juice containing potassium cyanide.

Fifty years ago most chemists believed the tarry messes, often obtained in the bottom of their flasks and called polymers, consisted of the simpler monomers held together by vague physical attractive forces. The German chemist Staudinger believed true chemical bonding was responsible.

It was while Carothers was investigating this theory of polymers that he discovered (in 1930) that the polymerisation of chloroprene yielded an interesting polymer called neoprene, which was later to save the war for the US.

Carothers then worked on polyamides. These are polymers made by stringing together pairs of two different monomers. If you think of a polymer as a string of paper clips, then for a polyamide the paper clips come in two different types where the length of each paper clip can be changed independently.

He found the best polyamide for a textile fibre in 1935, which was elastic, stronger than silk, inert to moisture and solvents. The two monomers had 5 and 10 carbon atoms respectively, so it was called Fibre 510. Du Pont looked for cheaper starting materials and so Fibre 66 came into being.

The commercial name for this product as used for stockings was suggested as *norun* because it was more resistant than silk to laddering. There were problems and so norun was spelt backwards to read *nuron*. However this was too close to neuron and could be construed as a nerve tonic. An Asian intonation to *nulon* ran into trade mark problems and a change was made to *nilon*. English speakers differed in their pronunciation of this, so, to remove ambiguity the name finally became *nylon*.

Come Pearl Harbour, Japan was in a position to strangle the American war effort by the loss of two vital natural products, rubber and silk (for parachutes). Japan was the major supplier of silk and rubber came from South East Asia.

Carothers had already invented neoprene or synthetic rubber (which was superior to the natural product in many applications, particularly if exposed to oil). Nylon was to replace silk for parachute cloth, 'Flakvests', tow ropes, tyre cords.

All nylon production was commandeered and nylon stockings were donated by thousands of patriotic women to the war effort.

Today about half the chemists in the US work on the preparation, characterisation, or application of polymers. And as for nylon, the objects in regular use that use this versatile material include hosiery, textiles, paint brushes, fishing lines, runners, nets, tennis racquets, upholstery, sewing thread, tyre cords, ropes, films for boil-in-the-bag, sails and rigging, gears, oil seals, casings, hoses, combs, zip fasteners, hinges, syringes, spectacle frames, ski bindings, and photoengraving printing plates for this book.

The versatility of nylon (and the later polyesters) is due to the fact that it is made from two paired monomers. By varying the length of the two monomers the properties can be changed from a reasonably water-friendly fabric, suitable for body contact, to a tough engineering material replacing metal in gears.

Source: From Ben Selinger, 'Millions made from a mess' *Canberra Times*, 29 January 1989

Mixed fibres

In a blended fabric, two or more fibres are blended before spinning them into yarns. In a combination (or union) fabric, individual yarns composed of one fibre are combined during weaving with yarns composed entirely of another fibre.

Cotton and rayon, for example, are combined with other fibres to increase absorbency and comfort, decrease static build-up, improve dyeability, and reduce production costs. Acrylics improve softness and warmth without adding weight. Nylon adds strength. Polyester contributes several properties to blends, including wash-and-wear qualities of abrasion resistance, wrinkle resistance, and dimensional stability. Acetate improves drapability and texture.

The quantities of blend needed vary considerably: 15% nylon improves the utility of wool; 10% elastic fibre gives strength properties to clothes; the addition of 30% modacrylic reduces the flammability of acrylic carpets.

Textile technology

The traditional weaving process has been replaced by machine knitting techniques which are simple, fast and flexible. Machine knitting is particularly suitable to synthetic fibres. Knitting has had its great impact in apparel for women and children and increasingly for men. Second-grade fibres (not up to specification) have been exported to Eastern Europe and Third World countries for processing.

FABRICS AND FLAMMABILITY

The large sheets of butcher's paper that are often used for children's paintings can be a fire hazard. Paper can be rendered non-flammable by soaking it in alum, $K_2SO_4 \cdot Al_2(SO_4)_3 \cdot 24H_2O$. The protection afforded by the alum can be simply demonstrated. If you write on paper with alum solution and allow it to dry and then heat the paper carefully (over a warm stove), the dry invisible alum letters become visible as dark carbonised areas. The alum dehydrates the (polyalcohol) cellulose by acting as a proton donor to form H_2O from the—OH groups of the cellulose. Alum is representative of substances that are inert at normal temperatures but become active when heated. However, high retention of flame-retardant properties on washing and dry-cleaning is nowadays required of flame-retardant treatments.

The serious nature of clothing fires was aptly summed up in a National Bureau of Standards (1973) report: 'If your house catches fire you will probably escape with your life and your skin. If your pyjamas catch fire you will probably lose your skin and possibly your life.'

A study of burned garments received through the Burns Research Unit at the Royal Children's Hospital, Melbourne, allowed Tom Pressley of the CSIRO Protein Chemistry Division to devise sensible tests for real-life flammability. The most important single property that makes a garment dangerous is ready ignition by momentary exposure to a flame. A special case concerns fabrics with a flammable pile of cotton or rayon (i.e. cellulose). It can be demonstrated that when a dried (in a desiccator) cotton chenille strip is exposed to a flame, the flame rapidly travels over the whole surface, gradually penetrating and eventually lighting the bare fabric over a wide area, even when the bare material is heavy and slow to ignite. The most dangerous garments are those that ignite readily and become a mass of flame in 30 seconds or less. Much summer clothing is in this category.

Garment flammability depends on the *fibre*, the *fabric* and the *fashion*. Most burns are caused by cellulose fabrics, either cotton or rayon, which ignite easily and burn rapidly. Acrylic fibre (e.g. Acrilon, Courtelle, Orlon) is harder to light than cotton, but, once alight, burns freely. Polyamides (e.g. nylon) and polyesters (e.g. Terylene and Dacron) are rarely involved when worn as outerwear because their ignition temperatures are above their melting points. When exposed to a flame they melt and retreat away without lighting. However, in a blend with another, flammable fibre, the melt can burn fiercely. (A nylon-rayon garment is highly flammable.) These materials are less suitable for underwear; if the outer garment burns, the underwear melts and sticks to the skin. Even before melting, the fibres shrink, causing close contact with the skin, and can then efficiently conduct heat from the flame to the skin.

Availability of air has a large influence on a fire, so that fabric weight and structure have an important effect on burning characteristics. Loosely spun yarn or lightweight or loosely woven fabric ignites more easily and burns more rapidly than tightly constructed or heavyweight material. A raised or pile surface is an extreme case of loose structure: a cotton chenille dressing gown is one of the most dangerous garments in common use. Conversely, heavy weight and tight structure bring safety: cotton jeans of tightly spun and woven drill are relatively safe. It has been found repeatedly in burned garments that the seams remain unburned and act as a sort of 'fire break' to the spread of the burning.

Garment shape also plays a part. A flowing nightdress, loose pyjamas, ski pyjamas, all from cotton, gives a range from dangerous to relatively safe. Data collected by the Burns Research Unit have shown that British and American figures for burns to children are not applicable in Australia, and also that the local situation is continually changing. A report in 1964 showed a girl : boy ratio of accidents of 3 : 1. In 1975 the ratio was 1 : 2 (see Table 7.1).

Two points to note from Table 7.1 are (1) that the widely held notion that nightclothes are most involved in burning accidents is not valid; and (2) that, whereas the domestic fires represent a decreasing hazard, the number of accidents involving flammable solvents such as petrol and kerosene is large (adults too are victims of these accidents). Australian standards have been set to deal with the problem and they are now being made mandatory under State laws. The first standard is AS1176, 1976, which defines *methods of test for combustion characteristics of textile materials*. These methods are divided into three parts:

Part 1. Method for the determination of ease of ignition.

Part 2. Method for the determination of burning time and heat output.

Part 3. Method for the determination of surface burning properties.

Parts 1 and 3, in effect, measure how easily a fabric can catch fire, while Part 2 indicates the time available to deal with an accident. (These methods can form the basis of some experiments or demonstrations.) Although the individual tests are not very stringent, their combination provides a stiff test for a material to pass.

TABLE 7.1
Summary of statistics on clothing fires

	Percentage
Sex	
Boys, 1½–14 yrs	66.1
Girls, 1½–14 yrs	28.4
Babies, 0–1½ yrs	5.5
Ratio of boys to girls—2.3 : 1	
Clothing	
Daywear	68.3
Nightwear (including bathrobes)	26.2
Babywear	5.5
Ratio of daywear to nightwear—2.6 : 1	
Fabric	
Cellulose or cellulose blend	81.4
Fire source	
Solvent (e.g. petrol, kerosene)	45.0
Matches	15.0
Domestic fires	8.1

Sources: T. A. Pressley, results from the Burns Research
Unit, Royal Children's Hospital, Melbourne.

The next standard, AS1248, 1976, is entitled *Fabrics for domestic apparel of the low fire hazard type.* This standard sets the performance requirement of a material when subjected to the test methods of AS1176 in order that it might qualify as a low fire hazard material.

Having dealt with the scientific tests and how a safe material should perform, the standards also deal with the actual clothes—in AS1249, 1983, *Children's night clothes having reduced fire hazard.*

A flow diagram for evaluating garments is given in AS1249. It is not a simple matter. Further information on the burning behaviour of fibres can be found in Appendix III.

Natural fibres, (except cotton) and skins are low in flammability. Wool burns only in an upward direction and needs plenty of oxygen, so local depletion of oxygen causes the flame to be extinguished. In fact, wool is so sensitive to oxygen availability that skiers are safe—it fails to burn above 1000 metres.

CARPETS

The criterion of a hard-wearing carpet is simple—it should stand up to wear without looking shabby. Where you put a carpet and how much use it gets is very important in assessing its life. Underlay also helps a carpet survive. The two most important questions to ask when assessing the quality of a carpet are 'how much pile?' and 'what is it made of?'

The pile *density*, *height* and *weight* will tell you how much pile there is. Density is the thing to look for first. Fibres wear more if you tread on the sides rather than the ends. The more tightly the fibres are packed, the more likely they are to stay upright and the better the carpet will wear. Provided the pile is equally dense, the higher the pile the better. The type of fibre used is also important, and is discussed in greater detail in the box 'Carpet materials and construction' below.

To test a carpet:

- Bend the carpet sample back on itself. You can compare samples by seeing how much they gape. The more easily you can see the backing through the pile, the less dense the pile is.
- Tug at a few tufts to see if they are firmly anchored.
- Look at the backing. A closer weave lasts longer and the threads should be straight and at right angles. A rigid backing will help to keep the carpet in shape.

Underlay is used for the following reasons:

- It forms a cushion between the carpet and the floor, so that the carpet wears evenly. This is particularly important if the floor is uneven.
- It protects the backing from rubbing and rotting.
- It stops any dirt that comes through the floor boards from soiling or abrading the pile.
- It provides extra insulation.
- It makes the carpet feel softer and thicker.

The greater the likely traffic, the heavier the underlay needs to be. Do not use foam with underfloor heating as it can disintegrate. Carpets with heavy foam secondary backing (like many tufteds) do not need a separate underlay. This seems to save money, but of course you cannot use the underlay again if you want to change the carpet. There is also the danger that if the underlay disintegrates before the carpet, or if it separates from the carpet backing, the carpet will wear unevenly and more quickly.

Standards Australia has prepared an excellent standard for the classification and terminology of textile floor coverings (AS2454, 1981) in which the various terms are also defined pictorially.

The UK Consumers' Association produced a consumer handbook in 1976 for its Advice Centre Servicing Unit, entitled *Carpets* and edited by G. Clegg and G. Davies. Parts of the first two sections and the last part of Section III are reproduced here with permission.

CARPET MATERIALS AND CONSTRUCTION

Materials

The properties considered desirable in a carpet—durability, appearance retention, resilience, dirt and stain resistance, ease of cleaning—depend to a large extent on what kind of fibre the carpet is made of. But the characteristics of a fibre are not always constant. They depend on the way the fibre is used. Wool, for example, is generally considered to be a resilient fibre, but if a carpet is constructed from an inadequate amount of loosely-spun woollen yarn, it will flatten relatively quickly. So, although fibre content is an important consideration, there is a danger in making definite assumptions on this basis alone.

FIBRES

Fibres are either:

staple—short fibres from 12 mm to several cm long; or
continuous filament—fibres of indefinite length.

TABLE 7.2
Carpet fibres

Natural			Man-made	
Animal	Vegetable	Mineral	Regenerated	Synthetic
Wool	Cotton	Metal	Rayon (viscose and various modified versions of viscose is the type of rayon normally used in carpet production)	Polyamide (nylon)
Hair	Jute			Acrylic (e.g. Acrilan)
Silk	Sisal			Polyester (e.g. Terylene)
	Coir			Polypropylene (e.g. Meraklon)
				Poly(vinyl chloride) (PVC) (e.g. Fibravyl)

All natural fibres, except silk, come in staple form. Man-made fibres are produced in continuous filament form and then chopped into required lengths. Fibres are used either on their own or blended together in various combinations.

Natural fibres are derived mainly from animal and vegetable sources.

Man-made fibres are either:

regenerated fibres—made from natural fibre-forming materials such as cellulose; or
synthetic fibres—made from raw materials such as coal and petroleum which are not natural fibre-forming materials. Fibres are produced from these by forcing the liquid base through fine holes in a 'spinneret' which hardens into filaments by chemical action, evaporation or cooling.

With the man-made process it is possible, also, to produce a fibre containing more than one fibre type or two different types of the same fibre, called *bi-component fibres*. The two components can be arranged side by side or 'wrapped' around each other. Using bi-component fibres enables a manufacturer to make a carpet directly from fibre without converting into yarn.

YARNS

To produce a continuous thread of adequate strength for carpet manufacture, fibres are generally converted into yarn (apart from exceptions such as bi-component fibres) by spinning and/or twisting. Extra twist may be put into a yarn to give it better resilience or a more textured appearance. *Staple fibre yarns* are produced by a spinning process. The type of staple fibre and the spinning process used determine the texture of the yarn. In a *woollen spun* yarn the fibres lie in random directions and this increases the bulk of the yarn. With a *worsted spun* yarn the fibres are combed before being twisted. This gives a finer, smoother yarn and is used for more expensive, patterned carpets to give clarity to the design. The yarn formed as a result of the spinning process is referred to as a *single* yarn. Two or more of these can be twisted together to form *two-fold, three-fold, four-fold,* and so on.

Continuous filament yarns consist of filaments combined with a small amount of twist to prevent them separating. Variations can be achieved by crimping (processing by heat, steam or pressure) individual filaments to give a fuller texture. Fibres treated in this way are called *bulked continuous filament* (bcf). Sometimes continuous filaments are chopped into staple fibres to give a more 'woolly' appearance. The thickness of continuous filament yarns can be varied according to the size and number of filaments twisted together.

Blends of one or more different fibres in a yarn are produced either for economy or to combine the advantages of different fibres. Generally, at least 20% of one fibre is required to affect the carpet's overall characteristics but the amount differs depending on the fibres involved—for example adding as little as 5% of nylon to a viscose rayon carpet improves its wearing qualities. A common 'economy' blend is 45% wool 45% viscose 10% nylon. The large proportion of the cheaper fibre, viscose, gives the carpet bulk whilst keeping the price low. Another popular blend is a combination of 80% wool and 20% nylon. The introduction of nylon gives greater durability than wool alone yet the carpet retains the appearance and feel of wool. Some other commonly used blends are: acrylic/nylon, nylon/viscose, acrylic/viscose, acrylic/viscose/nylon.

Fibre characteristics

ANIMAL FIBRES

Wool Although expensive, wool is considered nearest to an ideal fibre for carpets. It retains its appearance well, doesn't soil easily, is soft, warm, resilient and has good resistance to burning and static. Wool is moderately hard wearing if used at a high enough density. *Blends:* Nylon is often combined with wool to improve durability. Wool, nylon and viscose are blended to produce a lower-priced carpet.

Hair (from animals other than sheep) Hair fibre is usually tough, rather coarse and fairly cheap. It is used mainly in the production of cord carpets and felt underlays. *Blends:* it is sometimes combined with viscose rayon and wool to give a softer feel.

Silk A luxurious and hard-wearing fibre, which, because of its cost, is normally found only in intricately designed Eastern carpets.

VEGETABLE FIBRES

Cotton   A hard-wearing fibre which is cheap but compresses and loses its appearance quickly. It is most suitable for bathroom rugs which can be washed easily.

Jute   Used mainly as a backing material but sometimes as a cheap fibre in cord carpets. It is hard wearing but loses strength when wet. It is difficult to clean and suffers a rapid loss of appearance. *Blends:* It is frequently blended with viscose rayon for cord carpets to combine durability with softness.

Sisal   Used to make cord carpets and matting. It is fairly cheap and very hard wearning but feels harsh.

Coir   A very cheap, harsh but hard-wearing fibre used for matting.

MINERAL FIBRES

Stainless steel   These fibres are sometimes blended in very small quantities with other fibres to counteract static charges. These are used mainly for contract (commercial) carpets because of the high cost.

MAN-MADE FIBRES: REGENERATED

Viscose rayon (e.g. Darelle, Evlan, Evlan M, Fibro). Ordinary viscose is a cheap, low-performance fibre which soils and flattens easily. It is easily flammable but a flame-retardant version—Darelle—is available. Modified versions with improved resilience and higher resistance to abrasion and soiling are now being produced for carpeting—Evlan, Evlan M. Both ordinary and modified viscose are used in blends with wool and/or nylon for medium to low price carpets.

MAN-MADE FIBRES: SYNTHETIC

Nylon (e.g. Antron, Bri-Nylon, Du Pont Nylon, Enkalon, Monsanto Nylon, Timbrelle). Basically an expensive fibre but because of its good durability it can be used in low density to produce inexpensive carpet. Nylon is not easily flammable but will melt if exposed to flame. Traditionally, nylon's main problem has been its retention of static electricity. Consequently it attracts and shows dirt more than most other fibres and needs frequent cleaning. (This is fairly easy, but it can be difficult to remove animal hairs from a nylon carpet.) Recently developed nylons are considerably improved. Nylon is often blended with wool and other fibres to add greater durability.

Acrylic (e.g. Acrilan, Courtelle). The nearest of the synthetic fibres in appearance to wool, with a somewhat similar feel. One of the more expensive synthetics but cheaper than wool and suffers a greater loss of appearance. It has good resistance to abrasion and wears well. Although generally less liable to retain static than nylon, it still has a tendency to attract dirt. However, it has good stain resistance and can be cleaned easily. It is fairly flammable. Acrylic is often blended with nylon and/or wool.

Modacrylic   (e.g. Dynel, Teklan, Verel). Modified acrylics have better flame resistance and, it is claimed, better wear and soil resistance than ordinary acrylics. Consequently they are more expensive.

Polyester   (e.g. Dacron, Terylene). Has similar properties to nylon but is less resilient. However it is not used to the same extent in carpets because it is both more difficult and expensive to produce; but its soft handling qualities make it particularly suitable for shag-style carpets (carpets with a pile height over 25 mm). *Blends:* Sometimes blended with nylon.

Polypropylene   (e.g. Fibrite, Meraklon). A fairly cheap fibre used mainly as a backing material but sometimes as a pile fibre. As such, it is very hard wearing and has good resistance to soiling and staining. It is particularly suitable for bathrooms and kitchens as it absorbs virtually no moisture and usually has a plastic backing. *Blends:* Occasionally with nylon.

Poly(vinyl chloride) normal   (e.g. Clevyl, Fibravyl, Movil). Because PVC fibres are non-flammable they are sometimes blended in small quantities with other fibres to improve flame resistance.

Carpet backing materials

There are two methods of carpet construction:
    where the backing is woven at the same time as the pile (woven and some cord carpets).
    where the pile is inserted into or stuck onto an existing backing (tufted and non-woven carpets).
Carpets may have one or two layers of backing:
    primary backing—the backing in or to which the pile is anchored.
    secondary backing—sometimes added to give a carpet extra stability and resilience and perhaps to act as an integral underlay.

PRIMARY BACKING MATERIALS

Jute   Woven into hessian, it is the traditional fibre used for both methods of carpet construction. It gives a carpet a thick, stable and durable backing. It is also strong unless weakened by water. This makes it susceptible to mildew (though it can be proofed against this). The yarn usually has a high oil content, which can work its way into the carpet pile and cause soiling, unless controlled during the yarn-making process. High-density carpets are usually backed with jute to give adequate dimensional stability.

Polypropylene   Was developed as an alternative to jute and is now the material used most frequently. Its advantages over jute are that it is lighter, cleaner, more consistent in quality and resistant to water. A disadvantage of polypropylene is that it does not pick up colour well when the carpet is dyed. With low-density pile carpets, this can sometimes cause the backing to show through the pile (referred to as 'grinning'). This can be overcome by adding a thin web of nylon—which will take dye—to the backing. This is called *needleweave.* Polypropylene is available in two forms:

Woven. Granules of polypropylene are heated and extruded as a film, slit into strips and woven.

Non-woven. Usually 'spun-bonded', where filaments are compressed by heat into sheet-form. Non-woven polypropylene will not unravel, which prevents the tufts separating, making it particularly suitable for backing carpet tiles and for closely tufted carpets.

Some manufacturers now combine the advantages of both jute and polypropylene in a woven *poly-jute* backing fabric.

Cotton   Is sometimes used as a primary backing material for woven carpets. It is stronger than jute and less affected by water.

ADHESIVE LAYER
On almost all tufted and non-woven carpets a coating of latex or PVC is added to the primary backing to strengthen tuft anchorage. Some woven carpets are coated with a starch or modified starch solution, sometimes combined with synthetic resin or latex. Besides improving tuft anchorage this gives woven carpets additional stiffness and prevents fraying.

SECONDARY BACKING MATERIALS
Most woven carpets do not structurally need a secondary backing (though some Axminsters are now made with a sheet of foam bonded to the back of the carpet to act as an integral underlay). Most tufted carpets need a secondary backing for added stability and resilience. The most common form of secondary backing is rubber foam or crumb, which also acts as an integral underlay. This saves consumers money in that they do not need to buy separate underlay and will probably be able to fit their carpets themselves.

Foam backings were originally used to give support to cheaper, lightweight carpets and were mainly of mediocre quality. Since then much improved and heavier grades of foam have been produced and quite expensive carpets are now foam backed.

Foam   Latex foam is now made mainly from synthetic rubber combined with fillers and additives. The density is controlled by the amount of air introduced into the foam latex mixture. The mixture is then applied to the back of the carpet, 'set' under infra-red heaters and dried to remove excess moisture.

Crumb (or sponge)   Usually black, often made from reclaimed rubber (e.g. old tyres) which is broken up into granular form then pressed together under heat with a binding agent. It is usually cheaper than foam.

Polyester and PVC   Are also used occasionally for foam backings. They absorb less water and are stronger than latex but more expensive. They are sometimes used for backing carpet tiles, where greater dimensional stability is required.

Polypropylene Is only used as a secondary backing where the primary backing is also polypropylene because of the problems of adhering polypropylene to jute. However, some manufacturers now highlight the fact that their carpets are entirely polypropylene-backed and therefore suitable for potentially wet areas like bathrooms.

Jute Is sometimes used as a secondary backing material but this is increasingly less common.

Construction

The way a carpet is constructed is no longer such an accurate guide by which to assess it. At one time 'woven' was synonymous with 'quality' carpets, and other types with 'cheapness' but this is not particularly true today. The price and performance of a carpet now relate more to the type and amount of fibre used than to the method of construction. However, construction still has some bearing on a carpet's durability and appearance retention.

The three main methods of carpet construction are: *woven, tufted* and *non-woven.*

WOVEN CARPETS

Woven carpets are those where the backing is constructed at the same time as the pile. *Wilton* and *Axminster* are the names given to the two main weaving processes and are *not* brand names as is commonly believed. While many high-quality carpets are Wiltons or Axminsters, there are others of mediocre quality made by these methods. Although the majority of woven carpets are made from wool, all-synthetic fibre wovens are now available.

A woven carpet is made of warp and weft yarns intertwined.

Warp yarns run lengthwise and are of three types:

Chain warps, which go over and under the weft yarns to bind them. These may be made of cotton or viscose rayon but are now increasingly polypropylene.

Stuffers, which run through the centre of the carpet between the chain warps and the two layers of weft to add bulk and to strengthen the carpet. Stuffers are normally made from jute but sometimes cotton or wool.

Pile warps form the carpet surface and are made from all types of fibre used in carpets.

Weft yarns run from side to side of the fabric. Normally there is both an upper and lower weft. They are usually made from jute or cotton.

Wilton

This process can produce both plain and patterned carpets. They can be *cut pile* (this gives the smooth, level surface most commonly associated with the name Wilton); *loop pile* (Brussels is the name given to loop pile Wiltons); low loop pile (cord carpets) and mixtures of cut and looped pile.

Most Wilton carpets are made on wire looms. The pile warp yarn is carried under the weft and then over a flat wire which raises it above the backing yarns. The width (or height) of the wire determines the depth of the pile. The yarn forms loops which are left uncut for Brussels and cord carpets (a round rather than a flat wire is used for the latter). A small blade, attached to the wire, cuts the loops as the wire is withdrawn to give a cut pile surface.

Figured (patterned) Wilton These are produced by running several pile yarns of different colours (up to five) through the weave. The carpet design is transferred to a punched card operated mechanism called a *jacquard,* which selects the right coloured yarn for each tuft position, lifting it to the surface and leaving the other yarns hidden in the carpet (these yarns are called unused or dead yarn).

The pile yarn is drawn from *bobbins* arranged on *frames.* Each colour comes off a separate frame. A carpet is referred to as a 2, 3, 4 or 5 frame Wilton depending on the number of colours used.

Five colours are normally the maximum but more can be incorporated by *planting* (substituting) one coloured yarn for another in parts of the design. As this is an expensive process it is only used where both the heavier quality of a top-grade Wilton and a wide variety of colours are required.

The dead yarn in a patterned Wilton gives it a more solid (and more expensive) construction than a plain Wilton. Occasionally, though, plain Wiltons are made by the jacquard process to produce a heavier quality carpet.

Sculptured (sometimes called *embossed or carved*) *pile* Wiltons can also be produced. These combine straight and twisted yarn, cut and loop pile or pile of differing heights (by using varying heights of wire). 'Shag-style' Wiltons (carpets with a pile about 25 mm long) are made by a variation in the kind of wires used.

Face-to-face weave is a more economic way of producing a Wilton. Two carpets are woven at the same time by the same pile yarn crossing between two sets of backing yarns. A blade, the width of the carpet, cuts through the pile as it forms, so separating the two carpets. The face to face method is mainly used for lower quality, plain Wilton.

Axminster

This process produces only a cut-pile surface as each tuft is inserted separately. There is no unused yarn running through the carpet, which makes it cheaper than the Wilton method. The construction process, also, allows for much greater variety of colour and design, so Axminster carpets are often highly coloured and intricately designed.

Gripper Axminster is the most common method of making Axminsters. The carpet design is transferred to jacquard cards which program the colour

TABLE 7.3
Carpet fibres compared

Pile fibre	Basic cost	Durability	Resistance to flattening	Soil and stain resistance	Ease of cleaning	Handle	Flame resistance	Resistance to static
Acrylic	Expensive	Good	Very good	Shows dirt easily. Good stain resistance.	Easy; needs to be done frequently. Doesn't soak up liquids.	Resembles wool in appearance and handle.	Is ignited fairly easily but modified versions (modacrylics) are available which have a high level of flame resistance.	Good
Cotton	Cheap	Good	Poor	Fair for stains, poor for dirt.	Easy; needs to be done frequently, colours may run.	Soft, smooth pile.	Fairly flammable	Good
Nylon	Fairly expensive	Very good	Good	Shows dirt easily (although there are some nylons—like *Antron*—which conceal dirt more effectively). Good stain resistance.	Easy; needs to be done frequently.	Rather harsh but nylon for carpets is now being developed with a more wool-like feel.	Moderate. Will melt and may continue to burn slowly.	Poor
Polyester	Expensive	Good	Fairly good	Shows dirt easily. Good stain resistance, except to oil.	Easy; needs to be done frequently. Doesn't soak up liquids.	Fairly soft texture.	Moderate. Will melt and may continue to burn.	Poor

Polypropylene[a]	Fairly cheap	Very good	Fair	Good resistance to dirt and stains.	Very easy. Doesn't soak up liquids.	Fairly harsh.	Moderate. Melts and may continue to burn slowly.	Fairly good
Sisal	Cheap	Good	Good	Fair; dirt builds up underneath.	Fair; difficult to vacuum clean. Dirt can lodge in or under the pile.	Harsh.	Fairly flammable.	Good
Viscose[b]	Cheap	Fair	Poor	Shows dirt easily. Fair stain resistance.	Easy, but should not be saturated with shampoo.	Fairly soft and warm.	Burns readily.	Good
Modified viscose[c]	Cheap	Fairly good	Fair	Shows dirt easily. Fair stain resistance.	Easy, but should not be saturated with shampoo.	Fairly soft and warm.	Burns readily but *Darelle* is flame-resistant.	Good
Wool[d]	Expensive	Good	Very good	Good	Easy, but may felt or shrink if made too wet.	Soft, warm.	Good. Does not burn easily, largely self-extinguishing.	Good

[a] Good for kitchens and bathrooms.
[b] Used on its own, only suitable for light traffic areas.
[c] Has better wear resilience and soil resistance than ordinary viscose.
[d] Susceptible to insect attack but usually moth proofed. Can felt under damp conditions.

selection and the yarn is mounted on frames behind the loom (as in figured Wilton). Eight is the most common number of frames and although 12 and 16 frame looms are available additional colours are still usually introduced by planting. The different coloured yarns are threaded through carriers, which extend across the loom, and are controlled by the jacquard mechanism. A row of beak-like pincers (grippers) moves up and takes hold of the free ends of the pre-selected yarns, which are then severed at the required length from the carrier. The grippers, carrying the tufts, move down to the weaving position, where the tufts are released and bound in place by the backing yarns.

Spool Axminster.   Each colour of yarn used in the pattern is wound onto 70 cm spools in the order it appears in the design. The required number of spools are placed side by side to make up a complete row of the pattern. As many different rows of spools as are needed to form the pattern repeat are assembled and then mounted between a pair of chains above the loom, which carries the rows one after the other to be woven into the carpet. At each stage, the required spools are lowered from their chains and the yarn is drawn down through the backing threads and held there while the spool is withdrawn. The pile yarn is severed from the spool, which is returned to the chain. The next row of spools is then lowered to repeat the process.

Spool Axminster is the most flexible weaving method for patterning as, theoretically, each length of yarn wound on to the spool can be a different colour. In practice economic considerations limit the number of colours used, although some floral chintz or Persian designs incorporate over 30 or 40 different shades. Because of the cost of setting up the design, spool Axminster is normally used only for long production runs.

Spool Axminster carpets are less tightly woven than gripper Axminster and need a final coating of modified starch or latex to give them added stiffness. Carpets requiring the colour potential of the spool process and the tighter structure of the gripper process can be produced on a *Spool/Gripper* loom. The yarn is wound on spools and inserted into the carpet by a gripper mechanism.

If it isn't obvious from the colours in the design, the backing will show which process has been used. With spool Axminster, the backing is smooth and the pattern shows through; with a gripper Axminster the backing is heavily ridged and the pattern does not emerge.

Tuft density

The quality of a carpet is largely determined by the amount of pile it contains (pile density) and its weight (the higher the density the better). In *Axminster* carpets the density can be calculated by multiplying the *pitch*—the number of pile tufts measured across the width of the carpet (fixed at 24 or 28 tufts/100 mm) by the *number of rows*/100 mm (which varies between 16 and 40). Therefore, with six-pitch Axminster the tuft density ranges from 384 to 960 tufts/dm^2 and with 28 pitch from 448 to 1120 tufts/dm^2.

The *weight* is calculated from the tuft density (number of tufts per dm^2), the pile height and the count of the pile yarn.

With *Wilton* carpets, the tuft density and weight are calculated in a similar way but the pitch is not fixed as in an Axminster construction. The number of *frames* used in making the carpet is also important in determining quality as the dead yarn contributes to the strength and resilience of the product.

Identification

Wiltons and Axminsters differ mainly in the colours and designs which can be produced. As an aid to identification, Wiltons have a more closely woven backing and, when patterned, dead yarn shows among the backing material. Only one weft thread normally shows in the backing of a Wilton. In an Axminster double weft threads are visible.

Kara-loc weave

Consumers may see Kara-loc given as a method of construction on some carpet labels. It is another (but rather different) type of weaving process which is used solely by one manufacturer (Crossleys). Kara-loc, it is claimed, allows for a greater variety of patterning, particularly in achieving sculptured and textured effects.

TUFTED CARPETS

Tufted carpets weren't introduced until the mid-1950s and early production problems gave them a reputation of cheapness and inferior quality. They are quicker and more economical to produce than woven carpets and therefore predominate in the lower price ranges. However, expensive, high-performance tufted carpets are also now available.

Tufted carpets are produced by inserting the pile yarn into a prepared backing material. The yarns are threaded through a row of needles across the machine. A downward movement of the needles forces the yarn through the backing. Loopers (hooks) hold back the length of yarn required to make the loop while the needles are withdrawn. The loopers then release the loop to pick up the next insertion of yarn. Where a cut pile carpet is being produced, blades sever the loops as they are formed.

When the stitching process is complete, the carpet backing is coated with adhesive to secure the tufts or loops in position. Usually a secondary backing material is also applied.

Colour and design

A major drawback in tufted carpet production has been the limitations in colour variation and design. However, recent developments with differential-dyeing yarns and colour printing techniques have increased their patterning potential. Sculptured pile effects can be achieved by adding patterning attachments to the tufting machine. Moving the position of the needles creates zig-zag designs. Coloured patterns can be produced by threading the needles with different coloured yarns and programming them to tuft at varying heights.

Tuft density

Slightly different terms are used to express the tuft density of a tufted carpet. Instead of pitch (as in woven carpets), the distance between the tufts widthways is referred to as *gauge* (the smaller the gauge the better). The most commonly used gauges are 5 mm, 4 mm and 3 mm, though carpets are also produced in gauges of 2.5 mm, 2.1 mm or 2 mm. The number of rows 100 mm down the length of the carpet is referred to as *stitches/100 mm* and varies between 16 and 48.

NON-WOVEN CARPETS

This term refers to the miscellaneous collection of carpets which do not involve *weaving* (woven carpets), *stitching* (tufted carpets), *knotting* (the method used for making carpets by hand) or *knitting* a popular method of carpet construction in the US and Europe but mainly used only for rugs in the UK).

The two main manufacturing processes for non-woven carpets are *needlepunch* (also referred to as needlefelt or needleloom) and *bonding*. Neither of these is used to the same extent as weaving or stitching. However, because of their relative cheapness, they are becoming increasingly popular.

Needlepunch

'Endura' (Gilt Edge) is a well known carpet made by this method. Needlepunch carpet is cheap because it uses loose fibre rather than yarn. This means there is no pile but a fibrous surface which is punched into the backing by a series of barbed needles. The needles move up and down very rapidly, entangling the fibres with the backing material. The backing is coated with an adhesive to secure the fibres.

The fibre used can be of one type or a mixture. Some needlepunch carpets have a top layer (wear surface) of hard-wearing fibres—like nylon—and a substrate of jute or polypropylene.

Bonded

Bonded is the technique of using adhesive to stick pile fibres onto a pre-manufactured backing or support fabric. Some of the more common uses of this process are:

Non-woven cord carpets (e.g. Criterion Cord)   Loose fibre is pressed into a ridged fibrous web, coated with adhesive and bonded to a similarly coated backing material.

Vernier method (e.g. Stoddard's Telsax)   The pile yarn is folded between, and bonded to, two parallel surfaces of backing material and then cut through the centre to form two cut pile carpets. This is a cheap method of producing a carpet which is similar in appearance to a plain Wilton but lacking a Wilton's high degree of tuft anchorage.

Flocked carpet The surface pile is composed of short lengths of fibre (usually nylon) which are implanted vertically into an adhesive coated backing fabric by means of an electrostatic attraction between the fibres and the backing. At present, this type of carpet is used mainly in the contract field and for carpets in cars.

Source: Reprinted from G. Clegg and G. Davies (eds), *Carpets,* UK Consumers' Association, 1976.

LEATHER

Leather-making is an ancient art, originating in the Mediterranean region about 3500 years ago. It is a process used to preserve hides, skins and furs. The word 'tan' derives from the crushed bark of an oak tree (called tan), which was one of the first materials found to be capable of converting skin to leather. It was not till the late nineteenth century that it was realised that the tanning was brought about by chemical agents called 'tannins' present within the oak bark.

Today, the leather industry is a large international concern. Until recently there was no substitute for leather products, but the development of cheap, high-quality plastics over the past decade is damaging the industry. Much research is being done into the production process, and there are still many fundamental points to the chemistry of leather production that are not understood. (A good introduction to leather and its manufacture has been given by Kirk and Othmer.[1])

The tanning process is in fact a chemical reaction that takes place within skin. All skins contain a protein called *collagen.* (This is present in other parts of the body as well as the skin; e.g. the cornea is almost pure collagen). Collagen is the structural support for the skin. Collagen contains large sections of the amino acid sequence glycine–proline–3-hydroxyproline. This accounts for about 50% of the molecule. A further 20% contains the amino acids glutamic, arginine and asparagine. Unlike the first three amino acids these are highly polar and contain large, bulky side-chains. The two types of amino acids differ greatly in their physical properties and tend to be found together in different parts of the molecule.

Collagen is not highly branched, and the amino acids form into strands of length about 250 pm (picometres). Each molecule contains four identical sections, of length 64 pm. Each section contains one region of the polar amino acids and one region of the apolar amino acids. Three such strands usually intertwine and the collagen appears as a right-handed triple helix to form a fibril ('little fibre'). In these fibrils the small non-polar regions pack closely together, resulting in a highly ordered crystalline region. The large, bulky polar regions cannot pack so closely, resulting in regions of disorder within the fibril. These open regions are suspect to attack and subsequent decomposition by biological materials, which destroys the collagen and the skin. Tanning makes these regions resistant to bacterial attack and thus preserves the skin.

In skin, the fibrils are intertwined, forming fibres 29–50 μm in diameter which can be seen under an optical microscope. It is these fibres of collagen that give structural support to the skin. Gelatin is just denatured collagen. Gelatin forms large three-dimensional matrices which give structural support to jellies, etc.

Vegetable tannins are mixtures of large polyphenols (molecular masses 500–3000). They are extracted from the bark, roots, wood, leaves and fruit of trees and plants. Important sources of tannins are the barks of wattle, mangrove, oak, eucalyptus, pine and willow trees and the woods of oak, chestnut, poplar and quebracho trees. Tea leaves contain a tannin, tannic acid, as the brown colouring material.

All tannins possess several properties that are necessary for their action as tanning agents. They are all water soluble, and they precipitate proteins from solution.

In adding milk to tea, the tannins and milk proteins react and this results in neutralisation of the astringent taste of the tea. Tannins appear in beer from the barley and in wine from grape skins. The tannins react with protein to form a (non-biological) haze, as occurs in home brewed beer and wine. Wines and beers are clarified with a 'fining' agent such as protein (e.g. gelatin) or a colloid such as bentonite.

Being polyphenols, tannins have many phenolic groups which can form hydrogen bonds. Because of their length they can form hydrogen bonds to places far apart within the collagen. Finally, they are small enough to diffuse through skins and enter into the collagen fibres.

Vegetable tanning

If a skin is placed in dilute acid, the non-crystalline, polar regions of the collagen will be penetrated by the acid. The peptide linkages within this region will be protonated, resulting in the breakage of the hydrogen bonds within this region. The collagen then swells, allowing the tannins to penetrate. The phenyl groups within the tannins are protonated at a much lower pH than the peptide linkages and so the pH is controlled to a point where the peptide linkages are protonated and the phenyl groups are not. The tannins thus form strong hydrogen bonds to the collagen, cross-linking the fibrils in many places. The process is complete when about 50% by mass of tannin has been absorbed. The leather is then resistant to bacterial attack, is much stronger and will repel water. No resistance to heat is achieved.

Chrome tanning

Chrome tanning is performed in a basic solution (pH > 7). The processes involved are still uncertain, but the key step is believed to be the production *in situ* of some insoluble chromium (III) hydrated hyroxides. These either form hydrogen bonds to the collagen fibrils or react with them in some way. The final result is again cross-linking of the fibrils. The process provides resistance to bacterial attack and shrink-resistance to boiling water, but provides no resistance to water penetration. Water resistance is achieved by adding vast quantities of waxes and other substances after tanning is complete.

Synthetic tanning (syntans)

The most important syntans are condensation products of formaldehyde with one of the following classes: naphthalene sulfonic acids, phenols, sulfonated phenols, diaryl sulfones, urea, melamine and dicyanodiamine. They were introduced around the 1920s and were developed during the Second World War. They may act as partial or complete replacements for natural tannins.

The leather shoe—the use of leather

The shoe is a very good example of the use of leather. There are four different pieces of leather within a shoe—the sole, the shoe upper, the innersole and the laces.

The shoe sole must be strong and stiff, with high resistance to abrasion and water. A heavy leather is used here, made from vegetable-tanned cowhide. The shoe upper, on the other hand, must be light and flexible but still strong and water repellent. These are generally made from calfhide that has been chrome tanned. Often a plastic 'patent' paint is applied to the outside surface to improve the finish, change the colour and add water resistance. An innersole leather must be light and spongy, being able to absorb perspiration and not be destroyed by it. A combination tannage is often used here, where the leather is half tanned by vegetable tanning and then tanning is completed by chrome tanning. Lastly, a shoelace must be strong and very flexible and have resistance to water. Chrome tanning is invariably used to tan either calf or kangaroo skin for this purpose.

Because dry collagen will contract rapidly and irreversibly when heated to above about 70°C, the dry-shrinkage temperature of all leathers is about 70°C. When wet, vegetable-tanned leather is heated, the same occurs. However, when wet, chrome-tanned leather is heated, shrinkage does not occur until the temperature reaches 95° to 100°C or the leather dries, in which case shrinkage occurs at 70°C again. This goes some way to explain what happens when wet shoes are placed near a fire or oven to dry. At 90°C the wet shoes may dry properly but, as soon as they do, they start to shrink because they are above 70°C (greater than the dry-shrinking temperature). Because the thin uppers dry and shrink fastest, the shoe curls upwards. Even at 70°C the sole leather will shrink long before the shoe dries, and the shoe will curl down. In either case the result is not a pretty sight.

Some experiments with leather chemistry

The tanning reaction
As mentioned earlier, tanning is brought about by the cross-linking of the large protein collagen by active tanning agents. If collagen is in solution and some tanning agent is added, the collagen will be precipitated from solution.

An easily available form of collagen is animal glue, although its concentration is low. Gelatin is a protein of similar structure to collagen. A concentrated solution of glue can be made using a minimum amount of water. A teaspoon of gelatin in 50 mL of water, heated gently to aid dissolving, provides a suitable alternative solution.

For the tannins, two teaspoons of tea leaves brewed and stewed in half a cup of boiling water and then strained provides an excellent solution. Tannins may be extracted from wattle bark by breaking the bark into small (~5 mm) pieces and boiling it in a beaker of water.

On mixing tannin and protein a bright orange precipitate should result (the glue–bark combination is poorest.)

Shrinkage tests

The shrinkage of different leathers under wet and dry conditions can be readily observed. A 0°–100°C thermometer is required. Take strips of different types of leathers about 100–130 mm long and 5–10 mm wide. The wet-shrinkage temperature can be measured by tying the leather to the thermometer and lowering the thermometer into warm water or by pushing a long nail through the leather and letting it rest on top of the container (see Fig. 7.1).

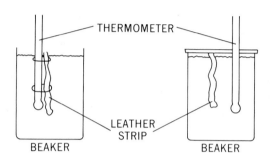

Fig. 7.1 Wet-shrinkage experiment

The leather should be close to the thermometer so that the thermometer measures the temperature of the water near the leather. Heat the water gently whilst stirring to keep the water at an even temperature throughout. Vegetable-tanned leathers will curl up very dramatically at around 70°C. Chrome-tanned leathers will curl up much more slowly, at around 95°–100°C. It may be necessary to leave the chrome-tanned leathers in the water for 30 minutes to an hour before noticeable contraction occurs.

Dry shrinkage can be determined by placing leather strips in an oven with a thermometer. The effect is not as dramatic as wet shrinkage in water: the leather does not shrivel but only contracts in length. All leathers shrink at about 70°C dry. Measure accurately the length of the leather strips and place them in an oven at 75°C for five to 10 minutes. Take them out and measure their length. Return them to the oven for 12 to 24 hours at 100°C. Again measure their length. What does this tell you about the speed of the reaction?

Some model results for three different leathers are given in Table 7.4. The vegetable-tanned cattle hide shrivelled rapidly in water at 67°C. The water became discoloured and was able to precipitate gelatin. This indicates that tannins were freed from the leather. The length and width of the strip shrunk about 25%. The vegetable-tanned kangaroo thong also shrivelled rapidly when the water temperature reached 77°C. The chrome-tanned sheepskin began to curl very slowly at 90°C and was held at 97°C for 30 minutes before the deformation was high.

The only noticeable sign of curling came from the vegetable-tanned cattle hide. The contractions were all about 5% of the length of the leather and occurred within the first few minutes of heating.

TABLE 7.4
Lengths (in mm) of leather strips heated in an oven

Leather	Start	10 min @ 75°C	30 min @ 75°C	11 hours @ 106°C
Chrome-tanned sheepskin	124	119	118	117
Veg.-tanned cattle hide	126	121	120	120
Veg.-tanned kangaroo thong	124	119	118	118

The Australian leather industry[2]

In 1794 an attempt was made to tan kangaroo skins but failed. By 1803 the first two successful tanneries tanning cattle hides were established in Sydney. Demand increased steadily, so that in 1945-46 Australia had 152 tanneries or related establishments employing 5022 people. In that year, the industry treated 3.8 million cattle and calf hides, 2.5 million sheep skins and 0.8 million goat hides. However the industry soon declined. Synthetics moved into traditional leather areas, such as shoe soles, suitcases, and motor vehicle upholstery. Imports increased and by 1978 employment in the industry had been halved.

In 1978-79 (the last year for which figures are available), 90% to 95% of cattle and sheep hides were exported. France was the major buyer of woolly sheep skins, while cattle hides went to Japan and Poland. Hardly any goat skin leather is now produced in Australia and imports have increased particularly from India. Interestingly, India provides a substantial air freight subsidy for its export of finished leather products and does not allow the export of raw hides and skins (except furskin and skins of 'stray' dogs), to keep the price to its own manufacturers below world parity. However, many more of our exports are now only semi-processed. The total value of raw hide and skin exports was $373 million in 1978-79.

Home tanning kits

Quite often one sees tanning recipes appearing in homecraft magazines. These methods tend to be very cheap and although they undoubtedly can be used to tan skins, the quality and life of the finished product are not guaranteed. Treat these recipes with care, especially the shorter ones, as effective tanning is not a simple two-step process.

Glossary of terms

Chamois leather—leather that can absorb and desorb large volumes of water. It is oil-tanned sheepskin.
Chrome tanning—conversion of skin to leather through the application of chromium salts.
Collagen—large protein that gives structural support to the skin.
Fatliquoring—lubrication of a tanned piece of leather.
Leather—skin that has been chemically treated to make it resistant to bacterial attack.

Oil tanning—method of tanning sheepskin to produce chamois leather.
Suede—leather that has been buffed on an emery wheel so that some of the fibres are loosened at one end.
Tannins—chemicals extracted from plants that can tan skin to leather.
Vegetable tanning—conversion of skin to leather through the application of vegetable-extracted chemicals.

Types of leather

The type of leather produced by the tanning process depends on four main factors:
- The *type* of skin to be processed and its pretreatment.
- The *agent used* to tan the skin. There are four main types of tanning process: vegetable tanning; inorganic tanning (e.g. chrome tanning); combination tanning (vegetable and chrome tanning); and tanning by synthetic organic molecules called syntans.
- The *conditions*, such as the temperature and pH of the tanning mixture, and the time that the skins are in the mixture.
- The *coatings, dyes* and other substances applied to the leather after tanning.

Heavy leather. This has high tensile strength and is used for soles, belts, harnesses etc. It is generally made from ox hides, being tanned either by vegetable or combination tanning.

Soft leather. This is flexible and light and is usually tanned by chrome tanning. Upholstery is made from calfskin, whilst sheep, goat and reptile skins are usually made into clothing or ornaments.

Chamois leather. This is leather with an ability to absorb and desorb large volumes of water. It is made by oil-tanning sheepskin.

Furs. These are tanned the same way as leather is tanned except that the hair is not removed. Usually chrome tanning is used.

REFERENCES

1. Grayson, M. (ed.) *Kirk-Othmer encyclopedia of chemical technology,* 3rd edn. Wiley, New York, 1983.
2. This section is based on *Tanned and finished leather: dressed fur,* IAC report, no. 245, 17 June 1980.

BIBLIOGRAPHY

General
Blandford, J. M. and Gurel, L. M. *Fibres and fabrics.* A consumer's guide from the US National Bureau of Standards, 1970 NBS CIS 1.
Buying floor covering. South Australian Department of Consumer Affairs.

'Children's sleepwear', *Consumer Research*, November 1973, 19.

'Core labelling of clothing, household textiles, furnishings, upholstered furniture, bedding, piece goods and yarns'. AS1957, Standards Association of Australia, Sydney, 1982.

Gohl, E. P. G. and Vilensky, L. D. *Textiles for modern living*, 3rd edn. Longman, Sydney, 1983.

'How to buy a carpet'. *Consumer Research Magazine* **23**, March 1974, Consumer Research Inc., Washington, NJ.

'How to choose your carpet'. *Choice* **30**, ACA, May 1986.

Identification of textile materials, 6th edn. Textile Institute, Manchester, UK, 1970.

Industrial production of leather. CIBA Review 1970(1).

Knitted fabrics. CIBA Review 1971(4).

Nordic carpet classification (NCC) marking instructions. Danish Institute for Informative Labelling, Copenhagen, 1976.

Reactive dyes. ICI Educational Publications, 1979. Beautiful production.

Ridley, A. and Williams, D. *Simple experiments in textile science*. Heinemann, 1974. The experiments cover the topics of (1) comfort and safety, (2) appearance, (3) resistance to wear, (4) fibre content. Most experiments call for nothing more than normal laboratory equipment.

'Textile floor coverings—terminology, definition and structure classification'. AS2454, Standards Association of Australia, Sydney, 1982.

Flammability

Australian Standards AS1249(1983); AS1248(1976) and AS1176(1982). Standards Association of Australia, Sydney.

'Children's nightwear. The burning question'. Consumer leaflet no. 4, Standards Association of Australia, Sydney, 1977.

'Fire research: a special issue'. *Chem. Brit.* 23(3), 1987.

'Flaming dangerous'. *Choice,* ACA, July 1982.

Flammability of fabrics. CIBA Review 1969/4. A technical discussion with excellent colour illustrations.

Hall, N. Stop home fires burning. *New Scientist,* 21 January 1988, 30.

Paul, L. 'Behaviour of plastic and rubber foams in fire'. *Chemistry and Industry*, 17 January 1983, 63.

Pressley, T. A. 'Apparel flammability'. *Control* (National Safety Council of Australia, Queensland Division) **1**(1), 1974, 23–33. Very readable, and the basis for the section on flammability.

Pressley, T. A. *et al.* 'Textile flammability testing'. Textile Journal of Australia **48**(2), 1973, 17.

Sacks, T. 'Fire Detectors with brains'. *New Scientist*, 23 October 1986, 42.

Chemistry of hardware and software (III)

8

PAINTS

Before the First World War, practically all pigments, linseed oil, turpentine and varnishes were imported for bulk sale to tradesmen, who mixed their own paint as required. The mid-1920s saw the development of ready-to-use hard gloss paints and nitrocellulose lacquers, and the whole marketing pattern changed—the decorator carried 'colour cards' to arrange colour schemes with his customers and ordered ready-to-use paints from the manufacturers for immediate delivery. The mass-media advertising of the 1950s and 1960s, together with the development of paints that were easier to apply and improved painting techniques, has led to the situation today where 75% of homes are painted by their owners.

Until the early 1950s the *vehicles* used in paints were principally natural 'oils'—tung, fish, linseed—which are to a greater or lesser degree polyunsaturated. Linseed oil was thinned with turpentine and pigmented with white lead (i.e. basic lead carbonate, $Pb(OH)_2 \cdot PbCO_3$) and tinted with one of a small range of colouring agents.

Alkyd resins were introduced into the industry during the 1940s and have since become the basis of nearly all 'oil'-based paints (e.g. house paints, alkyd enamels, undercoats and primers). Water-based paints, known as latex paints, plastic paints etc., based on poly(vinyl acetate) (PVA) or acrylic resins, were introduced during the 1950s and have become increasingly popular. At first these water-based paints were recommended only for interior use but in recent years water-based paints for exterior use have become available. The introduction of easily dispersible pigments ('universal stainers') during the 1960s meant that the retailer need only keep white paint bases and add small quantities of universal stainers to obtain a wide range of colours.

Function

The basic function of a paint, the protecting of a surface from the action of light, water and air, is achieved by the application of a thin, resistant, impervious, flexible film to the surface. The film usually contains pigments to hide and decorate the surface. Thus paints have two basic components:

- The *vehicle*—the liquid part of the paint, which polymerises in some way to provide the bonding and protective film.
- The *pigment*—a solid, suspended in the vehicle, which is opaque, scatters light and colours the film.

PAINT QUALITY IMPORTANT TO ARTISTS

Leonardo da Vinci may not have been the greatest painter of his day, but his superiority in the technology of paintmaking ensured that his works survived while those of many of his contemporaries did not. His technical competence was of course shown in many other spheres.

The protection of a film of paint is remarkable considering it is only about 25 to 50 μm thick. Compare this to the thickness of newsprint, which is about 75 μm.

Basic terminology

Chalking. Loose pigment powder on the surface of a weathered film left by erosion of the outer layer of binder under action of ultraviolet light. Some chalking is desirable to give a self-cleaning surface.

Checking. Slight fine breaks in the surface of a film visible to the eye or under $\times$ 10 magnification.

Cracking. Breaks in the paint film that extend from the surface to the underlying material.

Extender pigments. Pigments which, in themselves, have little hiding power but which enhance the properties of the paint by filling voids in the film, spreading the main pigment, improving brushing quality etc. They are usually natural carbonates and silicates (clay, talc, gypsum, silica).

Flooding. Also known as floating. A defect sometimes occurring in paint films involving the separation of individual pigment particles, thus giving a non-uniform colour.

Flow. The ability of an applied paint film to level out and produce a smooth coat.

Hiding power. The ability of a pigment to hide a surface depends on its refractive index (see below) and on the particle size. The measure of hiding power is the area of a black-and-white check design obscured completely by a pound of pigment.

Mineral spirits (turpentine). A petroleum fraction boiling between 150°C and 200°C, containing aliphatic hydrocarbons. Evaporates from paint as it dries.

Pigment. Provides colour and opacity to dried film. Properties depend on chemical composition, refractive index and size of particles.

Primer. Paint intended as the first coat on a surface.

Refractive index. Ability of a material to bend a ray of light: the larger the value, the greater the refraction (e.g. water, $n = 1.33$; crown glass, $n = 1.5$; diamond, $n = 2.42$). It is related to the density of the material.

Thinner. Usually a volatile solvent added to facilitate application of the paint.

Thixotropy. The property of a liquid or gel to lose viscosity under stress and regain the gel state when the stress is removed. Thixotropic paints flow when the brush is moving but should not drip from the brush. (See also box 'Fun with fluid flow'.)

Varnish. A solution of a natural or synthetic resin in a solvent, sometimes with the addition of a drying oil.

The protective film

There are two ways of establishing the flexible film on the surface to be protected:
- Apply the monomer and allow it to polymerise *in situ.* The old oil-drying paints operated on this principle but curing times tended to be long.
- Apply a suspension of a high molecular mass polymer in a solvent and allow the solvent to evaporate. This method, which is faster, is utilised in the modern plastic paints. Both oils and alkyds cure (convert from liquid to solid form) by the same mechanism.

Oil-drying paint

The monomer is linseed oil or some other unsaturated oil (tung, soya, castor, menhaden etc.) These oils are mixtures of long-chain, unsaturated fatty acids such as linoleic acid (see also Fig. 3.3).

$$CH_3—(CH_2)_4—CH=CH—CH_2—CH=CH—(CH_2)_7—COOH$$

The double bonds can react with oxygen on exposure of the oil to air, leading to the formation of peroxides and eventually to polymerisation by cross-linking at the active oxygen sites. The cross-linking can continue until there are two links for every double bond in the original oil, which is far beyond the point required for good film properties. In addition, such materials tend to degrade under the influence of the ultraviolet component of sunlight. Yellowing and a tendency to become brittle are two principal ageing deficiencies. The natural oxidation is slow but the rate is enhanced by the use of blown linseed oil (air is blown through the oil heated to 130°C) or boiled linseed oil (oil is heated to 260°C in the presence of litharge, PbO, giving a lead soap, which acts as a drier). Paints that contain lead soaps (at less than 1%) but no other lead compounds, such as pigments, are legally non-lead paints. (Lead and other metal soaps with variable valency (e.g. cobalt and manganese) catalyse the uptake of oxygen and, with the exception of lead, the decomposition to free radicals.)

The old oil-based paints had the advantage of being cheap and adhering well to porous surfaces such as wood, paper and plaster. In addition, the film was flexible enough to cope with the swelling and shrinking of wood with changing humidity. The films have a degree of porosity, which allows the escape of a certain amount of moisture through the walls. However they are slow drying (several days between coats) and tend to weather because the surface continues to react, which leads to checking and gives a surface unsatisfactory for repainting. Paints with a linseed oil base were not suitable for metals.

THE ROMANS HAD A WORD FOR IT

On some tubs of polyunsaturated margarine sitting on our breakfast table, one of the ingredients listed will be linoleic acid. This is one of the polyunsaturated fatty acids we eat in the hope of reducing the cholesterol deposited in our arteries (see Chapter 3). Linseed oil and linoleic—can you see the connection? The only oil the Romans knew about was olive oil. The Latin for oil, *oleum* (we use oleum for oily fuming sulfuric acid), gives us oleic acid, the main component of olive oil. So linoleic is the oil of linseeds. What about the word linoleum? Good old 'lino' was made by pouring linseed oil over hessian (with cork and a few rosins etc thrown in) and allowing it to harden in air (like putty and oil paints, also made from linseed oil). Today we use plastics like PVC and, until very recently, asbestos matting. What about the linen cupboard? Unless you have very expensive tastes you should really call it a cotton cupboard. Linen comes from the flax plant (genus *Linum*), whose seeds are linseeds! And as for lint? This is just the linen (cotton) scrapings.

Plastic or latex paints

As mentioned above, plastic or latex paints contain highly polymerised resins. Solutions of such polymers are too viscous to be readily applied and so the paints are formulated as emulsions of the resin solution in water. These emulsions are known in the industry as *latexes*, after the natural, milky, emulsified juices of certain plants (e.g. rubber and dandelions). Latex paints have several advantages over oil-based paints: they dry quickly (several coats can be applied in rapid succession without having to dismantle and re-erect scaffolding); they lack a persistent odour; they are water-soluble (hence spills can be cleaned up easily and brushes and rollers can be washed in water); the vehicle at least is non-flammable and non-toxic; and they can be used to paint on damp surfaces. On the other hand, they tend to be low in gloss and rather soft; hardness is achieved by increasing the amount of pigment or other additives. Their composition varies with the intended use (exterior or interior, nature of surface) and the degree of gloss required. Exterior paints may contain poly(vinyl acetate)(PVA), based on the monomer vinyl acetate, or acrylic polymers based on the monomer methyl methacrylate (Fig. 8.1).

Fig. 8.1 Vinyl acetate and methyl methacrylate

Latex paints generally contain a high proportion of resin to provide a film that is stable to weathering. Gloss paints also have higher resin : pigment ratios.

Interior paints may also contain PVA or styrene-butadiene resins (Fig. 8.2), and they have high proportions of pigment.

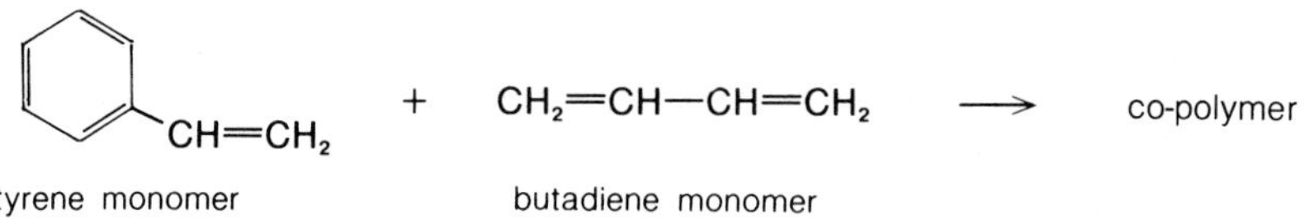

$$\text{styrene} + CH_2{=}CH{-}CH{=}CH_2 \longrightarrow \text{co-polymer}$$

styrene monomer   butadiene monomer

Fig. 8.2   Styrene and butadiene monomers

To improve washability and adhesion, up to 15% of drying oil or alkyd resin is added. One of the alkyd resins used for this is Glyptal (trade name of a glue), which is formed by the reaction shown in Figure 8.3.

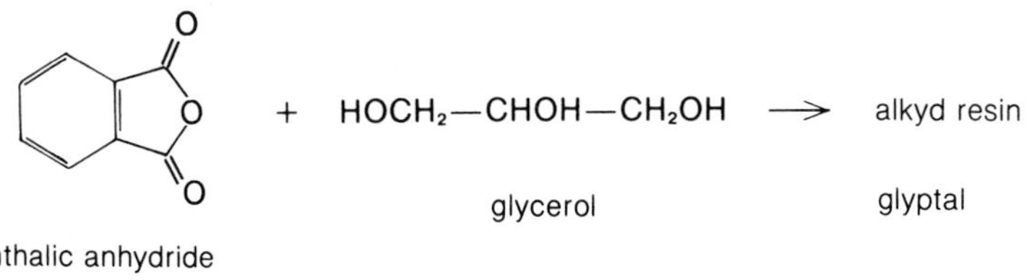

$$\text{phthalic anhydride} + HOCH_2{-}CHOH{-}CH_2OH \longrightarrow \text{alkyd resin}$$

glycerol   glyptal

phthalic anhydride

Fig. 8.3   Formation of Glyptal

Plastic paints contain a number of additives to stabilise and thicken the emulsion (sodium methacrylates, carboxymethyl cellulose, clays, gum arabic), to assist in dispersion of the pigment (tetrasodium pyrophosphate, lecithin), to reduce foaming, and to preserve the paint. In more humid areas, anti-mould agents become essential. (This is true also for products such as fibro-cement.)

Polymerisation of alkyd resins with polyamides gives *thixotropic* (non-drip) paints, which have a gel-like consistency when standing but which flow under stress when applied with a brush. The degree of thixotropy must be modified so that the brush marks can level out before the paint re-gels. The opposite effect, namely an *in*crease in resistance to flow with stress is called *dilatant.*

There is a paint made by ICI called Tempro which was developed as a temporary cover to ensure that cars arrive in the showroom without marks or scratches. It is then washed off in a car wash using mildly alkaline solutions. This paint has found other uses such as changing the camouflage of military vehicles temporarily from green and black to white during winter exercises, or painting temporary tracks for city races on roads.

An advantage of water-based paints is that the carboxylic acid binder is negatively charged and the paint heads for the anode under an electric field. (Ammonium salts are actually used.) This electrodeposition process is used extensively for car bodies. More recently positively charged paints are used giving cathodic deposition and better corrosion resistance.

FUN WITH FLUID FLOW

The change in flow properties with applied pressure can be exploited in making fun materials. The examples described here originally came from David A. Katz of the Community College of Philadelphia, Pennsylvania, USA.

Slime

This delightful material is a reversible cross-linked gel made from guar gum. The cross-linking occurs through the addition of borax. Guar gum is described in Merck (9th edn) as a vegetable gum (mol mass about 220 000) used as a protective colloid, stabiliser, thickener and film-forming agent for cheese, salad-dressing, ice-cream and soups (see *Model Food Regulations*, NH & MRC, Canberra, 1983, entries under Additives, Vegetable gums; see also Chapter 11) and as a binding and disintegrating agent in tablet formulations, creams, toothpastes etc.

To prepare Slime, try to obtain 'viscosity builder' grade gum from a food or cosmetic company.

Disperse 0.75–1.0 g of guar gum in 20 mL of water. Dissolve 0.75 g of borax in 20 mL of water. Also boil 60 mL of water and then remove it from heat. Then add the 20 mL suspension of gum to the hot water while stirring. Now add borax solution while stirring and allow the mixture to cool. The natural colour of the solution is green, the intensity of which depends on the grade of guar. The quantity of guar needed depends on the quality of the gum. Store the thickened Slime in a tightly closed container to prevent it from drying out.

An alternative recipe involves using a 3% solution of poly(vinyl acetate) with a molecular mass (MM) of 70 000. As PVA degrades at the boiling point of water (i.e. it breaks down to give a lower MM polymer) the mixture is dissolved in boiling water quickly and then cooled rapidly. To this is added 10 mL of a saturated borax solution (plus food colourings). The mixture is stirred slowly. For polymers of lower MM (e.g. 10 000), higher concentrations (e.g. 5%) are needed; however these lower MM do not work as well.

Slime belongs to a class of materials that do not obey the usual laws of viscosity and are called non-Newtonian fluids. Another example is Silly Putty, which is a silicone polymer.

A low stress, such as slow pulling, allows Slime to flow and stretch and you may even be able to form a thin film; a high stress such as pulling sharply will cause a break. If Slime is poured from its container and the container is then tipped upwards slightly, the gel will self-siphon. Try cutting the pouring stream with scissors. Hitting a small piece of Slime with a light hammer will not cause splashing or spattering and the material will even bounce to a small extent. If pushed through a tube, Slime will emerge with a swell (known as die swell in the plastics extrusion trade).[1]

Other fun materials

There are a number of other toys based on this type of behaviour. For example, a Bad Case of Worms (Mattel Toy Co.) is made from a Shell isoprene polymer which is a thermoplastic elastomer that is plasticised and tackified. When the worms are thrown against a smooth, clear surface such as a wall they stick, but after a while the worms start to 'crawl' down the wall. The rate of progress depends on the cleanness of worm and wall. The worm is soap-washable.

A Magic Octopus works the same way and is probably a styrene-butadiene co-polymer. Excess plasticiser can leave an 'oily' residue that is difficult to remove.

Shrinky Dinky

These are plastic films on which you draw pictures. They are then placed in an oven for four minutes at about 165°C. The film shrinks to about one-third all round, thereby becoming nine times thicker. The films are made from a polystyrene film that has been extruded under stress to become oriented in both planes (directions). You can check the composition of the film with in-frared spectroscopy (see Experiment 13.15).

Viscosity

To understand a little more about viscosity consider the following experiment. Take a tall glass with a small amount of honey in the bottom. If the glass is tipped on its side with the open end facing slightly downwards, then the honey will begin to flow. The molecules closest to the wall will be attracted to the surface and will be held back from flowing. These molecules will in turn slow down the molecules in the next layer trying to flow past. The process of each layer slowing down the next layer extends throughout the whole fluid. The fluid moves in layers like a pack of cards being pushed along a table by a flat hand on top of the pack. The card in contact with the table hardly moves at all (see Fig. 8.4).

Fig. 8.4   Pushing a pack of cards

If the forces between the molecules are weak then this drag does not extend very far from the surface, but for strong forces the drag is transmitted further into the liquid.

Increasing temperature reduces the attraction between molecules and hence reduces the viscosity of liquids.

For non-Newtonian fluids the viscosity also depends on the rate of shear, sometimes increasing with shear (dilatant), sometimes decreasing (thixotropic), sometimes behaving in a complex combination of both, depending on the way the force is applied and for how long. The mechanisms responsible for the different behaviours can be quite diverse.

Have you ever experienced the 'shear' pleasure of running along the wet sand near the water's edge at a beach? The sand feels as solid as a footpath and your feet hardly mark it at all. If you stand for even a short time you will sink in a little leave an obvious temporarily dry footprint. Running on dry sand is more difficult, because the sand collapses rapidly under your feet. When water is added to dry sand, it acts as a lubricant and the sand grains pack together much more efficiently.

However the sudden shear force of the feet of a runner on wet sand forces the grains past each other, forming cavities which do not have time to be filled with lubricating water. The resultant friction hinders movement of the grains. If you stand still on the other hand, the water can move in, lubricate, and hence allow the sand beneath your foot to reshape.

If you are in the kitchen, rather than at the beach, you can demonstrate the same effect with a mixture of 2 : 1 (by volume) of cornflour and water. This slurry has a low viscosity to slow stirring with a spoon, but will instantly thicken to almost a solid paste if stirred quickly, becoming fluid again when the stirring slows.

Although Slime behaves in a similar manner, the mechanism relates to the behaviour of the polymer *molecules* in solution rather than *macroscopic* solid particles, as is the case for sand grains and flour.

The natural shape of polymer molecules is like a random coil. This shape results from an entropy effect. Random has a special meaning here. It means there are many more possibilities for throwing down a chain that looks tangled in roughly the same way than putting it down in a stretched out manner. So being coiled in a variety of ways is the most probable shape a polymer molecule will take (see Appendix VIII, 'The Entropy game').

For example the molecules in rubber are randomly coiled. When rubber is stretched the molecules are unwound. The elasticity of rubber comes about from

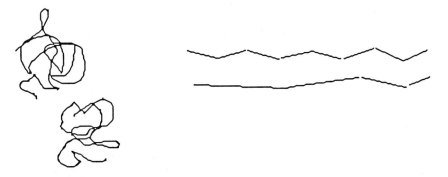

Fig. 8.5 Randomly coiled and stretched rubber molecules

the tendency of the molecules to return to the random coil shape—the condition of maximum entropy. In Slime, a rubber-like polymer molecule is dissolved in a solvent. Shear forces can unwind the coils and the molecules then have a greater chance of entwining and this increases the viscosity of the solution temporarily. With the shear force removed, the molecules move back to random coil, and this moving back is manifest macroscopically as the elastic behaviour seen in the experiments on Slime. A stream of Slime pouring out of a container will be stretched under its own weight and, when you cut it, the upper cut portion *re-coils* (literally) elastically. You cannot cut a stream of honey.

Thixotropic fluids are easier to understand. They all have some loose secondary structure which binds them sufficiently to give the initial high viscosity.

You don't have to get exotic paint formulations to see the effect; just take out your tomato sauce bottle from the pantry. If you tap or shake the bottle, the shear forces break the secondary structure and lower the viscosity to the point where the contents may flow so well that they spread themselves around the kitchen. It is the weak hydrogen bonds in the starch gel that provide the secondary binding in tomato sauce.

Shear thinning is used in many food and cosmetic products. When making a sandwich by spreading a solid onto bread, the viscosity at the shear rate of about 100 mm per second (appropriate for a hurried early morning lunch maker) must not be much higher than 10–30 poise (the units in which viscosity is measured, not the composure of the hurried lunch maker), if the bread is not to tear. Some idea of the magnitude of viscosities (in poise) are given by the following: marshmallow creme 10 000; honey 100; motor oil 1: water 0.01 and air 0.0001.[2] Peanut paste (butter) is much harder to spread than margarine, because its viscosity is much higher at the shear rate of spreading. Other products with similar shear thinning properties to margarine are mayonnaise and mustard.

Next time you are stirring a bowl of tomato soup, watch what happens when you suddenly stop stirring. It shows Slime-like elasticity and may re-coil with a slow reverse spin.

Imagine toothpaste with the properties of honey! It would be too viscous to squeeze from the tube, but not viscous enough to stay on the brush. Ink held in a ballpoint pen is much more viscous than honey, but on shearing with the ball in the nib it flows on to the page, setting when the pen has passed by. Higher temperatures lower the viscosity and ballpoints can leak under these conditions. More seriously, higher temperatures make engine oils less viscous and the friction protection they offer falls dramatically. Polymers are added to increase the viscosity and so compensate for this change with temperature (see 'Chemistry of the car—Lubricating oils', Chapter 10).

The children's fun material Silly Putty is a silicone polymer called a polyborosiloxane and has more complex time-dependent viscosity properties. Its preparation has been described,[3] but is not recommended except by competent chemists.

A silicone consists of chains of Si-O-Si-O-Si-O- with methyl groups attached to the silicon (see Fig. 6.12). However in this polymer about 5% of the silicon atoms are replaced by boron, and these boron atoms can cross-link weakly with oxygen atoms in other chains. Silly Putty can be made into a ball as if it were plasticine because the cross-linking bonds readjust. However when such a ball is dropped on the floor, the bonds have no time to break, and the material is elastic.

When the material is pulled apart rapidly, or hit with a hammer, the shear force is much greater than on bouncing and now all the bonds break at once and the material behaves as if it were brittle. When the material is left on a table for a long time, the weak long-term shear exerted by merely sitting under its own weight is sufficient to cause the bonds to rearrange and the material to flow.

When cooled in liquid nitrogen ($-197°C$),[1] all the bonding effects are locked and a Silly Putty ball will bounce and sound like glass. As it starts to warm up, all the shear is concentrated on the thin warmed layer on the outside of the ball, which is sufficient to cause bond breaking and energy absorption. It behaves like a golf ball soaked in honey and stops bouncing. As the entire ball warms up, the shear force is distributed over the whole mass again and the force per unit mass of material drops. We are in the elastic range again. Quite a complicated piece of fun.

Solid paints

Paints consisting of 100% solids (such as epoxies and polyesters) are sprayed onto metal surfaces and held electrostatically. They are then heated to cause the material to flow and coalesce into a continuous film, at the same time causing chemical cross-linking in three dimensions. These products find use in coating aluminium frames and heavy-duty machinery.

Chalking

Under the influence of sunlight, oxygen and water, paints tend to degrade by the long chain of the polymeric binder breaking down into shorter chains. This process continues until pigments at the surface of the paint film become liberated as 'chalky' residue (Fig. 8.6). This erosion continues until the entire paint film has weathered away, a process requiring anything from one to 20 years. Although this erosion is advantageous in white paints as it provides a self-cleaning attribute, it is most undesirable in coloured paints, because it gives the impression of fading. (Coloured pigments are lost from the chalky layer more rapidly than white ones, resulting in a gradual fading). Resistance to chalking can be improved by changing the monomer used to produce the polymer.

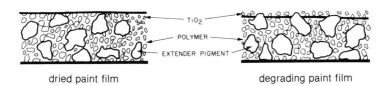

dried paint film   degrading paint film

Fig. 8.6 Schematic representation of a dried paint film. The paint film degrades by exposure to sunlight, water and air. The pigment particles are liberated at the surface giving the appearance of a chalky residue (From F. L. Floyd, 'Emulsion polymers in coatings', *Chem. Tech.*, August 1973, 484, with permission of the American Chemical Society)

Flexibility

Polymeric substances are glass-like in their mechanical behaviour as a function of temperature. Like glass (silica), an inorganic polymer with SiO_2 repeat units, these polymers do not 'melt' in the sense that crystalline materials do. Rather, they exhibit a characteristic 'glass transition' region, in which they pass from a hard, brittle substance to a soft, rubbery, elastic substance. Hardness is at a maximum below the glass region and flexibility at its greatest above. These two properties must be balanced.

Hiding power

The hiding power is associated with the difference in refractive index of pigment and polymer. The reflectivity F is an approximation for the hiding power:

$$F = \frac{(n_1 - n_2)^2}{(n_1 + n_2)^2}$$

where n_1 is the index of refraction of pigment and n_2 is the index of refraction of polymer. Rutile (titanium dioxide, (TiO_2) is the most commonly used white pigment because of its high index of refraction (2.73). Hiding is therefore maximised by minimising the refraction index of the polymer. This is accomplished using branching side chains in the monomer so that the polymer density is decreased through poorer packing. A polymer of refraction index $n = 1.0$ would provide a 40% increase in opacity over an acrylic polymer/rutile system (acrylic, $n = 1.48$). In other words, the expensive TiO_2 can be correspondingly reduced while still providing equal hiding power, thus generating an enticing increase in profits. The manufacturer of one major brand of paint has been using small polymer spheres as the 'pigment' to give a white paint. This saves on the use of titanium dioxide. However, the lower density polymer may have other undesirable properties. Contrast the process of maximising hiding power by maximising the difference in refractive indices with the design of translucent toothpastes (Chapter 4) and translucent urea-formaldehyde filled resins (Chapter 6), where the refractive index difference is minimised.

The traditional white pigments added to paints were white lead (basic lead carbonate, $Pb(OH)_2 \cdot PbCO_3$) and lithopone (30% ZnS + 70% $BaSO_4$). Today, the major white pigment is titanium dioxide (TiO_2), which exists in two forms: *rutile* and *anatase*. (Rutile has the higher refractive index and hence the higher obliterating power but reflects a light that is slightly yellow.) Besides its other advantages, TiO_2 also promotes the self-cleaning of the paint by chalking. Paints sometimes contain mixtures of the two forms in varying proportions to achieve a compromise of their best features. Examples of coloured pigments are given in Table 8.1.

Particle size

Levelling (disappearance of brush marks after paint application) improves with increasing particle size. If the particles are large they are encapsulated by the polymer rather than being stuck together. This improves levelling because the flow properties are improved. Gloss, on the other hand, declines with increasing particle size.

DISAPPEARING TRICKS

The index of refraction of Corning 7740 (Pyrex™) glass is 1.474 while that of pure glycerol is 1.4746. When two pieces of glass tubing, one soda and the other pyrex, are placed in pure glycerol the soda tube is seen shifted by parallax, while the pyrex 'disappears' completely from view (see Plate VII). Glass blowers use this effect to distinguish glass types. (See also the discussion on hiding abrasives in clear toothpaste, 'Chemistry of cosmetics—Toothpaste' in Chapter 4.)

Another experiment is to pour two immiscible liquids of different densities on either side of a U-tube. The system stabilises with the liquids on each side at a different height. By matching the refractive indices, the interface cannot be seen. Plate VIII shows such an experiment with carbon tetrachloride (care, toxic vapour!) in one arm and glycerol (to which ~6–7% water has been added) in the other arm. (You may need to measure the refractive index of the glycerol mixture and adjust the amount of water added to obtain an exact match.) The interface is completely invisible. Adding some iodine crystals, which dissolve only in the CCl_4, shows the two liquids. When the iodine was completely dissolved, I was surprised to see a short column of virtually uncoloured liquid on top of the carbon tetrachloride. What had happened was that during the demonstration I had shaken the tube backwards and forwards in response to a suggestion from the audience to prove that there was no careful layering. Because of the invisible interface, I had not noticed some glycerol moving across to the other arm. Actually it is a good idea that it is there, because it prevents the evaporation of toxic carbon tetrachloride vapours. For this reason, the arms of the U-tube are stoppered when not in use.

A close inspection of the interface shows that it is curved and seems to slope away from you, no matter from what angle you look at it.

Note the lack of correspondence between density and viscosity. Carbon tetrachloride is more dense but less viscous than glycerol. A safer set of liquid is provided by glycerol and paraffin oil (with about 8% decane added to match the refractive index).

A grease spot (butter, for example) on paper becomes transparent because the refractive index of the grease matches that of the paper fibres. What could be simpler?

Paint procedures

Many consumer products are made from prepainted steel strip, called 'coil coating' because the strip is transported as a giant, painted coil. When refrigerators, freezers and washing machines are stamped and pressed out of this steel, there is a tremendous stress applied to the paint. Because different parts of each item come from different parts of the coil, there has to be a high level of consistency in paint colour. In the older process of painting after manufacture, it

TABLE 8.1
Coloured pigments

Colour	Inorganic	Organic
Black	carbon black copper chromate/MnO_2	
Yellow	Pb, Zn, Ba chromates cadmium sulfide iron oxides (ferrite yellow)	azo (arylamide) yellows nickel azo yellow
Blue/Violet	ultramarine Prussian blue cobalt blue	phthalocyanin blue indanthrone blue carbazole violet
Green	chromium oxide mixtures of yellow and blue	phthalocyanin green
Red	red iron oxides cadmium selenide red lead chrome red	azo pigment dyestuffs and toners (e.g. toluidine red—fades badly in tints) quinacridones, perylenes

Source: Based on W. M. Morgans, 'Paints and plastic coatings', *Education in Chemistry* **10**(1), 1973, 12.

was easy to keep each item looking uniform and it did not matter that no two looked exactly the same!

Paint removers

Paint removers must act to remove the paint film, or to swell and soften it so that it can be easily removed by scraping or flushing with water. Dichloromethane is the most common organic-based paint remover. It is a small, polar molecule and can penetrate the coatings rapidly. Co-solvents are sometimes used to extend the range of usefulness on coatings that are little affected by dichloromethane alone. For example ethanol/dichloromethane enables shellac coatings to be removed. To inhibit the decomposition of chlorinated hydrocarbon solvents various additives are used (amines, epoxides and aliphatic alcohols).

Dichloromethane and many other solvents are volatile. A wax (usually stearin or ceresin) is added at a maximum concentration of 1–3%. The rapid evaporation of the solvent, and consequent cooling, causes the wax to be thrown out of solution to form a continuous surface film that keeps the solvent on the paint. Thickeners are added so that the remover can be used on vertical surfaces, and to provide a solvent reservoir above the paint film (metal and amine soaps, proteins, starch, clay etc.). Wetting agents are used in flush-off strippers. For low-resistance coatings, cheaper solvents can be used (e.g. for lacquers, methyl ethyl ketone, which incidentally will dissolve epoxy resins such as Araldite). Caustic soda can be used as a paint remover but is not recommended.

Additives that increase stripping rates in caustic alkali solutions are sequestering agents such as gluconates. They speed up the removal of the paint film by helping to dissolve oxide pigments, thus creating holes in the coating. Once the coating has been breached, the sequesterant helps to remove the coating of oxide, phosphate etc. from a metal surface, thus loosening the film from beneath.

Surfactants are also added. Paint removers of this type are restricted to use on steel substrates, but removers based on sodium metasilicate can be used on aluminium and other non-ferrous alloys. The alkali helps to hydrolyse the ester linkages present in many polymers.

The composition of various *frypan and oven cleaners* is very similar to the paint strippers described here, as they carry out the same sort of job. Paint strippers should *not* be used on frypans because some of the residues may be toxic.

Toxicity

The toxicity of a paint depends on two factors:

- The chemical composition. For example, compounds of metals such as lead, cadmium, antimony and barium are generally poisonous.
- Solubility. A substance that is *insoluble* in the body fluids, even though it is a compound of a metal whose soluble salts are toxic, will pass through without injury. For example, barium sulfate is used to provide contrast in X-ray pictures, and is taken as a barium meal. However, white lead is *soluble* in the body fluids, rendering it highly toxic. Legislation was passed after the Second World War to prohibit the use of 'soluble' lead in paints.

Red lead and zinc chromate are used in primer coats; they protect metal surfaces against corrosion. Primers are always thin, because the thinner the layer the more the adhesive forces overcome the cohesive forces. The molecular contact must be close; hence the need for a very clean surface and, preferably, a large surface area. The latter is achieved by sand-blasting to provide ridges and valleys. However, multiple coating increases the labour and the need for dilute, low solid-content paints increases solvent costs and pollution.

Solvents

In Los Angeles County, the largest single source of air pollution is the automobile. However, in 1966, solvents released from surface coating operations (i.e. all kinds of painting) constituted the largest single *stationary* source of air pollution. Photochemical smog is formed by a sequence of reactions in which sunlight, oxides of nitrogen and oxygen, and reactive organic chemicals are the chief ingredients (see also Fig. 10.16). A regulation (of Los Angeles County) classifies organic solvents according to their degree of photochemical activity, and controls their emission.

Special-purpose paints and coatings

Heat-resistant paints

Ordinary paints blister, then char and disintegrate at relatively low temperatures. Coatings for ovens, heaters, stills, engines, turbines etc. must withstand much greater heat. Paints with silicone vehicles, metallic powders (Al, Zn, Sn to reflect and conduct heat away) and heat-resistant pigments (Cr_2O_3, Ti_2O_3, C, TiO_2, CdSe) are widely used for specialist applications. Polymers of alkyds with saturated fatty acids give non-drying paints that can be baked on and are called oven or baked enamels. These must be contrasted with vitreous or porcelain enamel (see 'Vitreous enamel' later in this chapter).

Fire-retardant paints

The addition of a variety of compounds to paint renders it less flammable. These compounds (phosphates, tungstates, borates and carbonates) decompose on heating to give gases that do not support combustion and hence tend to extinguish the flames.

Alternatively, a substance that fuses on heating to give a glass-like layer on the surface can be added to the paint.

A third approach is to use non-flammable constituents—silicones, chlorinated resins, mineral powders. Water-based plants are usually non-flammable before they are applied, but become flammable when dried out.

Anti-fouling and insecticidal paints

Anti-fouling paints have much application in marine construction. They usually contain inorganic poisons (mainly Cu and Hg salts) or organic molecules (such as pentachlorophenol). More recently organo-tin groups have been directly incorporated into the polymer. These have been found to be very effective against marine organisms while at the same time rapidly decomposing to non-toxic inorganic tin on release into the sea water. DDT and other insecticides have been added to paints to control insects.

Luminous paints

Luminous paints fall into two groups:

- *Fluorescent* paints absorb ultraviolet radiation and re-emit it as visible light only while being irradiated. They contain zinc and cadmium sulfides, together with organic dyes.
- *Phosphorescent* paints are irradiated with ultraviolet light and continue to glow in the dark for some hours after the irradiation has ceased. Phosphors include ZnS (green, yellow, orange) or CuS and SrS (bluish). Additional salts can be used to change the colour.

EXPERIMENT 8.1

Preparation of ferric tannate

Equipment

Tea bags, saucepan, two glass jars, a pad of steel wool, a small quantity (100 mL) of vinegar, about 1 mL of a 3% hydrogen peroxide solution, filter funnel, cotton wool.

Procedure

Prepare a solution containing tannic acid by adding boiling water to fresh tea leaves or a tea bag. Make the solution equivalent in concentration to strong black tea. One cup of solution will be sufficient.

To prepare a solution containing iron (III), boil a small quantity (~100 mL) of vinegar to which a pad of steel wool has been added (the steel wool should not contain any soap). Allow the solution to simmer for 5 to 10 minutes, and then strain the solution through a filter funnel containing a loosely fitting plug of cotton wool.

When the solution is cool, add about 1 mL of hydrogen peroxide solution. The colour of the solution should now be a dark brownish-red, indicating the presence of iron (III).

To produce ferric tannate add a quantity of your tannic acid solution (say 10 mL) to a roughly equal amount of your solution of ferric ion. The solution should turn black as a result of ferric tannate being produced. Ferric tannate is the major ingredient of many black inks.

Reactions

$$2H^+ + Fe \rightarrow Fe^{2+} + H_2$$

$$2H^+ + 2Fe^{2+} + H_2O_2 \rightarrow 2Fe^{3+} + 2H_2O$$

$$Fe^{3+} + \text{tannic acid} \rightarrow \text{ferric tannate}$$

Notes

1. Hydrogen peroxide is available from pharmacies and is used for cleaning wounds. It is a strong oxidising agent and should be handled with care. It is also poisonous if swallowed, so it must be used carefully. Before you add the peroxide to your solution, make sure the solution has cooled to near room temperature. Don't spill the peroxide, because it is a powerful bleach.
2. Ferric tannate may stain, so be careful to avoid contact with your clothing.

(See also 'Leather' in Chapter 7 for an experiment using tannic acid and gelatine.)

ADHESIVES

Nebuchadnezzar used bitumen in the construction of buildings in Babylon about 3500 years ago, and it may have been used on the Tower of Babel. The ancient Egyptians used gum Arabic (and hence the name), egg white and animal glues for furniture. Papyrus sheets were joined into folios with flour and water paste. Many skills were lost in the Dark Ages but by the sixteenth century animal glue was used for furniture again. Apart from some improvements, there was little further development until the twentieth century. In the First World War, plywood bonded with casein and blood albumen was important for boat and aircraft construction. The Second World War saw the rapid development of synthetic adhesives. By 1981 the total western European consumption of adhesives was estimated to have reached two million tonnes dry weight, at a value of 1300 pounds sterling per tonne. An example shows why. Boots were once held together by a combination of sewing and nailing. Today, 90% of the construction is by adhesives.

The term 'adhesive bonding' can be used to refer to a variety of operations— sealing envelopes, applying bandages and repairing torn paper with cellophane tape are but a few common examples. Adhesion involves the fastening together of solid materials by a thin, generally continuous, intermediate layer.

Today's hardware shops provide a bewildering assortment of packages and types of adhesives which involve different types of adhesive bonding (see Plate IX). More elaborate surface preparation is required than for mechanical fasteners, which function quite well with dirty surfaces. The mechanical strength of most adhesives takes time to develop and pressure is generally required to ensure the necessary close contact. Adhesive bonding can be 'irreversible' in the sense that the structure can be disfigured or destroyed if pulled apart.

Adhesives make simple, strong and cost-effective joints. They hold attractions for both children and engineers. For fixing together the components of the European Airbus, for example, phenolic and epoxy adhesives were chosen because of the cost savings when many simple parts are fixed together simultaneously rather than consecutively. A complete adhesive-bonded section of a McDonnell Douglas YC15 aircraft 12 metres long was constructed and tested for 120 000 hours without the catastrophic failure that would have occurred with a riveted structure. However, unlike traditional ways of making joints, such as welding, riveting, bolting or brazing, there is no established method of predicting the strength of a bonded joint, although the aerospace industry has obviously developed some confidence in them for complex structures involving many materials.

Let us attempt to define what we want in an adhesive.

There are many sticky materials that we would not consider to be good adhesives—chewing gum, for instance. To be predictable and consistent, an adhesive must stick more strongly to the substrate(s) than to itself, i.e. it must be stronger in adhesion than cohesion. Thus a bonded joint should fail only in the breaking apart within the adhesive or within the components, but not at the interface between the component and adhesive. Epoxies are brittle and 30 times stronger than two-part acrylics, which are weaker, more flexible and stretch before failing.

The strength of a joint depends on a number of factors, including the thickness and strength of the components. The strength is not simply proportional to the overlap and a point is often reached where there is no increase in strength with further overlap. There are expert system programs for designing joints, e.g. ADJOIN from Pera's Material Engineering Group in the UK.

Adhesives have been found to differ over a factor of 30 in strength, 180 in stretch at fracture, and 300 in modulus of elasticity. The descriptions in the literature can be misleading. Data sheets sometimes quote 'shear modulus' as an indication of strength, whereas it is only a measure of stiffness. The toughness of an adhesive, on the other hand, is probably less crucial than other properties such as yield strength, and the toughness of the components, which occupy a greater volume than the adhesive, has a greater effect on the behaviour of the joint. Lap joints are stronger when they are bent than when they are subject to shear stresses.

As aircraft are not subject to very rough treatment, the brittleness of epoxy adhesives is not a critical factor. However for wider uses incorporation of a little elastomeric material acts to absorb energy and stop crack propagation. This matrix polymer is often acrylate and is applied as two separate components, one to each side of the joint. This has the added advantage of dissolving grease from an inproperly cleaned surface, which would otherwise have hindered adhesion.

For less demanding situations (footwear, laminating, floor and roof tiles, upholstery and trim of motor vehicles, for example), solvent-based adhesives are used. Natural rubber in an organic solvent has now been all but replaced by a range of polymers such as polychloroprene. As with paints, solvent-based adhesives are being replaced because of the cost of the lost solvent and because of its environmental and occupational health effects. Hot-melt adhesives have been successful. The co-polymers of ethylene and either vinyl acetate or ethyl acrylate are used, e.g. in the DIY handguns for the handyman. However they melt again at high temperatures and so are not recommended for the kitchen table.

Adhesive failure occurs when the surfaces have not been prepared properly or when there is a large difference in surface energy between the surface of the substrate and the adhesive. Liquids wet only surfaces that have higher surface energies (the solid equivalent of liquid surface tension). Polar adhesives with molecules that attract each other strongly and are thus liquids of high surface energy do not adhere well to Teflon or polyethylene. Cohesive failure often occurs along weaknesses caused by gas bubbles left behind from volatile by-products of the solidification process. A thinner glue line, obtained by using less adhesive, can be the answer here.

The process of bonding can be described as falling into two types—physical and chemical. Physical bonding occurs when the process is one of cooling or solvent evaporation. Starch and animal glues and poly(vinyl acetate) (PVA) emulsions (white glues) all solidify by the evaporation of water. They cannot be used where exposure to water is likely to occur. Chemical bonding results from a reaction that changes the nature of the adhesive. An example is the cyanoacrylate systems, which cure rapidly with traces of moisture as catalyst.

It can be seen from Table 8.2 that some adhesives can be multi-purpose.

Poly(vinyl acetate) glues are milky liquids generally sold in squeeze bottles. They consist of a latex of polymer, fillers and plasticisers in water and are analogues of rubber latexes. They keep very well in a sealed container. Their setting time is short for porous substrates such as paper and cardboard (two to three minutes) but longer for non-porous ones such as polished woodwork and ceramics (12 hours). Surplus glue can be wiped off with a wet rag before it has set. The glue line is transparent and the set glue is resistant to hydrocarbon solvents (oil, grease etc). Formulations with an acid hardener have better water-resistance. The joint softens if heated and normally has poor creep resistance.

Plastic glues are glues made to join plastics and are generally clear solutions. They set by loss of solvent and the joint is not very strong. They are suitable for light-weight applications such as for models, books, ceramic ornaments and leather.

Some plastics require special treatment. Vinyl plastics (e.g. inflatable toys, swimming pool liners, some upholstery) can be joined with vinyl kits. Polystyrene cements are available for polystyrene toys. Polyethylene is very non-polar and so some polarity has to be introduced by heating it in air with a flame (but not so as to melt it). The surface has become suitably polar if water will spread on the plastic. It can then be bonded with most flexible adhesives, except water-based ones (e.g. not PVA). Casein and starch dextrose glues work well on polyethylene and polypropylene bottles. Incidentally, a 3% caustic soda solution will remove

TABLE 8.2
Adhesives

Substrate	Animal	Casein	Vegetable	Urea-formaldehyde	Phenol-formaldehyde	Resorcinol-formaldehyde	PVA	PVA water-resistant	Natural rubber	Synthetic rubber	Plastic	Epoxy	Hot melt
Cardboard	X	X	X				X	X					X
Ceramics							X	X		X		X	
Glass										X		X	
Leather	X						X	X	X	X		X	X
Metal					X					X		X	X
Paper	X		X				X	X	X	X	X		X
Plastics							X	X	X	X	X	X	X
Rubber									X	X		X	X
Textiles	X						X	X	X	X	X		
Tiles										X		X	
Vinyl											X		
Wood (exterior)					X	X		X				X	
Wood (interior)	X	X		X	X	X	X	X				X	X

Note: Look for a cross against *both* substances you intend to join.
Source: 'Adhesive', *Consumer* **128**, 1976, 110. New Zealand Consumer Council.

wine labels. A strong adhesive is used to prevent the labels washing off in wet Eskys.

Synthetic rubber glues include the elastomeric (contact, pressure sensitive) adhesives and silicone rubber cements. They are stronger than natural rubber. Whereas neoprene rubber adhesives lose solvents in the setting process, the others absorb moisture to set.

Epoxy resins

Epoxy resins were introduced commercially just over 20 years ago. The term is applied to a whole family of resins and combinations of resins and curing agents whose properties can vary widely. An epoxy resin is defined as a molecule that contains more than one epoxy group (Fig. 8.7) and is capable of being converted into a thermosetting plastic by the use of curing agents or catalysts.

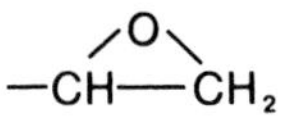

$$-CH-CH_2$$ with epoxy O bridge

Fig. 8.7   The epoxy group

Epoxy resins provide good adhesion to a wide range of materials, including metals, wood, concrete, glass, ceramics and many of the plastics. This is because polar groups are present in the cured resin. Since no water or other by-products are liberated during the curing of epoxy resins, they exhibit very low shrinkage. They can be formulated to withstand very high temperatures, and they are chemically very resistant. The most commonly used curing agents are polyfunctional amines in the form of an amine adduct (to reduce the volatility of the objectionable amines).

Ultrafast-setting adhesive

An interesting glue sold for home use was first introduced by Kodak for industrial purposes (Eastman 910). It is a one-component system and depends on the polymerisation reaction of a monomer, *methyl 2-cyanoacrylate* (Superglue), shown in Figure 8.8.

methyl 2-cyanoacrylate

methyl 2-methacrylate
(forms perspex)

Fig. 8.8   Acrylates

It is an exceedingly strong adhesive, very fast setting (10 seconds to two minutes) and bonds a wide variety of materials. It can be based on the methyl, ethyl or butyl cyanoacrylate monomer with the addition of a suitable stabiliser. A weak base such as adsorbed water causes rapid anionic polymerisation. It gives very strong bonding, provided the gap is less than 0.5 mm. The bonding is brittle and has poor resistance to warmth and moisture, but this adhesive is suitable for electronic and light electrical units. It has also found favour with the police for 'developing' latent fingerprints on certain materials. Doubts have been expressed about its safety as a consumer product because of the toxic vapour (tolerance 2 ppm in air) and fast setting time. The glue adheres very strongly to the skin and is difficult to remove. In the USA a boy's eye was once glued shut by a squirt of the liquid.

Formaldehyde resins

The group of formaldehyde resins are related to bakelite—the original plastic. They have a number of specialist applications. Urea-formaldehyde is used in low-stress veneer applications such as joining Formica to wood and in particle board, while resorcinol and phenol-formaldehyde are used in marine applications and outdoor furniture. Another use of phenol-formaldehyde is in brake linings and clutch facings, where a dry abrasive material such as asbestos or its more recent replacements, which must withstand high temperatures, is bonded to a substrate.

VITREOUS ENAMEL

Vitreous enamels (called porcelain enamels in North America) are alkali borosilicate glasses formulated to have temperature expansion coefficients slightly lower than the metal base to which they are applied. (The temperature expansion coefficient is the amount by which a material becomes larger when it is heated.) When the hot ceramic is applied to the hot metal and then cooled, the metal contracts more and keeps the enamel compressed. This takes advantage of the high compression strengths that glasses possess. Normally the base metal is covered with two layers—a dark 'ground coat' next to the metal, covered with a decorative coating. (Lately, though, with the special grades of steel now available, the ground coat can be eliminated.) A ceramic glaze is similar to a vitreous enamel, except that the substrate for the ceramic glaze is non-metallic (e.g. clay), generally has a low expansion coefficient, is weaker under tension and is brittle.

Vitreous enamelling and ceramic glazing are basically the same process. The basic operations are:
1. manufacture of the glass;
2. melting of the glass to a slip (frit or flake);
3. application of the slip to the substrate;
4. drying;
5. firing at around 830°C.

Contrary to what one might expect, vitreous enamel is not a thermal insulator but is a relatively good heat conductor when applied in thin coats and is thus useful in ovens.

Pyrolytic, self-clean ovens literally burn off foods at temperatures of 430°C–540°C. They are put through the cleaning cycle periodically and the higher temperature is held for about one and a half hours.

Continuously cleaning ovens work on an entirely different principle. The enamels inside the oven have a porous structure (like pumice) that has the effect of increasing the surface area of the material by a factor of 40 to 80 times the apparent surface coating area. The surface catalyses the burning of fats and oils and so removes them at cooking temperatures. They are less effective with other soils such as sugar.

Paint enamel is a completely different material. It is basically a normal paint that is dried by baking at a relatively low temperature to remove traces of solvent.

DENTAL AMALGAM

Tooth enamel is composed mostly of a hydroxyapatite mineral, $Ca_{10}(PO_4)_6(OH)_2$, with many other elements in small amounts (3% in total). Bacteria in the plaque on teeth produce (predominantly) lactic and acetic acids, which are largely prevented from attacking the outer enamel by protein adsorbed on the surface, but they can diffuse through to attack inner surfaces. Fluoride can replace in part the hydroxide in the enamel and form a much tougher mineral.

When a tooth develops caries (a 'cavity'), the area is drilled to remove decaying matter and filled. The material used to fill teeth must be malleable, must adhere tightly to the tooth and thus exclude air and saliva, and must be able to withstand abrasion, corrosion, discolouration and the force of chewing. The material used must therefore satisfy quite a variety of requirements.

The search for suitable materials is fascinating and today's standard dental alloy is made up of the following metals:

silver	66.7–74.5%
tin	25.3–27.0%
copper	0.0–6.0%
zinc	0.0–1.9%

The major component has the approximate formula Ag_3Sn. To save money, copper used to be used as a replacement for silver. Zinc serves as an oxygen scavenger during manufacture. The dental alloy is mixed with mercury (1:1 weight ratio) just before packing into the cavity:

$$Ag_3Sn + Hg \rightarrow Ag_2Hg_3 + Sn_8Hg + \text{unreacted } Ag_3Sn$$

The phase most readily attacked by saliva is Sn_8Hg. Saliva acts on the Sn_8Hg and releases Sn^{2+}. The process is accelerated if precipitating or complexing reagents are present, such as sulfide (from eggs) or citrate (from citrus fruit—see Experiment 13.5). Such corrosion is accelerated if there is contact with a less re-active metal such as gold from an inlay on an adjoining tooth. A piece of aluminium foil placed on a mercury amalgam filling can give a galvanic shock (at greater than 30 μA passing through the tooth).

Health problems from the mercury would not appear to be great although a certain amount vapourises during the first few hours of a filling and mercury can be detected in the urine. The older copper amalgam released vapour for many months after a filling and was a very likely health hazard (see 'Metals in food—Essential elements—Mercury', Chapter 12).

PORTLAND CEMENT AND CONCRETE

Ever since people first started to build, they have felt the need for some cementing material to bind stones and other aggregates to form strong walls and floors and to give them a smooth appearance. The Assyrians and Babylonians are known to have used moistened clay for this purpose. The Egyptians used the mineral gypsum, which had been calcined or 'burned', mixed with sand to make the mortar used in constructing the pyramids. Gypsum is calcium sulfate dihydrate ($CaSO_4 \cdot 2H_2O$). When heated to 121°–132°C, 1.5 moles of water per mole of gypsum are lost, resulting in the hemihydrate $CaSO_4 \cdot \frac{1}{2}H_2O$. On mixing with a small quantity of water, the hemihydrate is converted slowly to the dihydrate, serving as a binder in the process. The hemihydrate is known as plaster of Paris, having received its name from work carried out by French chemists in the late eighteenth century on the mechanism of the setting of calcined gypsum by reaction with water.

The Greeks produced lime by 'burning' limestone and found that the action of the lime admixed with sand was enhanced if some volcanic ash was also used in the mixture. The Romans improved such mixtures by utilising pozzuolana, a volcanic deposit in Italy, which had proportions of alumina and silica nearly optimum for use with lime to make an effective cement.

The quality of building cements declined with the decline of Rome, and the art of cement making was practically lost until the middle of the eighteenth century. It was not until 1824 that Joseph Aspdin, an English bricklayer, found that the addition of volcanic ash to the lime could be avoided if a limestone containing a relatively high proportion of clay was used and the calcination was carried to incipient fusion. The product was named Portland cement because after hardening it resembled a natural limestone quarried at Portland, England.

Many limestone deposits contain appreciable proportions of clay, such 'clayey limestones' being known as cement rock. Since in such deposits the limestone and clay are intimately mixed, the early practice was to use cement rock itself if it had the approximate proper proportions of ingredients to make a good cement. The process for making Portland cement is relatively simple but the chemistry of cement manufacture is highly complex. An example of the overall chemical composition of a high-quality Portland cement is given in Table 8.3.

TABLE 8.3
The overall chemical composition of Portland cement

Mineral	Formula	Per cent
Lime	CaO	62.0
Silica	SiO_2	22.0
Alumina	Al_2O_3	7.5
Magnesia	MgO	2.5
Iron oxide	Fe_2O_3	2.5
Sulfur dioxide	SO_3	1.5
Other	—	2.0

Although a cement can have the correct overall *chemical composition* (as listed in Table 8.3), unless the chemicals are present as the *proper compounds,* it will be worthless as a cementing material. The proper compounds in a good cement are tricalcium silicate, dicalcium silicate, tricalcium aluminate and tetracalcium alumino-ferrate. After the calcination step, the magnesium oxide remains largely as such, there is some free lime, and the sulfur is present as calcium sulfate. The basic calcium oxide reacts mostly with the acidic silica and sulfur trioxide, and with the amphoteric alumina and ferric (iron) oxide.

Tricalcium silicate is the essential ingredient in cement. The dicalcium salt hydrates much more slowly in the mortar than the tricalcium salt but contributes greatly to ultimate maximum strength. The bonding properties of a cement in a sand-gravel-water aggregate are the result of gels, formed by the hydration of the silicates, plus reactions of the other ingredients with water. The anhydrous calcium sulfate is converted to gypsum as in plaster of Paris, and free calcium and magnesium oxides are converted to their hydroxides as in regular plaster. The gels are largely colloidal, and their fibrous nature accounts, in part, for the strong bonding strength. It requires months for a concrete prepared from Portland cement, sand and crushed stone or gravel to reach maximum strength.

The first theories of cement hardening were put forward by Le Chatelier (1893) and Michaelis (1893), both famous for other contributions to physical chemistry.

The system containing cement, water, sand, aggregate (gravel) and air is called *concrete*. Concrete's most important engineering property is its *compressive* strength (i.e. strength against compression). Its *tensile* strength (i.e. strength against stretching) is only about one-tenth of its compressive strength. Both the aggregate and the cement paste have higher tensile strengths than the concrete they form—suggesting that the weakest part of a hardened concrete is at the inter-face between cement paste and aggregate and sand. It is because of these properties that steel mesh is used to reinforce concrete and why some structures are *prestressed* to ensure that the concrete is compressed rather than stretched.

High alumina cement called *cement fondu* is particularly interesting because it provides high performance and fast setting (traffic can run on it the next morning). However it lacks durability.

Calcium chloride can be added to cement to obtain rapid hardening. It accelerates the hydration and the heat of hydration causes the concrete to become very hot. Limestone is often used as an aggregate in concrete for sewerage pipes. The acid waste attacks the aggregate preferentially and increases the life of the concrete. The compressive strength of concrete is normally 'specified'. A minimum value of 20 MPa is specified, but with better water control, a value of 30 MPa is obtainable. The more water used, the weaker the concrete, but the easier it is to work. The amount of water needed increases as the grind of concrete used be-comes finer, because of the larger surface area that needs to be wetted. The finer grind creates more pores in the concrete and the concrete is weaker.

Why does concrete crack?

Concrete changes its volume for a number of reasons. Expansion can be caused by the presence of undesirable compounds (e.g. MgO and CaO). Alkalis can dissolve some of the aggregate containing amorphous silica or carbonate minerals. Shrinkage can be caused by carbonation (reaction with carbon dioxide in the air).

The problem of changes in volume caused by freezing of water (ice has 9% greater volume than water) is overcome by the use of surfactants to entrap air in the concrete, thus making room for expansion.

A survey of buildings carried out in 1979 in Sydney indicated that 70% of build-ings less than 15 years old displayed noticeable durability problems and that those less than five years old were in even greater trouble. Expansive forces generated by rusting steel reinforcement chip pieces of concrete from the surface, disfiguring it and ultimately creating a structural weakness. Carbonation converts calcium hydroxide into calcium carbonate and lowers the pH. This in turn increases the corrosion potential of the steel (see 'Experiment 8.3'). It was estimated that, in 1979, the cost of repairs amounted to $50 million p.a., which could rise fourfold in the same dollars by 1990.

Concrete mixtures

The proportioning of the ingredients is usually in the order of cement, sand and gravel by volume; for example 1:3:6 is a lean mixture, 1:2:4 is used for stronger structures, reinforced cement or when concrete is used under water, and 1:1:2 is used when the concrete is exposed to sea water (and the cement must be free of lime and alumina). The amount of water required to combine chemically with the cement is about 16% by weight, but, for efficient mixing, a greater amount than this must be used.

An effective waterproof concrete developed by the US Geological Survey admixes a heavy residual mineral oil (density 0.93—engine oil, new or old, seems to work well) with Portland cement and sand in the ratio of 1:3, and oil not more than 10% by weight of the cement. In fact a 50:50 mixture of oil and water works well. This concrete takes half as long again to set but its compressive strength is only slightly reduced. The grip on steel is greatly decreased but on reinforcing bars, wire mesh or expanded metal it is satisfactory.

Sugar inhibits the setting of concrete and also weakens it. A tanker of sugar syrup is often kept on hand for dealing with large-scale spillage of concrete. The exact mechanism of this process is unknown, although some sugars (e.g. sucrose and glucose) are particularly effective. Sugars are weak acids, as shown by their tendency to migrate to the anode in an electrophoresis experiment, but some migrate to the cathode in salt solutions at pH 7, indicating that there is coordination with the cation. Divalent actions form the strongest complexes, with Ca^{2+} being the most effective.

EXPERIMENT 8.2

Strength of cement with changing water content

Equipment

Moulds (from milk cartons), 100 mL measuring cylinder, 400 mL beakers, cement, water, plastic wrap, steel balls, hammer.

Procedure

Prepare five moulds by cutting off the tops of some cardboard milk cartons. Label them 1,2,3,4,5. Half-fill a 400 mL beaker with dry cement sand mixture. Fill a 100 mL measuring cylinder with water. Slowly add water a little at a time, with stirring, to the initially dry mixture. Continue adding water until the mixture becomes a thick paste. If you add too much water the mixture will suddenly become very thin and you will need to start again. When you have a thick mixture with no pockets of dry material, pour it into the first cardboard mould and scrape as much of the material out of the beaker as possible. Smooth the surface of the cement in the mould. Wipe the inside of the beaker with a paper towel and rinse it well with water.

How much water is left in the cylinder? Record the amount of water you used.

Dry the beaker and repeat the experiment, but this time with 20% less water. Repeat the experiment again, but with another 20% reduction in water. Repeat again with 20% *more* water than the first mixture. Cover all four moulds with plastic wrap to prevent evaporation. Prepare one more sample identical to the first one you made, but leave this one *un*covered. Leave all the moulds in a warm place for two days.

After two days examine the samples. Do the surfaces look different? Scratch the surfaces with your fingernail, a nail or point of a file. Note any differences. Try dropping steel ball or marble from the same height onto the surface of each sample. *Wear safety glasses.* The harder the surface, the greater will be the bounce height. Tear the cardboard off the samples and try a 'reproducible' hit with a hammer, of increasing intensity.

Record the order of surface hardness by both methods and the resistance to breaking. What effect does the amount of water have? What difference does keeping cement under wraps make to the surface?

There are other parameters you might wish to explore, e.g. the ratio of sand to cement. Try mixtures varying from 50 mL sand plus 150 mL Portland cement, to 50 mL of cement plus 150 mL of sand. What is the effect of adding 1% sugar to an otherwise strong setting mixture?

CAUTION

Dispose of waste cement 'thoughtfully' (as they say on the wrappers), by wrapping it up in thick layers of newspapers before placing it in waste containers.

EXPERIMENT 8.3

Alkalinity of concrete

Equipment

Steel reinforcing (or nails) jars.

Procedure

Concrete is cement plus gravel and is often reinforced with steel rods. The steel is prevented from corroding because of the alkaline nature of the cement. Demonstrate this protection by placing pieces of steel reinforcing which has been cleaned of rust with sand paper, into two half filled jars of water. Into one jar also place some broken pieces of concrete. Seal the jars and leave for a week. The rod without the concrete should corrode faster.

Reactions

In exposed buildings, carbon dioxide from the atmosphere slowly penetrates the surface of concrete and reacts with lime, $Ca(OH)_2$, converting it to limestone,

$CaCO_3$, the reverse of the process by which cement is made. This reduces the alkalinity of the surroundings of the steel which then can rust. The oxides and hydroxides of iron have a larger volume than the iron and this expansion cracks the concrete (compare to the reduction of volume when magnesium oxidises to magnesium oxide page 235).

Obtain a broken piece of concrete where the surface has been exposed to the atmosphere for a while, e.g. from a building site. Break the piece open, and wet the whole new surface with some phenolphthalein indicator solution. A pink colouration will show the area of high alkalinity inside, with a rim of untinted concrete around the edge. The extent to which the gas penetrates depends on the size of the pores in the concrete; the more water originally used, the larger the pores.

REFERENCES

1. J. Walker, 'It is fun with Polyox, Silly Putty, Slime and other non-Newtonian fluids', *Scientific American* **239**, November 1978, 142, as used in S. Kohlagen, 'slime show', *Austr. Sci. Mag.* **57**.
2. J. L. Sutterby, 'Viscosity at home', *Chem. Tech.*, 1985, 416.
3. D. A. Armitage *et al.*, 'The preparation of "Bouncing Putty"' *J. Chem. Ed.* **50** (1973), 434.

BIBLIOGRAPHY

Adhesives

Allen, K. W. Adhesion and adhesives. *Chem. in Brit.*, May 1986, 451.
Findlater, D. 'New clues to glues in action'. *New Scientist,* 21 November 1985.
Foster, Van R. 'Polymers in caulking and sealant materials'. *J. Chem. Ed.*, **64**(10), 1987, 861.

Cement

Dandy, A. J. 'Chemistry of cement'. *Chemistry,* March 1978, 13.
Daugherty, K. E. and Robertson, L. D. 'Practical problems in the cement industry solved by modern research techniques'. *J. Chem. Ed.,* **49**(8), 1972, 522.
Figg, J. 'Chloride and sulfate attack on concrete'. *Chemistry and Industry,* 17 October 1983, 770.
Hall, C. 'On the history of Portland cement after 150 years'. *J. Chem. Ed.* **53**(4), 1976, 222.
'Has concrete changed?'. *Rebuild,* CSIRO, **7**(1), 1982.
'Investigating concrete's durability'. *Ecos* **35,** Autumn 1983, 30.
Lea, F. M. *Chemistry of cement and concrete,* 3rd edn. Edward Arnold, London, 1970.
'Protection of steel in concrete', in *CSIROPRAC: a source book for science teachers and students.* Australian Science Teachers' Association and CSIRO, 1985.
Taylor, H. 'Modern chemistry of cements'. *Chemistry and Industry,* 1981, 620.

Ceramics

Bell, J. *et al.* 'The ceramic age dawns'. New Scientist, 26 Jan. 1984, 10.
Bowen, H. K. 'Advanced ceramics'. *Sci. Am.,* October 1986, 147.

Emsley, J. 'Glass: past elegant, future thin'. *New Scientist,* 8 Dec. 1983, 728.

Hicks, J. F. G. 'Glass formation and crystal structure'. *J. Chem. Ed.,* **51**(1), 1974, 29.

Koleske, J. V. and Faucher, J. A. 'Demonstration of the glass transition'. *J. Chem. Ed.,* **43**(5), 1966, 254.

Michalske, T. A. and Bunker, B. C. 'The fracturing of glass'. *Sci. Am.,* December 1987, 78.

Milberg, M. E. 'Ceramics for cars'. *Chem. Tech.,* September 1987, 552.

Marquis, P. M. and Wilson, H. J. 'A tooth for a tooth: ceramics in modern dentistry'. *Chemistry and Industry,* 6 October 1986, 657.

Real, M. 'Ceramics in surgery: a materials and design challenge'. *Chemistry and Industry,* 6 October, 1986, 661.

Treptow, R. S. 'Amalgam dental fillings'. *Chemistry,* April 1978, 17; May 1978, 15.

Paint

British Paints. *Product manual and painting specifications.* Similar guides issued by Dulux, Taubmans, etc.

'Exterior house paints'. *Choice,* ACA, April 1984.

Floyd, F. L. 'Emulsion polymers in coatings'. *Chem. Tech.,* August 1973, 484.

Griffith, J. R. 'Chemistry of coatings'. *J. Chem. Educ.* **58**(11), 1981, 956.

Hill, L. A. *Some aspects of weathering of paints in Australia,* Technical Note 172, June 1971, Department of Supply (now Materials Research Laboratories), Melbourne, unclassified.

Nicholson, J. 'Paint is only skin deep'. *New Scientist,* 29 May, 1986, 40.

Nylen, P. and Sunderland, E. *Modern surface coatings.* Wiley Interscience, New York, 1965.

Oil Colour Chemists' Association Australia and Australian Paint Manufacturers' Federation. *Surface coatings,* 2nd rev. edn. NSW University Press, Sydney, 1983, 2 vols. Text for a post-certificate course in surface coatings technology; basic and very clear.

Turner, G. P. A. *Introduction to paint chemistry,* Chapman and Hall, London, 1967.

Chemistry in the medicine cabinet

9

DEVELOPMENT OF MEDICATION

One of the earliest records of human medication is to be found in an Egyptian papyrus dating from 1550 BC. During the nineteenth century the isolation and examination of the active principles of plant drugs were developed, while the twentieth century has heralded the manufacture and distribution of potent synthetic drugs. Early medicaments were mainly alkaloids (complex plant chemicals containing nitrogen); the toxic properties of mandrake, opium poppy and nux vomica seed are due in each case to alkaloids—atropine, morphine and strychnine, respectively.

The first synthetic drugs were ether, chloroform, chloral hydrate (1869), phenacetin, acetanilide (1887), aspirin (1899) and procaine (1905). From this point there was an increase in synthetic drug production—an increase strongly reinforced by the discovery of the organic arsenicals as a treatment for syphilis and of the antibacterial action of certain dyestuffs. This second discovery led in turn to the development of the sulfonamides, which encouraged firms already engaged in the dye industry (the main chemical industry at the time) to move into pharmaceutical preparations. Table 9.1 shows this development. The first pharmacopoeia was published in London in 1618, but the first to regulate standards for drug purity was published in 1846. In 1960 the first international edition was prepared by the World Health Organisation (WHO).

Drugs were traditionally taken by mouth, but in 1853 the hypodermic syringe was invented. It allowed drugs to be injected directly into the bloodstream and so take rapid effect (e.g. morphine). Later, the hypodermic syringe was used to administer drugs that are destroyed in the gut, such as insulin. Anaesthesia also became possible by injection, removing the limitation of having to use gases. Preparations for injection had to be sterile and preferably made up in single-dose containers, and fabrication moved out of the hands of pharmacists to the drug manufacturer. The cost of advertising drugs to physicians is responsible for about 75% of the price of the drug. Another important step was the development of the compressed *tablet* (or 'tabloid' as it was first known).

The international pharmaceutical industry started after the First World War and grew quickly to multinational proportions. By the 1950s painstaking research was giving way to a new era of commercialism, aimed at achieving the greatest possible consumption of pills and medicines.

TABLE 9.1
The rise of potent drugs

1803	Sertürner isolated first 'active principle' (morphine)
1846	Ether first used as an anaesthetic in surgery
1876	Salicylates found to be pain killers
1884	Cocaine found to be a local anaesthetic
1899	Aspirin synthesised
1902	Adrenaline isolated from the adrenal gland
1903	Veronal (barbitone—the first barbiturate)
1905	Organic arsenic compounds used for the treatment of syphilis
1907	Ergot alkaloids found to counteract adrenaline
1911	Vitamin studies started
1912	Phenobarbital used as an anti-epileptic
1916	Heparin used as an anticoagulant
1921	Insulin discovered and used to treat diabetics
1929	Penicillin discovered—no medical use because of crudeness of product
1935	Activity of the sulfonamides discovered and used on humans
1936	Pethidine, the first synthetic narcotic, synthesised
1937	Curare introduced as a muscular relaxant
1941	Penicillin first used on an Oxford policeman who was suffering from multiple suppurating abscesses and osteomyelitis and was in the terminal stages of a generalised infection. Although after four days of the treatment his condition showed an astonishing improvement, supplies of the drug had run out (in spite of recycling it from his urine) and he eventually died.
1943	Diphenhydramine—the first practical antihistamine
1950s	Development of tranquillisers
	Synthetic anti-inflammatory steroids
1960	Oral drugs for mature onset diabetes

By the end of the 1950s difficulties were becoming apparent. Really new drugs were becoming harder to find. The areas of therapeutics which remained were less easy to find drugs for. Some diseases afflicted only a comparatively small number of people, who were generally in hospital and were not likely to prove a source of great income to manufacturers even if better curative agents were found.

In 1976 the chemical and allied products industries in the USA had sales of about $100 billion, of which the pharmaceutical industry contributed about $15 billion and employed about 30% of the total technical staff. In research and development the pharmaceutical industry spent over $1 billion out of a total of $2.85 billion. Companies can spend about 10% of sales revenue on research and up to 30% on advertising and promotion. The after-tax profits of the pharmaceutical industry were 11.2%, compared to 7.5% for the chemical industry overall. In spite of declining profits, the pharmaceutical industry is still one of the most profitable and research intensive areas of the chemical industry.

The production of pharmaceuticals can be classifed into five groups. Fermentation, used for antibiotics such as penicillin, streptomycins, tetracylines and modification of steroids, is probably the most important in terms of dollar earnings, but, on a weight basis, synthetic chemistry is dominant because it produces the heavy-use products such as aspirin, tranquillisers and antihistamines. Animal extracts provide insulin and hormones, while biological sources lead to vaccines and other serums. Vegetable extracts provide alkaloids such as opium, quinine, atropine and the precursors of steroid drugs.

An interesting technological comparison can be made between the pharmaceutical and petrochemical industry. In the petrochemical industry, sophisticated catalysts but simple processes are the rule. On the other hand, in the pharmaceutical industry, multistage processes and expensive reagents are more common. The petroleum industry invariably uses hydrogen as a reducing agent, whereas the pharmaceutical industry can afford to use the sodium metal–liquid ammonia Birch reagent (for a crucial step in producing the contraceptive pill).

Because smaller quantities are involved and high levels of purity are required, the pharmaceutical industry generally uses batch production rather than continuous processes. The exception is aspirin!

WHAT'S IN A NAME?

The unnecessary multiplicity of commercial names for drugs of the same kind can lead to confusion and wastage.

Drugs have three sorts of names (see 'Language and labels—Nomenclature' in Chapter 1). A precise name for a chemical substance in an internationally agreed system has been devised to provide a *systematic* name. Because this is unwieldy, subject to typographical error and highly forgettable, there is a simple, uniform 'official' or 'non-proprietary' name, often called the *generic* name (INN), approved on an international basis for a chemical likely to prove worthwhile as a drug. The third name, the *trade* name, is a creation of the manufacturer of the product. When a trade name is specified in a medical prescription (89.7% of the time according to the industry), *only that particular brand may be dispensed.*

The arguments for generic labelling are basically that it provides a simple, unambiguous, international label for drugs, and doctors are more likely to know exactly what substance they are dealing with. A trade name effectively disguises the constituents of a combined preparation, thus increasing the chances of a patient receiving something that contains ingredients which (to that person) are potentially hazardous.

Many preparations are marketed under different names in different countries (for trademark reasons). The same name may refer to different drugs in different countries—which can have unfortunate results for travellers. Some examples are Anotox, Avlon, Benol, Bilagen, Bilitrast, Cedrox (which can be vitamin C or aspirin!), and we are only up to C in the alphabet. The pharmaceutical industry's reply to this is that preparations may differ in their bio-availability (the rate of release of active ingredient). It is true that the manufacture of the pill in which the drug is contained plays a critical role in the stability of the drug and the rate of its release. In Australia, poor quality control in this aspect led to the development of standard testing methods by the National Biological Standards Laboratory (initially for digoxin, 1966).

An interesting aspect of bio-availability is that of *polymorphism* (different solid-state structures for the same chemical). Chloramphenicol palmitate exists in two forms, only one of which is biologically active. In the other, the long hydrocarbon chain is possibly 'wrapped around' the outside of the crystal preventing it from dissolving. There are other cases as well.

MIMS NO. 1

The first Australian edition of this pharmaceutical handbook appeared in January 1977. Beautifully produced and bound, quarto size, with full-colour reproduction, the annual is 4.5 cm thick (including advertisements). It starts off with a product identification section that illustrates in colour most of the solid, identifiable dosage forms (tablets, capsules etc.) available for prescription (white, unmarked tablets are not shown). This is followed by a section on drugs used in pregnancy and lactation; a guide to clinically significant drug interactions; the GP's library; a guide to poisoning by therapeutic substances; a directory of manufacturers; a guide to adjustment of dosages in renal failure; and tables of Normal Values. Then there is a large section on prescription by therapeutic class giving composition actions, indications, contraindications, precautions or warnings, adverse reactions, dosage and administration, and presentation. Finally there is a series of indexes. Subsidiary booklets come out every second month.

Australian Prescriber was first published in late 1975 as an Australian Government independent review of therapeutics. It was discontinued in 1982 as an economy measure! In 1983 it resumed publication four times a year and is distributed free of charge to all doctors, dentists, pharmacists and other health professionals who request it. However, forms for the reporting of adverse drug reactions are no longer included. Our free-enterprising medical profession finds the collection of such vital statistics burdensome (see also Appendix V, 'Child poisoning').

The Australian Drug Evaluation Committee approves approximately 40 new drugs each year and about 40 new presentations (see Table 9.1). A neutral, critical review of these developments is essential. The cost to the taxpayer of this publication is 0.03% of the cost of drugs prescribed under the Pharmaceutical Benefits Scheme.

The industry is deliberately confusing science with commerce. In 1971 there were about 40 000 trade names used internationally for 5000 distinct pharmaceuticals. Aspirin had 198 international synonyms and, as well, there were its salts: ammonium (one name), lithium (three), sodium (four), calcium (the soluble aspirin, 36), magnesium (20) and aluminium (20). All the trade names are different—a triumph in marketing linguistics. The number of drug approvals in Australia over the last few years is given in Table 9.2. A distinction is made between new drugs and new presentations. Even the new drugs often represent only small variations on existing ones.

Generic labelling ensures that molecular manipulations are obvious because the names will be related. Compare the lists of generic and trade names for some tricyclic antidepressants in Table 9.3. A set of similar sounding trade names may be coincidental and does not necessarily mean that there is a relationship between the drugs.

TABLE 9.2
Approvals recommended by the Australian Drug Evaluation Committee (ADEC)

		75–76	76–77	77–78	78–79	79–80	80–81	81–82	82–83	83–84	84–85	85–86
Approval for general marketing of new drugs	Approval	48	45	63	46	38	69	48	41	31	41	40
	Rejection	11	4	19	22	25	20	15	22	16	9	19
	Deferral	10	15	6	5	7	3	1	2	4	4	6
	Application	69	64	88	73	70	92	64	65	51	54	65
Extend therapeutic indications or amend dosage regimens for drugs already on the market	Approval	4	7	23	16	8	24	19	22	19	31	41
	Rejection	3	3	3	7	4	6	6	11	14	7	12
	Deferral	—	—	—	—	—	—	—	6	—	1	2
	Application	7	10	26	23	12	30	25	39	33	39	55

TABLE 9.3
Generic and trade names for tricyclic antidepressants

Generic name	Trade names
Imipramine 10 mg	Imiprin, Tofranil
Imipramine 25 mg	Imiprin, Tofranil, Melipramine
Desipramine	Pertofran
Trimipramine	Surmontil
Chlomipramine	Anafranil
Amitriptyline	Laroxyl, Saroten, Tryptanol, Amitrip, (Endep)
Nortriptyline	Allegron, Nortab

Many have about 20 other trade names throughout the world.
Source: Pharmaceutical benefits, Australia, December 1987.

An interesting illustration of the confusion resulting from industry practices has been devised by a former Professor of Pharmacology at Albany (New York) Medical College, Dr Solomon Garb. Dr Garb describes what would happen if drug manufacturers were responsible for the manufacture and marketing of baked beans.

> They would all stop using the word 'beans' and each would give the product a new, coined name . . . Picture the confusion in the grocery store if beans were no longer named 'beans' but if each maker gave a completely new name to his product. Further, try to imagine what would happen if there were 300–500 additional new names of this type in the grocery store each year. This is approximately what is happening in medicine, and it is becoming exceedingly difficult for physicians to keep things clear.

Some estimates in the UK in 1983 found that profits on top-selling drugs are so great that switching from a single-brand-name heart drug to its unbranded equivalent would save £14.5 million. There are about 4000 brand-name drugs on the market, which are protected by patent but can be substituted, except for about 30 for which there is no good substitute. Patents on products developed during the 1960s are now running out and within five years about 90 of the top 100 drugs will be out of patent. It is fascinating to watch the trade advertising change for a drug that is reaching this point. Suddenly it is no longer so wonderful, as its owners try to move sales to another drug with a longer patent protection and simultaneously spoil the pitch for competitors waiting for the current protection to lapse. In the UK, drug companies are entitled to compensation for losses on branded products that are out of patent by raising the price of protected drugs. The Pharmaceutical Price Regulation Scheme actually regulates profits and not prices, which is bad for consumers but does have a certain economic rationale.

In September 1984 the US president signed into law the Drug Price Competition and Patent Restoration Act. This Act permits simplified approval for generic versions of post-1962 products that are no longer protected by patents and whose safety and effectiveness have already been demonstrated. The increased competition brought about price reductions. Valium was approved by the FDA in November 1963 and the patent held by Hoffmann–La Roche ran out in early 1985.

In September 1985 Valium acquired three generic duplicates to join the other top-selling eight prescription drugs with generic competitors.

As shown in Table 9.4, sales of nine of the 10 top-selling prescription drugs in the USA in 1984 added up to US$2.1 billion. Forty million Americans also spend something like $500 million annually on antihistamines for hay fever (allergic rhinitis). This thus affects one in six citizens (and the author as he was trying to type this section!).

THERAPEUTICS IN THE USA[1]

A scant 50 years ago, American drug manufacturers could produce drugs and sell them—without trying them out first on animals or people, or clearing the products with the US federal government. Until 1938, the onus was on the government to show that a particular drug was adulterated or misbranded. The US Food and Drug Administration (FDA) sought new legislation via a 'chamber of horrors' presentation of cases, such as that of the top executive who died in agony from a non-prescription drug then widely available: radioactive radium water.

None of the horror stories prevailed over the manufacturing-advertising lobby until children started dying. In October 1937, physicians reported to the American Medical Association the deaths of six patients from a liquefied version of a new formulation of the then-new wonder drug, sulfanilamide. Because pills are hard to swallow, a small company found a good solvent for the drug—and put it on the market without testing. Elixir of sulfanilamide killed 107 people, mainly children, before supplies were recalled. Sulfanilamide was such a new drug that no-one was sure whether it was the drug or the solvent that was toxic. Dr Francis Kelsey (then a graduate) found it was the solvent, diethylene glycol, also used as an antifreeze. (Ethylene glycol was also the adulterant used more recently in the Austrian wine scandal in 1986.)

The tragedy of the elixir of sulfanilamide led to the 1938 Food, Drug and Cosmetic Act, requiring drugs to be cleared for safety before they go on the market. Later, as an employee of the FDA, Francis Kelsey stopped thalidomide from being sold in the USA in contrast to what happened in Australia and most other countries (see Appendix IV). It took the thalidomide disaster to make Australia pass corresponding legislation 25 years after it was passed in the USA. The procedure now used in the USA is outlined in Table 9.5 and Figure 9.1.

THERAPEUTICS IN AUSTRALIA

There are over 200 000 therapeutic goods on the Australian market, the majority of which are imports. This is a multibillion dollar business. The drugs and therapeutic devices for human use selected for inclusion as pharmaceutical benefits alone cost Australians a total of $839 566 038 in 1986 (patient plus

TABLE 9.4
The US top ten prescription drugs in 1984

Brand name	Generic name	1984 brand name sales US$ million	Use
1. Dyazide	hydrochlorthiazide-triamterene	337	A diuretic/antihypertensive drug, for which there is an alternative with a different bioavailability and potency for which the physician must adjust.
2. Inderal	propanolol HCl	413	For hypertension, migraine headaches and various heart problems. Three generic version now available.
3. Lanoxin	digoxin	101	A heart drug for which generics have been available for some time.
4. Valium	diazapam	354	Sedative, tranquilliser and muscle relaxant.
5. Tylenol with codeine	acetaminophen-codeine	115	A pain killer with generic versions since the 1970s.
6. Amoxil	amoxicillin	229	For bacterial infections.
7. Tagamet	cimetidine	Not available	For ulcers etc. FDA approved in 1983 and still under patent. No generics.
8. Lasix	furosemide	146	A diuretic with uses including mild to moderate hypertension. Generic approved in 1983.
9. Motrin	ibufren	214	Pain killer used in arthritis.
10. Darvocet-N 100	acetaminophen/propoxyphene	189	Pain killer with three generic competitors.

Source: HHS News, US Department of Health and Human Services

TABLE 9.5
How experimental drugs are tested in humans

	Number of patients	Length	Purpose	Per cent of drugs successfully completing[a] the phase
Phase 1	20–100	Several months	Mainly safety	70
Phase 2	Up to several hundred	Several months to 2 years	Some short-term safety, but mainly effectiveness	33
Phase 3	Several hundred to several thousand	1–4 years	Safety, effectiveness, dosage	25–30

[a] For example, of 100 drugs for which investigational new drug applications are submitted to FDA, about 70 will successfully complete phase 1 trials and go on to phase 2; about 33 will complete phase 2 and go to phase 3; 25 to 30 will clear phase 3 (and, on average, about 20 of the original 100 will ultimately be approved for marketing).

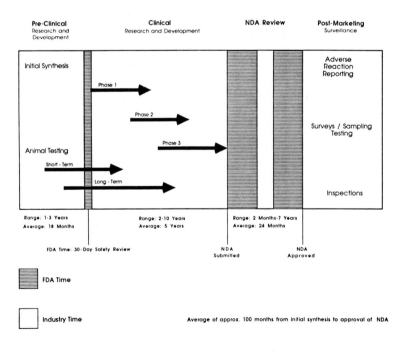

Fig. 9.1 New drug development in the USA (From *From test tube to patient: new drug development in the United States*, an FDA consumer special report, January 1988)

government contribution, see Appendix VI). When non-benefit and non-prescription items are included, as well as drugs for treating animals, the drug business is seen to be big business.

The control of this area falls on the States, and Commonwealth Department of Community Services and Health. A national register of all products on sale in Australia since 1984 that are claimed to be of therapeutic use in humans is now compiled as a computer data bank. It contains details of over 17 000 medical devices and 800 suppliers. The drug information held includes the active components, so that when a problem arises all the products containing that material can be located. When for example a 'natural' appetite suppressant containing the Japanese substance glucomannan was found to swell and block the oesophagus, all other products with that ingredient could be identified.

The 'approved product information' included with a drug contains details of a drug's chemistry, pharmacology and clinical use. The text is negotiated between the pharmaceutical company and the Department of Health. The document includes the purposes for which the drug can be used (called *indications*), when it should not be used (*contraindications*), precautions, warnings, adverse reactions, and safety for use during pregnancy and lactation. The full text is often published in the MIMS Annual and it must be inserted in the package of any drug that is administered parenterally (i.e. *not* via digestion).

Drugs for use in Australia are evaluated by the Australian Drug Evaluation Committee (ADEC) and its many expert subcommittees, set up after the thalidomide tragedy. Drugs are evaluated according to a number of criteria making use of worldwide data. However, virtually no independent toxicology research is done in Australia, because of a lack of trained staff and research support. Even when a drug is approved for marketing, the approval is based on studies under close clinical control (i.e. the administration of the drug to the patients is carefully controlled) and only a few, perhaps unrepresentative, patients take part in the study. Often larger-scale trials take place in the 'less demanding' environment of Third World countries. In any event, less common adverse effects of the treatment are not revealed, particularly if they take some time to appear. Thus close post-treatment monitoring is essential. Chloramphenicol, for example, causes aplastic anaemia, a result brought to the attention of the world community by the international consumer movement. It took an estimated 1.5 million observations of the drug use to confirm this adverse effect, which only happens in one of approximately 50 000 patients.

Another committee determines which drugs should become pharmaceutical benefits and thus be available free to pensioners and at subsidised cost to other patients. Pensioners make up just under 15% of the Australian population but use over half of the total prescriptions at about 70% of the total cost of the scheme. The inclusion of drugs as pharmaceutical benefits greatly influences what doctors will prescribe, and so acts as a second filter in the evaluation of drugs, particularly now in regard to efficacy, adverse reactions (drugs will not be considered for benefits until they have been in use for some time), and cost in relation to competing products. The price of a drug in the scheme is subject to negotiations between the company and (yet another committee) the Pharmaceutical Benefits Pricing Authority. A major factor in setting prices (if at first you don't agree, try, try again) is the level of activity being undertaken in Australia including investment, production, research and, in particular, export activity. The drug you take is determined in part by its contribution to the Australian economy! Appendixes VI and VII give some statistical information on drug sales. Market share information for individual brand names is no longer released because of its alleged commercial value. Appendix VI thus only provides an example of a brand name for the generic drug. Appendix VII gives a set of data collected by the industry, which includes hospital prescription and non-benefits items. The over-the-counter (OTC) products arise from the traditional distinction between proprietary or patent medicines and prescription medicines, once called *ethical* drugs. There are few restrictions on the sale and advertising of the OTC products and there is fierce competition with supermarkets in this area. The uses of a few of the less obvious products are included.

Schedules

The NH&MRC *Standard for uniform scheduling of drugs and poisons* has nine schedules for drugs and chemicals which the State governments have largely adopted into their appropriate Acts.

In the second edition of *Chemistry in the marketplace*, criticism was levelled at the quite unclear distinctions between some of the schedules. This situation has now been rectified to some extent and new definitions have been introduced. These are listed below.

Chemicals are scheduled on the basis of use, toxicity, potential for abuse and need for the substance. Chemicals for therapeutic use (drugs) fill schedules 2, 3, 4 and 8 in a progression of increasing strictness of control. Likewise schedules 5, 6 and 7 provide increasing control on agricultural, domestic and isolated industrial chemicals. Most industrial chemicals are to be controlled by Worksafe Australia (National Occupational Health and Safety Commission), if it ever transfers the hundreds of millions of dollars spent on Australia's fastest growing bureaucracy to something useful. Schedule 9 contains the prohibited substances, available only for special research projects. Schedule 1 keeps us all in touch with the fact that modern medicine arose out of the dark Middle Ages of ignorant alchemy.

The following is extracted from the *Standard for uniform scheduling of drugs and poisons.*

Poisons are classified according to the schedules in which they are included, as follows:

Schedule 1. Poisons of such danger to life as to warrant their being available only from medical practitioners, dentists, veterinary surgeons, pharmacists or persons licensed to sell Schedule 1 poisons.

Schedule 2. Poisons for therapeutic use that should be available to the public only from pharmacies; or where there is no pharmacy service available, from persons licensed to sell Schedule 2 poisons.

Schedule 3. Poisons for therapeutic use that are dangerous or are so liable to abuse as to warrant their availability to the public being restricted to supply by pharmacists or medical, dental or veterinary practitioners.

Schedule 4. Poisons that should, in the public interest, be restricted to medical, dental or veterinary prescription or supply, together with substances or preparations intended for therapeutic use, the safety or efficacy of which requires further evaluation.

Schedule 5. Poisons of a hazardous nature that must be readily available to the public but require caution in handling, storage and use.

Schedule 6. Poisons that must be available to the public but are of a more hazardous or poisonous nature than those classified in Schedule 5.

Schedule 7. Poisons which require special precautions in manufacture, handling, storage or use, or special individual regulations regarding labelling or availability because of exceptional danger or deficiencies in toxicological information.

Schedule 8. Poisons to which the restrictions recommended for drugs of dependence by the 1980 Australian Royal Commission of Inquiry into Drugs should apply.

Schedule 9. Poisons which are drugs of abuse, the manufacture, possession, sale or use of which should be prohibited by law, except for amounts which may be necessary for medical and scientific research conducted with approval.

ADVERTISING
A person shall not include any reference to a poison included in Schedule 3, 4 or 8 of this Standard in any advertisement except in genuine professional or trade journals, or other publications intended for circulation only within the medical, veterinary, dental or pharmaceutical professions or the wholesale drug industry.
A person shall not include any reference to a substance included in Schedule 9 of this standard in any advertisement.

SALE OR SUPPLY
A person other than a pharmacist (or an assistant under the direction of a pharmacist), a medical, dental or veterinary practitioner in the lawful practice of their professions or a licensed person shall not sell or supply a Schedule 1 or 2 poison. A person may apply for a licence to sell a prescribed range of Schedule 1 or 2 poisons if
1.   he keeps open shop for the sale of goods more than 25 kilometres by the shortest practicable road from the nearest pharmacy;
2.   he produces such evidence as may be required that he is a fit and proper person to be so licensed.
   A person other than a pharmacist or a medical, dental or veterinary practitioner in the lawful practice of their professions, shall not sell or supply a Schedule 3 poison. The person who sells or supplies a Schedule 3 poison shall
1.   provide adequate instructions for use, either written or verbal, at the time of supply or sale;
2.   label the container with his name or the name of the pharmacy and the address from which it was supplied;
3.   where provided in regulations make a record of the transaction, in a prescription book or other approved recording system.

[*Note.* An earlier requirement that 'Any person who supplies or sells Schedule 3 poisons shall keep such substances in a separate part of the premises to which customers do not have access' has been dropped. Now we can search the shelves for our favourite poison.]

A pharmacist may sell or supply without prescription any Schedule 4 poison, other than a poison excepted by Regulation from this provision, provided that

1. the patient is under medical treatment with the poison and continuation of medication is essential;
2. the quantity sold or supplied does not exceed three days medication;
3. the pharmacist is satisfied that an emergency exists.

A scheduled substance has specified labelling requirements, including warnings, statements of contents, strengths etc. and must include the manufacturer's name and address.

Summary

S1   These are dangerous substances *not* normally required for any usual purpose and so do not occur in medicine or commerce (except where exempted); e.g. bromine and, more recently, comfrey, which joins other dangerous herbs in this schedule.

S2   These are dangerous substances which *are* normally required for usual purposes and are sold only by pharmacists; e.g. silver nitrate, mercury (except in scientific instruments).

S3   all S3 drugs are basically those which are potent and are needed so regularly by sufferers of specific ailments that requiring a prescription each time is considered unwarranted.

S4   The S4 classification is the one generally seen on prescription-only drugs, and it contains one of the largest lists of chemicals.

It is sometimes difficult to fathom the logic used in deciding entries into the S2, S3 and S4 schedules.

The NH & MRC lists change regularly. In 1973 Enterovioform (clioquinol), once a common treatment for travellers' diarrhoea, was scheduled S3. Agitation by consumer organisations led to discrediting of this drug and it was rescheduled S4 in 1975. More agitation followed and so as a drug for internal human use it was scheduled as a *prohibited substance* (S9 in the new scheme). Various antihistamines in various preparations spread through S2, S3 and S4.

S5   Schedule 5 can be regarded as the 'hardware' schedule. It contains a variety of chemicals used around the home—from epoxy resins, turpentine and kerosene to moth balls, antifungal agents in wallpaper paste and some pesticides.

S6   For the consumer this represents the pesticide and OTC (over-the-counter) veterinary products.

S7   Awkward chemicals needing special rules; e.g. vinyl chloride and acrylonitrile monomers (see Chapter 6) and *o*-tolidine (see Chapter 5).

S8   Addictive drugs—morphine, pethidine etc. that are legal.

S9   In the second edition of this book, I noted the odd bedfellows here: cannabis, heroin and thalidomide. In spite of its criminal history, thalidomide has legitimate use and in the third edition we noted its removal from this list and its rescheduling in S7 for use in treating leprosy. We now see one of its optical isomers, (+) thalidomide scheduled as S4 (restricted in one of the appendixes of the NH & MRC Standard to treating leprosy). The chemical preparation of a substance in general provides equal mixtures of optical isomers and no attempt is made to separate them unless the use demands it. It has now turned out that only the (−) isomer causes birth abnormalities. The original British patent 768 821 obtained by Chemie Grünenthal describes the preparation and uses without reference to isomers. Experimental work with thalidomide indicates that it is effective in preventing a fatal complication in bone marrow transplants (graft versus host disease). In the third edition I noted that dioxins and clioquinol, that 'cure' for 'Bangkok belly', had been added.

Clioquinol, as predicted, has been promoted for good behaviour to S7 (and so has dioxin). The S9 schedule has some of the more recent illicit designer drugs added, but there is great difficulty in defining them adequately from a chemical-legal point of view.

Exemptions

There is also a list of *exemptions* from the *Standard for uniform scheduling of drugs and poisons*: timber and wallboard, ceramics, electrical components and electric lamps, vitreous enamels, explosives, glazed pottery, matches, motor fuels, lubricants, paper, photographic paper and film, inorganic pigments unless S6, paints as defined in the paint standard, blankets moth-proofed with dieldrin in the mill during finishing.

These are followed by a series of appendixes which exempt individual chemicals for undisclosed reasons. They may be covered under other jurisdictions, or they may require a warning without other restriction, or they have just been placed in the too hard basket. There is now quite an elaborate system of individualised warning and first-aid instructions for poisons. The cartoon created for earlier editions of *Chemistry in the marketplace* must have had some effect. It showed a frightened little girl looking at a large elephant labelled 'if swallowed seek medical advice'.

Aspirin and analgesics

The following are unscheduled and may be sold freely (e.g. in supermarkets): *plain* aspirin in tablets of less than 325 mg in packets of not more than 25, or powders of less than 650 mg in packets of not more than 12; *plain* paracetamol (or *plain* salicylamide) in tablets of less than 500 mg in packets of not more than 25, or powders of less than 1 g in packets of not more than 12. Dosages or packs greater than these are scheduled S2 and are available only in pharmacies. The plain analgesic may contain not more than 10 mg of codeine per dosage or not more than 0.5% in individual preparations. However, if the pack size exceeds the restrictions for S2 (that is, if you want one big pack rather than lots of small packs), the preparation is scheduled S3 and you should not be able to find it in the accessible part of the pharmacy, but will need to ask for it.

Mixtures of analgesics (including mixtures with caffeine) are scheduled S4 and require a doctor's prescription. These may contain not more than 30 mg of codeine per dose or not more than 1% in undivided preparations. Codeine in mixtures with concentrations in excess of this are scheduled S8.

TABLE 9.6
An aspirin by any other name . . .

Australian approved name: Aspirin
Synonyms: Acetylsalicylic acid, ASA

TRADE NAMES (August 1987)

Unscheduled (products containing aspirin only as an active ingredient)

1. Plain formulations
   Aspirin BP (DHA)
   Asprin BP (Fawns & McAllan)
   Aspro (Nicholas)
   Bayer Aspirin (Bayer)
   Bex (Beckers Pty Ltd)
   Vincent's Powders (Nicholas)
2. Soluble formulations
   Aspro Clear (Nicholas)
   Aspro Soluble (Nicholas)
   Dispersible aspirin (Fawns & McAllan)
   Disprin (Reckitt & Colman)
   Nyal Soluble Aspirin (Nyal)
   Solvin (Reckitt & Colman)
3. Soluble, with excess antacid
   Alka Seltzer (Miles)
4. Soluble, buffered formulations
   Bufferin (Astra)
   Bufferin 500 (Astra)
5. Enteric coated formulations
   ASA Arthritic Strength Aspirin (SK & F)
   Ecotrin (SK & F)
   Rhusal (GP)
   Winsprin (Winthrop)
6. Sustained release formulations
   Bi-Prin (Boots)
   SRA (Boots)

Scheduled (products containing aspirin in combination with other active ingredients)

Asco-Tin (Faulding)
Aspalgin (Fawns & McAllan)
Asprodeine (Nicholas)
Bex (Beckers Pty Ltd)
Code-Co (Hamilton)
Codiphen (GP)
Codis (Reckitt & Colman)
Codox (Glaxo)
Codral Blue (Wellcome)
Codral Cold Tablets (Wellcome)
Codral Forte (Wellcome)
Decrin (Nicholas)

Doloxene Co. (Eli Lilly)
Equagesic (Wyeth)
Hycoden (Faulding)
Morphalgin (Fawns & McAllan)
Orthoxicol Cold and Flu Caps (Upjohn)
Percodan (Endo)
Solcode (Reckitt & Colman)
Solusal-Co. (Hamilton)
Veganin (Warner)

SOME FOREIGN TRADE NAMES

Canada
Acetophen
Acetyl-Sal
Ancasal
Asadrine
Cetasal
Entrophen
Monasalyl
Neopirine-25
Nova-Phase
Novasen
Sal-Adult
Sal-Infant
Supasa
Triaphen-10

France
Aspirisucre
Aspro E
Claragine
Ivepirine
Juvepirine
Seclopyrine

Germany
Acetyline
Colfarit
Godamed

Italy
Cemerit

Sweden
Albyl-Selters
Apernyl
Damyl
Dispril
Instantine
Premaspin

United Kingdom
Aspro
Aspro Clear
Claradin
Safapryn
Safapryn Co.

USA
ASA
Measurin

Fig. 9.2   A headache! (Courtesy of Alan Foley Pty Ltd)

The contention put by pharmacy guilds that restricting the sale of all drugs to pharmacies on the grounds of long-term toxicity even of drugs like aspirin is weakened by the advertisements that appear in their trade journals offering inducements 'to boost your analgesic sales'. As a result of discussion in the first edition of this book, the ACA carried out a survey on the behaviour of pharmacists when approached with a request for analgesics in large quantities. The results, published in *Choice* (February 1976), were alarming. The industry was disturbed, as is shown by the following extract from the *Australian Journal of Pharmacy*, March 1976, p. 121:

> The Pharmaceutical Society of Australia believes 'strong intra-professional action' needs to be taken to ensure that pharmacists exercise personal supervision over sales of minor analgesics.
> ... in 97 randomly chosen pharmacies in all capital cities *Choice* buying panel members purchased unusually large quantities (12 packs of large sizes) of five common analgesic products. On 73 occasions sales were made by a pharmacy assistant, and buyers were refused the dozen packs asked for at only three pharmacies ... [Pharmacies], far from taking action to question the unusual size of the request, encouraged and hoped to increase the size of the sale.

Australia has the world's highest rate of kidney failure, at least 20% of which has been attributed to analgesic abuse (we use 50% more than Britons and 85% more than Americans).

Pharmacies

Australia has one pharmacy for every 3000 people, compared to one per 4500 in the USA, one per 5000 in Britain and one per 15 000 in Holland (where the system is to 'allocate' people to a pharmacy). The number of wholesale outlets per pharmacy is about one in 17 (Australia), one in 53 (Britain), and one in 31 (Holland).

At your local pharmacy you can probably buy footwear, sunglasses, health and leisure equipment, toys and electronic goods, photographic equipment and processing, cosmetics, perfumes, toiletries and toothpaste, as well as OTC medicines and, of course, prescriptions. What is the history behind this situation?

In the 1930s, pharmacists enjoyed a virtual monopoly over drugs and patent medicines. Suddenly department stores began marketing more cheaply items such as toothpaste and analgesics, which had previously been available only at the pharmacy. At the same time the pharmacist's professional skills in compounding preparations from basic materials became less and less necessary as prepackaged medicines began to dominate the field, particularly when a completely new generation of therapeutically potent drugs (the sulfonamides and penicillin) appeared in the 1940s.

At this time the British chain pharmacy Boots wanted to enter the Australian market. With their advantage of scale and efficiency, the threat to the local community pharmacy could not be ignored. The argument that standards would drop was rejected by a NSW government inquiry headed by Justice Browne. Nevertheless laws were passed prohibiting the *establishment* of chains (thereby allowing the continuing existence of the Soul Pattinson chain), limiting the ownership of pharmacies to one per person (who soon had to be a pharmacist), setting up licensing boards and so on. Some controls have since been relaxed and in NSW there can now be up to three pharmacies in partnership, while in other States a pharmacist can own more than one pharmacy. Anyone can own a medical practice and hire a doctor to operate in it, but pharmacies cannot be owned by non-pharmacists. The argument for this restriction is that the more commercial nature of the pharmacy business makes owner interference and coercion into non-ethical practices more likely.

The intervention of government in the pharmaceutical distribution system both as a monopoly buyer and third party subsidiser of pharmaceutical consumption has helped turn pharmacists into shopkeepers and encouraged a certain style of practice, of high-volume dispensing with little professional intervention. In a typical pharmacy, prescription medicines typically account for 45% of turnover, non-prescription for 15% to 20%, and approximately 40% of pharmacy turnover comes from cosmetics and toiletries etc. for which they enjoy about 60% of the market. These average levels fluctuate wildly with the location of the pharmacy. The turnover from pharmaceuticals can vary from as little as 20% to as much as 80%. Within the non-prescription area, pharmacists hold 25% of the photo-processing market, but competition is increasing. In more health related products, they do better. The Trade Practices Commission has outlawed formal agreements on 'chemists only' on items outside the Poisons Schedules, although some companies still have such a 'policy'. Many items are now available through super-markets, such as antiseptics, bandages, toothpaste and analgesics, which pre-viously were not.

Do the increasing professional qualifications required of pharmacists (over the last 20 years it has gone from an apprenticeship with some technical training to a degree course at university) make sense when less than 3% of prescriptions are now compounded by the community pharmacist? On the other hand the hospital pharmacy has increased in importance (representing about 15% of pharmacists) and the need for individualised medication makes greater demands on skills.

The community pharmacy is moving into the health promotion arena with programs such as *Self-Care*, which is an information and referral service on minor health matters. As the supermarkets encroach on the pharmacy, the pharmacy encroaches back on the medical practitioner (to call such information 'advice' would involve a demarcation dispute with the medicos). In the 1950s people used the community pharmacy extensively for minor medical matters and treatments, rather than worry the busy doctor (a visit would normally turn out to be more expensive, anyway). The structure of our medical and pharmaceutical benefits in the 1980s makes a visit to the doctor with a consequent script a much cheaper option. Recent modifications to the scheme may tend to reverse this trend to some extent. It is estimated that pharmacies have about 40 million non-prescription consultations every year compared to 60 million GP consultations.

Another interesting development has come about through hi-tech developments in clinical screening kits. The legal semantics here are interesting. A *diagnostic* kit is something used by medical practitioners and they are jealous of the term. Use of a *testing* kit in some States requires a licence (~$1000). So the pharmacist is left with a *screening* kit. They are chemically the same. The fingerprick devices allow blood glucose, cholesterol, triglycerides, uric acid etc. to be measured (with a reflectance meter) without the need for a syringe sample. Blood sugars, for example, provide much more useful information than urine levels. One such instrument called Reflotron™ (Böhringen) employs a complex reagent strip onto which a drop of the blood is placed. The strip is placed into the machine which reads a magnetic strip to calibrate for that particular batch of strips and then gives a readout of the result in a few minutes (see Plate X). The results are of clinically useful accuracy. Pregnancy tests (screens) are also now much more user friendly.

In the USA, pharmacies have moved into the hi-tech home-care services, providing dialysis, cytotoxic drug delivery etc., which would otherwise be carried out in a hospital. A high-level laboratory in the pharmacy is required and in Australia the move in this direction has not been followed. However computers have moved in and as well as the usual management tools (such as stock control, accounting, invoicing, all aspects of the dispensing routine, autobilling by disk or modem of benefit payments), a database warns of drug allergies, drug–disease interactions, and scaled drug–drug interaction. Because of the confusion in drug naming there are warnings when dispensing errors are likely—a so-called PDL message (PDL is Pharmaceutical Defense Ltd, which handles the indemnity insurance).

ADVERTISING DRUGS

Prescription drugs cannot be advertised on radio or television. As of 1 June 1977 a new voluntary code for the advertising of proprietary medicines and appliances came into being through a joint committee of newspapers, radio and television broadcasters, advertisers, pharmaceutical manufacturers and pharmacists, under the guidance of the then Commonwealth Department of Health. The code was reissued in 1981. Prohibited are claims for relief from complaints that should be dealt with by a medical or dental practitioner. These prohibited claims form an interesting list—including development of the bust, cures for baldness or impotence, and raising of a person's height.

The code includes a long-overdue section on analgesics. Analgesics advertisements must include the approved name of the drugs contained in the preparation and a warning against prolonged use. The warnings on radio and television are to be spoken as part of the advertising message, using the same vocal expressions as for the main message. An advertisement for analgesics shall not contain:

- any claim that analgesic consumption is safe;
- any claim that a preparation will relax, relieve tension, sedate or stimulate;
- unsubstantiated claims that one preparation is appreciably less irritant to the stomach, more rapidly absorbed, faster in action, or more effective or less harmful than another.

Conditions for vitamin are also spelled out advertisements in the code:

- No suggestions of food lacking nutrients through soil depletion.
- Vitamin therapy can only help if there is a 'deficiency' (undefined).
- No claims or dramatisation of benefits in dealing with irritability, sexual activity, nervousness, or of stimulation of appetite or growth, or providing nutritional insurance.
- No claim for good looks, good health and long life can necessarily be attributed to the use of vitamins.

There are further sections on claims, treatment, professional recommendations and testimonials. A very interesting clause is the one on disparagements:

> An advertisement relating to goods for therapeutic use shall not contain claims intended to disparage other medication *or the medical or allied professions*. [Italics added.]

Copies of the code are available from the Commonwealth Department of Community Services and Health. Breaches are best reported to the Minister for Health (Parliament House, Canberra) for an assured reply. For media other than radio and TV the State Acts or the *Trade Practices Act* 1974 apply (Section 52, Part V, Consumer Requirements). It says *inter alia*: 'A corporation shall not, in trade or commerce, engage in conduct that is misleading or deceptive' (see Chapter 14, 'Consumer information', for information circulars of the Trade Practices Commission that deal with breaches of this Act).

ASPIRIN

Recent experiments have shown that both salicylic acid and acetylsalicylic acid (aspirin) can breach the protective barrier in the stomach and cause stomach bleeding. For most people the bleeding produced is trivial—from half to two millilitres after two tablets; however, for some people it can be hundreds of millilitres, with such people requiring emergency hospitalisation. In acid solution aspirin is un-ionised (Fig. 9.3a) and is fat-soluble and can diffuse through the stomach's protective barrier. Once through it is in a neutral environment—it ionises and then cannot pass back again. (Compare with the effectiveness of HOCl versus OCl$^-$ as a bactericide in swimming pools: see 'Swimming pool chemistry' in Chapter 5.) The rate of diffusion is enhanced by alcohol even when the contents of the stomach have a low acidity. The cocktail party story of aspirin and alcohol being potent is seen to be well founded. Such cooperative action is often called *synergism*.

a. acid solution   b. neutral solution

Fig. 9.3 Aspirin in acid and neutral solutions

The related methyl salicylate (Fig. 9.4), which has the common name oil of wintergreen, is used externally to ease the pain from rheumatism and strained muscles. Aspirin that is kept too long begins to hydrolyse to salicylic acid, which is not well tolerated by the human body.

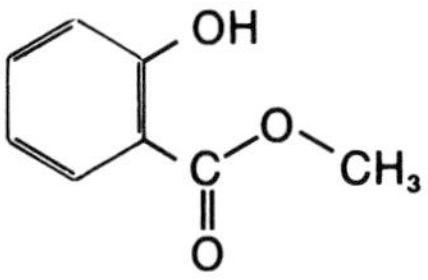

Fig. 9.4 Methyl salicylate

Soluble aspirin is either the sodium or the calcium salt of normal aspirin. These salts immediately form aspirin in the acid stomach in the form of *fine* crystals and possibly cause less gastric distress. Some analgesics, such as Panadol, contain *p*-acetylaminophenol (4-hydroxyacetanilide, paracetamol), which is comparable to aspirin as a pain-reliever (Fig. 9.5), but is gentler to the stomach.

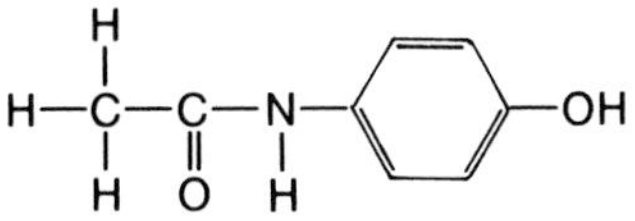

Fig. 9.5 Paracetamol

CHEMISTRY OF DRUGS

Introduction

A discussion of the chemistry of drugs could easily fill a book on its own. It has a fascinating history for a start. Modern research on the mode of action of drugs is exciting and potentially very useful in paving the way for a more scientific selection of effective substances. It is perhaps not realised how much we still rely on naturally occurring substances—either directly or with some modification—for our pharmaceuticals. Even where it is possible to synthesise a drug such as morphine, it is often more economical to produce it from a plant and then perhaps add a few trimmings. A substance as complex as insulin is only synthesised once! This is done because synthesis is the ultimate proof of the correctness of the structure of the compound. However, as new and better chemistry is developed, the synthetic route to a drug can again become interesting and competitive.

I have selected only a section of this huge subject and will concentrate particularly on the drugs that affect behaviour. These have become a consumer item in the sense that their use is widespread and some of the preparations are available without a medico's script. They also have the greatest potential for abuse.

SENSITIVITY TO DRUGS DETERMINED BY GENES

Individual people fall into two genetic groups; those who acetylate drugs like sulfonamides fast and those who do so slowly. The amount of this enzyme produced is caused by a single difference in one chromosome. While 90% of Asians (Japanese and Chinese) are fast acetylators, only 40% of Americans (both black and white) acetylate drugs fast. Acetylation is often the first step in metabolising and thus deactivating a drug and so *slow* acetylators are exposed to *higher* levels of a drug given at the *same* dose. The acetyl derivative of sulfathiazole is not very soluble. It tended to block kidney tubules and lead to many deaths. It was replaced by sulfadiazine.

The same acetylating enzyme deactivates some carcinogens, e.g. aromatic amines such as benzidene and *o*-tolidine, used in dyestuff manufacture and as analytical reagents in the detection of blood (and chlorine levels) in water. Slow acetylators are at higher risk of bladder cancer from these chemicals. In a study at ICI in the UK (*New Scientist*, 5 June 1986, p. 24) 23 former dyestuff workers with bladder cancer were examined. Twenty-two were slow acetylators. Genetic probes allow the detection of such differences (other enzymes are related to the detoxification of some pesticides) and genetic screening will have profound effects on our attitudes to occupational health and the grounds on which legal compensation claims are liable to be based.

A parallel genetic difference exists in regard to alcohol dehydrogenase, which converts alcohol in the liver to acetaldehyde. Fast acetylators can be embarrassed by the flushing that occurs on imbibing, because of the sudden release of acetaldehyde. The further oxidation to acetic acid occurs at the same rate in both fast and slow acetylators.

Drugs can be classified under headings of what they do; for example, analgesics (pain deadening), sedatives and tranquillisers (reduce anxiety), stimulants, anti-depressants, hallucinogens etc. But from the point of view of understanding, the relation between structure and activity is more useful. Let us consider the very first of the antibacterial drugs—the *sulfonamides*, which were found to be effective against the 'cocci infections' caused by the bacteria streptococci, gonococci and pneumococci.

The basic compound is called sulfanilamide (Fig. 9.6).

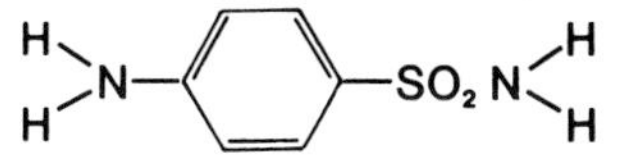

Fig. 9.6 *p*-aminobenzenesulfonamide

A whole *family* of derivatives can be built up on this compound by modifying the molecule in a manner which either changes its potency or reduces side-effects or toxicity. Thus, if we write a general formula for a sulfonamide as

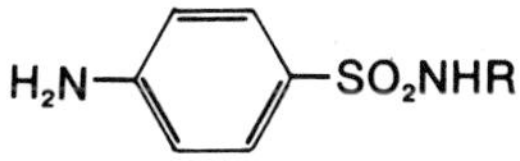

then, in sulfathiazole

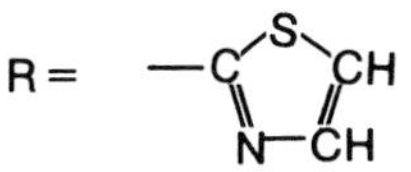

and in suffadiazine

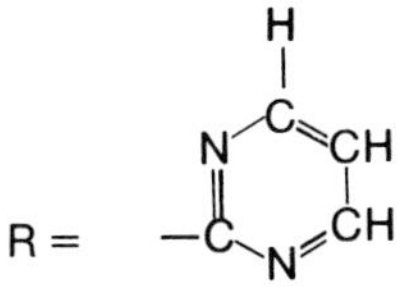

These are then members of the sulfonamide family or generic group. It appears that the effectiveness of these drugs depends on maintaining the basic structure and shape of the molecule, and you might wonder why. One of the essential growth compounds for most bacteria that are susceptible to the sulfonamides is *p-aminobenzoic acid.*

The fact is that bacteria absorb a sulfonamide 'by mistake' because their shape and charge distribution are similar to *p*-aminobenzoic acid (Fig. 9.7), and then they cannot metabolise it. It fits into the cell machinery, but doesn't come out; that is, it *blocks* the active sites of the cell.

sulfanilamide

p-aminobenzoic acid

Fig. 9.7 Sulfanilamide and *p*-aminobenzoic acid

The sulfonamides as chemicals have been known for a long time but their medical value was discovered in 1932 by Domagk only by accident while looking at a series of azo-dyes he used for attenuating virulence of experimentally used bacteria. Why are the sulfonamides active against bacteria and not against people? Well, p-aminobenzoic acid is used by bacteria to produce folic acid, which they need, just as we do. By blocking the enzyme that carries out the first step the bacteria get no folic acid, and bacteria cannot absorb folic acid from their food. Humans do not synthesise folic acid but can obtain it from their food, and so the sulfonamides cannot deprive them of it.

DISCOVERY OF SULFA DRUGS

The story of the first sulfonamide is interesting. Gerhard Domagk, working for IG Farbenindustrie patented an azo-dye, Prontosil, on Christmas Day, 1932. However his experiments with the dye on infected mice and safety tests on rabbits and cats were not published until they appeared in the *Deutsche Medizinische Wochenschrift* of 15 February 1935. The paper describes how 12 mice infected with streptococcal bacteria stayed lively for eight days, while of 14 controls, 11 survived for one day only, 2 for two days and only 1 survived three days. This paper was followed immediately by another from the local hospital with enthusiastic descriptions of clinical tests on human patients over a period of more than two years. Domagk had also used Prontosil to cure his daughter of a bacterial infection. Domagk noted that the chemical was ineffective *in vitro* and 'only works as a true chemotherapeutic agent in a living organism'. This was a problem because the antibacterial properties of the separate sulfanilamide entity were recognised back in 1919 (although not in animals), and therefore prevented the patenting of Prontosil (or, for that matter, the sulfanilamides themselves). He may have delayed publication in the hope of finding another compound that could be patented. In November 1935, a publication by researchers at the Pasteur Institute in Paris showed that the sulfanilamide entity alone was found effective in humans (as well as *in vitro*).

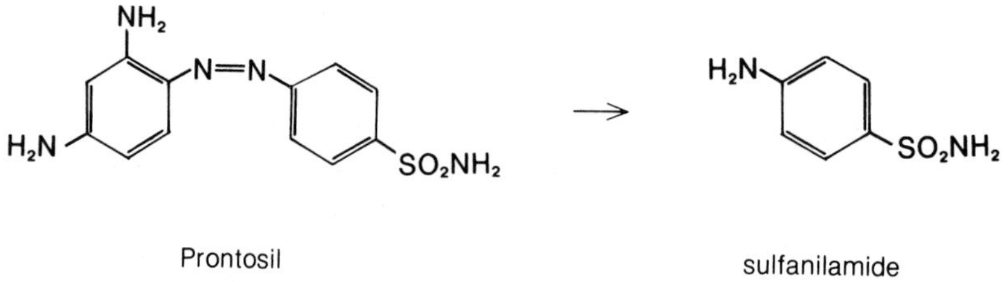

Prontosil   sulfanilamide

Fig. 9.8 The action of Prontosil

I vividly remember taking the huge 'M & B' wonder tablets (May & Baker's sulfa drugs) as a child in the 1940s and being required to drink copious quantities of water at the same time. The kidney damage caused by earlier products (see box 'Sensitivity to drugs determined by genes', in this chapter) meant that cautious doctors still insisted patients should drink many pints of water per day while on the tablets.

See also *New Scientist*, 18 July 1985, and the section on azo food dyes in 'Food additives—Colouring matter' in Chapter 11.

Antibiotics are substances produced by micro-organisms within themselves, which, when excreted, interfere with the growth or metabolism of other micro-organisms—a sort of chemical warfare on the microbe scale. In 1929, Fleming discovered a mould of the *Penicillium* genus that inhibited the growth of certain bacteria. The active compound, which he was unable to isolate, was called *penicillin* (Fig. 9.9). Florey and Sharp later performed the decisive clinical tests and were responsible for the success of the drug.

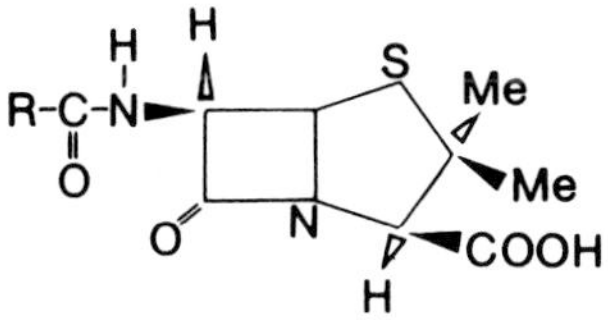

Fig. 9.9 Penicillin

In the original penicillin, the R-group was a mixture. The R-group is varied often by adding molecules to the nutrient solution in which the mould is growing to produce many different penicillins. The mode of action of penicillin was determined in 1962. Penicillin interferes with the building up of the cell wall (which is continuously being digested and rebuilt) and the cells of certain bacteria are very much more sensitive to this interference. Penicillin is effective against a series of bacteria called Gram-positive (which take up and hold a certain stain or dye) but not against Gram-negative bacteria (in which the stain is washed out). In addition, many bacteria have developed or can develop the *enzyme* or biological catalyst *penicillinase,* which can destroy *penicillin* (*-ase* often means an enzyme which is related to the compound or chemical reaction immediately preceding it).

In order to balance the argument on safety testing of drugs it should be pointed out that penicillin is quite toxic to guinea pigs. During the war there was no time to carry through an adequate testing program on animals. Just as well! Mind you, virtually everything is bad for guinea pigs, poor defenceless creatures. Their standard use today is as a test for tuberculosis bacterium.

Now I will discuss the behavioural drugs. In order to see how sensitive structure is in relation to pharmacological activity, consider the alkaloids associated with opium. The alkaloid morphine was first isolated from the latex of the opium poppy (*Papaver somniferum*) by a German tinker, Sertürner, in 1803, although the ancient Babylonians probably used crude opium to relieve pain 5000 years ago. Its addictive properties were known from early times. In 1832 another alkaloid, codeine, was isolated from opium. Although codeine has only about one-tenth of the potency of morphine, its prolonged use in low doses can cause physical dependence. In 1898 morphine was acetylated to produce diacetylmorphine or *heroin,* which was quickly realised to be even more addictive than morphine. Figure 9.10 illustrates the structure of morphine and some of its derivatives.

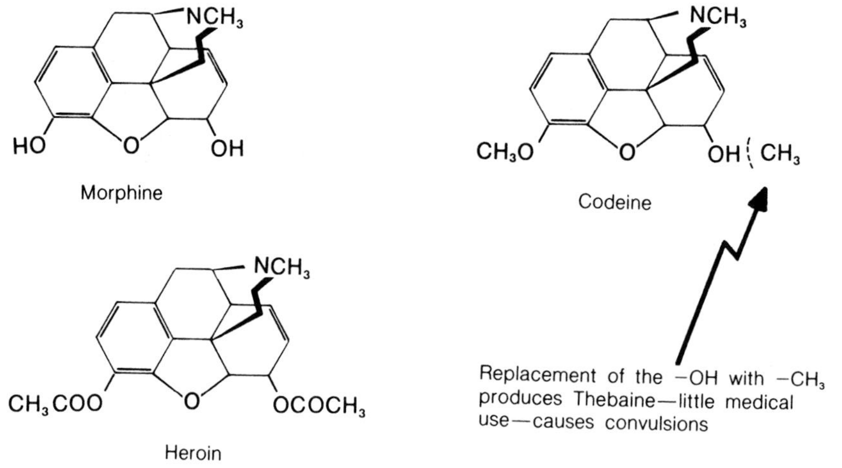

Fig. 9.10   Morphine and its derivatives

The first potent analgesic to be prepared that did not depend upon opium for its prime source was discovered quite by chance in 1939 by Eisleb and Schaumann during a search for atropine-like activity. The substance, *pethidine*, seems only vaguely related to morphine but, if the molecule is drawn to show a particular *conformation*, the relationship becomes apparent (Fig. 9.11).

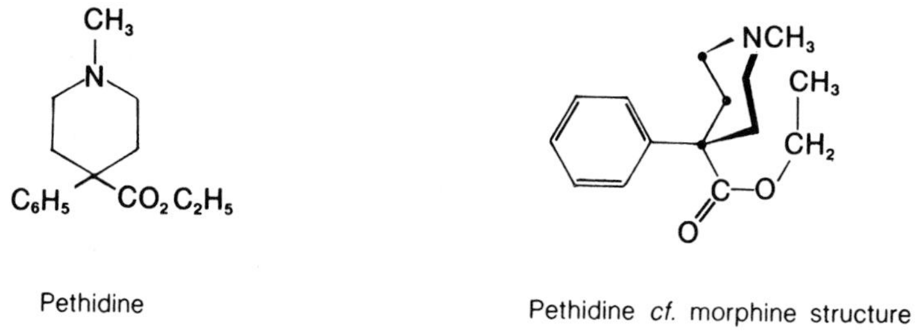

Fig. 9.11   Pethidine

In 1946 the first member of an important new group of synthetic analgesics (based on 3,3-diphenylpropylamine) was introduced under the name *methadone*. Again the structural relation to morphine can be detected when the flexible methadone molecule is rearranged (Fig. 9.12).

As long ago as 1915 a simple derivative of codeine was prepared in which the methyl group attached to the nitrogen ring was replaced by another alkyl group. Although this compound seemed itself devoid of any analgesic properties, it was noted that it *antagonised* the properties of codeine. In 1941 a corresponding transformation was effected on morphine to give a substance that was named *nalorphine* (Fig. 9.13). The first and obvious use of nalorphine was therefore to treat cases of poisoning by morphine and its derivatives. However, it has also proved very useful for diagnosing cases of addiction. When nalorphine is given to a

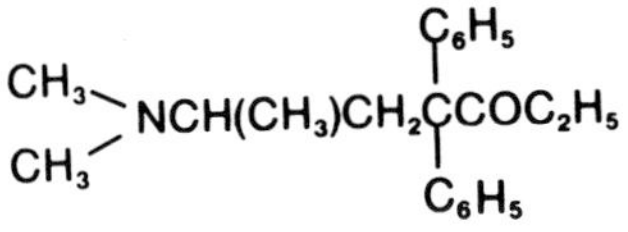

methadone
only the laevo form is active

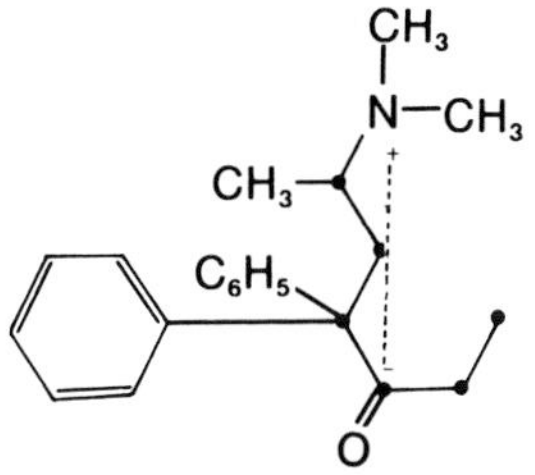

methadone *cf.* morphine structure

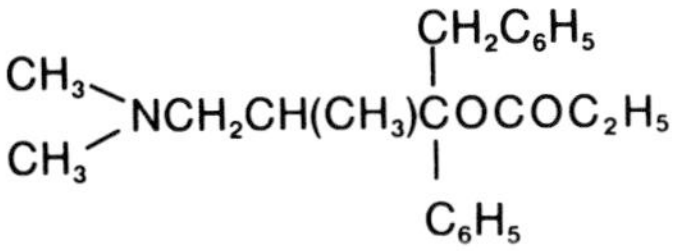

dextropropoxyphene

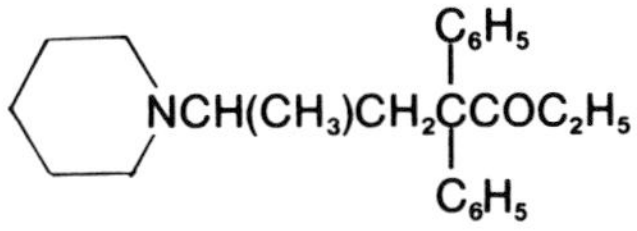

dipipanone

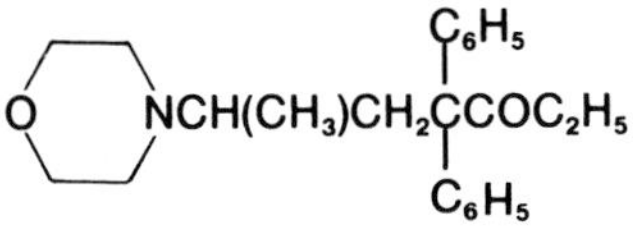

phenadoxone

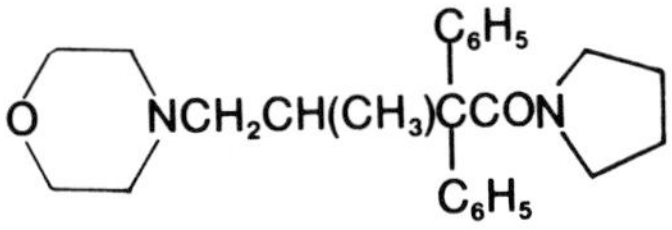

dextromoramide

Fig. 9.12 Methadone and structurally related compounds

person addicted to morphine or any of its derivatives, it brings about a rapid and conspicuous onset of withdrawal symptoms. Also, if a new drug is given over a period of some weeks, followed by an injection of nalorphine, and this leads to the onset of withdrawal symptoms, this can be taken to indicate that the particular drug is liable to cause dependence.

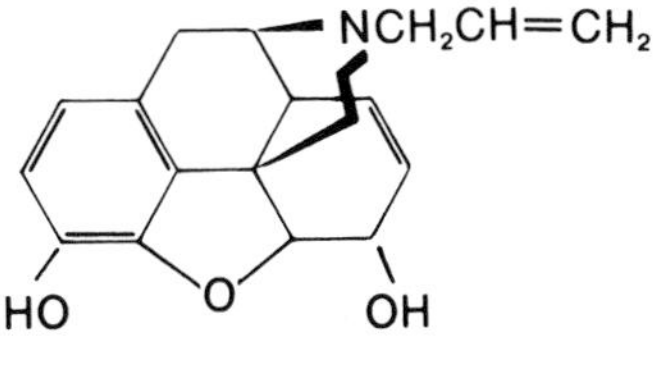

Fig. 9.13 Nalorphine

The time of onset of physical addiction of the opiates used to be characterised as follows:

heroin—4-5 days
morphine—1 week
pethidine—10 days-2 weeks
methadone—1 month

This can be compared with the barbiturates, where addiction takes about six months.

Comparison of potency is done in two ways. Chemists consider a mole-for-mole effect between drugs. Physicians compare them on a dose-for-dose basis—where in fact the number of molecules in a dose will be different for different drugs. In order to maintain a steady level of the drug in the body, the dosing must take into account the half-life of the drug in the body (see Table 9.7). Both these approaches have validity in their particular usages, but the distinction must be kept in mind.

TABLE 9.7
Half-lives of various drugs in the human body

Drug	Time(hours)
Acetylsalicylic acid (aspirin)	0.25 ± 0.03
Amitriptyline	16 ± 6
Amoxicillin	1.0 ± 0.1
Bromide ion	168
Caffeine	4.9 ± 1.8
Chlorpromazine	30 ± 7
Cimetidine	1.9 ± 0.3
Clonazepam	23 ± 5
Cocaine	0.71 ± 0.26
Dapsone	28 ± 3
Desipramine	18 ± 6
Diazepam (Valium)[a]	18 ± 6
Erythromycin	1.8 ± 0.6
Flurazepam	1.5
Imipramine	18 ± 7
Isoniazid	
— fast acetylators	1.1 ± 0.1
— slow acetylators	3.1 ± 1.1
Methadone	35 ± 12
Morphine	3.0 ± 1.2
Nicotine	2.0 ± 0.7
Nitrazepam	26 ± 3
Nitrogylcerine	2.3 ± 0.6 min.
Nortriptyline	31 ± 13
Phenobarbital	99 ± 18
Phenylbutazone	56 ± 8
Protriptyline	78 ± 11
Streptomycin	5.3 ± 2.2
Sulfadiazine	7.0 ± 3.9
Tetracycline	10.6 ± 1.5
Theophylline	9.0 ± 2.1
Thiopental	9.0 ± 1.6

Source: A. G. Gilman, L. S. Goodman, T. W. Ball and F. Murad, *Goodman and Gilman's pharmacological basis of therapeutics*, 7th edn, Macmillan, New York, 1985, Table A–11–1.

[a] You must also consider the active metabolites, e.g. desmethyldiazepam (62 ± 16) and oxazepam (7.6 ± 2.2).

Barbiturates

Barbiturates were once the major ingredients used in sleeping pills and provided valuable adjuncts to anaesthetics. Phenobarbital, one of the earliest barbiturates to be synthesised, has distinctive anticonvulsant properties useful in the treatment of epilepsy. Barbiturates are derivates of barbituric acid, first synthesised more than 100 years ago by the German chemist von Bäyer, by the reaction shown in Figure 9.14. Barbituric acid is not pharmacologically active. Replacement of the hydrogen atoms at the fifth carbon position with alkyl or aryl substituents yields drugs with sedative or hypnotic properties.

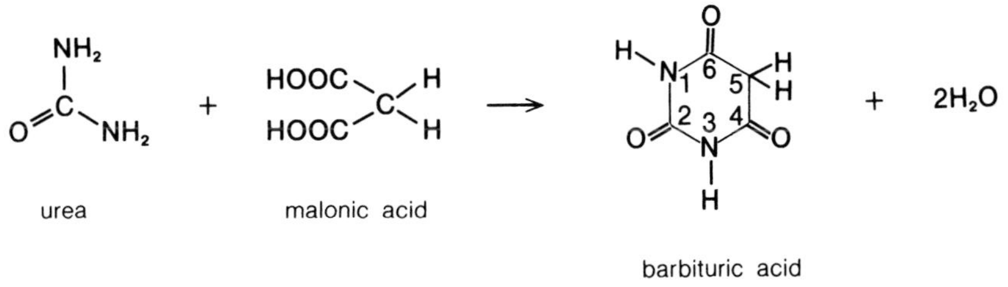

Fig. 9.14   Preparation of barbituric acid

Since 1900, well over 2000 barbiturates have been synthesised; some are illustrated in Figure 9.15. A mere half-dozen became widely adopted in medicine, but now chronic insomnia is no longer treated with these drugs. Thiopental is the standard injectable general anaesthetic. The chain length of the substituents at the C-5 position plays an important role in establishing the action of a particular barbiturate. Phenobarbital, with phenyl group at C-5, was more valued for its selective anticonvulsant activity than for its sedative properties. Alkyl groups at C-5 also provide anticonvulsant activity. However, if the chain is too long or if alkyl groups are placed on the two nitrogen atoms at positions one and three, *convulsants* are produced.

If thiourea, $S=C(NH_2)_2$, is used in place of urea in the synthesising reaction, thiobarbiturates are obtained; an example is thiopental (Fig. 9.15).

Generally, the more fat-soluble a barbiturate is made, by having non-polar groups, the more rapid the onset of the action is. Hypnotic properties may often be increased by increasing fat solubility and may be abolished by introducing polar groups on the side chains. Barbiturates and alcohol are both metabolised in the liver and, even when both are taken at non-toxic dosages, the combination can be toxic because the alcohol retards the excretion of the barbiturate. This is one example of a drug interaction.

The problem with all *chemically* effective hypnotics is that, when they are first taken, they reduce dreaming. As a certain amount of dreaming is essential to the brain, the amount of dreaming experienced on coming off the barbiturate is believed to increase to make up the loss. This can have the effect of apparently interrupting sleep (although it is often an illusion) so that the haggard patient will feel like taking more drugs. It may take many days or several weeks to re-establish

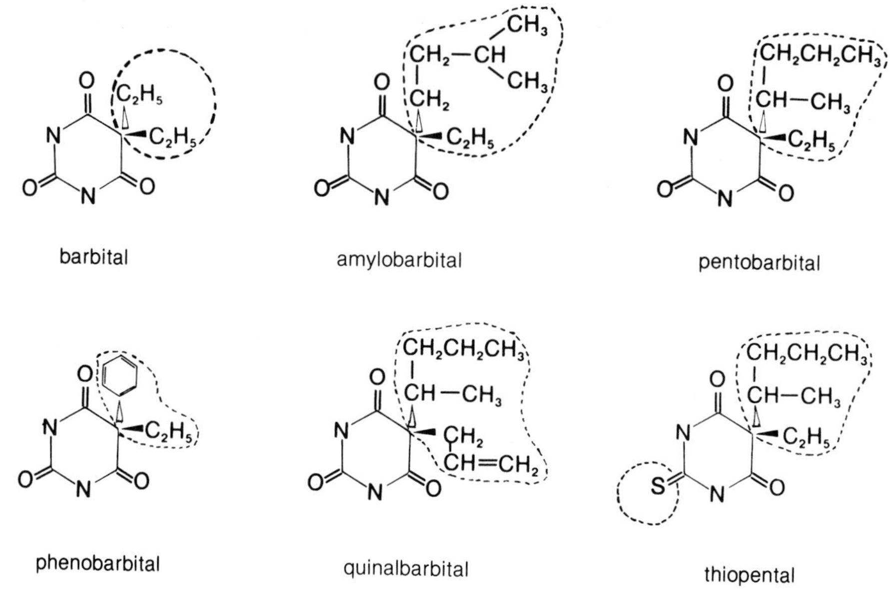

Fig. 9.15 Barbiturates

a normal sleep pattern. (Some rats treated with barbiturates were found to suddenly respond less to a given dose than previously. The cause was exposure to DDT, which activates the same liver detoxification enzymes that destroy barbiturates.) The word *chemically* is emphasised because, in many cases, sleep is induced indirectly; for example, by relieving anxiety with a tranquilliser or a placebo (used because of the major interaction between expectation and effect).

Barbiturates are sometimes classified according to length of action (Table 9.8), although this method is discounted by some authorities.

TABLE 9.8
Classification of barbiturates according to action time (see also Table 9.6)

Long-acting
   phenobarbital[a]; methylphenobarbital (Prominal[b])[a]
Intermediate duration
   amylobarbital (Amytal)[a]; butobarbital (Sonabarb), butethal (Neonal); hexethal (Oral); vinbarbital (Delvinal)
Short-acting
   cyclobarbital (Amnosed); pentobarbital (Nembutal, Petab, Sommital, Penbon, Sodepent, Pentone, Pentobeta); secobarbital (Seconal)
Ultrashort-acting
   hexobarbital sodium (Evipal); thiamylal sodium (Surital);[c] thiopentonesodium (Pentothal)[c]

[a] Available under the Pharmaceutical Benefits Scheme. International spelling (not yet adopted in Australia).
[b] Names in parentheses are trade names.
[c] The thiobarbiturates (Pentothal and Surital) are inactive by mouth and can be administered only by intravenous or rectal routes—they belong to the group of infamous truth drugs.

Both *tolerance* (increasing quantities needed for an effect) and physical dependence occur with high doses. The barbiturates stimulate enzymes in the liver that break down the drug, thus reducing its effect. Although tolerance develops to the sedative effect of the drugs, the lethal dose remains essentially constant. As tolerance increases, therefore, the margin of safety decreases, and accidental poisoning may occur at doses that no longer provide sedation. The *therapeutic index* is the ratio of the toxic dose to the effective dose. The larger this factor is, the greater the safety in the use of the drug. The therapeutic index is dependent on two types of drug tolerance:

- *pharmokinetic tolerance*—a tolerance due to changes in the concentration of the drug in the body caused by changes in liver activity.
- *pharmodynamic tolerance*—a tolerance caused by the receptor (where the drug acts) requiring more drug while the concentration for receptor poisoning may not change.

Amphetamines

Amphetamines were once used to treat obesity, mild depression and narcolepsy (a tendency to fall asleep at any time), and certain behavioural disorders in children. The last is the only current use. In Australia, they cannot be prescribed without the consent of the State health authorities. Amphetamines are pep pills. Ordinary doses of 10 to 30 mg per day provide a feeling of well-being and increased alertness. Amphetamines are structurally similar to the naturally occurring *biogenic amines*, such as *ephedrine*, which act as stimulants of the central nervous system, in a similar manner to epinephrine.

Amphetamine and epinephrine (see Fig. 9.16) are *optically active*. This means there are two compounds with *exactly the same formula* but whose structures are mirror images of each other and cannot be superimposed. If you look at your hands they are pretty much the same shape—I hope!—but you can't place one hand in an identical position on top of the other. However, if you hold them parallel, one acts as the image of the other in an imaginary mirror placed in between. In fact the two pairs of chemicals related in this way are called left and right handed (!) or, using the Latin *laevus* and *dexter*, *l-* and *d-* for short. It is actually a bit more complicated than this—but isn't it all? Strange as it may seem, compounds differing only in this way can be biologically very different in their activity. *Benzedrine* is a 50 : 50 mixture (racemic) of the *d-* and *l*-amphetamine but, as the *l-* is less active on the central nervous system, pure *d-* or *dexedrine* is obviously nearly twice as potent.

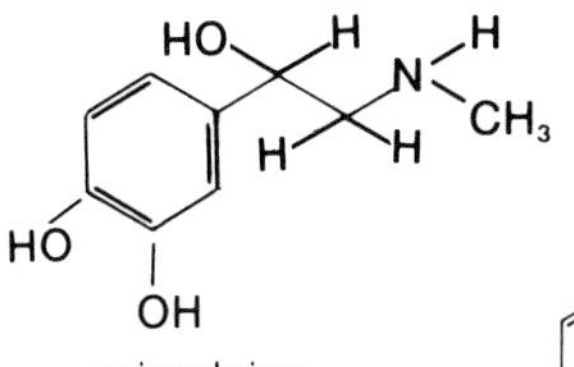

epinephrine
(adrenaline)

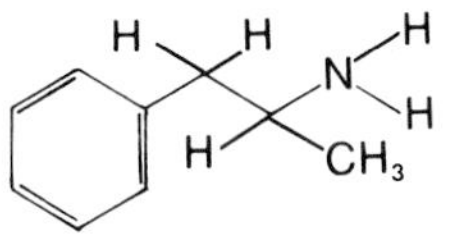

'amphetamine'

Fig. 9.16 Biogenic amines

Amphetamines and barbiturates were often used in conjunction. Thus amphetamines may be consumed in the morning to alleviate the symptoms of a barbiturate hangover, whilst the barbiturates may be necessary to counteract the stimulant properties of amphetamines and allow the user to sleep. In case of overdose they were also used as mutual antidotes. The deeply held belief by the public in antidotes is somewhat dangerous, because although two substances may be antidote in *one* aspect, they can reinforce each other (*synergism*) in other side-effects. The death rate can be very high. The amphetamines also form a family of drugs although the pattern is somewhat difficult to see and tends to overlap other categories of drugs.

Tranquillisers

These are drugs that sedate without inducing sleep. The *major* tranquillisers are used in the treatment of schizophrenia by blocking dopamine receptors in the brain. Many of them are based on a compound called *phenothiazine* (Fig. 9.17) and its derivatives (Table 9.9).

watch this point

phenothiazine

chlorpromazine

promethazine

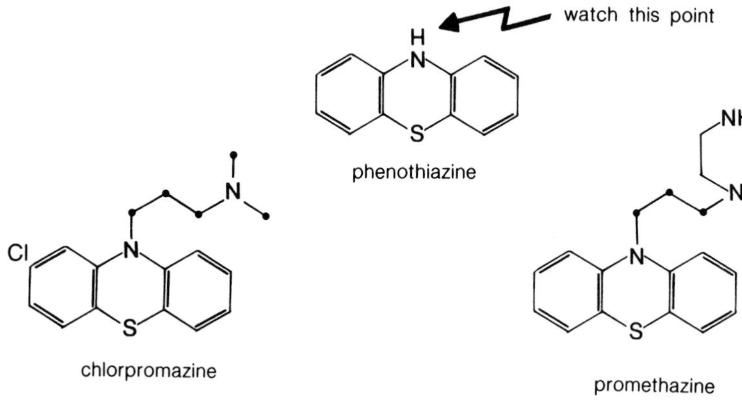

Fig. 9.17 Phenothiazine and two of its derivatives

TABLE 9.9
Series of phenothiazine tranquillisers with different types of substituent

Generic name	Trade name
Aliphatic series	
Chlorpromazine	Largactil, Protran, Promacid
Promethazine	Phenergan[a], Meth-Zin[a], Progan[a], Prothazine[a], Avomine[a]
Piperidine series	
Thioridazine	Melleril, Aldazine
Pericyazine	Neulactil
Piperazine series	
Prochlorperazine	Stemetil, Compazine, Anti-Naus
Thiopropazate	Dartalan
Fluphenazine	Anatensol
Fluphenazine decanoate	Modecate
Trifluoperazine	Stelazine, Calmazine

[a] Not available on Pharmaceutical Benefits Scheme.

Derivatives of phenothiazine that retain the sulfur atom but not the nitrogen are the thioxanthine tranquillisers (Table 9.10).

TABLE 9.10
Thioxanthine tranquillisers

Generic name	Trade name
Chlorprothizene	Taractan
Clopenthixol	Sordinol
Flupenthixol	Fluanxol
Thiothixene	Navane

If the sulfur and the nitrogen atoms of phenothiazine are replaced by —CH=CH— and —CH— respectively, one of the derivatives is protriptyline (Fig. 9.18). Compare it with the tricyclic antidepressants in Figure 9.20.

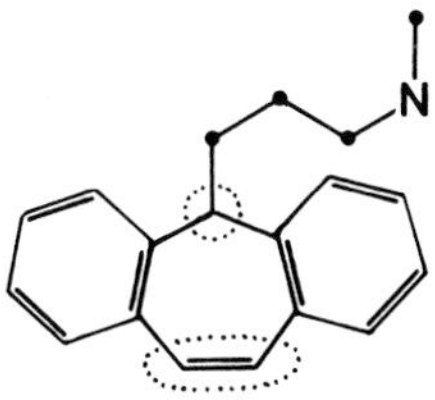

Fig. 9.18   Protriptyline

All these compounds are used to relieve the symptoms of schizophrenia and reduce the likelihood of relapse, and they affect the brain stem rather than the cortex. Their use has profoundly modified the problems of the mental hospital, but they do carry a high incidence of adverse reactions.

Minor tranquillisers
The most common of the minor tranquillisers are built up on a benzodiazepine nucleus. Four are illustrated in Figure 9.19.

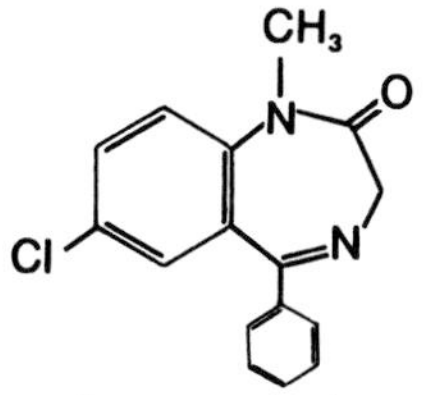

diazepam, trade name: Valium

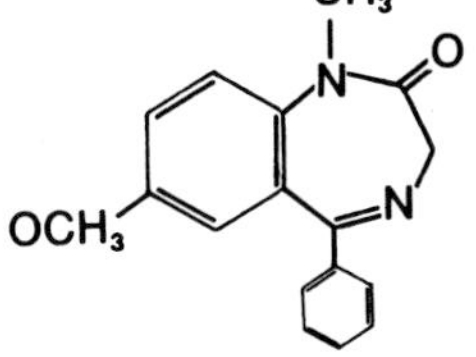

oxazepam, trade name: Serenid

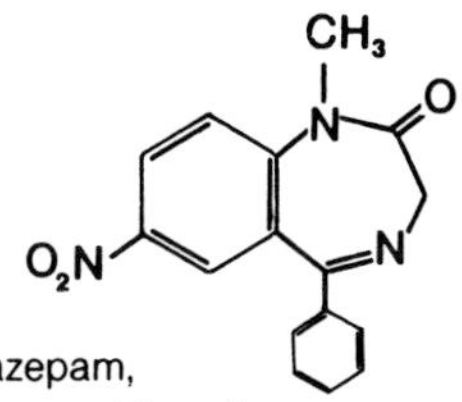

nitrazepam,
trade name: Mogadon
(sleeping pill)

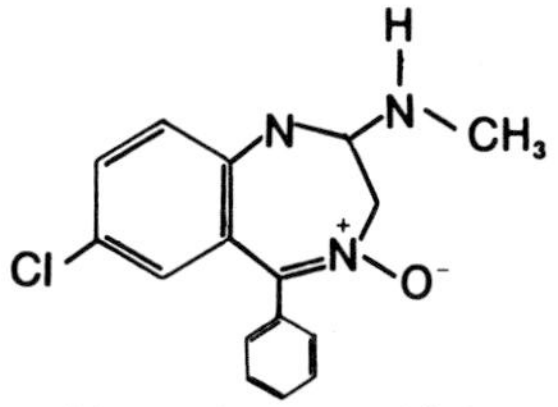

chlordiazepoxide, trade name: Librium

Fig. 9.19   Minor tranquillisers

Librium was used in the treatment of neuroses, behaviour disturbances, alcoholism and as premedication for anaesthesia. Valium is used to reduce symptoms of anxiety. The differences relate to how fast they metabolise to the fast-acting actual drug—nordazepam. Valium loses the 1-methyl group, while Librium hydrolyses the 2-methylamino group to an oxygen.

Tricyclic antidepressants

Depression is a problem that faces many people and the 'tricyclics' (Fig. 9.20), usually derived from *dibenzazepine*, form a popular family of antidepressants. These drugs have as many side-effects as the tranquillisers. They present a particular problem of overdose abuse, and most are on the Pharmaceutical Benefits Scheme list (Table 9.11).

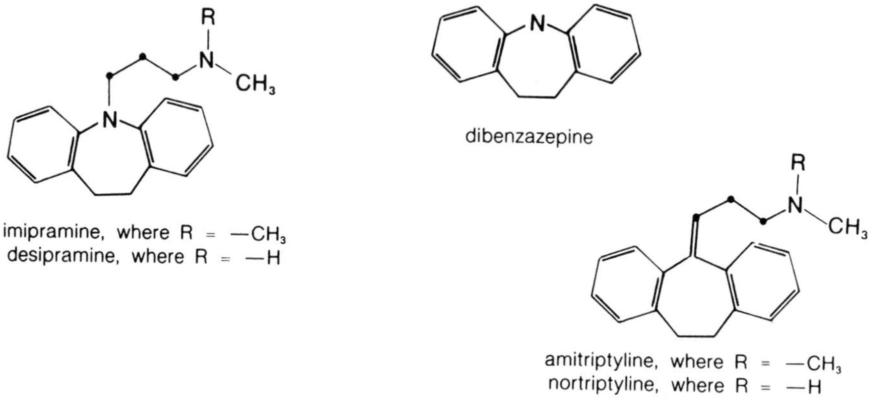

imipramine, where R = —CH₃
desipramine, where R = —H

dibenzazepine

amitriptyline, where R = —CH₃
nortriptyline, where R = —H

Fig. 9.20   Tricyclic antidepressants

TABLE 9.11
Tricyclic antidepressants (available on Pharmaceutical Benefits Scheme)

Generic name	Trade name
Imipramine	Tofranil, Imiprin, Melipramine
Desipramine	Pertofran
Amitriptyline	Tryptanol, Laroxyl, Saroten, Amitrip, Endep
Nortriptyline	Allegron, Nortab
Trimipramine	Surmontil
Doxepin	Sinequan, Quitaxon, Deptran

An interesting series of drugs that are still occasionally used to treat depression are the so-called *monoamine oxidase inhibitors* (or MAO inhibitors). The name means that they have the capacity to inhibit an enzyme that is normally responsible for removing certain substances (those *biogenic amines* again!) such as norepinephrine (noradrenaline) and serotonin from the body. Currently there is considerable evidence that depressive illnesses are associated with a *decrease* in the level of these amines in certain parts of the central nervous system, so that by inhibiting their destruction, their level is increased. An example of a biogenic amine is phenelzine (Fig. 9.21). Notice its close similarity to amphetamine (Fig. 9.22). Other MAO inhibitors are listed in Table 9.12.

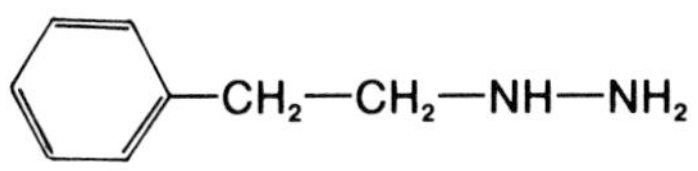

Fig. 9.21   Phenelzine

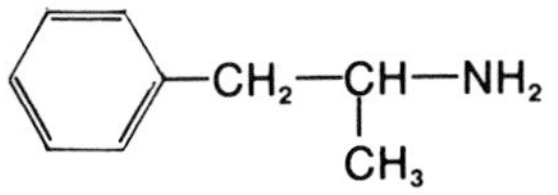

Fig. 9.22   Amphetamine

TABLE 9.12
MAO inhibitors

Generic name	Trade name
Iproniazid	Marsilid (5% rate of liver damage) (deleted from PBS August 1987)
Phenelzine [a]	Nardil
Nialamide	Niamid (less effective than a placebo)(deleted from PBS)
Isocarboxazid	Marplan
Tranylcypromine	Parnate [a], (strong 'cheese' effect, see below)
Mebanazine	Actomol (deleted from PBS)

[a] Available on Pharmaceutical Benefits Scheme (PBS)

Patients treated with these drugs have to be warned to avoid eating cheese, red wine, certain beers, piquant foodstuffs such as Marmite and Bovril, and must not take any other medication without consulting their doctor. The reason for this is that these foodstuffs contain *tyramine*, which is normally broken down in the alimentary canal. When MAO-inhibiting drugs are used, the enzymes that carry out the breakdown are inhibited, allowing tyramine into the bloodstream. This causes a massive release of norepinephrine, which in turn causes a sudden fluctuation in blood pressure, which produces intense headache—and sometimes death.

Antihistamines

Many people are allergic to pollen, stings, dust etc. An *allergen* is a substance that initiates the allergic response. It is usually a protein but is sometimes a polysaccharide. For a person with pollen allergy, a pollen grain enters the nose and clings to the mucous membrane. The nasal secretions acting on the pollen grain release the grain's allergens and other soluble components, which penetrate the outer layer of the mucous membrane. By a series of events that are not well understood, the allergen forms a complex with an antibody of a type that is present in unusually high concentrations in allergic persons. The complex is responsible for the release of the allergy mediators, one of the most potent being histamine (Fig. 9.23). Histamine is formed by the breakdown of the amino acid histidine; it accounts for many of the symptoms of hay fever and other allergies.

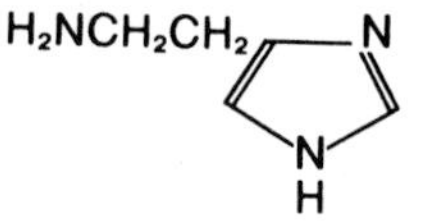

Fig. 9.23   Histamine

Antihistamines are most widely used for treating allergies, and there are more than 50 types available. Many of these contain, as does histamine, an ethylamine group, $-CH_2CH_2N=$. These drugs compete with histamine for the receptor sites normally occupied by it on cells and thus prevent it from causing allergic reactions. An example of a well-known antihistamine is Polaramine (Fig. 9.24). Note that some of the tricyclic antidepressants have antihistamine effects as well, because they also contain the ethylamine group.

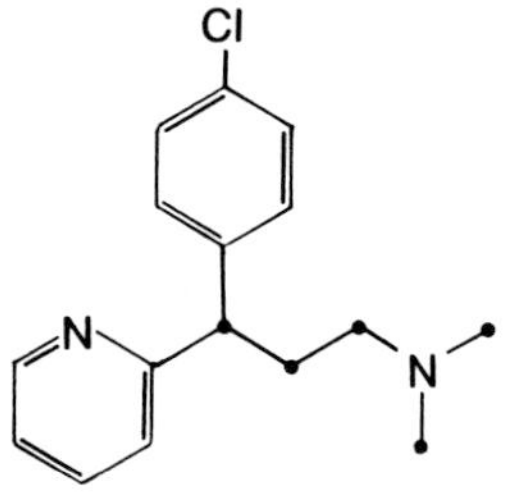

Fig. 9.24   Dexchlorpheniramine (Polaramine)

Hallucinogenic drugs

Hallucinogenic (or related psychotomimetic) drugs derived from various plants and fungi have been used from time immemorial. They were used for religious purposes and for festivals and orgies. The use of the emetic toadstool *Amanita muscaria* extends over thousands of years. The Aztec and Mayan cultures used the peyote cactus, from which *mescaline* (Fig. 9.25) is derived. They also used the psilocybe mushroom (or sacred mushroom Teonanacatl), the active principle of which is *psilocybin* (Fig. 9.26), which is about 30 times as potent as mescaline. Similar mushrooms are found in Australia. From a plant called ipomoea (morning glory) the Mexican Indians 'obtained' a substance similar to lysergic acid (Fig. 9.26), and from the plant *Datura stramonium* (thorn apple) they obtained *scopolamine (hyoscine)* and *atropine*. Other plants used in Central and South America contained *cocaine* (Fig. 9.27), and there is at least one hallucinogenic animal, a caterpillar found in bamboo stems. An interesting fact is that tannin in tea contains gallic acid, which can be converted to mescaline by complex

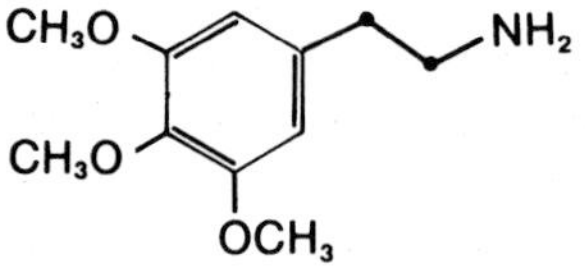

Fig. 9.25   Mescaline

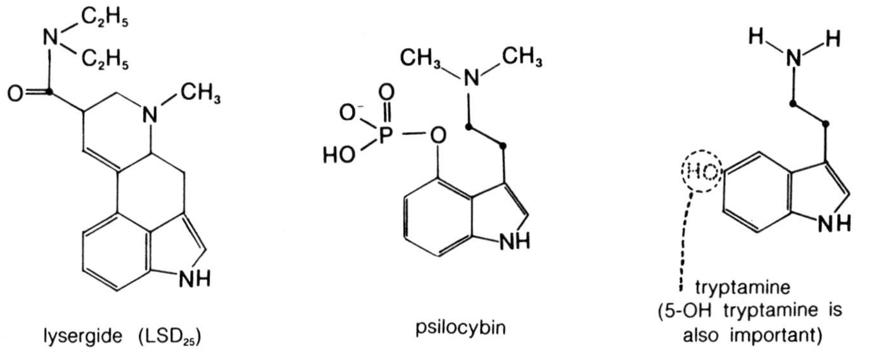

lysergide (LSD₂₅)   psilocybin   tryptamine
(5-OH tryptamine is
also important)

Fig. 9.26 Some hallucinogens based on tryptamine

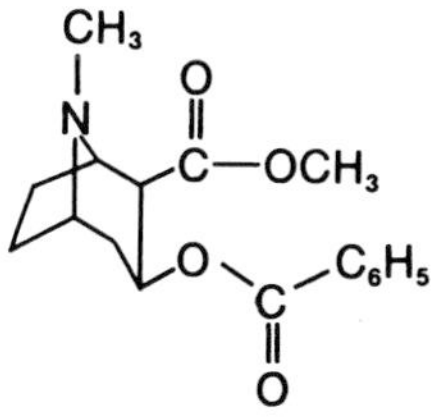

Fig. 9.27 Cocaine (blocks nerve transmissions)

chemistry. Mescaline was made famous by Aldous Huxley. It is classed as a *catecholamine*, along with amphetamines, to which it is structurally related (compare the structures).

Lysergic acid diethylamide (LSD)

LSD is classed as an *indoleamine*. It is one of the most potent drugs known and doses as low as 20 to 25 μg (1 μg = 10^{-6} g, 1 millionth of a gram) are capable of causing marked effects in susceptible individuals. It was at one stage believed to cause chromosome damage when taken in large doses but this is now disputed. Lysergide (Fig. 9.26) was discovered, in a chemical sense, by the Swiss chemist Albert Hoffmann in 1938, when he accidentally ingested some of the compound while investigating a modified ergotamine as an improved drug for childbirth. Clandestine manufacture is usually from ergot alkaloids to yield lysergic acid, to which the diethyl groups are easily added. Ergot itself is found on many plants, particularly rye. An ergot alkaloid is used to induce uterine contractions. In ergotamine, the diethylamino group is replaced by a peptide (a mini-protein).

Cannabis

The use of cannabis (marijuana, hashish, Indian hemp) has a long history. The most active ingredient in the extract is *tetrahydrocannabinol*, THC (Fig. 9.28). THC can be obtained from the fruiting or flowering tops of the cannabis plant, the cultivation of which is banned in Australia. There are many ingredients in cannabis other than THC, and their long-term effects are unknown. There is evidence to suggest that, chemically if not socially, it is less harmful than tobacco, except when driving a car.

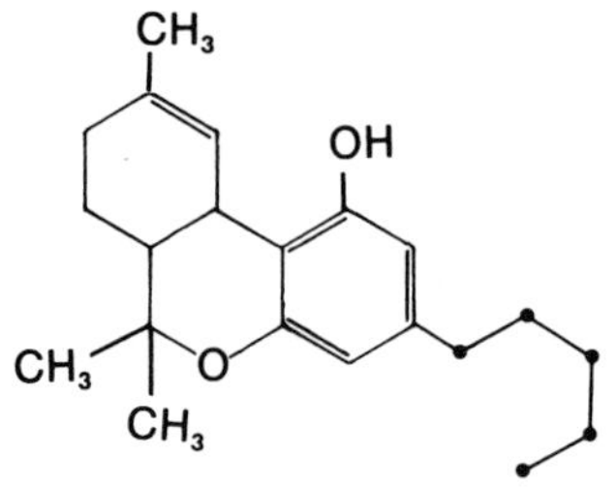

Fig. 9.28 Tetrahydrocannabinol (THC)

Mode of action[2]

To understand the effect of drugs on us, we naturally have to explore the chemical and pharmacological action of the chemicals involved. However the effect of drugs is strongly influenced by the personal and social environment. Traditional drugs used in traditional ways often cause few problems. Opium and cocaine are good examples. In a different legal and social climate their effects can be disastrous. Physical dependence on opiates is a complex issue, which I will not attempt to treat here. It is worth noting the chemistry. Opiates reduce pain, aggression and sexual drive, hardly the stuff to make violent criminals. (They do, however, make zonked car drivers.) The Chinese opium smokers were blissful and peaceful. Our social structure has made these drugs illegal. This provides a certain social attraction to some and also sets a high price for them. This in turn sets the scene for crime and corruption. One way of attempting to separate out the strictly biochemical from additional social effects is to look at animal (especially mammal) experiments.

People's inclination to take drugs is shared with that of other mammals, which show patterns of self-administration that are strikingly similar to those found in human users of the same drug. This seems to rule out any specific mental or physical weakness in the human addict. Animals will press a lever more than 4000 times to receive repeated injections of cocaine. When given free access they will self-administer high daily doses that cause severe toxic effects and even cause the animals to mutilate themselves. With cocaine (and amphetamines), the animals alternate between periods of self-imposed abstinence and periods of drug taking, and they generally die from starvation and poisoning. If saline solution is substituted, there is a burst of rapid lever pulling for several hours, then abruptly all responses cease and are not resumed. When morphine is provided the behaviour is very different. The animal will raise the dose it is giving itself gradually over a period of weeks, then keep the rate at a steady level that avoids excess toxic effects on the one hand, and withdrawal symptoms on the other. When saline solution is *now* substituted, the animal will continue to press the supply lever at a slow but steady rate (except during the peak of withdrawal) for weeks on end, showing that the addiction remains. Although the behavioural patterns in humans are influenced by social and psychiatric factors as well, the basic similarity is worth bearing in mind.

There are aspects of drug taking that have a definite chemical basis. One of these is tolerance.

PLATE I Structures of (left to right) water, ammonia and methane. The geometry of methane is a tetrahedron. In ammonia one hydrogen is replaced by a lone pair of electrons. In water two hydrogens are replaced. The angles between the remaining atoms change a little as a consequence

PLATE II Stick and space-filling models for aspirin. Each model has its uses for visualising the shape of this molecule and the way the atoms in the molecule can move

PLATE III Hats are best to protect from the sun: the author with (from left to right) his sons, Adam and Michael, and his wife Veronica

PLATE IV Space-filling model of DDT showing the shape of the molecule. This molecule acts by wedging open a channel in the cell membrane and causing leaks. Overall shape is more important than the exact nature of the atoms in the molecule

PLATE V Chlorine meter. It looks impressive, but is there any reason why it *should* work?

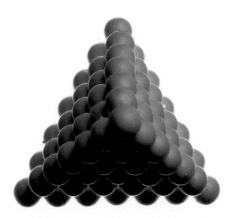

PLATE VI a. Cubic close packing: *abc* layering of spheres forms a tetrahedron with four layers of spheres. No channels are seen from the top

b. Hexagonal close packing: *abab* layering of spheres. Note the channel of light through the unfilled *c* position. The hexagonal shape is not seen unless the triangular layers are of equal size

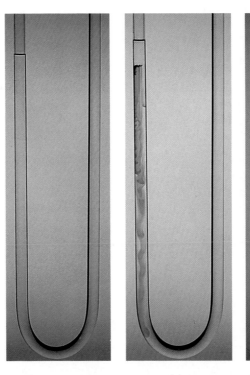

PLATE VIII Unbalanced U-tube. The liquid poured in the two arms of the U-tube settles at different heights. The reason for this is that there is a different liquid on each side! However they are matched for refractive index so the meniscus between them is not visible. The liquids have different densities. The two liquids are revealed by adding iodine crystals, which dissolve in one liquid but not in the other. (See Chapter 8 for further details.)

PLATE VII Disappearing glass tubing. The liquid matches the refractive index of one type of glass but not the other. No light is scattered when there is no refractive index difference

PLATE IX A selection of different adhesives for different tasks. How interchangeable are they?

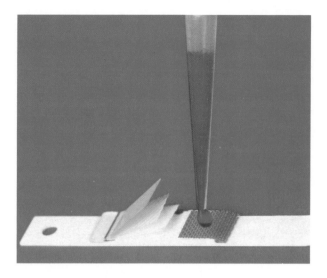

PLATE X Automated blood testing
(Courtesy of Boehringer-Mannheim)

a. Placing the blood sample on the
reagent carrier

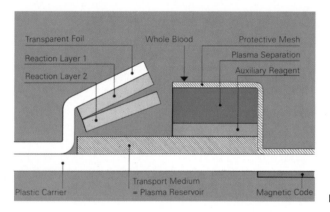

Transparent Foil | Whole Blood | Protective Mesh
Reaction Layer 1 | | Plasma Separation
Reaction Layer 2 | | Auxiliary Reagent

Plastic Carrier | Transport Medium = Plasma Reservoir | Magnetic Code

b. Reagent carrier before reaction

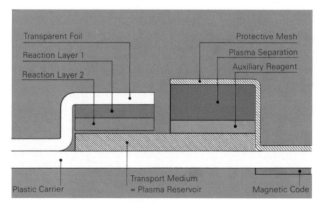

Transparent Foil | | Protective Mesh
Reaction Layer 1 | | Plasma Separation
Reaction Layer 2 | | Auxiliary Reagent

Plastic Carrier | Transport Medium = Plasma Reservoir | Magnetic Code

c. Reagent carrier during reaction

PLATE XI The big bang—the combustion chamber designed for a reproducible explosion of hydrogen and oxygen to illustrate flammability limits

PLATE XII The copper metal suspended over a trace of liquid acetone is kept red hot by the heat released at the surface of the copper from catalysing the oxidation of acetone vapour.

PLATE XIII   In ten years the level of hidden sugar in processed foods stayed high. We eat more sugar indirectly in this way, than directly by deliberate consumption of sugar

PLATE XIV   Space-filling models for some artificial sweeteners

a. Sucrose

b. Saccharin

c. Cyclamate

d. Acesuflam K

e. Dulcin

PLATE XV (Courtesy of Fritz Sondereggar, Urambi Hills Wholemeal Bakery, ACT)

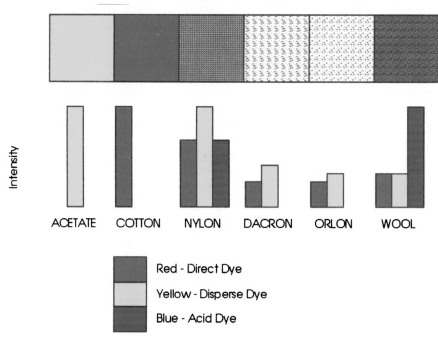

PLATE XVI Dyes and how they interact with different fibres

Tolerance

Drug users often develop tolerance to the chemical they are using. This can occur for a number of reasons. In the case of the opiates most of the tolerance comes from the adaptation of the cells in the nervous system to the drug's action.

In the case of alcohol, barbiturates and related hypnotics (a group of *depressants* of the central nervous system, CNS), chronic use causes the capacity of the enzymes that metabolise the drugs to increase, i.e. you can remove the alcohol faster. Social drinkers can metabolise ethanol only with the liver's slow acting enzyme, alcohol dehydrogenase. (Police sometimes use this information to calculate from the level they measure on a breathalyser back to some earlier period of time.) Chronic drinkers are no more efficient with this enzyme, but they induce a new alcohol-destroying enzyme of the P-450 (cytochrome mono-oxygenase) type in the liver. Such persons can perform well on difficult tasks at blood alcohol levels above 0.2 mg/mL (the legal limit varies; 0.05, 0.08 to 0.1). After several weeks of abstinence the capacity of this enzyme declines, so that the abstinent alcoholic and normal individual metabolise alcohol at the same low rate. Chronic use often means a higher blood concentration is needed to produce the same effects, i.e. it produces *pharmacodynamic* tolerance. This in turn means that to obtain the same effects, the person will consume more of the drug. However the fatal dose of the drug does not change. The result is often death by overdose.

When the same enzymes are involved, tolerance to one drug can cause cross-tolerance to another. An example is the cross-tolerance of alcohol with benzodiazepines. Thus chronic drinkers will deal not only with alcohol more effectively, but also with Valium. If they consume both drugs at the same time then the following happens. The alcohol will monopolise the enzyme which then is not free to deal with the Valium. The effect of the Valium is enhanced and prolonged (and of course added to that of the alcohol).

There are other aspects, such as linking behaviour to drug level, which are much harder to correlate chemically. The point at which marked intoxication is caused by drinking alcohol can be monitored by measuring the blood alcohol (or breath) levels. However there is an interesting difference between the level which is found to correspond to intoxication 'on the way up', i.e. while drinking, compared to 'on the way down', i.e. after drinking has ceased. The effect on behaviour (in particular for alcohol) appears to be far less on the way down than on the way up, and this has meant that drivers caught by a breathalyser test the morning after a heavy drinking bout have no idea that their level is still high. Owing to a genetic defect, Westerners oxidise ethanol only slowly in the first stage to acetaldehyde. The Japanese and Chinese, on the other hand, generally have a gene that codes for an enzyme that oxidises it faster. As a result, a few sips of ethanolic beverage bring a deep red colour to their cheeks and an unpleasant tingling.[3]

Cocaine and amphetamines belong to a group of drugs that mimic the natural substances that *stimulate* the CNS. They cause an elevation of mood, a sense of increased strength and mental capacity, and less need for sleep or food. (The natives living high in the Andes chewed the leaves of the coca bush for generations for just this purpose.) The cocaine is converted from the hydrochloride salt to the free base with alkali and extraction with organic solvents. Absorption from the lungs is then increased dramatically. The drug is highly addictive. Given a choice of cocaine and food, monkeys select cocaine (over a period of eight days). Continuous access causes weight loss, self-mutilation and death within about two weeks. Whereas the effects of cocaine fade quickly, because the esterase enzymes in the bloodstream quickly hydrolyse it, the effects of amphetamines last for hours. This has been demonstrated in double-blind experiments, where subjects could not tell with which drug they had been injected until some time had passed.

Nicotine and tobacco

Columbus discovered tobacco smoking as well as America, and the name of the plant, *Nicotiana tabacum*, derives from the entrepreneur who promoted its sale, Jean Nicot. The active ingredient, nicotine, was isolated in 1828. New varieties, better methods of curing the leaf, coupled with technology for mass production, allowed the introduction in the mid-nineteenth century of the cigarette—cheaper and neater than the cigar, with a smoke so mild it could be inhaled! About 4000 compounds have been found in cigarette smoke. No other drug of dependence causes cancer, and tobacco is the only environmental cause of cancer that is on the increase.

Classification for health purposes has concentrated on the levels of nicotine, tar (which contains the potent carcinogens) and carbon monoxide. The carbon monoxide reacts preferentially with the red corpuscles in the blood. On removal of the source of carbon monoxide, the equilibrium with oxygen is gradually restored.

The levels of these compounds are determined by using smoking machines. However smokers do not conform to smoking machines, and manufacturers can design cigarettes to perform well on the machines, while dosing smokers at high levels. (A common ploy is to include fine holes just up from the filter, which lower the machine reading through dilution of the smoke with air. However when smoked for real, these holes are covered by the smokers' lips.) The nicotine content of tobaccos can vary from 0.2% to 5% and provides from 0.05 to 2.0 mg (1982 average 1.0 mg) per cigarette to a smoking machine. In cigarettes the nicotine is nearly always present in a protonated form in which it is less readily absorbed through the mouth (hence the need to breathe the smoke into the lungs), in contrast to the basic form found in cigars and pipe tobaccos (smoke pH 8.5). Note the analogy to cocaine and 'free base' cocaine. The nicotine is suspended on the minute particles of tar and absorption from the lung occurs in *seconds* and is almost as fast as intravenous injection. Peak concentrations found in the blood are typically 25 to 50 ng/mL.

A fascinating aspect of smoking is the way smokers titrate their nicotine needs. When heavy smokers are unknowingly given cigarettes with a higher content of nicotine they subconsciously reduce the number smoked and alter their puffing pattern to maintain about their usual level of nicotine. Conversely when given low-nicotine cigarettes they increase the number smoked and/or puff more efficiently. Thus another example of counterproductive social engineering is to insist on low-nicotine cigarettes, because the smoker then takes in more tar and more carbon monoxide for the same level of nicotine. This is in stark contrast to marijuana smokers, who do not titrate their desire for the active drug because its absorption and effect are much delayed. It takes seven to 10 minutes for the blood plasma concentration of the active ingredient, Δ^9-THC, to reach its peak, and the physiological and subjective effects do not reach a maximum for 20 to 30 minutes. Titration is thus not possible. (Taking the drug through the mouth delays onset of effects for a half to one hour.) The drug is slowly destroyed in the liver and gradually excreted in the urine and faeces. This long delay makes it difficult for law enforcement authorities to determine whether the drug was in use while, for example, driving, if this is (as it should be) an offence. In fact the legal difficulties of enforcement of non-drug use while performing activities such as driving or operating machinery is a major technical obstacle to decriminalisation, if this were otherwise desired.

Drug interactions

The interaction of drugs is an extremely important area of study. One drug can change the pharmacological effect of another—using the term *drug* in the evident sense to include such substances as alcohol, some foods and food additives. Some drugs even precipitate in the bottle (e.g. tetracycline and calcium ion).

Drugs that are taken orally have to be absorbed through the gut, and this can be influenced by other material present. By using suitable coatings a drug can be absorbed either in the acidic stomach or the alkaline duodenum. Once the drug is in the plasma it can become bound to protein and only a small percentage remains free and active. This percentage can be drastically altered by another drug, which kicks the first one off its protein site. Often use is made of this method to boost the efficiency of a drug. Also, a drug can affect the efficiency of an enzyme and hence influence the rate at which a second drug is broken down by that enzyme. The MAO inhibitor drugs and the consequences of eating cheese while they are being taken is a classic example.

The way and speed with which drugs are metabolised by the body can also depend on *genetic factors*, so that comparisons between animals and humans, and between individuals, can be misleading. They also depend on *physiological factors,* such as age, diet, hormones (including the effects of pregnancy) and disease states—especially if the liver is involved. The old are particularly liable to be treated with several drugs simultaneously and they, in particular, will have impaired metabolism, which will affect the drugs' effects on them.

Very often a drug is changed in the body to another compound. Sometimes the new compound is inactive or it may be less active or more active than the original. The original may even be completely inactive and it is the new compound (metabolite) that is the 'real' drug. Some examples of this are shown in Figure 9.29.

In fact the body can be used as a chemical factory. Note the tremendous importance of the liver in the metabolism of drugs. You may begin to realise what immense problems this opens up. Not only do we have to have information about the effect of a new drug we may want to introduce—but also about the effect it has on other drugs and the effect they have on it. We have to know what other compounds it forms in the body and what their properties are. The rat is still one of the most popular animals used in toxicity testing, but it has active micro-organisms present in its stomach, so that orally administered drugs may be extensively metabolised by bacteria even before absorption, giving a markedly different metabolic pattern from that obtained in humans. On the other hand,

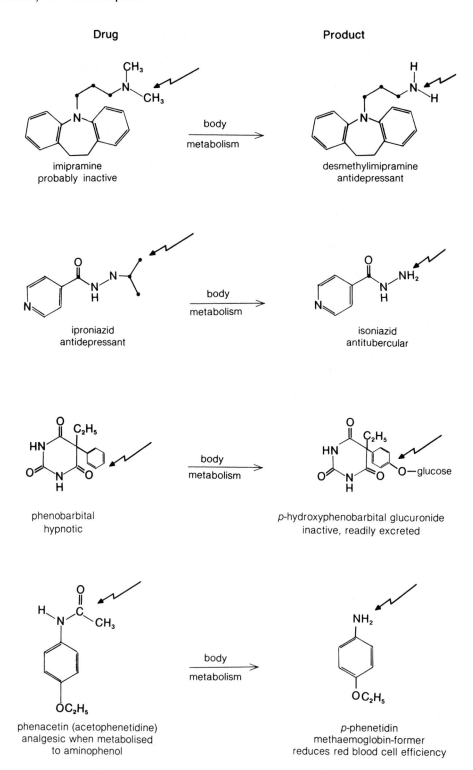

Fig. 9.29   Effects of metabolism on the pharmacological activity of drugs

many drugs used in treatment of illness are of high molecular mass (greater than 400) and, as a consequence, are excreted in the bile as well as the urine, so that they are frequently subject to bacterial metabolism in the intestines. These products can be reabsorbed and further metabolised by the liver—a cycle of absorption, metabolism by the body, excretion, bacterial metabolism, reabsorption and metabolism by the body.

To top it all, there is also the time factor to be considered—which we will deal with more fully under another heading (see 'Food additives—Colouring matter', Chapter 11). The chemical β-naphthylamine was a very important intermediate in the dyestuff industry. It was found, however, to be a very potent carcinogen—it causes cancer of the bladder. It took a long time to realise this because there is a time lag of about 30 years between contact and cancer. It was only because a large number of ex-employees of the German chemical firm died of the same disease at about the same time that the link with the past could be established. The same story has repeated itself more recently with the chemical vinyl chloride, used in the manufacture of the plastic poly(vinylchloride).

REFERENCES

1. This section is based on material in *HHS News*, 4 January 1988, US Department of Health and Human Services.
2. The information in this section is based on material in J. H. Jaffe, 'Drug addiction and drug abuse', in Gilman *et al.*, *Goodman and Gilman's pharmacological basis of therapeutics*, 7th edn, Macmillan, New York, 1985.
3. P. Propping, *Rev. Physiol. Biochem. Pharmacol.* **83**, 1978, 123.

BIBLIOGRAPHY

Essential
Commonwealth Department of Health. *Australian approved names and other names for therapeutic substances.* AGPS, Canberra, 1986.
Commonwealth Department of Health. *Therapeutics in Australia* (undated). Government printer, Tasmania. The official viewpoint.
Gilman, A. G., Goodman, L. S., Ball, T. W., Murad, F. (eds) *Goodman and Gilman's pharmacological basis of therapeutics.* 7th edn, Macmillan, New York. 1985.
MIMS. Intercontinental Medical Statistics (Australasia) Pty Ltd, Sydney. Published bimonthly with an annual.
NH & MRC. *Standard for the uniform scheduling of drugs and poisons, no. 2.* AGPS, Canberra, 1987.
Pharmaceutical Benefits (anon.), December 1987. One version for doctors and another for pharmacists.

General
'Brand-name drug prescriptions costing NHS millions'. *Guardian* (UK) 6 Feb. 1983.
Chemicals and Health. President's (US) Science Advisory Committee, September 1973. Science and Technology Policy Office, National Science Foundation.
Diesendorf, M. (ed.). *The magic bullet.* Society for Social Responsibility in Science, Canberra, 1975.

Drug problems in Australia: an intoxicated society? Report of the Standing Committee on Social Welfare, AGPS, Canberra, 1977.

Gould R. F. (ed.). *Drug discovery: science and development in a changing society. (Advances in Chemistry no. 108).* American Chemical Society, Washington DC, 1971. The proceedings of a symposium, this work includes a lot of interesting chemistry but also gives the points of view of the drug companies and their relations to the testing agency, the FDA.

Negwer, Martin. *Organische-Chemische Arzneimittel und ihre Synonyma* [Organic chemical drugs and their synonyms] Academie-Verlag Berlin, 4th edn, 1971. This is a standard reference work on pharmaceutical nomenclature. It lists 40 000 synonyms for 5000 distinct pharmaceuticals. (Compare the 3rd edition 1966: 26 000 synonyms for 4000 substances.) Volume 1 contains the structural formulae of the 4000 drugs and a running index number. The systematic chemical name of the compound (IUPAC rules), proposed and recommended WHO non-proprietary or generic names (in italics), and all the trade names are listed. Brief reference to therapeutic uses is also given. Volume 2 contains a group index to facilitate the finding of drugs related to each other chemically or pharmacologically with 1500 keywords. It also contains the 40 000 synonyms with index numbers corresponding to the structural formulae of Volume 1.

Many preparations are marked under different names in different countries (for trade mark reasons). The *same name* may refer to *different drugs* in different countries. It is hard to see that, for example, the antibiotics *chloramphenicol* and *tetracycline* need 203 and 190 synonyms respectively.

Drugs

Albert, A. *Selective toxicity,* 7th edn. Chapman and Hall, London, 1985.

Albert, A. *Xenobiosis: food, drugs and poisons in the human body.* Chapman and Hall, London, 1987.

Chem. Brit. **8**, 1972:

McDonald, A. D., 'Drugs that influence behaviour', 98.

Parke, D. V., 'Metabolism of drugs', 102.

Phillips, G. F., 'Controlling drugs of abuse', 123.

Rees, W. L., 'Modern developments in psychopharmacology', 109.

Robinson, A. E., 'Forensic toxicology of psycho-active drugs', 118.

Stockley, I. H., 'Basic principles of drug interaction', 114.

Craig, P. *Penicillin, the first half century. Chem. Brit.* **15**(8), 1979, 392. A careful analysis of what (probably) really happened in Fleming's laboratory.

Rainsford, K. D. *Aspirin and the salicylates.* Butterworths, London, 1984. A good historical and chemical coverage.

Kihlman, B. A., *Caffeine and chromosomes.* Elsevier, Amsterdam, 1977. Read the conclusion first: 'Caffeine: a chemical hazard in the environment of man?'

Chemicals causing cancer

Blackburn, G. and Kellard, B. 'Chemical Carcinogens I, II, III'. *Chemistry and Industry,* 1986, 15 September 607–779, 20 October 687–695, 17 November 770–779. A series of review articles directed to the molecular basis of the action of chemical carcinogens.

Doll, R. and Peto, R. *The causes of cancer.* Oxford University Press, Oxford, 1981. The classic epidemiological study.

IARC monographs programme on the evaluation of the carcinogenic risk of chemicals to humans. WHO, Geneva (from 1984 available from OUP). A multi-volume set of publications produced over many years by WHO. Most government regulations are based on its recommendations.

Kingman, S. 'Computers point the finger at cancer triggers'. *New Scientist*, 26 February 1987, 23. Computers are used instead of animals. The shape of a chemical is calculated, to see whether it can react with a particular activating enzyme which would then turn it into a carcinogen.

Metcalf, D. (ed.). *Cancer, causes and control.* Australian Academy of Science, Canberra, 1980.

Searle, C. E. 'Chemical carcinogens and prevention' *Chem. Brit.* **22**(3), 1986, 211.

Wong, J. L. 'Cancer and chemicals . . . and vegetables'. *Chem. Tech*, February 1986, 100. Better than most on similar topics.

Thalidomide

Sjöström, H. and Nilsson, R. *Thalidomide and the power of the drug companies.* Penguin, Harmondsworth, Middlesex, 1972.

Sunday Times Insight Team. *Suffer the children: the story of thalidomide.* André Deutsch, London, 1979.

Third World

Forty-four problem drugs: a consumer action and resource kit on pharmaceuticals, IOCU, 1981.

Prescription for change, IOCU, 1983.

CHEMICAL INDUSTRY ASSOCIATIONS

1. APMA, Australian Pharmaceutical Manufacturers Association Inc., Level 2, 77 Berry St, North Sydney, NSW 2060, Tel.: (02) 922 2699. Telex: APMAS 26953, Fax: (02) 959 4860. This organisation represents the manufacturers of prescription drugs.

2. AMDADA. Australian Medical Devices and Diagnostics Association, PO Box 338, North Sydney, NSW 2060. Same fax, telex and tel: (02) 922 1157.

3. AVCA. The Agricultural and Veterinary Chemicals Association of Australia, 12th Floor, 65 Berry St, North Sydney, NSW 2060. Tel.: (02) 957 5792. Telex: (02) 957 5792. Fax: (02) 929 0213.

4. NCCPI. The National Council of Chemical and Pharmaceutical Industries, address as for no. 3.

5. PSA, Pharmaceutical Society of Australia, 44 Thesiger Crt, Deakin, ACT 2605; PO Box 21, Curtin, ACT 2605. Tel: (062) 81 1366. This organisation represents the professional pharmacists: MPS = member of PSA.

6. PGA. The Pharmacy Guild of Australia, 14 Thesiger Crt, Deakin, ACT 2605. This organisation is registered under the Conciliation and Arbitration Act as an employer organisation of pharmacy owners, who must be trained pharmacists.

7. PAA. The Proprietary Association of Australia Inc., address as for no. 6. This organisation represents the manufacturers of OTC products.

8. NPDA. National Pharmaceutical Distributors Association, c/o API, 102 Briens Rd, Northmead 2152. Tel: (02) 683 0683.

Chemistry of energy

10

We have become concerned about conserving energy and we ponder the alternatives. We disapprove of the 'wasteful' production of packaging materials and the need to dispose of them. We worry about using large quantities of irreplaceable fossil fuels in our cars and electricity generators.

How wasteful it is to produce an aluminium can, fill it with beer, empty it and throw it away! But is it just as wasteful to eat a breakfast cereal largely provided by fossil fuel (with a small solar contribution) and then proceed to turn it all into waste heat by jogging mindlessly around the block or by belting a ball against a wall?

We exist as highly complex organisms by ensuring that we waste energy continuously. That's the price of being improbable. What we need to understand is the manner in which energy can be wasted most sensibly.

WHAT IS ENERGY?

What is energy? What do we use? What do we need? What about the sun? What do we pay for it?

The purpose of this section is to examine the topic of energy from the point of view of an individual. What does energy mean to a domestic consumer? What things can we do and ought we to do individually to optimise our own situations?

The first point is to understand exactly what energy is. This concept is not as easy as it seems.

Energy is a means of doing work or producing heat and is measured in *joules* (the SI unit) and many other common units as well (calorie, watt-hour, BTU, therm). All the units are interconvertible. Thus 'all types of energy are equal—but some are more equal than others' (with apologies to Orwell). Just as there are hard and soft currencies, there are forms of energy that are desirable (because they are easily converted to other forms) and others that are less desirable.

Conversion of energy from one form to another *always* involves losses. The size of these losses is set not only by the limits of our technology, but also by limits set by the nature of energy itself.

You buy energy to use it, but in fact you don't use it. You *convert* it to work and (finally) to a less useful form—heat. You extract *usefulness* from energy. Does energy come in a hierarchy of usefulness? It certainly does. If you were offered a

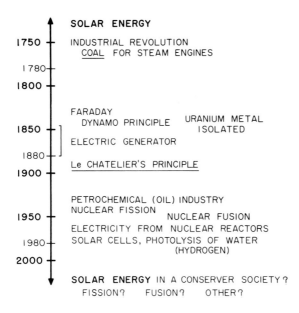

Fig. 10.1   Energy time-scale

certain amount of energy in the form of waste heat from a power station and, at the same price, you were offered the *same* amount of energy in the form of electricity or oil you would *not* accept the heat (and it is not even straightforward what 'same' means here). The reason is that the electricity and oil are more readily converted to other forms of energy, whereas heat is not. But using the electricity to drive a heat pump, in say an air-conditioner, you have an efficiency (heat pumped divided by electrical energy used) of much greater than 100%. (We shall return to this point.) In fact, you should be offered the waste heat from a power station at a cheap rate because, if the power authority can't sell it, the heat is lost, Fossil fuel can be stored, however.

The concept that the energy you 'use' is never destroyed but is still around afterwards in the form of heat is very important from an environmental point of view. All the fossil and nuclear energy we produce and use remains around as heat—to be radiated back out into space, hopefully.

Degrading the usefulness of energy is measured by a function called *entropy.* The more useless energy becomes, the higher its entropy. It is the entropy which is remorselessly increasing and is a measure of how we run down our energy sources.

Let us look at an analogy. Status in human society is linked to rarity. The least probable situations are the most desirable. You only have to look at 'situation' in a real estate sense to see that there are fewer more desirable blocks of land than less desirable ones. In fact, this should be stated inversely. Desirability is evaluated by rarity. Perhaps a better analogy comes from a consideration of the distribution of shareholdings in a public company. If a change occurs from a few large holders to a large number of small holdings, the effective control of the company by the shareholders is reduced drastically. This follows from the difficulty of obtaining coordinated, cooperative action if the number of shareholders is large. (See

Appendix VIII, 'The entropy game', for further discussion and exercises on this approach.)

Because heat is so low in status in the energy hierarchy, consumers converting electricity directly into heat in an electric heater are selling themselves short— even though the conversion occurs with 100% efficiency! (Because heat represents about half the end *use* of all our energy, its importance from *this* point of view should not be underestimated. Obtaining heat more efficiently *is* important.)

Instead of *converting* electricity directly into heat, electrical energy can be used more efficiently if it is used to drive a heat *pumping* system. A refrigerator is a good example. Electrical energy is used to pump heat from the inside of the fridge to the outside coils. If the unit is set in a wall with a rear heat-exchange coil outside the house (and the door is left open to the room) we have an air-conditioner. Turn it around (or more conveniently, reverse the cycle of inside and outside heat-exchange coils) and we have a reverse-cycle air-conditioner which heats the house and cools the outside. The electricity is converted to mechanical work to *pump* heat. It does this far more efficiently than if the electricity is *converted* directly into heat in a radiator. The efficiencies are greater than the 100% of a direct conversion and can be quite high for small temperature differences.

Instead of using electricity to create a temperature difference between two places, we can use an existing temperature difference to generate electricity. The temperature difference between the surface and depth of the ocean can be used to provide electricity by means of an air-conditioner type device being used in the opposite way, whereby heat going from the hot region to the cold region via the machine produces electricity.

Most of the 'higher' forms of energy such as electrical, mechanical, nuclear, solar, and chemical energy (with reservations) are about equally useful. Heat is the interesting one. Although it is the least useful form of energy, its usefulness is variable. The 'hotter' heat is, the more useful it is. To be more correct, the greater the difference between the temperature at which heat goes into an engine and that at which it comes out of the engine, the more useful it is. If you think about it, all combustion engines have a hot in, and a colder out (in a car, it is the hot cylinder and colder exhaust; in a steam engine, it is the hot boiler and colder condenser). All such engines are called *heat engines*. The *theoretical* (without any losses) efficiency with which heat can be converted into other forms of energy such as mechanical is given by the following formula:

$$\text{efficiency} = \frac{(t + 273)_{\text{hot input}} - (t + 273)_{\text{colder exhaust}}}{(t + 273)_{\text{hot input}}}$$

where t is the temperature in degrees Celsius. The 273 converts Celsius degrees into absolute degrees of temperature. The ultimate in coldness, or the absolute zero of temperature, is $-273°C$.

The efficiency is always less than one. The bigger the temperature difference is, the larger the efficiency (Fig. 10.2). For a refrigerator the formula is turned upside down and the efficiency is greater than one. Now the larger the temperature difference is, the *smaller* the efficiency. If the inside of a refrigerator is $-18°C$ (in the freezer) and the outside temperature around the coils is 40°C, the theoretical efficiency is

$$\frac{313}{(313 - 255)} = 5.4.$$

The ratio is greater than unity—you are *transferring* (not producing) about five times as much heat as the mechanical energy you are putting in.

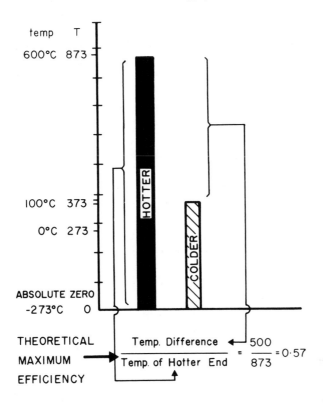

Fig. 10.2 Thermal efficiency

THEORY OF HEAT ENGINES

An ideal heat engine operates by extracting heat (reversibly) from a hot reservoir at temperature T_h and using it to expand a gas in a piston to do mechanical work (reversibly)—step 1. The transfer of the piston to a cold reservoir T_c (step 2) followed by injection of heat (reversibly) to that reservoir allows more work to be done—step 3. Transfer of the piston back to the hot reservoir completes the cycle—step 4 (see Fig. 10.3).

If the engine is regarded as a black box, then it can be regarded as cycling repeatedly, extracting heat q_h from the hot reservoir at temperature T_h and ejecting heat q_c to the cold reservoir at temperature T_c. In the process the engine (ideally) does an amount of work w determined by the first law of thermodynamics (which states that energy can neither be created nor destroyed) given by equation A in Figure 10.3:

$$w = q_h - q_c$$

Entropy is a concept derived from the second law of thermodynamics. The entropy associated with the transfer of heat is determined from the situation where the heat is transferred reversibly. (What 'reversible' means is a bit subtle for our present circumstances. Suffice it to say we are calculating an unattainable ideal which gives us a measure of relative performance.) This entropy is just the heat transferred divided by the absolute temperature (°C + 273). The cyclic engine cannot store or lose entropy and this gives us equation B in Figure 10.3:

$$\frac{q_h}{T_h} = \frac{q_c}{T_c}$$

A little algebraic rearrangement gives us the efficiency, which is the work obtained divided by the heat into the engine or w/q_h in terms of the temperature of the hot and cold reservoir (equation C in Fig. 10.3).

$$\frac{w}{q_h} = \text{efficiency} = \frac{T_h - T_c}{T_h}$$

So it is all as simple as the ABC!

For enthusiasts, the cycle of the heat engine can be shown on an indicator diagram, which plots temperature T versus entropy S. Step 1, the gas expansion, gives an entropy increase at temperature T_h, while step 3 gives the same entropy decrease at temperature T_c. In each case the (reversible) work done is the product of the temperature and change in entropy, $T\Delta S$, and the work output per (ideal) cycle is the area of the rectangle which the cycle encloses (see Fig. 10.3).

Problem: Assume that the total cost of operating an air-conditioner as a heat pump (pumping heat from the cold outside into a house) is four times the theoretical power cost for perfect efficiency, whereas the cost of direct electrical heating is just the power cost. If the room temperature is 27°C, at what outdoor temperature would the two systems yield equal cost?

When generating electricity from a temperature difference such as an ocean temperature gradient, the heat arrows in Figure 10.4 are reversed and heat is taken in at a higher temperature and ejected at a lower temperature. The motor is driven by this and becomes a generator. We are back to a heat engine.

The absolute levels of energy used by consumers in their homes varies with climate. For the year 1980–81 in Australia, it varied from 36 300 MJ in Queensland to 96 000 MJ in Tasmania. In Australia, 40% of the electricity generated is used in the home, and electricity represents about 40% of the total home use of energy.

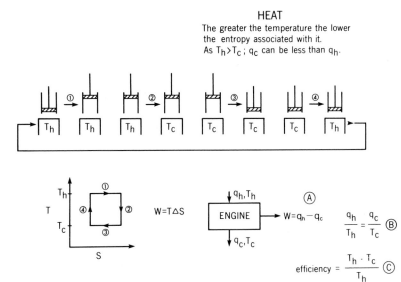

Fig. 10.3 Heat engine

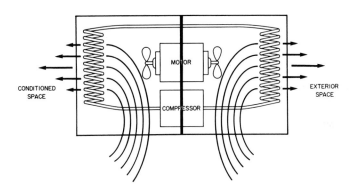

Fig. 10.4 Reverse-cycle air-conditioner

About one-third of home energy is used for space heating, one-third for water heating and one-third for all other uses (Table 10.1). Domestic appliances are the main component of the other uses.The efficiency of *similar* appliances using the *same* fuel and performing the *same* tasks can vary markedly. For example, over the period 1978–1982, tests conducted by the ACA on refrigerators of comparable size showed a spread of energy consumption from 4.1 to 14.1 watt-hours/per litre W h/L of storage space per day.

The cold facts

Household appliances are now given an energy rating label (see Fig. 10.5). This label rates the energy efficiency of appliances on a scale of one star to six, and tells

TABLE 10.1
Typical end use for home energy in Australia

End use (%)	Sydney	South Australia	Victoria	New South Wales
Water heating	37	30.3	31	34
Space heating	18	30.1	46	26
Cooking	18	10.9	9	16
Lighting	10	—[a]	2	6
Refrigeration	17	7.7[b]	5	11
Other appliances	—	13.2	7	7

[a] Included in other appliances.
[b] Estimated at 426×10^6 MJ.

Source: Consolidated papers on energy labelling, National Energy Conservation Program, July 1983.

you how much energy they use in units of kilowatt-hours per year, when tested in accordance with AS 2575.2. More stars mean a more energy-efficient refrigerator. That is, it uses less electricity to achieve the same level of cooling. An average household refrigerator (300–400L) can cost you between $38 (a five-star fridge) and $126(a one-star fridge) to run every year. The $88 per year saving adds up to $1300 over the 15-year average lifespan of a fridge. If this saving accrues interest

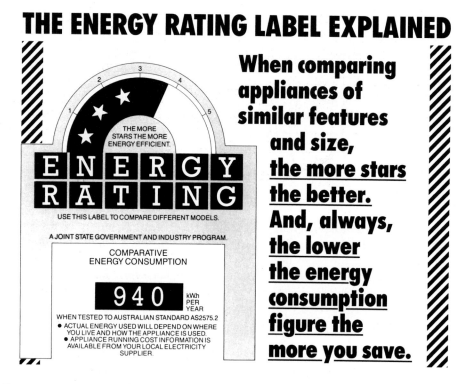

Fig. 10.5 The energy rating label helps the consumer to choose an energy-efficient appliance (Courtesy of the NSW Department of Energy)

to balance rising prices, then an efficient refrigerator can save you the cost of a new one. The NSW Department of Energy publishes lists of current models of refrigerators, grouped according to size, with their energy ratings. The total amount of power used will depend on the volume of storage space which is cooled and also on the efficiency. For refrigerators of comparable size, the more efficient will use less energy.

Appliance ownership by households is an important factor in total consumer demand. Over 99% of Australian homes have a refrigerator and so most new refrigerators are replacements. Although the replacements tend to be larger, if they are more energy efficient, the energy demand may decrease. The ownership of freezers has increased from 34% in 1976 to 43% in 1980.

What energy do we use?

How much energy is used in obtaining energy and how do we use it? The points to note as domestic consumers are:

1. The energy needed to *obtain* the primary energy (wood, coal, oil, natural gas) is very low. In fact, we have been lulled by a historical progression in which each successive primary source has been obtained more easily; that is, with *less* energy. The new sources—uranium, oil shale and coal liquefaction—on the other hand, need more energy for their provision.
2. We *use* our natural gas and oil (after refining) directly, but we use our coal via *electricity* generators, which results in an enormous conversion (and distribution) loss for coal. The reason for this enormous loss is that the production of electricity from coal proceeds via the production of heat (that lowly energy form). The hotter the heat (with respect to the temperature of the cooling water supplied to the condensers), the higher its status and the thermodynamic efficiency of conversion. That is why coal-burning power stations are noticeably more efficient in winter than in summer.

   We also export considerable quantities of coal, both directly, and indirectly (in the form of, say, aluminium, produced with electricity generated from coal).
3. For transport we use oil. Energy content comparisons are therefore not very helpful because having a scoop of coal in a car is like having an Australian dollar in Brazil—the value is there but not the conversion facility. Hence the interest in coal to oil conversion. In fact, so critical are liquid fuels for transport that they will set the 'parity' price for any other use (e.g. space heating).

   Some oil will continue to be a feedstock for the chemical industry, but even this will be replaced by coal. Even for transport, oil will become too valuable. Engines powered by hydrogen gas are being developed by Boeing—based on providing the hydrogen from electricity (electrolysis of water).

How much energy do we consume? The total consumption of primary energy in Australia in 1977-78 was 2.86×10^{18} J for 13 million Australians, which represented a continuous 7 kW per person(about 20 times the use of Third World countries).

An interesting twist to the conservation argument is to suggest that the accumulation of fossil fuels such as oil and coal can be seen as a failure of the ecological system to recycle materials. By burning fossil fuels, humans are helping nature retain the balance on earth! This line of argument is unlikely to be very convincing.

A comparison of energy from various sources is given in Table 10.2, including those sources humans use as food.

TABLE 10.2
The cost of energy to the Australian consumer

Type of energy storage	Inherent energy value (MJ)	Cost per kilo ($)	Cost other ($)	Cost per hundred MJ ($)
1 kg of petrol (1.2 L)	42	0.55		1.3
1 kg of heating oil	42	0.45		1.1
1 kg of butter (1 L of fat, 200 mL of water)	31^a	4.0		12.9
1 kg of sugar	17^a	0.80		4.7
car battery				
12 V			4.0	
50 Ah	2.16		(charging fee)	185
1 kW radiator	3.6		0.01	1.0
	(per hour)		per MJ	

^{a} This is the energy released on burning the material, not its availability to humans on eating.

What do we need?

The basal metabolic power used by humans (that is, the rate at which we use energy to keep alive), is about 100 W (a light globe's worth). This can be supplied by 20 g of sugar (or 10 g of fat) per hour—a very modest amount indeed.

The recommended daily allowance of usable energy for people is:

 9 MJ   12 MJ

An average woman weighing 60 kg and 160 cm tall uses energy as follows:
Basal metabolism   6 MJ/day
Normal activities   3 MJ/day
Increasing activity by 10% uses up 0.3 MJ, supplied to her by one slice of bread.
Rowing   4 MJ/hour for a 15 minute race   1 MJ
Gardening   0.8 MJ/hour for 3 hours   2.4 MJ
Even when we do work, like riding a bicycle, the performance is about 700 W. (The longer-term average is about one-third of this.) We are not going to run too many household appliances on these puny efforts. In our Western society, human energy is negligible.

We need to eat—so a study of the energy requirement in food production is an important undertaking.

700
Watts

Fig. 10.6   Peak performance

Energy down on the farm

In energy terms, our modern food system is expensive. Processing and distribution use up a lot of energy—far more than is actually contained in the food itself. At present much of that energy comes from oil—a resource that is becoming scarce and expensive.

Food and fuel are interchangeable. Farm produce can be chemically converted into fuel, and microbes can convert petroleum into food. However, a point that arose in our introductory comparison of costs of consumer energy must now be clarified. It is important to distinguish between the fuel value of food when it is burnt and the nutritional energy value when it is eaten. Only about three-quarters of the fuel energy in our food is absorbed when we eat. The figures to be used here refer to *fuel* energy. Of course the indigestible cellulose can be converted to sewer gas.

The arrival of a perfectly cooked meal on the dinner table marks the end of a long chain of energy-devouring processes. These begin not on the farm but in an earlier primary industry: mining. Before we can have mechanised agriculture we must have machines. All of this takes a great deal of energy, as does the production of fertilisers and pesticides, and the provision of irrigation. Even after it passes through the farm gate, most food still requires more energy expenditure for some kind of processing and packaging before it is transported to the supermarket. From there it is usually driven home in the back of a car. In the kitchen, it is likely to spend some time in a refrigerator before, finally, it is cooked and becomes a meal.

Two Canberra scientists at CSIRO (Dr Roger Gifford and Dr Richard Millington) studied this energy usage. The energy budget was a very broad one and dealt only with the aggregate of all Australian agricultural production over the years 1965–69. Individual products were not isolated because the web of energy inputs for each product cannot be clearly traced. Plant material harvested by humans and their livestock represents a mere 0.01% of the sunshine (falling on rural land) that could be usefully absorbed by plants. Nevertheless, this plant matter has a fuel value about 1.5 times greater than all the fuel energy burned by people in Australia. Only about 15% of this massive amount of plant tissue is harvested directly—the rest is eaten by livestock and converted into wool and animal products.

The energy inputs to farm products before they leave the farm are set out in Table 10.3. Many of the less tangible inputs to farming have been omitted from the budget, such as agricultural research and extension. The energy inputs to farm products after they leave the farm are set out in Table 10.4.

Food processing and distribution use much more energy than food production. Omitted from the table are such indirect items as the energy used for building food factories, refrigerators and stoves. About 14% of the energy in food produce

TABLE 10.3
Energy input on the farm

	$\times 10^{15}$ J per annum	
Direct farm use		
fuel	46.2	
electricity	8.4	54.6
Fertiliser		
mining	3.5	
shipping	9.8	
manufacture	5.5	18.8
Farm machinery (mining and manufacture)		6.8
Agricultural chemicals		4.4
Road transport not included elsewhere		1.0
Farm labour		1.2
		86.8

TABLE 10.4
Energy input from farm to dinner table

	$\times 10^{15}$ J per annum	
Transport from farm		
rail	2.1	
road	5.0	
grain handling	0.3	7.4
Factory processing		
bagasse as fuel	9.7	
other fuel	45.6	55.3
Food and drink packaging		
steel cans	10.8	
paper	8.5	
fuel value of paper packaging	4.3	
glass	5.5	29.1
Road transport from factory		7.7
Subtotal to retail store		99.5
Transport from store to home	33.0	
Domestic refrigeration	46.0	
Domestic cooking	42.0	
Subtotal store to dinner table		121
Grand total		220

Fig. 10.7 Where does all the energy go?

leaving the farm comes from animals, and it costs 3 J of grain (at least) to produce 1 J of meat. Losses during processing, retailing, kitchen preparation and digestion are difficult to estimate, but it seems that we absorb into our bodies only about half the fuel value that leaves the farm (Fig. 10.7).

In summary, what the final balance showed was that for each joule of *digestible* food energy eaten in Australia at least 5 J are expended in making it avaliable, of which

- 0.6 J is used in getting it to the farm gate—11%;
- 2 J is used in taking it from the farm to the retail store—38%;
- 2.8 J is used in getting it from the store to the dinner table—51%.

Work on converting the world's valuable fossil fuel reserves directly into food by growing microbes on the petroleum and harvesting them as food for livestock (or humans) has shown this to be feasible and economic. However it does not solve the problem of the much larger energy consumption past the farm gate. In the food chain, the internal combustion engine guzzles at least 40% of the total energy consumed in the total process. While shopping uses only 2% of the fuel going into petrol tanks of private cars, it represents 10% of the total energy cost of the food. Any improvement in transport would represent a substantial saving.

Our method of agriculture has been compared to that of the Tsembaga tribe living in the New Guinea highlands in 8.3 km² of tropical rainforest. The comparison is difficult because of different cultural values, but a very important difference is that their energy comes from renewable resources (trees), whereas ours comes from non-renewable fossil fuels almost exclusively.

A similar energy analysis can be done on uranium. If you assume that there is *no* inherent energy content in uranium, then nuclear technology can convert about 1 MJ of oil into 1 MJ of electricity (where the oil is used to mine, extract, transport and upgrade the uranium). This shows an energy advantage over a conventional oil-fuel generating station, which requires 4 MJ of oil (in this case burnt directly) to produce 1 MJ of electricity. Thus uranium is a component in a new technology that allows our oil reserves to last four times as long. If the uranium input to nuclear reactors is counted as a fuel input, then nuclear reactors operate as poor energy converters (1% conversion). Breeder reactors complicate the issue because they require 'burner' reactors to produce the plutonium.

We *need shelter*—and if we want to cost insulation, we need to understand what is meant by an energy analysis. Before insulation may be said to save energy, it

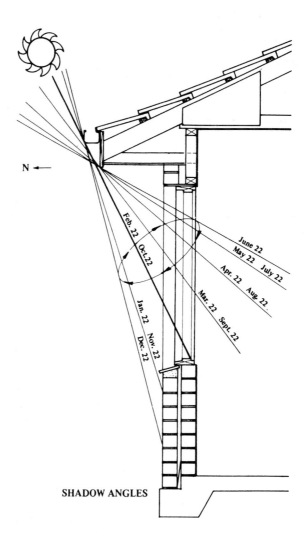

Fig. 10.8 Eaves designed for the Southern Hemisphere at the latitude of Sydney exclude the summer sun but allow winter sunlight into the house (From *Rebuild*, CSIRO, October 1978)

must first pay back the energy used in its manufacture and installation. It turns out that *energy* payback time can be less than a single heating season.

Along with insulation come all those handy hints for preventing heat loss by convection; for example, close up all the cracks and crevices around windows and doors. A report from the USA shows that some houses are now so well sealed that they are also trapping moisture and air pollutants generated inside the house. Amongst these are fumes from natural gas, household sprays, radon released from radioactive building materials, formaldehyde and other vapours from particle board, plywood and the insulation itself.

The presence of residual formaldehyde from insulation is a problem. The long-term stability of urea-formaldehyde foam has been questioned—even where it has been correctly installed—and its use as insulation is being discontinued (see also Chapter 6).

There are of course many other shelter questions—better design for passive use of solar energy, layout, materials and other heating and cooling sources—which are dealt with in an expanding popular literature. Simple care with the position of windows and eaves can make an enormous difference (Fig. 10.8).

What about the sun?

Australian houses—even new ones—are mostly designed with no thought given to the fact that, on the coldest day, trapped sunlight gives warmth and this warmth can be stored for the night. It's high time we changed our thinking.

If you are building a house, it generally costs no more to design and site the house to be warm in winter and cool in summer. If you own an older house and plan to remodel, there may be quite a lot you can do to make it more energy efficient. Figure 10.9 shows some of the design features to use. Ideally, the long

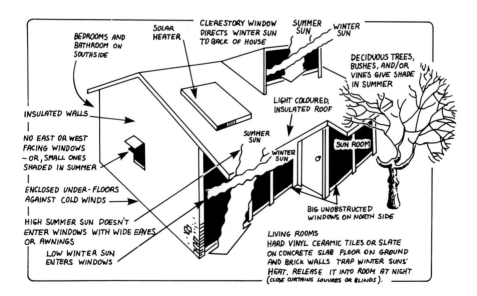

Fig. 10.9 Energy-efficient house design and siting for the Southern Hemisphere (From *Choice*, May 1983)

axis of the house should run east-west and few windows should be sited in the short east and west walls which get so much morning and afternoon sun during summer. Most windows should face north (south in the northern hemisphere) or within about 20° of it; and the window area should be about 20% of the floor area of the space you want to heat. With eaves of the right width, you can exclude summer sun but get maximum sunlight during winter. (The exact width of eave or awning needed for your latitude, window height and slope of roof can be worked out). This winter sunlight will provide a lot of daytime warmth, whatever type of floor you have. And if your house is of concrete slab construction and you choose the right flooring, the floor will store daytime heat and release it slowly into the house overnight.

Remember to check surroundings, too. Is the house protected from prevailing winter winds? Will it be free of shadow in winter? In southern latitudes (e.g. Melbourne), in mid-winter, something north of your house can cast a shadow twice as long as its height. What about your garden? Strategically placed deciduous trees and vines will shade walls and windows in summer and obligingly drop their leaves in winter.

What do we pay for energy?

Until recently, electricity and gas have been relatively cheap. But now that the costs of all forms of energy are rising, there's more motivation to use less and to use the cheapest. How much heat you need depends on your sensitivity to cold and on what you're wearing, what you're doing and when and where you do it.

Is the house empty during the day? What jobs do you do when you come home? During active periods, you provide your own central heating. It's when you are sedentary that you need extra heat. If you habitually relax in just one room in the evening, you're unlikely to need more than a single heater sufficient for the size of the room. Since you'll arrive home to a cold house and want warmth quickly, you should think about radiant heaters, fan heaters and radiant convectors.

On the other hand, if yours is a large house, occupied all day and used by a large family with divergent interests, you will need to maintain comfortable conditions most of the time in several areas. You will probably need convection heating, controlled by a thermostat, to minimise power consumption. Depending on which fuel works out cheapest in your area, you might end up settling for electric oil-filled heaters or heat banks in rooms you want heated, a large oil or gas space heater or even whole-house central heating. First though, you need to work out how much heat is needed and which fuel is cheapest in your area.

The heating estimator in the following box provides a method worked out by the CSIRO which you can use to estimate the heat output a heater must have to heat your room or open-plan area properly.

The most important parameter in the heating calculations is the heat value of the fuel—the amount of heat produced when a quantity of fuel is completely burnt, or when any energy source is completely converted to heat. Different heat values for different fuels result from the difference in the individual values for carbon and hydrogen (H = 121 MJ/kg, C = 32.8 MJ/kg). Other factors are the

WHAT SIZE HEATER?

Having worked out (from the heating estimator in Table 10.5) the number of kilowatts you need for heating, you may have to do a bit of calculating to see what rating of heater you need.

For an imaginary uninsulated room, suppose the estimator calculation gives 3.7 kW. We can use this as a basis for calculating relative heating costs for that room. Electric heaters are 100% efficient, whereas air-conditioners pump heat with over 200% efficiency. The efficiencies apply to optimum operating conditions, which are very often near maximum running.

TABLE 10.5
Heating estimator

Conditions (heat losses)	Calculation	Heat needed (watts)
1. *Heat loss through ceiling*		
(A = room area in square metres)		
Uninsulated roof	A × 40 =	
Insulated roof	A × 20 =	
2. *Heat loss through floor*		
Open or unheated space below floor	A × 20 =	
Slab floor on ground or enclosed crawl space beneath floor	A × 10 =	
3. *Heat loss through walls*		
(L = total wall length in metres)		
Internal walls		
When adjacent space is not heated	L × 40 =	
External walls		
With foil insulation	L × 25 =	
Uninsulated cavity brick	L × 55 =	
Uninsulated brick veneer or weatherboard	L × 65 =	
Uninsulated cement sheets	L × 80 =	
4. *Heat loss through windows*		
(A = window area in square metres)		
With double glazing or drapes from ceiling to floor	A × 40 =	
All other conditions	A × 80 =	_____
Watts (approx.) needed for continuous use	Total:	
Watts (approx.) needed for fast warm up	+ 50%[a]	_____
	Total:	_____

To turn watts into kilowatts, divide by 1000; this gives you the heat output capacity per hour you need from a heater.

[a] If you plan on intermittent use, CSIRO advises installing heaters with a capacity of about 50% greater than the estimated heat load, so that the room will heat up reasonably quickly after you first turn the heat on, for example early in the morning or when you arrive home from work.

Source: Choice, May 1983; derived from CSIRO Division of Mechanical Engineering Pamphlet no. 12/B/3, August 1977.

density of the material when costs are given per litre, and non-burning impurities. The heat values for all hydrocarbons lie between the values for hydrogen and carbon, but are closer to the carbon value because carbon is heavier than hydrogen.

Tables of thermodynamic properties of chemicals list values such as heats of combustion as a value per mole. For example, for methane, ΔH_c° is 889 kJ/mol, which divided by the molecular mass, 16, gives 55.56 MJ/kg; for butane, ΔH_c° is 3509 kJ/mol, which divided by 72 (molecular mass) gives 48.74 MJ/kg. Note the value for hydrogen, 285.2 kJ/mol (142.6 MJ/kg), which differs from the value given above because all the tabulated values assume combustion to carbon dioxide and water *as a liquid* (i.e. giving up its latent heat of vaporisation) and are thus *gross* rather than *net* calorific values. In practice, water formed during combustion is vented as a gas, taking the 21.6 MJ/kg (of water) latent heat with it to the atmosphere. This means a loss of about 8% of the heat.

The heats of combustion of compounds that are not listed in thermochemical tables can be calculated from a much wider tabulation of heats of *formation* because heats of chemical reactions are additive (Hess's Law).

A second important parameter is the efficiency of the heating device—that is, the heat delivered to the room as a percentage of the total heat available from the heat source. Just how efficiently we are warmed by the heat delivered to the room is discussed in Appendix IX. To make the calculation we also need to know the relationships between the various energy units (see Table 10.6). As well as the calorie, older units of energy include the British Thermal Unit (BTU) and the therm; the unit of electrical energy is the kilowatt-hour, kW h. The prices used in Table 10.2 are used only for illustrative purposes.

Oil

The heating value of liquid fuels varies little (kerosene 36.7 MJ/L, heating oil 37.7 MJ/L, distillate 38 MJ/L, diesel fuel 38 MJ/L). The efficiency of an unflued oil heater is about 95%, whereas the efficiency of flued oil heaters can vary from 60% to 75%. For comparison we shall use 75% as the optimum efficiency for a properly serviced heater.

TABLE 10.6
Energy conversion factors

1 therm	= 100 000 BTU	1 kJ = 0.948 BTU
	= 106 MJ	1 MJ = 948 BTU
1 kW h	= 3412 BTU	
	= 3.6 MJ	1 MJ = 0.28 kW h
1 calorie	= 4.187 J	1 J = 0.239 calories
1 horsepower hour	= 2.69 MJ	1 MJ = 0.37 horsepower hours

Solid fuel

The heating value of coke and coal is 27 MJ/kg and for briquettes it is 24.75 MJ/kg. For split wood it is 12.4 MJ/kg, but it drops to 6.7 MJ/kg for green wood.

Comparative heating costs do not take into account any fixed costs (supply or hire of gas cylinders, installation of oil tanks etc.). The householder must consider capital costs and maintenance.

All currently available commercial sources of energy produce pollution. Coal, briquettes and wood are the worst for air pollution, followed by diesel oil, heating oil, and then liquefied petroleum gas (LPG). Although electricity is non-polluting where it is used, its production causes pollution—air pollution (when coal is used), nuclear waste (from nuclear reactors) or the flooding of river valleys (to generate hydroelectricity).

Gas

An unflued gas heater is about 90% efficient. For flued liquid petroleum heaters the maximum and minimum efficiencies are given by the British Standard (BS1250 Part 4) as 78% and 50%. The United Kingdom Ministry of Housing gives an average of 60%. Bottled gas is generally sold by weight, but occasionally also by volume which of course depends on the temperature. If you buy by volume make sure you buy only on cold days. The heat value of LPG gas is 49 MJ/kg or 25.5 MJ/L (Australian Liquefied Petroleum Gas Association).

Town gas (made from coal) is decreasing in usage. Its composition is somewhat variable and so it is sold not on a weight or volume basis but on a heat supply basis measured in *therms*. Town gas is made by heating coal to a sufficiently high temperature (350°-1000°C) in the absence of air (a process called destructive distillation). What remains is *coke*, which is mainly carbon. When the volatile products cool to ambient temperature, a portion condenses to a black viscous liquid known as coal tar, which is a source of a wide variety of chemicals (hence coal tar colours—see 'Food additives—Colouring matter', Chapter 11). The non-condensable gases are known as coal gas. This gas varies in composition but consists mainly of hydrogen and methane in about equal volumes, along with some carbon monoxide (which is why it is so poisonous) and lots of nasty components such as hydrogen cyanide and hydrogen sulfide (which gives it its smell). Most of the nasty components are removed by scrubbing with water before sale as town gas. On the basis of 50:50 hydrogen (H_2) and methane (CH_4) the heat value of town gas is $(121 + 55.6)/2 = 88$ MJ/kg, which is over half again of that of natural gas, which is mainly pure methane (55.6 MJ/kg).

Exercise: Why do the burners on gas appliances have to be changed when converting to natural gas? What is the full combustion ratio of air to gas for the two types of gas? (Hint: write down the combustion equations of the two gases H_2 and CH_4 and work out the amount of oxygen (air) each needs. Remember that equal *volumes* of gas under the same conditions contain equal numbers of *molecules*.) If hydrogen gas was really H and not H_2, would it make any difference?

Table 10.7 gives some comparative heating costs for a typical situation (the State of Victoria in 1983).

TABLE 10.7
Comparative heating costs in Australia. Sample calculation for a room requiring a heater with an output of 3.7 kW (13.32 MJ/h)

Type of heater	Heater efficiency (%)	Input required	Cost of energy	Cost of energy at maximum efficiency (cents/hour)[a]
Electric	100	3.7 kW	6.76c/kW h[b]	25.0
Gas				
flued	70	19.0 MJ/h	0.424c/MJ	8.1
unflued	100	13.3 MJ/h	0.424c/MJ	5.6
Oil	75	17.8 MJ/h	40c/L	18.8
Kerosene	95	14.0 MJ/h	40c/L	15.3
LPG				
flued	70	19.0 MJ/h	48.8c/kg	18.9
unflued	100	13.3 MJ/h	48.8c/kg	13.2
Solid fuel (open fire)				
wood	25	53.3 MJ/h	9c/kg	30.0
briquettes	25	53.2 MJ/h	10.4c/kg	25.2
Slow fuel (slow-combustion stove)				
wood	60	22.2 MJ/h	9c/kg	12.5
briquettes	60	22.2 MJ/h	10.4c/kg	10.5

[a] Efficiencies of fuel burners drop considerably at less than maximum burning rates.
[b] Price for 900 kW h after the first 120 kW h is 6.76c/kW h, the middle of a three-tiered price structure. There is a cheaper off-peak rate for storage space heating of 2.70c/kWh

Source: Choice, May 1983.

To make a running-cost comparison table like this, for yourself, follow this procedure:

1. Work out the heater output required from the heating estimator, using the formulae below. MJ per hour = heat output required (kW) × 3.6.
2. Write down the heater efficiencies.
3. To find the input required, divide the heat output required by the heater efficiency.
    For example: heat output required = 15 MJ/h
             heater efficiency   = 75%
             input required (MJ/h) = 15 ÷ 0.75
                              = 20 MJ/h
4. Find the cost of energy where you live, from your local suppliers.
5. Calculate running costs.

Natural heat?

The hot-water service and space heating represent the major domestic uses of energy. Everytime I discuss energy, the emphasis has changed. My first lecture notes in Canberra 25 years ago discussed the replacement of wood heaters and stoves with oil—the wonder fuel. Next time round, gas was moving in. With the oil crisis everyone was converting to electricity. Recent increases in electricity tariffs have caused a shift back to wood and solid fuel again. So why not discuss wood.

About 70% of the mass of dry wood is carbohydrate, of which 40% to 50% is cellulose. The calorific value of carbohydrate is about half that of oil (both as a fuel and as a food). The remainder of the wood is lignin, a complex substance whose major claim to fame is as a source of natural(?) vanillin. Softwoods are higher in lignin. One kilogram of firewood has an energy content of about 16 MJ (equivalent to 0.5 litre of oil or 4 kW h of electricity).

When wood burns, a number of processes occur at the same time. The wood is being dried, volatile material is driven off and ignited charcoal is being formed and burnt. As a rough estimate, 70% of the energy released by the combustion of wood comes from burning gases while the rest comes from the glowing coals, the burning of charcoal. The temperature needed to ignite the gases differs for different gases, with some being quite high (carbon monoxide, 600°C; methane, 650°C; acetic acid, 540°C; and hydrogen, 540°C) and so these gases must be kept in the high-temperature zone with sufficient air for long enough to burn completely. Wood stoves with catalytic burners are available, which lower the ignition temperature (see continuously cleaning ovens, under 'Vitreous enamel' in Chapter 8.) The *more* creosote the wood produces, the better it burns. The more water in the wood, the more efficiently creosote will be 'steam distilled' out of the wood. As this occurs at just below the boiling point of water, the creosote will distil away before being reached by the burning edge. The creosote will deposit from the distillate in the cooler parts of the flue. (For more on steam distillation, see Experiment 13.12.) Catalysts are not used up but can be 'poisoned' by impurities (such as sulfur) which are found in wood.

In order to achieve high efficiencies, a normal heater must burn the wood at high temperatures (to make sure all the gases burn) with just the right amount of air. (Insufficient air means some of the gas will not burn, and too much means hot air is lost up the chimney.) The air and the wood gases have to be well mixed and the heat must be transferred from the heater to the room.

Open fires are very pleasant but can funnel a whole room (40 m^3) of air through the chimney every five minutes. As well as losing hot air, the fire is cooled, which means gases are not burnt efficiently. An efficiency of about 10% is reported.

Non-airtight stoves include Franklin, pot-belly, parlour and the fancy glass-enclosed heaters. They have an efficiency of about 30% to 40%. They are not easy to control and rely mainly on radiation from hot surfaces.

Controlled combustion (airtight) heaters are easily controlled and are the most efficient if not most attractive heaters, with efficiencies ranging from 40% to 50% at high heat output. When there is a slow-smouldering fire, the efficiency drops off and more creosote reaches the flue. If the area around the fire is insulated, the wood gases will stay hot enough to burn but will need extra preheated air added *above* the fire so that the gases will burn without increasing the burning rate of the wood.

The efficiencies given above are somewhat less than those quoted in the sales brochures but are the results obtained by research workers in Tasmania. My own choice was a compromise between visual display and efficiency.

SOLAR ENERGY

There has been a resurgence of interest in solar energy in recent years because of the realisation of the limitations of our present energy resources, and an immense amount of information on solar energy has been published. From the point of view of the consumer, a useful report is that of the Australian Senate Standing Committee on National Resources, *Solar energy*, which states:

> For 1973–1974 the percentage of primary energy consumption of the various sectors of the Australian community was as follows:
>
> | | % |
> |---|---|
> | manufacturing | 33 |
> | electric utilities | 28 |
> | transport | 27 |
> | domestic/commercial | 5 |
> | mining | 3 |
> | agriculture | 2 |
> | other | 2 |
>
> These figures suggest that the thrust of Australia's solar energy and development strategy should be aimed at manufacturing and liquid fuel applications. The more important applications in manufacturing are the generation of heat for industrial uses where process hot water and low pressure steam are required.

After examining the evidence, one of the conclusions of the committee was:

> At the scientific level, debate indicates that some experts are more concerned with the promotion of their own projects than with the overall development of solar energy utilisation.

The use of collected solar energy in Australia represents less than 1% of total energy consumption, with domestic water heating and heating of swimming pools pools being the major uses. Photovoltaic cells are being used in some remote areas to provide low-power requirements for communications purposes.

Solar hot-water systems consist of panels of collectors, generally copper pipes, installed to face north (south in the northern hemisphere). The water heats up and circulates through collectors into a storage tank, which generally has an electric booster for overcast days. The lifetime of commercial plastic collectors is believed to be about ten years. The plastic used in pipes is a black acrylonitrile-butadiene-styrene (ABS) co-polymer (see Chapter 6). It has much lower thermal conductivity (heat transferring properties) than metals.

As solar heating generally competes with electricity, it is important to consider how electricity is costed. The electricity authorities assume that coal (used for most generators) has *no intrinsic value* when they determine the rates. To them the cost of coal is simply the cost of digging it out of the ground and, because the big power stations are now located on coal fields, the cost of digging it out of the ground is very low. The export price of coal varies by a factor of 10 from steaming coal from mines tied to electricity authority power stations to export high-grade coking coal. If export parity prices for coal were used, the electricity tariffs could be expected to be much higher and would also provide an incentive to preserve coal rather than offer off-peak tariffs.

There are many interesting physical aspects of solar heating technology. One of these involves studying selective surfaces which have high absorbance ($\sim 90\%$) for sunlight but low emissivity for heat radiation (3% to 8%) so that the heat that is obtained from the sun is not lost again. How does it work?

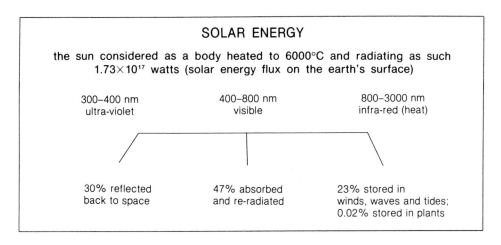

Fig. 10.10 The fate of the solar energy that reaches the earth

There is some interesting chemistry in the design of solar panels. After all, we learn in physics that a good absorber of radiation (such as black paint) is also a good emitter. We need a good absorber of light (in which half the sun's energy is concentrated) and a poor emitter of heat radiation (from the hot panel) and chemistry provides us with just the right surface. Something like copper black has a polished metal surface (good reflector, poor radiator), covered with a layer (2 μm thick) of non-metallic copper oxide, which strongly absorbs visible radiation (it is thick compared to the wavelength of visible light) but is transparent to heat radiation (which has a long wavelength compared to the thickness of the film). The wavelength of maximum emission depends on the temperature of the source and is given approximately from Planck's law. From this law we obtain the relation that $\lambda_{max} = 2.9 \times 10^{-3}/T$ (for the wavelength in metres). At the temperature of the sun (6000 K), λ_{max} is 483 nm, while on the roof of a house the temperature is about 350 K and λ_{max} of the heat radiated back is 8286 nm. The thickness of the oxide layer is 2000 nm. Thus the oxide layer is thick compared to the wavelength of the light coming from the sun, but thin compared to the wavelength of the heat radiating from the copper underneath. The solar absorber thus acts in exactly the opposite manner to the glass cover on the solar panel, which is transparent to visible light but absorbs and re-emits heat radiation (the greenhouse effect).

A second type of panel surface is also used in which a dark non-metal is covered with finely divided metal. Finely divided metals appear black. An example is the silver grain in a developed photographic film). Chrome black (Cr in Cr_2O_3) and iron in iron carbide are examples.

Silicon solar cells

Silicon makes up 25.7% of the Earth's crust, by mass, and is the second most abundant element, being exceeded only by oxygen. Sand is just silicon combined with oxygen. However, to have silicon pure enough for use in solar cells, it must have impurity atoms of vanadium, titanium and zirconium below 10^{13} atoms cm^{-3}; that is, one impurity atom for every 10 billion silicon atoms is allowed. For iron, manganese and chromium we can be 10 times more generous, whereas for copper and nickel one in a million is permitted.

Purification of chemical substances to such a degree requires the art to be taken to its ultimate perfection. It is also highly energy intensive.

The solar car race

On 1 November, 1987, a solar car race set off from Darwin to Adelaide, 3200 km away. A description of the vehicles was given in Australian Geographic, November 1987. The cars were made as lightweight and aerodynamically efficient as possible. The other critical features were the solar panels, the storage batteries and the electric motors.

The solar cells are manufactured from pure silicon (valency 4). This is doped with a tiny amount of phosphorus (valency 5) to provide an extra electron and give an n-type (negative) semiconductor. The silicon can also be doped with boron (valency 3) to provide the absence of an electron (called a positive hole) and give a p-type (positive) semiconductor. These two types of silicon are layered in a single cell, forming an n-p junction. Diffusion of electrons across the junction sets up a voltage difference which then opposes further diffusion. However, the photons of the blue and ultraviolet section of sunlight, which cause sunburn, are also energetic enough to 'promote' an electron in the n-p junction to a higher level where it can move freely (a conduction band) at the same time leaving a hole behind, which is now also free to move (in the valence band). This movement apart constitutes the flow of an electric current, which can operate a motor. A silicon solar cell generates about half a volt, but the current depends on its surface area, and the amount of sunlight. By connecting several cells in series or parallel, the voltage and current respectively can be increased. The more expensive single crystal silicon cells are more efficient in direct sunlight while the cheaper amorphous silicon cells seem to work better in diffuse light. The Japanese entrant in the race, 'Nippon TV Leyton Solar Japan', used both types to advantage. A typical output of a set of amorphous solar panels of area 6.9 m^2 (384 cells), was 600 W (0.9 hp) at 1000 W sunlight irradiation per m^2 of cells.

Fig. 10.11 Powered by the sun—the vehicle entered by the Australian Geographic team, Marsupial, in the solar car race from Darwin to Adelaide in November 1987 (Courtesy of *Australian Geographic*)

The *Australian Geographic* team, Marsupial, used rare earth permanent magnet motors on each rear wheel operating at a nominal 100 V through mosfet switching dc/dc converters, giving an overall conversion efficiency of 90%. On the dashboard, a 200 A digital ammeter was flanked on either side by 30 A motor current meters.

NUCLEAR ENERGY

By 1990 the installed nuclear generating capacity of the USA will be about 200 GWe (gigawatts electrical). Each 1-GWe light–water reactor (LWR) burns most of the ^{235}U and a little of the ^{238}U from about 5600 tonnes of U_3O_8 during its 30-year lifetime. Natural uranium is 99.27% ^{238}U and only 0.72% ^{235}U. Thus the present reactors are only using a small fraction of the uranium.

Since zero-energy neutrons can cause fission of ^{235}U, this material is said to be *fissile*. Only neutrons with energies in excess of 0.6 MeV (million electron-volts) can cause fission of ^{238}U. As these fast neutrons are not produced during fission, it is not possible to produce a chain reaction with ^{238}U alone. However, with a source of fast neutrons, neutron absorption in ^{238}U can result in the formation of fissile ^{239}Pu, a fuel suitable for fission reactors (and of course, if of sufficient quality, nuclear bombs). (See also 'Radioactivity' in Chapter 12.) The reactions are:

$$^{238}U(n, \gamma) \xrightarrow[t_{\frac{1}{2}} = 23 \text{ min}]{\beta} {}^{239}Np \xrightarrow[t_{\frac{1}{2}} = 2.3 \text{ days}]{\beta} {}^{239}Pu$$

where β represents beta emission with the given half-life, and γ indicates gamma-ray emission. Since fissile fuel can be bred from ^{238}U by neutron absorption, ^{238}U said to be *fertile*. Another fissile uranium isotope, ^{233}U, with negligible natural abundance can be produced from the thorium ^{232}Th in beach sands, by the reactions:

$$^{232}\text{Th}(n, \gamma) \, ^{233}\text{Th} \xrightarrow[t_{\frac{1}{2}} = 22 \text{ min}]{\beta} \, ^{233}\text{Pa} \xrightarrow[t_{\frac{1}{2}} = 27.4 \text{ days}]{\beta} \, ^{233}\text{U}$$

Although the amount of ^{235}U is limited, the reserves of fertile ^{238}U and ^{232}Th are vast.

Efficient conversion of fertile isotopes into fissile fuels requires high flux sources of high-energy neutrons. Such sources are the liquid metal fast-breeder reactors (LMFBR), fusion–fission hybrid reactors, and electronuclear reactors. A fusion-fission hybrid is a fusion reactor with fertile isotopes in the blanket around it. The point here is that the requirements for a fusion–fission hybrid are far less demanding than for pure fusion reactors (Q ~ 2 compared to Q ~ 10; see Glossary). A fusion–fission hybrid has advantages over the fast-breeder in that a critical mass is not needed, and no fissile fuel is needed for start-up. Fuel doubling times are months (compared to 15 years) and a hybrid reactor can fuel many light-water reactors. Loss-of-coolant accidents are less likely and, as the hybrids produce *fuel* rather than *electricity*, they can be operated intermittently. Hybrids do have large requirements for handling tritium for the fusion element and design, and maintenance is more difficult. They produce a virtually inexhaustible supply of nuclear fuel. The Soviet Union is understood to be moving into this technology.

CHEMISTRY OF THE CAR

Combustion engines

All combustion engines convert heat energy into mechanical energy and the heat is produced from fossil fuels such as petrol, diesel fuel, coal etc. During the conversion, there are further energy losses from friction in the mechanical parts of the engine and, for cars, additional frictional losses on the road and against the air. Quite apart from these 'mechanical' losses there is the much more fundamental loss of all conversions of heat into mechanical work we discussed earlier.

If the gas in the cylinder of a car engine has an *average* temperature of about 600°C (the peak temperature is much higher—see later in the section on gas turbines) and the exhaust temperature is 100°C, the maximum possible efficiency is (873 − 373)/873 = 57%. Transmission losses cut this back to 20% to 25% in real life. A brief description of some combustion engines follows.

Steam engine

The steam engine is an external combustion engine: the fuel is burnt separately from the motor. Hence a wide variety of fuels can be used, and can be burned efficiently without additives. However the power to weight ratio is unfavourable compared with the later internal combustion engines, and acceleration is lower. It also requires a driving fluid (water) that has to be carted around as well as the fuel.

Diesel engine

The diesel engine is an internal combustion engine which explodes its charge by the heating caused by the compression. Both four-stroke and two-stroke engines are used. The four-stroke diesel cycle is:

Down . . . charge with air
Up . . . . . . compression with heating (compression ratio varies from 16:1 to 23:1)
Down . . . fuel injection: burning and expansion with a smaller rise in cylinder pressure than for petrol engines
Up . . . . . . exhaust

The diesel engine exhausts only about a tenth of the amount of carbon monoxide exhausted by the petrol engine. Hydrocarbon emission is about the same. Blowby is negligible in the diesel engine, since the cylinders contain only air on the compression stroke. Evaporation emissions are also low because the diesel uses a closed fuel-injection system and because diesel fuel is less volatile than petrol. The levels of nitrogen oxides in the exhaust gases are probably worse than for petrol engines. The major problem is smoke and odour. The smoke can be reduced by about 50% by adding about 0.25% barium additives, which probably work by inhibiting the hydrogenation of hydrocarbons to carbon particles or by promoting their oxidation, or both. The sulfur in the fuel helps to convert about 75% of the barium to insoluble barium sulfate.

Diesels, with their higher compression ratios and higher temperatures, are more efficient than petrol engines, running up to about 35% thermal efficiency.

Petrol engine

The petrol engine is the most common internal combustion engine for family cars. Like the diesel, both two-stroke and four-stroke cycles are possible, with the latter being the more common and more efficient (but more complex mechanically). The differences between petrol and diesel engines are: petrol engines work on lower compression (usually 7–10:1) and thus need a spark to ignite the charge; carburation is more common than fuel injection; a more volatile, more highly flammable fuel (petrol) is used and this normally contains a range of additives (see later this chapter) to 'improve' its performance. The basic four-stroke petrol cycle is:

Down . . . air *and fuel* intake
Up . . . . . . compression and *spark ignition* near completion of compression
Down . . . power stroke: expansion caused by increased volume of gas and rise in temperature
Up . . . . . . exhaust stroke

A major problem with petrol engines is atmospheric pollution, caused by the following:

1. Lead compounds from lead tetra-ethyl ($Pb(C_2H_5)_4$—anti-knock additives (see later this chapter)
2. Carbon monoxide (CO) from incomplete combustion
3. Nitrogen oxides (NO, NO_2, etc.) formed from air during the high temperatures of the explosion
4. Hydrocarbons—unburnt or incompletely burnt fuel: 65% in exhaust, 15% from evaporation, 20% from blowby (gases that escape past the piston rings into the oil sump)

As increasing the compression ratio increases the thermal efficiency and performance of the motor, this has been a standard 'advance' in automotive design. The increased compression, however, requires higher octane fuel (see later this chapter and Appendix X), which earlier on had meant increased lead tetra-alkyl content and hence more pollution. Also, more nitrogen oxides are formed at the higher temperatures of the high-compression engines. With the current emphasis on pollution problems the compression ratio in modern cars has begun to drop. The decrease in power output is usually made up by a larger capacity (size of cylinders), which leads to lower economy, more carbon monoxide and hydrocarbon pollution. You can't win! (See Fig. 10.14.)

Some general facts about the energy consumption in petrol cars are shown in Table 10.8.

TABLE 10.8
Some facts about petrol cars

The heat value of the fuel distributed as	%
Useful work	24
Cooling water	33
Exhaust	36
Friction	7

Variations in temperature in the engine[a]	°C
Near the spark plug	1000–1650
Central electrode of plug	500–900
Cylinder wall (water cooled)	80–150
Cylinder wall (air cooled)	95–220
Exhaust value	up to 860
Piston base	300–500

[a] At the point of ignition the piston, plugs, etc., reach temperatures of 2000–2500°C and withstand pressures 3–6 MPa (1 atm ≈ 0.1 MPa).

Methane

Sewage gas is 95% methane, 4% CO_2. The production from four London sewage works in 1939 amounted to over 28 000 m³ per day (equivalent to about 5000 litres of petrol). Methane has a high heating value and it does not dilute or destroy the oil film on the cylinder walls.

While on the topic of car engines here are a few handy definitions.

Compression ratio

A compression ratio of 10:1 means the mixture of petrol and air is compressed to one-tenth of its volume at normal outside air pressure and then ignited. Methanol, CH_3OH, is often used as a racing fuel. This is not because it has more 'power' (44 MJ/kg for all hydrocarbon fuels such as petrol, diesel oil, kerosene, heating oil etc. and only 19 MJ/kg for alcohols), but because of its higher octane rating (160). It can therefore be used at higher compression ratios (19:1 or higher).

The internal combustion engine is in fact a heat engine. Gas is the working medium, which is heated internally by the ignition of the mixture (see box 'Theory of heat engines' at the beginning of this chapter). If we can get a cool and therefore more concentrated charge into the engine, we will get a higher power output. For example, alcohol needs more heat to convert it from liquid to vapour (0.5 MJ compared to 0.14 MJ for petrol), allowing it to act as a refrigerant and cool the gas entering the cylinder between cycles. The latent heat is lost from the point of view of heat value (compare the earlier discussion on space-heating fuels in regard to the latent heat of water). The alcohol fuel is usually ethyl alcohol with 0.5% of vile-tasting pyridine. It costs about 1.5 times the cost of top-grade petrol and about double the amount is needed because of the heat-value difference.

Air : fuel ratio

A petrol engine has an air to fuel ratio of 14:1 to 15:1. For alcohol as a fuel the ratio is 7:1 to 9:1, which means carburettor jets for alcohol are $\sqrt{2}$ (1.4) times in diameter (not twice!). (Carburettor = burette used in cars.)

Nitromethane (CH_3NO_2) is also used in bike racing. It supplies its own oxygen for burning, but it produces nitric acid in the exhaust and requires the driver to use a gas mask. It increases power but *reduces* the allowable compression ratio.

Safety aspects

The *flash point* is the temperature to which a fuel must be heated before vapours will ignite by a free flame in the presence of air (Table 10.9).

The *ignition temperature* is the temperature at which a combustible mixture with air ignites in the absence of a flame (Table 10.10).

TABLE 10.9
Flash points of some common liquids

Liquid	Temp. (°C)	Liquid	Temp. (°C)
Methanol	+11	*n*-hexane	−21
Ethanol	+13	Benzene	−11
Cyclohexanol	+68	*n*-heptane	− 4
Ethyleneglycol	+111	*n*-octane	+13
		n-decane	+46
Diethyl ether	−45		
1,4-dioxan	+12	nitromethane	+35
Cyclohexanone	+44	carbon tetrachloride	nf
		chloroform	nf
Acetone	−20	dichloromethane	nf

nf = non-flammable

TABLE 10.10
Ignition temperatures of gases

Gas	Temp. (°C)	Gas	Temp. (°C)
Hydrogen	580	n-Pentane vapour	309
Petrol	550	Carbon disulfide	100
Town gas	600–650	Acetylene	335
Methanol	...	Acetone	538

The *flammability range* is the range in composition of vapour with air between which explosions can occur. In fuel-rich mixtures there is insufficient oxygen to sustain combustion. In air-rich mixtures there is insufficient fuel.

Table 10.11 shows the flammability limits of various gases. Note:

- the wide limits for acetylene, hydrogen sulfide, and in particular carbon monoxide and hydrogen;
- acetone has a smaller range than methanol, a point that becomes important for an experiment on catalysis described later;
- demonstration instructions, for use by a very competent chemist (*never by an unskilled individual*), of the explosion of hydrogen in air are described in the box 'The big bang'.

TABLE 10.11
Flammability limits in air at ambient temperature (% by volume)

Methane	5.00–15.00
Propane	2.12–9.35
Acetylene	2.50–80.0
Acetone	2.55–12.80
Methanol	6.72–36.50
(Diethyl) ether	2.0–10.00
Hydrogen sulfide	4.30–45.50
Ethanol	3.28–18.95
Carbon monoxide	12.50–74.20
Hydrogen	4–74.20

Note: Flames or explosions can occur when the gas concentration is *between* these limits.

THE BIG BANG

There are many descriptions of *the* chemical experiment:

$$2H_2 + O_2 \rightarrow 2H_2O$$

The combustion of hydrogen and oxygen can be done in many ways. One is to use stoichiometric ratios (i.e. mole ratios to ensure complete use of reactants) of the gases in soap foam.

A bunsen burner (or better an oxy-acetylene torch) is adjusted to give a good flame and then the mains gas control is used to turn the gas off and on, so that the burner adjustments are not changed. This gas mixture is then used to produce a foam by bubbling the gas into a soap solution contained in a large shallow plastic bowl. The foam filled with the right combustion mixture can be set alight with a *long* taper, and the bubbles will explode. This method was used by the British army to detonate land mines on the Falkland Islands after the war with Argentina. The foam can absorb up to 90% of the pressure of a detonation. Incidentally, firefighters use foams for fire control, particularly in enclosed spaces, because people can still breathe even when completely immersed in a foam. The foam traps the products of the fire, such as noxious chemicals, coal dusts or radioactive products.

A more interesting approach is to allow the composition of the gas mixture to change with time, while the hydrogen is burning. The following demonstration must be done with great care and is included here only because so much can be learnt from it. While tin cans with detachable lids and gas-filled balloons are variously suggested for this experiment, the results can be erratic. I have found the following custom-built unit to be reasonably safe and reliable.[1]

The combustion chamber consists of two inverted cones (0.5 mm copper), soldered together around their rims to give a chamber of length about 20 cm and diameter about 15 cm (Plate XI). At the top is a small hole (about 2 to 3 mm) and at the bottom is a short neck of copper tubing about 2 cm long with an internal diameter of about 15 mm.

Seal the top hole temporarily with a piece of tape and flush the chamber thoroughly with hydrogen gas long enough to ensure that all air is removed. Then close the bottom hole with a rubber stopper, and place the chamber, large hole down, in a ring or metal tripod, with a piece of scrap wood underneath.

Preferably carry out the demonstration outdoors, or otherwise in a very large lecture theatre, with the audience at least 7 metres away.

Remove the tape and light the hydrogen gas at the small hole at the top. *Stand clear*! The hydrogen gas burns with an almost invisible flame and may appear to go out. Do not be deceived. After a while you will hear a soft whistle, which starts at a high frequency (your dog will hear it earlier while it is still out of your hearing range) and rapidly lowers in pitch, ending in an explosion with flames shooting out of the bottom. Why does this happen?

The speed of sound depends on the average molecular mass of the gas through which it is propagating. As the hydrogen burns at the top and air moves in at the bottom, the average molecular mass of the gas in the container goes up and the speed of sound goes down. The chamber acts as a resonant cavity like an organ pipe (or your own voice box), so the wavelength of the resonating sound waves is fixed. If the speed of sound drops, so must the pitch. (The same effect occurs when deep sea divers replace air for breathing by a mixture of helium and oxygen to avoid the (nitrogen) bends. The average molecular mass of the gas goes down and the pitch goes up, to give the diver a squeaky voice.)

While the reaction stoichiometry is simple (two hydrogen molecules and one oxygen molecule to give two water molecules), the mechanism through which it occurs is a complex chain process, similar to the combustion of other fuels. There are certain ranges for the ratios of hydrogen to oxygen which give quiet burning, and others which cause explosions. These are defined by flammability limits (see Table 10.11). Thus a further lesson to be learnt here is that a flame near an 'empty' container of solvent with air and solvent vapour still in it (i.e. an empty petrol can) can cause an explosion, while a full container on spilling may burn quietly.

The same type of chain reaction occurs in atmospheric processes (see 'The ozone layer' later in this chapter) and in the polymerisation of monomers to form polymers (see 'Chemistry of plastics' in Chapter 6).

Flywheels

A flywheel is a very compact way of storing energy, coming after nuclear and chemical fuel, and well ahead of storage batteries. Flywheel buses were used in Switzerland, where the routes are up and down. Whereas a petrol engine will not pour petrol back in to the tank on a downhill coast, the flywheel will take energy back and also store enough to drive a bus for half a day. A high-tensile steel flywheel spun up to its bursting speed can store the same amount of energy as water of the same volume as the flywheel, raised to 30 km. A single decker Leyland bus in the UK is being tested with a glass fibre flywheel (developed by BP) which is lighter and safer than the steel. The bulk of the flywheel is in the rim, which is made of glass and aramid fibres, bound by epoxy resin. It spins at a maximum of 16 000 rev/min, with the rim moving at 1300 km/hr, faster than the speed of sound. At this speed the energy is equivalent to taking a 16 tonne bus from start to 48 km/h, while braking returns most of this energy to the flywheel. A flywheel can repeatedly discharge 75% of its energy, something which would quickly destroy a battery.

Other engines

Gas turbine engines
Gas turbine engines have large acceleration lags (1.5 to 7 seconds) and their high continuous operating temperature of 3500°C means a high production of nitrogen oxides (NO_x). NO_x is produced at temperatures above 2900°C, which is only intermittently reached in a piston engine.

Electrical motors
The electrical motor is simplicity itself. There is only one major moving part—the rotor—and the modern electrical motors are remarkably efficient in operation. Some 80% of the electrical power is commonly converted into mechanical energy, the efficiency depending on the loading. The rest, as usual, is 'degraded' to heat. The theoretical efficiency is 100% because the electric motor is not a heat engine—it does not use heat as the driving energy. The main problem for transportation purposes is that electrical engines have a much lower power-to-

mass ratio than the more usual petrol engine (by about five times when batteries are included). The standard batteries (see later in this chapter) are heavy and have a low power capacity. In an electric car the battery weight constitutes about half the weight of the car, to give a total of about 2 tonnes (for a converted standard sedan). About 14 kW continuous power can be achieved. The high efficiency and particularly the very low pollution aspects of electrical motors are currently drawing much research in an effort to overcome weight and capacity problems.

All other common portable power sources involve combustion processes—either internal (within a closed system) or external (the burning fuel is open to the air).

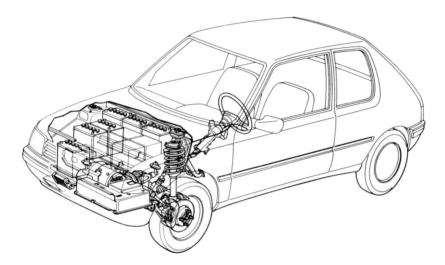

Fig. 10.12 Peugeot 205E electric car

Solar-powered air engine

In the early nineteenth century Robert and James Stirling devised a method of driving an engine using air in the cylinder and heat supplied by a furnace through a heating surface. In Stirling's engine, the heating effect was slow and eventually the heater surface burned out, and so the engine was abandoned.

Importantly, Stirling's idea was to use a *regenerator*, which in essence is a heat storage device, so that some of the heat rejected during the engine cycle of operation was stored and re-used in another part of the cycle of operation. When Stirling built his engine his regenerator was a massive construction using iron plates maintained at high temperature at one end by the furnace and at low temperature at the other end by a water cooler. With the inclusion of the regenerator in the engine cycle of operation the engine had a higher thermal efficiency than any other form of engine using heat energy, and this is still true today.

There has been a revival of interest in the Stirling cycle with its process of regeneration recently, and modern research has shown that, with improved maufacturing techniques and by using gases other than air (in particular, helium and hydrogen), and by operating the engine at higher cylinder pressures, engines with thermal efficiencies as high as 40% with good power outputs are possible.

The principle followed in this engine has been applied in satellites as a power source using solar energy as the external heat source. Also, some large industrial organisations are developing this type of engine. Some of these new forms of the engine are being used in German buses, burning gas or petroleum products to provide the heat energy.

Petrol and diesoline

With the demise of the steam engine, Australia, following the path of most Western, so-called developed, nations, has switched almost exclusively to *petroleum* as the fuel for its vital transportation industry. Current world usage is in excess of 10 million litres of crude petroleum oil and 3000 million cubic metres of petroleum gas *a day*. At this steady rate, not allowing for increases, the known reserves will be gone by the year 2000.

Petroleum is the fossilised organic remains of minute marine plants and animals that settled to the sea floor millions of years ago. Crude petroleum consists of a complex mixture of compounds, mainly hydrocarbons, but also of smaller amounts of organic molecules containing oxygen, sulfur, nitrogen and even metals. The percentage volume composition of West Texas crude oil is shown in Table 10.12.

TABLE 10.12
Composition of West Texas crude oil

	%
Butanes	2
Petrol	11
Naphthas (diesel, kerosene)	14
Furnace oil	17
Gas oil (heating oil)	39
Residue (lubricating oil, asphalt)	17

The standard 'barrel' is 42 US gallons, or 160 L.

The usual first step in refining petroleum is to separate the crude oil into fractions on the basis of their boiling points. Fractions of a typical crude petroleum, arranged in order of increasing boiling point, can be seen in Table 10.13. The petrol obtained in this separation is known as *straight-run gasoline* and is of too low a quality to be used directly in today's automobiles. The naphthas yield kerosene, and solvents for paints, varnishes and lacquers. Furnace and gas oils are burned in oil heaters and diesel engines or are used to make more petrol (see later this chapter in the section on cracking). The residues furnish a great variety of common products, ranging from waxes, mineral oils and paraffin to asphalt for paving.

TABLE 10.13
Petroleum oil fractionation

Fraction	Composition	Boiling range, °C	Principal use
Gas	C_1–C_4	Below 20	Heating fuel
Petroleum ether	C_5–C_6	20–70	Solvents, petrol additive for cold weather
Petrol (straight-run)	C_6–C_{10}	70–200	Motor fuel, solvent
Kerosene	C_{10}–C_{18}	175–320	Jet and diesel fuel
Gas-oil	C_{12}–C_{18}	Above 275	Diesel fuel, heating fuel oil, cracking stock
Lubricating oils	Above C_{18}	Distil under vacuum	Lubrication
Asphalt	Above C_{18}	Non-volatile liquid	Roofing and road materials

Petroleum oil fractionation

Before the advent of electricity as a utility in about 1900, the most useful fraction was kerosene, which was used for home lighting. The advent of electricity and the automobile then made gasoline the most important fraction. Straight-run gasoline (octane rating is about 70) consists mainly of straight-chain hydrocarbons, which cause engine knock and are relatively poor fuels for today's modern car. To get a smooth, constant push on the piston you need a *slow* explosion. When detonation occurs you have *pinging* or *knocking*. Furthermore, until the advent of jet engines, the kerosene fraction had much less use than gasoline. Over the years the petroleum industry has made high-octane gasoline by *reforming* the structure of the C_4–C_{18} fractions to give a highly branched C_6–C_{10} fraction with octane ratings of 90–110.

Alkanes vary greatly in their ability to burn in an engine without knocking. The *octane ratings* of a number of alkanes are presented in Appendix X. These are relative values of the ability of a fuel to burn smoothly and not cause knocking or pinging; they are determined under carefully defined engine conditions. It was discovered that 2,2,4-trimethylpentane (called industrially *iso-octane*) caused little knocking, whereas *n*-heptane caused a great deal (Fig. 10.13)

Fig. 10.13 Iso-octane and *n*-heptane

The measure that later became the octane rating was devised by Sir Henry Tizard (1885–1959), a wartime science administrator who advocated the development of radar during the Second World War (He was the first scientist in recent times to become head of an Oxford college.) The octane rating of any hydrocarbon was defined as being equal to the proportion of iso-octane in a mixture of iso-octane and *n*-heptane that knocked under the same conditions. Methylcyclohexane, for example, knocks under the same conditions as a mixture of 75% iso-octane and 25% *n*-heptane, and hence has an octane rating of 75. A similar system is used for diesel fuel, with an index called the *cetane* number, which is set by comparing a fuel with a mixture of cetane (100) and α-methylnaphthalene (0). A high cetane number indicates a greater ability of the molecules to continue a burning process.

The twin problems of low quality and low quantity of petrol from direct distillation of crude petroleum were solved by a variety of processes known as cracking, alkylation, reforming and isomerisation, and by using additives.

Cracking (thermal, catalytic, hydro-)

By heating the high-boiling fractions (C_{10} and above), with a catalyst, to 400°–500°C, the larger molecules 'crack' into the smaller hydrocarbons and at the same time tend to rearrange into the branched-chain hydrocarbons. (The reason for this transformation lies in the fact that the branched-chain products are more 'disordered'—that is, they have a higher entropy than the straight-chain ones.) This not only increases the amount of petrol that can be obtained from crude oil but also improves the octane rating (see Appendix X), typically into the 85–95 range. For example:

$$CH_3(CH_2)_8CH_3 \quad \xrightarrow{cracking} \quad CH_2 = CH_2 \quad + \quad CH_3 - \overset{\overset{\displaystyle CH_3}{|}}{CH} - (CH_2)_4 - CH_3 \text{ etc}$$

Alkylation

Alkylation involves heating isobutane with the low-boiling alkenes (C_3–C_6) under acid conditions. This causes addition of the isobutane to the alkene—leading to a larger branched alkane.

This process converts some of the lower-boiling gas fractions into high-octane fractions (typically 90–95). For example:

$$CH_2{=}CH{-}CH_3 \quad + \quad \overset{CH_3}{\underset{CH_3}{>}}C\overset{H}{\underset{CH_3}{<}} \quad \xrightarrow{H^+} \quad \overset{CH_3}{\underset{CH_3}{\overset{|}{>}}}C{-}CH_2{-}CH_2{-}CH_3$$

an alkene   isobutane   branched chain alkane

Isomerisation

Heating at relatively low temperatures, with special catalysts, causes rearrangement or isomerisation of straight-chain hydrocarbons into their branched-chain isomers. For example:

$$CH_3(CH_2)_6CH_3 \quad \xrightarrow[\text{heat}]{\text{catalyst}} \quad CH_3 - \overset{\overset{\displaystyle CH_3}{|}}{CH} - \underset{\underset{\displaystyle CH_2CH_3}{|}}{CH} - CH_2 - CH_3$$

This improves the octane rating to the mid-80s and is used primarily on the straight-run petrol fraction.

CATALYSIS

A catalyst is a material that offers an alternative, easier pathway for a chemical reaction to go along. If one describes a chemical reaction as having to go from a valley of starting materials over a mountain range to a valley of product materials, then a catalyst offers a new, lower mountain pass through the range. Of course the path makes it easier to come backwards from the product valley to the starting material valley and so the catalyst does not change the position of equilibrium for the reaction. Catalysts certainly take part in the reaction but are not used up and are chemically, if not physically, the same at the end of the reaction as they were at the beginning.

Catalysts are extremely important in making a particular reaction go faster by offering that reaction rather than competing ones an easier way. Catalysts in biological systems consist of proteins called enzymes. For chemical engineering purposes, metals (particularly transition metals) and their oxides are more often used.

Experiment

A very simple but instructive experiment involves the catalysis of the oxidation of acetone vapour. A catalyst for this reaction is a mixture of the two oxides of copper.

Materials
A 100–250 mL beaker; copper (1 or 2 cent) or alloy (e.g. 10 cent) coins suspended by a paper clip, or copper wire wound in a coil around a pencil as a mould; enough acetone to just cover the bottom of the beaker; and a gas burner.

CAUTION
Acetone is very flammable.
bp: 56.2°C. Flash point: −20°C
Ignition temperature: 484–538°C
Flammability limits in air: 2.55–12.8%
by vol.

Procedure
The copper is heated to red heat in the flame and then supported from a wire to hang just above the acetone in the beaker (Plate XII). The shimmering colours of the copper surface are caused by the heat being maintained in the copper coin from the heat of the chemical reaction taking place on its surface, as well as the colours of copper(II) oxide (black) and copper(I) oxide (red) and copper(0) metal (salmon pink). It may be necessary to warm the beaker in warm water to produce sufficient acetone vapour.

After a while, you may notice a smell of squashed ants, from the product of the reaction. The red heat is maintained as long as some acetone remains.

The flammability limits as a percentage of air can be converted to vapour pressure as follows:

$$\text{Lower: } \frac{2.55}{100} \times 1.013 \times 10^5 = 2.6 \text{ kPa (22 torr)}$$

$$\text{Upper: } \frac{12.8}{100} \times 1.013 \times 10^5 = 13 \text{ kPa (97 torr)}$$

The actual equilibrium vapour pressure of acetone at 25°C is 26 kPa, well *above the upper* flammability limit. Starting with warm acetone, which then keeps warm through the heat of the reaction, sets up this safe situation, except at the top of the beaker where air dilutes the vapour. (Will the same be true for methanol? See Table 10.11.) A flame at the top of the beaker can be avoided by moving the hot copper quickly out. Should a flame occur, it is easily extinguished.

With care, drop the red hot coin or wire into a beaker which has a few millimetres of water or acetone at the bottom. Note that the sizzling sound caused by the cooling of the copper maintains a steady volume until a crescendo heralds the end of the cooling. The same effect occurs when you pour liquid nitrogen into a dewar. Why? (Gases are very much less efficient conductors of heat than liquids. When the copper is very hot, the acetone in contact with it is in the vapour phase and cooling is slow. As the copper cools, it reaches a point where liquid acetone comes in contact and the rate of cooling increases rapidly, as does the sizzling sound.)

Catalytic reactions are extremely important industrially; for example, the Haber Process for producing ammonia from nitrogen and hydrogen, the decomposition of nitrogen oxides by the catalysts used in car anti-pollution mufflers, and the catalytic converters (cats) used in producing petrol from crude oil. The exothermic (heat given out) nature of the reaction shown is exemplified by the fact that, once heated, the copper is kept hot by the reaction.

Reformation

Reformation is by far the most important change process in refining and involves heating with special catalysts similar to those used in cracking and isomerisation. By careful choice of conditions, however, C_6 and above hydrocarbons are converted into aromatic compounds, mainly benzene, toluene and xylenes. The improvement in octane rating is marked—into the mid-to-high nineties. However, the aromatic compounds on incomplete combustion in an engine can produce larger molecules, which form known carcinogens in the exhaust.

With all this refining of crude oil the average petrol produced today has the broad composition:

butane	10%
paraffins	60–65%
aromatics	25% (up to 40%)
alkenes (olefins)	small

and the average octane rating is in the high eighties for the 'pure' hydrocarbon mixture. Since the modern car usually has a compression ratio of 8.5–9.5:1, and this requires 93–96 octane fuel, lead compounds were added to upgrade the octane rating. Non-lead petrol requires more branched chain and aromatic components.

Measurement of fuel consumption

The design of car engines, and thus the fuel used and the gases exhausted, depends on the expected driving conditions. A comparison of the driving cycles of Europe and the USA (both done on dynamometers, not on the road) is quite fascinating. Europe takes London, Paris, Turin and the Ruhr Valley as its norm, with a stretch of 4.05 km, a maximum speed of 50 km/h and average speed of 19 km/h, and the use of only first, second and third gears (never top or overdrive). The exhaust gases rarely rise above 400°C. The US driving cycle is based on Los Angeles, with a 'route' of 12 km, maximum speed of 93 km/h and average speed of 31.5 km/h. This gives exhaust temperatures of up to 600°C.

The US catalytic converter systems do not work below 400°C and so would be inoperative under European conditions. On the other hand, the US tests are just above the legal speed limit and a US catalyst driven at high, legal, autobahn speeds in Europe (which are not reflected in the European test conditions) would burn out. One common, untraceable way to disable a catalyst is to drive at high speeds for a few kilometres and thus melt the catalyst and improve a car's performance.

Lead additives cannot be used in cars with catalytic converters because lead poisons the platinum group catalysts, which are the only ones capable of converting nitrogen oxide and carbon monoxide into nitrogen, oxygen and carbon dioxide, which are thermodynamically more stable at lower temperatures.

An alternative to catalytic converters for reducing emission is to operate with a higher air:fuel ratio or weaker mixture (Fig. 10.14) with a 'fast burn' and 'lean burn' technology.

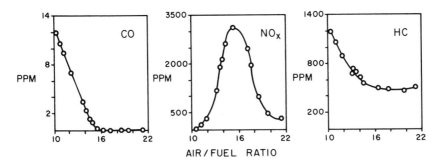

Fig. 10.14 Pollution of exhaust materials as a function of fuel mixture setting. Best ignition occurs in the range A/F = 10 to 14.5 and the overall ratio must allow for variations within the cylinder and from cylinder to cylinder. Uncontrolled engines are thus usually tuned to an A/F of about 13. The stoichiometric ratio (where in the theory the exact amount of air is present for the fuel) is 15.5

In rough terms, fast burn refers to an air : fuel ratio of around 16 : 1 and lean burn can be anywhere between 18:1 and 22:1. Operating with these lean mixtures requires exquisite engine design to provide super-efficient mixing and individual timing for each cylinder, which is set to provide the best economy and power under the driving conditions demanded by the driver. Fast combustion of a lean mixture is the key to low emissions while catalysts try to correct the emission pollutants after the event.

The catalyst is coated onto a honeycomb filter of sintered aluminium alloy and comes in two sections, the first to deal with oxides of nitrogen and the second with carbon monoxide. (Both catalysts use platinum and rhodium, but in different proportions.) To keep the NO_x catalyst hot it must have NO_x on which to operate and the stoichiometric fuel ratio is used to provide maximum NO_x (maximum power but low efficiency and high emissions). To activate the second catalyst for the oxidation of carbon monoxide, extra air must be fed in at the junction of the two honeycombs to provide oxygen in correct proportion to the unburnt gases. This sytem requires oxygen sensors and feedback control.

Lubricating oils

Typically, if a car engine is running at 5000 revolutions per minute (rpm), each piston is slamming up and down inside the cylinder 83 times every second. Because the piston has to go from being stopped at the top to being stopped at the bottom, the acceleration can be as great as 1500 times gravity.

Against these high stresses and the temperatures generated, the engine's only protection is a thin layer of oil between the moving metal surfaces. If this oil coverage fails and the metals do come in contact, the high points on both surfaces are welded together and then the weaker side is torn off as one surface slides past the other. This creates a higher peak which is more readily in contact with the other side and the process escalates. Finally the drag of one metal surface upon another becomes greater than the power produced by the engine and the engine locks solid. This is known as seizing. The pistons and the bearings which hold the spinning crankshaft are the most likely places for seizing to occur. The engine coming to a grinding halt is the most extreme example, but *any* metal-to-metal contact will cause wear.

The effectiveness of modern oils largely eliminates accelerated wear in a well-maintained engine. But inside a worn engine, the forces produced may become beyond the protective capacities of even the best oil.

A much more common wear situation arises when a car has been parked for some time, the engine is cold and the oil has had time to drain back into the sump. When the car is started, there may be some metal-to-metal contact in the top of the engine before the oil is pumped back around it. This is not critical provided the engine is not revved up. One should always wait at least a few seconds before driving off after a cold start. Waiting until the engine has warmed up so that the parts have heated to the optimum operating clearances is not really necessary and just adds more pollution to the atmosphere.

Modern engine oil also acts as a coolant and a washing agent. As the oil circulates it spends some time in the hot operating areas, and then some time in the sump, where some of the collected heat can be radiated out through the metal walls of the sump, which are cooled by the outside air flow.

The combustion process is not a particularly clean one and some of the by-products accumulate in the engine oil. Most of these are forced past worn or badly fitting piston rings. How the oil is designed to cope with all this waste matter is discussed later.

How to read labels

It is difficult to check visually the viscosity of an oil and impossible to gauge its quality by casual inspection. You must rely on the information provided by the manufacturer and written on the oil container. If you are using oil from refillable glass containers at the service station, you must also rely on the honesty of the proprietor in filling the container.

SAE viscosity grading

The numbers you see on all car oil containers are the United States SAE (Society of Automotive Engineers) ratings. The numbers indicate viscosity (see Fig. 10.15). Viscosity is a measure of a substance's resistance to flow. Honey, for instance, is more viscous than water. With engine oils, the thicker (more viscous) the oil is, the higher the SAE number. Oils with an SAE rating over 80 are the thicker gearbox and transmission oils.

If your oil has only one SAE number (engine oils in Australia normally range from 10 to 50 grade) it is a monograde. Monograde oils do not keep a constant

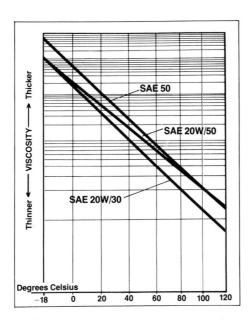

Fig. 10.15 Temperature dependence of oil viscosity. The graph illustrates how, although all crankcase oils get thinner as the temperature rises, a multigrade with a large SAE range can be thinner when cold and thicker when hot than a comparable monograde (*Choice*, July 1981)

viscosity throughout the temperature range: as the engine heats up, a monograde oil becomes thinner. This is not always ideal and, in many cases, an oil is required that is thin when cold but does not get a lot thinner as the engine temperature rises. To meet this need multigrade oils were developed. A 20–40W multigrade has the same viscosity as a monograde 20W when cold but is as thick as a 40 grade oil when both are measured at 100°C. The W after the 20 shows it is the viscosity measurement at colder temperatures (−18°C) .

It may seem logical to suppose that a thicker oil will give better lubrication than a thin one. However, provided an oil is within the correct viscosity range for the machinery to be lubricated, what matters is its performance rating. In some more extreme performance situations a thinner oil may perform better because it is 'wetter' and will tend to stay on the surface to be lubricated instead of flowing off and exposing the metal.

The reason for using a thick oil is to use less. Thick oil is scraped off the cylinder walls by the piston rings and returned to the sump, whereas, under the same conditions, a thinner oil would become mixed with the fuel and burned.

Although the American Standards are now used worldwide, they were devised for North American conditions. Much of the USA and Canada suffers extreme winter conditions and a thick engine oil can become so viscous at temperatures below −20°C that the starter motor cannot turn the engine over. Conditions in most of Australia are very different; therefore the requirements of the different SAE gradings have been widened recently, and the oil industry claims that this makes them far more relevant for Australian conditions.

What is in a modern oil?

First, there is a base oil to which other ingredients are added. For engine oils this base oil is normally a mineral oil from an oil refinery. It is widely accepted that re-refined sump oil also may be an acceptable base oil. However, it must be properly processed—and to do this can produce considerable pollution. There is one re-refining plant in Australia, and some of the oil produced is sold to the oil companies for blending with other base oils. So it may end up back in your sump.

Although some of the base oil is destroyed as it is used in the car engine, the chains of molecules that give oil its lubricating properties are not readily broken down. The main reason the oil in your car has to be changed every so often is that the level of contamination by combustion by-products becomes too high, and the additives can no longer assist the base oil to do its job.

The list of additives includes detergents, dispersants, corrosion inhibitors, anti-oxidants, viscosity index improvers, pour point depressants, extreme pressure additives and friction modifiers. Without delving too deeply into the chemistry behind these, their functions can be summarised as follows.

Detergents reduce or prevent deposits forming in your engine, particularly carbon build-up, which would otherwise cause the piston rings to stick.

Dispersants stop sludge (made up of fuel combustion products, unburned fuel, carbon, lead anti-knock residues and water) from precipitating out of suspension in lumps that can coat parts, block oil-filters and eventually interfere with engine operation. Sludge is mainly formed in stop-go, city, driving conditions. By keeping sludge in fine suspension, dispersants keep the oil and oil-filter effective longer and allow the detergents to do their job more effectively.

Corrosion inhibitors and rust inhibitors coat metal components that are susceptible to corrosion or rust. The most important function of a corrosion inhibitor is to help prevent corrosion of bearings and valve lifters by the acidic residues of combustion and of the oxidation of the oil itself.

Antioxidants prevent oil from oxidising when it comes in contact with the air inside the engine. Oxidation reduces oil life as it greatly increases oil viscosity and encourages the formation of carbon and acidic contaminants.

Viscosity index improvers help the oil maintain an even viscosity so that it does not thicken at low temperatures, or thin at high ones. These additives are the basis of multigrade oils. The way some are said to work is quite fascinating. They are normally polymers (see Chapter 6). As the temperature increases the polymer unravels, resulting in more surface area and a thicker oil. When the temperature drops again, the polymer returns to its original state.

Pour point depressants. After refining, vestiges of wax remain in the base oil. Pour point depressants stop these vestiges forming into a network, which would thicken the oil greatly at low temperatures.

Extreme pressure (EP) additives are not as necessary in engine oil as they are in many gear oils. When the pressure between two moving surfaces becomes so high that most of the film of lubricating oil is squeezed out (a 'boundary lubrication' situation), some other kind of lubrication is required. An EP additive reacts with the surface of the metals to create a film of metal salts that has a lower shear strength (that is, it is more likely to shear off) than the metal surfaces. This film acts as a solid lubricant. EP additives only come into effect in high-pressure and high-temperature situations.

Friction modified (FM) oils are quite controversial at present. In America it seems they are on the verge of being accepted and guidelines are being drawn up for the consumer. Basically, they are additives that are claimed to reduce the level of friction inside an engine and, consequently, to increase efficiency and the potential for fuel saving. Savings of about 5% in fuel consumption are being predicted. Some preliminary tests suggest that many motorists may simply change their driving style, thus using any increase in efficiency to provide greater power, not better fuel economy.

Miscellaneous additives may be required to deal with undesirable side-effects of other additives. Anti-foaming agents are one of the more common extra additives in engine oil.

The automotive oil market is naturally tied closely to the stated requirements of the vehicle manufacturers. Most of the upgraded standards for car oils have been in response to the lubrication requirements of the automotive industry. The American Petroleum Institute (API) and the American Society for Testing and Materials (ASTM) combined to work out an open-ended grading system that could be added to as required.

Hydraulic brake fluids

The hydraulic brake fluids used until just after the Second World War were essentially a solution of castor oil in *n*-butanol or diacetone alcohol. The availability of synthetic lubricants at a stable price helped the change to higher-performance fluids based on ethylene oxide derivatives. These new fluids had a

TABLE 10.14
Composition of brake fluids

Ingredient	Percentage	Description
Solvent	60	A mixture of mono alkyls usually either methyl, ethyl, or n-butyl, or mono-, di- and triethylene glycol.
Anti-swell	20	A mixture of mono-, di- and triethylene glycol.
Lubricant	20	A condensate of n-butanol and a 50:50 mixture of ethylene oxide and propylene oxide.
Additives	minor amounts	Corrosion inhibitors and antioxidants.

boiling point of about 190°C in contrast to 140°C for the earlier fluids. A typical formulation for the new brake fluids is shown in Table 10.14.

ICI Australia developed its own brake fluid technology and commenced manufacturing in 1964.

When the trend was for cars to become more powerful and heavier and to travel faster, and when these factors were combined with the increasingly widespread use of automatic transmission (which provides less engine braking), the demand for more efficient braking systems became more pressing. However, these trends resulted in more energy having to be dissipated as heat in the braking system. The introduction of disc brakes, with intrinsically higher operating temperatures, accentuated the effect of vapour formation and resulted in car manufacturers finding that more of their vehicles were experiencing brake failures because of vapour locking. Vapour locking is the effect produced by the local boiling of the brake fluid producing sufficient vapour in the system to allow the brake pedal to be depressed, thereby compressing the vapour without activating the brake.

At the request of some car manufacturers in the early 1960s, ICI Australia developed brake fluids with boiling points of 260° and 290°C, which overcame this problem. These compounds are similar to some of the non-ionic surfactants but are of much lower molecular mass. One type is a co-polymer of ethylene and propylene oxides (compare non-ionic surfactants, Chapter 2) and belongs to a group called polyglycols. However, these new brake fluids absorbed water from the atmosphere more readily, thus lowering the boiling point, so that after about one year in service the boiling point in many cases would reach the danger level where vapour locking might occur. In one experimental exposure of a simulated braking system to 80% relative humidity at 32°C for 141 days, the water concentration reached 3.85% in the front wheel cylinders and 1.65% in the diaphragm-protected master cylinder. Much of the water enters *through* the brake hoses (see Physics of plastics—Permeability of plastic films' in Chapter 6, and Figure 6.17).

The effect of water on the vapour-locking temperature of a brake fluid as measured by two different procedures is shown in Figure 10.16 (see also Experiment 13.14). In interpreting this graph it should be noted that trials in the Rocky Mountains in the USA showed that the highest operating temperatures brake fluids reached were 148°C, whereas in high-speed pursuits by London police temperatures of 188°C have been reached. A safe operating temperature of 155°C has been nominated by the US Department of Transportation in their federal specification, and this roughly corresponds to about 1.5% water in the brake fluid.

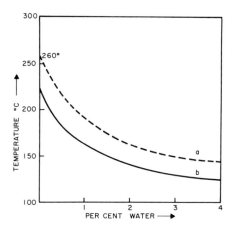

Fig. 10.16 Vapour locking temperature of brake fluids:
a. completely vapour;
b. first vapour produced (*Stop*, ICI)

Field trials with fleets of vehicles in Britain, as well as laboratory experiments, suggest that water uptake by brake fluids will reach this level in about 12 to 18 months, so the brake fluid should be changed before this danger zone is reached.

ICI Australia has also developed a brake fluid that is tolerant of moisture absorption, as it is able to react with the water chemically and the products of the reaction are normal brake fluid components. Australia was one of the first countries in the world to use this new, safer brake fluid.

Antifreeze

Water freezes at 0°C and in so doing it expands. (Ice is less dense than water and floats on it. There are very few materials that have this strange property.) The expansion can damage the chambers in which radiator water flows. When water has other materials dissolved in it, the freezing point is lowered and the boiling point is raised. The amount by which this happens depends only (to a first approximation) on the *number* of molecules of the added material—not on its mass or the type of molecule. Thus, for a given *mass* of material, light molecules are more efficient than heavy molecules. For this reason methanol (CH_3OH; MM 32) was commonly used, but it tends to boil away, so that ethylene glycol (HO— HC_2—CH_2—OH; MM 62) is generally used, but it is only half as effective.

Antifreeze solutions tend to be quite corrosive and so corrosion inhibitors are added that protect, to a certain extent, the metals found in an engine's cooling system (copper, solder, brass, steel, cast iron and cast aluminium). The inhibitors used are generally sodium nitrite with sodium benzoate, although other materials are occasionally quoted.

Ethylene glycol is sweet (see 'Food additives—Artificial sweeteners', Chapter 11)—and poisonous! Death on swallowing probably occurs, however, from nitrite poisoning. This is unfortunate because the antidote for ethylene glycol is alcohol!

The reason? The enzymes that would otherwise convert the glycol into the toxic oxalic acid, which is the real poison, are kept busy by the alcohol.

An Australian standard (AS2108, 1977) recommends a depression of freezing point of $-12°C$ when used according to the manufacturer's recommendations. The British standard also calls for a depression of freezing of approximately $-12°C$ from a 25% v/v antifreeze solution. This makes the quality of the product independent of the manufacturer's recommendation.

Accumulators (batteries)

Normally, car engines rely on electricity stored in the battery to start them. During running, the electricity drawn from the battery is returned by the generator (or alternator) and the battery fails only through mechanical breakdown. Batteries of this type are called secondary or storage cells; electricity is stored by driving a set of chemical reactions in one direction by connecting the battery to a charger. Current may then be drawn from the battery when the reactions proceed in the reverse direction.

Lead–acid accumulator

The familiar 12-volt car battery contains six cells, each providing 2 volts. Each cell contains two series of plates—positive and negative—immersed in sulfuric acid. The position plates are made by forcing a paste of lead dioxide (PbO_2) into a grid made of lead alloy, while the negative plates contain lead in a highly active spongy form. The overall reactions that occur are:

at the positive plates—

$$\underset{\substack{\text{lead}\\\text{dioxide}}}{PbO_2} + 4H^+ + SO_4{}^{2-} + 2\varepsilon^- \underset{\text{charge}}{\overset{\text{discharge}}{\rightleftarrows}} 2H_2O + \underset{\text{insoluble}}{PbSO_4 \downarrow}$$

and at the negative plates—

$$\underset{\substack{\text{spongy}\\\text{lead}}}{Pb} + SO_4{}^{2-} \underset{\text{charge}}{\overset{\text{discharge}}{\rightleftarrows}} PbSO_4 \downarrow + 2\varepsilon^-$$

Ideally these reactions can be recycled indefinitely, but, in practice, material is gradually lost from the plates and falls to the bottom of the cell, where eventually it may short circuit the plates, rendering that cell useless. The state of charge of a lead battery can be estimated readily by a hydrometer, which measures the density of the sulfuric acid. As can be seen from the reactions above, sulfuric acid is consumed by both reactions during discharge and thus the density of the electrolyte falls from about 1.28 (about 37% sulfuric acid) in the fully charged cell to about 1.15 (about 21% sulfuric acid) in the flat battery. If the water is allowed to evaporate from the battery, the concentration of the electrolyte rises to a value that can adversely affect the performance. Thus the electrolyte should be topped up with pure water as often as necessary. Distilled water is preferable, but *some* domestic water supplies are sufficiently pure.

Batteries can be tested against a standard set of parameters (Australian Standard AS2149, 1980). The tests depend on the size of the battery and they reproduce the various tasks a battery has to perform.

The rapid discharge current test assesses the capacity of the battery to supply a high current (160 A) under cold conditions ($-7°C$). (Readers in colder climates will laugh at this 'cold' temperature and no doubt their standards will require conditions that match their local environment.)

The chemical reactions in a battery producing electricity for the starter motor are sensitive to temperature and starting a car on a cold morning places heavy demand on the battery. Not only does the battery not function as well but the engine oil is more viscous and so the engine turns over with greater difficulty. The starter motor draws over 160 A for a short period; therefore we need to define a parameter for this aspect as well.

The standard requires the battery to maintain a voltage of more than 8.4 V for the first five to seven seconds and at least three minutes should elapse before the voltage drops to half the battery's value (6 V). The maintenance of an adequate voltage is normally more important than the length of time the battery will crank the engine: if the car doesn't start in one minute (never mind three!) you're in trouble. The test applies to batteries initially fully charged.

Another use of the battery is to supply a small current for a long time to the car lights. The storage capacity of the battery under these undemanding conditions is measured in electrical energy units (ampere hours, A.h). Batteries are tested by drawing for 20 hours a uniform current of one-twentieth of the 20-hour capacity. That is, after 20 hours the battery has reached the end of its rated capacity and its voltage should at that stage not have dropped below 10.5 V. A rating of 53 A.h for a battery should thus allow it to supply 2.65 A for 20 hours.

Another situation occurs when the alternator or generator on your car fails. How long can you keep driving before your battery goes flat? This requirement is measured by the reserve capacity, which is the period in minutes that a battery is capable of delivering a moderately high current (25 A, enough to run the ignition and headlights) at 25°C before the voltage drops to 10.5 V. The standard sets this period at 70 minutes.

The battery must also *accept* an adequate charging rate from the alternator, otherwise lots of short trips will soon flatten the battery.

The life endurance of batteries is tested by a cycle of constantly overcharging and discharging. Such a test is unsuitable for 'maintenance-free' batteries because they require a well-adjusted charging circuit or topping-up will be necessary, which is often difficult. Maintenance-free batteries use calcium–lead grids rather than antimony–lead alloys. This delays self-discharge and, together with immobilised electrolytes and other measures, removes the need for water replacement. The standard sets a minimum of seven cycles of overcharging and discharging.

It is interesting to cut open batteries after such severe testing. The most common sign is corrosion of the positive grids (see 'Taking care of your battery'.)

The first electric car appeared in the USA in 1847, built by Moses Farmer. It

TAKING CARE OF YOUR BATTERY

There are a few points to remember to gain the maximum life from your car battery.

- Check the level of acid in the battery regularly (if it is a low-maintenance battery) and if it needs topping up add distilled water. Although it is best to use distilled water, tap water will do if the water in your area is soft. The minerals present in hard water will shorten battery life. (Make sure you don't get any acid on you or your clothes.) Using tap water may void some warranties.
- Keep the battery firmly clamped down—vibration is one of the principal causes of car battery failure. However, don't overtighten the clamps or you can damage the battery casing.
- Check the alternator belt regularly—if it is too loose it can slip; too tight it will wear out the alternator bearings. You should be able to move the belt about one centimetre up or down, midway between the pulleys.
- Don't let the battery go flat. If you don't use the car very much put the battery on a charger every now and again. If you are leaving the car unused for any length of time, it may be best to take the battery out or, at least, disconnect the cables.
- Make sure that you never accidentally lay a spanner or other tool across the two terminals of the battery—the extremely rapid discharge can buckle the plates of the battery. It won't do the spanner any good either—or you, if you try to retrieve it!
- Remember that the gases formed while a battery is charging are highly explosive. Never use a flame to examine battery electrolyte levels. Do not smoke near batteries.
- Clean and tighten the battery terminals and cable clamps regularly. They can be cleaned with a wire brush and any corrosive deposits washed off with a tablespoonful of baking soda (sodium bicarbonate) dissolved in a cup of water. Greasing the terminals and clamps with petroleum jelly will inhibit future corrosion.
- Have the alternator and regulator checked and adjusted by a competent auto-electrician. Consistent undercharging or overcharging will shorten the life of the battery.

Source: Choice, August 1983.

ran on a non-rechargeable battery. Gaston Plante developed the rechargeable lead–acid battery in 1859 and Thomas Edison the nickel–iron battery in 1901. By 1912 there were 34 000 electric cars in the USA. Two developments killed the electric car. One was the price drop of Henry Ford's Model T from $850 in 1904 to $265 in 1925. The second was the introduction of the electric starter motor in 1911 (but not in the Model T). This removed the major objection of women drivers to hand crank-starting the petrol-driven cars.

In 1899 an electric car held the world land speed record at 105 km/h, but the low energy density of conventional batteries meant that this was never repeated. A small sedan might require 10 kW h of energy to travel 100 km. This would be supplied by, say, 10 L (8.6 kg) of petrol. The energy density of a lead–acid battery is about 25 W/kg, and so a 400 kg battery occupying 250 L would be needed to drive the car the same distance. However, this size of battery may not be needed if chemical systems based on solid electrolytes are used.

There are two major causes for the deterioration of a lead accumulator. The first is physical breakdown of the positive plates, which is accelerated to a certain extent by the movement of the car. The second cause is a chemical one. On standing, the lead sulfate produced by the discharge reactions tends to recrystallise into a form that is less soluble, making it difficult to recharge the battery properly. It has often been suggested that the addition of magnesium sulfate (Epsom salts) reduces the crystallisation of the lead sulfate, but there is no real evidence for this and it is unlikely that any of the commerical battery 'dopes' will improve the life and performance of the battery. What may help resuscitate a battery with a shorted cell is to empty out the electrolyte, flush the battery, and then replace fresh acid.

Other accumulators

The lead accumulator has a number of advantages. It is cheap, it is easy to manufacture, it uses a cheap electrolyte, the density of which serves to indicate the state of charge, and it maintains a relatively constant voltage over most of the discharge range. However it is heavy (therefore the energy density in watt-hours per kilogram is consequently low), the electrolyte is corrosive and the battery cannot be sealed, as gas is generated during the charging process from the electrolysis of water.

Several other systems are known, and some of these have special applications. In particular, the nickel-cadmium accumulator, which can be sealed, is now commonly used for rechargeable domestic devices—electric toothbrushes, portable razors and radios, and so on. The simplified overall cell reaction is

$$2NiOOH + Cd + 2H_2O \underset{\text{charge}}{\overset{\text{discharge}}{\rightleftarrows}} 2Ni(OH)_2 + Cd(OH)_2$$

with the cadmium plate being the negative one. Because the electrolyte, potassium hydroxide, does not enter into the cell reaction, its density does not vary with the state of charge. Methods for determining the charge are accordingly somewhat more complex and so these accumulators are usually recharged after a set period of use or are continually 'trickle' charged.

The specific energy of a source, the amount of energy it holds per kilogram mass, depends to some extent on the rate at which the energy is being drawn. This is illustrated in Figure 10.17. Of the batteries, the lead accumulator falls off most markedly at higher power. An interesting comparison is that of a cyclist. The racing cyclist is quite competitive at low power.

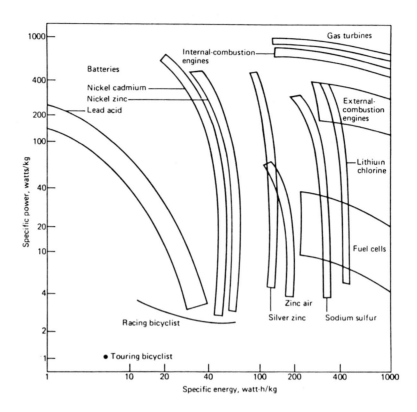

Fig. 10.17 The relationship between power and energy from various sources (From F. R. White and D. C. Wilson, *Bicycling science: ergonomics and mechanics*, 2nd edn, MIT Press, Cambridge, Mass., 1982)

Sodium sulfur accumulator

This accumulator has been around for about 80 years with a chequered history, but with improving technology it could be at an interesting stage. It is about one-fifth the weight of an equivalent nickel–cadmium accumulator, and at least five times the specific energy of a lead–acid accumulator (see Fig. 10.17). However with a sodium anode, alumina electrolyte and sulfur cathode, this cell needs to operate between 270°C and 410°C. If it is fully charged and in operation for at least 80% of the time, then it maintains the temperature range by heat released in its internal resistance (the battery is thermally insulated), but this limits it to commercial and military (submarines and satellites) operation.

The cell reaction consists of the reversible formation of sodium sulfide, with the ions travelling through the solid electrolyte. By using large numbers of small batteries in parallel (250 parallel strings of four batteries in series), the proneness to failure which has plagued these batteries in the past can be minimised in its effect. Chloride Silent Power is one of the main players in the field and is setting up pilot production to produce 6000 batteries per annum, rated at 60 kW h, which will give the trial Bedford vans a range of 170 km; 85 kW h boosts this to 250 km.

The market is there—in the UK there are some 40 000 electric cars of various types out of a total vehicle population of 21 million, one of the highest ratios in the world.

Primary cells

About five dry-cell batteries (the zinc and manganese dioxide Leclanché cell) are used every year for every man, woman and child in the Western world. The cell is cheap and easy to manufacture. In spite of developments such as the alkaline cell and more active manganese dioxide (see 'Chemistry of surfaces', Chapter 4), its energy density is low and its discharge characteristics are unsuitable for many applications. The voltage of the cell drops almost from the start of use (see Fig. 10.18).

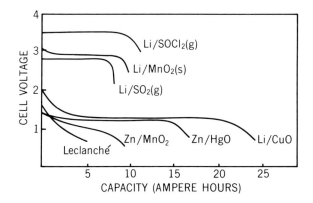

Fig. 10.18 Primary-cell discharge characteristics

The mercury cell developed by Ruben-Mallory in the 1930s has the advantage of maintaining a constant voltage of 1.3 V up to 95% of its total capacity. Although its energy per weight is only average, its energy per *volume* of 0.53 W h/cm^3 is as high as that of almost any other system. For consumer applications in watches and calculators, weight is hardly a problem. The disposal of mercury on the other hand is an environmental problem that is causing concern. Zinc–air batteries, where air replaces the mercury (or silver oxide), give twice the capacity of a mercury cell of equal volume and are used in hearing aids.

Lithium cells have had the most extensive development. As lithium reacts with water, solvents such as propylene carbonate are used to hold the electrolyte. When used with copper oxide, the cell voltage is about 1.5 V, and the cell can be a replacement for conventional batteries. Other systems give much higher voltages (see Fig. 10.18). For low-current applications in watches and calculators, these batteries are cheaper and most suitable. The military are exploiting lithium–sulfur dioxide cells, which have energy densities of up to 330 W h/kg and can operate in extreme temperature ranges and have an excellent shelf-life. Because abuse causes explosions, their use in consumer applications at present is prevented.

The most promising area of development is in the all-solid cell, which has resulted from the discovery in 1967 of solid electrolyte phases that allow ionic conductance at normal temperatures. The lithium–poly(vinyl pyridine)/iodine battery which generates solid lithium iodide as electrolyte has a high internal resistance, which makes it suitable only for low-current applications. The cell is used, for example, in automatic cameras, small computers and cardiac pacemakers, and is gradually replacing other batteries.

Metal-air batteries

If you can throw away one of the reacting electrodes of a battery and not have to carry it around with you, you can save a lot of weight. Why not use the oxygen in air as the oxidising material, and just blow in air as you go. For a travelling lecture series, some years ago I decided to make air–aluminium batteries out of beer cans and not to recycle the cans for 1 cent per can, less the greater cost of fuel to and from the recycling centre.

The first onerous task is to empty out the beer and replace it with the less drinkable caustic potash. Then I used a graphite electrode from a discarded dry cell to act as an anode. When the circuit is closed and air is bubbled over the anode, it is reduced to hydroxide, while the aluminium metal is oxidised to aluminium ion, producing electricity:

$$Al(s) \,|\, K^+ \, OH^-(aq) \,|\, O_2(g) \,|\, C\,(s)$$

which has a cell reaction

$$2Al + {}^3/_2\,O_2 + 3H_2O \rightarrow 2Al(OH)_3 \downarrow$$

The aluminium hydroxide does dissolve to a certain extent in the caustic.

This type of battery has been around for a long time, but now looks commercial because the price of aluminium fell from more than \$1600 a tonne in 1983 to less than \$800 a tonne in 1986. In the US alone, if 1% of cars used aluminium batteries, this would consume 600 000 t of metal per annum. On the basis of the masses of chemically active material (i.e. excluding packing), the battery is said to have a power density of 175 W/kg, and a specific energy density of 400 W h/kg. Zinc–air batteries have a longer history and some of the problems are similar. Both kinds of batteries form oxide layers, which prevent further reaction. For zinc this was overcome (in 1801) by amalgamating (alloying) the surface with mercury. However aluminium is more reactive, so reactive in fact that amalgamation causes whiskers to grow out of the surface of the aluminium, which makes the basis of an interesting overhead projection experiment. A piece of aluminium foil, if rubbed on the edges with mercury or a mercury salt such as mercuric chloride, will grow small whiskers, which can be enlarged with an overhead projector. **Care**, mercury salts are poisonous!

A practical solution is provided by alloying with metals such as gallium ($\sim 0.04\%$), indium or tin, plus the addition of sodium stannate to the electrolyte to reduce corrosion (by almost three orders of magnitude). Examination of the cell under an electron microscope reveals that the current comes from only very small areas of the anode (diameter $\sim 1\,\mu$m), and with enormous current density (hundreds of amperes per cm^2). This is enough to melt the metal added to the aluminium and allow the aluminium to form an amalgam with it.

Another problem concerns the formation of aluminium hydroxide gel at the anode, which requires agitation (by the bubbles or air) to cause it to precipitate.

A US company in Illinois called Voltek has powered a car on such a 10-cell battery, which develops 4.5 kW and keeps a pair of heavy-duty lead–acid batteries charged (when cruising at 64 km/h). The KOH needs to be replaced every 250 km or so, while the aluminium lasts for 1200 km. The main market will be for smaller disposable batteries for golf buggies and the like.

Fuel cells

As a contribution to the fight against pollution by car exhaust, much research is currently being devoted to the production of primary batteries to drive cars. These primary cells produce electricity from the reaction of chemical substances—they are not rechargeable. Many different types have been developed for special purposes, such as military use and space flight.

One of the most promising types is the fuel cell. Like the aluminium–air cell, most of these rely on the controlled reaction between oxygen and a fuel substance—that is, controlled combustion—although other mixtures have been investigated. Thus, in the hydrogen fuel cell, gaseous hydrogen and oxygen are bubbled over electrodes (carbon may be used, as well as certain metals) immersed in potassium hydroxide solution. The cell reaction is:

$$2H_2 \;+\; 4OH^- \;\longrightarrow\; 4H_2O \;+\; 4\varepsilon^-$$
$$O_2 \;+\; 2H_2O \;+\; 4\varepsilon^- \;\longrightarrow\; 4OH^-$$
$$\overline{2H_2 \;+\; O_2 \;\longrightarrow\; 2H_2O}$$

The maximum theoretical voltage is 1.23 V. One advantage of the fuel cell is that it generates electricity directly from chemical reactions without going through the intermediate phase of a (heat) engine. The theoretical limitations on efficiency of conversion thus do not apply. Other cells use alcohol or hydrocarbons in place of hydrogen. The products are water only, or water and carbon dioxide, and so the cells do not produce the pollutants (carbon monoxide and oxides of nitrogen) associated with the internal combustion engine. However, these cells have certain difficulties (e.g. the water must be removed or the electrolyte will become too dilute) and none has yet been developed to the stage where, whilst still being of a reasonable size, it can provide sufficient power to drive a car.

EXPERIMENT 10.1

Electrolysis of acetic acid solution

Requirements

One margarine container, 2 old batteries (1.5 V), 1 new battery (6 or 9 V), 2 small pill bottles, and a small quantity of vinegar (~ 50 mL).

Method

Prepare an electrolysis container by drilling two holes (~10 mm diameter) in the bottom of a margarine container. Insert a clean carbon rod from a 1.5 V dry cell battery through each hole and seal around the rod with silicon sealer so the container is watertight. Attach wires to each carbon rod as shown in Figure 10.19.

Fill your container with water to cover the carbon rods and add two teaspoons of vinegar. Fill the pill bottles with solution and mount a pill bottle over each carbon rod by attaching the bottles to a stick with rubber bands. Attach the carbon electrodes to a dry cell battery (6 or 9 V). After a few minutes bubbles should begin to form on the electrodes and then be trapped in the pill bottles.

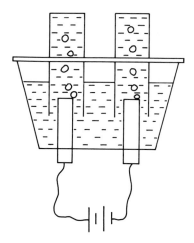

Fig. 10.19 Electrolysis apparatus

Reactions

1. At the electrode attached to the negative battery terminal:

$$2H^+ + 2\varepsilon^- \rightarrow H_2(g)$$

2. At the electrode attached to the positive battery terminal:

$$4OH^- \rightarrow 2H_2O + O_2 + 4\varepsilon^-$$

Notes

1. Don't trap more than a few millilitres of the gases because hydrogen gas is very flammable. It is explosive when mixed with oxygen (see box 'The big bang', p. 372).
2. Hydrogen and oxygen can also be produced from solutions of sulfuric acid (battery acid), or sodium sulfate (a laxative called Glauber's salt available from pharmacies) in water.
3. Try repeating your experiment with copper wire electrodes dipping into your acetic acid solution.

THE OZONE LAYER

The concerns of the fifties and sixties with E–zones (E for erogenous) has given away to those of the seventies and eighties with O–zone!

Ozone prevents radiation below 290 nm from reaching the ground. Radiation around 265 nm is most dangerous to living things, including the plants on which we are ultimately dependent. Ozone also stops a great deal of radiation in the 290 nm to 320 nm range, which causes skin cancer (see 'Chemistry of cosmetics', Chapter 4, on sunscreens). Concentrations of ozone in the stratosphere fluctuate with natural changes in rate of production and destruction; in any one year, the maximum concentrations in the spring can be half as high again as the minimum in the autumn. Production rates of ozone appear to be out of our control, but the compounds we add to the atmosphere will effect the destruction. The oxides of nitrogen, both natural and from car pollution account for perhaps two-thirds of the destruction.

One could not wish for a better example of the complexity of chemical feedback loops in the environment than the reactions of the atmospheric gases and the unpredicted effects of quite small perturbations.

Ozone is produced when ultraviolet (UV) light splits an oxygen molecule, O_2, into two atoms of oxygen, O.

$$O_2 + UV \rightarrow O + O$$

The oxygen atoms can react with other oxygen molecules to produce ozone,

$$O + O_2 \rightarrow O_3,$$

or react with ozone to produce oxygen molecules again:

$$O + O_3 \rightarrow O_2 + O_2$$

When ozone absorbs UV light it breaks up into an oxygen molecule and an oxygen atom again.

$$O_3 + UV \rightarrow O_2 + O$$

A delicate balance is set up which regulates the concentration of ozone at about 10 ppm. If all the ozone in the atmosphere were concentrated at the earth's surface, it would form a layer only about 3 mm thick.

In addition to being a filter for ultraviolet light, ozone is an important stabiliser of our climate. The lowest layer of the atmosphere, the troposphere, is warmed by the heat reradiated at the earth's surface. The rising of this heated air, its cooling high up and then its falling is what gives us winds and clouds. The next layer up, the stratosphere is heated from space by a greenhouse effect. The ozone absorbs UV, and emits infra-red radiation as heat. There is no convection and mixing, as occurs in the troposphere below. With less ozone and hence less absorption, the ultraviolet penetrates deeper and gives a warmer troposphere and colder stratosphere, with consequences for our weather. On the other hand, the increase in atmospheric carbon dioxide and methane (which is increasing three times as fast as carbon dioxide) *increases* the greenhouse heating effect. It is all very complex. So how did consumers come into the play?

Fluorocarbons, chlorofluorocarbons and chlorocarbons

Refrigerators used to be driven with ammonia gas and aerosols with butane. Ammonia stinks and is poisonous, and butane burns. Two fluorocarbons, CFC–11, CCl_3F and CFC–12, CCl_2F_2 (see Chapter 1, 'Language and labels', for nomenclature), were introduced in the 1930s as refrigerants (freons^{TM}), and after the Second World War as aerosol propellants. In 1975, something like 3000 million aerosol cans ejected more than 500 000 tonnes (t) of fluorocarbons into the atmosphere. Non-flammable, odourless, non-toxic except at very high concentrations and chemically inert, so that they do not react with the can contents— what more could we ask for? Because of their different boiling points and vapour pressures, CFC–12 gives the type of high-pressure spray needed for insecticides and paints, while CFC–11 gives a more gentle spray for spraying hair and even more delicate areas. In 1951 another use was found for these fluorocarbons: a non-

Fig. 10.20   Save the ozone layer (Courtesy of the *Canberra Times*)

flammable replacement for pentane as the foaming agent in the production of the foam plastics polyurethane and polystyrene.

The major use of another fluorocarbon, CFC–13 (CCl_2FCClF_2), is in the electronics industry, dry-cleaning and aircraft maintenance. It is one of the few solvents that is non-toxic and does not attack electronic components. In Australia, the defence force is the greatest user.

In 1985, approximately 12 000 t of CFC–11 and –12 were produced and consumed in Australia after a peak of 16 000 t in 1974. Of this, 45% was used for refrigeration and air-conditioning, 33% for aerosols, and 22% for foam manufacture. The total represents about 1.5% of the world usage.

Just as with the story of DDT, the stability of the fluorocarbons has led to our current atmospheric problems. The CSIRO Division of Atmospheric Research has published a study showing the following annual increase in concentration in the atmosphere. The half-life of the materials in the atmosphere is shown in brackets.

5% for CFC–11 (75 years) and CFC–12 (110 years);

13% for CFC–13 (CCl_2FCClF_2) (90 years);

8% for CFC–22 ($CHClF_2$) (20 years);

and for the chlorocarbons,

5% for CH_3CCl_3 (6.5 years);

1–2% for CCl_4 (50 years).

The fluorocarbons in aerosols have been replaced to some extent by trichloroethane, CCl_3CH_3, which is also used in vapour degreasing. However, as seen above, it is also somewhat persistent in the atmosphere. Each year 4000 t of dichloromethane, CH_2Cl_2, is used in Australia, of which 1500 t is used in paint strippers, 2500 t as solvent, the rest in aerosols and foam production.

Halon–1211 (bromochlorodifluoromethane, $CBrClF_2$) and Halon–1301 (bromotrifluoromethane, $CBrF_3$) are used in fire extinguishers. Halon–1301 is more effective than carbon dioxide because it does not require evacuation, while Halon–1211 is used in portable extinguishers.

Reactions

In the upper atmosphere the unfiltered UV irradiation breaks up these compounds to produce chlorine atoms with varying efficiency. A chlorine atom, Cl, from a chlorofluorocarbon can combine with ozone, O_3, to form ClO and an oxygen molecule, O_2. Then ClO and an oxygen atom, O, combine to produce another O_2 and a free chlorine atom, Cl, again.

$$Cl + O_3 \rightarrow ClO + O_2$$

$$ClO + O \rightarrow O_2 + Cl$$

The initial ozone is lost, and since the chlorine atom is regenerated, it can go on and repeat the process in a so-called chain reaction (see box 'Catalysis', p. 379). The chlorine atom is eventually removed from the chain by reacting with some other atmospheric impurity such as methane, forming hydrogen chloride, which in turn contributes to acid rain.

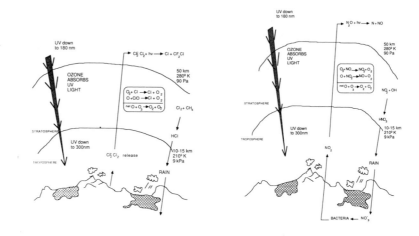

a. Initiated by chlorine   **b. Initiated by nitrogen oxides**

Fig. 10.21   Atmospheric chemical and photochemical reactions (From G. Pimentel *et al.*, *Opportunities in chemistry*, National Academy Press, Washington DC, 1985, pp. 199–200.)

The relative effect that these compounds have on the ozone layer is thus a function of their reactivity as well as their persistence. A table of relative ozone depletion potential, RODP, based on CFC–11 as unity, can be devised for the other compounds assumed to be at the same concentration (Table 10.15).

Ironically, one of the major pollutants caused by cars through their emission of hydrocarbons and oxides of nitrogen in sunlight is the ozone these emissions produce! In the stratosphere, the oxides of nitrogen are part of the natural cycle which *destroys* ozone.

TABLE 10.15
Relative ozone depletion potential

Compound	RODP
CFC–11	1.00
CFC–12	0.86
CFC–22	0.05
CFC–113	0.80
CFC–114	0.60
1,1,1–Trichloroethane	0.15
Carbon tetrachloride	1.11
Halon–1211	10.00
Halon–1301	10.00

BIBLIOGRAPHY

Bio-energy

Gifford, R. M. 'Energy in agriculture'. *Search* **7**(10), 1976, 411.
Rogers, P. L. 'Biological sources of energy'. *Aust. Natural History*, **19**(1), 372.

Electrical energy

'Air conditioners'. *Choice* ACA, February 1983.

Arreamides, J. 'Cars, kilowatts and chemistry'. *Chem. Aust.* **51**(10), 1984, 253.

Breeze, P. 'The electrochemistry of life'. *New Scientist*, 15 December 1983.

'Car batteries'. *Choice*, ACA, August 1983.

Ingram M. D. and Vincent, C. A. 'Solid state ionics'. *Chem. Brit.*, March 1984, 235. This article is more technical than the one below by Vincent.

Lead–acid starter batteries. Aust. Standard 2149—1980, SAA, Sydney.

Vincent, C. A. 'Battery research is back in business'. *New Scientist* **29**, March 1984.

Bennetto, P. 'Microbes come to power'. *New Scientist*, 16 April, 1987, 36.

Brooks, W. N. 'The chloralkali cell; from mercury to membrane'. *Chem. Brit.*, December 1986, 1095.

Fitzpatrick, N., and Scamans, G. 'Aluminium is a fuel for tomorrow'. *New Scientist*, 17 July, 1986 .

Fox, B. 'Terminal confusion over batteries'. *New Scientist*, 18 June, 1987.

Hamer, M. 'Batteries for the van about town'. *New Scientist*, 17 July, 1986, 37.

Haynes, M. 'Recent advances in primary batteries'. *Chem. Brit.*, December 1986, 1101.

Fuel and oil

Allaby, M. and Lovelock, J. 'Wood stoves: the trendy pollutant'. *New Scientist*, 13 Nov. 1980.

Australian energy outlook. Esso Australia Ltd., November 1984, 19 pp.

Cracker: The large-scale manufacture of organic chemicals. ICI Educational Publications, 1980.

'Firewood Forum'. The Institute of Foresters of Australia Inc., Canberra, April 1983.

Gibson, J. 'Chemicals from coal'. *Chem. Brit.* **16**(1), 1980, 26.

Methods of test for fuel consumption: of passenger cars, their derivatives and multi-purpose passenger cars. Australian Standard 2077-1982, SAA, Sydney.

'Motor oils'. *Choice*. ACA, June 1987.

Opportunities in chemistry, US National Academy Press, Washington DC, 1985, 34–41.

Spinks, A. 'Alternatives to fossil petrol'. *Chem. Brit.*, Feb. 1982, 99.

Standard for domestic solid fuel appliances. 1. Installation (AS2918, 1987). 2. Specifications (DR 88014, 1988). 3. Test methods (in committee).

Wittcoff, H. 'Nonleaded gasoline: its impact on the chemical industry'. *J. Chem. Ed.*, **64**(9), 773.

Mechanical energy

'Lightweight flywheels save fuel on buses'. *New Scientist*, 28 November 1985.

Moretti, P. M. and Divone, L. V. 'Modern wind mills'. *Sci. Am.*, June 1986.

Other

Brake fluids. Australian Standard AS1960-1976, SAA, Sydney.

Case-studies in chemical engineering. ICI Educational Publications, 1974.

Dunstan, I. 'Chemistry in the technology of explosives and propellants'. *Chem. Brit.* **7**, 1971, 62.

Pollution

Commoner, B. 'The environmental cost of economic growth', *Chem. Brit.* **8**, 1972, 52.

'It rains formic acid in the top end'. *Ecos* **50**, Summer 1986/7, 28.

Leichnitz, K. *Dräger: Handbook on gas sampling tubes*, 3rd edn. Drägerwerk A.G., Lübeck, 1976. Gives the reaction principle of all the gas testing tubes.

Mathews, R. 'Pollution begins at home'. *New Scientist*, 5 December 1985.

'Melbourne smog and the new breed of cars'. *Ecos* **50**, Summer 1986/7, 11.

Pearce, F. 'Acid rain'. *New Scientist*, 5 November 1987.

Ozone

Australian Industry position paper: chlorofluorocarbons, other chlorinated organic compounds and the ozone depletion issue, Association of Fluorocarbons Consumers and Manufacturers, 15 April 1987.

Ember, L. R., *et al.* 'Tending the global commons'. *C & EN*, November 24, 186, 14–64.

McKenzie, D. 'Anyone want to save the ozone layer?' *New Scientist*, 15 November 1984, 10.

'Mysteries of the Antarctic "ozone hole"'. *Ecos* **52**, Winter 1987, 7.

'Spray cans and the ozone layer'. *Ecos*, 14 November 1977, 3–9.

'When the air's carbon dioxide doubles'. *Ecos*, May 1981, 3.

The second law

Bent, H. A. *The second law: an introduction to classical and statistical thermodynamics*, Oxford University Press, 1965. Certainly one of the more interesting approaches to a difficult topic.

Ogborn, J. 'The teaching of thermodynamics: a teaching problem and an opportunity'. *School Science Review* **57**(210), 1976, 654.

'A Picture of Shuffling Quanta'. In *Nuffield Advanced Science Chemistry, Teachers' Guide 1*, Topics 1–11, pp. 75–79, Longman, England, 1984.

Porter, Sir George. *The laws of disorder*. BBC Films (National Library).

Chemistry in the dining room

11

INTRODUCTION

There is nothing new in the idea of food additives. The person who first smoked a herring was putting an additive in food. Thousands of years ago the Chinese used ethylene and propene (produced by the incomplete combustion of oil shale) to ripen bananas and peas. Incidentally, the earliest published literature on the use of oil shale would appear to be British Patent 330, issued in 1694 to 'make great quantityes of pitch, tarr and oyle out of a sort of stone'. Pliny the Elder (AD 23–79) records that wines from Gaul were artificially coloured and flavoured. Pickling in salt and fermentation processes resulting in the production of lactic acid, alcohol or acetic acid are methods of food preservation that date from ancient times. In salt caves near the Dead Sea there is evidence of even prehistoric preservative processes. Human and animal droppings containing protein and hence nitrogen on the salt floor produced both sodium nitrate and sodium nitrite, which preserve meat. This accidental (but perhaps socially undesirable) piece of chemistry was apparently put to good use. Curing meat with brines containing nitrites and nitrates is thus a long-established process. The nitrites apparently fulfil an essential role in the curing process; they certainly give ham a pleasing colour, which has been attributed to the conversion of myoglobin and haemoglobin into their nitroso-derivatives. The nitrite also prevents the development of the bacterium *Clostridium botulinum*, which causes botulism, the most deadly form of food poisoning. However, over the years there has been some concern about the possibility of a reaction in processed food, or in the digestive system, between nitrite and secondary amines to form nitrosamines, which are highly carcinogenic (i.e. cancer inducing). Small concentrations of nitrosamines (parts per billion) have been found in raw and particularly in smoked fish. It has also been found in red meat, but to a lesser extent.

Flavouring and seasoning were arts in many ancient civilisations, with the result that spices and condiments were important items in commerce. The spices were originally added as preservatives when refrigeration was not available. They also contain antioxidants, and these helped to preserve food longer. No doubt many of us would prefer to eat food straight from the farm, orchard or sea. But, in societies with large, heavily populated, urban areas that import food produced, perhaps, on the other side of the world, some form of processing, if only for preservation, is necessary. However, people do feel uneasy about 'chemicals' in their food because they are worried about the possible effects of eating substances that we do not eat 'naturally', or simply because some additives make some foods taste

Fig. 11.1   Junk food (© 1975 United Feature Syndicate Inc.)

different. All food, of course, consists of chemicals. The chemicals about which people are concerned are traces of substances that are not themselves food, and that may not be present in foods in their natural or traditional state. The presence of these substances may be either accidental or deliberate. Before allowing ourselves to become carried away by this concern, let us consider a few 'natural' culinary disasters.

In 400 BC an army of Greek mercenaries 10 000 strong became intoxicated and finally unconscious after eating honey in some villages on the shore of the Black Sea. In 1596 the members of an expedition to Novaya Zembla in the polar wastes became ill after eating bear's liver, and three of them lost their skin. Much the same thing happened again in 1913 to members of an Antarctic expedition who ate dog's liver. In 1816 Abraham Lincoln's mother died after drinking milk from a cow that had fed on snakeweed.

What happened in the first case was that the bees had fed on the nectar of rhododendrons, which contained a poison that had been deposited in the honey. Mrs Lincoln's cow had acquired a poison from the pasture. The case of the bears demonstrates that a necessary component in our diet, essential to our health, when eaten in excess can be disastrous. In this case it was vitamin A, stored in the bear's liver. These instances of normal foods being toxic are rare; that of Lincoln's mother is the only one recorded involving snakeweed, but outbreaks of honey poisoning do occur from time to time.

However, there are foods that appear to lead to disease in the long term. In Nigeria the starchy root *cassava* is widely eaten and is believed to be responsible for a nervous disease resulting in, amongst other things, deafness and difficulty in walking. The cassava root contains compounds that produce *cyanide* when the root is prepared as food. Although it is customary to wash the food well, and the toxic substances pass out of the root on soaking, it appears likely that enough cyanide remains to cause disease over a period of years. It is relevant to note that we have evolved culturally to avoid the dangers from natural poisons (e.g. the Nigerians wash their cassava without ever having done chemistry or having heard of cyanide!). On the other hand, it is hardly possible to pick up a popular magazine or weekly journal without finding some horrific story of the additives we are all consuming.

FOOD TECHNOLOGY AND THE LAW

The first modern processes for preservation were developed empirically some half a century before the true cause of food spoilage was known. In 1795 the embryonic revolutionary republic of France was beset by enemies on all sides and the government was desperate to seek ways of preserving food for its troops. A prize of

12 000 francs was offered, and it was won, after many years of patient experimentation, by a Parisian confectioner, Nicolas Appert. He published, in 1810, *The book for all households on the art of preserving animal substances for many years.* That's how it all started.

The legal situation in Australia has been described by Madgwick (then Chief Food Inspector with the New South Wales Health Commission) in an article in *Food Technology in Australia.*[1] In the early 1900s in New South Wales, people were rather pleased with their rich, creamy milk, which seemed to keep so well. The colour, it appears, was by courtesy of the milkman rather than the cow, and formalin acted as a preservative. The State of Victoria enacted the first general food legislation in 1905, followed by New South Wales in 1908. Other States followed shortly after, except the Australian Capital Territory. The Australian States adopted a system of prohibition (in contrast to a system of abuse). Under this system, all substances that are not expressly authorised are prohibited in food—in contrast to allowing what it did not expressly prohibit. Food products thus have to be defined by 'specifications of identity' known as food standards. Additives have to be defined separately as they are not foods. Although the 1908 New South Wales Act prohibits preservatives unless specifically allowed, it does not define *preservative* except to state that common salt, sugar, spices, wood smoke, vinegar and acetic acid are *not* included in the term. Other additives such as flavours, colours, essences and spices were originally defined as foods. After describing these first Australian food laws, Madgwick traces the legal development of food laws through to the early 1970s.

There are still areas of confusion. The labels *artificially flavoured* and *artificially coloured* mean that the colour or flavour is not normally found in the product to which they have been added and not necessarily, that the flavour or colour is not of natural origin, as many appear to believe. The allowed list of colours includes chemical dyes as well as 'natural' vegetable colours, but the same legal requirements apply to the whole set. 'Natural' dairy products such as butter and cheese are exempt (provided harmless vegetable colouring matter only is used). The definition of *food* in the 1908 New South Wales Pure Food Act is:

> 'Food' or 'article of food' means article used for food or drink by man, and includes confectionery, and any article that enters into or is used in the composition or preparation of food, and *any spices, flavouring substances, essences, and colouring matters so used* and any substance or article used for consumption by man which the Governor may by proclamation declare to be food or any article of food. [Italics added.]

Madgwick then explains that the *additives* that I have italicised can be used as food when they are added to a food for which there is no standard. A food product for which no standard has been defined by regulation, and which does not occur in nature, cannot be said to be artificially coloured and/or flavoured. The colours and flavours then used are considered to be an integral part of the food. Function defines legal status. Ascorbic acid (Vitamin C) can be used as an antioxidant (as in beer) or as a vitamin (as in orange juice). Sorbitol can be used as a humectant or as a sweetener. Sucrose (cane sugar), when added to sweeten a product, should, according to Madgwick, have the same status as other sweeteners such as cyclamate and saccharin when a product in its standard state does not normally contain sucrose. Thus *additive* is not defined directly in Australian food law.

By now you should be completely disoriented—thinking either of giving up food altogether or not worrying at all about these things and assuming that the 'authorities' will look after us. The first alternative is of no use in the long term and the second, unfortunately, is just not good enough because those authorities need to be continually prodded to ensure action within a reasonable time. The problem, like many in our lives, is one of cost versus benefit. We must, however, be careful about exactly what we mean. Is it consumer *health* benefits weighed against consumer *health* risks? Or is it consumer *economic* benefits against consumer *health* risks? Or consumer *convenience* benefits against consumer *health* risks? And, on the other hand, when manufacturers use the term *cost–benefit*, does it mean *industry economic* benefit weighed against *consumer health* risk? These very different factors are often covered by the same expression.

FOOD ADDITIVES

Types of food additives

In this section the different types of food additives are defined, and then are described in the major groups in greater detail. Classes of additives I have not covered include antibiotics, release agents in confectionery, starch-modifying agents, and water-correcting agents (used in brewing to give a uniform mineral content).

GRAS. This is a group of additives included in the list of additives of the US Food and Drug Administration (FDA) that are 'generally regarded as safe' (hence GRAS), including a large group of natural flavours and oils that are not specified.[2] To be on this list, an additive must have been in use before 1958 and have met certain specifications of safety. Additives brought into use since 1958 must be approved individually. Occasionally, substances are removed from the list by the FDA in the light of new evidence, a recent example being the cyclamate sweeteners.

Preservatives. Substances used to prevent spoilage caused by bacterial activity, fungus and mould, and which thus prolong the keeping quality of foods.

Antioxidants. Substances used to inhibit the oxidation of fats during storage (i.e. stop them becoming rancid).

Fig. 11.2   Food additives (Courtesy Brian Foley, Alan Foley Pty Ltd, Narrabeen, NSW)

Colouring matter. Substances used to colour foods.

Flavouring agents. Aromatic substances, both natural and synthetic, used as components of food flavours or directly in foods, and artificial sweetening agents.

Sweeteners. Substances to make food taste sweet.

Sequesterants. Substances that react with traces of metal ions, tying them up in a manner that prevents their normal reactions, such as catalysing the decomposition of food. Sequesterants such as phosphates are used in detergent formulations to tie up metal ions in water.

Gelling agents, stabilisers and emulsifiers. Substances used to produce or maintain a certain consistency in foods.

Acids and bases. Acids are used to impart a certain tartness to foods or to alter the acidity of the medium (i.e. to lower the pH in canned products or to prevent the crystallisation of jams and jellies). Bases are used as ingredients of baking powders used in pastry production and in powders for effervescent beverages.

Improving agents. This group includes chemical compounds that enhance one or more of the quality criteria of foods (flavour, consistency) and substances used for polishing and glazing confectionery products.

A summary of proposed food additive classifications is given in Table 11.1.

LD$_{50}$—A HISTORICAL PERSPECTIVE

Until about 50 years ago, the toxicity of a substance was usually expressed as the lowest dose which had been observed to kill an animal, even though it was realised that another animal might survive after a much larger dose. In 1926, however, it was shown that the individual minimal lethal doses of digitalis for 573 cats followed a log normal distribution, and in 1927 Trevan, who obtained similar results using an enormous number of frogs, plotted per cent mortality at each dose and interpolated a value for the dose that would kill 50% of the animals. This dose he designated as the LD$_{50}$. After another quarter century, in the early 1950s, the value of this measure of toxicity had been completely accepted and the methods necessary for obtaining it had been well worked out. The practical value of this approach is now being questioned, again on grounds of biological variability. It has been shown, for example, that the LD$_{50}$ for a drug on a single species of laboratory animal varies with strain, sex, age, diet, litter, season, social factors, and temperature. It seems hardly worth standardising all these variables when LD$_{50}$ variation between species is so great, sometimes more than a hundredfold between rats and mice.

Humans differ very greatly from the rat and other experimental animals in their response to some toxins. Some species of animals used in routine laboratory toxicity testing can live happily all their natural days on the seeds of a vetch that, when consumed by humans in amounts of a few hundred grams a day, produces irreversible paralysis of the legs. Much concern was recently aroused by the demonstration that lysinoalanine, a compound formed in foodstuffs treated with alkali, as in the preparation of protein isolates or more traditionally in the primitive use of maize, is severely toxic to rats, producing kidney damage when fed at levels of 100 ppm in their diet. It has now been shown that 10 times this level in the diet of quail, mice, hamsters, rabbits or monkeys has no discernible effect. Thus it appears that the rat is quite unusual in its susceptibility, and that in retrospect the primitive Middle Americans who developed maize as a staple of the diet, using alkali treatments in its preparation, were not foolhardy. Unfortunately there is no convenient escape from the truism that the proof of the pudding is in the eating, and that must be by humans, for it is clear that lack of effect of a new food on experimental animals does not guarantee that it is harmless to us, nor does toxicity in an animal species mean that we must at all costs avoid it.

The variability of response between animals of the same species occurs in humans too, and ignorance of this principle has led to many unsuccessful murder and suicide attempts, in which the presumed lethal dose of a toxic substance was not enough for the purpose envisaged. With very toxic substances the lethal dose resulting in death in 99% of a population may be several times the LD_{50}, but it is still a very small amount, and the distinction is perhaps not very important. It is interesting to speculate on the application of this biological phenomenon to the long-term deleterious effects of particular diets on individuals. Let us suppose, for example, that the consumption of a total of 2.5 tonnes of saturated fatty acid could be shown to result in the death from heart disease of 50% of those consuming it. It would take about 100 years to consume that much on a normal Australian diet, and we would perhaps be entitled to say that therefore it could not be regarded as toxic, as no one (or *almost* no one) would have an opportunity to consume that much. We might, however, be wrong to make this judgment, for in a highly variable population such as ours, one could expect death to result in *some* individuals in half or a third of the time. What I am suggesting is that, with the articles of diet present in major amounts, it is not possible to test for indications of long-term human toxicity using animals, simply because to increase the dosage of the suspect material in the diet enough to give reasonably clear-cut experimental results in animals is impossible, for to do this the component would have to amount to the total weight of the diet or more. Still less is it possible to make such tests on humans, for the institution of slavery was abolished in advanced societies half a century before the emergence of human nutrition as a science, which is one reason why we know so much less about human nutrition than about the nutrition of domestic animals.

Source: M. V. Tracey, 'The price of making our foods safe and suitable', *Food quality in Australia* (Academy Report no. 22), Australian Academy of Science, Canberra 1977, pp. 69–71. See also box 'Toxicity', in Chapter 5.

TABLE 11.1
Classification of food additives under the NH & MRC Model Food Code.

1. *Food additives*
   Addition of food additives to foods is forbidden unless specifically allowed. Foods and the additives allowed, and the levels allowed in that food, are listed.

2. *Preservatives*
   Benzoic acid and its salts, sulfur dioxide (HSO_3^-, etc.), nisin, sorbic acid and salts, propionic acid and salts. Other allowed additives which are not defined as preservatives are: antioxidants, salt, saltpetre (nitrates), nitrites, sugars, acetic acid, alcohol, glycerol, herbs, hop extract, spices, essential oils, materials formed during process smoking.

3. *Colourings*
   A list of natural colours, plus caramel, carbon black and synthetic colouring substances (see Appendix XII).

4. *Flavourings (flavour enhancers)*
   Monosodium glutamate, disodium 5'-guanylate and disodium 5'-inosinate, ethyl maltol, thaumatin.

5. *Antioxidants*
   Esters of gallic acid, BHA, BHT, phospholipids, vitamin C, tocopherols (vitamin E), TBHQ.

6. *Artificial sweetening substances*
   Saccharin, cyclamate, aspartame, acesulfame, thaumatin plus a variety of bases for the sweeteners.

7. *Vitamins and minerals*
   A, B_1, B_2, B_{12}, niacin (nicotinamide), C, D (D_2, D_3). Calcium, iodine, iron, phosphorus.

8. *Modifying agents*
   Group I—vegetable gums
   Agar, alginic acid (salts and esters), gums of larch, carrageenan, acacia, guar, karaya, locust bean, tragacanth. Hydroxypropylmethylcellulose, methylcellulose, pectin, sodium carboxymethylcellulose, sodium actenyl succinate starch, xanthan gum.

   Group II—mineral salts
   CO_3^{2-}, HCO_3^- of (Na^+, K^+, NH_4^+, Ca^{2+} and Mg^{2+}), $CaCl_2$, CaO, KH tartrate, PO_4^{3+} (Na^+, K^+, Ca^{2+}) including meta, poly and pyrophosphates.

   Group III—food acids
   Acetic, citric, fumaric, lactic, malic, tartaric (and NH_4^+, Ca^{2+}, K^+, Na^+ salts).

   Group IV—emulsifiers
   NH_4^+ salts of phosphatidic acids, diacetyltartaric acid ester of mono- and diglycerides, gylcerol lactostearate. Partially esterified fatty acids (with glycerol or sucrose), various polysorbates, sorbitan monostearate, phospholipids from natural sources.

   Group V—humectants
   Glycerol, sorbitol, polydextrose.

   Group VI—thickeners
   Starch and a variety of chemically, physically and enzymatically modified starches (with some chemical restrictions on the product formed).

As of June 1984, detailed specifications of *purity* of food additives have been laid down replacing the vague descriptions introduced in 1969.

Maximum tolerable daily intake (MTDI)

Originally known as the 'acceptable daily intake (ADI)' and now more thought-fully renamed 'maximum tolerable daily intake' (MTDI), this is the amount of a food additive, calculated on the basis of body mass, that can be eaten daily for a lifetime without adverse effect. If a particular MTDI is x mg/kg body weight, an average 70 kg man can afford to eat $70x$ mg/day for a lifetime. Where there is some doubt about the safety of an additive, a time limit on its use is set (e.g. five years) in which further work is to be done, and a conditional (or temporary) MTDI is set. The method of establishing the MTDI is to carry out experiments on animals—increasing the quantity of the additive to establish at which level acute and chronic toxicity occurs in any animal (that is, the level at which an immediate poisonous effect is observed and the one at which long-term effects occur). At least two species of animals are used and the most sensitive species is taken for determining the level. The quantity so determined is divided by 100—the usual safety factor—to set the final MTDI. Sometimes such a factor cannot be afforded. In the case of mercury, the MTDI based on *human* studies was set with a safety factor of only 10, because the naturally occurring levels in food are relatively high. If we consider a natural poison such as *solanine*, which occurs in the green patches of sprouting potatoes, we find that concentrations of solanine of 380–480 mg/kg have been obtained from potatoes implicated in fatal poisoning, whereas the normal level in potatoes is 30–60 mg/kg. Solanine is not destroyed by cooking but only washed out to a certain extent. It is a glycoalkaloid and inhibits cholinesterase enzyme (see 'Pesticides and alternatives—Organophosphorus insecticides', Chapter 5).

Later in this chapter we discuss irradiation of food. Potatoes can be irradiated to delay sprouting and prevent greening. While the green chlorophyll is stopped, the production of solanine is not, and so we are left without a warning sign.

The final task is to look at all the foods in which the particular additive under consideration occurs and to calculate the likely consumption of the various foods. It is then possible to calculate what the actual daily consumption of the additive might be. This knowledge allows a variety of decisions to be made.

1. In the case of an intentional additive (e.g. a preservative), should it be allowed in more types of foods or not? Should the present level allowed in foods be reduced? Is there a chance of an abnormal diet? For example, is someone who drinks a couple of litres of, say, Glugga each day getting too much of an additive such as SAIB (sucrose acetate isobutyrate), which is a replacement for the more dangerous brominated vegetable oils. Should technology be improved to reduce the need for the additive?
2. In the case of unintentional additives (e.g. pesticide residues) should the *withholding* period (see Chapter 5) be increased in order to reduce the amount of additive in the food when eaten?

It is very difficult to calculate the safe levels of some substances used in a wide range of food. An example is the preservative sorbic acid. Most toxicological data are based on laboratory mammals. Unfortunately such results are not necessarily transferable to humans. There are some compounds that are very toxic to some

AVOID BRUISED OR GREEN POTATOES

Glycoalkaloids are potentially toxic compounds found especially in plants from the Solanaceae family. There are many different glycoalkaloids produced by the various members of this family, i.e. potatoes, tomatoes, capsicum, tobacco. However, the only glycoalkaloids recorded as actually causing human death are those produced by potatoes (α-solanine and α-chaconine). Up to 30 deaths and over 2000 documented cases of glyco-alkaloid poisoning involving potato have been recorded as well as numerous cases of livestock losses. A large number of livestock have been killed both by potato glycoalkaloids and other glycoalkaloids produced by the various members of the Solanaceae family.

The toxicity of potato glycoalkaloids is far greater in humans than in other animals studied, with levels of between 3 and 6 mg/kg being reported as lethal, levels comparable to that of strychnine (5 mg/kg). There appears to be a considerable variation in response between people to high levels of glycoalkaloids. Toxicity is due both to the anticholinesterase activity on the central nervous system, and membrane disruption activity which affects the digestive system and also general body metabolism.

Potatoes affected by late blight fungus were suggested . . . as being a major cause of the greatly disabling conditions of anencephaly and spina bifida, through teratogenic effects on the early stages of the fetus. Such a hypothesis appears incorrect; however, laboratory studies, in which potatoes with high glycoalkaloid content have been fed to animals, have indicated that other less incapacitating abnormalities can be produced. Furthermore, studies have repeatedly shown that high fetal mortality and reabsorption occur even with only one high dose of glycoalkaloids given to pregnant animals. Because of this danger, it is suggested that any slightly green or damaged potatoes be especially avoided by women who are pregnant or likely to become pregnant.

Source: S.C. Morris and T. H. Lee, *Food Tech. Aust.*, **36**(3), 1984, 118.

animals but not to humans and *vice versa*. Infants and children cannot be considered simply as small adults, but reliable clinical data are generally unavailable. For a start, the energy intake of children per body mass is about three times that of adults. Very young infants are especially vulnerable to chemicals because the mechanisms that provide protection against these substances in adults are absent or not fully developed. Although the evidence for this derives mainly from studies with drugs rather than with food additives, it is likely that very young infants are less efficient than older children in metabolising some food additives and may therefore accumulate them to excessive levels. If this occurs at a time when sensitivity to toxic effects is critical, because of delicately balanced growth and differentiation processes, there may be deleterious consequences that may not appear until much later in the child's development. The physiological barriers protecting sensitive tissues, such as the blood–brain barrier or the protective barriers for retinal or lens tissue are not as effective in very young infants. In the case of food additives the immediate danger of poisoning is not nearly as important as the long-term chronic, carcinogenic and mutagenic effects. Even less information is available on these effects.

Preservatives or antimicrobials

Benzoic acid

Benzoic acid (Fig. 11.3) and its sodium salts are among the bacteriostatic or germicidal agents most widely used in foods. Many berries (e.g. raspberries) contain appreciable amounts (~0.05%) of benzoic acid. Benzoic acid is included on the permitted lists of at least 30 countries for a great variety of foods, particularly soft drinks. Benzoic acid preserves food by inhibiting the growth of bacteria. As it is only the free acid that is effective, it can be used only in foods of pH less than 4.5. (See also HOCl/OCl in swimming pools, in question 6, 'Swimming pool chemistry', Chapter 5; and ionisation of aspirin in 'A final word on aspirin', Chapter 9.)

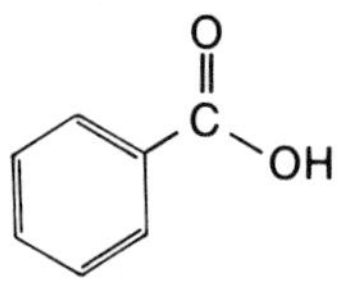

Fig. 11.3   Benzoic acid

At the international level, the Joint FAO/WHO Expert Committee on Food Additives considered benzoic acid in a report in 1962. It stated that benzoic acid is rapidly and completely excreted in the urine. Long-term tests in rats showed that no accumulation in the body occurs. The body excretes benzoic acid as hippuric acid within 9 to 15 hours of eating food containing it.

Sulfur dioxide and sulfites

Sulfur dioxide and sulfites were used by the ancient Egyptians and the Romans. Sulfur dioxide is unique in being the most effective inhibitor of the deterioration of dried fruits and fruit juices. It is used widely in the fermentation industry to prevent spoilage by microbes and as a selective inhibitor. It is also used as an antioxidant and antibrowning agent (for *casse brune*) in winemaking. Sulfur dioxide destroys thiamine (a vitamin); therefore its use is restricted to foods that are not important sources of thiamine. Thus it is forbidden in meat, except for manufactured meat (such as salami), where it is allowed because it protects against the danger of bacterial contamination during processing at elevated temperatures.

Propionates, CH_3CH_2COOH

Flour contains the spores of the bacterium *Bacillus subtilis*, which are not likely to be killed by the baking temperature. Under summer conditions, these bacteria become active and produce a condition called *rope*, which renders bread inedible. The calcium and sodium salts of propionic acid are used in bread (0.2%) to inhibit the growth of micro-organisms.

Sorbic acid

Sorbic acid (2,4-hexadienoic acid, CH_3—CH=$CHCH$=$CHCOOH$) is naturally present in some fruits. It is a selective growth inhibitor for certain moulds, yeasts and bacteria. It is used in cheese, pickles, fish products, cordials and carbonated drinks.

Nitrates and nitrites

As indicated in the introduction to this chapter, nitrates and nitrites occur naturally in many foods, particularly vegetables. The human infant is extremely sensitive to nitrites because of low ability to deal with a modification of blood haemoglobin caused by nitrites. Additional nitrite is derived from nitrate by bacterial activity in the gut, which is particularly efficient in the very young infant because of the inadequacy of acid production in the stomach.

Diethyl pyrocarbonate, $C_2H_5OCOOCOOC_2H_5$

Diethyl pyrocarbonate is a preservative that *decomposes* rapidly in water to form ethyl alcohol and carbon dioxide. It is thus useful in the preparation of wines, beer and other drinks. However, because it has been alleged to produce a urethane with some food products, it was removed from the NH & MRC permitted list. (See also section on wine in this chapter.)

Miscellaneous

Sodium diacetate is another rope inhibitor used in baked foods (e.g. bread).
Biphenyl is a fungistatic that migrates from the wrapping material to inhibit growth of mould causing decay of citrus fruits.
Hexamethylenetetramine is a preservative for certain fish products (but is not used in Australia).

In our cost–benefit decision-making process, preservatives would probably score well on the benefit side. Whenever there is a choice, however (price and convenience being comparable—such as for tomato sauce), it would seem sensible to opt for the unpreserved product. No case can be made for the indiscriminate use of preservatives. It is often better to encourage the use of refrigeration and good handling techniques.

Antioxidants

Preservatives for fatty products and oils are called antioxidants. They prevent the occurrence of oxidation, which is the cause of *rancidity*. Vitamin C (ascorbic acid) is commonly used for water-soluble products, but the most common antioxidants are the fat-soluble BHA (butylated hydroxyanisole) and BHT (butylated hydroxytoluene). They have similar properties to the 'natural' oxidant vitamin E (α-tocopherol). The word *butylated* is not widely used in chemical nomenclature. It is applied here because the usual names for BHA and BHT include the words *cresol* or *phenol*, which generally have connotations of toxicity. To avoid consumer rejection of these 'safe' compounds, the names were made to sound safe as well. Various esters (propyl, acetyl and dodecyl) of gallic acid (3,4,5-trihydroxy-benzoic acid) are used in margarine, oils, cream cheese and instant mashed potatoes.

Antioxidants are additives we shall probably have to accept if we want convenience and reasonable shelf-life. On the other hand, we should expect producers to use the maximum technological skill to reduce the required amount to a minimum.

Colouring matter

Colours are put in food mainly for aesthetic reasons. The way food looks has an effect on its palatability. Both natural and synthetic colours are used. The synthetic colours are mainly coal-tar dyes, many of which have been found to be carcinogenic. The list of permitted red dyes has halved during the past 30 years. No two countries seem to agree on which colours are safe.[3] The USSR puts its faith in natural colours and allows only three synthetic ones (although the distinction is questionable when a natural dye is synthesised). When the British had to conform to the EEC norm (1 January 1977), the 'gold' put in the kippers and 'pink' in the sausage had to be changed.

A typical example of the type of compound used as a food dye is allura red which was added to allowed food colours in Australia in 1977 (Fig. 11.4).

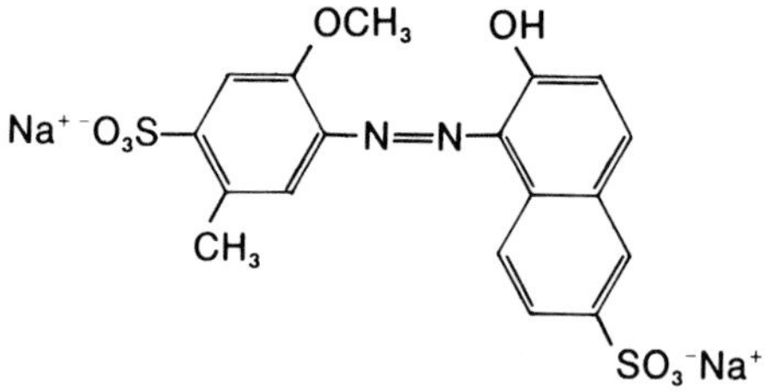

Fig. 11.4   Allura red

The coal-tar designation comes from the presence of aromatic rings, mainly benzene and naphthalene (Fig. 11.5). The colour is introduced by one or more *diazo* (dinitrogen) groups: —N=N—. To make the dyes soluble in water, one or more sulfonic acid (SO_3^-) groups are attached, with Na^+ or NH_4^+ being the other ion. The dyes are generally made from two halves, which are brought together by joining the nitrogens in a diazo coupling reaction.

A case was reported of a young boy who was passing red stools. Unfortunately, he was subjected to extensive tests in hospital before it was found that the source of the red was *erythrosine*, which was used to colour the cereal he was fond of. Apparently very little of the colour is absorbed: most is excreted in the faeces, which explains its lack of toxicity.

benzene   naphthalene

Fig. 11.5 Aromatic hydrocarbons

This metabolic inertness is not found with the food azo dyes described above. The azo linkage is split by the bacteria in the gut of humans and animals and it is probable that the products are the problem. The process has been known since the discovering in 1935 of the first antibacterial sulfa drug—Prontosil—which was later found to form the active drug sulfanilamide in the bowel (Fig. 11.6). (See also section on sulfa drugs in 'Chemistry of drugs', Chapter 9.)

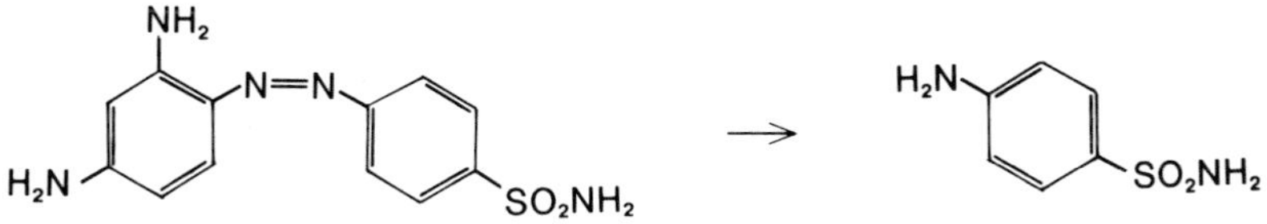

Prontosil   sulfanilamide

Fig. 11.6 The action of Prontosil

The less toxic members of the azo series are sulfonated on *both* aromatic partners of the azo link (see Fig. 11.4). When split, both bits will still be sulfonated and hence poorly absorbed. The more toxic members split with one part not sulfonated and hence are probably more easily absorbed. The triphenylmethane colours (these are chemically a different family—see Appendix XII) are all sulfonated and highly water-soluble. They do not break up metabolically and are poorly absorbed.

Prior to the Australian 1955 list, some 40 coal-tar dyes were approved as food colours. The reduced list (22) excludes all *fat-soluble* dyes (and other suspect dyes) including the notorious carcinogen butter yellow 11020—p-dimethylaminoazobenzene (see Fig. 11.7), which had already been deleted (along with the colour sudan I) by Canada in 1934.

The $SO_3^-Na^+$ group, which makes these dyes soluble in water, also reduces their physiological activity. The Australian 1955 list was 'known to include some dyes which were, to some extent, suspect but which were included for the time being to prevent undue embarrassment to the food industry, it being understood that they were subject to removal at any time'. In 1966 this number was still 22— eight red, one orange, six yellow, one green, two blue, one violet, two brown (of mixed composition) and one black. In 1969 two yellow dyes were removed. Now another eight colours have been deleted because of lack of technological need (see Appendix XII). A new red colour has been added (allura red, see Fig. 11.4).

In 1960 the USA had 14 permitted food dyes and the UK had 30. But nine of the permitted US dyes were banned in the UK. Because the chemical structures of all azo dyes are very similar, one might imagine that there is not much to choose

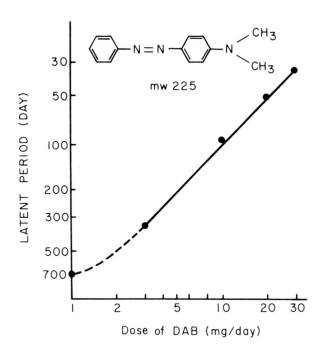

Fig. 11.7 Log-log plot of latent period for the appearance of liver cancer in rats fed *para*-dimethylaminoazobenzene (butter yellow) against dose levels. The larger the dose is, the shorter the time for the cancer to appear. Curves of this type are used to justify the use of high dosages in testing for carcinogens which at the levels in normal use take long times to show an effect. It was as a result of such tests that this colour was removed from the list of allowed food colours. (International Agency for Research on Cancer, *Evaluation of carcinogenic risk of chemicals to man*, vol. 8, 1975. WHO)

between them from the point of view of danger to health. However, two of the basic materials in the preparation of some of these dyes are the naphthylamines (Fig. 11.8). One of these, 2-naphthylamine, is a very potent carcinogen and causes cancer of the bladder, with an induction period between intake and disease of about 20 to 30 years. On the other hand, 1-naphthylamine is less potent (i.e. compared to the 2-) and its greatest danger is that it may be contaminated by the 2-compound. All this, of course, raises another point—most toxicological testing is done with massive overdoses of the material over a relatively short time. Commercial enterprises are unhappy at waiting 30 years before releasing their latest find to the public. There is little reason to believe that dose can replace time as an experimental variable. In fact, if high concentrations of a possibly potent test chemical are used, the effect can be to *kill* the exposed cells—instead of revealing mutagenic or carcinogenic responses. This also applies to the effects of radiation. Large concentrations can also act as physical irritants and lead to tumour formation. On the other hand, large concentrations can reveal the presence of a potent impurity, even if present in very low concentration, as in the herbicide 2,4,5-T and in saccharin.

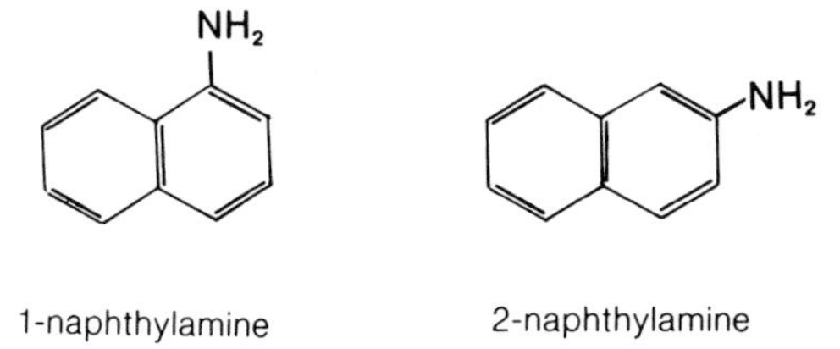

1-naphthylamine   2-naphthylamine

Fig. 11.8 Naphthylamines

The use of healthy animals to test the toxicity of chemicals to humans of various ages and states of health (liver and kidney function etc.) is also questionable. This explains why, with three exceptions (4-aminobiphenyl, nitrogen mustards and vinyl chloride monomer), the carcinogenic action of various industrial compounds was detected primarily by exact medical observations of humans, and not in animal experiments. Confirmation by *animal* experiments has often lagged several decades behind the medical observation—over 40 years in the case of 2-naphthylamine (humans, 1895; animals, 1938). Other examples are tar (1775, 1918), asbestos (1930, 1941), chromates (1912, 1958) and benzidine (1940, 1946).

Although it can be argued that preservatives are essential under modern conditions, the case for adding colours is not strong. Although the reasons for using colouring agents are to restore colour that may have been lost in processing or to standardise the final look of a product that may be made from varying ingredients, it is really a question of what you are used to. Dutch people faced with a can of green peas would be most suspicious: they are used to their natural cooked

SOFT DRINKS

Arguments based on what children drink suggest a limit on the average daily consumption of soft drink at about 25 mL of beverage per kilogram body mass. For a 50 kg reference individual, 25 mL/kg corresponds to 1250 mL/day or 456 L/year. The Australian average for consumption of soft drinks is 70 L/year. The question remaining is what is the average consumption for those consumers who drink a lot of soft drinks. This does not influence the question of what level of food colour should be allowed in soft drink, given an accepted rate of consumption and an accepted no-response dose.

For the food colour amaranth, a temporary maximum tolerable daily intake has been set at 0.75 mg/kg body mass.[4] With the usual assumption that 50% of intake will occur in food other than soft drinks, this level fixes a maximum level for amaranth in soft drinks at 15 mg/kg of soft drink.

The current (1988) allowed level in Australia is 70 mg/kg of soft drink.

colour of bluish-grey. There can be no justification for exposing the most sensitive section of our community—our children—to relatively large concentrations of dyes in soft drinks, ice confectionery and sweets, where no natural colouring is being restored. In 1976, Australians drank 70 L of soft drinks per head of population, about one-third of which were cola-type drinks (and beer consumption was about 140 L a head). The permitted concentration of dyestuff in soft drinks as consumed (NH & MRC Food Standards) is 70 mg/kg (double in cordials and four times in solid foods). Because it may take decades for these substances to be cancer-producing, protecting our children should be a minimum objective. In addition, drinks of the cola variety can contain caffeine (a stimulant) and phosphoric acid, the use of which has never really been examined or justified.

Colour as flavour

The confusion between colour and flavour is not entirely fanciful. A carefully designed survey was carried out by CSIRO[5] in an ice-cream parlour in a Sydney beach suburb, for four weeks in summer. The aim of the survey was to determine the influence of artificial colour in what consumers choose. Four flavours were surveyed in a 'cross-over' mode. The flavours are normally sold with and without colour.

Week 1. Uncoloured passionfruit and coloured butterscotch, plus controls.
Week 2. Coloured passionfruit and uncoloured butterscotch plus controls.
Week 3. Uncoloured rockmelon (cantaloupe) and coloured peppermint plus controls.
Week 4. Coloured rockmelon and uncoloured peppermint plus controls.
Controls. Ten other flavours of ice-cream—vanilla (white), chocolate (brown), vanilla chocolate chip (white), orange chocolate chip (orange), peach mango (white), boysenberry (purple), honeycomb (white), caramel toffee (brown), rum raisin (white) and coffee walnut (brown)—that is five coloured and five uncoloured flavours were used as an unchanging 'background' for the full four weeks.

For the surveyed samples, coloured and uncoloured ice-creams of the same flavour were never presented together, but for the controls they were. The colours used were orange/yellow (beta carotene), brown (caramel) and green (chlorophyll). For each of the surveyed flavours, the time taken for the purchase of about 12 litres (121 ice-creams) was recorded as an inverse measure of popularity (Table 11.2).

TABLE 11.2
Time taken (hrs) for the purchase of 121
ice-creams (the longer the time, the less popular)

Flavour	Uncoloured	Coloured
Passionfruit	44	15
Rockmelon	40	16
Butterscotch	27	9
Peppermint chip	36	9

All flavours were clearly labelled, but no reference was made to colour (i.e. both coloured and uncoloured passionfruit were labelled 'passionfruit'). The coloured outsold the uncoloured by approximately 3:1! While there is clearly a difference in popularity of flavours, it is not as large as the effect of colour.

Exercise: Note how carefully the above experiment was designed to avoid answering the wrong question or biasing the result. Think of other interesting topics, such as artificial versus natural colourings. Devise your own survey to test out a marketing hypothesis.

Flavours

Flavours constitute the largest class of food additives. It is estimated that 1100 to 1400 natural or synthetic flavours are available. This represents a tremendous task of checking for dangerous effects, which, needless to say, has hardly been attempted. (Many synthetic flavours are identical to the natural ones.) Some countries publish lists of permitted and prohibited flavours; others have a short list of prohibited flavours, many of which are natural; and others allow flavourings that are found only in the aromatic oils of edible plants.

The NH & MRC has a model standard for essences that covers the major flavour chemicals. It is still a very limited coverage.

The International Organisation of the Flavour Industry divides flavourings into five groups:

- Aromatic raw materials of vegetable and animal origin, such as pepper or meat extract.
- Natural flavours such as concentrates prepared from aromatic raw materials by extraction and concentration. For example, the flavour in tarragon vinegar is extracted from tarragon.
- Natural flavouring substances that are isolated from aromatic raw materials by physical means. Lemon oil bought in a bottle, for example, contains the same chemicals as the zest of lemon, which can be collected by rubbing a sugar cube over a lemon.
- Natural substances that are isolated by chemical processes but are identical with the natural substance. For example, monosodium glutamate prepared synthetically is the same as monosodium glutamate found naturally in tomatoes, mushrooms, parmesan cheese and sweet corn.
- Artificial flavouring substances that have not been identified in natural products but that simulate natural flavours.

Most flavours fall into the first four categories and are therefore natural products normally found in food. However, not all are. Furthermore, in Australia, there is only a very limited list of regulated flavourings. Manufacturers can add any other flavouring they like. Solvents permitted in flavourings are regulated.

Flavours are often complex chemical mixtures. Usually, only small amounts are added to foods or drinks and it is difficult to analyse for many of them. In practice, in Australia, virtually no checking is done to ensure that all the substances used are safe.

In Australia, the act of adding colour or flavouring was previously classed as artificial. This was confusing because, if a natural flavour such as natural vanilla

Fig. 11.9 Former regulations produced apparent contradictions in labelling, as shown by this ice-cream package (August 1977). The ice-cream is claimed to be made only from natural ingredients, yet is labelled 'artificially flavoured'.

was added to ice-cream, the label stated 'artificially flavoured' and yet also stated 'made only from natural ingredients' (see, for example, Fig. 11.9). The regulation made it more difficult for consumers to distinguish between products with truly artificial ingredients and those with natural ones. The labelling rules have changed somewhat, and reliance is now placed on ingredient labelling (see Fig. 11.10). If synthetic vanillin is used, it must be labelled 'imitation vanillin essence' and not 'vanilla'. However, vanillin can be produced from wood lignin, which is natural.

Gas chromatography is a very sensitive physical chemical technique used to separate components in a chemical mixture, particularly if they are fairly volatile. (The method is an extension of thin-layer chromatography used in Experiment 13.3)

Chromatograms of two varieties of passionfruit, *Passiflora edulis* (Sims) and its mutant *P. edulis f. flavicarpa* (Degener) are shown in Figure 11.11.

Fig. 11.10 The detailed list of ingredients on an ice-cream label now distinguishes between natural and artificial flavours

It is possible to detect compounds present at concentrations as low as 10 μg/kg (1 in 10^8). The differences in composition shown in Figure 11.11 lead to quite a different taste. During the 1950s, Australian plant breeders set out to combine the flavour of *edulis* with the hardiness and juice yield of *flavicarpa*. Today *edulis* is no longer grown commercially; it has been replaced by four purple-skinned hybrids, but some of these lack the ionone-related compounds that gave *edulis* the flavour we remember in earlier days. (So it is not just that your taste buds have dulled!)

Amongst the important flavours are the 2-methoxypyrazines, one of which, the 3-isopropyl derivative, is responsible for the characteristic aroma and flavour of green capsicums. It also occurs in pea *pods* but far less in the pea seed. The aroma of crushed pea pods that emanated from kitchens in the days when peas were still shelled came from the pods, and the practice of adding a few pods during cooking enhanced the pea flavour.

Avocados must be eaten at just the right time to obtain the best flavour. Left too long, the bland, slightly nutty flavour becomes fatty–tallow and putty-like. The flavour chemicals have not changed but the gas chromatograph reveals that the *relative* amounts are quite different. The avocado is very rich in unsaturated fatty acids (see Chapter 3), and these oxidise during ripening to carbonyl compounds. Too much oxidation and the taste becomes rancid; too little and there is no taste at all.

There can be little that is more nauseating than a rotting potato. The cause is a bacterium that breaks down two amino acids, tryptophan and tyrosine, to produce

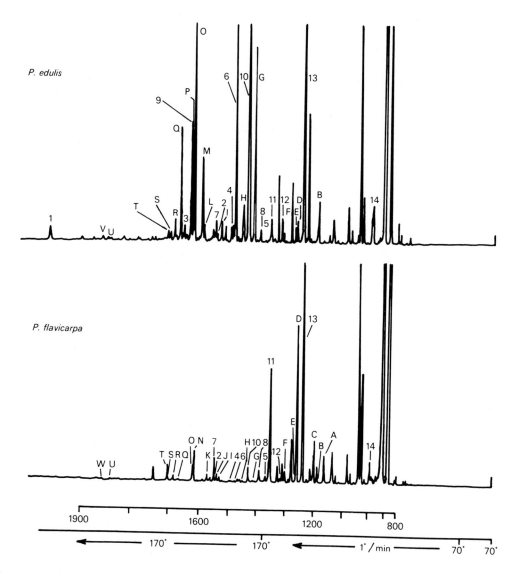

Fig. 11.11 Chromatograms of passionfruit. Gas chromatograms of headspace vapours from purple (upper) and yellow (lower) passionfruit juices. Conditions: Juice sample 10 g, temperature 40°C, collection time 1 hour. Stainless-steel column (150 m, 0.75 mm internal diameter) wall-coated with Carbowax 20 M, programmed 70°C for 16 minutes. 70° to 170°C at 1°C/min, 170°C for 2 hours. Peaks (the named compounds are the major components in each peak).
A, Myrcene. B, Heptan-2-one. C, Limonene. D, *trans*-β-Ocimene. E, 2-Heptyl acetate. F, α-Terpinolene. G, 2-Heptyl butanoate. H, Ethyl octanoate. I, Ethyl 3-hydroxybutanoate. J and K, the Theaspiranes. L, *cis*-Hexa-3,5-dienyl butanoate. M, 2-Heptyl hexanoate. N, Unknown M.W. 154. O, Hexyl hexanoate. P, Octyl butanoate. Q, *cis*-Hexa-3-enyl hexanoate. R and S, the 2-(3-Hydroxybutyl)-butanoates. T, α-Terpineol. U, Dihydro-β-ionone. V, Dihydro-ionone. W, Geranyl acetone. 1, β-Ionone. 2, Ethyl *cis*-octa-4,7-dienoate. 3, Megastigma-4,6,8-triene. 4, Ethyl *cis*-oct-4-enoate. 5, Rose oxide. 6, *cis*-Hexa-3-enyl butanoate. 7, Linalool. 8, *cis*-Hexa-3-enol. 9, Edulan I. 10, Hexyl butanoate. 11, Hexanol. 12, Heptan-2-ol. 13, Ethyl hexanoate. 14, Methyl butanoate (F.B. Whitfield, 'The chemistry of food acceptance', *Food Res. Q.* **42**, 1982, 52.)

compounds found in faeces and horse manure. These compounds can be absorbed by other potatoes nearby and render the lot unacceptable.

We are incredibly sensitive to some compounds used in fungicides and other protective chemicals. Chlorophenols and their precursors are now environmentally ubiquitous and the related chloroanisoles can be detected in food by their musty–mouldy flavours at concentrations well below 1 μg/kg (1 in 10^9). These compounds are formed when micro-organisms detoxify chlorophenol-based antimould preparations. One of these compounds can be detected in water at a concentration of 3×10^{-7} μg/kg (3 in 10^{16}). As you can see, regulating 'flavours' is not an easy task. At such low concentrations we are concerned with aesthetics and not safety.

Monosodium glutamate

Related to flavours are the additives known as *flavour enhancers*. The commonest of them is monosodium glutamate, MSG, which is the monosodium salt of glutamic acid, one of the natural amino acids (Fig. 11.12). In excess, this compound can cause the unpleasant complaint known as *Kwok's disease* or, more commonly, *Chinese restaurant syndrome*, because of the generous application of MSG in the dishes of many of these establishments. It was also once used in baby foods, but this has been banned.

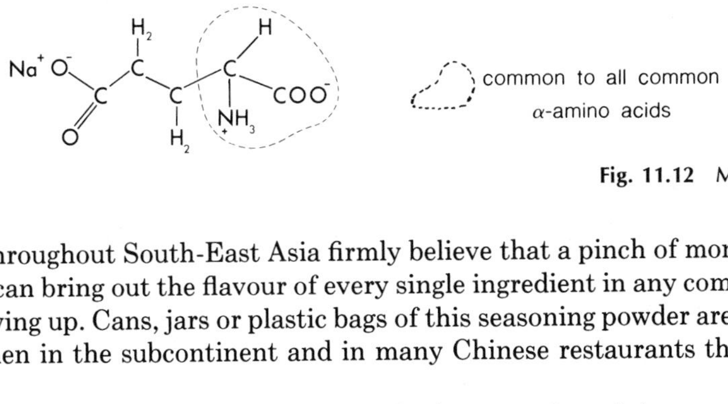

Fig. 11.12 Monosodium glutamate

Cooks throughout South-East Asia firmly believe that a pinch of monosodium glutamate can bring out the flavour of every single ingredient in any complex dish they are frying up. Cans, jars or plastic bags of this seasoning powder are found in every kitchen in the subcontinent and in many Chinese restaurants throughout the world.

The most efficient form of production is the fermentation of glucose or sucrose produced by the acid hydrolysis of any cheaply available carbohydrate, such as (in Thailand) molasses and tapioca, in a suitable nutrient medium containing nitrogen (e.g. urea). The organism involved is *Corynebacterium glutanicum*. The product comes as short, needle-like crystals that look dull. They smell like sauerkraut and taste sweet and a bit salty. The reason for this detailed description is that adulteration by shopkeepers in the region is widespread. In spite of its relative cheapness (about $1 per kilo in Thailand in 1973), similar looking crystals of borax or sodium metaphosphate (and sugar and salt) are often added, with disastrous results for the consumer in the case of borax. Borax is such a common illegal additive in the region (in meatballs, fish etc.) that methods for its detection suitable for local use must be publicised. One method involves the preparation of turmeric paper from turmeric tubers, which are readily available in the market-place. Ground turmeric is extracted into ethanol (or methylated spirits), and filter paper (or newspaper edges) are soaked in the solution and dried. The suspected crystals can be scattered on the paper and then drops of 1 M hydrochloric acid (HCl) added. Alternatively, the paper can be added to the borax acidified with HCl. In either case a pink colouration develops. When then placed in ammonia vapour turmeric paper turns from yellow to pink and the pink borax stain turns blue.

Sugar

Sugar was known in India 5000 years ago and one of the earliest forms of Indian sweet, *khandi*, has its name preserved in a modern American equivalent. Sweets join jewellery and perfume as one of the earliest forms of present. However sugar was not known to the Romans, who had to make do with honey for confectionery. They did however invent the first artificial sweetener, *sapa* (hence sapor). Sir Edward Barry, an historian of the ancient art of making wine, wrote a book in 1775 called *Observations historical, critical and medicinal on the wines of the ancients*. He commented that the Romans boiled down grape juice in lead pans to give a concentrated sweet syrup. (When other fruits were used it was called *defrutum*.) The syrup contained lead salts such as lead acetate (called sugar of lead), which had about the same sweetness as sucrose. This both sweetened and preserved the wine (lead kills microbes). As lead affects the brain (see also Chapter 12), it is alleged that this sweetener (along with many other lead products used), finally 'lead' to the decline of the Roman Empire.

Sugar cane (*Saccharum officinarum*) was cultivated in southern Europe only around AD 800, and sugar beet (*Beta vulgaris*) much later, in around AD 1800. Sugar infiltrated European consciousness because it was sweet and expensive and thus prompted early entrepreneurs to establish slave-based empires in the tropics.

Sugar is used as a preservative for meat and fruit. It is used in soft drinks, because it makes water feel more 'refreshing', and in tomato sauce and peanut paste because of its 'go-away' properties—the ability to degrease the palate after eating fat. The average Western person's daily consumption of 170 g provides about 1.8 MJ of energy. The percentage of sugar deliberately added to food in the home has declined but this has been more than compensated by the increase in the amount of sugar now added directly to manufactured foods (see Plate XI); hence the interest in non-joule sweeteners.[7]

Nevertheless, we must never forget that sugar is a very important food, and it is only our overindulgence that makes artificial sweeteners attractive. Raw sugar is a cheap source of minerals and some protein as well as carbohydrate, and keeps indefinitely and without refrigeration. Australia exports almost 80% of its sugar as raw sugar. However sugar is also exported in other forms[8], as shown in Table 11.3.

Over 12 million children a year die from the effects of diarrhoea, mainly in the Third World. A simple treatment consists in feeding them with a water containing an 8:1 ratio of sugar and salt.

The joint FAO/WHO report *Carbohydrates in human nutrition* found that 'there is no conclusive evidence that the consumption of simple sugars is of

TABLE 11.3
Export of sugar in other forms, by tonnage,
1985–86

Category	Tonnage
Preserved foods	11 155
Confectionery	5 373
Dairy foods	1 899
Non-alcoholic beverages	1 583
Alcoholic beverages	1 057
Bakery products	937
Other groceries	1 086
Total	23 090

aetiological significance in diabetes mellitus'. This common form of diabetes occurs in obese adults and control of total kilojoule intake is required.

Sugar *is* implicated in dental caries because it acts as food for the bacterium *Spectrococcus mutans*, which converts it into acid. Sticky forms of sugar (and starch etc.) are more effective because they stay on the teeth.

Sugar is a most natural food, consumed with relish in the form of honey by our hunter-gatherer forebears and produced in all green plants. The nutritional problem with sugar is that it replaces other sources of energy which provide additional nutrients not present in sugar. It is for this reason that controlled intake is recommended.

It is interesting to compare the relative sweetness of sugars. The 'inversion' (splitting) of sucrose to form glucose and fructose increases the sweetness by 24% (see Fig. 11.13).

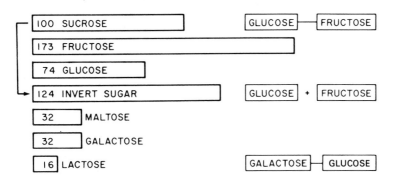

Fig. 11.13   Relative sweetness of sugars. Sucrose is a chemical combination of glucose and fructose, and lactose is a chemical combination of galactose and maltose. When sucrose is 'inverted' it breaks up into glucose plus fructose, which is thus sweeter than the sucrose it came from.

Artificial sweeteners

As a chemical educator I never cease to emphasise the need for care and cleanliness in a chemistry laboratory. To my shame the story of the discovery of modern artificial sweeteners is one story after another of unhygienic serendipity (see Table 11.4).

TABLE 11.4
The discovery of artificial sweeteners

Sweetener	Date	Discoverer and laboratory	Comments
Saccharin	1878	I. Remsen and C. Fahlberg (Johns Hopkins University, US)	Tasting of new compounds was accepted practice at the time.
Dulcin	1884	J. Berlinerblau (Bern University, Switzerland)	Tasting of new compounds was accepted practice at the time. Introduced 1890s; withdrawn in US, 1951.
Cyclamate	1937	M. Sveda (E. I. DuPont de Nemours, US)	The chemical contaminated a cigaratte he was smoking in the lab. Banned in US and UK, 1970; approved in Europe (production and sales, 2000 tonnes p.a.) Canada and Australia.
Aspartame	1965	J. Schlatter (G. D. Searle, US)	Accidentally discovered sweet taste while working on anti-ulcer drugs. Approved in US, 1974; banned, 1975; ban lifted, 1981. Monsanto took over Searle on basis of this one patent; annual sales now over US$1 billion. Patent expired 1987.
Acesufam K	1967	K. Claus (Hoechst, Germany)	While in lab., licked fingers to pick up piece of paper.
Sucralose	1970s	S. Phadnis (Queen Elizabeth College, UK)	During the planning of a project, a foreign research student misinterpreted a request for testing a compound (from a sugar company) as a request to taste, which he did. The lab. immediately applied for the patent.

The sweet-tasting property of an ortho-toluenesulfonamide was discovered during a routine research project at a time when the taste of a new chemical was considered one its noteworthy properties.[9] Fahlberg had been working in the US. He returned to Europe and carried out tests for potential toxicity, then applied for patents in 1894 and began to manufacture. Saccharin was widely adopted, although concerns about its safety did surface at the beginning of its production and again in the US at the time of the 1906 Pure Food and Drug Act. Saccharin's main attraction was that it is excreted unchanged and hence not metabolised. Recent research shows it may be a very weak carcinogen in rats (which concentrate urine very highly in their bladders), but it is unlikely to be more dangerous than cyclamate (sodium cyclohexylsulfamate), which dominated the US market in the mid–1960s until it was banned. Unlike saccharin, cyclamate survives cooking.

Joseph Berlinerblau was involved in synthetic chemistry at the University of Bern when he made and tasted a compound, now called (4-ethoxyphenyl)urea (CAS Registry number 150-69-6). He patented the compound seven years later. It was called Dulcin[TM] (sweet) in 1893 by one of the workers testing it for toxicity. The animal tests gave mixed results, but it was decided that the benefits outweighed the risks. Dulcin had one major advantage over saccharin in that it did not leave a bitter aftertaste. However its marketing was nowhere near as effective.

A US FDA study in 1951 declared it to be unsafe, and it was removed from the US market.

Aspartame (NutraSweet™-US) is something quite different chemically. It is just two amino-acids joined together to give a dipeptide, aspartyl phenylalanine (methyl ester) (Fig. 11.14). Phenylalanine is an essential amino-acid, which however must be avoided by the one person in 15 000 who has the genetic condition called phenylketonuria (PKU). Aspartame decomposes at the rate of about 10 per cent per month at ambient temperature, so it can be used only in foods with a fast turnover, such as soft drinks and fruit yoghurts. At higher temperatures the rate increases so that it cannot be used in cooked foods. There are some anecdotal reports that aspartame has an effect on mental behaviour,[10] allegedly through the formation of methanol, although the quantities seem insufficient for this to be the cause (it contains less methanol than pineapple juice, for instance).

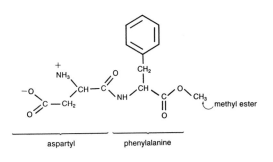

Fig. 11.14   Aspartame (*N*-L-α-aspartyl-L-phenylalanine 1-methyl ester)— sweeteners from a pair of amino acids (Courtesy Seale and *New Scientist*)

Aspartame (NutraSweet™ and Equal™) is permitted in a number of foods in Australia and must be included in the ingredient list. Up to 10 g/kg may be present in low-joule chewing gum, other low-joule foods may contain up to 5 g/kg, and brewed soft drinks may have up to 1 g/kg. Products containing aspartame must carry a warning of risk to PKU sufferers. One Equal™ tablet (equivalent in sweetness to a level teaspoon of sugar) contains 18 mg of aspartame; one Equal™ sachet (equivalent to two level teaspoons of sugar) contains 38 mg. A cup of tea (250 mL) with a sachet contains more than twice the dose recommended for pregnant women.

A derivative of aspartame (methyl fenchyl L-aspartylaminomalonate) was found to have a relative sweetness of 25 000 (25 000 times sweeter than sugar). It is still very much in the experimental stage.

A larger peptide molecule (molecular mass, 22 000) provides another sweetener, thaumatin (Talin™), which comes from a west African plant ketemfe (*Thaumatococcus danielli*). It does not provide instant sweetness but does provide a long, lingering effect 3000 times sweeter than sugar. It probably has greater potential as a flavour enhancer. Pet food manufacturers have found it a very effective additive and farmers have found pigs eat more and gain up to 10 per cent extra weight with thaumatin added to feed. Thaumatin is an approved additive in Australia.

Acesulfam K has a structure similar to saccharin and cyclamate, and a sweetness comparable to aspartame. (See Plate XIV) Unlike aspartame, it is stable in water and heat. It is excreted and not metabolised. However, it has not had the same commercial success.

The relative sweetness of some artificial sweeteners is given in Table 11.5.

TABLE 11.5
Relative sweetness of some artificial sweeteners

Sweetener	Relative sweetness[a] (mass for mass)
Sucrose (cane sugar)[b]	1.0
Lead acetate (sugar of lead)	1.0
Glycerol (glycerine)[b]	0.6
Ethylene glycol (anti-freeze)	1.3
D-tryptophan	35
Cyclamate[b]	30–80
Acesulfam-K	150
Aspartame[b]	100–200
Dulcin (sucrol)	250
Saccharin[b]	500–700
Sucralose	650
Hernandulcin	1000
Thaumatin[b]	∼3000

[a] Sweetness is measured by comparing the taste in water to a 4% solution of cane sugar. The results vary with the people in the taste panel. Some find saccharin 700 times as sweet as sugar.
[b] Approved in Australia.

As chemists have been able to pin down more closely the section of molecules that are responsible for taste, molecular design has become more logical (Fig. 11.15).

Another approach has been to modify the sugar molecule, although this can have the opposite effect. One sugar derivative, sucrose octoacetate, is so bitter it is used as a harmless denaturant for alcohol to make methylated spirits undrinkable. (In this, it replaces methanol, which causes blindness in determined drinkers.) Adding chlorine atoms to sucrose has a variety of effects,[11] but a tetrachloro substitution in the correct place gives an increase in sweetness of 2200. With a substitution in the wrong place the chemical is as bitter as quinine. Sucralose™ is 4,1′,6′-trichloro-4,1′,6′-trideoxygalactosucrose, a trichloro-derivative 650 times as sweet as sucrose. It is currently being tested in the UK.

Sequesterants

Metals such as copper, iron and nickel get into food from processing machinery or because of chemical reactions with the container. A sequesterant such as citric

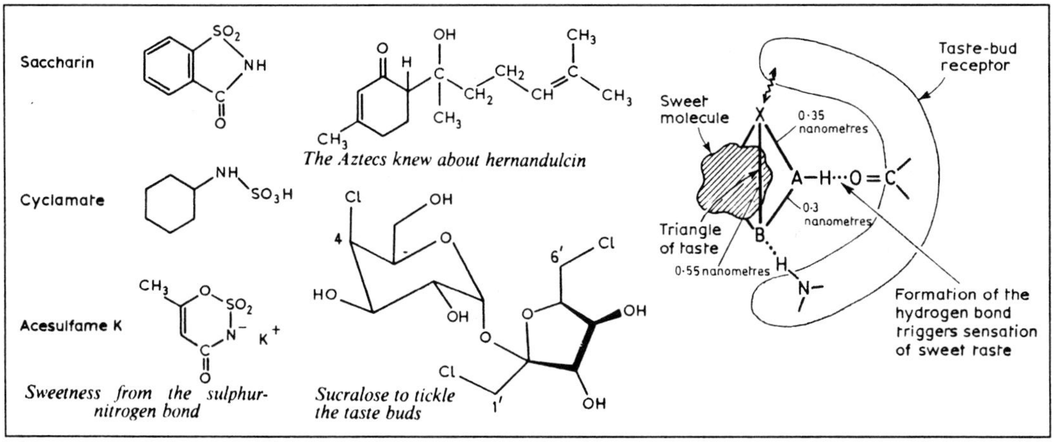

Fig. 11.15 Triangle of sweetness (From L. Hough and J. Emsley, 'The shape of sweeteners to come', *New Scientist*, 19 June 1986, p. 50.)

acid (chief acid in citrus fruit, 6–7% in lemon juice, Fig. 11.16) thus acts as a synergist (a helping agent) for antioxidants. Sequesterants are used in shortenings, mayonnaise, lard, soup, margarine, cheese etc.

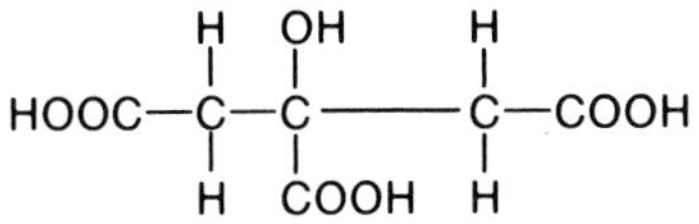

Fig. 11.16 Citric acid

Stabilisers and thickeners

Stabilisers and thickeners are added to improve the texture and blends of foods. An example is *carrageenan* (a polymer from edible seaweed), which belongs to a group of chemicals called polysaccharides (carbohydrates of high molecular mass, including sugars, cellulose and starch). They are particularly effective in icings, frozen desserts, salad dressing, whipped cream, confectionery and cheeses.

Emulsifying (surface-active) agents

The food 'soaps' are used to stabilise emulsions of oil and water components in foods.

Polyhydric alcohols

Polyhydric alcohols are additives used as humectants (to prevent foods from drying out). Tobacco is also kept moist by them. An added feature of these compounds is their sweetness. Some particularly effective alcohols added to sweeten sugarless chewing gum are mannitol, sorbitol, glycerol and xylitol, the latter two being recent additions. These polyhydric alcohols have the same energy (calorific) value as cane sugar—16.5 kJ/g.

Water-retention agents

Polyphosphates are increasingly used in the processing of poultry and mammalian meats to bind water and to minimise 'drip', and as an aid to further processing. They are also used to a large extent in processed fish. Phosphates are also used in soft drinks and in the production of modified starches. Because of these many uses, phosphate, although an essential mineral, is likely to be consumed in larger amounts than would be the case if the diet consisted of unprocessed foods. There are indications that an excessive daily intake can lead to the premature cessation of bone growth in children, with a consequent significant reduction in adult height.

The NH & MRC Food Additive standards allow the following usages:

> Sodium phosphate (soluble) including ortho-, pyro-, and metaphosphates and low molecular weight polyphosphates (analysis as P).
> Canned and processed meat products—1.3 g/kg; cheese and cheese products—3.0% total emulsifying agents, including phosphates; fish and fish products— 2.2 g/kg; frozen fish and poultry—1.3 g/kg.

As a member of Group II modifying agents (mineral salts), sodium phosphate is allowed in all foods in which these agents are allowed. There is a specified list, the last member of which is 'Foods not elsewhere standardised'. Permitted concentrations are not set for most of the specified list and, in particular, not for its open-ended final member.

It is an interesting exercise to calculate the potential daily amount of phosphate in the diet of a typical Australian child. It is also fairly straightforward to analyse for phosphates in food, although some special reagents are required.

Food irradiation[12]

A method for killing parasites in meat with x-rays appeared in a patent more than 60 years ago. In the 1950s Eisenhower's 'Atoms for Peace' initiative experimented with irradiated food for combat and space rations but could not overcome the problems of changing taste. Interest renewed in the 1980s and extension of shelf life by irradiation is again on the agenda. About 20 countries have approved irradiation for about 30 foods.

The core of an irradiation unit is either a radioactive source, usually cobalt-60 (gamma rays of energy 1.17 and 1.33 MeV), or caesium-137 (gamma rays of energy 0.66 MeV), or an electron beam.

Radiation breaks up molecules, and big molecules, such as the DNA heredity material are more susceptible. Micro-organisms are destroyed in this way. Plant development and ripening can also be inhibited. The physical process involves the gamma ray hitting a molecule which ejects an electron and ionises. Each ejected electron may ionise further molecules. As electrons are slowed down, they emit their excess kinetic energy as weak x-rays called *Bremsstrahlung*, which can cause nuclear transformations. The maximum energy levels of irradiation are set to prevent food becoming radioactive. The susceptible atoms are the common carbon-12 and oxygen-16.

In the UK limits have been set to ensure that these atoms are not made radioactive and these levels are as follows

Max. overall dose	10 kilograys
Max. energy (gamma or x-rays)	5 MeV
Max. energy (electrons)	10 MeV

The *dose* is measured in a unit (now) called a gray and this gives the amount of energy that the radiation deposits in the irradiated material. The other limits set how energetic (million electronvolts) the different types of radiation are allowed to be (see Chapter 12 for an explanation of radioactivity.) The threshold for making carbon radioactive is 18.6 MeV, and for oxygen it is 15 MeV.

The distinction between dose and energy is exactly the same as for sunlight. The bulk of the sun's energy is in visible light, but it is not energetic enough to cause sunburn or tanning. The ultraviolet light is energetic enough to cause these changes, but the amount present in sunlight is very small. (See 'Chemistry of cosmetics—Sunlight on skin' in Chapter 4 for more on sunlight's effects.) However any increase in the level of ultraviolet light, for instance as a result of the destruction of the ozone layer, is critical (see Chapter 10).

While a dose of 5 grays can kill a person, a dose of 1 kilogray is needed to reduce the population of most microbes to 10 per cent of their initial value, the D_{10} value. *Clostridium botulinum* (an organism responsible for botulism) is not very susceptible and it can withstand up to 50 kGy. Irradiation can thus kill its competitors and leave it better off to multiply.

Irradiation does affect food components, splitting carbohydrates into sugars and causing some fruits and vegetables to become soft, while others, like strawberries, become less acid. Irradiation splits fats into a variety of smaller molecules, and unless meats are irradiated in vacuum packs, oxidation is accelerated. A taste variously described as 'wet dog' or 'goaty' occurs in meat that has not been irradiated at sub-zero temperatures. In the presence of oxygen, irradiation turns meat bright red and then brown as myoglobin is oxidised. Lobsters and shrimps turn black from enzyme damage. Eggs and cheeses develop bad flavours. One successful use of irradiation is in preventing the sprouting of the surface buds of potatoes. However with onions the germ cells are in the interior, and if irradiated sufficiently to prevent sprouting, the unsuspecting consumer finds the interior has turned brown. Dry foods require smaller doses, but a change in taste in (boiled) rice irradiated even at a level of 1 kGy, can be detected by Australian consumers, while the more discerning Japanese can detect 0.5 kGy. South Africans have had success with irradiation of mangoes to kill the seed weevil, but similar experiments in Australia have shown the local varieties are very susceptible to damage.

There are some rather subtle aspects of the effect of irradiation. The greening of potatoes on exposure to light is a significant problem; losses of up to half a million dollars per annum are reported in New South Wales alone. The greening due to chlorophyll formation is in itself quite harmless, but the greening is accompanied by the simultaneous synthesis of the poisonous substance, solanine, which may be dangerous for pregnant women (see box 'Avoid bruised or green potatoes'). Solanine is not inhibited by irradiation. Is irradiation of potatoes likely to remove the visible green warning signal of its presence?

Mobility of sources is important in a large country with widely separated agricultural areas and short harvesting seasons for many products. While cobalt sources are static, electron beam sources can be mobile. One of the largest electron beam machines is found in Odessa in the Soviet Union, where it is used to kill insects in (imported) grain. Another factor is that irradiation kills only while the source is on and cannot prevent reinfestation. In contrast, pesticides have a withholding period during which their level drops and they are still effective.

While prolonging shelf life is the obvious commercial advantage, the main use of food irradiation is likely to be for quarantine purposes. As little as 0.1 kGy will sterilise insects, while 1 kGy is needed for a kill. However, this poses a problem. On seeing an insect infestation, how does a customs officer decide whether the insect is fertile or sterile? The higher kill doses will inevitably be used.

At the moment, food irradiation is very much a technique looking for an application. While it does have limited possibilities, consumers must be wary that it is not used to cover slack food storage and processing habits.

Packaging and other accidental additives

Finally, there is the matter of accidental additives that enter food without any wish for their presence, for example residues of pesticides. An NH & MRC standard sets a tolerance for various pesticides and substances derived from pesticides. The amount specified is different for different chemicals and for the same chemical may be different in different foods. For example, dichlorvos (the active constituent in Shelltox insect strips) is allowed up to 0.1 mg/kg in fruit, 0.5 mg/kg in vegetables and cereal products, 0.5 mg/kg in milled cereal products and 2 mg/kg in raw cereals. The reason for this ought to be related to the average daily intake of these foods but could possibly be related to the *rate of use* of the pesticide or the difficulty of its removal. Incidentally, there is no level set for citrus fruit. Analyses of foodstuffs (particularly meat) have regularly shown levels in excess of the standard. Thus, its enforcement will be politically embarrassing.

Another source of accidental additives comes from the material used for packaging, and this has been causing increasing concern. Until recently there were no regulations that specifically controlled the composition of packaging materials, nor were there regulations that allowed for the migration of substances from packaging materials to foods. Since the second edition of *Chemistry in the marketplace*, standards and regulations have been introduced, partly as a result of revelations made therein.

Despite past lack of regulatory permission, a variety of materials are known to migrate from packages to foods. Paper and cardboard packages can be the source of inorganic and organic migrants from inks, pigments, dyes and preservatives if the paper and cardboard have been produced using wastepaper recovery processes. Perhaps the most famous of the migrants from such sources were the celebrated PCBs (polychlorinated biphenyls), which have toxicities similar to chlorinated pesticides (DDT, hexachlor etc.). The use of virgin woodpulps presumably would preclude any migration risks in paper and cardboard intended for food-contact use, but the economics of such processes may not always be acceptable to industry.

Glass containers, despite statements to the contrary, can frequently contribute to food quantities of contaminants substantially in excess of those derived from plastic containers, but it must be admitted that, as yet, there has been no evidence of toxicity from this source. Nevertheless, resaleable glass containers that are subject to sterilising chemicals and detergent processes could be a significant source of these materials in foods unless strict controls are observed.

Metal containers, apart from metallic contamination derived from corrosion, may also contribute cadmium from lacquers and can linings, as well as the surface lubricants that are used in the rolling of metallic foil and sheet stock. The US FDA has set limits not only on the amount of such materials on metal sheets intended for food containers but also on the nature of the lubricants. The use of electroplating to give a layer of tin on a steel can has produced a cheaper but less satisfactory product than the older dip process, as there are greater problems of ensuring a continuous cover.

It is generally conceded that the major area of regulatory interest centres upon plastics intended for food contact use. This is because plastics are highly complex mixtures and the problems that they pose appear to be of much greater significance than those of any other packaging material. The number of different packaging materials is very large, their individual characteristics differ widely, and a very large variety of substances may be used in their manufacture. New ingredients, new formulations and new packaging uses are introduced frequently. It has been estimated that from 5000 to 15 000 different chemical compounds may be available to the plastics manufacturer who supplies materials for both food and non-food packaging applications. Migrants may take the form of unpolymerised monomers, catalysts, surface active agents and release agents; additives, such as antioxidants and plasticisers; lubricants, such as fatty acids and fatty alcohols; and antioxidants that do not necessarily figure in the permitted lists of additives for food. A survey in 1974 showed that margarine generally contains about 0.2 mg/kg of cadmium, while a margarine packed in a container with about 1.4% cadmium content gave a value of about 0.7 mg/kg cadmium. Despite the impressive list of migrants from packages that one could produce, it has been argued that the hazard to health they constitute is small. Where migration and adulteration do occur and the migrant is toxic, the extent of migration is generally small in relation to the total diet and (it is hoped) is below the maximum tolerable daily intake. However the discovery that vinyl chloride monomer (for which there is no recommended MTDI) on inhalation is a potent liver carcinogen has raised doubts about such a complacent attitude. (For further details see 'Chemistry of plastics—Vinyl polymers', Chapter 6.) A limit of 0.05 mg/kg has been set for any food.

Contaminants of microbiological origin are even more critical. *Aflatoxin* is one of a number of naturally occurring toxic products (poisons) found in various moulds. There is usually no visual evidence of its presence. It has been found in greatest amounts in peanuts and other nuts, in corn and products manufactured from these commodities. It has been shown to cause liver cancer in some test animals and has been a suspected, but unproven, cause of liver cancer in certain African and Asian countries, where high amounts of these toxins have been detected in foods normally consumed. Not all moulds produce dangerous toxins, and some are useful in food processing, such as the moulds used to produce Roquefort cheese. Other moulds are used to produce antibiotics. Precautions to help control aflatoxins involve the prevention of mould formation by proper drying and storage of crops, removing damaged material before storage or processing, and providing adequate moisture and humidity control of stored foodstuffs. A limit of 5 μg/kg has been set for all foods except peanut paste (peanut butter), where, for 'practical reasons', the limit is 15 μg/kg (see later).

Labelling

Everyone is an expert on food labelling (ingredient, nutrition etc.), and there are as many systems as there are experts. Anyone who has bought Australian food products overseas will know that a lot of the technical problems held up against labelling food adequately in Australia have been overcome on export labels.

On the question of additives, the EEC developed a simple rational system in the early 1970s. Each additive is given a code number: E 100 to E 199 are colours; E 200 to E 299 are preservatives; E 300 to E 399 are antioxidants; E400 to E 499 are texture modification agents. Flavours have not been dealt with. The colours are further subdivided: E 100 to E 109, yellow; E110 to E119, orange; E 120 to E 129, red; E 130 to E 139, blue; E 140 to E 149, green; E 150 to E 159, brown and black; E 160 to E 170, unclassified; E 170 to E 189, unique surface colourants. Some specific examples are: ammoniacal caramel, E 150; tartrazine, E 102; benzoic acid (sodium benzoate, potassium benzoate, calcium benzoate), E 210 (to E 213); propylene glycol alginate, E 405. This system makes it possible to search for specific allergies caused by food additives and to select a diet that avoids them. Technological improvements resulting in the use of fewer additives can then be followed by the consumer who studies the labels. I introduced the EEC system to the Food Standards Committee of the NH & MRC in 1974. In 1983 the NH & MRC finally recommended this system for Australia (the same numbers, without the 'E'), after years of consumer agitation. The list is given in Appendix XII.

In the summer of 1976 the French Federal Consumer Organisation declared a boycott on food additives they regarded as unsafe and unnecessary, particularly food colours. I was in Paris at the time and spoke to François Lamy, who was responsible for the assault. Contrary to popular belief, the French take food much more seriously than sex, and the boycott was an outstanding success. Unfortunately, the accuracy of the eivdence on which the boycott was based was very poor in places. However, the consumers had been subject to so much delay and procrastination by the food industry that the industry's scientific protestations made no impact at all (see Fig. 11.17).

To give readers some idea of how food standards are prepared in Australia it should be mentioned that the original draft is generally prepared by the industry wishing to introduce the food or additive. This is not so much a Machiavellian plot as a lack of any alternative source of expertise or concern. Many of the changes in food standards are designed to increase the benefit to the manufacturer, with some spin-off (perhaps) to the consumer.

Fig. 11.17 Cover of *Que Choisir?*, journal of the French Federal Consumer Organisation, recommending a boycott of food colourings

ALCOHOLIC PRODUCTS

Beer—more than barley and hops

The most powerful 'food' industry in Australia must be the brewing industry. Its product is both revered and suspected. This attitude is best illustrated in the article reproduced here. We are all experts on beer, and we resent technological interference with our 'natural' products to an irrational extent. Or is it irrational?

Richard Boston (the Guardian, 1 February 1975) examines the ingredients of British beer.

> In 1516 Count William IV of Bavaria issued a Reinheitsgebot, or Purity Law, which allowed the use in brewing of only barley, hops, and water. This law remains in force to this day, not only in Bavaria but throughout Germany.
>
> There are some who would have you believe that the situation here is not totally dissimilar, and will quote to you the old rhyme:
>
> > He that buys land buys many stones,
> > He that buys flesh buys many bones,

> He that buys eggs buys many shells,
> But he that buys good beer buys nothing else.

This doggerel, it must be obvious to anyone experienced in these things, was written by a brewer. Apart from the feebleness of the verse, it is completely misleading, this being an unmistakable hallmark of brewers' utterances. Unless we are to be told what the ingredients of good beer are considered to be, the last line is simply tautologous.

The brewers of today are more prosaic, but their statements have the same qualities. A few months ago the Brewers' Society announced that 'British beer today is brewed from the same basic ingredients as it has been for the past 500 years or more—barley and hops'. It depends, as the late Professor Joad might have said, on what you mean by British beer, and what you mean by basic.

Apart from the presumably accidental omission from the ingredients of yeast and water, anyone who believed that particular statement was doubtless surprised to learn from recent newspaper reports that a shortage of seaweed was threatening the foaming head of his pint.

What, then, are the ingredients of British beer? Since the wicked W.E. Gladstone's Inland Revenue Act of 1880 brewers have been allowed to use not only sugar but virtually whatever they like provided it is not actually injurious to health. That was no small provision, considering some of the things that used to go into the beer. That neglected Tudor poet John Skelton, the Rector of Diss, who kept falcons in the church, amongst other things, penned the greatest ever denunciation of a brewer. This was the ill-favoured Eleanour Rummyng.

> Droupy and drowsy,
> Scurvy and lowsy:
> Her face all bowsy.
> Comely crynklyd,
> Woundersly wrinklyd,
> Like a roast pygges eare
> Brystled with hare.

Apart from being physically repulsive, one of Eleanour's less endearing habits was that of allowing the hens to roost above her mash-vat, so that the droppings fell into the beer.

> And somtymes she blennes
> The dung of her hens
> And the ale together.

As far as I know no modern brewer follows this malpractice. However, in the eighteenth and early nineteenth centuries still worse things were used, such as a poisonous drug called *Cocculus indicus* and what H. Jackson described in 1758 as 'green vitriol called copperas or salt of iron'. When mixed with alum this was guaranteed to give a 'head like a collyflower' (to the beer, I assume, but possibly to the consumer).

Cobbett frequently denounced 'beer-doctors' and 'beer-druggists' and in *Cottage economy* (1821) he quotes from a book on brewing a recipe for porter [a heavy, dark brown beer], of which the ingredients include 'one quarter of high-coloured malt, eight pounds of hops, nine pounds of treacle, eight pounds of colour, eight pounds of sliced liquorice-root, two drams of salt of tartar, two ounces of Spanish-liquorice, and half an ounce of capsicum'.

Complaints of adulteration were by no means rare in that period as is shown by a ballad of about 1825.

The brewer's a chemist, and that is quite clear,
We soon find no hops have hopped into his beer:
'Stead of malt he from drugs brews his porter and swipes.
No wonder so oft we all get the gripes.

We find Dickens in 1856 also writing about the 'brewhous-chemist' and similar complaints continue well into this century. For example, Beachcomber in his 1930s' *Dictionary for today* was defining beer as 'a drink made of various chemicals in various proportions'. (Fig. 11.18)

Fig. 11.18   The pub with unreal beer

Of course that's all in the distant past. What of today? The answer is that the consumer has no way of knowing. If you buy a tin of soup the ingredients are listed on the label. If you buy a bottle of beer there's no way of knowing what's in it. This is not simply a matter of curiosity. Quite recently cobalt sulfate was used in brewing in the United States: it gave the beer a lovely head, but it gave the consumers lousy hearts, and after more than 40 people had died its use was discontinued.

Many of the independent brewers could echo the words of Mr John Young, chairman of Young's of Wandsworth, in his company report this year: 'Our definition of beer is that it is brewed from malted barley and hops, and we have no use for wheat flour, rice, or potato starch.'

These last ingredients, along with such things as maize grits, and flaked rice, are called adjuncts. They are harmful to flavour rather than health and, though some are used for specific properties, are mostly cheaper than malt. I have even heard of one large brewer who produces a so-called beer which contains no malt at all.

Many people feel that beer-drinkers should be allowed to know what they are consuming, and last May the Food Standards Committee invited interested parties to submit evidence on the way beer should be made, defined, and labelled. The Consumers' Association, which for at least 15 years has kept a vigilant eye on beer, proposed among other things that adjuncts should be limited to 30%—or, to put it the other way round, that the 'malt fraction' should be at least 70%.

Fig. 11.19   Beer labels do not list ingredients
(Courtesy of Dr Simon Brooke-Taylor, Head Brewer,
Eagle Hawk Hill Brewing Co., Sutton, NSW)

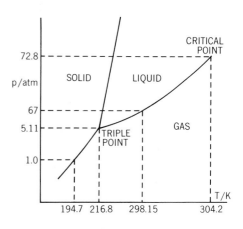

Fig. 11.20   Phase diagram for carbon dioxide

They point out that the average malt fraction has declined from 81.1% in 1953 to
76.6% in 1973. The Campaign for Real Ale has also submitted a report suggesting
that the malt fraction should be at least 70%. I don't know what the Brewers' Society
said in their evidence, because when I asked them they wouldn't tell me.

Whatever the Food Standards Committee finally recommends [this report has
now appeared and has accepted to a large extent the submissions by the Consumers'
Association], there are the EEC regulations to consider. A good row could well
develop here. While the Germans will resist any regulation that allows the use of any
adjuncts and chemicals, the British brewers (at least the big British brewers) will
doubtless resist anything that inhibits their ancient and venerable right to use them.
In this particular argument it will be hard for even someone as fond of the big British
brewers as I am to cast them in the role of the good guys. [The Germans lost this
argument.]

In the traditional method of making beer, hops (or hop extract) were added to
malted grain and then fermented. The yeast converts glucose to alcohol, but it also
destroys some of the bitter flavouring substances. By extracting the hops with
liquid carbon dioxide, the flavour can be obtained and added to the beer *after*
fermentation. Solvent extraction of foods to remove certain components has
always had the problem of residual solvent. By using a material that is normally a
harmless gas, this problem is solved. Carbon dioxide is normally a gas or a solid
(dry ice) but, under pressure in a cylinder, it can be a liquid and is then a
marvellous solvent. Above a critical temperature the distinction between a liquid

and a gas disappears. This is the critical point and beyond this the CO_2 is supercritical (see Fig. 11.20). For carbon dioxide this occurs at 31°C and 73 atmospheres pressure.

SUPERCRITICAL SOLVENTS

Supercritical CO_2 is a very selective solvent and is used, for example, in extracting caffeine from coffee beans while leaving their flavour and aroma virtually unchanged. This gives us decaffeinated coffee and surplus caffeine, which goes into analgesics and cola drinks.

Nicotine can be extracted from tobacco to give a cigarette that still causes cancer but has not the kick—not much demand for this! Nicotine in chewing gum has been introduced as an alternative product—to give the kick without the cancer. Nicotine affects you very quickly because it reaches the brain in just eight seconds from the lungs. (See 'Chemistry of drugs—Nicotine and tobacco', Chapter 9, for further details.)

Oilseeds (such as soya and sunflower) can be extracted with supercritical CO_2. This results in less gum being extracted than with conventional solvents. The oil separates as the CO_2 pressure drops and so a very neat cyclic process is possible. The natural insecticide pyrethrin (see Chapter 5) is now extracted from plants this way and essential oils for perfumes are being studied.

For water, the critical constants are 374°C and 218 atmospheres pressure. Supercritical water is used in modern steam-turbine electricity generation because it simplifies the design calculations for heat transfer (there is only one phase not two) and the higher temperature provides higher efficiency (see Fig. 10.2).

Wine

> And much as Wine has play'd the Infidel
> And robb'd me of my hale of honour—well
> I often wonder what the Vinters buy
> One half so precious as the Goods they sell

White or red wine? The process of manufacture differs from country to country. In Australia the shifting popularity from red to white can possibly be attributed to the cleaner chemistry in the production of white wines (see box 'Red or white').

The effects of wine

The chemistry of wine-induced migraine is generally associated with histamine. In Australia red wines are believed to be more potent in this regard. In France the opposite is true. It reminds me of the scientific connoisseur of spirits who had a hangover every time he drank scotch and soda. When he switched to gin and soda without improvement he realised immediately it was the soda. There are headaches and headaches, and careful neurological study quite clearly dis-

RED OR WHITE?

Red wine	*White wine*
Skin contact	No skin contact
Phenol/colour extraction from skins 700–2000 mg/kg.	Juice phenol only present at levels of 200–500 mg/kg.
Extraction of acid salt from skin higher at pH 3.5–4.0.	Limited extraction of acid salt at low pH 3.0–3.5.
Unavoidable presence of wild micro-organisms, yeast and bacteria, leading to off odour formation.	Juice can be clarified prior to fermentation; microbiological control high; no off odours.
Extraction of growth activators from skins and stimulation of fermentation both primary and malolactic.	Controlled fermentation because of absence of solids and growth stimulators.
Believed by winemakers to require high temperatures—fermentation conditions conducive to high temperature loss of fruit volatiles.	Low-temperature fermentation: retention of fruit volatiles.
Addition of SO_2 is limited because of its effect on colour and it is less effective because of high pH and binding by phenol. Malolactic fermentation all but inevitable; causes loss of acid and fruit.	Addition of SO_2 at low pH leads to free levels of SO_2 effective in controlling malolactic fermentation and oxidation.
Promotion of belief in requirement for oxidation and extraction of wood flavour to complete wine complexity.	Promotion of simple, identifiable fruit flavours.
Promotion of belief in requirement for bottle ageing to complete complexity.	Fresh fruit flavours promoted; product consumed early.
Taste incompatibility of high tannin and high extraction with presence of sugar.	

tinguishes migraine, cluster headaches and tension headaches, and the mechanism of the effect of wine on each is quite distinct. What does seem to be clear is that it is highly unlikely that any of the amines (e.g. histamine) in alcoholic beverages can be directly connected with the onset of a migraine attack. Clarets and port are often associated with the attack and yet are very low in amines. This brings us back to the scotch and soda story—it now appears that it is the *alcohol* that is important. Alcohol is known to be able to release histamine from the liver into the bloodstream in dogs and so possibly also in humans. Histamine release is also known to be the cause of headaches similar to migraine. Any carbohydrate meal rapidly increases serotonin levels in the brain, which probably accounts for the attraction of sugar. The Bedouin, forbidden alcohol, sucked sherberts. The level of serotonin is known to fall during a migraine attack. This is interesting in that it lends some credence to the belief that sugar acts as a prophylactic in forestalling a hangover.

Lowering the alcohol content of wine is of course prevented by law. The NH & MRC 1974 standards set the level for wine at between 8% and 17% v/v (above which the wine is fortified). 'Non-intoxicating' is set by Federal Customs at below 1.15% (2° proof spirit). Any 'wine' containing between 1.15% and 8% alcohol is presumably a grape-based beverage.

The more permanent effects of alcohol show that the blanket generalisations of social drinkers and alcoholics as distinct breeds are no longer acceptable. Our habit of applauding drinking yet stigmatising the alcoholic is inconsistent with the facts. Careful studies show that there are no distinct groups. Alcohol consumption is now seen from a pharmacological point of view—a given intake provides a given risk and arguments of the increase in consumption being the result only of social drinking have been statistically damned. The risk of alcohol damage to the fetus (particularly in the first trimester) has been rediscovered as it was known in ancient Carthage and in Biblical times.

When have you had enough? As one champagne specialist told me—the time to stop is when you can no longer read the label on a Wolf Blass 'Renepo Gel' backwards.

Champagne

In a true champagne the bubbles continue to rise steadily for a considerable period of time, compared to a sparkling wine where there is normally a short, sharp burst. In the mid-fifties two Russian chemists, Parfentov and Kovalenko, claimed that when carbon dioxide is produced during secondary fermentation in the wine, it is combined chemically with the alcohol as diethyl pyrocarbonate ($C_2H_5OCOOCOOC_2H_5$), from which it is slowly released. When carbon dioxide is simply added to wine under pressure it just dissolves and does not react. Interestingly, diethyl pyrocarbonate was once an allowed preservative in wine (see also 'Food additives'). It was useful precisely because it rapidly decomposed into carbon dioxide and alcohol (quantitative hydrolysis in distilled water in 150 minutes). However, as its main purpose was to offer an easy way of avoiding more careful processing, it was prohibited when it became suspected of forming carcinogenic urethanes with food amine components. Carbonic esters, the best known of which is diethylcarbonate, ($C_2H_5OCOOC_2H_5$), play a significant role in the formation of wine flavour. They release carbon dioxide on hydrolysis and champagnes contain the highest concentrations.

Labelling

Even the question of accurate labelling of wine raises highly technical discussion. It apparently wasn't until we had a visit from a French ampelographer that we began to know something about the origin and nature of our vines in Australia. The nature versus nurture argument in regard to human intelligence is one with which we are all familiar, but there is a parallel argument applied to 'wines from vines'.

It is often assumed that any variety will give a specific flavour in its wine, which can be recognised by those who have developed enough skill. Often it is the *process* that is characteristic, such as low-temperature fermentation giving very flowery scented aromas. A characteristic flavour is produced by initially fermenting the grapes *before* crushing, whereby the first alcohol is produced in the grape without added yeasts. Of course, all this has legal connotations in labelling. Wines that are subsequently blended are distinct from wines made from grapes that are initially blended. The sugar content at harvest of a Rutherglen muscat may be much more important than the grape variety. There is general agreement that the French agricultural trade-mark system of 'Appellation Controlé' has little relevance in Australia and even in France it might appear as a method of protecting 'inferior' technology. It depends whether you regard the 'malolactic fermentation' which often gives the 'unique character' appellation of wines as secondary fermentation or bacterial spoilage.

What is the place of new technology in European winemaking? The German consumers are is interested in what is done to their grapes and reserve judgment on what is good for themselves. French consumers are more concerned with the pedigree and adjust their taste accordingly. On the other hand if you want to improve your Latin, then Chaucer's Summoner[14] had the answer:

> And whan that he wel dronken hadde the wyn,
> Thanne wolde he speke no word but Latyn.

Try the experiment on wine analysis in Chapter 13 (Experiment 13.15).

Legal aspects of alcoholic products

Beer was included in the rations of workers building the pyramids of Egypt and since ancient times alcoholic products have been a lucrative source of revenue for the tax-gatherer. Taxes are based on alcohol content. *Proof* spirit was defined as 'that percent of pure alcohol which when mixed with water and poured onto gunpowder of specified composition just permits the gunpower to be ignited'. In the UK this is 57.155% by volume at 20°C. The Americans earlier had adopted 50% by volume. Since 1972 *proof* has not been used in Australia and the measure is percentage by volume, and taxes vary for different products.[15]

Brandy is distilled from grape wine and matured in wood for not less than two years. Apple brandy is distilled from grape cider. Whisky is distilled wholly from barley malt, whereas Australian blended whisky is distilled partly from barley malt and partly from other grains. Rum is distilled wholly from sugar, sugar syrup, molasses or the refuse of sugar cane. Gin is distilled from barley malt, grain, grape wine, apples or other approved fruit. No spirits can contain more than 83% by volume of alcohol.

Various products are protected by law. Under French law *Cognac* may be manufactured only from grapes grown in the Cognac region of France. *Scotch* is regarded as being a product of Scotland, but has become a generic name for whisky whether the Scots like it or not. Hogarth's London was full of gin mills, but by 1960 there was not one gin distillery left in the city of London proper as known to Hogarth, and 'London gin' imported into Australia is now manufactured in Singapore. Spain has lost the legal battle to protect the name *Sherry*.

In Australia today there are very few real liqueurs, in spite of legal definitions in some States. Most are compounded and are thus purely synthetic. The excise definition requires alcohol of not less than 22.484% by volume (except Advocaat— 17.136% v/v). Rye and corn whisky are defined—but the definition is analytically unenforceable! French brandies with names such as *Napoleon, VSOP* etc., terms which have no legal standing in Australia, are misleadingly regarded as being of necessity products of high quality. The NH & MRC introduced a standard for spirits and liqueurs in 1983 that sets out to protect names of origin such as Cognac, Armagnac, Scotch, Bourbon etc. Ratios of amyl, propyl, and butyl alcohols give indications of the origin of the spirit and can determine that some French brandies are not legal brandies in Australia.

Australian wines exported to the EEC must be tested for density, alcohol content, total dry weight, total acidity, volatile acidity, total SO_2, and citric acid. There are no similar requirements for wine imported into Australia.

The *excise* definition of beer had to be changed some years ago to meet the changing technology of the brewing industry. The definition is:

> any fermented liquor that—
> 1. is brewed from a mash, whether or not the mash contains malt, and
> 2. contains hops (including any substance prepared from hops) or other bitters, whether or not the liquor contains sugars or glucose or other substance, but does not include liquor that does not contain more than 1.15% alcohol v/v.

This rather extended definition now permits the use of potato mash, sugars, and of enzymes and lupinones, some of which were not previously permitted. The NH & MRC standard (1982) specifies *yeast* fermentation but also allows any source of carbohydrate.

Home brewing and home winemaking are legal in Australia provided you do not produce more than 1800 L per member of the family in any year and you do not sell the product. It seems an adequate allowance!

CAFFEINE

Coffee is consumed in one form or another by about one third of the world's population. The two most important commercial varieties are *Coffea robusta* and *Coffea arabia*. *C. robusta*, grown in West Africa and Indonesia, has a higher caffeine content. *C. arabia*, grown in East Africa, Central and South America, the Caribbean and New Guinea, yields a stronger flavour. Blending and roasting change the character of the crop considerably.

Instant coffee was popularised through the US armed forces in 1938, and it is now the most widely used form of coffee in Australia. Australian food regulations set a minimum level of 3% for caffeine in instant coffee, which requires a greater use of the milder *C. robusta* beans.

Decaffeinated coffee is even older, having been introduced by the German firm Kaffee HAG in the early 1900s, but has only now attracted a loyal and growing public. There was initially some concern with the safety of the solvent, trichlorethylene, used to extract the caffeine from coffee. However the safer trichlorethane, and liquid carbon dioxide are now used (see box 'Supercritical solvents', p. 439).

Worldwide consumption of caffeine has been estimated at 120 000 tonnes per annum, which works out at 70 mg/per person per day. Approximately 54% of this is from coffee, 43% from tea and 3% from other sources. In spite of the image of the USA as a heavy coffee-consuming country, Scandinavians consume almost three times as much caffeine (340 mg/day) from coffee as Americans, while the British match the Scandinavians in their caffeine intake from tea (320 mg/day). Only the USA and Canada show significant caffeine intake from soft drinks (35 and 16 mg/day respectively). Caffeine is allowed in non-cola drinks in these countries (but not in Australia). The total caffeine intakes of USA and Canada (211 and 238 mg/day respectively) is well below that of Sweden (425 mg/day) and the UK (444 mg/day). The estimated level of consumption of caffeine in Australia is 240 mg/day from all sources.

The approximate caffeine content of beverages is given in Table 11.6.

TABLE 11.6
Approximate caffeine content of beverages

Beverage	Container	Volume (mL)	Content (mg)
Percolated coffee	cup	150	100
Drip coffee	cup	150	80–200
Instant coffee	cup	150	60–70
Cocoa	cup	150	5
Chocolate bars	bar	100 g	20
Cola drink	can	375	35–55[a]
Tea (av.)	cup	150	50
Tea (weak)	cup	150	30
Tea (instant)	cup	150	30
Tea (Chinese)	cup	75	15
Decaffeinated tea	cup	150	2
Coffee beans (green)		0.8–1.8%	
Tea leaves (undried)		0.68–2.1%	

[a] The NH & MRC max. 145 mg L^{-1}; see also Experiment 13.4 on caffeine in cola, Chapter 13.

In addition to caffeine, tea contains about 1 mg per cup of the much more active alkaloid theophylline, while cocoa contains about 250 mg per cup of the much less active alkaloid theobromine (which does not contain bromine; see section on chocolate in Chapter 3). The relation between the chemical structures of xanthine, caffeine, theophylline and theobromine can be seen in Figure 11.21.

Modern medicine[16] makes use of caffeine as a respiratory stimulant, and the related theophylline for bronchial asthma (doses for adults are in the 250 mg range). Conversely, there have been many studies attempting to link coffee with disease. A long-term study involved 1910 men at a large engineering company in Chicago over 20 years[17] and showed that the risk of heart disease for both smokers and nonsmokers increased in proportion to the amount of coffee drunk (70% higher for six cups per day compared to one). The survey made no distinction between normal and decaffeinated coffee. There are several other biological active ingredients in coffee apart from caffeine. In fact, plants store these chemicals in their leaves as a natural insecticide.[18] Tea and coffee grounds are therefore also effective as insecticides or repellants.

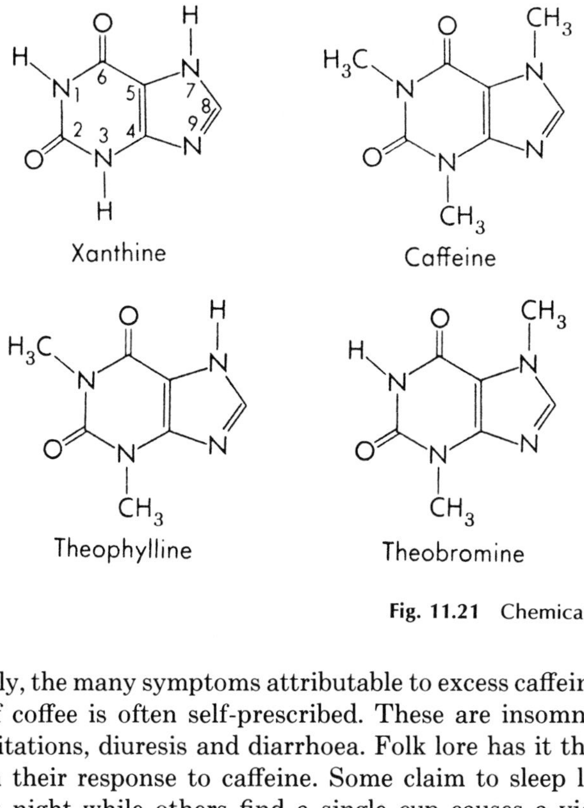

Xanthine

Caffeine

Theophylline

Theobromine

Fig. 11.21 Chemical structure of some xanthines

Unfortunately, the many symptoms attributable to excess caffeine are those for which a cup of coffee is often self-prescribed. These are insomnia, irritability, headache, palpitations, diuresis and diarrhoea. Folk lore has it that people vary considerably in their response to caffeine. Some claim to sleep like a log after several cups at night while others find a single cup causes a violent reaction. Closely controlled experiments do not bear this out however, and effects such as tolerance play a large part in perceived differences to caffeine intake.

The symptoms of caffeine withdrawal are usually mild and seldom last for more than seven days. They usually start within 18 hours and consist of diffuse throbbing headache made worse by exercise (like a 'tension' headache), and other varied effects.

Caffeine is metabolised in the liver and eliminated in a first order process (i.e. the level decreases exponentially like radioactive decay). The half-life (the time to reduce the level in the body to half its initial value) is three to four hours. Caffeine does not accumulate. The rate is increased by smoking and other causes that increase the activity of 'mixed function oxidases', and is greatly slowed in the later stages of pregnancy, an effect related to endocrine function. It is well established that women automatically reduce their caffeine intake during pregnancy. Consumption of caffeine, unlike many other drugs, is self-regulating. Caffeine and the benzodiazepam drugs (e.g. Valium) are directly antagonistic.

The question of caffeine intake by children is a vexed one. An 18 kg child consuming a 375 mL can of soft drink with a legal caffeine level of 145 mg/L consumes 55 mg of caffeine which represents 3 mg/kg body mass. (Actual levels found in soft drinks in the USA are given in Table 11.7.) As a level of 1.5 mg/kg is already effective in postponing sleep, one wonders about the 'technological need' for such a high legal level.

TABLE 11.7
Caffeine content of various soft drinks (US)

Soft drinks containing caffeine[a]	Caffeine content (mg/375 mL)
Jolt	76.1
Sugar-Free Mr PIBB	62.1
Mountain Dew	57.1
Mello Yello	55.8
TAB	49.5
Coca-Cola	48.2
Diet Coke	48.2
Shasta Cola	46.9
Shasta Cherry Cola	46.9
Shasta Diet Cola	46.0
Shasta Diet Cherry Cola	46.4
Mr PIBB	43.1
Dr Pepper	41.8
Sugar-Free Dr Pepper	41.8
Big Red	40.6
Sugar-Free Big Red	40.6
Pepsi-Cola	40.6
Aspen	38.0
Diet Pepsi	38.0
Pepsi Light	38.0
RC Cola	38.0
Diet Rite	38.0
Kick	33.0
Canada Dry Jamaica Cola	31.7
Canada Dry Diet Cola	1.3

[a] There are many types of soft drinks, manufactured by the leading bottlers, that contain no caffeine.

Australian limit for caffeine content, 55 mg/375 mL.

Source: Food Tech. Aust. **40**(3) 1988, 106

CAFFEINE AND OTHER MEDICAL CONTROVERSIES

When considering the more serious accusations made against coffee it is interesting to see the problem from the point of view of a publisher of scientific research. Dr Arnold Relman, editor of the prestigious *New England Journal of Medicine* was interviewed on his approach by David Dale in the *Sydney Morning Herald*, 30 August 1986 (p. 41)

'About five years ago we got a paper which said that there was a statistical association between drinking more than five cups of coffee a day and cancer of the pancreas. Now cancer of the pancreas is an increasingly common cancer, not as common as cancer of the bowel or breast, but it is getting there, and there's no treatment for it. It's usually rapidly fatal.

'So this correlation with coffee drinking was a very interesting finding. It didn't prove that drinking coffee *causes* cancer of the pancreas—there are many possible explanations—but it's a piece of evidence; it may provoke further research.

'We debated a long time whether to publish. We could just see the headlines. We sent the paper to our statistical referees; they said the methodology was OK; the conclusions were justified. It was a close call, but we went ahead.

'Oh boy, did that hit the fan. Headlines all over the world. Reuters called me up and said that the bottom had dropped out of the coffee market. It was terrible.

'There was a lot of criticism that we published it. And now it turns out that it's probably not going to hold up. People have done studies since—we have published them—which suggest that the initial association is probably not true [the effect may have been due to the solvent used for decaffeinated coffee, see above].

'The problem is in conveying to journalists and to the public that science works in small steps.

'*We publish on biodegradable material, not stone.* [Italics added.]

'We publish in the full knowledge that tomorrow we may get a paper in the mail which gives a completely opposite explanation of the facts . . . All we can promise is that in the judgment of competent, hardworking people with no axe to grind, the report you are reading is interesting and original and important.'

In the last ten years the *New England Journal of Medicine* has brought to our attention the following:

- The first description of Legionnaire's disease as a new type of pneumonia, and its probable cause.
- The first description of toxic shock syndrome and its probable cause.
- The first report that death from heart disease can be prevented by the use of drugs called 'beta blockers'
- The first demonstration that for certain types of heart disease, coronary bypass surgery does *not* prolong life.
- The first report on a new treatment for advanced cancer using a drug called Interleukin II.
- The first major evidence that complete removal of the breast is not always necessary to prevent spread of breast cancer, and that removal of the lump can in some cases be as effective as removing the whole breast.

The 'mastectomy versus lumpectomy' was controversial because the journal took a year to publish it—the editor felt the data were insufficient.

We felt the follow-up time on the cases was too short—only three and a half years. There are a few unfortunate women who seem to be cured initially and then five years later have trouble with a distant spread of the tumour. We told the authors we wanted at least another year of follow-up time.

Eighty-five per cent of papers rejected by the journal are published elsewhere, however in this case the authors agreed to wait. The prestige given by a journal's rigorous refereeing of submitted papers itself gives credence to what is published.

These then are the issues that must be balanced. Total avoidance of caffeine seems unwarranted and an intake of less than 300 mg/day seems acceptable by all the evidence. What is disturbing is the attitude that for children, coffee is undesirable, but cola drinks are acceptable. The lower mass of children means that their corresponding dose per kilogram is higher.

NUTRITION

Ever lost a friend after serving chilli con carne? Red kidney beans have been found to be a cause of several outbreaks of serious food poisoning. The beans contain a poison that is normally destroyed during cooking. If soaked raw beans are eaten in a salad or after slow cooking in a casserole, then occasionally vomiting and diarrhoea may occur, with subsequent dehydration. Interestingly, slow cooking increases the amount of poison, a five-fold increase has been recorded at 80°C, whereas boiling for ten minutes renders the beans harmless.

Red kidney beans contain traces of a protein poison belonging to a class called lectins which coagulate red blood cells and attach themselves to the carbohydrate in the cell walls of the intestine. The Bulgarian émigré George Markov was murdered in London by such a poison injected from an umbrella. That poison has a close relative (ricin) that comes from the castor oil bean. Other beans (runner, broad and French) also contain small quantities of lectins, which form during the ripening process.

Our knowledge of the chemistry of cooking has come from the experiments of our forebears with various foods and methods of preparation. Those whose experiments worked survived—and led to us. No wonder that food habits tend to be quite conservative. Yet food habits vary enormously from one culture to another.

Chemistry has shown that all staple diets supply basic nutrients of proteins, fats and carbohydrate. Foods do not have 'special' attributes; expressions such as 'fish is good for brain' are without foundation. The food is broken down and we build up the components 'in our image' again. The plants and animals we eat didn't design their body composition to suit us as food. Much of what we eat, we break down and remake, but some substances we ingest are active individual chemicals and not just 'generic' food (see box, 'The myth of pure natural foods').

Our level of understanding is not impressive. Until recently, we believed that protein deficiency (kwashiorkor) was the major nutritional disease of the Third

World, whereas the problem is simply starvation. The role of essential minor food components such as vitamins and minerals is under continuous scientific review. Fact and fiction are difficult to separate. An apple a day was supposed to keep the doctor away. But apples are a very poor source of vitamin C, their roughage is less effective than cereals and bran, and they make a poor toothbrush. So what?

Vitamins

Our bodies are metabolic factories and, as for all well-organised manufacturers, a decision must be made about which chemicals should be obtained from outside and which should be synthesised internally. The bulk materials (fats, carbo-hydrates, proteins and minerals) are imported, degraded into simpler units and then reconstituted. Minor essential chemicals that occur with the major ones are obtained at no extra cost, so there is no need to make them. We call such a group of unrelated substances vitamins. Their only common characteristic is that they are *not* produced in the body. Different 'bodies' have made different decisions. We share with the fruit bat, guinea pig, anthropoid ape and red-vented bulbul (a bird) the decision not to produce vitamin C (ascorbic acid) but to obtain it from fruit. This evolutionary decision can also be seen as a genetic deficiency disease which, unlike other such diseases, happens to be common to the whole human species (and four other species). Because of the wide availability of vitamin C in the tropical climates where these five species evolved, such a deficiency disease had no evolutionary disadvantage and, as extra production capacity of the body was freed to produce some other vital material, its decision gave a metabolic edge that ensured the survival of that decision.

How much vitamin C do we need? Can excess do us good? Can excess do us harm? Chemistry can suggest some answers. Our bodies can use only up to 60 mg per day: the rest is excreted. In the body the vitamin is converted to a number of products, including oxalate. Although oxalate is a component of kidney stones, the ingestion of excess vitamin C cannot lead to this complication because the amount the body can convert is sharply limited. The amount of vitamin C *intake* needed to reach the maximum rate of *use* of the vitamin in the body varies. Since the 1940s it has been known that smokers, for example, need a much higher level—perhaps over twice as much. Megadose levels on the other hand have been suggested as palliatives for anything from the common cold to cancer.

Unlike the other vitamins, which seem to have very specialised roles to play, the range of activities of vitamin C is very wide. Vitamin C also has an important role in food technology. There are sound chemical reasons for adding it to diverse products such as beer and bacon. On the other hand, it decolourises the legal synthetic coal-tar food colours (both azo- and triphenylmethane types), particularly in sunlight, and so is not added to food when these are used. There are also patents specifying the use of ascorbic acid for the removal of chlorine from drinking water.

Food processing

It is in food processing that much of food chemistry has been learnt. Prehistorical methods of food preservation had a common mechanism for making food inedible for micro-organisms. This involved either removing the water component by drying or rendering the water 'unavailable' to the microbe by increasing its salt content. This meant the use, not only of common salt (NaCl) but of other 'salts' (such as in curing with sodium nitrate and nitrite). Such concentrated brines are said to have a high osmotic pressure and microbes in them dehydrate as the water is extracted from their cells by osmosis. Bacteria, moulds and yeasts have different sensitivity and thus need different levels of dissolved material to prevent their growth. (See Fig. 3.9.)

The chemical nature of the dissolved material is not important, only the amount dissolved, so sugar, for example, can be used instead, which accounts for the longevity of honey and jam. Actually, chemists talk of a colligative property, which is a property of the number of molecules etc. present and not their nature. *Precisely* the same chemical argument is used in selecting antifreeze materials, except that there are other good reasons for not using salt or sugar. Likewise antifreeze is not a recommended food preservative.

The next level of sophistication involves the adjustment of the acidity of the food, because many dangerous microbes (e.g. *C. botulinum*) are sensitive to acid. Fermentation of food to produce acid is common to all cultures and cuisines and the list of foods is enormous. Some examples are dill pickles, sauerkraut, coffee beans, vanilla, kimchi, tarhana, sajur asin, kishk, salami, yoghurt, cheese, sour dough bread, soy sauce.

Bacteria, yeasts and moulds are used to produce mainly lactic acid or acetic acid (or both). Fermentation to alcohol produces a pleasant, but not stable, product unless the concentration of alcohol is high enough.

While curry and spices are alleged to have minor preserving action, their original use was to cover the bad taste of already decomposing food.

Modern food technology has expanded the types of preservation techniques and reduced the hazard of traditional methods. Fat soluble antioxidants mimic the action of vitamin E, while vitamin C is added to cured meats to reduce the production of cancer-causing nitrosamines. Canning followed by (commercial) sterilisation has given us products with long shelf-lives and has made modern urban living possible. On the other hand, we are able to feed ourselves bizarre diets equally well from the supermarket or the health-food store.

We are subjected to bewildering propaganda from the processed-food industry, nutritionists, the medical profession and consumer organisations. A paper entitled 'Processed food: a pain in the belly' (presented to the Chemistry Section of the 1982 ANZAAS Conference in Sydney by the Australian Consumers' Association) was launched, with criticism, as an educational publication by the Commonwealth Minister of Health at the time, Mr Carlton. The paper dealt with the contrasting and conflicting messages contained in the advertisements of the food industry and the pamphlets of the Department of Health nutrition experts.

Natural versus artificial

Perhaps the best place to start is with the concepts of natural and artificial. There is no doubt that natural is by no means best. Our introductory discussion on beans should have dispelled that misconception. Moulds produce both penicillin and aflatoxin (probably one of the most carcinogenic chemicals yet found either in nature or in the laboratory). Natural sugar contains, in concentrated form, a great variety of material of unknown composition or effect, and is certainly less desirable than purified sugar, which would seem a very natural food. Taken unknowingly (in processed food) and in excess it causes problems. Is vanillin produced from softwood less natural than vanillin from a vanilla bean? Is a nature-identical synthetic vanillin produced in the laboratory from a coal-tar precursor unnatural? Ascorbic acid (vitamin C) is almost entirely produced synthetically from glucose, although a crucial step in the process requires a microbiological reaction performed by courtesy of *Acetobacter suboxydans*.

THE MYTH OF PURE NATURAL FOODS

In the introduction to his talk 'The myth of pure natural foods' which was given at a public seminar on Nutrition, Health and the Consumer, organised by the Queensland Consumer Affairs Council *et al.* in July 1977, Bill McCray, the director of the Biochemistry Animal Research Institute of the Queensland Department of Primary Industry, said:

> To the nutritionist, and I quote, food is 'nutritive material taken into an organism for growth, work, repair, or the maintenance of vital processes'.—An objective view?—Agree?—It is not. It is the highly subjective view of the nutritionist. The truth is that food, for all holozoic creatures, like ourselves, consists of almost any readily available, independently living fellow-creature and in our case often the carefully preserved remains of such fellow creatures. Further, these unfortunate victims of the gratification of their fellow creatures are complex mixtures of chemicals, each victim containing a host of chemicals that are inimical to the 'growth, work, repair' etc. quoted in the so-called objective definition.
>
> George Bernard Shaw, the Irish wit and playwright of my youth said it for us. He said he refused to make his stomach a graveyard for dead animals. He was a confirmed vegetarian. Hence his stated wish was that his funeral cortege include little lambs, calves and chickens to represent those dead animals for whom he had not made his stomach a graveyard. Though he did not admit it, he was forced to make his stomach a compost heap for dead plants.
>
> It is easy to accept the predator–prey concept for the tiger or the eagle but even the gentle lamb devours the living grass. For ourselves, we can readily accept this thesis for meat and fish, but, consider the staff of life—bread. Here we take the grain into which a plant, our fellow creature, has poured all its reproductive energy; its hopes and aspirations for the future generation lie in the germ, the energy for whose early struggle is in the starch, and we crush both to flour, snuffing out the life in the grain. Then into this bleached and whitened corpse we place our fellow creature, yeast, and when it has grown, reproduced, and prospered on the energy of the grain, we put it living into an oven at 200°C to die to make our daily bread.
>
> In the multimillennium that man existed as a food gatherer, he existed as a fringe species under constant threat of the extinction that overtook many homonoid species, not the least threat was that posed by his food. Then, as now, eating provided the greatest exposure to exotic chemicals that most of us ever receive. Man evolved in an environment that provided many poisons

produced by micro-organisms, plants and other animals and is clearly not without biochemical defence mechanisms to protect himself. In his role as a sort of metabolic crematorium for his fellow creatures he manages to burn up on most occasions the harmful and the helpful with equal facility. Usually it is only when his defence mechanisms are overwhelmed by the sheer numbers of toxic molecules—the total amount of the toxin rather than its toxicity—that man is adversely affected: the dose makes the poison.

That man's fellow creatures are often far from wholesome is a matter for history or prehistory; primitive men developed not only an extraordinary lore about the use and hazard of foods, particularly plants, but also an extraordinary technology to render useful the more recalcitrant of their plant brothers. No need to go to prehistory to study food gatherers—just go to the museum to learn how our own local Aborigines worked the highly toxic seeds of the Moreton Bay Chestnut to make food of it in the absence of more prolific or less seasonal other foods.

While I have concentrated in this chapter on food additives, a proper balance requires a closer look at 'natural' foods. Some of the problems were discussed in the introduction and also when explaining maximum tolerable daily intake (MTDI), but the list can be extended greatly. Bananas contain serotonin (one of the biologically active amines); chick peas, while highly nutritious, are also toxic. So fad diets are to be avoided. One nutmeg is nice, two can cause an abortion, while three at one sitting could be fatal. Avocados will poison animals, while onions cause anaemias in animals. Half a kilo of horseradish (containing isocyanates) can kill a pig in three hours. Broad beans cause a disease (favism) in people with a particular genetic inclination. Aflatoxin is the most potent known carcinogen (in animals) and is formed by a fungus that attacks peanuts and other nuts, and cereals such as corn, rice and wheat, when the humidity and temperature are right for it. It is believed that it causes liver cancer, which is prevalent in humans in parts of Africa and Asia. Along with other pesticides, the fungicides used to control fungi are becoming increasingly less effective, so problems will arise in protecting our 'natural' food. Dairy herds in Tasmania were being fed on kale (1955), which transferred an oxazolidine to the milk. This increased benign goitre amongst children in spite of the use of iodised salt. Today the problem is one of excess iodine in milk from iodophor disinfectants used in the dairies (NH & MRC limit is 0.5 mg/L). I have quietly slipped back to an 'unnatural' additive. The difference in definition is marginal, and the only sensible approach appears to be to spread any risk by spreading the diet to cover a large selection of foods.

Flour and bread

Magnus Pyke, in his book *Food and society* acknowledges that even those with 'knowledge and facts' fall short of objectivity on this topic: 'Few people are able to discuss the composition of bread temperately'.[19]

On the whole the British seem to prefer white bread. At one time the demand was met by treating flour with *nitrogen trichloride*, but this agene process was abandoned some years ago after dogs fed large amounts of agenised bread developed running fits or canine hysteria. The toxic compound in the treated flour was later shown to be methionine sulfoximine, which is formed by the action of nitrogen trichloride on methionine, an essential amino acid of proteins. Today over half the total flour production in the UK is treated with chlorine dioxide as a maturing and bleaching agent. Additional whiteness is achieved, if necessary, by bleaching with *benzoyl peroxide*, giving benzoic acid as a final product. Various other improving agents are added—vitamins, buffers to control the acidity, surfactants to keep suspensions in the baking process, sequestrants to inactivate tiny amounts of possibly harmful metals—as well as preservatives such as propionic acid and sulfites to prevent spoilage.

The NH & MRC (1983) lays down standards for flours, meals and bread. Bleaching of flour can only be carried out by ozone and/or oxides of nitrogen and chlorine, or benzoyl peroxide. The flour can contain various additives (see Table 11.8).

TABLE 11.8
Additive levels in flour

Additive	mg/kg
Calcium acid phosphate , $Ca(H_2PO_4)_2$	7000
Ammonium chloride, NH_4Cl	600
Bromates, $KBrO_3$	30
Iodates, KIO_3	— (deleted 1982)
Sodium and calcium stearoyl-lactylates	4000
Calcium sulfate, $CaSO_4$	800
Sodium metabisulfite, NaH_5O_3	60
L-cysteine	75 (added 1978)

Bread[20] represents nearly half the Australian domestic use of wheat. Consumption of bread peaked in 1960 at about 60 kg per person per year, then levelled off, and now appears to be on the increase again. Over the last 15 years the proportion of all breads accounted for by combined wholemeal breads has risen from 20% to nearly 50%. Cereals are currently our most important source of complex carbohydrates, thiamin, iron and energy; the second most important source of protein, and the third most important source of calcium. Cereal grains exceed in value all other food exports combined and the three-year mean to 1985–86 was nearly A\$4000 million; unfortunately 95% of this was in unprocessed form. We produce enough wheat to provide each Australian every day with 10 plates piled with about 300 g of grain. One plate is eaten here and the other nine are exported.

Gluten

Gluten is the most important protein component in wheat and gives flour its unique physical properties for bread making. Gluten can be isolated by kneading dough under running water, which removes most of the starch and other water-soluble components. The dry matter of gluten consists of two main groups of proteins, *glutenin* and *gliadin*, plus some bound fats and carbohydrates.

The addition of water to flour hydrates the protein. Kneading, or high speed mixing in bakeries, 'works' the dough. This breaks the disulfide bonds between protein chains, allowing the chains to move. Then the bonds reform in this new position. The protein then changes shape to accommodate the new bonding positions (Fig. 11.22). The process is thus analogous to what happens in setting hair. (See also 'Chemistry of cosmetics—Hair'; 'Leather', Chapter 7.)

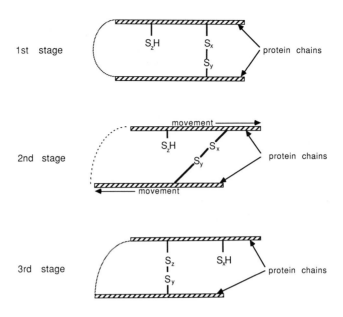

Fig. 11.22   Making and breaking of disulfide linkages during kneading of dough (Courtesy of Bread Research Institute of Australia)

Bakeries sometimes use a gluten-softening agent such as L-cysteine or sodium metabisulfite to help break bonds and aid mixing, and gluten-strengthening agents such as ascorbic acid or potassium bromate to help set the new bonds.

During 'proofing', starch breakdown and fermentation occur. The carbon dioxide produced opens up the gluten network giving it a cellular structure. Baking denatures the protein, gelatinises the starch granules with water from the gluten and sets the bread.

'You pays your money and you gets your dough'

Protein content and hardness are two of the most important physical characteristics of dough. Hard wheats with strong doughs are suitable for bread flours. Soft wheats are cultivated to give weaker doughs and low protein. As such, they are ideal for biscuits. The different requirements of wheats for different products are shown in the two-dimensional plot in Figure 11.23.

In practice a somewhat more complex set of tests and parameters is used.

Pasta is made from durum wheat (protein content 14% to 15%), milled to produce semolina. A minimum of 12% protein is needed to make pasta that remains firm on cooking (some hard bread wheat is sometimes mixed in). The semolina is blended with water, extruded through variously shaped nozzles and dried to 12% to 13% moisture level.

Noodles require flours of between 9.5% and 12% protein level. Higher levels giving stronger gluten are liable to cause tearing during processing. With lower levels the noodles are liable to fall apart during processing. The flour is also required to be very free from bran specks, as these are responsible for discolouration.

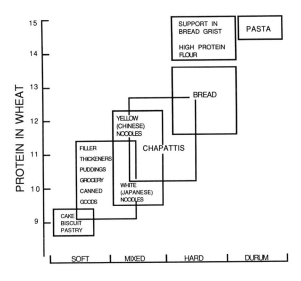

Fig. 11.23 Requirements of certain wheat products in terms of protein percentage and hardness (From H. J. Moss, *J. Aust. Inst. Ag. Sci.* 1973, 109; courtesy of Bread Research Institute of Australia)

Flat breads (such as chapatti, pita, mafrood, Lebanese) rely for their quality on steam produced when the dough is heated. Flour with a high degree of mechanical damage to the starch grains absorbs more water. Such flour is best made from hard bread wheats. Relatively low strength is required so the dough will not distort during preparation.

Biscuit doughs are expected to be weak, as tough texture is undesirable. Low-protein flours from soft wheats are used.

Fig. 11.24 From wheat to bread (and money!). The old Australian $2 note shows William Farrer, who pioneered the development of wheat in Australia.

　　　The number of parts by mass of flour that is extracted per 100 parts of wheat is called the extraction rate. Milling to produce white flour results in the removal of varying amounts of bran, pollard and germ which together constitute 12% to 15% of the mass of the grain. The limit of white flour extraction is 75% to 79%. Figure 11.25 shows the influence of extraction rate on nutrient retention. Australian white flour has increased in extraction rate from 71% in 1940 to 77% in 1985, and is expected to plateau at about 79%.

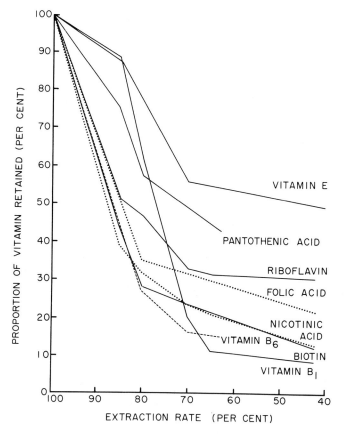

Fig. 11.25　Milling extraction rate and proportion of total vitamins retained in the grain in flour (From Magnus Pyke, *Man and food*, World University Library, London, 1970)

White or brown bread?

The small difference in nutritional value of white (70% extraction) versus wholemeal (100% extraction) flour in the context of a Western diet today is unimportant. However, the selling point of wholemeal bread is strong (see Plate XV). The advantages of white flour are as follows:

* Bread made from it is lighter, finer and is white.
* It contains less fat. It is thus less prone to rancidity and keeps longer.

- It contains less phytic acid and fibre, which tie up minerals such as calcium, iron, magnesium and zinc. Thus minerals in white bread are more available.

The disadvantages of white flour are:
- It contains less of the B-group vitamins.
- It contains less of certain minerals and trace elements.
- It contains less fibre.

The limiting amino-acid in wheat flour is lysine, which in the normal diet is compensated for by other sources of protein. The enormous effort to determine whether one can be properly nourished 'by bread alone' seems to be quite pointless for the normal sane-eating human being. Where undernourishment occurs in the world, it is the lack of sufficient food in total, rather than its balance, that is critical.

Bleaching makes use of benzoyl peroxide (maximum 40 mg/kg flour).

Wholemeal flour (wholewheat, wheatmeal) contains all the constituents of the wheat. Wholemeal bread is made from at least 90% wholemeal flour. It cannot be stored for long periods because the oil from the wheatgerm may become rancid. This makes it unpopular with millers and bakers. The bread must contain no more than 48% water. Brown bread must be made from at least 50% wholemeal flour, and brown colouring (in the form of malt) can be added. Brown bread is *not* wholemeal bread. Rye bread must be made from at least 30% rye flour. Milk bread has at least 4% skim milk powder (non-fat milk solids) added, which makes the loaf bulkier and helps retard the staling process. Protein-increased bread must have at least 15% protein (2.7% nitrogen $\times$ 5.5). (In the last two cases the percentage is calculated on a *moisture-free* basis.) Extra protein can be achieved by adding skim milk powder, soy flour or gluten to the dough. Special grain breads such as soy contain a high proportion of wheat or rye flour because only these rise satisfactorily. There is no guaranteed minimum quantity of the flour named on the loaf. Preservatives in bread (generally propionates and sodium acetate/acetic acid) now need to be declared on the label, but for a long time this was not the case.

Orange juice

The NH & MRC standard for fruit juices and fruit drinks was gazetted in all States. Fruit juices shall not contain added water (except concentrates diluted to original). They may contain preservatives and added vitamins and minerals. Unless labelled 'sweetened', they must contain no more than 4% of added sugar. A concentrated fruit juice has had at least 50% of the water removed. The minimum levels of vitamin C are shown in Table 11.9.

Note that apple juice is not considered a source of vitamin C. Concentrated juices must have the required vitamin level on dilution and must have dilution instructions on the label. The preservatives used are sulfur dioxide (maximum 115 mg/kg) or benzoic acid (maximum 400 mg/kg); sorbic acid can replace some of the benzoic acid. No other additives are allowed in fruit juices; neither colour nor flavouring should be added.

The Trade Practices Commission (TPC) issued Information Circular no. 16 on 25 June 1976 in response to complaints from consumers and the fruit juice industry about shady practices in the industry. The ruling was as follows. Unless a product complies with the standard (as described above) it must not be sold as

TABLE 11.9
Required levels of vitamin C
in fruit juices

Juice	Vitamin C (mg/L)
Orange	400
Lemon	350
Grapefruit	300
Guava	1000
Pawpaw	400
Mango	250
Blackcurrant	700
Pineapple	100

'juice'. It may (depending on State law) be called a 'fresh fruit drink', a 'fruit juice drink', a 'fruit squash drink', a 'fruit drink', or a 'fruit flavoured drink'. A product reconstituted from concentrates should (according to the TPC) clearly state that it has been reconstituted. The word *pure* can only be used for a fruit juice without preservatives, artificial colourings, flavouring, or the like.

The word *fresh* should not be used for reconstituted juice. The word should convey a time lapse between production and consumption of not more than two or three days. Suppliers must be able to provide backing for their claims. The terms *fresh picked, fresh chilled* and *fresh delivered* should not be used to indicate non-existent freshness. Trade names should not be used to mislead.

DIGESTION[21]

'Full of sound and fury signifying nothing' said Macbeth after consuming Jerusalem artichoke soup. (Well not really! In fact the Jerusalem artichoke is a form of sunflower, *Helianthus tuberosus*. In Italian the characteristic 'turning towards the sun' is 'girasole', which the fine English ear for language turned into Jerusalem.)

The scientist-cum-statesman Benjamin Franklin proposed the following project to the Royal Academy of Brussels in the 1770s as a problem for their proposed prize for 'useful science':

> It is universally well known, that in digesting our common food, there is created or produced in the Bowels of human creatures, a great quantity of Wind. That the permitting of this Air to escape and mix with the Atmosphere, is usually offensive to the Company, from the fetid smell that accompanies it. That all well bred People therefore, to avoid giving offence, forcibly restrain the Efforts of Nature to discharge that Wind. That so retained contrary to Nature, it not only gives frequently great present pain, but occasions future Diseases such as habitual Cholics, Ruptures, Tympanies, &c., often destructive of the Constitution and sometimes Life itself.
>
> Were it not for the odiously offensive smell accompanying such escapes, polite People would probably be under no more restraint in discharging such Wind in Company, than they are in spitting or in blowing their Noses.
>
> My Prize Question therefore should be: To discover some Drug, wholesome and not disagreeable, to be mixed with our common Food, or Sauces, that shall render the

natural discharges of Wind from our Bodies not only inoffensive, but agreeable as Perfumes.

The solution is *not* to try to disguise the output but to modify the input. The human gut cannot digest certain medium-sized sugar molecules such as raffinose and stachyose, found particularly in beans. Not so the bacteria in the colon, which attack them with vigour, producing a variety of gases and vapours, both inoffensive (carbon dioxide, hydrogen and methane), and offensive (possibly hydrogen sulfide or rotten egg gas, indole, skatole and ammonia). Bacteria are responsible for the production of about one-third of the gas, while the other two-thirds comes from swallowing. Both hydrogen and methane are flammable and have caused explosions during operations involving electrocautery on the colon. (The explosion limits in air for H_2 are 4% to 74%, and for CH_4 are 5% to 13%.) Normal people pass about half a litre of gas by the anus per day, although amounts as high as 168 mL per hour have been recorded in volunteers when half their diet was baked beans. Gastric gas causing a bloated feeling is also released upwards and it is interesting that certain substances relax the sphincter muscle at the *top* of the stomach, releasing the gas in that direction. These include nutmeg, ginger, caraway, cinnamon and peppermint. Taking peppermints after meals (as a sweet or liqueur) is obviously a tradition with a purpose.

It is only today, using genetic engineering, that this problem is being solved. Because soya beans are used in so many processed foods, research in the US has concentrated on locating the genes responsible for producing the indigestible sugars and replacing these with genes that will produce more acceptable sugars.

ALLERGIES[22]

While all people react in identical ways to some poisonous substances, there are other substances to which individuals react quite differently. These are called *idiosyncratic* reactions. Idiosyncratic reactions to food were known to Hippo-crates, but systematic studies have really been documented only since the early 1940s. There are believed to be three mechanisms of adverse reaction to food. The first of these reactions results from the body's *metabolism* and relates to a genetic fault that affects the way an individual metabolises certain food components. Examples of such genetic faults are diabetes, phenylketonuria (PKU), lactase deficiency and favism. The second and most common mechanism is *pharmaco-logical* (see Table 11.10) and is really an adverse 'drug' reaction in which the 'drug' is a natural or synthetic food component. The component is a relatively small molecule that is biologically active, just like a drug. For example in Chapter 9 we saw that the *drug* sulfonamide is modelled on the *natural* food component para-aminobenzoic acid and the molecule is the same size. Far less common than drug reactions is the third mechanism, which is a true *allergic* reaction to food (see Table 11.10). Allergic reactions involve the immunoglobulins, the carriers of specific antibodies. Some of these, the gammaglobulins, carry well-identified antibodies such as tetanus antitoxin and Rh antibodies, whereas a more recently discovered member, IgE (immunoglobulin E), was found responsible for forming antibodies against food proteins. Note that the chemical causing the production of

TABLE 11.10
Clinical features of adverse reaction to food

Pharmacological	Allergic
Common	Uncommon
All ages	Mostly children
Non-atopic	Atopic[a]
Many foods	Specific food
Delayed reaction	Immediate reaction
Difficult diagnosis	Easy diagnosis

[a] Atopic is a description of the type of person who has the constitutional make-up that responds with allergic reactions to certain stimuli; that is, the sort of person who gets dermatitis, hay fever, hives, etc. People who have other allergic reactions are the sort who get allergic reactions to food.

the antibody in this case—namely a protein—is very large. The *allergic* reaction to a bee-sting is a reaction to a foreign *protein*. The end response to the drug reaction and allergic reaction may come together to a common final path and so produce similar symptoms in the victim.

Allergic reactions in any one individual generally involve only a couple of foods (e.g. eggs, nuts, milk, seafood) and, in particular, their protein components. Local swelling, itching and burning around the mouth are common, and may be followed by nausea, vomiting, abdominal cramps and diarrhoea.

Pharmacological food reaction is much more prevalent. The reaction is to a chemical that may be common to a number of foods (and drugs). The time between exposure and reaction may be delayed for periods varying from hours to days, which makes linkage between cause and effect quite difficult. Detection is complicated by the fact that the amount of chemical needed to cause any adverse reaction does not stay the same but can depend on how much of it has recently been eaten. Just as antihistamines can make you very drowsy initially, and continuous use reduces this effect, the same masking can occur with pharmacological reactions to food.

Sometimes pharmacological and allergic reactions are combined in an individual to cause severe symptoms, with the former increasing the severity of the latter.

Skin tests detecting specific IgE may provide useful information on true food allergy, although the reverse is not true. High levels of IgE occur in normal blood donors and are of no significance in the absence of symptoms. Testing for pharmacological reaction is best done with the purified individual components given as a food (in the absence of other foods that could also be supplying that or other chemicals).

Other methods are still in the development stage and these methods yield high levels of false positive and false negative results. The patients may be encouraged to remove the wrong foods from their diet or leave the wrong foods in their diet.

Natural and synthetic food chemicals to which there are common reactions are salicylates (including aspirin), benzoates, metals and synthetic food colours such as tartrazine (colours include azo and non-azo dyes), and sulfur dioxide. Sulfur dioxide is quite soluble in water and forms a clathrate (compare chlorine in 'Swimming-pool chemistry', Chapter 5). The solution is acidic but sulfurous acid as such does not exist. Just as with chlorine, the species present in water depend on the acidity, yielding sulfur dioxide, sulfite and bisulfite. The only common solids are sodium and potassium sulfite and bisulfite. (Sulfur oxides are common air pollutants as well and form the acid in 'acid rain'.) Excess of the normal protein containing amino acid, glutamic acid (in the form of monosodium glutamate; see earlier in this chapter) is reported as causing reactions. For further details see the excellent series in the issue of *Chemistry and Industry* devoted to food intolerance (6 February 1984).

CONCLUSION

This chapter on food additives and the chemical aspects of nutrition has grown to become a dominant part of the book, extending its tentacles into other chapters. Food is the most intimate chemical contact we have with our environment and our reaction to it is a peculiar one.

Pharmacological reactions to foods are quite common. Because of this individual response, the responsibility for avoiding these problems is thrown much more on the consumer. Without adequate information there is no way in which this responsibility can be assumed. In the first edition I suggested that, where labelling created problems, detailed information should be made available by the manufacturer or producer so that particular products known to cause problems could be avoided. This was just a natural extension in the provision of specified foods for rather larger sections of the population, such as diabetics (no sugar), coeliac (gluten free), phenylketonuria (low in the amino acid phenyl-alanine). The plea in the second edition of this book for a coding system based on the EEC method was accepted (96th session, NH & MRC, October 1983). The presence of food additives must be declared in the list of ingredients by their functional class names followed in brackets by either their code number (see Appendix XII), their prescribed names or appropriate designations.

Some mechanism must also be introduced into the scheme to provide incentives for manufacturers to use a minimum amount of additive and not necessarily the legally allowed maximum. The work of the Food Standards Committee must move much more into the area of providing information on the components (and their variability) in 'natural' foods consumed in Australia.

At the moment the NH & MRC has the dual function of funding and fostering medical research and building national agreement on public health legislation, standards and administration. While Australia undertakes less than 2% of the world's medical research, that 2% is highly regarded, so that the first function of the NH & MRC would appear to be well fulfilled.

Our policy makers do very badly at making use of the other 98%. As the previous discussion shows, our history in regard to regulating food additives has shown a very slow use of current knowledge, with the food colours being a classic example. Australia has not ratified a single international food standard under the Codex Alimentarius Commission. Other areas of public health have suffered similar neglect (e.g. see urea-formaldehyde foam in 'Chemistry of plastics', Chapter 6).

Finally, consumers who found this chapter interesting should extend their reading (as suggested in the extensive further reading list) and keep it and their eating habits as broad as possible. Eating a narrow selection of foods in excess or reading only articles with which you agree are both undesirable.

REFERENCES

1. W. J. Madgwick, 'Food additives and the law in Australia', *Food Tech. Aust.* **26**(12), 1974, 541.
2. G. O. Kermode, 'Food additives', *Sci. Am.* **226**, March 1972, 15.
3. FAO publications (available from Australian Government bookshops). Food additive control in—no. 1, Canada; no. 6, France; no. 7, the Federal Republic of Germany; no. 8, USSR, 1969.
4. World Health Organization, *Evaluation of mercury, lead, cadmium, and the food additives amaranth, diethyl pyrocarbonate, and octyl gallate* (WHO Food Additive Series no. 4), Geneva, 1972.
5. G. G. Severino, R. L. McBride and B. M. Cox, 'People prefer coloured flavours', Food Research Quarterly **47**, CSIRO, 1987, 64–65.
6. R. H. M. Kwok, 'Chinese-restaurant syndrome', *New England Journal of Medicine* **278**, 1968, 796 (letter to the editor).
7. L. Hough and J. Emsley, 'The shape of sweeteners to come', *New Scientist*, 15 June 1986.
8. A. S. Truswell, 'Sugar and health: a review', *Food Tech. Aust.* **39**(4), 1987, 134.
9. C. Fahlberg and I. Remsen, *Berichte* **12**, 1879, 469.
10. 'Sweetener blamed for mental illness', *New Scientist*, 18 February, 1988.
11. L. Hough, 'Haworth Memorial Lecture: the sweeter side of chemistry', *Chem. Soc. Rev.* **14**, 1985, 357–374.
12. This section is based on S. Sonsino, 'Radiation meets the public's taste', *New Scientist*, 19 February 1987; and S. C. Morris, 'The practical and economic benefits of ionising radiation for the post-harvest treatment of fruit and vegetables: an evaluation', *Food Tech. Aust.* **39**(7), 1987, 336.
13. R. A. Edwards, 'Food contamination from packaging materials: Regulatory aspects', *Food Tech. Aust.*, **26**(6), 1974, 242.
14. Geoffrey Chaucer, *The Canterbury Tales*, the General Prologue, lines 637–638.
15. F. E. Peter, 'Legal aspects of alcoholic products', *Aust. Wine, Brewing and Spirit Review*, 18 July 1975, 10.
16. J. and I. Couper-Smartt, 'Caffeine consumption: a review of its use, intake, clinical effects and hazards', *Food Tech. Aust.*, **36**(3), 1984, 131.
17. See *New Scientist*, 11 February, 1988, 31.
18. 'Caffeine: a cuppa is a killer for insect pests', *Chem. Brit.*, February 1985, 126.
19. M. Pyke, *Food and society*, John Murray, London, 1968. I particularly like the chapter 'If it's poisonous, why eat it?' Chapter 14 also provides an excellent discussion of the Codex Alimentarius.
20. Most of the information on bread and wheat came from Colin Wrigley, Officer-in-Charge, CSIRO Wheat Research Unit, Sydney, and John Moss, Bread Research Institute of Australia, Epping Road, North Ryde NSW 2113.
21. This section is based on Stephen Young, 'A Christmas digest', *New Scientist*, 19/20 December 1985; Bernard Dixon 'Begone with the wind', *New Scientist*, 24 April 1986.
22. D. H. Allen, *et al.* 'Adverse reactions to foods', *Med. J. Aust.*, 1 September 1984, 537.

BIBLIOGRAPHY

Additives

Caffeine: a scientific view, meeting International Life Sciences Institute, Sydney, 15 July 1986. Published as a supplement, *Food Tech. Aust.* **40**(1), 1988.

Evaluation of caffeine safety, a scientific status study by the Institute of Food Technologists' expert panel on food safety and nutrition. *Food Tech. Aust.* **40**(3), 1988, 106.

Food chemicals codex, 2nd edn. Committee on Specifications, Food Chemicals Codex of the Committee on Food Protection, National Research Council, National Academy of Sciences, Washington DC, 1972.

Kihlman, B. A. *Caffeine and chromosomes*, Elsevier Amsterdam 1977. Read the conclusion first, 'Caffeine: a chemical hazard in the environment of man?'

Lu, F. C. 'Toxicological evaluation of food additives and pesticide residues and their "acceptable daily intake" for man: the role of WHO in conjunction with FAO', *Residue Reviews* **45**, 1973, 81.

Nitrate, nitrite and nitroso compounds in foods, a scientific status study by the Institute of Food Technologists' expert panel on food safety and nutrition. *Food Tech. Aust.* **40**(3), 1988, 100.

Roberts H. R. (ed.) *Food safety*. Wiley Interscience, New York, 1981.

Sharp, D. H., and West, T. F. (eds), *The Chemical industry*, Society of chemical Engineers, Ellis Harwood Ltd UK, 1982, ch. 39.

Sulfites as food ingredients: a scientific status study by the Institute of Food Technologists' expert panel on food safety and nutrition, *Food Tech. Aust.* **39**(11), 1987, 532.

Advertising

Processed food: a pain in the belly, Sect. 2 (Chemistry). ANZAAS, 1982. ACA, Sydney (its first educational publication).

Allergies

Friedman, J. 'Naturopathic treatment and childhood behavioural disturbances: a pilot study', Appendix 16 in *Report of the Committee of Inquiry into Chiropractic, Osteopathy, Homeopathy and Naturopathy* AGPS, Canberra, 1977. This study appears to be statistically more significant than the committee is comfortable to acknowledge. If it had been a clinical test of a conventional drug there would have been little hesitancy about its efficacy.

Kauffman, G. B. 'Asthma, anti-allergens and aerosols: the discovery and development of disodium chromoglycate (Intal)' *Education in Chemistry*, March 1984, 42.

Larkin, T. 'Food additives and hyperactive children'. *FDA Consumer* (Washington DC), March 1977, 19.

Michaelsson, G., and Juklin I. 'Urticaria induced by preservatives and dye additives in food and drugs. *Dermatology* **88**, 1973, 525.

Randolph, T. G. *Human ecology and susceptibility to the chemical environment*. Charles C. Thomas, 1962. This book predates the fashionable interest in the environment and ecology.

Thune, P., and Gramholt, A. 'Provocation tests with antiphlogistica and food additives in recurrent urticaria'. *Dermatologica* **151**, 1975, 360.

Tracey, M. V. 'The price of making our foods safe and suitable'. *Food quality in Australia*, (Academy of Science Report no. 22), Australian Academy of Science, Canberra, 1977.

Analysis

Heavy metal contamination in seafoods, cadmium and zinc. NH and MRC, Australian
  Department of Health, 1973.

Kefford, J. K. 'Analytical problems with fruit products'. *Aust.* CSIRO *Food Preservation
  Quarterly* **29**(4), 1969, 65.

Liston, J. 'Health and safety of seafoods'. *Food Tech. Aust.* **32**(9), 1980.

Methyl mercury in fish. NH & MRC, Australian Department of Health, 1972.

Pearson, D. *The chemical analysis of food*, 6th edn, Churchill, 1970.

Carcinogenesis

Cancer: causes and control. Australian Academy of Science, Canberra, 1980.

Doll, R., and Peto, R. *The causes of cancer.* Oxford Medical Publications, Oxford, 1981.

Evaluation of carcinogenic risk of chemicals to man. WHO, International Agency for
  Research of Cancer, IARC Monograph, Lyon, no. 1, 1972; no. 8 'Some aromatic azo
  compounds', 1975.

Colours

Bonin, A. M., and Baker, R. S. U. 'Mutagenicity testing of some approved additives with
  the salmonella microsome assay'. *Food Tech. Aust.* **32**(12), 1980, 608.

Bonin, A. M., Farquharson, J. B., and Baker, R. S. U. 'Mutagenicity of arylmethane dyes in
  salmonella'. *Mutation Research*, 1982.

Boycottez les colorants', *Que Choisir?*, Journal of the French Consumer Association
  (Union Fédérale des Consommateurs), 7 rue Leonce Raymond 75781, Paris, Cedex
  16, April 1976.

Carpenter, T. I., *et al.* 'Synergistic effect of flourescein on rose Bengal, light-dependent toxicity'. *Environmental Entomology* **10**(6), 1981, 953.

Yoho, T. P., Butler, L., and Weaver, J. E. 'Photodynamic killing of house flies fed food, drug
  and cosmetic dye additives'. *Envir. Entomology* **5**, 1976, 203.

Flavour

Boudreau, J. C. (ed.). *Food taste chemistry* (ACS Symposium Series 115). ACS, Washington DC, 1979.

'Chemistry and flavour'. *Chemical Society Reviews* **7**(2), 1978, 167–219.

Hornstein, I., and Teranishi, R. 'The chemistry of flavour'. *C & EN*, 3 April 1967, 92.

Johnson, A. E., Newston, H. E., and Williams, A. A. 'Vegetable volatiles: a survey of
  components identified' (Parts 1 and 2), *Chemistry in Industry*, 22 May 1971, 556;
  23 October 1971, 1212.

Laing, D. G. 'On the measurement of smell'. *Food Res. Q.* **35**, 1975, 88.

Murray, K. E., Shipton, J., and Whitfield, F. B. 'The chemistry of food flavour'. *Aust. J.
  Chem.*, **25**, 1972, 1921.

Schlegel, W. 'Nutrition and flavour legislation'. *Food Tech. Aust.* **27**(1), 1975, 10.

Theimer, E. T. (ed.). *Fragrance Chemistry: the science of the sense of smell.* Academic
  Press, New York, 1982.

Vodoz, C. A. 'Flavour legislation: world trends'. *Food Tech. Aust.* **29**(10), 1977, 393.

Whitfield, F. B. 'The chemistry of food acceptance', *Food Res. Q.* **42**(3) 1982, 52.

Whitfield, F. B., Last, J. H., and Tindale, C. R. 'Skatole, indole and p-cresol: components in
  off-flavoured frozen French fibres'. *Chemistry and Industry*, 4 September 1982, 662.

Food

Chandler, B. V. 'Quality of Australian honeys'. *Food Res. Q.* **37**, 1977, 1

Farrer, K. T. H. 'Adulterations of all descriptions: F. M. Bird Memorial Lecture'. *Food
  Tech. Aust.*, August 1979, 340.

Farrer, K. T. H. ' Nutrition as a factor in marketing'. *Food Tech. Aust.* **32**(5), 1980, 236.
   Food Preservation. Unilever booklet, revised no. 3, 1970.
Hall, E. G., and Scott, K. J. *Storage and market diseases of fruit.* CSIRO. 101 colour plates
   of diseased or damaged fruit—just the thing to take with you to the market.
Jeffries D. *The role of the Australian food industry in disseminating nutritional
   information.* CRES, Australian National University, Canberra, 1978.
Mabey, R. *Food: the impact of food technology on everyday life..* Penguin Education, 1972.
   Part of a series of topic books for students in schools and colleges.
Potter, N. N., *Food Science,* 2nd edn. Avi Publishing Co., Connecticut, 1973. An excellent
   general textbook.
Ruello, J. H. 'Prawns: fresh and frozen', *Food Res. Q.,* **36**, 1976, 13.
Tracey, M. V., 'Foods and heart disease', *Food Res. Q.,* **34**, 1974, 61.
Specialist courses for the Food Industry. No. 1 'Instrumental Techniques', 1971. No. 6
   'Food borne microorganisms of public health significance', 1975. CSIRO and
   Department of Food Technology, University of NSW.

Food composition—Australia

Composition of Australian Foods, research papers from the Department of Food
Technology, University of New South Wales; written by H. Greenfield, R. B. H. Wills and
others; published in *Food Technology in Australia*, vol. 31, 1979 to vol. 39, 1987.

   1.   'Tables of food composition and the need for comprehensive Australian tables', **31**,
        1979, 458.
   2.   'Methods of analysis', **32**, 1980.
   3.   'Foods from a major fast food chain', **32**, 1980, 363.
   4.   'Lebanese foods and meals', **32**, 1980, 578.
   5.   'Fried take-away foods', **33**, 1981, 26.
   6.   'Chinese foods', **33**, 1981, 176.
   7.   'Minerals in Lebanese, Chinese and fried take-away foods', **33**, 1981, 274.
   8.   'Fortification of McDonald's foods', **33**, 1981, 378.
   9.   'Meat pies, sausage rolls and pasties', **33**, 1981, 450.
  10.   'Health food confectionery', **33**, 1981, 476.
  11.   'Mueslis', **33**, 1981, 564.
  12.   'Hamburgers', **33**, 1981, 619.
  13.   'Rice', **34**, 1982, 66.
  14.   'Margarines and cooking fats', **34**, 1982, 240.
  15.   'Pizza', **34**, 1982, 340.
  16.   'Foods from Pizza Hut restaurants', **34**, 1982, 364.
  17.   'Snack foods', **34**, 1982, 452.
  18.   'Foods from Kentucky Fried Chicken', **34**, 1982, 566.
  19.   'Greek foods', **35**, 1983, 84.
  20.   'Salad dressings and blended vegetable oils', **35**, 1983, 467.
  21.   'Mayonnaises and salad creams', **35**, 1983, 562.
  22.   'Tomato', **36**, 1984, 78.
  23.   'Brassica vegetables', **36**, 1984, 176.
  24.   'Italian foods', **36**, 1984, 469.
  25.   'Peas and beans', **36**, 1984, 512.
  26.   'Luncheon meats and continental sausages', **37**, 1985, 114.
  27.   'Vitamins in take-away foods', **37**, 1985, 162.
  28.   'Citrus fruit', **37**, 1985, 308.
  29.   'Beers', **37**, 1985, 450, 468.
  30.   'Apples and pears', **38**, 1986, 77.

31.   'Tropical and sub-tropical fruit', **38**, 1986, 118.
32.   'Leaf, stem and other vegetables', **38**, 1986, 416, 421.
33.   'Lamb', **39**, 1987, 202.
34.   'Beef and veal', **39**, 1987, 208, 227.
35.   'Pork', **39**, 1987, 216.
36.   'Beef, lamb and veal offal', **39**, 1987, 223.
37.   'Manufactured meat', **39**, 1987, 234.
38.   'Tuber, root and bulb vegetables', **39**, 1987, 384.
39.   'Vegetable fruits', **39**, 1987, 488.
40.   'Temperate fruits', **39**, 1987, 520, 530.

Other papers, written by G. I. Hutchison and A. S. Truswell
1.   'Nutrient composition of Australian chicken', **39**, 1987, 196.
2.   'Nutrient composition of Australian beef', **39**, 1987, 199.

Foods (native)

'Aboriginal bushfoods study arouses wide interest'. *Syd. Uni. Gazette*, September 1984.
Anderson, M. 'Some mushrooms to enjoy'. *RSC News* **17**(5), May 1986, 4.
Larkin, T. 'Herbs are often more toxic than magical'. *FDA Consumer*, October 1983, 5.
Warren, R. G. 'The quality control of medicinal herbs'. *Chem. Aust.*, January/ February, 1987, 38.

Packaging

Brown, I. A. 'Food contamination from packaging materials'. *Food Tech. Aust.* **27**(2), 1975, 65.
Crane, W. J. 'Food contamination from packaging materials: biscuits and snack foods'. *Food Tech. Aust.* **27**(3), 1975, 104.
Davis, E. G., McBean, D. McG., and Rooney, M. L. 'Packaging foods that contain sulphur dioxide. *Food Res. Q.,* **35**, 1975, 57. Discussion of permeability of plastic films to gases.
Davis, E. G., and Huntingdon, J. N. 'New cell for measuring the permeability of film materials'. *Food Res. Q.,* **37**, 1077. Methods used on testing wine casks where up to 50% of the oxygen leakage into the wine is contributed by the taps and valves; see Chapter 6.
Richardson, K. C. 'Shelf life of packaged food', *Food Res. Q* **36**, 1976, 1.

Sweeteners

Fahlberg, C., and Remsen, I. *Berichte* **12**, 1879, 469.
Hough, L., 'Haworth Memorial Lecture: the sweeter side of chemistry', *Chem. Soc. Rev.* **14**, 1985, 357-374.
Hough, L. and Emsley, J. 'The shape of sweeteners to come'. *New Scientist*, 15 June 1986.

Vitamins

Cameron, E., and Pauling, L. *Cancer and vitamin C.*, Warner Books, New York, 1981.
Detlman, G. C., and Kalokerinos, A. 'Estimation of vitamin C with a dip stick', *ACBS Journal*, September 1973, 55.
Seib, P. A., and Tolbert, B. M. (eds). 'Ascorbic acid: chemistry, metabolism, and uses', *Advances in Chemistry, Series 200*, American Chemical Society, 1982.
Stone, I. 'Humans: the mammalian mutants'. *American Laboratory*, April 1974, 32.

Water

Duckworth, R. B., Oswin, C. R., Johnson, D. S., Lamb, J., Aitkin, A., Rankin, M. D. and Goodall, J. B. 'The role of water in food', *Chemistry and Industry*, 18 December 1976, 1039. A series of papers on what water does in food, how to keep more of it there and why (other than profits!). The UK Food Standards Committee is reviewing whether foods are being sold with significantly different water contents from those accepted as normal in the past.

The heavy metals and radiation

12

INTRODUCTION

The contamination of our environment by heavy metals is a constant and disturbing problem. Heavy metals are among the most dangerous and least understood of contaminants. Because they exist naturally as part of the earth's crust they occur in all soils, rivers and oceans. In the right quantities some are essential to life. Elements that have a function in our bodies are H, O, C, N, S, P, K, Na, Cl, Ca, Mg, Fe, Cr, Mn, Zn, Cu, Al, Co, Se, I, F and Br. Selenium has only recently been shown to be significant. Many we require only in minute (trace) quantities, and they are, like selenium and fluorine, very toxic indeed at higher concentrations.

The heavy metals are in widespread industrial use and, when released into the air or into rivers, they distort the naturally occurring distribution of metals with which we have come to terms in evolving in our 'natural' environment. This distribution is very critical because our need for traces of some elements, such as selenium, is very low. It is the most toxic element known to be essential to mammals. The uptake and need for copper depends on the level of zinc and molybdenum to which we are exposed and *vice versa*. It is also known from

Fig. 12.1

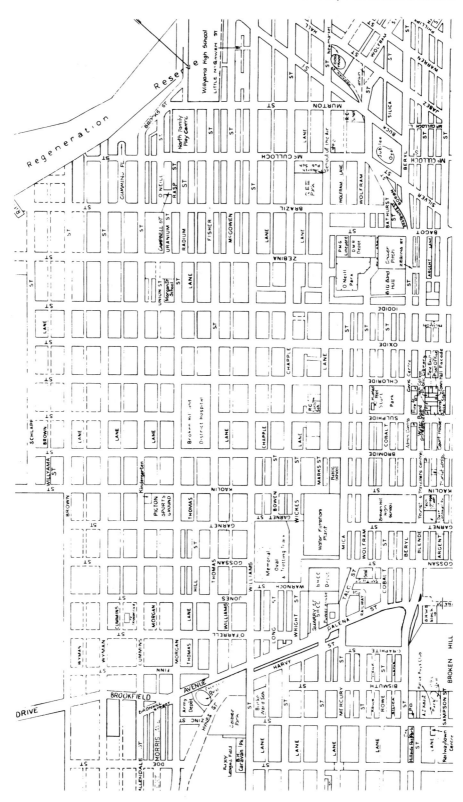

Fig. 12.2   Section of map of Broken Hill, NSW. How many 'chemical' street names can you find?

laboratory experiments that selenium compounds have a protective effect against the toxic action of mercury. Seals and dolphins from many parts of the world show a correlation between the accumulation of mercury and selenium (e.g. Hg = 2.53, Se = 10.2, in mg/kg), which is not found in the fish on which they feed. This seems to indicate that the ratio is metabolically established in the marine mammals. We are only just beginning to understand some of these subtle interactions.

Our modern technology is heavily dependent on the use of these materials, but they are by no means indispensable in all their uses. Where there is a known problem and a reasonable alternative solution, consumer groups and environmentalists have good reason to believe that their case for restriction and control is justified. To develop an understanding of the complexity of the argument, a number of factors have to be considered.

What materials are involved?

The term *heavy metals* in this context signifies the metals mercury, cadmium and lead in the first instance. In a broader sense, other toxic elements are included. Some of these, such as arsenic and barium, have no part in our personal biochemistry; others, like copper, zinc and selenium, are essential at the correct level.

In what form are they dangerous?

Depending on the physical and chemical form of the metal, the substance can either be very poisonous or completely harmless. *Liquid* mercury, as in a thermometer bulb or in a mercury dental amalgam, is harmless, but long exposure to the small amount of *gaseous* mercury given off by the liquid (but not the amalgam) is readily absorbed into the lungs and can lead to poisoning. The toxicity of *inorganic* mercury compounds depends on how soluble they are in the body fluids. The use of *soluble* mercuric nitrate in the felt-making process led to the traditional notions of 'hatters' shakes' and 'mad as a hatter' and also accounts for the use of the term 'hatter' for solitary, acutely shy goldminers who recovered their gold by distillation of the gold-mercury amalgam in crude stills.

The most toxic forms of mercury are now recognised to be the organo-mercury compounds—in particular, alkyl-mercury, which is produced from inorganic mercury by micro-organisms present in the bottom of waterways. These compounds, while not soluble in water, are very soluble in fat and hence are stored in the body with an average residence of about 70 days, in contrast to inorganic mercury, the residence time of which is about six days. Thus, with continuous exposure of low levels, it is possible to build up concentrations in the body of 10 times the amount of organic mercury compared to inorganic mercury at the same level of exposure. A publication of the Australian Department of Health (1973) shows how the calculation of a maximum allowable level of alkyl-mercury in fish of 0.5 mg/kg was arrived at. This level was set with a safety factor of only 10, whereas the normal practice involves a factor of 100. (Further details are given in Table 12.2, p. 484.)

While soluble barium compounds are very poisonous, barium sulfate, which is used for increasing the contrast of internal organs for x-rays, is so very insoluble as to be harmless. Metallic lead is not considered particularly dangerous in itself, although water collected from lead-lined roofs once was a considerable problem. Some forms of lead used in paint are very soluble in body fluids and, hence, are toxic. Copper is an essential metal, but water passing through copper plumbing in Australian hospitals was found to produce up to 1 mg/kg of copper compound in dialysis fluid, which could add about 10 times the dietary intake of copper to the copper stored in the liver. Domestic copper water-pipes normally present no problems.

Cadmium compounds are used as pigments (lemon, yellow, orange, maroon) in ceramic glazes, paints and plastics. Cadmium is absorbed from food and water slowly so that something like 98% of the cadmium in food is excreted within 48 hours. However, cadmium that is absorbed remains, so that cadmium levels build up from zero at birth. The absorption from the air is much more efficient (10%–15%). Cigarette smoke contains cadmium—20 cigarettes contain an average of 30 μg, of which about 70% is extracted into the smoke. There is a good correlation between stored cadmium levels and high blood-pressure, and injection of cadmium chloride into animals can apparently cause irreversible damage to the testes. More details on cadmium are given later in this chapter.

At what levels are they dangerous?

Medical tradition divides poisoning into three categories—acute, chronic and subclinical.

Acute toxicity means that you show a rapid poisoning response and need immediate treatment. Generally, this is the result of an accidental intake.

Chronic toxicity involves the effect of long-term low-level exposure to a material which causes a slow and steady poisoning process. Again this is less likely to be a common consumer problem than an occupational hazard, although children eating lead paint come into this category.

Subclinical toxicity. This is the region in which the levels are so low that any abnormality cannot easily be detected or associated with the particular substance. The main concern here is the effect of such low levels on children because:
- the child's immature nervous system, including the brain, appears more susceptible to permanent damage than that of the adult;
- the absorption rate (for lead) through the gastrointestinal tract is about 50% for a child compared to 5% to 15% for an adult.

Grimbledon Down   Bill Tidy

Fig. 12.3 Misleading poison (From *New Scientist*, 9 January, 1975, reprinted with permission)

- children consume more food in relation to their body mass than adults do because they have a higher metabolic rate and also chew non-food objects that may contain additional heavy metals.

Standards

Obviously, we should set exposure to heavy metals as low as is reasonably feasible. It was therefore disturbing to find that, in 1974, glazes used by Canberra primary-school children that were allegedly lead free actually contained, in one instance, almost 5% lead. In 1975 teething rings were analysed and found to contain both lead and cadmium in one instance, and a popular form of plastic building kit commonly chewed by children used over 1% cadmium selenide as a pigment. An Australian Standard for children's toys and playthings (safety standards) AS1647 was first published in 1976.

Of course one of the questions that can be raised is to what extent these materials are extractable from the toy. The classic piece of work in this area has been carried out by G.W.A. Fowles.[1] He examined toys in the UK and found cadmium levels of up to 1% in the plastic (cadmium sulfide/selenide). Extraction of cadmium from shavings of plastic (type ABS) toys taken with a variety of hand tools that do not heat the plastic showed that, under conditions likely to be present in the human stomach (namely 37.5°C and 0.1 M hydrochloric acid), significant amounts of cadmium can be extracted from plastic shavings.

Practical gnawing experiments by Fowles, his children, and a number of 'guinea pigs' have shown that similar shavings can be obtained quite easily. The risk is not associated with sucking, because saliva will not favour extraction, but with the gnawing. Toys made from polypropylene or polystyrene type materials are less easily wetted and give lower extraction levels. Fowles was critical of the test methods used by industry, in which sawn cubes rather than small shavings are used, because any method involving heating can seal over exposed pigment particles and give falsely low results. To give some idea of a consumer chemist's dedication, the following is quoted from his manuscript:

> Toy 47 (red)—samples taken by teeth.
> Bulk sample gnawed by the teeth of the author and gnawings spat out. One gram of sample obtained in 30 minutes. Sample washed by decantation, so loss of very fine particles; remaining sample passed through 1 mm mesh with the exception of a few larger pieces—all used in standard extraction test (4 h, 25 mL 0.1 M HCl, 37.5°C). Cadmium extracted in HCl (simulates stomach juices): 128 micrograms. Cadmium in saliva: 3.5 micrograms.

In the first edition of this book (1975) the story ended at that point. However, since then there have been some chemically interesting developments. The factors that showed up were:

- The acid strength should correspond approximately to the stomach contents, ~0.1 M HCl, and should be checked after extraction to guard against loss of strength by neutralisation by a basic filter.
- Raising the temperature from ambient to 37.5°C (body heat) increases the extraction rate by about two to three times.

- There are differences between measurements made at different times during the day and at different seasons.

Light has a dramatic effect—a photoflood can increase rates of extraction by a factor of about 10 for selenide (red) and about 3 for sulfide (yellow). The effect is perhaps not too surprising since both cadmium sulfide and cadmium selenide are photosensitive. Most human stomachs can be assumed to be dark! Another trap lies in wait for those who don't read labels. Not only does most of our food come pre-packaged these days but so do many standard chemical solutions. The 0.1 M HCl concentrate (Volucon, see Fig. 12.4) is labelled *preservative added 13.5 mg HgCl$_2$ per litre* (even chemicals have to be preserved!). The presence of mercuric chloride in the hydrochloric acid used by Professor Fowles in his original extraction method would inhibit the cadmium extraction presumably by forming a *more* insoluble layer of mercuric sulfide over the grains of cadmium sulfide, thus preventing further extraction. The levels of cadmium extracted under these more rigorous conditions were lower.

Fowles has also examined copper enamelling powders for lead and used his gnawing talents to test whether comics contained lead or chromium compounds that could be ingested in the stomach of a child who chewed and swallowed part of a comic. He gives an issue by issue report of the levels of extractable metal in the comic series.

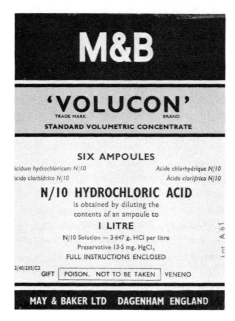

Fig. 12.4   Acid with preservative

HEAVY METAL ANALYSIS OF CONSUMER PRODUCTS

The following is a summary of some typical research results and the effects of their publication. The research was undertaken in conjunction with Microanalytical Laboratory, Research School of Chemistry, Australian National University.

Margarines, November 1973

Brand X: 38 mg/kg lead (material in contact with container). Same manufacturer, Brand Y: 11 mg/kg lead (material in contact with container). The problem resulted from incorrect plastic used in package and led to setting of standards for plastic packaging materials. Brands X and Y retested September 1974: no lead detected, no cadmium detected.

Glazes used by schoolchildren in pottery classes, September 1974

Brand X, labelled 'lead free':

	Pb (mg/kg)	Cd (mg/kg)
Type 1	2300	—
Type 2	700	—
Type 3	1000	—
Type 4	49200	—

This matter was soon rectified.

Babies' teething rings (imported), April 1975

	Pb (mg/kg)	Cd (mg/kg)
1. Round yellow	1295	1100
2. Animal motif	1060	340
3. Paint from (2)	810	267

Led to setting standards for children's toys. (This investigation was conducted in conjunction with the ACT Consumer Affairs Council.)

Wrapping paper for bread, October 1975

	Cr (mg/kg)	Pb (mg/kg)
Brand 1	—	—
Brand 2	540	5186
Brand 3	—	—
Brand 4	—	—
Brand 5		
White colour	—	—
Yellow colour	trace	130
Orange colour	391	2100
mixed	205	1000

Submission made to Department of Health. Retested March 1977: Brand 5—no detectable chromium or lead.

METALS IN FOOD

As well as the avoidable intakes of heavy metals, we ingest a certain amount with our food. Some work done on tea urns in Western Australia highlights this problem. Domestic utensils used for boiling water are of variable quality: some of them introduce significant amounts of cadmium into the water boiled to make beverages. The practice of connecting urns to boiling hot water supplies rather than using normal outlets should be avoided. 'Enrichment' (in one case × 197) generally occurs in hydrotherms. These are continuously heated, thermostatically controlled, sealed units in which a constant water level is maintained. Impurities in the water source concentrate on evaporation, particularly if the device has a low usage.

Food consists of chemical compounds mainly based on the elements carbon, hydrogen, oxygen and nitrogen, and to a minor extent such elements as sodium, calcium, potassium and phosphorus. Early investigations found a large number of other elements in quantities too small for accurate determination by available methods. These elements were said to be present in trace amounts; hence the term *trace elements*. Modern techniques, such as atomic absorption spectroscopy, anodic stripping polarography and neutron activation, have permitted the determination with reasonable accuracy of many trace metals at levels of about 10^{-7} to 10^{-9} g/kg.

Trace elements—essential, non-essential or toxic?

Trace elements were at first generally regarded as undesirable contaminants of food, but later it was realised that some of these elements, notably iron and copper, were essential for the health and well-being of humans. Trace elements were accordingly classified as essential, non-essential and toxic. Essential elements, like fluorine and selenium, were at first regarded as toxic, but at high intake almost all elements become toxic, and the margin between beneficial and harmful levels may be small. Hence the classification of essential, non-essential and toxic is unsatisfactory.

Research work has gradually added to the list of essential trace elements, and it is now believed that 14 trace elements are essential for animal life: iron, iodine, copper, zinc, manganese, cobalt, molybdenum, selenium, chromium, nickel, tin, silicon, fluorine and vanadium. Probably other elements will be added to this list as experimental techniques are further refined. Five of the 14 elements—nickel, tin, silicon, fluorine and vanadium—have emerged as essential nutrients in the diets of laboratory animals more recently, following the introduction of ultra-clean environments and the use of pure crystalline amino acids and vitamins. Much remains to be learned about the metabolic functions of the 'newer' trace elements and their practical significance in the health and nutrition of humans.

To state the obvious, all the elements in food come from the environment in which the food is produced. Plants extract elements from the soil in which they grow, and some of these elements are required for the plant's nutrition. We receive our trace elements from our food and from the air. Processing and packaging may increase the undesirable trace elements while depleting essential trace elements. It should not be surprising that minute quantities of some elements are essential for the health and well-being of people, for the human race evolved in an environment containing most of the known elements. The environmental relationship is illustrated by the fact that a person's body burden of sodium and potassium follows fairly closely the levels of these elements in the Earth's crust.

In the following more detailed discussion of some elements there is no significance in the selection of elements.

Zinc

Until recently, zinc deficiency was regarded as important only in the practical nutrition of pigs and poultry and therefore was only of indirect importance to humans. Zinc deficiency has now been demonstrated to be a public health problem in several countries. Inadequate zinc in the diet results in growth failure, in sexual infantilism in teenagers and in impaired wound healing. Zinc-responsive growth failure has been observed in Egypt and Iran, where a major constituent of the diet is unleavened bread prepared from high extraction wheat flour. In the USA the same phenomenon has been observed in young middle-class children if they consume little meat (less than 30 g/day). The average adult requires 15 mg or more of zinc per day and lactating women require about twice this quantity, but the biological availability of zinc is related to the type of food consumed. For example, zinc availability in cereals and vegetables appears to be lowered by the complexing action of some cereal components and this appears to be largely responsible for zinc deficiency in Egypt and Iran. Studies indicate that only about 20% to 40% of the zinc in a mixed Western diet is actually available for absorption.

The food that contains the most zinc is oysters, some of which contain more than 1000 mg/kg. Another good source of zinc is yeast, which may contain about 100 mg/kg, but meat is the most important source in a normal diet. Of course, high levels of zinc are undesirable in food. For example, a large number of Sydney schoolchildren were ill when they consumed a cordial that had been stored overnight in a galvanised container and had developed zinc levels of about 500 mg/kg.

Oysters and Zinc

In 1972 the CSIRO reported on the level of zinc in oysters collected from the river Derwent on which the city of Hobart, capital of Tasmania, is situated. Downstream from the Electrolytic Zinc Co. refinery, these oysters concentrated so much zinc from the water that consumption of as few as six oysters could cause vomiting from zinc poisoning. High cadmium and copper were also found, as well as fish with high levels of mercury, the latter probably from Australian Newsprints, further upstream. Things were not helped when on 5 January 1975 the ore-carrying ship, *Lake Illawarra*, on its way to the zinc refinery, crashed into the bridge across the Derwent and cut the bridge in two. Apart from the economic and social upheaval this severing of the city caused, thousands of tonnes of heavy-metal-containing ore then lay on the bottom of a deep part of the harbour from where it was not salvaged.

Claims by oyster growers against the Electrolytic Zinc Co. were settled out of court, and all the oysters were removed from Ralph's Bay, 15 km downstream from the zinc works and moved to nearby areas, Pipeclay Lagoon and Cygnet, which lie outside the estuary. While regulations at that time in Tasmania and elsewhere in Australia allowed 40 mg/kg of zinc (based on wet mass) of foods, the CSIRO found wet levels of 10 000 mg/kg. Even those oysters with the lowest levels contained eight times the limit. This was soon rectified in our regulatory tradition by raising the national legal limit (for oysters) to 1 000 mg/kg, where it still stands today (see Table 12.1). Tasmania, to be on the safe side, raised its limit to 1 500 mg/kg.

The situation with other heavy metals was worse from a health point of view. Analyses by the Australian Government Analytical Laboratories and the CSIRO found levels up to 35 mg/kg (as wet mass) for cadmium, 148 mg/kg for copper and 17 mg/kg for lead. High levels of mercury in fish were also found. In the meantime the Electrolytic Zinc Co. has taken steps to get its factory in order. Although it still needs and receives ministerial discretional exemptions from the State's *Environmental Protection Act 1973*, the pollution levels shown by experimental replants of oyster in the critical areas have dropped significantly.

In 1988 the metal levels of most oysters outside the estuary are within the legal Tasmanian limits, but these are set to ensure that most oysters are inside the limit. This leads us to ask what really are natural and safe levels.

Analysis of oysters at Port Davey in the remote southwest of the State, away from direct industrial and urban pollution, still showed mean levels of 1000 mg/kg of zinc, but mussels were mostly below the old standard of 40 mg/kg. This is not entirely explained on the basis of known bio-accumulation rates (Table 12.3), but may have to do with the high tissue level of calcium in the oyster, which could suppress the availability of zinc for its enzymes. The mussel has most of its calcium in the shell. The interaction of metals in different species raises the whole question of how best to use these organisms as biomonitors of pollution.

A final point raises the interesting question as to how the oyster itself survives such a massive load of toxic metals (it obviously has survival value in that the more inedible they are the better their chance of reproducing!). Oysters apparently have amoeba-like cells reminiscent of our own white cells, in which little membrane-covered bags (vesicles) store copper and zinc. These beasties scavenge amongst all the tissues of the oyster in much the same way as macrophages (also large amoeba-like cells), scavenge around our lungs for dust particles and ingest them to prevent damage to other cells. (This does not always work; for example, asbestos fibres of critical size kill these cells and they are again released—hence the danger of this commonly used mineral.) The oyster appears to remove some of the metal-loaded scavengers slowly through its gills.

Copper

Copper deficiency has not been reported in adults, even in areas where copper deficiency is severe in grazing animals. But copper deficiency is implicated in anaemia in infants from impoverished communities where the diet is based on cow's milk. Infants' diets containing less than 50 μg of copper per kilogram body mass per day have resulted in copper depletion and produced clinical lesions. (Copper is a component of several amine oxidases, and it is possible that, in some animal species, defects in the synthesis of vascular elastin and of collagen in the bones and connective tissues are the result of copper depletion causing a decline in amine oxidase activity.) Sulfur in the diet in the form of sulfides can markedly decrease copper absorption, and cadmium concentrations in the order of 3 mg/kg can adversely affect copper utilisation. Infants appear to require 50 to 100 μg/kg per day, and adults about 30 μg/kg per day. The condition commonly known as 'kinky hair' in infants is caused by a genetic defect causing inefficient copper absorption.

Liver, oysters, many species of fish and green vegetables are good sources of copper, but milk and cereal products are poor ones. Indeed, copper is an undesirable constituent of milk, fats and fatty foods as it acts as a catalyst and promotes rancidity of fats even at a very low concentration. Polyunsaturated milk produced by special feeding is even more sensitive to traces of copper and fairly high levels of antioxidants must be added.

Copper is frequently implicated in food poisoning and copper in water at a concentration of about 20 mg/L is often the problem. Under certain conditions, drinks from machines dispensing carbonated beverages may contain high concentrations of copper. Water from a copper water service may also contain high copper concentrations (up to about 70 mg/kg) if allowed to stand for some days without flushing. Following illness in children immediately after eating ice blocks, 285 samples (mostly of the offending brand) were analysed for copper. A small number had levels of 43 to 80 mg/kg—levels that are sufficient to cause vomiting. This is a recurring problem and arises because some manufacturers use tinned copper moulds. The acidic mixture used to produce ice blocks will dissolve copper from de-tinned areas of the moulds if the mixture is left in the moulds for excessive lengths of time.

Copper bracelets

Is there anything in the myth that copper bracelets have therapeutic value? This question has been investigated in Australia by W.R. Walker in the Department of Chemistry at the University of Newcastle. Through letters in newspapers about 300 sufferers from arthritis were contacted (half of whom previously wore 'copper bracelets') and were randomly allocated for a psychological study. This involved wearing copper bracelets and placebo bracelets (anodised aluminium resembling copper) alternately. The results showed that for a statistically significant number of subjects, the wearing of a copper bracelet appeared to bring some therapeutic benefits. Copper from the bracelet is found to dissolve in sweat and the amount lost was about 13 mg/month. If this were absorbed into the system it would amount (over 12 months) to more than the usual copper level in the body. The level in sweat was about 500 mg/kg from the bracelet, but, if sweat is left in contact with copper turnings for 24 hours, the concentration rises by a factor of 100 (turns blue). The skin is permeable to some materials and copper does move through. The perceived effectiveness of the bracelet was tested with carefully prepared psychological tests and found to be significant. It was also interesting that respondents to the questionnaire described over 100 different conventional medical treatments of their arthritic conditions, most of which were accompanied by undesirable side-effects.

"Could I 'ave a copper pair for me arthritis?"

Fig. 12.5 'For me arthritis' (From *Newcastle Morning Herald*, 18 September 1976)

Chromium

Chromium deficiency in humans appears to reduce tolerance to glucose. (Chromium is a co-factor of insulin, essential to proper glucose metabolism.) Chromium deficiency may result from a grossly deficient diet. The chromium requirement of humans is very difficult to estimate, because little is known of its form in food or its biological availability. Meat appears to be the best source of chromium in the diet, as it may contain several mg/kg. *Yeast*-leavened bread is also an important source of chromium.

Selenium

As discussed earlier, selenium may well prove to be one of the most important elements, since it has been shown to be essential in animal diets, although many toxic effects have been described. No pathological conditions in humans appear to have been identified as resulting from selenium deficiency, but selenium *reduces* the toxicity of methyl mercury, and selenium deficiency may reveal an underlying heavy-metal toxicity.

Selenium and sulfur can replace each other in certain chemical structures and reactions, but sulfur cannot replace selenium in its essential nutritive role. The selenium intake varies widely throughout the world and blood levels accordingly vary from about 0.8 μg/mL in Venezuela (selenium-rich area) to 0.07 μg/mL in Egypt (selenium-poor area).

Cobalt

Cobalt is thought to be unique because it is the only trace element shown to be physiologically active in one particular form only, namely cyano-cobalamin, or vitamin B$_{12}$. Cobalt nutrition in humans is thus primarily a question of source and supply of vitamin B_{12}. Ruminants, on the other hand, can utilise dietary cobalt directly because the microflora of the rumen convert cobalt into vitamin B_{12}. Hence sheep have been given metallic cobalt pellets, which remain in their rumen. All ordinary diets contain considerably more cobalt than can be accounted for as vitamin B_{12}. Cobalt intake is about 0.15 to 0.6 mg/day. Levels of 25 to 30 mg/day can be toxic to humans.

Cobalt was implicated as the cause of heart failure in beer drinkers consuming about 12 L a day. The cobalt had been added to the beer in concentrations of about 1.2 to 1.5 mg/L to improve foaming. At this level, the heavy drinkers ingested about 6 to 8 mg cobalt per day. This represents a quantity that can be ingested without ill-effects in normal diets, and it appears that the cardiac problem arose from the combination of poor quality diet, high alcohol consumption and high cobalt intake.

A number of other trace elements have been shown to be essential to animal nutrition, but there is no clear evidence that manganese, vanadium, tin or nickel are essential to humans. The concentration of many trace elements in food is high. These elements include mercury, cadmium, arsenic, antimony and lead. The levels of these elements naturally present in food are only very rarely sufficiently high to produce symptoms of acute toxicity, but chronic toxicity may arise because the intake of a trace element may exceed the body's ability to eliminate the element.

NATIONAL HEALTH & MEDICAL RESEARCH COUNCIL MODEL FOOD STANDARDS REGULATION
A12. Metals and contaminants in food

(Adopted by Council at the Ninety-Fifth Session in June 1983)

1. For the purposes of this regulation and save where the contrary intention appears—
   (a) 'metal' includes compounds of that metal;
   (b) where food contains a metal and any compound or compounds of that metal, that metal and compound or compounds shall be expressed as the metal;
   (c) antimony, arsenic and selenium are deemed to be metals;
   (d) maximum permitted concentration shall be determined on the edible content of the food that is ordinarily consumed and, in the case of food in a dried, dehydrated or concentrated form, shall be calculated with respect to the mass of the food after dilution or reconstitution;
   (e) 'beverages and other liquid foods' include fruit juices and beverages with a fruit juice content, milk, alcoholic beverages and frozen liquid foods, but do not include thick gels or other semi-solid foods.
2. Food specified in column 2 [of Table 12.1] shall not contain a metal specified in column 1 thereof in a concentration greater than the maximum permitted concentration specified opposite and in relation to that food in column 3 thereof.

TABLE 12.1

Column 1	Column 2	Column 3
		Maximum permitted concentration in food (mg/kg calculated as the metal)
Metal	Food	
Antimony	Beverages and other liquid foods	0.15
	All other foods	1.5
Arsenic	Beverages and other liquid foods	0.1
	Galline (chicken) livers	2.0
	Fish, crustaceans and molluscs (inorganic arsenic only)	1.0
	All other foods	1.0
Cadmium	Beverages and other liquid foods	0.05
	Bran	0.2
	Fish and fish content of products containing fish	0.2
	Edible offal other than liver	2.5
	Liver	1.25
	Meat muscle	0.2
	Molluscs and the mollusc content of products containing molluscs	2.0
	Wheat germ	0.2
	All other foods	0.05
Copper	Beverages and other liquid foods	5.0
	Cocoa and chocolate	50.0
	Edible offal other than ovine (sheep) livers	100.0
	Ovine livers	200.0
	Molluscs and the mollusc content of products containing molluscs	70.0^a
	All other foods	10.0
Lead	Beverages and other liquid foods	0.2
	Bran	2.5
	Fish in tinplate containers	2.5
	Fruit juices and fruit juice drinks	0.5
	Infants' foods in containers other than tinplate	0.3
	Infants' foods in tinplate containers	A mean level of 0.3 in 10 sample units. No sample unit shall exceed 0.8
	Meat in tinplate containers	2.5
	Milk, condensed milks and liquid milk products in tinplate containers	0.3
	Molluscs	2.5
	Tomato products, as specified in regulation F2, in tinplate containers	2.5
	Vegetables	2.0
	Wheat germ	2.5
	All other foods	1.5
Mercury	Fish, crustaceans, molluscs and fish content of products containing fish	A mean level of 0.5^b
	All other foods	0.03

TABLE 12.1 — *Continued*

Column 1	*Column 2*	*Column 3*
Metal	Food	Maximum permitted concentration in food (mg/kg calculated as the metal)
Selenium	Beverages and other liquid foods	0.2
	Edible offal	2.0
	All other foods	1.0
Tin	Canned—Asparagus	250.0[a]
	Fruits	250.0
	Fruit juices	250.0
	Green beans	250.0
	Tomato products as specified in regulation F2	250.0
	All other food packed in tinfoil or tinplate containers	150.0
	Foods not packed in tinfoil or tinplate containers	50.0
Zinc	Beverages and other liquid foods	5.0
	Oysters	1000.0[a]
	All other foods	150.0

[a]Considered too high by an expert committee (author's note).
[b]The mean level of mercury in fish, crustaceans, molluscs and the fish content of products containing fish in the prescribed number of sample units, as determined by the methods prescribed by sub-regulation (5) of this regulation.

Tin

Tin in its inorganic form is generally regarded as non-toxic, but the attachment of one or more organic groups to the tin atom produces biological activity against most species and this is greatest when the number of attached groups is three (R_3SnX). If the chain length of the *n*-alkyl group is steadily increased, the highest toxicity to mammals is attained when R = ethyl. Tributyltin compounds, on the other hand, show a high activity against fungi and are used as fungicides in wallpaper pastes. They are less dangerous to mammals. A combination of quaternary ammonium salts ($R_4N^+X^-$) with tributyltin oxide gives a water-soluble formulation. Organo-tin fungicides have also been incorporated into marine paints; they protect the surface from marine growth. Triphenyltin compounds are also toxic to fungi. Increasing the chain further reduces biological activity and the tri-*n*-octyltin compounds are of low toxicity.

The largest single application for organo-tin compounds is as stabilisers for poly (vinyl chloride) plastic (see Chapter 6) and sulfur-containing organo-tins are unsurpassed in their ability to confer heat resistance to the plastic. Several dioctyltin compounds are allowed in PVC used for food packaging.

Mercury

The first recorded mention of mercury was by Aristotle in the fourth century BC, when it was used for religious purposes. Earlier still, vermilion (cinnabar, HgS) is known to have been used as a decorative war-paint (cosmetic). Paracelsus (1493–1541) introduced the treatment of syphilis with mercury. In 1799 Howard prepared mercury fulminate as a detonator for explosives. Economically mercury often cannot be replaced by any other metal (nor can it replace other metals); therefore, its unique properties have proliferated its uses.

Fig. 12.6   Can of roast veal, 1824 (Courtesy of the International Tin Research Institute, UK)

With the exception of iron almost all other metals can be *amalgamated* (alloyed) with mercury. Sodium amalgams are formed in electrolysis cells used for producing chlorine and caustic soda. Many mercury compounds are used as industrial *catalysts*. Dental amalgam is also prepared with mercury (see Chapter 8). For this application, the mercury and alloy must amalgamate upon mixing to form a smooth paste within 90 secs. Within three to five minutes it should set to a carvable mass and remain so for 15 minutes. Within two hours it must develop sufficient strength, hardness and toughness to resist biting and chewing stresses. It must expand to maintain a good marginal seal but not so much as to overstress the tooth. It must not produce toxic or soluble salts or tarnish or produce significant amounts of mercury *vapour*.

The *equilibrium* vapour pressure of mercury (20°C, 13 mg/m^3 of air) is about 200 times the recommended atmospheric concentration. In an amalgam the equilibrium vapour pressure is greatly reduced and ventilation prevents build-up to equilibrium.

POISONING BY MERCURY VAPOUR

The suggestion that covering mercury spills with water 'will keep the mercury from vapourising'[2] is thermodynamically false,[3] although it can give short-term protection (see demonstration of immiscible liquids, 'Disappearing tricks' in Chapter 8, and Plate VIIIa).

Raoult's law predicts a vapour pressure for mercury proportional to its concentration in the water layer ($\sim$10^{-7} M or $\sim$0.05 mg/kg). It is however only applicable, at best, to completely *miscible* solutions. When positive deviations from Raoult's law occur, then in the limit where the liquids are *im*miscible, each component evaporates independently to give a total vapour that is the sum of the two pure component vapour pressures. This is exploited in steam distillation (see Experiment 13.12) to remove organic material from water, and its converse, the Dean and Starke distillation, to remove water from organic material (see Experiment 13.1).

The only limitation to the equilibrium evaporation of mercury is the rate of diffusion of the mercury through the water. A simple experiment confirms that this is not slow.

Introduce 1 mL of mercury carefully (via a hypodermic syringe) into a 1 L Erlenmeyer flask half-filled with distilled water. Purge the air inside the flask gently with compressed air and stopper the flask. With a UV absorption or atomic absorption spectrometer, using a 10 cm quartz cell, show that in a few hours the mercury vapour approaches the equilibrium value of the pure liquid (13 mg/m^3 at 20°C).

In 1854 Regnault H.V. did the first experiment on the vapour pressure of almost immiscible liquids.

TABLE 12.2
Vapour pressures (kPa) of H$_2$O and CS$_2$, and mixtures

Temp.°C	Pure H$_2$O	Pure CS$_2$	Sum	Mixture
12.07	1.40	28.89	30.29	30.11
26.87	3.51	51.81	55.32	54.96

Note that a three-phase, two-component system has only one degree of freedom. You are free to vary either the temperature or the pressure, but not both.

Inorganic mercury is swept through the body fairly rapidly by the body's natural detoxification system. This means that it concentrates rapidly in the liver and kidneys, where, if the dose is high enough, massive damage can occur, but the half-life in humans is six days (i.e. half the material ingested is excreted in six days). In contrast the half-life for organic mercurials is about 70 days (see Fig. 12.7). A daily intake of 2 mg of *inorganic* mercury will not reach a steady state level of 20 mg in the body, however long the exposure. On the other hand, a similar intake of *organic* mercury would top 200 mg within a year—although you would be dead long before that. Even worse is the fact that, although the best biologically protected areas such as the brain or fetus have a much slower uptake of organo-mercury, they have a much slower excretion rate as well.

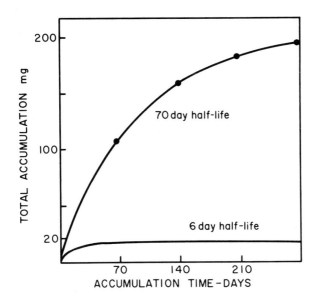

Fig. 12.7   Half-life accumulation of mercury in the body. Organic mercury reaches 10 times the level of inorganic mercury in nine months. At the 2 mg/day ingestion level shown here, symptoms of severe poisoning from organic mercury would appear about the third month (From A. Tucker, *The toxic metals*, Earth Island Ltd, London, 1972)

Figure 12.8 shows accumulation in the brain and liver, assuming a guessed 20% and 50% absorption respectively of a daily dose of 100 μg—and a longer brain half-life. A dose of 100 μg would be contained in a normal single meal of fish that contained 0.5 mg/kg mercury wet mass (the permitted maximum in Australia). Brain half-life is not known accurately but is believed to be considerably longer than that of the liver. Half-life in the testes is also long. According to some scientists, symptoms of damage to the brain first appear at concentrations as low as 3 μg/g of brain tissue—which would be reached in 50 days on this half-life basis. Other scientists believe that 20 μg/g of brain tissue is the damage symptom threshold. This explains the devastating effect of organo-mercury poisoning: the cells of the brain, which do not reproduce and have limited repair facilities, are destroyed.

It would seem that a sensible precaution would be to restrict severely the use of organo-mercury compounds. However, one of the biggest mercury poisoning disasters, at Minamata (Japan) in 1953, was the result of effluent of a poly(vinyl chloride) factory using inorganic mercury. The inorganic mercury waste was

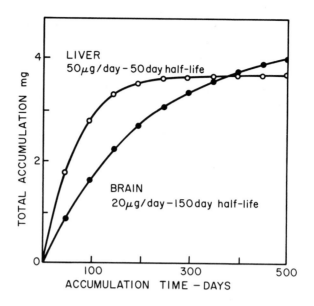

Fig. 12.8 Accumulation of organo-mercury in the brain and the liver (From A. Tucker, *The toxic metals*, Earth Island Ltd, London, 1972)

converted in the sludge at the bottom of a bay by anaerobic microbes (oxygen depleted conditions) to organic mercury and entered the food chain in its most deadly form. The preferential methylation of mercury by methane-producing anaerobes results in most aquatic animals containing monomethylmercury, the concentration being a rough guide to the species position in the food chain. Thus shrimps, which are low in the chain, generally contain less than 0.05 mg/kg mercury, and sharks, which are at the top of the chain, contain levels often in excess of 2 mg/kg. Marlin and swordfish can contain around 16 mg/kg.

Limits for metals such as mercury in fish are set in ppm wet mass (i.e. the fish mass) or, more correctly, mg/kg. The original Minamata data were in mg/kg dry mass (i.e. fish without water), and this yields a number about 10 times higher (because fish are ~90% water). Because this was not appreciated initially, danger levels were set too high. As things stand 15 to 30 μg a day is within the range of the normal daily intake from 'uncontaminated' food (limit 0.03 mg/kg) and this seems to be about the maximum level that the most sensitive section of the population (i.e. children) can tolerate without fear of damage, so that setting 'safe' limits for contaminated foods in the case of mercury may not be realistic. About 150 g of fish with the maximum of 0.5 mg/kg of mercury eaten by a 25 kg child *once a week* alone contributes the maximum allowable daily intake (30 μg per day per 75 kg body mass).

There is also a real danger in dealing with a population *average* daily intake. If it takes 40 mg of mercury to kill and I get the 40 mg while you get none, according to the average we are both safe. But I will be dead. It is the possible repetitive dose taken by an individual that is critical.

While the natural quantity of mercury in sea water is about 35×10^6 tonnes only, it takes tens of thousands of years for metallic marine pollutants to disperse *uniformly*; meanwhile high concentrations build up in discharge areas: estuaries, coastal areas, and shallow continental shelves—the food productive areas. The other factor that is very important is the ability of marine animals to concentrate heavy metals. An oyster can concentrate mercury by a factor of 100 000. If the *natural* water level is about 50 parts in 10^{12} (a million million), an oyster might be able to concentrate this to 5 mg/kg. The concentration factors of some marine organisms are given in Table 12.3.

TABLE 12.3
Average abundance of certain trace elements in the earth's crust and sea water, and enrichment factors and in selected sea organisms (all values are in mg/kg)

Element	Earth's crust	Ocean water	Oceanic residence time (years)	Enrichment factors		
				Scallop	Oyster	Mussel
Be	2.80	0.000 001	150	—	1 000 000	—
Ag	0.07	0.000 1	2×10^6	2 300	18 700	330
Cd	0.20	0.000 05[a]	5×10^5	2 300 000	320 000	100 000
Cr	100.00	0.000 6	350	200 000	60 000	320 000
Cu	55.00	0.003	5×10^4	3 000	14 000	3 000
Mn	950.00	0.002	1400	55 500	4 000	13 500
Mo	1.50	0.01	5×10^5	90	30	60
Ni	75.00	0.002	1.8×10^4	12 000	4 000	14 000
Pb	12.50	0.000 03	2×10^3	5 300	3 300	4 000
V	135.00	0.002	1×10^4	4 500	1 500	2 500
Zn	70.00	0.005	1.8×10^5	28 000	110 000	9 000
Hg	0.08	0.000 05	4.2×10^4	—	100 000	100 000

[a]0.000 02 to 0.000 8 ppm in some NSW waters.

Source: K. J. Doolan and L. E. Smythe, 'The cadmium content of some New South Wales waters', *Search* **4**(5), 1973, 162.

AMAZING! EIGHTEEN OYSTERS BEFORE CHUNDERING!

Fig. 12.9 Zinc Point Casino (The first legal casino in Australia was Wrest Point in Hobart.)

Cadmium

Cadmium is entirely a by-product metal extracted from zinc and lead. It looks like zinc but, on bending, the coarse-grained cadmium gives a crackling sound similar to that given by tin. It is used for plating, bearing alloys, soldering aluminium (Cd-Zn), orange-red glaze in ceramics, and in nuclear reactors as a shield against neutrons and as a control rod. It is used in nickel oxide cadmium accumulators. Organo-cadmium compounds are used as PVC stabilisers and as mould-release agents for plastic articles.

Oysters grown in unpolluted water may contain as little as 0.05 mg/kg of cadmium, whereas oysters grown in polluted water may contain cadmium in excess of 5 mg/kg. Obviously oysters should not become a staple diet item.

Large concentrations of cadmium have bizarre effects. In 1955 the Japanese reported an affliction called 'Itai-Itai Byo'. The name mimics the cries of sufferers and has been translated 'Ouchi-ouchi disease'. The disease, which takes a long course of increasing painfulness, begins with simple symptoms such as 'joint pains' and ends with total and agonising immobility as the result of skeletal collapse. Cadmium leads to bone porosity and to the total inhibition of bone repair, so that the load-bearing bones of the skeleton suffer deformation, fracture and collapse. In Japan the disease has been associated with rice and soya (in the range of 0.37–3.36 mg/kg dry weight) in the local diet, and was also known to occur in workers engaged in the preparation of cadmium-based paints.

Cadmium is chemically related to zinc (they are in the same column of the periodic table, see Fig.1.1) and it is found together with zinc in nature and in zinc products. The zinc/cadmium group is also related to the magnesium/calcium group (for reasons which can be found in elementary chemistry books) so that is why cadmium (and strontium) can interact so effectively with the calcium in the bones. The mutual replaceability of metals depends both on their *chemical similarity* and *the size of the atomic ion* (see Table 12.4).

TABLE 12.4
Ionic radii in picometres (10^{-12} metre)

Group IIA			Group IIB		Group IV	
Ca^{2+}	Sr^{2+}	Ba^{2+}	Zn^{2+}	Cd^{2+}	Hg^{2+}	Pb^{2+}
99	113	135	74	97	100	121

Zinc and cadmium are mutually inhibitory in both absorption and retention processes, perhaps because they compete for similar protein-binding sites. High zinc intake may reduce the toxicity of cadmium and conversely high cadmium intake with marginal zinc deficiency may aggravate the deficiency. Unfortunately, the residence time of cadmium in the body is very long.

Arsenic

Arsenic is a component of almost all soils and hence all food contains trace amounts af arsenic. Oysters, which concentrate arsenic, probably contain the highest levels. The arsenic in marine animals is organically combined; it does not appear to be very toxic and is rapidly excreted by man, apparently unchanged.

Organic arsenicals are widely used as additives in stock food because these compounds appear to stimulate growth and improve food utilisation. Animal diets typically contain 35–40 mg/kg arsenic, at which level there is some accumulation in edible tissues, usually less than 0.5 mg/kg in muscle and 2.0 mg/kg in liver. Before the advent of DDT, arsenic was one of the principal pesticides, being used as both an insecticide and herbicide. Therefore it is likely that arsenic levels in food were much higher than they are today, as very little arsenic is now used in agriculture.

There appears to be a level of intake above which arsenic accumulates in the system and below which the body appears to be able to excrete all the arsenic ingested. Normal intake is about 0.007–0.6 mg/kg body mass per day. It appears unlikely that any problem should arise with this element: arsenic poisoning is only likely in individuals occupationally at risk or who happen to be the victims of attempted murder.

Hg^{2+} and As^{3+} are poisonous, with a number of similar symptoms. Both cases can be treated with BAL (British Anti-Lewisite developed to combat the arsenical war gases; see Fig. 12.10).

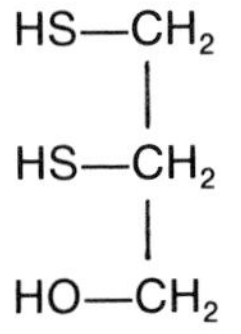

$$
\begin{array}{l}
HS—CH_2 \\
\quad | \\
HS—CH_2 \\
\quad | \\
HO—CH_2
\end{array}
$$

Fig. 12.10   BAL (British Anti-Lewisite)

This compound complexes with the metals via its SH groups. It thus competes with the ability of these metals to complex, via a sulfur atom, with a vital intermediate in the biological process by which an acetyl group is attached to co-enzyme A. Less than 1 millimole of Hg^{2+} is capable of disrupting this process, which is a key step in our ability to metabolise glucose. The BAL shifts the equilibrium in favour of its complex with the metal, which is excreted. Glucose metabolism is vital to nearly every cell in the body but the brain is particularly dependent, so nervous disorders are an early symptom of heavy-metal poisoning. The high content of sulfur-containing proteins in hair means that arsenic can be detected in victims of poisoning long after death (see 'Chemistry of cosmetics—Hair', in Chapter 4).

Lead

Lead was used in ancient times by the early Egyptians to glaze pottery. It was also used by the Romans to make water pipes and to store wines (this being a possible source of lead poisoning).

Lead tetra-ethyl additive for petrol accounted for about 10% of the world lead consumption in 1970, with the USA consuming about 0.26 megatonnes per year for this purpose. Most of the *bromine* produced from sea water is used to produce ethylene dibromide, which, along with lead tetra-ethyl, is added in order to exhaust the lead from the cylinder as volatile lead dibromide. (For each molecule of lead tetra-ethyl there are 0.5 molecules of ethylene dibromide and 1.0 molecule of ethylene dichloride in the usual antiknock additive; see also Chapter 10.)

The toxicology of lead is complicated. Inorganic (Pb^{2+}) lead is a general metabolic poison and is accumulated in humans. It inhibits enzyme systems necessary for the formation of haemoglobin (levels of urinary δ-aminolaevulinic acid are monitored to indicate this interference). In particular, children and young people appear liable to suffer permanent damage. Lead can replace calcium in bone, and so it tends to accumulate there. An unpleasant feature is that it may be remobilised long after the initial absorption; for example, under conditions where calcium is suddenly needed by the body, such as during feverish illness, cortisone therapy and also in old age. It can also cross the placental barrier, and thereby enter the fetus.

Lead alkyls (organo-lead) such as lead tetra-ethyl are even more poisonous than Pb^{2+}, and are handled quite differently in the body. Lead tetra-ethyl causes symptoms mimicking those of conventional psychosis. There is little or no elevation of blood lead, so correct diagnosis is difficult in the absence of suspected exposure. Yet leaded petrol has been almost a 0.1% hydrocarbon solution of this compound. The danger of using leaded petrol for degreasing or cleaning is generally not appreciated. Humans have evolved in the presence of a certain amount of lead—it averages about 10 mg/kg in the earth's crust. We know that over 2 megatonnes of lead is mined each year, in comparison with the 180 kilotonnes estimated to be naturally mobilised and discharged into the ocean and rivers.

Typically an adult living in a city who breathes air containing, say, 3 μg lead/m^3 and who respires about 15 m^3 of air per day will absorb 20–25 μg of lead per day (about 50% efficiency in absorption). A normal daily diet containing about 300 μg of lead will result in about the same amount of absorbed lead (5–10% efficiency in absorption). Fallout of airborne lead also adds to the lead present in food and water. Even the present average levels of lead in the blood of adults in industrial countries are not far below those that can lead to obvious clinical symptoms.

RADIOACTIVITY

CHANT OF THE RADIOACTIVE WORKERS

We're not afraid of the alpha ray,
A sheet of paper will keep it away!
A beta ray needs much more care,
Place sheets of metal here and there.
And as for the powerful gamma ray
(Pay careful heed to what we say)
Unless you wish to spend weeks in bed
Take cover behind thick slabs of lead!
Fast neutrons pass through everything.
Wax slabs remove their nasty sting.
These slow them down, and even a moron
Knows they can be absorbed by boron.
Remember, remember all that we've said,
Because it's no use remembering when you're dead.

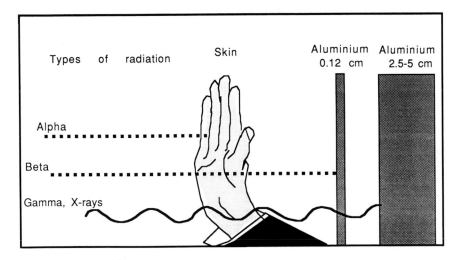

Fig. 12.11 How far radiation travels through matter. Alpha particles lose their energy first—a layer of skin is enough to stop them. Beta particles are halted by a relatively thin sheet of aluminium. Gamma and x-rays are the most penetrating, and will pass through relatively thick layers before being absorbed (From *New Scientist*, 11 February 1988)

Radiation

Radiation comes in two forms. Firstly, there is electromagnetic radiation like light, but with a range of frequencies from radio waves to gamma rays. All this radiation travels at the speed of light (3×10^8 m/s) and can also be regarded as a stream of massless particles called photons. When photons hit you, the *effect* depends on their *energy*, which increases—the shorter the wavelength (higher the frequency) of the radiation. The *amount* of radiation depends on the *intensity* (i.e. the number of photons). The small percentage of ultraviolet radiation in sunlight can cause burning in 20 minutes, while you can spend weeks behind glass (which cuts out the UV, but lets through the visible light) without a tan.

The second type of radiation consists of a stream of particles *with* rest mass travelling at various speeds, depending on their kinetic energy. Examples of these are alpha particles (helium nuclei), beta particles (electrons) and neutrons (discovered after the wave-particle duality was understood and so there is no special name for their radiation).

Radioactive disintegration

Radioactive decay is a random process. We do not know which atom in a sample will decay, nor do we know when. In fact we can use the model presented in the Entropy Game, (Appendix VIII) to describe radioactive decay. In Figure 12.12 the horizontal axis is time and the vertical axis is the average number of atoms emitting at that time. No matter how many atoms you postulate are emitted at particular times, you are obliged to play the exchange game to obtain the correct answer *because atoms of the same type are indistinguishable*. The general application of this simulation game to quite diverse areas takes a while to be

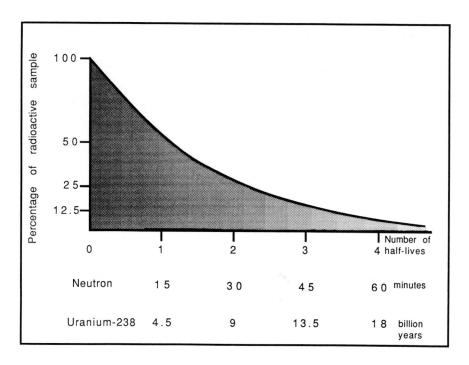

Fig. 12.12 Exponential decay. The decay pattern for all radioactive nuclei is the same, regardless of the half-life (From *New Scientist*, 11 February 1988)

appreciated. The fixed half-life of a radioactive species, which is the average time for half the atoms to emit radiation (i.e. to decay), corresponds to the fixed number of counters per square in a particular run of the game.

The rate at which a large collection of radioactive nuclei emits particles is called the *activity* of the source. One becquerel (Bq) is one radioactive disintegration per second. (The older curie is the activity of 1.02 g of radium, and thus 3.7×10^{10} Bq.)

Biological effect of radiation

The biological effect of radiation depends on the amount of energy that the radiation deposits in your body. This is basically the average energy of the ray as emitted, measured in joules, deposited per kilogram of body mass. The unit used is the gray (Gy) (corresponding to 100 rads [*radiation absorbed dose*] in older units). When β particles are emitted they are accompanied by an anti-neutrino (which is not absorbed) and hence the energies of β particles are spread over a wide range. The average energy for the β particle is about one-third of the maximum value quoted. Alpha emission has a sharply defined energy, and the energy value quoted is used directly. Note that the gray is a strict physical conversion of radiation energy to energy per kilogram of body mass.

We now have a measure of our source in terms of a rate of production of particles and a measure of biological damage in terms of an energy deposited. The difficulty is to connect these two. Consider the following. If I punch your nose, I am transmitting energy from my fist to your face. A hard punch can draw blood. The amount depends on how often I repeat the process. A soft blow has little effect, no matter how often it is repeated. In fact a whole series of soft blows may far exceed in energy, the energy of a single hard blow, but it is not absorbed in a manner that causes damage. As we saw, the same argument holds for the UV versus visible radiation in sunlight. For the same amount of energy deposited, alpha particles do approximately 20 times as much damage as beta particles or gamma rays, mainly because they are more effective at forming ions in the body. The designation *ionising* radiation comes from this property of these three emissions. Don't think from the poem that alpha rays are harmless because they are easily stopped. If you ingest or breathe in alpha emitters, the tissues stopping the alpha rays are severely damaged in the process.

The proper unit with which to assess the effects of radiation on health is the *dose equivalent*, measured in a unit called a sievert, corresponding to 100 rems (röntgen equivalent man) in older units. Its units are also joules/kg. The amount of energy absorbed (Gy) and the dose equivalent (Sv) are connected by a quality factor, which for example is 20 when comparing alpha with beta and gamma rays.

The average energy deposited by the radiation in joules/kg or grays is multiplied by the quality factor (see Table 12.5) to give the dose in sieverts.

TABLE 12.5

Type of radiation	Quality factor
X- and γ-rays and β^- particles	1.0
Thermal neutrons	3.0
Fast neutrons or protons	10
α-particles or ions heavier than $^4He^{2+}$	20

In summary, disintegrations/s (Bq) and average energy absorbed per disintegration, measured in millions of electronvolts (MeV), gives energy deposited per kg of body mass (Gy), which when modified by a quality factor gives dose equivalent (Sv). Later we shall see that because different organs have a different sensitivity to radiation, another weighting factor, W_t, is used to accommodate this and give an *effective* dose equivalent, also measured in Sv. No wonder it is hard to see what facts mean in this field.

Internal exposure to radiation

A naturally occurring radioactive element in the body, potassium-40 (^{40}K), has a half-life of over one billion (10^9) years. The average human (70 kg) contains about 140 g of potassium. Since the isotopic abundance of ^{40}K is 0.0117%, there are 16.4 mg of ^{40}K or 2.47×10^{20} atoms of ^{40}K in the average body. From the number of atoms, N, of ^{40}K and the half-life $t_{1/2}$ of 1.28×10^9 years, we can calculate the number of disintegrations per second :

$$\frac{dN}{dt} = kN = \left(\frac{0.6931}{t_{1/2}}\right) N$$

This yields 4250 Bq, i.e. 4250 radioactive disintegrations per second in the whole body (i.e. all 70 kg of it).

The average energy of the beta radiation from ^{40}K is 0.548 MeV (max. 1.35) and coupling this to the activity per kg body mass, we obtain a dose of 0.17 millisievert (mSv) per year. ^{40}K also emits high energy gamma rays, which contribute to the dose.

It is interesting to compare ^{40}K to another naturally occurring radioisotope in the body, ^{14}C. Assuming 1.6×10^4 g of carbon in the average 70 kg body, an isotopic abundance for ^{14}C of 1 in 10^{12}, and a half-life of 5730 years, we obtain an activity of 3000 Bq, which is roughly the same as for ^{40}K. However, because of the energy difference of their beta rays (average 0.0441 MeV, max. 0.156 for ^{14}C), the dose is 0.01 mSv/yr, almost 20 times less. ^{40}K is the dominant contributor to the 0.2 mSv/yr whole body exposure of beta and gamma radiation from internal sources.

The β-particle (electron) emitted by the ^{40}K has a large energy. It travels at less than the speed of light in a vacuum, but its speed in a medium such as water would be greater than the speed of light in water. Thus when say KCl is dissolved in water, the electron is slowed down instantly and this leads to a shock wave (like from a supersonic plane) and the emission of bluish white light called Cerenkov radiation. The light is far too weak even in a saturated solution of KCl to see with the naked eye, but it can be measured by very sensitive light detectors. (I once used a saturated KCl solution to provide an absolute calibration of sensitivity of photon counting instruments developed as part of a research project.) Cerenkov radiation is seen as the eerie glow that surrounds water-cooled nuclear reactor cores.

With a scintillation counter, you can use this effect to do an experiment in radioactivity without using special radioactive materials, such as to determine the percentage potassium in dietary salt substitute, where KCl replaces some or all the NaCl.

External exposure to radiation

Now let us consider some external sources of radioactivity and study the radioactive decay series, namely that from uranium-238 (^{238}U). (The other external sources are from thorium-232 and actinium-235.) A simplified radioactive decay series for ^{238}U is shown at the top of page 495.

The half-lives are in yr (years), d (days), m (minutes), and s (seconds). Eight elements in the series decay with the loss of an alpha particle (a helium nucleus) and their atomic number goes down by two while their atomic weight goes down by four. The other mode of decay is emission of a beta particle (an electron from the nucleus), which converts a neutron into a proton. This does not alter the mass appreciably, but raises the atomic number by one. This series can be shown on a graph of atomic number versus atomic mass (Fig. 12.13).

External exposure from rocks and soil comes from ^{40}K (clay), uranium, thorium and their descendants.

Cosmic rays contribute about 0.3 mSv per year at sea level. Living at 3000 m trebles the dose because there is less atmosphere to absorb the rays. Short periods

Simplified radioactive decay series for U^{238}

$$^{238}U \xrightarrow{\alpha} {}^{234}Th \xrightarrow{\beta} {}^{234}Pa \xrightarrow{\beta} {}^{234}U \xrightarrow{\alpha} {}^{230}Th$$

4.51×10^9yr 24.1 d 17 m 2.5×10^5yr 8×10^4yr

$^{230}Th \xrightarrow{\alpha}$

$$^{214}Bi \xleftarrow{\beta} {}^{214}Pb \xleftarrow{\alpha} {}^{218}Po \xleftarrow{\alpha} {}^{222}Rn \xleftarrow{\alpha} {}^{226}Ra$$

19.7 m 26.8 m 3.05 m 3.823 d 1620 yr

$^{214}Bi \xrightarrow{\beta}$

$$^{214}Po \xrightarrow{\alpha} {}^{210}Pb \xrightarrow{\beta} {}^{210}Bi \xrightarrow{\beta} {}^{210}Po \xrightarrow{\alpha} {}^{206}Pb$$

1.63×10^{-4}s 22 yr 5 d 138.4 d ∞

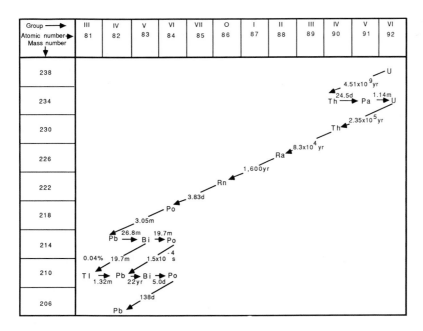

Fig. 12.13 Uranium-radium series (mass numbers $4n + 2$)

at high elevation, such as 10 trans-Atlantic plane trips, give an exposure equivalent to a whole year at sea level.

Rocks contribute radon gas and its descendants, which are usually considered as internal exposure sources. On the basis of existing measurements, and an average of one change of air per hour in buildings, the UK National Radiological Protection Board has estimated that the average annual dose of radon received in Britain is equivalent to a dose of nearly 1 mSv/yr, about half the annual exposure of members of the public from natural background sources.

Radon has a half-life of 3.8 days and decays to polonium-218, another alpha emitter with a half-life of three minutes, to give lead-214, with a half-life of 27 minutes. Two β emissions and we are at polonium-214, another short-lived alpha emitter, which produces lead-210, a β emitter of 22 years half-life. (Incidentally, this is one of the few isotopes that cannot be made artificially.) These solids attach to aerosols that can be trapped in the lungs and deliver their dose there. Many alpha emitters also concentrate in bone. The uranium-238 series contributes about 1 mSv (including 0.8 mSv from ^{222}Ra and descendants), and the thorium-232 series 0.33 mSv (including 0.17 mSv from ^{220}Ra and descendants) per year.

Problems occur if a granite used in home-building has been crushed to a gravel because the fracturing allows an easier escape for radon. In the case of housing built on rock fill, the deciding factor is the depth and permeability of the soil. Normally a metre or so of soil is sufficient to slow down the gas diffusion so that the radon has decayed to solid descendants (which are trapped) before escaping. Radon in the house wall construction materials is generally effectively trapped. The usual problem, particularly in cold climates, is that the urge to conserve heating, inevitably involves restricting ventilation. This leads to a build up of radon and its descendants.

Removal of uranium ore in mining *in*creases the radioactivity in the local environment, because of the break-up of the radioactive ores and release of radon. Diffusion of radon from the ground decreases when the ground is wet as the surface becomes sealed. Radon is quite soluble in water.

Radioactivity in food

Radioactivity in food comes mainly from the alpha emitter polonium-210. Other contributions are from the beta emission from lead-210 and bismuth-210. Relatively high concentrations of polonium are found in seafoods, such as in the muscles of fish and in molluscs. Because reindeer and caribou eat lichen and these accumulate ^{210}Pb and ^{210}Po, the peoples of the sub-Arctic that live off these animals take in 140 Bq and 1400 Bq per year respectively from these isotopes, compared to a normal 40. Whereas most alpha emitters such as lead accumulate in the bone minerals, polonium is the exception and distributes in the soft tissues after intake, settling in the bones after decay to lead. For the average citizen the yearly radiation intake from food amounts to about 0.37 mSv per year.

Let us carry out a typical calculation on polonium levels for non-smokers (smokers receive extra doses in their lungs), in areas of normal dietary intake. Because different organs in the body have different sensitivities and masses, the dose they receive is weighted with a factor W_t to compute an *effective dose equivalent* for each organ. This weighting factor is the first entry in Table 12.6. The next entry gives the average activity of ^{210}Po found in each organ. Then comes the absorbed dose, which is the energy deposited by the radiation. The final column is the effective dose equivalent. This is the absorbed dose in grays converted to the effective dose in sieverts by a factor of $Q=20$ because polonium is an alpha emitter. Then this is multiplied by the weighting factor W_t for each organ given in column 1, to give the effective dose *equivalent*, also in sieverts in the last column. That is the figure that is usually quoted.

TABLE 12.6

Tissue	W_t	^{210}Po Bq/kg	Absorbed dose μGy/yr	Effective Dose Equiv.μSv
Gonads	0.25	0.2	5.4	27
Breast	0.15	0.2	5.4	16
Red bone marrow	0.12	0.11	5.1	12
Lung	0.12	0.1	2.7	6
Thyroid	0.03	0.2	5.4	3
Bone lining	0.06	—	36	43
Remainder	0.24	0.2	5.4	26
Σ (sum)	1.00			133 μSv

Thus the dose from polonium-210, (0.13 mSv/yr), is about the same as we calculated for potassium-40 (0.17 mSv/yr).

When exposures are given for medical X-rays of a particular part of the body, it is the effective dose *equivalent* to the whole body that is quoted. The actual dose of the X-ray is higher by the inverse of the weighting factors.

A single chest X-ray deposits an effective dose *equivalent* of about 0.1 mSv. Among the higher exposures, X-rays of the small intestine, for example, vary from 3 mSv for radiography to 8 mSv for fluoroscopy per X-ray. For medical reasons, X-ray exposure is thus characterised by high dose rates and uneven distribution of dose in the population. They are by far the largest contribution from non-natural sources. However, when averaged over the whole world population, medical exposure contributes an effective dose equivalent of 0.4 mSv/yr.

This highlights how much individual exposure can vary. It can be very high for a citizen in the developed world having easy access to diagnostic X-rays and very low for a citizen in the Third World far from medical services. The low figure for the average 'dilutes' the high figure for the individual.

The same argument applies to an accident such as at Chernobyl, which for the following 12 months contributed only 0.03 mSv per person to the average world exposure, about the same as one person would obtain from a trans-Atlantic flight. However for people downwind of the accident or in the Third World eating cheap (contaminated) European food exports, the situation is different.

Although nuclear explosions in the atmosphere have diminished from the intensity of the mid-1950s and early 1960s, the fall-out effects continue to expose the world population. It is estimated that the exposure resulting from all these tests until 1980 is equivalent (for the present world population) to being exposed to the natural background for an additional four years. Each new atmospheric test commits future generations to exposure. The most significant exposure is through ^{14}C, ^{90}Sr and ^{137}Cs. The total committed dose varies from 3.1 in the Southern Hemisphere to 4.5 mSv in the Northern.

Risk from occupational exposure

For radiation workers, the whole-body exposure limit is set at 50 mSv/yr, while for the citizen it is set so that the exposure over a 70-year lifetime does not exceed, on average, 1 mSv/yr. Thus for the citizen, this additional exposure from industrial sources is equal to about one half of what is received from natural sources (1 mSv from radon descendants and 1 mSv other), not including medical exposure. It should be noted that the International Commission on Radiological Protection (ICRP) has insisted since 1977 that these levels should no longer be regarded as the boundary of acceptable practice, but rather the point at which practices become distinctly unacceptable. In 1977 the emphasis shifted to the so-called 'alara' principle, (*as low as reasonably achievable*, economic and social factors being taken into account).

Radiation effects are called 'somatic' if they become manifest in the exposed individuals themselves, and 'hereditary' (or genetic) if they affect the descendants of the exposed people. These effects are now categorised in two ways[5].

1. Non-stochastic: where the severity of the effect increases with the dose—e.g. skin reddening, cataract of the lenses of the eyes, cell depletion in bone marrow, and impaired fertility of the gonad cells.
2. Stochastic: where the *probability* of the effect, rather than its severity, is proportional to the dose. This generally refers to the induction of later cancer or genetic damage.

Crudely, the difference is determined by whether cells are killed and therefore have no long-term influence, or whether they are only damaged, in which case they have a latent ability to cause harm later on. Thus, ironically, the very heavily exposed population of Hiroshima and Nagasaki often showed less than expected cancers because of the high death rate in their cells.

The stochastic effect is of the same type as, say, between cigarette smoking and lung cancer, or the various risk factors associated with heart disease. You do not see a direct cause and effect relation in each individual but a statistical correlation in an exposed population.

As far as stochastic effects such as cancer are concerned, the critical factor is the radiation dose to cancer conversion rate. This conversion was set on data based on exposure to radiation of the fringe population of Hiroshima and Nagasaki. For technical reasons, it is now believed that the level to which these people had been exposed was lower by a factor of at least two than had been assumed and that, coupled with a longer period in which to study cancer rates, the risks are correspondingly higher by factors that range up to 15 for some organs, with an average of about 3 overall. No changes in the international standards have yet been made.

The international occupational level is based on an assumed acceptable level of fatalities in comparable industries, *namely 100 fatalities per year per million workers*. This magical figure is heavily quoted by occupational health and safety authorities but it is hard to find a really quantitative basis for it.

The data from Japan provided the figure of 12.5 excess delayed cancers per mSv radiation exposure, and 2.5 excess per mSv for genetic effects expressed in the next two generations of offspring. Thus one million people exposed to an average of 1 mSv would be expected to develop 12.5 excess cancers per year; one million workers exposed to 50 mSv would develop 625 excess cancers per year. This figure, already six times higher in comparison to other safe industries, now appears to be an underestimate by a factor of 2 as well. Obviously one should aim in practice for average exposure levels of about one-tenth of the legal maximum in order to be comparable to other safe industries, and this is generally achieved.

For different types of exposure there are comprehensive tables[6] where you can look up the value of dose per intake of radioactive material, i.e. Sv/Bq.

Radon

Risk factors for radon derive from the modelling of doses to the lung and radiological models for the radiation damage. They are also derived from epidemiological studies of uranium miners who were heavily exposed and who contracted lung cancer. The matter is very complex indeed. A consensus appears to be emerging that for radon 20 Bq/m^3 = 1 mSv. At 400 Bq/m^3 = 20 mSv, the ICRP suggests that a householder should contemplate corrective action (such as good ventilation, venting of basements to the outside air etc.). It has been suggested in the USA that the levels in homes be recorded with other information on the title deeds. The major entry of radon is through the underground basements. The UK national average level is 22 Bq, but as we said, averages lie. In Devon and Cornwall, a study of 400 houses found levels 15 times the national average, and levels of 1000 Bq have been recorded. The UK uses lower radon-to-cancer conversion factors. On their figures only 800 to 900 people a year in the UK die from the effects of radon (a striking 7000, if the USA conversions are used).

In the USA, a survey of a million homes is about to begin. Australia has completed a year-long survey of 3600 volunteer homes chosen randomly from the electoral roll in which radon monitors have been set. The results of preliminary measurements on very small samples are shown in Table 12.7.

TABLE 12.7
Average radon concentration for Australian homes

Location	No. of homes	Month	Radon concentration Bq/m^3	
			Geom. mean	Interquartile range
Melbourne	47	May	53	39–70
Melbourne	20	July	19	6–24
Sydney	45	July	21	5–45
Perth	45	May	15	7–32
Jabiru	43	May	19	9–40
Armidale	55	August	10	4–31

Jabiru is a town near the Ranger uranium mine in the Kakadu National Park in the Northern Territory, and Armidale is an inland NSW town in an area of granitic rock containing uranium and thorium in higher than average levels. The highest level measured in a home was 395 Bq/m^3. The apparatus used to collect the radon samples consisted of a container with activated carbon, which was left to allow the radionuclides to equilibrate and then measured. (See Fig. 12.14.)

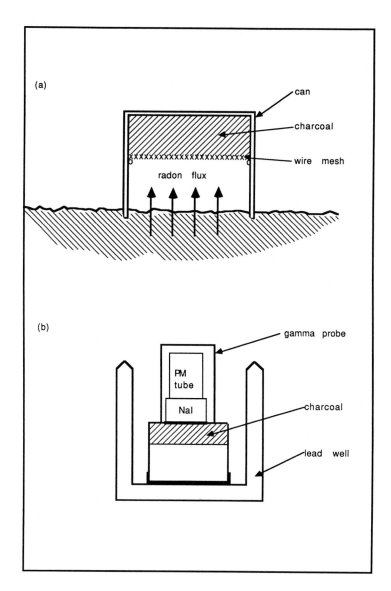

Fig. 12.14   Radon collection. (From S. B. Solomon, J. R. Peggie and K. N. Wise, 'Radon levels in Australian homes', IRPA Congress, April 1988)

While one might have thought that mining and processing uranium ores constituted the major industrial hazard in Australia, it appears that beneficiation of beach sand for the separation of rare earths is the real problem area. The decay of thorium-232 provides a series of hard gamma ray emitters with high energies. As these operations tend to be dusty, quite high levels of exposure are probable. The thorium decay series is given below in simplified form and in Figure 12.15.

Simplified radioactive decay series for Th²³²

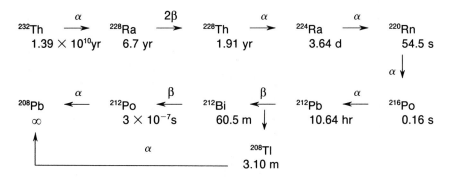

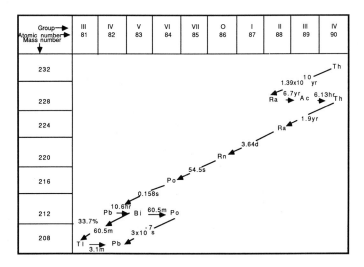

Fig. 12.15   Thorium series (mass numbers 4n)

Radiation-emitting consumer products[7]

The energy emitted during the radioactive decay of radium-226, promethium-147 and tritium (^3H) can be converted into light by use of a scintillator. Cancer of the lips was common amongst women painting dials for military instruments and watches during the Second World War, because of their habit of sharpening the paint brush between their lips. There are also long term dangers to users. Today tritium is used almost exclusively, such as gaseous tritium light sources for liquid crystal displays. Television screens once contributed about 0.005 mSv from soft x-rays, but modern colour sets under normal operation and servicing have negligible emission.

Starters for fluorescent lamps and trigger tubes in electrical appliances incorporate radioisotopes for faster and more reliable operation. These are kyrpton-85, promethium-147, and thorium-232. The level of activity is appreciable, but not a hazard except in the case of breakages or careless disposal. In some countries anti-static devices use polonium-210 to ionise the air. The only significant hazard is from the very small gamma component emitted. Smoke detectors usually contain americium-241. These devices have a useful life of about 10 years and their safe disposal is important. Uranium is used as a pigment in ceramics and glassware, while thorium is used in the incandescent mantles of kerosene lamps, and in some optical products. The beta emission from the decay products can be hazardous under special circumstances, e.g. from the optical lenses containing high levels of thorium. Uranium added to the porcelain used in dentistry to simulate the natural fluorescence of teeth could give an undesirably high dose to the gums.

Art forgery detection

In 1937 a Dr Abraham Bredius, a noted art expert, announced that he had discovered *the* masterpiece of Johannes Vermeer (1632–75), *Christ and His disciples at Emmaus*. Vermeer was perhaps one of the finest of the seventeenth century Dutch genre painters, whose work was wrongly attributed until 1866. Because there were never more than 36 authentic Vermeers catalogued, they are highly prized. The owner, a certain Hans van Meegeren, netted $280 520 from Boyman's Museum in the Netherlands for a painting he had painted himself. It was so well done that the museum even had to restore it before the official unveiling. A gifted artist in his own right, but with little recognition because of his old fashioned style, van Meegeren accepted that consumers bought on brand name and that art connoisseurs were no exception. He continued to produce what the public wanted by forgery and amassed the tidy sum of three million dollars. The buck(s) stopped in 1945 when the art collection of Hermann Goering was discovered in a salt mine at Alt-Ausée in Austria. Amongst the treasures was a Vermeer, *Woman taken in adultery*.

The Dutch rapidly traced the sale to van Meegeren, who was tried as a Nazi collaborator. This carried the death penalty at the time. He changed his tune rapidly and pleaded that his Vermeers were forged and that therefore he had thwarted the Nazis' art acquisition program, by selling them his own works. However, the experts disputed this (there were after all considerable investments threatened, not to mention reputations), and the art world closed ranks and rejected van Meegeren's claims as a reckless attempt to save his neck.

The chemical examinations at the time were ambiguous and a desperate van Meegeren set out to prove his skills by painting under police supervision, another 'Vermeer', *Jesus amongst the doctors*. This allowed him to escape execution but, many did not accept that the *Emmaus*, among other works, was forged. The controversy continued after his death and was not finally resolved until 1968, when radiochemical analysis of the lead paint was undertaken.

The manufacture of lead pigment involves the roasting of galena, lead sulfide ore. The ore always contains ^{238}U together with its descendants, all in equilibrium. (It is not strictly an equilibrium situation, although that term is used. It is a steady state.) In this situation, as fast as isotopes in the chain are emitting particles and converting to daughter products, so their concentrations are being replenished from further up the chain. A good analogy is to consider a series of dams of different sizes (amounts of isotopes). The time taken to half-empty an isolated dam is constant and fixed. This is its characteristic half-life. It can vary from a trickle to a flood depending on the nature of the dam (milliseconds to millennia). However the amount of water leaving an isolated dam at any time is not constant, but is proportional to the amount of water currently in that dam. As the amount in the dam drops, so does the outflow, giving the characteristic falling exponential decay.

Because our dams are linked, the outflowing water from one dam flows into the dam(s) below, giving a connected stream of water flow. Some dams will slowly fill and others empty, until a steady state is reached.

Now look at the radioactive series for ^{238}U again, at the point at which ^{226}Ra occurs. It has a half-life of decay of 1600 years, and its daughter ^{210}Pb has a half-life of only 22 years. Radium and lead behave quite differently on refining lead ore. While ^{210}Pb is carried through with the normal lead, most of the ^{226}Ra stays in the waste slag products. The ^{210}Pb, now without most of its parent, decays away with the 22-year half-life, and reaches a low level when it re-equilibrates with the remaining ^{226}Ra after about 200 years. The amount of ^{226}Ra is measured by its alpha emission (4.78 MeV), while the amount of ^{210}Pb is measured through the alpha emission of its descendant (^{210}Po), because the β emission of ^{210}Pb is too weak. The count rates for ^{210}Po and ^{226}Ra in the *Emmaus* painting were found to be 142 and 13.3 Bq/kg lead respectively. The radium level is low, indicating proper purification of the lead had been carried out. However to explain the current polonium level, the level back in the mid-seventeenth century (about 13 half-lives back) would have had to have been $142 \times 2^{13} = 1.2 \times 10^6$ Bq, which is unrealistically high. Using our analogy, if the upstream dam (radium) had really been emptied in the seventeenth century, then there is just too much water today in the (lead/polonium) dam below.

There are many other fascinating chemical aspects of art examination and you should consult some of the references for further reading.

For a discussion of irradiation used in food preservation, see 'Food additives' in Chapter 11.

EXPERIMENT 12.1

Tarnishing and corrosion

Equipment
Alfoil sheet, tarnished silver, sodium bicarbonate (baking soda), aluminium soft-drink cans.

Procedure—Part A
Air contains amounts of sulfur compounds as a result of natural and human activities. These compounds react with many materials, including silver, in which case the product is black silver sulfide, which adheres strongly to the silver surface. (This is particularly noticeable on silver spoons used for eating eggs.) The sulfide can be polished off, with resulting loss of silver, or dissolved off by solutions containing cyanide, but the best method uses an oxidation-reduction reaction to reverse the corrosion process.

Put some hot water in an enamel or pyrex pan and add a few teaspoons of sodium bicarbonate ($NaHCO_3$, baking soda) to the water. Wrap the silver to be cleaned in aluminium foil. Leave a small opening in the foil and put the foil and silver in the solution. Ensure the foil is completely covered with solution and that no air bubbles are trapped inside the foil (this ensures proper cleaning). All of the silver utensils must touch the foil at some point. The silver will be cleaned in about 1–5 hours, depending on the degree of tarnishing.

The redox processes which occur are:

$$Al(s) \rightarrow Al^{3+}(aq) + 3\varepsilon^-$$
$$2\varepsilon^- + Ag_2S(s) \rightarrow 2Ag(s) + S^{2-}(aq)$$

The overall reaction is:

$$3Ag_2S(s) + 2Al(s) \rightarrow 2Al^{3+}(aq) + 3S^{2-}(aq) + 6Ag(s)$$

Procedure—Part B

A common method used for packaging food and beverages is to use tin coated steel or aluminium cans. These forms of packaging are chosen because both tin and aluminium are readily oxidised, and a protective oxide coating develops over the metal surface.

Take two aluminium soft-drink cans and cut the tops off them. Fill both cans with water and to one can add a few teaspoons of sodium bicarbonate and to the other add a few teaspoons of acetic acid. Allow the cans to stand for a few days.

You will notice that the can containing the acetic acid solution shows no evidence of corrosion, but the can containing sodium bicarbonate has corroded. The aluminium oxide formed on the surface of the aluminium is stable in the acid solution, but in the basic solution formed by the sodium bicarbonate, the oxide dissolves. The aluminium surface is therefore unprotected in basic solution.

This explains why the sodium bicarbonate was added to the water in Part A— to keep the aluminium surface clean and free from the aluminium oxide. It also explains why soft-drinks can be stored in aluminium cans—soft drink solutions are acidic.

Note

Try the experiment in Part B using tin-coated steel. Be sure you choose a can which has a tin coating. Some cans are coated with a protective polymer coating instead of tin. Also, try a set of experiments in which you have scratched the tin surface to reveal the underlying steel.

EXPERIMENT 12.2

Exchanging metals

Equipment

Two clear glass containers, copper sulfate, clean iron nails, zinc metal from a torch battery.

Procedure

Make up 50 mL solutions of copper sulfate in water in two glass containers. Add sufficient copper sulfate to each till the colour is clearly blue (~1 teaspoon heaped).

To one container add some shiny iron nails and to the other add some small shiny pieces of zinc metal. Notice the change in colour of the solution and any precipitate that forms when the solutions are allowed to stand for several hours.

When the reactions are complete, add some iron nails to the solution containing the zinc and some shiny zinc to the solution containing the iron nails. Notice any further reactions that take place.

Keep the solutions you have prepared, since you can use them in other experiments. Be sure to label them carefully and store them in a safe place.

Reactions

$$CuSO_4 + Zn \rightarrow ZnSO_4 + Cu\downarrow$$
$$CuSO_4 + Fe \rightarrow FeSO_4 + Cu\downarrow$$
$$FeSO_4 + Zn \rightarrow ZnSO_4 + Fe\downarrow$$
$$ZnSO_4 + Fe \rightarrow \text{no reaction}$$

Notes

1. Try other metals in the place of iron or zinc (for example, lead, aluminium, nickel).
2. See if any reactions occur between the metals and ammonium sulfate solution. (Ammonium sulfate is a garden fertiliser.)
3. Copper sulfate should be available from some hardware shops or nurseries. If you cannot buy it there, it can be ordered from chemical supply houses.

CAUTION

Some solutions of salts are poisonous if they are swallowed. You should ensure that you carry out your experiments in a safe location, and either dispose of the solutions when your experiments are complete, or label them and store them in a safe place.

REFERENCES

1. G.W.A. Fowles, *The leaching of cadmium from ABS samples containing CdS and CdSe pigments.* Department of Chemistry, University of Reading, UK, 1975.
2. *Criteria for a recommended standard—occupational exposure to inorganic mercury.* HSM 73-11024, US National Institute for Occupational Health, 1973, 9, 83.
3. M. L. Sanders and R. R. Beckett, 'The mercury–water system : a deviation from Raoult's law'. *J.Chem.Ed.* **52**(2), 1975, 117.
4. This section is based on R. Beckmann, 'Oysters and zinc—the Derwent revisited'. *Ecos* **50**, Summer 1986-87.

5.   K.H. Lokan, *ICRP—history and developments*, and private communication, Australian Radiation Laboratory, Melbourne, Victoria, 1988.
6.   'Limits of intakes of radionuclides by workers'. *Annals of the ICRP*, ICRP publication 30 & supplements, Pergamon Press, Oxford, 1978.
7.   *Ionizing radiation: sources and biological effects*. UN Scientific Committee on the Effects of Atomic Radiation, 1982 Report to the General Assembly, 1985.

BIBLIOGRAPHY

General

Jaworski, J. F. (ed.). *Data sheets on selected toxic elements*. National Research Council Canada, 1982.

Neuhaus, J. W. G. 'Trace metals in foods'. *Food Tech. Aust.* **27**, 1975, 195. Used as a major source. *Recommended health based limits in occupational exposure to heavy metals*, WHO, Geneva, 1980.

Tucker, A. *The toxic metals*. Earth Island Ltd., 1972. This book provided the inspiration for, and some of the contents of, this chapter.

Underwood, E. J. *Trace elements in human and animal nutrition*, 4th edn. Academic Press, New York, 1977.

Waldron, H. A. (ed.). *Metals in the environment*. Academic Press, New York, 1980.

Cadmium

Carroll, D. M. and Halpin, M. K. 'The light sensitivity of cadmium release from glazed ceramic tableware'. *Nature* **247**, 1974, 197.

'Checking out a pollutant', *Rural Research* **79**, 1973, 19.

David, D. J. and Williams, C. H. 'Heavy metal contents of soils and plants adjacent to the Hume Highway', *Aust. J. Exp. Agric. and Animal Husb.*, **15**, 1975, 414; 'The accumulation in soil of cadmium residues from phosphate fertilizers and their effect on the cadmium content of plants', *Soil Science* **121**(2), 1976, 86; 'Some effects of the distribution of cadmium and phosphate in the root zone on the cadmium content of plants', *Aust. J. Soil Res.* **15**, 1977, 59.

IARC Monograph. *The evaluation of carcinogenic risk of chemicals to man*. Vol. II, *Cadmium*, [etc.]. WHO, Lyon, 1976.

Miller, G. J., Wylie, M. J. and McKeown, D. 'Cadmium exposure and renal accumulation in an Australian urban population'. *Med. J. Aust.* **1**, 1976, 20.

Ministry of Agriculture, Fisheries and Food (UK). *Survey of cadmium in food*. HMSO, London, 1973.

Page, A. L. and Bingham, F. T. 'Cadmium residues in the environment'. *Residue Reviews* **48**, 1973, 1.

Rosman, K. J. R., Hosie, D. J. and de Laeter, J. R. 'The cadmium content of drinking water in Western Australia'. *Search* **8**, 1977, 85.

Thrower, S. J. and Eustace, I. J. 'Heavy metal accumulation in oysters grown in Tasmanian waters'. *Food Tech. Aust.*, 1973, 546.

'Toxic metals in Tasmanian rivers'. *Ecos* **1**, 1974, 3.

Underwood, E. J. *Trace elements in human and animal nutrition*, 3rd edn. Academic Press, New York, 1971.

Waldron, H. A. 'Health standards for heavy metals'. *Chem. Brit.* **11**, 1975, 354.

'What's in Port Pirie's vegetables'. *Ecos* **12**, 1977, 17.

Copper

Health Commission of NSW. *Annual report 1975–76.* Division of Analytical Laboratories, Food and Water Branches.

Walker, W. R. 'An investigation of the therapeutic value of the "copper bracelet": dermal assimilation of copper in arthritic/rheumatoid conditions', *Agents and Actions* **6**(4), 1976, 454 (Birkhauser Verlag, Basel); 'The copper bracelet story: some aspects of the role of copper in inflammatory disease', *Chem. Aust.* **4**(10), 1977, 247.

Lead

Bloom, H. and Smythe, L. E. 'Environmental lead and its control in Australia'. *Search* **14**(11/12), 1984.

Dartnell, P. L. 'Lead in petrol, 1: Energy conservation'. *Chem. Brit.* **16**(6), 1980, 308.

Davies, D. R. L. *Lead in petrol: towards a cost benefit assessment,* CRES, Australian National University, Canberra, 1980.

Fell, G. S. 'Metals in the environment: health effects'. *Chem. Brit.* **16**(6), 1980, 323.

Harrison, R. M. and Loxen, D. P. H. 'Metals in the environment: chemistry', *Chem. Brit.* **16**(6), 1980, 316.

Health and environmental lead in Australia, Australian Academy of Science, Canberra, 1981.

Lead glazes for dinnerware, International Lead Zinc Research Organization Inc. (UK).

Proceedings of a conference on ambient lead and health. Division of Occupational Health and Radiation Control, Health Commission of New South Wales, 1974.

Turner, D. 'Lead in petrol, 2: Environmental health', *Chem. Brit.* **16**(6), 1980, 312.

Mercury

Methyl mercury in fish. NH & MRC, Australian Department of Health, AGPS Canberra, 1972.

'Mercury'. *Which,* UK Consumers' Association, June 1974, 180.

Report on mercury in fish and fish products, Department of Primary Industry, AGPS Canberra, 1980.

Tin

Smith, P. and Smith, L. 'Organo-tin compounds and applications'. *Chem. Brit.* **11**, 1975, 208.

Zinc

Beckmann, R. 'Oysters and zinc—the Derwent revisited', *Ecos* **50**, Summer 1986–87.

'Zinc deficiency and Aboriginal health,' *Ecos* **40**, 1984, 14.

Radiation

Bell, A., 'A new approach to entombing nuclear waste'. *Ecos,* Winter 1987.

Bodner, G. M. and Rhea, T. A. 'Natural sources of ionising radiation'. *J. Chem. Ed.* **61**(8), 1984, 687.

Feynmann, R. P., Leighton, R. B., and Sands, M. *The Feynmann lectures,* Addison-Wesley **1** (51), 1982.

Flemming, S. J. *Authenticity in art: the scientific detection of forgery.* The Institute of Physics, London 1975.

Fox, R. H. and Schuetzman, H. I. 'The infrared identification of microscopic samples of man-made fibres'. *J. Forensic Sciences* **1**, 1968, 397.

Gove, R. C., Hannah, R. W., Patlacini, S. C., and Porro, T. J. (Perkin Elmer Corp.). 'Infrared and ultraviolet spectra of seventy-six pesticides'. *J.A.O.A.C.* **54**, 1971, 1040.

Haslam, J., Willis, H. A. Squirrell, D. C. M. *The identification and analysis of plastics*, 2nd edn, and Iliffe Books, London, 1972, reprinted 1982, Heyden.

Hausdoff, H. H. 'Infrared applications to the analysis of cosmetics and essential oils'. *J. Cosmetic Chemists* **4**, 1953, 251.

Krause, A., Lange, A., and Eyrin, M. *Plastics analysis guide: chemical and instrumental methods.* Hanser Publ., Munich, 1983.

Murphy, J. E. and Schwemer, W. C., 'Infrared analysis of emulsion polishes'. *Anal. Chem.* **30**, 1958, 116.

Pearce, F. 'A deadly gas under the floorboards'. *New Scientist*, 5 February, 1987.

Perkin Elmer Infrared Applications Studies: *Polymer plastics industry*, 3 pp., 1968; *Paints and coatings* (reflectance methods) 3 pp., 1968; *Packaging and containers* (plastic films) 3 pp., 1968; *Waxes*, 3 pp. 1968; *Analysis of diesel and petroleum base lubricants*, 3 pp; *Drugs* (a very well prepared publication with a flow diagram system of identification of about 60 drugs), 34 pp., 1972.

Pullen, B. P. 'Cerenkov counting of ^{40}K in KCl using a liquid scintillation spectrometer'. *J. Chem. Ed.,* **63**(11), 1986.

Recommended radiation protection standards for individuals exposed to ionizing radiation, Commonwealth Department of Health, AGPS , Canberra, 1981.

Rogers, F. E. 'Chemistry in art: radiochemistry and forgery'. *J. Chem. Educ.* **49**(6), 1972, 419.

Rotblat, J. 'A tale of two cities: Hiroshima and Nagasaki'. *New Scientist*, 7 January, 1988.

Sutton, C. 'Radioactivity: inside science'. *New Scientist*, 11 February, 1988 (special).

Wilke, T. 'Radiation? It's as plain as the nose on your face'. *New Scientist*, 12 June, 1986.

Experiments in consumer chemistry

13

INTRODUCTION

In April 1937 Consumers' Research Inc., Washington NJ, produced a 40-page booklet *Consumers' test manual*, 'Comprising simple and readily applied tests of common household articles and supplies suitable for use by students of chemistry, physics, general science, household arts, and by consumers generally'.

The introductory note starts off with the following statement:

> It was expected that the task of compiling this *Test Manual* would be a fairly easy one and that it would be completed within a few months after work upon it was begun, but the amount of material available was so great and the necessity of careful selection and checking of diverse and conflicting methods so obvious as the work proceeded, that a far more ambitious job was done than was originally intended.

The price charged was a nominal 25 cents in order to make it generally available. Unfortunately the foreshadowed additional sections on 'the efficiency and safety of electrical appliances, performance of radio sets, strength and wear resistance of textiles and paper, accuracy of time pieces, efficiency of flashlights, cutlery, can openers, kerosene, gas and electric stoves' never appeared.

The manual gives a method for determining the water content of butter by distillation with xylene, on which the Dean and Starke experiment described in Experiment 13.1 of this chapter is based. It also gives the wool test for coal-tar dyes, the turmeric test for borax (and vice versa), and tests for starch in face powders, sulfides in depilatories, lead in petrol, lead in paints, water in oil paints (!), gas efficiency of baking powder, ammonia in household ammonia and lead in drinking water (dithizone test). There is a strong section on the analysis of soap, starting with the water content, and followed by carbonates, silicates, rosin, free alkali, sugar and starch. The quality of eggs is checked in detail. An interesting and simple test (the Preece test) for the uniformity of coating of zinc on galvanised sheet metal and wire fencing is given. One of the tests for bleached flour unfortunately involves a dangerous reagent, but this could be modified. There is an appendix on methods for producing standard solutions.

In 1939 Professor Römpp produced his *Chemie des Alltags (Chemistry of Everyday)*, which has gone through 22 editions and 127 000 copies (1975) and is now edited by Professor Raaf. Its 300 pages make little allowance for dilettantes and, although occasionally lighthearted (by German standards), it is a serious textbook organised in an alphabetical format ('Von Alkohol bis Zündholz'). It had an accompanying volume *Chemische Experimente die gelingen (Chemical Experiments Which Work)* expressing the same sentiment as Consumers' Research on the accuracy of the literature.

PRACTICAL SESSION

The aim of this chapter is to introduce you to the ways chemical analyses are carried out. Modern analytical techniques involve sophisticated machines as well as the traditional boiling flasks and test tubes. A tremendous amount of care, skill and patience is required to obtain accurate measurements of low concentrations of materials. Often the search for one substance will be inhibited by interference caused by the presence of another. I have chosen a number of (atypical) examples. They are chosen to be relevant and interesting as well as *safe* and *simple* to perform in the laboratory.

A note on units

The SI (Système International d'Unités) symbol for concentration of a substance B is c_B or [B], with units mol m^{-3}. Alternatively, concentrations in terms of molality, m_B in units of mol kg^{-1} are also acceptable. While these are accepted units for thermodynamic calculations, the former is not convenient for solutions. For most practical purposes, the latter is similar for aqueous solutions to the unit that is used almost exclusively, namely the molarity, M, with units of mol/L^{-1}. (One litre of water has a mass of one kilogram at 20°C. Once you make up a solution or the temperature varies, this equivalence disappears.)

The older term of normality, N, is based on a concept called 'equivalents' which was discontinued 30 odd years ago. While a N/10 solution of hydrochloric acid is also M/10; a N/10 solution of sulfuric acid is M/20 because it has two replaceable hydrogens when reacting as an acid. Normality is thus useful when specific titrations are involved and still has industrial usage (see Experiment 13.16 on wine analysis, item 6, Tannin content). Normality is confusing when the reaction type changes (from, say, acid–base to oxidation–reduction), and should be discouraged.

Care

Although these experiments have all been tested by the author, no responsibility is accepted for their safety. Unlike most of the simple experiments scattered throughout the other chapters, *these experiments must be carried out in a properly equipped laboratory under professional supervision.*

Safety

A chemical laboratory can be a hazardous place. The trained chemist learns to assess and deal with these hazards almost by instinct after years of contact. It is difficult, therefore, to condense advice on safety to a few lines. What follows is thus rather sketchy.

1.   You must wear adequate protective clothing—preferably a laboratory coat or old clothes—and 'sensible' shoes which cover the top of the feet.

2. You must *never* eat or smoke in the laboratory, and make sure you wash your hands thoroughly after working in the lab.
3. Heat solutions only electrically or via a hot-water bath, except for aqueous (water) solutions, for which a bunsen burner (flame) can be used. Practically all solvents are flammable (will burn).
4. Chemicals are manipulated with spatulas (nickel spoon-like objects). Fingers must *never* be allowed to come in direct contact with chemicals. Any spillage should be washed off immediately.
5. Carefully label all containers used for holding chemicals and solutions.

Fig. 13.1

Reports

You should make notes of what you are doing and, on the conclusion of your experiment, write a short report. This will be useful to you and to others attempting to repeat similar work.

EXPERIMENT 13.1

Determination of water (and fat) content using a Dean and Starke apparatus

Equipment

Dean and Starke apparatus	Heating mantle with simmerstat
Fume hood	Toluene or cyclohexane
Water bath	Weighted vials and beakers
250 mL round-bottomed flask	Vacuum pump

Principle

The Dean and Starke apparatus is shown in Figure 13.2. It is used to determine the amount of water present in such materials as detergents, oils, meat, cheese etc. It can also be used to determine the amount of fat or fat-like material present in a product.

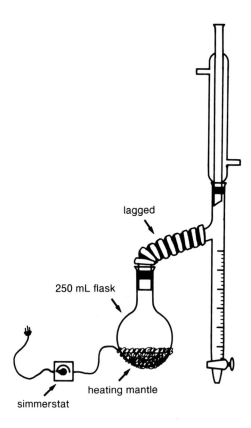

Fig. 13.2   Dean and Starke apparatus

The principle of the experiment is that, in a mixture of toluene and water, boiling occurs at 85°C and both water and toluene are given off in the vapour. On condensing, two layers of liquid are formed: a bottom layer of water (with 0.06% toluene dissolved) and a top layer of toluene (with 0.05% water dissolved). The relative volumes are 18% water and 82% toluene. The excess toluene flows back into the flask and distils back over with more water. (The sample size should be such as to provide no more water than will fit into the calibrated arm.)

The experiment should be carried out in a fume hood because toluene vapour can be harmful over a period of time. If a hood is not available, cyclohexane should be used. It is less efficient (carrying over only 8.5% water) but also less harmful.

However, if ethyl alcohol is present (as in some concentrated detergents), use *n*-hexane so that the bottom layer contains mainly water and ethyl alcohol and only a little of the carrier hydrocarbon. The efficiency of *n*-hexane is low (3.8% water) and so the process is much slower. In any event the concentrates are hygroscopic and should be measured immediately on opening or stored in sealed containers until they are tested.

If you have any doubts about the suitability of the method for a product, it is best to make up a standard sample of known composition.

TABLE 13.1
Fat and water analyses of ground meat[a]

Description[a]	Analyses	% fat	% water
Topside mince	2	16.0	66.5
	2	16.4	64.9
	2	19.1	63.1
Hamburger mince	4	8.0	71.8
	4	12.6	66.6
	2	13.2	67.7
	2	13.3	69.7
	2	13.5	67.4
	2	14.5	68.2
	2	20.2	63.9
Sausage mince	2	16.1	54.1
	2	17.5	50.9
	2	19.3	51.7
	4	23.7	55.1
	2	24.2	53.3
	2	26.6	47.6
	3	27.6	50.5
	2	28.5	47.8
	4	33.3	50.0

[a] Listed in order of most expensive (topside) to least expensive (sausage).
Source: Canberra Consumer, **50**, 1975, 9.

Method

1. Detergents

10 g of liquid dishwashing detergent is distilled with 90 mL of toluene in a Dean and Starke apparatus for about 20 minutes or until no more water is seen to be distilling over. (The distillation temperature is 110.6°C at this point.) The heating should initially be slow to avoid excessive frothing and carry-over of detergent. The amount of water distilled over (as the bottom layer) is measured approximately as the volume in the calibrated arm but is best collected in a pre-weighed vial and then weighed.

After the toluene has been poured out of the flask, any solid residue remaining should be noted. A white precipitate suggests a salt—common salt (NaCl), Na_2SO_4 etc. A hard white compound formed around the joints in the condenser suggests the presence of urea in the detergent.

2. Minced sausage meat (or cheese)

20 g of meat (or cheese) is distilled with 80 mL of toluene as in (1). The toluene solution remaining in the flask is carefully poured into a pre-weighed beaker and

several additions of toluene are made to the meat, the flask shaken up to extract the last traces of fat and this toluene is also poured into the beaker. The toluene is evaporated in a fume hood over a water bath and under a water pump vacuum until *all* the toluene has been removed. The beaker is reweighed to determine the weight of fat. (The approved method of analysis of fat content involves the use of diethylether, which is very volatile and flammable.)

Discussion

If powdered laundry detergents are being examined then there will be water of hydration tied up in some of the components (e.g. tripolyphosphate). A higher boiling solvent is needed to ensure the hydrate is broken down (xylene or petroleum ether, bp 140°C).

Urea is often used as a solubilising agent to keep fairly insoluble material in solution. This is true for some of the detergent concentrates. In dilute detergents it can be used as a thickening agent. Because it is alcohol-soluble, *initial* tests on a detergent would add it to the fraction of material classed as 'active', whereas it is not a surfactant.

While most students will be familiar with steam distillation, which has the double objective of separating the volatile from non-volatile components with a reduction in boiling point to reduce thermal decomposition, few meet the same experiment with the inverse emphasis, a volatile organic carrier to distil over water.

Figure 13.3 shows a phase diagram for the toluene–water system (with the mutual miscibility exaggerated). At a temperature below 85°C, the two liquid phases coexist. At 85°C the sum of the individual vapour pressures (of water and toluene) is equal to the external pressure (of one atmosphere, $\sim$100 kPa) and boiling takes place.

Since three phases and two components are present and the pressure is fixed, the phase rule $(P + F = C + 2)$ tells us that the system is invariant. The compo-

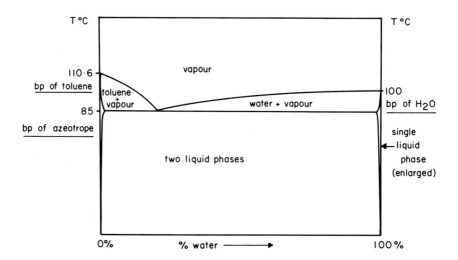

Fig. 13.3   Phase diagram of the toluene–water system (From B. K. Selinger, 'Water, water, everywhere', *Education in Chemistry*. **16**, 1979, 125)

sition of the vapour phase is fixed, and the mole fraction of each component is proportional to its vapour pressure at that temperature. At 85°C, the vapour pressure of water is 57.7 kPa and the vapour pressure of toluene is about 50.6 kPa, so that the composition (by mass) of water will be (MM H_2O/MM toluene) x (vp H_2O/vp toluene) = 0.22 or 22%, which is slightly higher than found in practice (entry 1 in Table 13.2).

TABLE 13.2
Properties of azeotropes with water

Components			Azeotrope				
			Per cent composition				
Compounds	bp (°C)	bp (°C)	In azeotrope vapour	Distillate Upper	Distillate Lower	Relative volume of layers at 20°C	Relative density of distillate layers
1. toluene	110.6	85.0	79.8	99.95	0.06	U 82.0	U 0.868
water	100.0		20.2	0.05	99.94	L 18.0	L 1.000
2. ethanol	78.5		37.0	15.6	54.8	U 46.5	U 0.849
toluene	110.6	74.4	51.0	81.3	24.5	L 53.5	O 0.855
water	100.0		12.0	3.1	20.7		
3. ethanol	78.5		21.2	20.6	32.0	U 95.5	U 0.722
vinylpropyl ether	65.1	57.0	73.7	77.8	0.2	L 4.5	L 0.923
water	100.0		5.1	1.6	67.8		
4. ethanol	78.5		12.0	3.0	75.0	U 90.0	U 0.672
hexane	69.0	56.0	85.0	96.5	6.0	L 10.0	L 0.833
water	100.0		3.0	0.5	19.0		
5. hexane	69.0	61.6	94.4			U 96.2	U 0.660
water	100.0		5.6			L 3.8	L 1.000

Source: After R. C. Weast, *Handbook of chemistry and physics,* 58th edn, CRC, Cleveland, Ohio, 1977. D180, D198.

Normally toluene is used as the carrier because of this favourable transfer rate for water. The azeotrope vapour consists of 20.2% water, and the composition of the lower layer of the distillate is 99.94% water. The toluene upper layer contains 0.06% water; this represents a small error, which is progressively reduced as the volume of water collected increases and the toluene in the upper layer is reduced by running back into the still. The distillation is finished when all the water has been transferred. The temperature of distillation rises as the system moves into the liquid toluene + vapour region in Figure 13.3.

Where a consumer product contains a volatile component other than water, such as ethanol found in some hand dishwashing detergents, a ternary azeotrope can form. Ethanol, toluene and water form a ternary azeotrope vapour (entry 2 in Table 13.2). Whereas the upper layer of the distillate has only 3.1% water, the lower layer has a 24.5% contribution from toluene, as long as there is enough ethanol. The upper and lower layers have very similar densities and so are hard to separate. Once all the ethanol has been carried over, the system reverts to a binary one. (Ethanol–toluene in the *absence* of water azeotropes to a *single* phase distillate.)

The alkyl ethers appear to be ideal carriers for ethanol plus water mixtures (entry 3 in Table 13.2), giving a lower layer virtually free of carrier. But these are notorious peroxide formers and so are undesirable in a student laboratory.

Hexane, however, represents a considerable improvement over toluene (entry 4 in Table 13.2) and works well in practice, although it is very slow because the azeotrope vapour contains only 3% water. The lower layer of the distillate contains only 6% while the ethanol lasts and then reverts to a binary system (entry 5 in Table 13.2).

EXPERIMENT 13.2

The use of oils in frying

100 mL glycerol
100 g stearic acid
100 g glyceryl monostearate
2 L lard
2 L beef dripping
2 L hydrogenated coconut oil
2 L peanut oil
2 L olive oil
2 L deep-frying oil
Nine 100 mL beakers
250 mL measuring cylinder
25 mL measuring cylinder
Six thermometers (0–360°C)
Bimetal thermometer
5 kg potatoes
Ruler
Potato cutter (40 mm diameter)

Equipment
 Sharp knife
 Chopping board
 Paper towel
 2 L 40–60°C bp light petroleum +
  solvent residue bottle
 Porous pot
 Four bunsen burners
 Deep-fryer with basket (e.g. Sunbeam)
 Six soxhlet extractors ('Quickfit'
  Cat. No. 200 RASX)
 6-hole steam bath with stands, clamps
  etc.
 Fume chamber
 Rotary vacuum evaporator
 Air oven at 100°C
 Analytical balance

Principle

Frying essentially involves placing a food into heated fat. When fats or oils are heated to high temperatures, decomposition occurs and a temperature, known as the smoke point, is reached where visible fumes are given off. The smoke point is primarily a function of the free fatty acid content of the oil (being lower in the case of high free fatty acid content fats) but other factors such as impurities (e.g. glycerol), heating rate and area of fat exposed to the air during heating are also involved. It has been suggested that fats and oils used for frying should have smoke points in excess of 185°C so that the odour and irritating effect of the fumes can be avoided. The foaming properties of the oil or fat are also an important consideration, with excessive foaming being a major drawback.

Samples of lard, beef dripping, hydrogenated coconut oil, peanut oil, olive oil and a specially prepared deep-drying oil will be examined for their suitability as frying fats.

Method

1. Examination

Note the colour, consistency and odour of the oil samples.

In a 100 mL beaker, heat about 20 mL of each sample to 100°C and note the presence of any 'off-odours'. Continue heating until the fat starts to decompose and again note any 'off-odours'. At what temperature does the smoke point of each sample occur?

2. Smoke point

Add each of the following hydrolysis products to separate samples of beef dripping:

100% glycerol

10% stearic acid

10% glyceryl monostearate

Determine the smoke point of each mixture and assess the effect of the hydrolysis products on smoke point.

3. Absorption of oils

(a) Effect of different oils

Half fill the deep-fryer with a sample of oil and set the thermostat at 150°C. When constant temperature is reached, determine the actual temperature with a bimetal thermometer.

Prepare standard potato slices by cutting scrubbed potatoes into circles 40 mm diameter using the potato cutter.

Deep fry to an even yellow-brown colour a minimum of 25 slices at a setting of 150°C, totally submerging all slices by means of the wire basket. After frying, drain for two minutes in the basket and then place the potato slices on paper and allow to cool.

Repeat this procedure for all the oils. Do the different fats show different degrees of foaming?

Weigh 15 circles fried in each fat. Place each lot in a soxhlet apparatus and extract with 40–60°C light petroleum (150 mL) for six hours. The flask (plus porous pot) is weighed accurately before commencing the extraction. At the end of the extraction remove most of the solvent by pouring from the soxhlet tube. Then remove the residual solvent on a rotary evaporator followed by one hour in a drying oven at 100°C. Express results as per cent of fat in the chips.

Comment on any correlation between fat properties and fat absorption by the potatoes.

(b) Effect of different drying temperatures

Use the above technique to establish whether fat absorption is influenced by frying temperature. Use beef dripping at 170°C and 190°C to fry chips until they are cooked to a similar degree to those fried at the lower temperature. Determine fat absorption and compare with the results at 150°C.

4. Selection of a suitable frying fat
Allow samples of the chips to equilibrate at ambient temperature. Give them a quality rating based on appearance and flavour, taking particular note of any difference in the appearance or 'greasy' taste of chips prepared from oils of different melting points.

Estimate the cost of potatoes and oil required to produce one tonne of chips, based on recovery of peeled potatoes, and the per cent of oil extracted from the chips. Obtain quotations of the current cost of potatoes and at least three of the oils (of suitable quality) used for frying.

From these experiments, what would your recommendations be for the commerical production of chips based on acceptability and cost?

Source: University of New South Wales, School of Food Technology.

EXPERIMENT 13.3

Separation of food and drug components using chromatography

Chromatography is a term used for a variety of separation techniques. The components of the mixture to be separated are distributed between two phases, one of which remains stationary while the other phase, the *mobile phase*, percolates through the interstices or over the surface of the *stationary phase*. The mobile phase can be a liquid or a gas, and the stationary phase can be a solid or a liquid. Several combinations (types of chromatography) are thus possible.

Equipment (for parts A and B)
500 mL polythene wash bottle
Pyrex measuring cylinder, 10 mL (unstoppered, graduated)
Pyrex measuring cylinder, 100 mL (unstoppered, graduated)
800 mL beaker or 1 L low form with spout
Disposa-gloves (medium, large)
Filter paper (1 sheet of 9 cm × 6 cm)
Test-tube rack, aluminium
Micro-spatula
Two square sheets of aluminium foil
Ten test-tubes, 8 × 1.2 cm (semi-micro)
Commercial Merck plastic-backed SiO_2 TLC plates will be used; cut into 4 × 6.7 cm size and stored in desiccator. (You will need four per student.)
2-propanol (isopropyl alcohol) (20 mL per student)

Solutions
1.5% hydrochloric acid solution. Dissolve 96 mL of conc. hydrochloric acid (36% w/v HCl) in distilled water and make up to 2 L. Use for preparation of amino acid standards.
2% ammonia solution. Dissolve 240 mL of conc. ammonia solution (density 0.88) in distilled water and make up to 4 L. Store in winchesters in fume cupboard.

Amino acid standards. Prepare in 500 mL volumetric flasks and transfer 50 mL to labelled reagent bottles for storage:

1. 0.05 M glycine. Dissolve 1.88 g of glycine in the 1.5% HCl solution and make up to 500 mL.
2. 0.05 M tyrosine. Dissolve 4.53 g of tyrosine in the 1.5% HCl solution and make up to 500 mL.
3. 0.05 M leucine. Dissolve 3.28 g of leucine in the 1.5% HCl solution and make up to 500 mL.
4. 0.05 M aspartic acid. Dissolve 3.32 g of aspartic acid in the 1.5% HCl solution and make up to 500 mL.

Make up about 5 × 50 mL quantities of the unknown amino acid mixtures. Store in 25 mL dropping bottles labelled as follows:

A	tyrosine and aspartic acid
B	glycine and tyrosine
C	tyrosine, glycine and leucine
D	aspartic acid and leucine
E	glycine and aspartic acid
F	aspartic acid, tyrosine and leucine
G	glycine, aspartic acid, and leucine
H	leucine and tyrosine

Solvent mixture (butyl acetate/dichloromethane/85% formic acid). Mix 1200 mL of butyl acetate, 800 mL of dichloromethane and 200 mL 85% formic acid and divide up into five equal portions. Store in Kipps dispensing apparatus labelled 'TLC eluent' in fume cupboard.

1:1 dichloromethane/acetone. Make up the mixture by mixing 1 L of each solvent and store equal portions of the mixture in Kipps apparatus. Store in fume cupboard.

Analgesic standards. Make up 250 mL at a time.

1. Caffeine in dichloromethane. Dissolve 5 g of caffeine per 100 mL of solvent.
2. Aspirin in acetone (prepare fresh every two to three days). Dissolve 5 g of aspirin per 100 mL of solvent.
3. Phenacetin in dichloromethane. Dissolve 5 g of phenacetin per 100 mL of solvent.
4. Paracetamol in acetone. Dissolve 3 g of the paracetamol per 100 mL of solvent.
5. A mixture in 1:1 dichloromethane/acetone. Dissolve 3 g of aspirin, 2 g of phenacetin, paracetamol and caffeine per 100 mL of the 1:1 solvent mixture (make fresh every two to three days).

Also make up five commercial analgesic samples; e.g. aspirin, Panadol, Vincents, Bex, APC. Grind up about five tablets of each and store in clearly labelled wide-necked reagent bottles (125 mL).

Other equipment

Labels (2 cm × 1 cm self-adhesive)

Whatman no. 1 filter paper or Whatman no. 1 chromatography paper. Cut up into
   12 × 22 cm sheets.

Paperclips

Stapler and staples

Small filter paper squares for practising spotting

Spare packets of melting point tubes

One desiccator of TLC plates

In the fume cupboard (one set per cupboard). (All fume cupboards must be
adequately protected from the ninhydrin by masking with Benchkote and
masking tape; i.e. backs and base.):

Two winchester bottles labelled

1.   butyl acetate/dichloromethane/formic acid residues for disposal
2.   dichloromethane/acetone residues for disposal

Ninhydrin aerosol pack (two cans per cupboard)

Paper chromatography rack and paper clips

Also provide the following:

UV light boxes in instrument room

Ovens set at 110–120°C

Chromatography tanks. Students use a 250 mL beaker for their TLC work.
   Therefore each is provided with a piece of aluminium foil of a size which covers
   the lip of the beaker with some overlap.

Paper chromatography. Each student is provided with an 800 mL or 1 L beaker
   for their chromatography tank. Paper for the tanks is cut to size (12 × 22 cm)
   using no. 1 Whatman 48 cm circles and one issued to each student. A
   commercial aluminium foil square is issued for the tank lid.

Rubber gloves of appropriate sizes are provided in a fume cupboard with the
   ninhydrin aerosol cans.

Ninhydrin aerosol pack. Two cans per fume cupboard in each section.

Spotting bench. Provide three semi-micro burners and a flint gun on large ceramic
   tiles with a supply of melting point tubes at each burner for drawing out
   spotters.

A maxi-bin labelled 'used spotters' will also be required.

A. Separation of amino acids by paper chromatography

Principle

In paper chromatography, one liquid phase is allowed to move along a strip or
sheet of cellulose (filter) paper. The paper and a second liquid phase, which is
strongly absorbed on the paper, constitute the stationary phase. The liquid phases
are referred to as the *solvent system*. One of the earliest applications of
chromatography was the separation of a mixture of pigments from leaves into
components of different *colours*; hence the name, which comes from the Greek
words *chroma* (colour), *graphe* (writing).

In this experiment, a mixture of amino acids is to be separated using aqueous 2-propanol as the solvent system. The components can be identified by comparison with pure samples of amino acids chromatographed in the same way. The sample solutions are applied to the paper as spots with any device that will transfer a very small amount of material, such as a capillary tube (spotter). The paper is then placed in a vessel containing a small volume of the chosen solvent system for development (see Fig. 13.6). The experimental procedure will be described in detail below. The vessel must be covered to maintain an atmosphere saturated with solvent vapour. The solvent mixture ascends the paper by capillary action and solutes will move at a rate that depends on their partition between the stationary and mobile phases and hence on their structure. The ratio of the distance travelled by a compound to that travelled by the solvent front is called the R_f value (or retardation factor) of that compound (see Fig. 13.4). For a given solvent mixture and paper, the R_f value is characteristic of the solute. The R_f values of the amino acids in the solvent system given above vary from 0.1 to 0.8. In any homologous series among the amino acids (such as glycine, alanine, α-aminobutyric acid) the R_f value increases with the size of the molecule, indicating a successively smaller tendency to absorb as the molecule gets bigger.

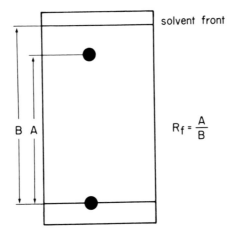

solvent front

$R_f = \dfrac{A}{B}$

B A

Fig. 13.4 Chromatography plate

Common methods of detecting the position of compounds on the paper after development are: (1) intrinsic visible colours when possible; (2) reactions with colour-producing reagents; (3) fluorescence; and (4) prevention of fluorescence of the stationary phase by absorption of ultraviolet light.

The amino acids (colourless) are detected using a ninhydrin spray. Ninhydrin reacts with α-amino acids to yield highly coloured products. Ninhydrin should be kept off the skin because it reacts with proteins in the skin to form a rather long-lasting purple discolouration. The spray reagent is prepared as a 2% solution of ninhydrin in ethyl alcohol. Fingerprints also yield ninhydrin reactive material. Therefore, to protect chromatograms as well as the skin, you should wear protective gloves when handling chromatographic paper and also when spraying.

Method

Place labels on five clean 8 × 1.2 mm test tubes, and place one spotter into each tube. From stock solutions (containing about 0.05 M solutions of the amino acids in 1.5% hydrochloric acid), transfer a few drops of the appropriate material to each of the first four test tubes. Obtain a few drops of an unknown from the instructor for the fifth tube. (The unknown will contain from one to four of these same amino acids at a concentration of about 0.05 M each in 1.5% HCl.)

Stationary phase. Obtain a clean sheet of Whatman no. 1 filter paper, about 12 × 22 cm, and draw a light pencil line parallel to the bottom and about 1.5 cm from the edge of the paper (Fig. 13.5). Along this line, at intervals of about 2 cm, place 10 light crosses. Above each cross place identifying marks, two for each known and two for the unknown.

Using the capillary tubes, place a small amount of each appropriate solution on its two positions along the line on the filter paper. Avoid getting the spot on the paper larger than about 2 mm in diameter. (Although your paper gives you two chances to make a proper addition, it is advisable first to practise transferring solution to an ordinary piece of filter paper.) Let the paper dry for a few minutes in air. Add a second portion of the unknown to one of its two positions to make certain that sufficient quantities of each component of the unknown will be present for good visual observation when the paper is developed. Ensure the spots on the paper are dry (it may be necessary to place the paper in an oven at about 100°C for a short time).

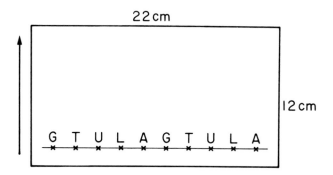

Fig. 13.5   Amino acids spotted on plate (U = unknown)

Roll the paper into a cylinder (avoid putting fingerprints on paper), and staple the ends together about a third of the way in to form the edge (Fig. 13.6). Staple the paper in such a fashion that the ends of the paper do not touch each other— otherwise the solvent will flow more rapidly at that point and form an uneven front.

Place it carefully in the beaker of solvent, and cover carefully and tightly with the aluminium foil (Fig. 13.6). Make certain that the paper does not touch the sides of the beaker, and use care in keeping solvent from splashing onto the paper.

Let the solvent rise up the paper for *at least* 1½ hours, being careful not to let it reach the top edge. If the time is shorter, the components may not be sufficiently separated for easy identification. Remove the paper and place it upside down in

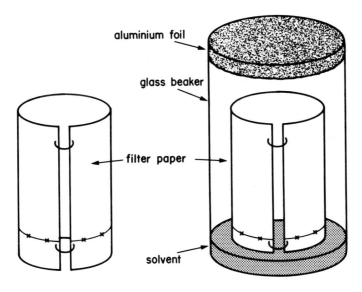

Fig. 13.6 Paper chromatography in process

the fume cupboard to dry. When most of the solvent has evaporated, open the cylinder by tearing it apart where it was stapled and hang it in a fume hood (Fig. 13.7). Spray the paper lightly but completely with a solution of ninhydrin, and leave the paper in the hood until the spray solution is dry. Place the paper in an oven at 100–110°C for about 10 minutes or until all the spots have developed.

Circle each spot with a pencil immediately, as the spots will fade. Now measure from the centre of each spot to the base line. Measure the distance the solvent has travelled at each position and calculate R_f values for each amino acid (Fig. 13.4). Determine the composition of the unknown by visual comparison of spot colours and by the relationships of R_f values.

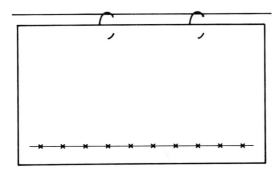

Fig. 13.7 Paper hanging up to dry

B. Separation of drugs by thin layer chromatography

Method

Analgesics contain a number of standard drugs in various formulations. It was common, until the early 1960s, for analgesics to contain phenacetin, until it was withdrawn from the market because of its adverse effect on the liver (see Fig. 9.29). It was replaced in many proprietary lines by the structurally related product paracetamol (Fig. 13.8) and many products still retain this drug. Other components are aspirin and caffeine (Fig. 13.9). Codeine is also a common ingredient (present as the phosphate) in pain-relieving preparations, available in Australia only on prescription. In the USA the alkaloid caffeine (Fig. 13.9) replaces codeine in analgesic preparations. Indeed many products in Australia also contain this drug. However, the most common and well accepted pain-relieving ingredient is aspirin, acetylsalicylic acid (Fig. 13.9).

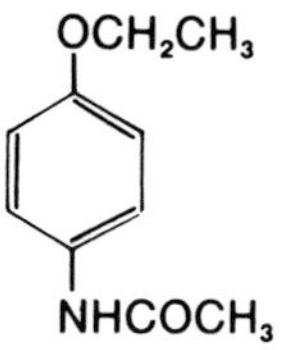

phenacetin

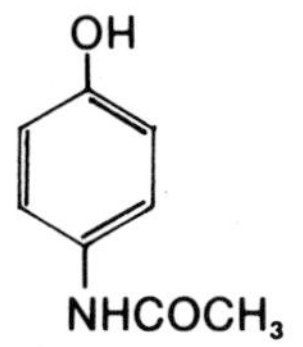

paracetamol

Fig. 13.8 Phenacetin and paracetamol

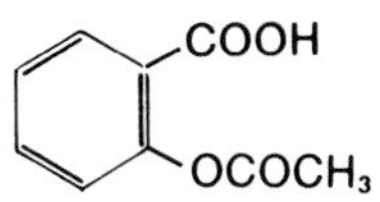

aspirin

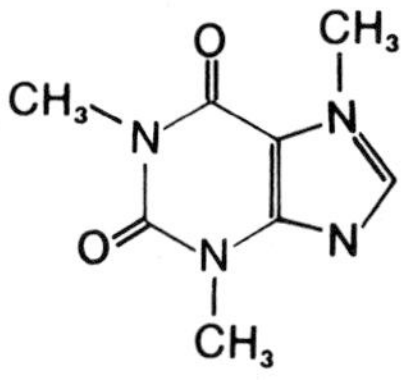

caffeine

Fig. 13.9 Aspirin and caffeine

The chromatographic technique used in this experiment is thin layer chromatography (TLC), which is similar to paper chromatography. Instead of paper, a thin layer of finely divided adsorbent (e.g. silica gel, alumina, powdered cellulose) supported on a glass plate is used. Commercial TLC plates, prepared by coating a plastic film with the adsorbent, are also suitable.

(a) Separation of aspirin, caffeine and paracetamol
Label four tubes A, B, C, M.

Collect a few drops of stock solution of authentic aspirin (A), caffeine (B), paracetamol (C) and a mixture consisting of equal volumes of all three (M).

Solvent: butyl acetate : chloroform : 85% formic acid (6:4:1 volume).

Spot samples of A, B and M near the bottom of the chromatographic plate (as shown in Fig. 13.10). Prepare a similar plate with C and M. (The chromatographic plates comprise a layer of silica gel as adsorbent, calcium sulfate as binder, together with a fluorescing agent to aid location of the sample by ultraviolet illumination.) Develop the plates using the solvent system described above. When the solvent front has reached almost to the top of the plates, remove, dry, and examine the plates under ultraviolet light (short wavelength, 254 nm; e.g. a germicidal lamp). Mark the position of the samples and of the solvent fronts.

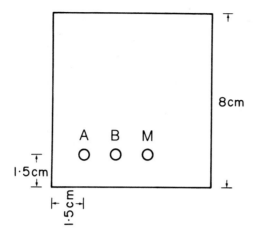

Fig. 13.10   Drug separation by plate chromatography

Record the R_f values (see experiment 13.3A). Repeat the experiment and compare duplicate R_f values. (Remember to clean spotter between samples.)

When satisfied that the reference samples can be detected clearly, continue with stage (b) of the experiment. If R_f values are not consistent, check the procedure with your instructor. Common errors include: spotting too close to edge of plate; spot too large in diameter; too much sample (this leads to streaking); spot not dry prior to commencing chromatography.

(b) Investigation of a selection of analgesic drugs
The same procedure is used for chromatography and detection in this section of the experiment. Samples are prepared from a tablet (or powder) of each product by grinding the dry tablet and then mixing with chloroform (15 mL) and using this as the test sample. Fill in Table 13.3 with the manufacturers' specifications for their analgesic preparations (from their labels).

TABLE 13.3
Analysis of analgesics

Product name	Sample	Aspirin	Caffeine	Paracetamol	Other
		Manufacturer's specification			
Blue Codral	E				
Codiphen	F				
Aspirin	G				
Veganin	H				
Panadeine	I				
Vincents	J				
Bex	K				
APC	L				

It is best to use the mixture M prepared previously as a reference sample.

C. Food colours in jelly beans

Equipment

Jelly beans (or Smarties)
White woollen thread (1 m for each
 sample)
Chromatography paper + melting
 point capillaries
Five chromatography tanks 20 × 4
 cm diameter
Twenty 100 mL beakers
Stapler and ruler
Hot plate

Water bath
Five 100 mL graduated cylinders
Five glass rods
Fume cupboard
Solvents:
 acetic acid (0.1 M)
 ammonia 2% solution
 developing solution—
 butanol:ethanol:2% ammonia =
 3:1:1

Method

White woollen thread is used in the extraction of food colours. It is prepared beforehand by boiling for 10 minutes in 2% ammonia solution, then rinsing well under the tap to remove the fluorescing dyes, and then drying.

Warm water (20–25 mL) is added to about three to six jelly beans of each colour. This is done separately for each colour. The number of jelly beans used depends on the depth of colour wanted. They are allowed to stand for about 10 minutes or until they look white. If they are allowed to remain too long in solution, or if the water is too hot, too much sugar is extracted. The solutions now contain both dye and sugar, which must be separated. This is done by dyeing the wool.

A length—about a metre—of prepared white wool is placed in each 100 mL beaker and a dye extract is added and the whole is acidified by adding a few drops of dilute acetic acid (0.1 M). The mixture is brought to the boil and simmered for five minutes, at which stage the wool can be seen to be dyed. The wool is now removed and washed well under the tap until no stickiness or jelliness remains.

The dye is now re-extracted from the wool by placing it in a beaker containing about 20 mL of 2% ammonia solution and simmering for about five minutes. The wool is then removed from the beaker and the extract is evaporated to dryness on a hot plate in a fume cupboard (ammonia has a nasty smell). The residue is stirred with the minimum number of drops of water to mix it completely (to maintain a high concentration).

Each solution is spotted on chromatography paper as in part A. The labels on the paper are now the colours of the beans used. The solvent system used is butanol : ethanol : 2% ammonia in a 3:1:1 ratio.

For more details see British Standard BS3210 (1960): *Analysis of water-soluble coal-tar dyes permitted for use in foods.*

This method has been criticised as being more complex than necessary and an alternative method is suggested using Smarties. Because the intense colour of Smarties is on the outside layer only, it is sufficient to lick the tablet and rub it directly on chromatography paper. (The wool method, however, ensures that only *acid* dyes are extracted.) The separation can also be simplified. If the paper is dipped in 1% aqueous sodium chloride solution, the dyestuff separates within 10–20 minutes. (The separations are not affected by pH in the range 4–9. The salt is essential to reduce the electrostatic attractive forces between the dye and paper molecules.)

EXPERIMENT 13.4

An experiment with dyeing

Equipment
Multifibre fabric or individual fabrics
Set of suitable dyes
250 mL beaker
Bunsen burner on a tripod with a wire gauze.

Introduction
Dyeing involves immersing fabrics in a bath of dye at a suitable temperature for a period of time. The dye molecules are attracted to the fabric partly as a result of their large size and partly because of the presence of particular molecular groups or charges which interact with the fabric. Hydrophilic (water-loving) fibres such as wool and cotton swell appreciably in water and allow entry of water-soluble dyes. The relatively few polar groups in synthetics make them hydrophobic (water-fearing) and allows little swelling. Polyester is thus difficult to dye with water-soluble dyes.

Method
A multifibre fabric woven from wool, orlon, dacron, nylon 66, cotton and acetate are dyed with a mixture of chlorazol scarlet (a direct dye—large molecule), dispersal yellow, (a disperse dye—non-ionic) and edicol blue (an acid dye—negatively charged). (The dyes suggested in the second edition have been modified in the light of the availability of a kit from Educational Services, PO Box 416, Cheltenham, Victoria 3192.) The dye bath contains

0.05 g chlorazol scarlet
0.03 g dispersal yellow
1.25 g edicol blue (40%)

in 100 mL of water. Other dyes of a similar nature will also work.

The temperature of the dye bath is brought to boiling and a small piece of fabric is added. After about 10 to 25 minutes, the hot liquid is poured off and the murky-coloured fabric is rinsed several times in a large volume of cold water, squeezing between washes, and then allowed to dry. Plate XV shows schematically how the dyes interact with different fibres.

Principle
Wet wool has NH_3^+ and COO^- groups. In acid solution the COO^- is neutralised to COOH and the fibre is positively charged and thus attracts mainly the negatively charged sulfonic acid dye. That is why wool is used to extract (legal) food colours in experiment 13.3C (Food colours in jelly beans), because these are all acid dyes.

Cotton is attracted to the direct dyes, which are similar to acid dyes but consist of much large planar molecules with groups that form strong hydrogen bonds with cellulose. Cotton fibres are cellulose and the OH–groups give a negative charge in water and so salt is often added to reduce repulsion between like charges.

Acetate is a cellulose-based polymer, sufficiently modified to warrant its designation as synthetic. This fabric serves directly as a solvent for the dye. It thus attracts the non-ionic yellow (disperse) dye. Such dyes are called disperse because they rely for attraction on so-called dispersion forces, which increase with the size of the molecule. (The same forces are responsible for polyethylene being solid, although ethylene itself is a gas). An alternative explanation for the name is that the dyes are insoluble and for use need to be dispersed by a detergent.

Like wool, nylon is positively charged in water, so it also takes up the blue acid dye. It takes longer for the larger red direct dye to be adsorbed (this is speeded up at a lower pH of 3, given by adding acetic acid). Nylon is less polar than wool and so it takes up the yellow non-ionic dye as well, and so becomes green.

Dacron is a trade name for polyester, while Orlon is a trade name for polyacrylonitrile (sometimes with 10% other monomers present to aid in dyeing).

Well, it is all a little more complicated than this and for a useful discussion, see the bibliography.

EXPERIMENT 13.5

Extraction of caffeine and benzoic acid from soft drinks

Introduction

An American study was undertaken in 1970 on the effect of food acids on human teeth. The teeth were soaked in acidic solutions of equivalent strength to those found in foods, cola drinks and orange juice. Colas turned the teeth brown but did not damage the enamel, but the enamel of the teeth soaked in orange juice could be flaked off with a fingernail. Experiments conducted with Australian students also confirmed that cola drinks do not dissolve baby teeth. It is important to remember that the contact time for drinks with teeth is in fact quite short. Of far more concern is the sugar content of these drinks, which can vary between six and nine teaspoons per 375 mL can.

Coca Cola comes from the cola nut, a native of West Africa and coca, a native shrub of Peru. At first the cocaine from the coca leaves was left in the drink, but later it was removed and sold to drug companies (e.g. Burroughs Wellcome) to be used in medicaments to treat ear, nose and throat complaints. Since the Second World War, the (cocaine free) coca component has been omitted altogether.

Equipment

Ten 1 L conical flasks	Cans of Coke (or cola)
Five 250 mL separating funnels	Cans of lemonade
Ten 100 mL beakers	Infra-red spectrometer, Nujol, ethanol,
Five watch glasses	sodium chloride plates, mull plates
Five 100 mL graduated cylinders	and razor
Ten funnels	Sodium carbonate
Five spatulas	Dichloromethane (methylene chloride)
Ten glass rods	Conc. HCl

Water bath and distillation
  apparatus
Pasteur pipettes
Five 250 mL beakers
Ten cotton-wool pads
Fume cupboard

Potassium chlorate
2 M ammonia
Benzoic acid (standard)
Caffeine (standard)
Magnesium sulfate (5 g)

Method

1. Isolation of caffeine

Each student takes a sample of cola (record brand name) and carefully measures 150 mL of this sample into a conical flask (1 L). To this is added 2 g of sodium carbonate (the amount is not critical provided the resultant solution is basic—test with pH paper), which is added to stop contamination of the extract with benzoic acid, a commonly added preservative. The benzoic acid, being acid, reacts with the basic sodium carbonate to form sodium benzoate, which is water-soluble and so not extracted into organic solvents.

Then 50 mL of dichloromethane (methylene chloride) is added to the conical flask and the flask is swirled gently for at least five minutes. (*Do not* shake vigorously; this often causes an emulsion to form, which hinders separation.)

The combined liquids are then transferred into a separating funnel and allowed to settle (5–10 minutes). The lower methylene chloride layer is drained through the tap into a clean 250 mL conical flask. A fresh sample of dichloromethane (50 mL) is added to the separating funnel and the flask stoppered. *Gently* invert the funnel a few times to allow the remaining caffeine to be extracted into the dichloromethane layer. The lower layer is separated and combined with the first extract.

The total extract is treated with anhydrous magnesium sulfate ($\sim$5 g) to remove the water (it forms an insoluble hydrated magnesium sulfate salt) and the dichloromethane (it should be clear) is filtered through a cotton-wool pad into a 250 mL beaker. Evaporate the dichloromethane on a water bath in a fume cupboard (it is better to distil if off and recover the solvent). The product should be weighed and then tested as follows:

a. Place a small amount of precipitate on a watch glass and mix with two to three drops of concentrated hydrochloric acid (care!). Add a *few* small crystals of potassium chlorate. Mix well with glass rod and evaporate to dryness on a water bath (boiling) in a fume cupboard with the front protector down. Cool the glass and moisten the residue with a drop or two of 2 M ammonia solution (NH_4OH). Residue turns purple.

b. A sample should be prepared for analysis by infra-red spectroscopy (if an instrument is available) by formation of a Nujol mull and the spectrum run using sodium chloride plates on the spectrometer.

2. Isolation of benzoic acid

Half a can of lemonade is poured into a 1 L conical flask and acidified with two drops of dilute hydrochloric acid. To this 50 mL dichloromethane is now added and the flask swirled gently for at least five minutes and then poured into a separating funnel and allowed to settle (five minutes). The solvent layer only is drained into a 100 mL beaker and allowed to evaporate, leaving a residue of benzoic acid. Confirm by running an infra-red spectrum of a mull of residue and compare with a standard sample (if a spectrometer is available; see Fig. 13.11).

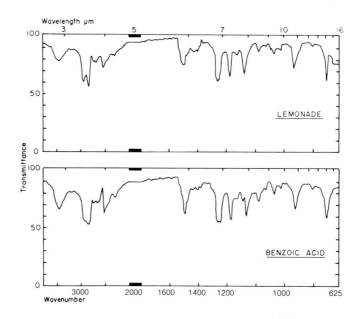

Fig. 13.11   An infra-red fingerprint

EXPERIMENT 13.6

Determination of vitamin C content in orange juice

Equipment

50 mL burette, clamps and stand
Separating funnel
100 mL beaker, low form with spout
250 mL beaker, low form with spout
500 mL polythene wash bottle
100 mL measuring cylinder
Two 250 mL conical flasks, narrow-necked
250 mL filter flask

250 mL volumetric flask, stoppered
Porcelain Buchner funnel, 5 cm diameter
Conical funnel, 55 mm
Pyrex weighing funnel
Three pipettes (5 mL, 10 mL, 25 mL)
Stirring rod, 7 mm diameter
Plastic spoon
Ceramic tile, white, 15 cm

Chemicals

L-ascorbic acid (50 mg/student), freshly opened bottle. Check before using.
Oxalic acid. Supply about 3 kg, repacked in screw-capped jars labelled 'Oxalic acid—Poisonous'.
Solvent ether. Provide two winchesters.

Solutions

0.002 M *N*-bromosuccinimide. Make up 5 L batches by dissolving 1.780 g
N-bromosuccinimide in distilled water and making up to the mark. Mix well.
Divide into two winchesters and standardise against L-ascorbic acid in distilled
water. Approximately 10 batches, made up as required, will be needed.

10% acetic acid. Make up a 5 L batch by diluting 500 mL glacial acetic acid to 5 L
with distilled water. Label as '10% acetic acid/water'. Store in 500 mL
quantities with a 2 mL Kipps dispenser.

4% w/v potassium iodide (KI). Dissolve 200 g KI in distilled water and make up to
5 L. Mix thoroughly and transfer to winchesters. Distribute in 500 mL portions
in a Kipps apparatus so that 5 mL can be delivered.

Starch solution. Add 10 g soluble starch to a small amount of water to form a
slurry.

Add 500 mL boiling distilled water, stirring until dissolved. Cool and transfer to
dropping bottles. Label as 'Starch'. Repeat as required.

Selection of orange juices (pure), orange concentrate, fresh oranges, grapefruit
etc., as required.

Celite Hyflosupercel filtering aid. Need about 500 g overall.

Notes

1. Cover the fume cupboard bench-top with Benchcote and supply 2 winchesters
with tops and funnels, and label them 'Oxalate residues'. When full the
residues can be treated by acidifying with H_2SO_4, heating to about 80°C and
mixing with potassium permanganate solution (~5% w/v) until a permanent
pink end-point is reached. Then discard down the sink.

2. Aspirettes: these must be checked daily for proper working order. They are
faulty if they deflate on squeezing without the valve operating or if they
contain some solution.

3. The orange juice concentrates are to be made up according to the packaging
instructions. Store in winchesters. Label 'Orange juice'. Keep the carton for
reference.

Introduction

Ascorbic acid (vitamin C) plays a part in many physiological functions in humans;
for example, the transport of iron between blood plasma and bone marrow, spleen
and liver. It has also been linked to prevention of the common cold and the slowing
of senility, although both these subjects are still under intense discussion
throughout the world.

Most species are able to synthesise their own ascorbic acid from glucose, but
humans, other primates, guinea pigs and Indian fruit-eating bats are unable to do
this and must obtain this vitamin from their diet. A large number of green
vegetables are excellent sources of vitamin C, particularly green capsicum, which
contains more ascorbic acid per 100 g than citrus fruits. Many people drink
commercial orange juice as a source of vitamin C and this experiment aims to
analyse the vitamin C content of some commercial orange juices.

Ascorbic acid is not a carboxylic acid as the name implies, but is a lactone (a
cyclic ester) and is very easily oxidised. The fact that ascorbic acid is a powerful

reducing agent (since it is easily oxidised) enables it to be determined using a redox titration procedure. In this experiment we will use N-bromosuccinimide as the oxidising agent. The reaction involves the oxidation of ascorbic acid to dehydroascorbic acid and the reduction of N-bromosuccinimide to hydrogen bromide. Hence one mole of N-bromosuccinimide (MM 177.99) reacts with one mole of ascorbic acid (MM 176.13) (MM = molecular mass).

The end point of this titration is detected by adding a little iodide ion and starch indicator to the ascorbic acid solution before titration. When all of the ascorbic acid has been consumed, any excess N-bromosuccinimide will oxidise iodide to iodine, thereby turning the starch blue. The appearance of the blue colour therefore marks the end point.

Method

The procedure may be conveniently divided into three stages: preparation of the sample; standardisation of oxidising agent; and titration procedure.

1. Preparation of sample

Whilst it is possible to perform the titration directly on orange juice, the colour of the juice can sometimes obscure the end joint. For this reason, it is advisable to either *filter* off the pulpy components of the juice or to *extract* the orange colouring matter using ether. Both procedures are described below. You should choose only one.

Orange juice is very difficult to filter because it quickly clogs the pores of the filter paper; use of a filter aid such as celite is therefore necessary. Use the following procedure.

Collect 40–50 mL (no more!) of orange juice (record its brand name) in a 100 mL beaker. To it add 0.2 g (weighed on top-pan balance) of solid oxalic acid, which acts as an antioxidant. Stir the mixture till all the solid has dissolved. Then add one to two teaspoonsful of celite into the mixture and make a slurry. Vacuum filter the slurry into a clean, *dry* Buchner flask. (All glassware has to be dry so that the orange juice is not diluted. Also, do not use too strong a suction for the filtration to avoid solvent evaporation, which will concentrate the orange juice.) The filtered solution should be pale yellow and is now ready for analysis.

The alternative method of sample preparation is to extract the orange colour into ether. Collect about 40–50 mL of orange juice in a 100 mL beaker and add 0.2 g of solid oxalic acid and swirl until the oxalic acid has dissolved. Transfer this solution to a dry separating funnel and add about 20 mL ether from a measuring cylinder. (*Be careful!*—ether is highly flammable. Ensure that there are no naked flames in the laboratory.) *Gently* shake the funnel—too much shaking will result in the formation of an emulsion, which will take a long time to separate. The orange colour should be extracted into the ether (upper) layer and most of the pulp will tend to collect at the phase boundary between the aqueous and organic layers. Drain off the aqueous phase leaving as much as possible of the pulp in the separating funnel. The aqueous phase is now ready for analysis.

2. Standardisation of oxidising agent

A solution of N-bromosuccinimide (approx. 0.001 M) is required but, because NBS in solution is unstable, it is necessary to standardise the solution before use.

It is possible to standardise the solution using many common primary standards; however it is more convenient to use pure ascorbic acid as standard. Weigh about 50 mg of ascorbic acid accurately and *quantitatively* transfer this acid to a 250 mL volumetric flask containing about 2 g solid oxalic acid. (Oxalic acid is *poisonous*! Do not handle with bare hands or ingest.) Fill the flask to the 250 mL mark and shake gently until all solid has dissolved. This solution is used to standardise the *N*-bromosuccinimide solution, using the procedures described below.

N-bromosuccinimide. Add 25 mL of the standard ascorbic acid solution prepared above (use a pipette filler since poisonous oxalic acid is present in the solution) to a conical flask containing 5 mL of 4% potassium iodide solution, 2 mL of 10% acetic acid solution and three drops of starch indicator. Dilute with about 30 mL of distilled water and titrate with the *N*-bromosuccinimide solution provided. The end point is marked by the appearance of a permanent blue colour.

Whilst it is conventional to express the concentration of *N*-bromosuccinimide in terms of molarity, it is easier in this experiment to find the amount of ascorbic acid titrated by 1 mL of the *N*-bromosuccinimide. From the weight of ascorbic acid used, calculate the number of mg contained in 25 mL of standard solution. Dividing this amount by the average titre on *N*-bromosuccinimide gives the number of mg of ascorbic acid titrated by 1 mL of oxidant solution. (See also the section on dipstick chemistry later in this chapter.)

3. Titration of orange juice
Using 10 mL of your prepared sample of orange juice, proceed as outlined above in 'standardisation of oxidising agent', except that the orange juice aliquot replaces the 25 mL aliquot of standard ascorbic acid. Repeat the titration until you are convinced of the validity of your results. Express the ascorbic acid content in terms of mass of ascorbic acid per 100 mL of juice.

TABLE 13.4
Ascorbic acid levels in fruit juices—1984

Juice	Brand	No. of determinations	Mean mg/L	NH & MRC min. requirements mg/L
Orange	Farmland	11	657	400
	Sunrise (conc.)	11	365	400
	Sunburst	11	481	400
	Sunburst (conc.)	12	391	400
	Mr Juicy	15	719	400
	Fresh fruit[a]	17	450	(400)
Grapefruit	Farmland	16	684	300
	Fresh fruit[a]	4	438	(300)
Lemon	Fresh fruit[a]	6	206[b]	(350)

[a] Fruit juices are normally fortified with ascorbic acid to meet the legal minimum, on the assumption that the natural vitamin in the juice has disappeared on processing. Hence the high levels found in juices compared to fresh fruit. Concentrates on dilution according to directions give lower levels than the straight juices.

[b] The minimum requirement for lemon juice is greater than that found naturally in the fruit—apparently to meet consumer expectations!

Question

How does the ascorbic acid content as analysed by you compare with that claimed by the manufacturer? Suggest plausible reasons for any variations. Some results from first-year chemistry students at the Australian National University are given in Table 13.4. (See also Table 11.9 for the legal levels for vitamin C in fruit juices.)

EXPERIMENT 13.7

Electrochemistry

Equipment

Overhead projector
Petri dishes
Platinum wire
$m/20$ sulfuric acid
DC power supply
1 M nitric acid
Mercury
Potassium dichromate crystals

3 M potassium dichromate solution
Iron nail (clean)
2 M stannous chloride solution
Electrician's solder (fluxless)
1% agar gel (plus phenolphthalein
    indicator) at pH 8
Bent iron nails
Freshly prepared potassium
    ferricyanide solution

From dust to dust is the story of our lives and also of many of the products we manufacture. In particular, we take metals in their natural state of positive charged ions (as oxides and sulfides mainly) and expend energy to reduce them to the metal. For the rest of the time we fight continuously against their return to their natural state (in the oxidising environment that life on earth has caused) by a process we call corrosion.

The most sensible way to study corrosion is through electrochemistry and the most spectacular electrochemistry demonstration is the mercury beating heart and the mercury amoeba. These experiments are reported repeatedly in the chemical education literature, but my research leaves little doubt that the honours go to Professor E. J. Hartung, who was professor of chemistry at Melbourne University in 1933. The following two demonstrations come from Hartung's book, published in 1953 by Melbourne University Press (unfortunately now out of print), and were brought from Melbourne to Canberra with Professor Arthur Hambly.

In Hartung's experiment 199, cathodic attraction of mercury globules, shows one circular platinum electrode (anode) around the (inside) circumference of a petri dish in which is located a mercury globule under $m/20$ sulfuric acid and the other (cathode) dipping into the acid. A potential 40 VDC shows the mercury repelled from the circular anode and chasing the moving cathode around the dish.

Experiment 200, lowering of surface tension of mercury by contamination, is taken directly from Hartung.

A globule of mercury about 1 cm diameter is introduced into a dish so that it is completely covered by 1 M nitric acid; the dichromate crystal is then dropped in about 2 cm from the globule. As soon as the advancing orange ring of the dissolved salt comes in contact with the globule, this is thrown into violent contortions due to the formation of mercury chromate, which lowers its surface tension. The solution soon becomes turbid with the red precipitated chromate, which gradually hides the moving mercury from view. If, before the dichromate crystal is introduced, the globule is touched with a clean steel needle, it quivers and appears to shrink momentarily. The surface tension is here raised . . .

The beating heart experiment is best done in sulfuric acid of about 3 M with dichromate *solution* added (until it turns pale yellow) and with a clean, sharp iron nail, held so as to form a make-and-break contact.

EXPERIMENT 13.8

Test for lead in petrol

Equipment

Ultraviolet lamp
Dilute acetic acid
Dithizone solution (2–5 mg in 100 mL chloroform, freshly prepared)

Filter paper
Potassium iodide solution (16.5 g/100 mL)

Method

A piece of filter paper is saturated with petrol and exposed to strong sunlight for several hours or, more quickly, under ultraviolet light. It is then moistened with dilute acetic acid and a few drops of freshly prepared potassium iodide solution (16.5 g/100 mL). The appearance of a yellow colouration after several minutes indicates the presence of lead as lead iodide.

Alternatively, an even more sensitive test involves dithizone (diphenylthio-carbazone) reagent (2–5 mg of dithizone in 100 mL of carbon tetrachloride or chloroform—use freshly prepared). The reagent changes from green to red if placed on a filter paper containing lead salts (concentration limit 1 mg/kg). See also 'Chemistry of the car—Petrol engines', in Chapter 10).

EXPERIMENT 13.9

Test for lead in glazed pottery

Equipment

Detergent
Distilled water
4% v/v acetic acid
Dithizone solution (as for Experiment 13.8)

Method

The surface of the sample of pottery must be clean and free from grease. It should be washed in water containing detergent, then rinsed with distilled water. The utensil should then be filled with acetic acid (4% v/v) and left covered at room temperature for 24 hours. The acid should be stirred thoroughly and tested for lead as in Experiment 13.8.

The British Standard allows 2 mg/kg lead and 0.2 mg/kg cadmium for large casseroles (holloware), 7 mg/kg lead and 0.7 mg/kg cadmium for small casseroles and 20 mg/kg lead and 2.0 mg/kg cadmium for plates.

A very pretty electrolysis experiment is Hartung's experiment 160, electrolysis of stannous chloride solution.

> A few mL of 0.5 to 2 M stannous chloride solution are poured into the flat cell (deep petri dish) and the parallel electrodes [two pieces of electrician's solder* work nicely and prevent complications of gas evolution] are sharply focused on the screen. On starting the current [3–6 V depending on distance apart of the electrodes] large flat blades and thin spears of tin grow rapidly from the cathode . . . on reversing the current, the crystals of tin rapidly dissolve while blades grow from the other electrode . . . It often happens that some of the crystals of tin become detached; these appear to crawl along the base of the cell towards the anode in a remarkable manner as the metal adds on in front and is dissolved at the rear . . . At the lower concentrations of $SnCl_2$ the crystals are more ferny.
>
> With silver nitrate (and inert electrodes) at higher concentration (1 M), a deposit appears on *both* electrodes, silver as expected on the cathode and 'argentic oxy-nitrate' on the anode. This salt is a good electrical conductor, so that on reversing the current, silver grows from the points of the spikes while the silver at the other electrode dissolves, and when it has disappeared, the oxy-nitrate begins to grow.

A corrosion experiment which should be done in all schools is the ferroxyl indicator (Hartung's experiment 178).

> Having been stressed at the head and point during manufacture, a steel nail shows differences in potential on its surface when immersed in an electrolyte, the stressed parts being positive. Corrosion at these parts tends to occur, with the formation of ferrous ions, if the electrons left behind in the metal are removed. Dissolved oxygen in the electrolyte may do this at the negative areas by the formation of hydroxyl ions. To demonstrate this effect, a 1 per cent solution of agar is made by soaking the material for some hours in water and then boiling until dissolved. To this solution, phenophthalein is added and the pH adjusted to 8. A few drops of freshly prepared potassium ferricyanide solution are then added, the warm solution is poured into a petri dish, the well-cleaned nail introduced and the preparation allowed to set. After a few hours, deep pink zones diffuse out from the shaft of the nail (hydroxide ion) and blue green zones form around the head and point (ferrous ion). For instant colour, apply a voltage from a battery across the 'electrodes'.

Hartung goes on to show that aluminium will not be attacked in copper sulfate solution, but is rapidly attacked in copper chloride. This can be used to illustrate the properties of the protective oxide layer of aluminium and the susceptibility of aluminium window frames near the ocean.

An interesting electrochemical question regarding a consumer product is to ask what the physicochemical difference is between the silver, the red and the gold (black) batteries. This takes you back to the section on the chemistry of surfaces (Chapter 4) and chemistry of the car (Chapter 10).

---

* Solder, an alloy of tin and lead, melts at a lower temperature than either of the components. The solid (eutectic) is the analogue of the azeotrope in the Dean and Starke experiment, Experiment 13.1 (see also Chapter 6).

EXPERIMENT 13.10

Rapid determination of lead in paints

Equipment

Solution A: 0.5% ammonium nitrate + 5% tartaric acid prepared in distilled water (shelf-life one month).

Solution B: 0.2% rhodozonic acid (Na salt) prepared in distilled water (shelf-life approx. one hour).

Distilled water

Filter paper

Tissue paper

The Continental Can Company of the USA has a patented analytical technique to ascertain the overspray of lead-based solders on the inside of beverage cans. This technique was modified by D. Mulcahy (South Australian Institute of Technology) for testing paint to provide faster and deeper penetration into the surface but was otherwise very similar.

Distilled water is first sprayed onto the paint surface and any dust removed using a tissue. It is then similarly sprayed with solution A. An 11 cm diameter no. 1 Whatman filter paper is sprayed with solution A to saturate the paper. Press the paper onto the prepared surface, taking care to make a good surface contact. Use pins if necessary. Allow to stand for at least one minute. Remove paper and air dry.

Spray solution B onto the dried filter paper. If the paper does not turn bright orange initially, solution B should be reprepared and the paper sprayed again.

Any lead is shown by a pink colouration. If any colouration is indicated then the paint is assumed to have a lead content in excess of 2%. The depth of pink colouration is an indication of lead content: the deeper the colour the higher the lead content. The area of colouration is an indication of the area of penetration of solution A and not an indication of the lead content.

Incidentally, the *hardness* of a paint surface can be measured by pushing forward across the paint surface at an acute angle a series of pencils of different hardness (6B to 6H) in a reproducible manner. This is the basis of an SAA standard.

EXPERIMENT 13.11

Permanent press for wool

Equipment

Permanent crease solution (3% sodium bisulfite or sodium metabisulfite solution containing a little detergent)

Two wool samples

Small sponge

Watch glass

Detergent

Steam iron

Method

Pour some of the permanent crease solution into a watch glass and sponge a line of solution down the centre of one of the wool samples. Then crease the sample along the sponged line and press the crease in with the steam iron for about 30 seconds. Using the steam iron again, press a similar crease in the untreated sample.

Immerse both samples in warm water (about 70°C) containing a small amount of detergent. Then check to see if the treatment has produced a 'permanent crease'.

Wool is a protein polymer, and human hair is a protein polymer very similar to wool. A 'permanent wave' in human hair is much the same as a permanent crease in a skirt or a pair of trousers.

EXPERIMENT 13.12

Steam distillation of eucalyptus leaves

Equipment

Steam distillation apparatus
Eucalyptus leaves

Method

The oil produced from eucalyptus leaves contains eucalyptol (1,8-cineole) in many cases as the main component in an oil yield of about 1–3%. The oil is useful in cough drops, mouthwashes, gargles, dental preparations, inhalants, room sprays and medicated soaps. It is an effective disinfectant. (See Fig. 13.12.)

eucalyptol   eugenol

Fig. 13.12   Essential oils

Because the oil decomposes when boiled (bp 176°C), the method used to extract it from the leaves and purify it is to distil the leaves with steam. The amount of oil carried over with the steam depends on how volatile the oil is. Using this method of steam distillation the temperature of the oil never exceeds 100°C and the oil is not destroyed. In fact this technique is identical to the one used in the Dean and Starke experiment (Experiment 13.1), except in this case we are interested in the oil phase rather than the water phase (see Fig. 13.14).

You might also like to produce oil of cloves (eugenol, bp 164°C) from cloves.

An interesting consumer aspect of steam distillation involves the use of the antioxidant BHA in vegetable oils used for frying (see Fig. 3.4). The frying of wet

Fig. 13.13 A bush eucalyptus still, Braidwood, NSW, c. 1900. Cooling pipes run into a creek (just visible). (Courtesy of Professor A. J. Birch)

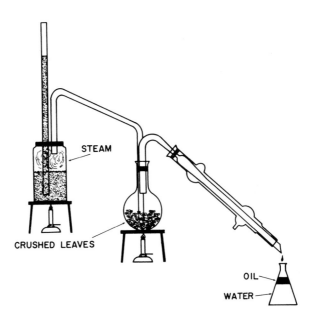

STEAM

CRUSHED LEAVES

OIL

WATER

Fig. 13.14 Steam distillation of eucalyptus leaves in the laboratory: the oil settles out as a separate phase (layer) on top of the water (compare with Dean and Starke distillation, Experiment 13.1).

foods involves the evolution of steam, which helps to distil out these (phenolic) antioxidants. It is interesting to note that the natural antioxidant α-tocopherol, because of its long chain, is less volatile and thus more useful. Actually, the chemical effects of frying are quite complex (see Experiment 13.2). Most of the products of oxidation at room temperature are not found. Peroxides are decomposed rapidly and volatile rancid compounds distil off. On the other hand, polymers form (which causes high viscosity), the colour darkens (oxy-compounds) and excessive foaming occurs (the polar oxygenated compounds formed act as surfactants).

EUCALYPTUS AND TEA-TREE OILS

Ian Rae of Monash University describes a marvellous eucalyptus distillery a couple of hundred kilometres out of Melbourne at Wedderburn[1].

'Leaves and twigs are collected into a large tank mounted on a truck. At the distillery, on goes the lid, steam is blown in at the bottom, and away she goes. Most of the boiler fuel consists of dried gum leaves and the whole thing has a Heath Robinson look about it, with the oil finally collecting in an old Hoover Twin Tub.

TABLE 13.5
Australian eucalyptus oil production

Year	Local production (in tonnes)	Imports (in tonnes)
1947–48	900	—
1948–49	560	—
1949–50	520	—
1950–51	780	—
1951–52	775	—
1952–53	540	—
1977	200	c. 75
1987	140–160	c. 270

Fig. 13.15   Australian tea-tree oil has many different uses.

Back in 1788, only 10 months after the establishment of the colony, the Surgeon-General John White sent a quarter of a gallon of eucalyptus oil to England for further testing. Unfortunately the thriving industry that built up over the years has declined, and today Australia is a net importer of eucalyptus oil (see Table 13.5).[2]

In contrast, exports of tea-tree oil are increasing.[3] Distillation of tea-tree oil (*Melaleuca alternifolia*) is mainly (+)terpinen-4-ol and related compounds. This oil has bactericidal and fungicidal uses, from dealing with *Salmonella typhi*, to treating tinea, acne and diabetic gangrene. It is used as a perfume toner and nutmeg substitute. It kills dry rot fungi, and is used in a variety of products ranging from water-gel fire blankets to oral contraceptives. It is also used to combat legionnaire's disease.

Today, the essential oil industry seems to be concentrated in Tasmania, where they turn out oils of fennel (exported to make Pernod), peppermint, lavender, caraway, parsley, boronia, blackcurrant and hops.

EXPERIMENT 13.13

Tests on plastics

Equipment

Samples of plastics
Copper wire in cork holder
Large beaker with water
Spatulas

Method

When burning plastics:

- Use very small pieces of your samples and hold them with tongs or a wooden peg.
- Experiment in a well ventilated place because of the fumes produced.
- Hold the burner or candle at an angle so that any drops of molten plastic fall onto an asbestos or other non-flammable mat.

1. Copper wire test

Heat a copper wire (stuck into a cork as a holder) in a gas flame until any yellow or green colour disappears. Press the heated wire into the plastic sample and then put the wire with a little molten plastic on it back into the flame. A green colour indicates that the plastic contains halogen—probably chlorine in poly (vinyl chloride) (PVC) or poly (vinylidene chloride) (PVDC). Before repeating the test with another plastic, again heat the copper wire until the green colour disappears.

It should be noted that an additive may contain a halogen (chlorine, bromine or iodine) that gives rise to a positive result. Also cyanide (from, say, Orlon) may give a positive result.

2. Density

Some polymers are less dense than water and hence will float. These are polyethylene, polypropylene, styrene-butadiene and nitrile (some types). It is essential for the sample (not a foam type) to be properly wetted and pushed below the surface, and then released. The presence of large amounts of additives can change the density.

3. Feel

Poly(ethylene) and poly(tetrafluoroethylene) have a waxy feel not possessed by other polymers. Clean the surface to remove grease or plasticisers.

4. Heating tests (in a laboratory)

A small piece (0.1 g) of the material is placed on a clean spatula (nickel spoon-like object), previously heated to remove traces of combustible material. It is then gently warmed, without ignition, over a small colourless gas flame until it begins to fume. The sample is removed from the flame and the fumes are tested with moist litmus paper (red and blue) to determine whether they are acidic, alkaline, or neutral. Check the odour as well. The sample is now moved to the hottest zone of the small gas flame and the following points noted:

1. Whether or not the material burns and if so, how easily.
2. The nature and colour of any flame (very sooty flame generally indicates an aromatic polymer, but may result from carbon black filler).
3. Whether or not the material continues to burn after removal from the flame.
4. The nature of any residue.

Table 13.6 considers some of the possibilities. It was compiled using *uncompounded* polymers. Thus highly plasticised PVC may *not* be self-extinguishing.

EXPERIMENT 13.14

Measurement of vapour-locking temperatures

Equipment

Gilpin dilatometer	Large beaker
High boiling silicone oil	Magnetic stirrer
Silicon carbide boiling chip	Thermometer (high temperature)

Method

Any hydraulically operated braking system must use a non-compressible fluid in the master and wheel cylinders. The formation of vapour (which is compressible) in the wheel cylinder, the result of boil off from the brake fluid, leads to 'vapour locking' and brake failure. The boiling point of fluid mixtures is seldom 'ideal' and so is not a good parameter for describing what will happen in real circumstances. (*Ideal* is an expression used in physical chemistry to describe systems that obey predictable, simple mathematical expressions. Real systems vary from almost ideal to completely non-ideal—just like people!)

TABLE 13.6
Burning test on plastic

Plastic	Flame colour	Odour	Other features
1. The material burns but extinguishes itself on removal from the flame			
Casein	yellow	resembles burnt milk	very difficult to ignite, alkaline fumes
Melamine-formaldehyde	pale yellow with light blue-green edge	formaldehyde and fish-like	
Nylon	blue with yellow tip	resembles burning vegetation	melts sharply to clear liquid which can be drawn into fibre
Phenol-formaldehyde	yellow	phenol and formaldehyde	very difficult to ignite
Poly(tetrafluorethylene)	yellow	none	burns with extreme difficulty, chars very slowly, acidic fumes
Poly(vinyl chloride) } Poly(vinylidene chloride) }	yellow with green base	acrid	acidic fumes
Urea-formaldehyde	pale yellow with light blue-green edge	formaldehyde and fish-like	very difficult to ignite—alkaline fumes
2. The material burns and continues to burn on removal from the flame			
Acrylonitrile-butadiene-styrene	yellow with blue base, smoky	styrene	
Alkyd	yellow, smoky	pungent, unpleasant	
Cellulose acetate	yellow	acetic acid (vinegar)	acidic fumes
Cellulose nitrate	yellow	possibly camphor on gentle warming	burns at a very fast rate, may explode
Epoxide	orange-yellow, smoky	acrid	
Ethyl cellulose	pale yellow with blue-green base	resembles burning wood	
Poly(acrylonitrile)	yellow	cyanide initially and then resembles burning wood	drips on igniton
Poly(carbonate)	yellow, smoky	phenolic	difficult to ignite initially
Poly(ethylene) } Poly(propylene) }	yellow with blue base	resembles burning candle wax	becomes clear when molten
Poly(methyl methacrylate)	yellow with blue base	methyl methacrylate	
Polystyrene	yellow with blue base, very smoky	styrene	
Polyurethane	yellow with blue base	acrid	
Poly(vinyl acetate)	yellow, smoky	vinyl acetate	black residue
Poly(vinyl alcohol)	yellow, smoky	unpleasant sweet	black residue

A Gilpin dilatometer (see Fig. 13.16) consists essentially of a U-tube, one arm of which has a calibrated tube having a volume of 5 mL. Enough brake cylinder fluid is placed in the dilatometer so as to fill completely the calibrated side-arm and then the other side of the U-tube is filled to the same height. The filled portion of the U-tube is completely immersed in a bath containing a high boiling silicone oil (which remains transparent at high temperature). A boiling chip of silicon carbide should be added to the oil bath. A metal shield surrounding the apparatus helps reduce temperature fluctuations.

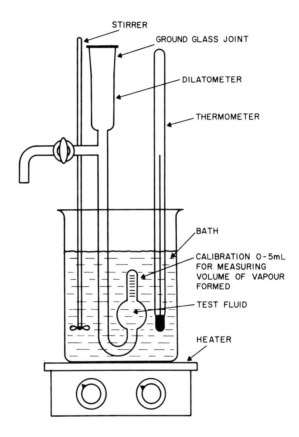

Fig. 13.16   Gilpin dilatometer apparatus for measuring vapour locking temperatures

The temperature of the stirred bath is raised slowly, when approaching the vapour formation temperature, and a plot of bath temperature versus volume of vapour produced is recorded. Once vapour begins to form there is a rapid increase in vapour volume for only a few degrees rise in temperature.

The experiment should be carried out on brake fluid from a sealed container and then on fluid with 1%, 2%, 3% and 4% added water. A plot of vapour formation temperature versus added water should be plotted (see Fig. 10.16). An 'unknown' sample from your car can then be measured. Alternatively, different brands of brake fluid can be compared.

When measuring used brake fluid from a car, a black precipitate is often formed on heating and this remains when the fluid is cooled. This could be because the heat causes flocculation of carbon black, which is leached out of the rubber caps during service. Another possibility is that, if the brake fluid has been in the braking system for more than about 12–18 months, the inhibitor system then becomes exhausted (depending on use, moisture uptake, etc.) and corrosion may occur. In this case, the colloidal ferric hydroxide would similarly flocculate on heating. If this precipitate does occur, filter it off (after cooling) and test it.

EXPERIMENT 13.15

Infra-red spectroscopy

Equipment

Infrared spectrometer
Plastic film used for storing food, oven bags, food storage bags, windows of window envelopes, computer disc box wrapping.

Method

If an infrared spectrometer is available, there are a number of quick experiments that are possible. Plastic film has just the right thickness for an infrared spectrum to be taken *au naturel*. A number of these are shown in Figures 13.17–13.19. The compositions can be obtained from authenticated standards, which also allow the other major additives to be assessed (e.g. plasticisers). In addition, the interpretation of the particular spectral bands can be discussed. The infrared spectrum for polystyrene is shown in Figure 13.20.

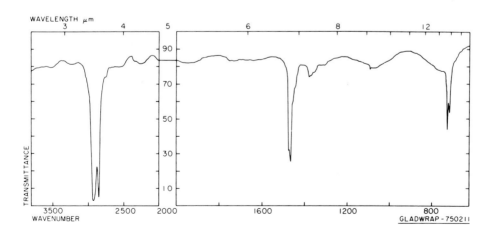

Fig. 13.17 Gladwrap, which shows the infra-red spectrum of polyethylene

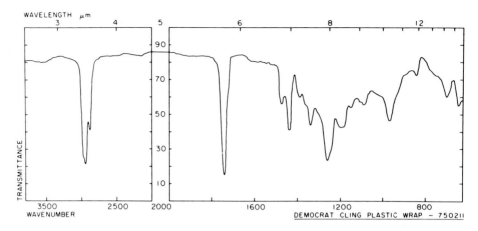

Fig. 13.18   Democrat Cling Plastic Wrap, which shows the infra-red spectrum of poly(vinyl chloride). The large peak at 1750 cm^{-1} is not present in pure PVC. It is due to an ester plasticiser

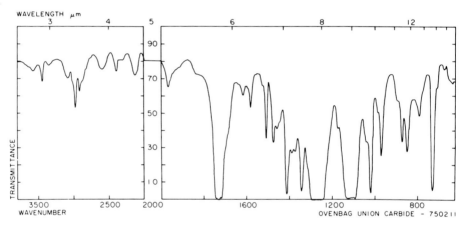

Fig. 13.19   Ovenbag (Union Carbide), which shows the infra-red spectrum of polyterephthalate (polyester). This material is also found as a laminate on the cardboard bases of frozen pizza packs

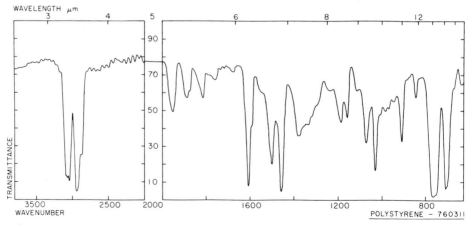

Fig. 13.20   Infra-red spectrum of polystyrene, used for window envelopes and meat packs, also to calibrate infra-red spectrometers.

Whenever a beam of radiation traverses a boundary in which there is a change in refractive index, part of the radiation is reflected and interference effects, which show up in the spectrum as *fringes*, occur (see Fig. 13.21). Interference fringes can usually be recognised because of their uniform, evenly spaced (frequency scale) appearance. The thickness of the film d is given by

$$d = \frac{\triangle m}{2n(\tilde{v}_1 - \tilde{v}_2)}$$

where $\triangle m$ is the number of fringes (complete cycles) in the region $\tilde{v}_1$ to $\tilde{v}_2$ (units of cm^{-1}), and n is the refractive index. For example, for polyethylene, $n = 1.51$, and using Figure 13.22 we can calculate

$$d = \frac{4}{2(1.51)\,(1380-800)}$$
$$= 2 \times 10^{-3}\,cm$$
$$= 0.8\,mil\,(thousandths\,of\,an\,inch)$$

When making the calculation, choose an area free from absorption bands.

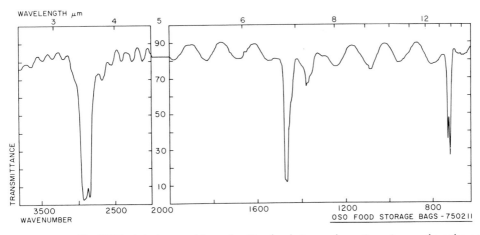

Fig. 13.21 Interference fringes for Oso food-storage bags. Superimposed on the polyethylene spectrum are interference fringes, from which the thickness of the film can be calculated

EXPERIMENT 13.16

Some aspects of wine analysis

Equipment

Burettes, 50 mL
Stirrers
Beakers, plastic cups
Evaporating dishes (small)
Semi-micro test-tubes and racks

Aluminium foil for chromatography
   tank covers
Filter paper cut to fit 250 mL beaker
   (9 × 6 cm)
Distillation kit

Volumetric flasks, 50 mL

Bulb pipettes, 1 mL

Heating mantles, hot plate, water bath

Water bath 60°C, in which to suspend 250 mL Erlenmeyer flasks

Desiccator with chromatography plates (Merck ART 5749) cut to 5.5 × 7.0 cm, reactivated

1 pear-shaped flask, 100 mL, B14

1 round bottom flask, 100 mL, B14

1 still head, B14

1 condenser, B14

1 recovery bend, B14

1 thermometer adapter, screw capped, with washer, B14

Chemicals

Acidified $K_2Cr_2O_7$, 0.1 M, 1 L

Ferrous ammonium sulfate, 0.3 M, 5 L

Iodine 0.1 M, 5 L

NaOH, 0.1 M, 30 L, labelled 'For pH titrations'

NaOH, 1 M, 5 L, labelled 'For distillation'

$Na_2S_2O_3$, 0.050 M, Volucon 500 mL

Cs^+(CsCl) and K^+(KCl) 10 000 mg/kg, 500 mL

H_2SO_4, 25% v/v, 4 L

HCl, conc., 500 mL

Indicators (shelf-life one week):

Starch 2%

N-phenylanthranilic acid, labelled 'Redox indicator'

Make up the following standard solutions:

20 mg/kg K^+ + 1000 mg/kg Cs^+ 500 mL (50 mL Cs^+ + 10 mL K^+)

40 mg/kg K^+ + 1000 mg/kg Cs^+ 500 mL (50 mL Cs^+ + 20 mL K^+)

80 mg/kg K^+ + 1000 mg/kg Cs^+ 500 mL (50 mL Cs^+ + 40 mL K^+)

100 mg/kg K^+ + 1000 mg/kg Cs^+ 500 mL (50 mL Cs^+ + 50 mL K^+)

120 mg/kg K^+ + 1000 mg/kg Cs^+ 500 mL (50 mL Cs^+ + 60 mL K^+)

150 mg/kg K^+ + 1000 mg/kg Cs^+ 500 mL (50 mL Cs^+ + 75 mL K^+)

Use the stated volumes of the two standard solutions and make up to 500 mL with distilled H_2O.

A separate laboratory area is needed as the spotting area.

Other equipment

Fireproof squares, 30 × 30 cm

Microburners

Maxi-bins, labelled 'Used spotters'

500 mL bottle with 10 mL Kipps dispenser labelled 'TLC solvent'

Melting-point tubes

Dropper bottles labelled:

0.5 M Lactic acid (MM 90.08)

0.5 M Succinic acid (MM 118.09)

0.5 M Citric acid (MM 192.12)

0.5 M Malic acid (MM 134.09)

0.5 M *d*-Tartaric acid (MM 150.09)

In the fume cupboard

Line the whole cupboard with Benchcote to protect from bromocresol green staining. Double-line the area where bases will be used.

The bromocresol green is applied to the plates via a Pyrex B24 spraying unit using a compressed air line. Provide a paper-towel-lined plastic basin for spraying.

Technical notes

1. Approximately 30 minutes before the class commences, the pH meters are calibrated against a solution of pH 7 buffer solution.

   The electrodes need to be checked very carefully from day to day to ensure that the chamber inside the electrode has sufficient saturated KCl solution and that there are no air bubbles trapped inside the glass bulb.

   The best results for chromatography are obtained by drying the plates in the oven overnight (80°C) after chromatographic separation has occurred, and then spraying with bromocresol green the next morning.

2. Ferrous ammonium sulfate, 0.3 M (acidified), 5 L.

   Dissolve approximately 588 g of ferrous ammonium sulfate in a sulfuric acid solution, which has been prepared previously by dissolving 250 mL of concentrated sulfuric acid cautiously in water (2.5 L). Always add the acid to the water, *never* vice-versa. Use ice-cooled distilled water and make up to 5 L. Label: '0.3 M Ferrous ammonium sulfate' (acidified).

3. Iodine solution, 0.01 M (stabilised with potassium iodide—KI), 5 L.

   Grind solid iodine crystals (mortar and pestle). Weigh 12.70 g of this material, dissolve in distilled water, add 100 g KI and make up to 5 L.

4. Sodium hydroxide, 0.1 M, 30 L.

   Dissolve 120 g of NaOH pellets in about 3 L of distilled water. Cool, then make up to 30 L. Mix the container thoroughly before standardising against Volucon 0.01 M HCl. Label: 'For pH titrations'.

5. Sodium hydroxide, 1 M, 15 L.

   Dissolve 200 g of NaOH pellets in distilled water. Cool and make up to 5 L. Store in a plastic container. Label: 'For distillation'.

6. Starch indicator.

   Dissolve 8 g of starch slurry in 400 mL of boiling distilled water. Cool and store in 50 mL dropping bottles.

7. Indicators (as per Vogel if required).

   N-phenylanthranilic acid: dissolve 0.25 g of the crystals in about 12 mL of 0.1 M NaOH and dilute to 250 mL with water. Label as 'Redox indicator: N-phenylanthranilic acid'.

8. Spotter solutions, 200 mL of each solution.

   Provide in the spare laboratory 0.5 M solutions of lactic acid (MM 90.08), succinic acid (MM 118.09), citric acid (MM 192.12), malic acid (MM 134.09), d-tartaric acid (MM 150.09) in 1:1 H_2O absolute ethanol.

9. Eluent solution, 2.5 L.

   Saturate n-butanol with water. Mix 10 parts of this solution with one part of 85% formic acid.

10. 25% v/v sulfuric acid, 4 L.

    1 L of concentrated sulfuric acid is made up to 4 L with distilled water whilst being cooled.

11. Caesium solution.

   25.334 g AR (Analytical Reagent) grade CsCl is dissolved in 100 mL distilled water and make up 2 L. Each student requires 10 mL.

12. Standard potassium solutions.

   Prepare 2 L of 100 mg/kg K^+ stock solution by dissolving 200 mg AR grade KCl in distilled water and make up to 2 L.

IN SEARCH OF A QUALITY RED

In young red wines, there is wide variation in the state of equilibria between coloured and colourless forms of the anthocyanin pigments. Also, because of the reactive nature of these and other phenolic constituents, there is progressive decrease in concentration of these monomeric forms, along with the formation of more stable polymeric pigment structures. Thus wine colour composition changes continuously during conservation and ageing.

Strong correlations have been shown recently between aspects of wine colour and quality rating in young red wines from the Southern Vales district of South Australia. Encouraging support has come from observations on premium French wines for the hypothesis that a high degree of ionisation of anthocyanins is a common feature of young red wines with superior quality.

Of the variable factors involved in anthocyanin equilibria, with consequent influences upon wine colour density and colour composition, pH is most important. High pH is also associated with organoleptic defects, as well as susceptibility to biological and oxidative spoilage.

In many Australian dry red wines, excessively high levels of potassium appear to be responsible for high pH. On these bases, it is proposed that wine quality in future vintages can be improved by adopting several modifications to current viticultural and oenological practice.

Source: T. C. Somers, In search of quality for red wines.

Introduction

Chemically, the major components of wines are water (up to 91%), ethanol (9–14%), various weak organic acids, sugars, naturally occurring pigments (such as anthocyanins in red wines), tannins, various metal ions (e.g. K^+, Mg^{2+}, Na^+) and additives acting as antioxidants and bactericides (e.g. sulfur dioxide, SO_2). Wines (with apologies to the oenophiles) are dilute, acidic, aqueous, ethanol solutions that may sometimes be coloured. With such a multitude of components (any one of which can alter the quality of a wine), it is clear that careful analysis of the various components of a wine is of critical importance during the manufacturing process. In this experiment, you are introduced to some aspects of wine analysis that are based on standard oenological analytical techniques used currently in the wine industry in Australia.

The parameters analysed are: (1) the various 'types' of wine acidity (including a qualitative identification of the different types of organic acids present); (2) the ethanol content; (3) the sulfur dioxide content; (4) the concentration of potassium; and (5) the approximate concentration of tannins.

Wine acidity

Acidity in wines is caused largely by the presence of various carboxylic acids in solution. Without such acids, wine would spoil easily and taste flat, whereas too much gives an acid taste. The main acids present in wines are tartaric, malic, lactic and acetic acids, and small amounts of citric and succinic acids are also present. The molecular formulae and some physical properties of these acids are given in Table 13.7.

TABLE 13.7
Acids present in wines

Acid		Dissociation constants			mp (°C)
		k_1	k_2	k_3	
Tartaric	COOH | (CHOH)$_2$ | COOH	1.0×10^{-3}	4.5×10^{-5}		170
Malic	COOH | CHOH | CH$_2$COOH	3.9×10^{-4}	7.8×10^{-6}		100
Lactic	COOH | CHOH | CH$_3$	8.4×10^{-4}			18
Succinic	COOH | (CH$_2$)$_2$ | COOH	6.9×10^{-5}	2.5×10^{-6}		183
Citric	CH$_2$—COOH | OH—C—COOH | CH$_2$—COOH	8.4×10^{-4}	1.8×10^{-5}	4.0×10^{-7}	153
Acetic	CH$_3$COOH	1.8×10^{-5}			16.6

There is no direct relation between pH and total titratable acidity in table wines. The concentration of hydrogen ions $[H_3O^+]$ may vary tenfold, whereas the titres generally vary no more than two or threefold. This is because of variations in concentration of the cations, principally potassium. The mineral content is mainly in the skin of the grape, so white wines (in which skins are minimally extracted) have lower potassium. The high potassium leads to high pH, unsatisfactory colour density and composition, and other deficiencies.

Tartaric acid usually constitutes half or more of the total acid content of a wine. Part of the tartaric acid is present as potassium bitartrate; that is, the tartaric acid is partially neutralised. Potassium tartrate is much less soluble in alcohol than in water and, during the fermentation of wine, the increasing alcohol content may cause the bitartrate to crystallise out as a deposit on corks and walls of vessels ('winestone', cream of tartar).

A dipstick wine-test kit is available for measuring the level of tartaric acid in fruit juices, musts and wines (see section on dipstick chemistry later in this chapter).

The amount of *malic acid* in grapes decreases during maturation. In wine, malic acid may be converted to *lactic acid* in a process known as *malolactic fermentation*. Lactic acid tastes less 'acidic' than malic acid. Too much conversion may give the wine a flat taste.

Acetic acid in small quantities is also formed during normal alcoholic fermentation (<0.03 g/100 mL). Much larger amounts are produced when wine goes 'sour' (i.e. on air-oxidation of alcohol in the presence of micro-organisms).

The *total acidity* of a wine is a measure of the total hydrogen ion content in solution, both as free H_3O^+ and undissociated species, and should therefore be expressed as amounts of hydrogen ions per unit volume. It is, however, common oenological practice to express total acidity as the mass of tartaric acid (in grams) per 100 mL of wine (i.e. to convert the measured $[H_3O^+]$ to an equivalent amount of tartaric acid). It can also be expressed in terms of the volume of 0.1 M base required to neutralise 100 mL of wine (e.g. in Germany).

Acetic acid and some other carboxylic acids comprise the *volatile acidity* of wine. The volatile acids can be separated from the wine by steam distillation (see Experiment 13.12) and estimated quantitatively by titrating the distillate with base. Alternatively, the volatile acids can be removed by repeated evaporation of the wine sample and the residue analysed for *non-volatile acids* or *fixed acidity*. The determination of volatile acidity is very important in the quality control of wines. Most countries have regulations governing maximum permissible limits. Volatile acidity is usually expressed as the mass of acetic acid (in grams) per 100 mL of wine. The determination of both total and volatile acidity can be done using standard methods of volumetric analysis. Qualitative identification or even a semi-quantitative estimation of the individual acids can be achieved by paper or thin-layer chromatography.

The process of malo-lactic fermentation mentioned above (i.e. the change in relative amounts of malic and lactic acids during fermentation) can also be followed in a qualitative or semi-quantitative way by paper or thin-layer chromatography. However, most of the analytical methods required for accurate quantitative determination of the individual acids are complicated and time-consuming.

The *pH of a wine* is the most important chemical parameter influencing the chemical and microbiological stability of the wine. The pH relates, among other factors, to colour, taste, redox potential, resistance to micro-organisms and the action of sulfur dioxide in preventing spoilage. At the usual pH of wine, sulfur dioxide exists in solution predominantly as bisulfite, HSO_3^-, but the lower the pH, the higher is the proportion of the species sulfur dioxide, $SO_2(aq)$, which is the active antimicrobial agent. (Compare to $HOCl/OCl^-$ in swimming pools, Chapter 5; and ionisation of aspirin, Fig. 9.3.) pH has an effect on the colour of red wines. Low pH favours a more desirable colour because of the effect on ionisation of anthocyanins, which are the pigments responsible for the colour. There is no direct and predictable relationship between pH and total acidity, although an empirical relationship between pH and the ratio of potassium hydrogen tartrate to tartaric acid has been proposed.

The *ethanol content* of wines is governed by the amount of sugar in the grapes and by the yeast cells, most of which are inhibited when the concentration of ethanol reaches about 15% (v/v). Ethanol content may be determined in several ways. The two methods that can be undertaken using standard laboratory equipment are determination of density following distillation and quantitative oxidation of C_2H_5OH with potassium dichromate. Both methods are used in winemaking quality-control work and research. More efficient and accurate instrumental methods of analysis are also available.

Sulfur dioxide is an essential additive to all wines; it acts both as an antioxidant and bactericide. If too little sulfur dioxide is used, the wine may suffer bacterial attack and may develop the characteristic taste of an oxidised wine. Too much sulfur dioxide, on the other hand, will be very evident to the palate. Most of the added sulfur dioxide combines with various aldehydes and ketones, especially acetaldehyde (from ethanal), the remaining uncombined SO_2 being available to protect the wine. There are legal limits to the total SO_2 permitted, and this varies from country to country, 250 mg/L (parts per million) being a commonly accepted value. About 20–40 mg/L of free SO_2 will safeguard a wine without affecting its taste, whereas levels below 10 mg/L in a white wine show that it is in imminent danger of going bad.

Alkali metals, notably potassium, are of particular concern in Australian red wines, which tend to have an excessively high potassium content, apparently leading to high pH, which has an adverse effect on the quality of the wine.

Method

Some of the procedures provided below are in outline only, since it is assumed that you are already familiar with them.

Before beginning any analysis, select a particular brand and type of wine and stay with this brand throughout the project.

1. Analysis of wine acidity
(a) pH of wine, pH titration and determination of total acidity
Collect about 120 mL (no more!) of the chosen wine and measures its pH. Then carry out a detailed pH titration of 25 mL of the wine with standard 0.1 M NaOH. To do this, pipette 25 mL of wine into a beaker, immerse the electrode and add NaOH from a burette, initially 1 mL at a time, noting the pH after each addition. The mixture must be swirled *gently* after each addition of base *before* taking any pH reading. When the pH rise begins to steepen, reduce base addition to 0.2 mL at a time. Continue base addition till the pH reaches beyond 10. Plot the titration curve of pH versus cumulative volume of NaOH added and determine from the

curve the volume of NaOH required for the pH to reach 8.3. This pH is taken to be the wine pH at equivalence point for the purpose of this titration (in the USA, pH 8.2 is used). Observe and describe any colour changes during the titration.

Repeat the titration to obtain duplicate results, but this time simply add NaOH at the *same rate* as above till the pH reaches 8.3 and note the volume added. The average titre is used to calculate the *total acidity* of the wine in terms of:

Number of moles of OH$^-$ required per 100 mL of wine

Mass of tartaric acid present per 100 mL of wine.

(b) Determination of fixed or non-volatile acidity

Evaporate two samples of wine (25 mL in 100 mL Erlenmeyer flasks) on a hot plate to about 5–10 mL. Use boiling chips to avoid excessive spattering. Add about 25 mL of water (using a measuring cylinder) and evaporate again. Repeat the process. Cool the final solution and dilute it to about 30 mL. Titrate with 0.1 M NaOH solution as in (b); that is by *slow* addition of NaOH to pH = 8.3, using the pH meter. The volume of NaOH used corresponds to the *fixed acidity* of the wine; the difference between the total and fixed acidity gives the *volatile acidity*.

Hence, calculate:

- the fixed acidity of your wine sample expressed as the number of moles of OH$^-$ required per 100 mL wine.
- its volatile acidity in terms of:
  (i)   no. of moles of OH$^-$ required/100 mL wine;
  (ii) mass of acetic acid (in grams)/100 mL wine.

2. Ethanol content analysis

Two different methods will be used in this part of the experiment. The density method involves distilling the ethanol from the wine as an aqueous ethanol solution and determining its density. A calibration curve of density versus composition is then used to determine the ethanol content. The dichromate oxidation method involves chemically oxidising all the ethanol in the wine sample to acetic acid with *excess* potassium dichromate ($K_2Cr_2O_7$) and then back titrating with a standard solution of a reducing agent.

(a) Alcohol content by the density method

Pipette a 50 mL sample of wine into a dry 100 mL round-bottomed, pear-shaped distillation flask and make it just alkaline with NaOH (use pH paper). Distil until the temperature reaches 100°C or until half the original volume has distilled over (be careful of foaming, especially with red wines). Collect the distillate in a clean, dry 50 mL Erlenmeyer flask.

Transfer the distillate *quantitatively* into a clean, dry and accurately pre-weighed 50 mL volumetric flask and top up to the mark with distilled water. Reweigh to calculate the density of the ethanol solution. Keep this solution for the next experiment.

The density-composition calibration curve is obtained by preparing five aqueous ethanol solutions of *accurately* known compositions in the range 5% (v/v) to 25% (v/v). Prepare these solutions by pipetting *appropriate* volumes of absolute ethanol into dry, clean and accurately pre-weighed 50 mL volumetric flasks and top up with distilled water and reweigh to determine their densities.

Plot the density-composition curve and estimate from it the ethanol content of your wine sample.

(b) Alcohol content by dichromate oxidation

The stoichiometric equation of the dichromate oxidation of ethanol to acetic acid is:

$$2K_2Cr_2O_7 + 3CH_3CH_2OH + 8H_2SO_4 \rightarrow 3CH_3COOH + 2K_2SO_4 + 2Cr_2(SO_4)_3 + 11H_2O$$

As stated earlier, an excess of dichromate is always used and a back titration is necessary to determine the amount of unreacted dichromate. For this purpose, a solution of ferrous ammonium sulfate, $Fe(NH_4)_2(SO_4)_2 \cdot 6H_2O$, is used. The reaction equation, expressed in ionic form, is:

$$Cr_2O_7^{2-} + 6Fe^{2+} + 14H^+ \rightarrow 2Cr^{3+} + 6Fe^{3+} + 7H_2O$$

Pipette 25 mL of the standard 0.1 M solution of potassium dichromate into *each* of *two* 100 mL Erlenmeyer flasks. To each, add 1 mL of the ethanol solution obtained above (use pipette), stopper the flask loosely and place it in a thermostat bath at 60°C for 30 minutes for oxidation to occur. Then transfer the mixture quantitatively to each of two 250 mL titration flasks, dilute to about 150 mL with distilled water, add 10 drops of the indicator solution and titrate with a solution of ferrous ammonium sulfate until the solution changes from an initial darkish brown colour to a clear dark green. (The solution assumes a violet-red colour just prior to the end-point.)

The ferrous ammonium sulfate solution, however, must first of all be standardised against the standard potassium dichromate solution used and this is best done with an ethanol-dichromate mixture maintained in the thermostatically controlled bath. Use the procedure below.

Pipette 10 mL of the dichromate solution into a 250 mL titration flask, add 100 mL distilled water (measuring cylinder) followed by 10 drops of the redox indicator solution. It should become a darkish brown colour. Titrate with the ferrous ammonium sulfate solution until one drop of titrant changes the solution to dark green. Repeat to obtain reproducible results and, from the given reaction equation, calculate the concentration of the ferrous ammonium sulfate solution.

From the back-titration results, calculate the amount of $Cr_2O_7^{2-}$ used in the oxidation of 1 mL of the ethanol solution and hence the concentration of the ethanol (wine) solution in terms of moles per litre. Convert this to % (v/v). Use the following ethanol density data for the conversion.

Temp.°C	18	19	20	21	22	23	23	25
Density g/mL	0.79114	0.79029	0.78945	0.78860	0.78775	0.78691	0.78606	0.78522

3. Detection of non-volatile acids by thin-layer chromatography

Relatively non-volatile acidic components of a wine can be detected by the technique of thin layer chromatography since each acid has its own unique R_f value (Experiment 13.3, part A). Detection of spots is by spraying the developed plate with an acid–base indicator solution. It is also of interest to compare the relative amount of each acid in different *types* of wines (e.g. red wines vs. white wines). Carry out the TLC analysis on both red and white wines.

Because of their pigments, the analysis of acids in red wines is likely to present problems. This is overcome by absorbing the pigments onto activated charcoal before TLC analysis. Collect about 10 mL of red wine in a measuring cylinder, add 3–4 spatulas of activated charcoal, swirl for a few minutes and filter through a piece of fluted filter paper into an evaporating dish. Concentrate down to 1 mL by heating on a hot-water bath. The white wine sample (10 mL) is similarly treated but *without* the pigment decolourisation step. Apply your sample (multiple applications will be needed for your wine sample) on the TLC plate, along with the following standard acid solutions: tartaric, lactic, malic, succinic and citric. Develop the plate in the eluent mixture (n-butanol : formic acid : water, in the ratio 8:1:2.5 by volume). The plate is then dried in an oven for at least two hours to remove all traces of formic acid before spraying with bromocresol green, which turns yellow when acidic and blue when basic.

Sketch your chromatogram, record all R_f values and identify all the non-volatile acid components in each of the wine samples (Table 13.8). Comment on the *relative amounts* of each acid in the different types of wines examined by comparing the colour intensities of the spots.

TABLE 13.8
Non-volatile acids in wine

Acid	R_f value	Present in white wine	Present in red wine
Succinic	0.53	yes	yes
Lactic	0.45	no	yes
Malic	0.23	yes	yes
Citric	0.16	no	no
Tartaric	0.14	yes	yes

4. Sulfur dioxide (SO$_2$) analysis

Sulfur dioxide in wine is present both in the free form, $SO_2(aq)$, HSO_3^- and SO_3^{2-}, and in combined form as bisulfite complexes with certain organic substances. Therefore analysis can provide either free SO_2 content or total SO_2 content, depending on the method of analysis used. In this experiment, both analyses will be done. The determination is based on a redox reaction between SO_2 (a reducing agent) and iodine, I_2 (an oxidising agent):

$$SO_2 + I_2 + 2H_2O \rightarrow 4H^+ + SO_4^{2-} + 2I^-$$

For free SO_2, the wine is acidified and titrated directly, whereas for total SO_2, it must first be made alkaline with sodium hydroxide to break down the bisulfite complexes, and then acidified and titrated. With red wines there is some masking of the end-point, and this analysis is difficult to perform.

(a) Free SO_2 content

Pipette 50 mL of the chosen wine into a 250 mL conical flask, add approximately

5 mL of 25% H_2SO_4 and 2-3 mL of starch solution. Titrate with 0.01 M I_2 to the first appearance of a dark blue colour, which persists for about two minutes. One *careful* titration ought to be sufficient. Dipsticks for determining sulfur dioxide can be used instead.

(b) Total (free and combined) SO_2

Using a measuring cylinder, place about 25 mL of 1 M NaOH in a 250 mL conical flask and pipette in 50 mL of wine. Shake the flask and then let it stand for 15 minutes. Add 10 mL of 25% H_2SO_4 and 2-3 mL of starch and titrate with 0.01 M I_2 as in part (a).

Calculate the amounts of free and total SO_2 in the wine as moles per litre, as % (w/v) and as mg/L.

5. *Determination of potassium ion (K^+) content*

As mentioned earlier, Australian red wines tend to have a high K^+ content (on average, 1.1-2.8 g/L), whereas other wines have contents like 0.65-0.94 g/L. This has implications in *wine quality* as well as in process control (see T.C. Somers reference). The K^+ content of white wines is generally lower (250-700 mg/L). Dipsticks are available for measuring the concentration of potassium.

6. *Tannin content*

Tannins are discussed in the section on leather (Chapter 7). The tannins are responsible for the 'bite' that is obvious in red wine (and strong tea). The tannin concentration is measured by oxidation with potassium permanganate, but this reagent also oxidises other species such as the alcohol and even the indicator. Two titrations are therefore carried out. We assume that only tannins and other colouring matter are adsorbed preferentially onto charcoal.

Solution required: 0.004 M $KMnO_4$ and 0.5% indigo carmine indicator. The indigo carmine indicator is made up by dissolving 0.5 g of the dyestuff in 60 mL of warm distilled water. The solution is cooled, 4 mL of conc. H_2SO_4 is added, and the volume made up to 100 mL with distilled water. The solution is filtered through a no. 42 Whatman paper.

Pipette 5 mL of wine into a conical flask and add 10 mL of distilled water. Place funnel into neck of conical flask (to reduce losses by spilling) and heat the flask gently until the volume of wine and water has been reduced to 5-7 mL. (By now the alcohol will have boiled off.) Add about 250 mL of cold distilled water and pipette in 2 mL of the indigo carmine indicator. Titrate with 0.004 M $KMnO_4$ until a golden yellow colour appears (titre A).

Take 20-25 mL of the wine and add 1 g of activated charcoal. Stir thoroughly. Filter. Pipette 5 mL of the decolourised wine into the conical flask and continue as above. Ensure you use the same volume of indicator. This blank determination allows for the indicator and any oxidisable compounds (other than tannins, anthocyanins, etc.) in the wine (titre B).

The amount of permanganate needed to oxidise the tannins is thus A − B = C.

Tannins have a variable composition and the titration is referred to a standard 'purified' tannin for which 1 mL of 0.1 N $KMnO_4$ = 4.11 kg of tannin. Thus 1 mL of 0.004 M $KMnO_4$ = 0.832 mg of tannin. (Note the difference between normality N and molarity M.)

$$\text{Thus \% tannin in wine} = \frac{0.832 \times C}{50} = 0.01664\ C$$

Source: This experiment was matured and refined over many years (like a good red) by many people at ANU (see Haddad *et al.* in bibliography). For further information contact the Australian Wine Research Institute, Waite Road, Urrbrae, SA 5064.

Fig. 13.22

DIPSTICK CHEMISTRY

Take a dipstick, like the sort your doctor gives you to test your urine, only this one is for ascorbic acid (vitamin C) not glucose. (Dipsticks are available from chemical supply houses.) In fruit juices we would expect to find vitamin C and in fact the law requires that you should find minimum amounts (see Table 11.9). You can use the dipsticks to measure vitamin C content in fruit juices (see Experiment 13.6). You normally find more than in the original fruit because the processor adds the legal minimum anyway to replace any vitamin naturally present that has not survived processing and storage.

Just how stable is the vitamin? There is considerable information available but why not do your own experiments, changing the following variables: (a) the effect of boiling in water; (b) cooking at different pH (e.g. by adding bicarbonate of soda); (c) the effect of boiling in the absence of oxygen. If you blend vegetables and measure vitamin C content before and after boiling, you will find a large *increase*, because the boiling extracts the water-soluble vitamin from the food. That gives us some hints about good cooking techniques (see Chapter 3).

Legally, a chemical is defined on the basis of what it is used for. When vitamin C is added to beer, it is there as an antioxidant (not as a vitamin), so it cannot be mentioned as a vitamin on the label. If you are keen on vitamin C, you may be better off eating ham, because vitamin C is added to the nitrate preservative to prevent formation of nitrosamines. Incidentally, the dipstick was developed for Dr Archie Kalokerinos, who asked the Ames company of Indiana, USA, for a quick, cheap diagnostic test that would enable him to test his theory that a great deal of ill health in the Australian Aborigines was caused by a lack of sufficient vitamin C in their diet.

Kalokerinos quotes that ascorbic acid appears in the urine when the concentration in the plasma reaches 1 mg/100 mL and suggests a value of 25 mg/mL in the urine should be established. Why not test your own urine and establish how you excrete a large dose (1–2 g) over a period of one day. An example is given in Figure 13.23. You will need to measure both the volume of urine and the concentration of vitamin C. You can also plot the amount of the original vitamin remaining and also the rate of excretion during the day (which peaks rapidly). Every person appears to have a different profile. The intake of excess vitamin C has been criticised by those who have learnt enough biochemistry to know of its metabolism to oxalic acid (calcium oxalate is the main component of kidney stones) but not enough to realise how slow this is, in comparison to the rate of excretions.

This high proportion excreted may encourage the conservationists among you to evaporate your urine and recycle the vitamin C. But perhaps not.

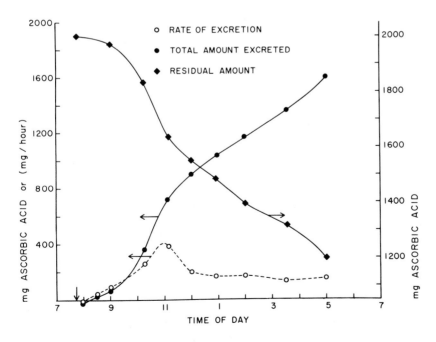

Fig. 13.23 Excretion of vitamin C

Having used the dipsticks in good faith, we might ask, 'How reliable are they?' You might be fired if you publish false results. By making up a set of standard solutions of ascorbic acid by carefully weighing out the pure compound and calibrating the dipsticks, we can breathe a sigh of relief if they correspond correctly to the colours.

If you then test some antifreeze you'll find it contains 200 mg/L of ascorbic acid. How very odd! You begin to wonder how the dipstick actually works and what might be interfering. Suddenly, there is a reason for wanting to know the chemistry of the dipstick. Ascorbic acid is a reducing agent, but so are lots of other things. Actually, as radiators rust, it would not be stupid if antifreeze contained an antioxidant—that is, a reducing agent. So you can't just test something with a dipstick without thinking about what's going on.

You can build up a story of chemistry, home economics and food law around this dipstick. The following box summarises the use of a number of other dipsticks.

COPPER Cu^+/Cu^{2+}

The test is suitable also for the detection of metallic copper. A colour change is produced on the reagent zone by as little as 0.5 μg of copper.
1. Copper in water
2. Copper in ice confectionery
Sensitivity: 10–5000 mg/L
Legal limits for food (mg/kg), see Chapter 12.

GLUCOSE

Reaction is not given by fructose or galactose nor by the non-reducing disaccharides, sucrose and lactose. Maltose does react. (For more details on sugars see Fig. 11.13.)
1. Hydrolysis of sucrose to glucose (invertase or H^+)
2. Formation of glucose in germinating seeds (~1 minute for halved barley grains against wetting)
See references for further details.

NITRATE/NITRITE

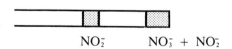

$$NO_2^-\qquad NO_3^- + NO_2^-$$

Interference from nitrite removed by adding aminosulfonic acid. Separate nitrite strip not needed.
1. Oxides of nitrogen in air. Sensitivity 1 mL of NO_2/m^3 of air.
2. Nitrite in saliva, average 7 mg/L, except after foods with high nitrate level (celery, beets etc.), where you obtain elevated levels for 24 hours.

3. Fermented raw meat (salami type). Legal limit 500 mg/kg, nitrate.
   Cured meat (corned beef). Legal limit 125 mg/kg, nitrite.
   Canned ham. Legal limit 50 mg/kg, nitrite.
4. Vegetable (nitrite):
     conventional carrots 40–100 mg/kg
     organically grown carrots 200–400 mg/kg
     fresh spinach ~5 mg/kg (if refrigerated for two weeks ~300 mg/kg).
5. Denitrification in waterlogged soils
   Soil + nitrate + glucose → N_2O
   Sensitivity:
     nitrate 10–500 mg/L
     nitrite 1–50 mg/L
   See references for further details.

SULFITE

1. Air pollution. Sensitivity: 5 mL (13 mg) of SO_2/m^3 air
2. Preservative (legal limits)

Fruit juices	115 mg/L; concentrated 600 mg/kg
Gelatine	1000 mg/kg
Dehydrated carrots	1000 mg/kg
Cheese	300 mg/kg
Sausages	500 mg/kg
Wine	300 mg/kg

Sensitivity: 10–500 mg/L

TARTARIC ACID

Grape juice and wine (added acetic acid ensures total tartrate is measured)
Less than 1 g/L indicates very poor quality.
See also Experiment 13.16.
Sensitivity: 0.5–10 g/L

You can also make up your own dipsticks. When I was in Thailand working on a consumer chemistry project in 1973, I discovered that one of the big problems there was the adulteration of food with borax to give it that tangy flavour. So it was a matter of buying turmeric rhizomes (bulbs) in the marketplace and making 'turmeric paper' with a piece of newspaper and then sending out teachers to test the more salubrious street restaurants for the local equivalent of 'bangers' with borax.

The tartaric acid test strip is basically used to test wine. I tried it out in a very expensive restaurant in Canberra. An elegant wine waiter arrived with the ordered vintage and poured some in a glass for me to taste. Out came the test-kit and the wine was right on the border-line of acceptability. I slowly shook my head, more in sadness than in anger. The wine-waiter's face had to be seen to be believed. How good the test strip is for wine I am not sure, but for testing the *sangfroid* of a wine

Fig. 13.24 'If this was made from grapes, why no tartaric acid?'

waiter it is absolutely superb. The fact that I dare not return to that restaurant can only be a saving on the family budget—so I have no regrets.

CHEMICALS AVAILABLE IN THE MARKETPLACE

It should be noted that some of the chemicals listed in Table 13.9 as readily available are *very poisonous* (e.g. cadmium salts) and many are *very dangerous* in mixtures (e.g. strong oxidising agents, potassium dichromate, calcium hypochlorite 'chlorine', etc. with oxidisable material: see 'Swimming pools', Chapter 5).

TABLE 13.9
Chemicals readily available in the marketplace

Chemical name	Common name	Availability
Acetic acid (dilute)	Vinegar	Supermarket
Acetone	Nail polish remover	Hardware
Aluminium	Aluminium (foil, cans)	Supermarket
Aluminium oxide (hydrated)	Alumina hydrate	Craft
Ammonia solution	Cloudy ammonia	Supermarket
Ammonium chloride	Sal ammoniac	Hardware
Ammonium nitrate	Ammonium nitrate	Garden supplies
Ammonium sulfate	Sulfate of ammonia	Garden supplies
Ascorbic acid	Vitamin C	Pharmacy, supermarket
Beeswax	Beeswax	Craft
Barium carbonate	Barium oxide flux	Craft
Barium chromate	Barium chromate	Craft
Benzoic acid	Preservative	Health food
Boric acid	Boracic acid	Pharmacy
Cadmium selenide	Red overglaze, cadmium red, pigment red 108	Craft

TABLE 13.9
continued

Chemical name	Common name	Availability
Cadmium sulfide	Yellow overglaze, cadmium yellow, pigment yellow 35, 36	Craft
Calcium carbide	Carbide	Speleologist's supplies
Calcium carbonate	Marble chips, egg shells, snail shells, sea shells, white wash, calcamine	Garden supplies, craft
Calcium fluoride	Fluorspar	Craft
Calcium hexametaphosphate	Calgon	Supermarket
Calcium hydroxide	Hydrated or slaked lime	Garden supplies
Calcium oxychloride	'Solid chlorine' 70%	Swimming pool supplies
Calcium phosphate	Bone ash	Craft
Calcium sulfate	Gypsum, modelling powder	Craft
Camphor	Some mothballs	Supermarket
Carbon	Coke, charcoal	'Lead' pencil, craft
Citric acid	Citric acid	Supermarket, health food
Chromic oxide	Chromium green oxide	Craft
Cobalt carbonate	Blue glaze	Craft
Cobalt oxide	Blue glaze	Craft
Collagen	Sausage casings	Butcher
Copper	Copper wire, foil	Electrician, hardware, craft
Cryolite (sodium aluminium fluoride)	Cryolite	Craft
Cupric carbonate	Green glaze	Craft
Cupric oxide	Black copper oxide	Craft
Cupric sulfate	Bluestone	Garden supplies, hardware
Cuprous oxide	Red copper oxide	Craft
Cyanuric acid	Cyanuric acid	Swimming pool water purifying tablets
Dolomite (calcium magnesium carbonate)	Dolomite	Craft
Ethanol	'Metho', methylated spirits[a]	Supermarket, hardware
Ethylene glycol	Antifreeze (contains additives)	Garage
Feldspar (aluminium silicates)	Feldspar	Craft
Fluorspar (calcium fluoride)	Fluorspar	Craft
Gelatin	Gelatin	Supermarket, pharmacy
Glucose	Glucose	Supermarket
Glycerol	Glycerine	Pharmacy, hardware
Gum Arabic	Gum Arabic	Health food, craft
Hydrochloric acid	Muriatic acid	Pharmacy, hardware, pool supplies
Hydrogen peroxide	Peroxide	Supermarket, pharmacy
Iron	Nails, steel wool	Hardware, supermarket
Iron chromite	Chromite	Craft
Iron (ferric) oxide	Rust, rouge, red iron oxide	Craft
Iron (ferrous) sulfate	Iron sulfate	Garden supplies
Lead	Lead 'sinkers'	Hardware, sports stores
Lead antimonate	Antimonate of lead, Naples yellow	Craft
Lead carbonate, basic	White lead flux	Craft
Lead dioxide	Car battery plates	Old batteries
Lead oxide, mono	Litharge, flux	Craft
Lead oxide, red	Red lead, flux	Craft
Lead sulfide	Galena	Mineral, rock shop

TABLE 13.9
continued

Chemical name	Common name	Availability
Lecithin	Lecithin	Health food
Linseed oil	Linseed oil	Hardware, craft
Lithium carbonate	Flux	Craft
Magnesium metal	Torch bulbs, fire starters	Camping supplies
Magnesium carbonate	Carbonate of magnesia, magnesite	Pharmacy, craft
Magnesium sulfate	Epsom salts	Pharmacy
Manganese carbonate	Manganese carbonate	Craft
Manganese dioxide	Manganese oxide	Used dry cells (see Experiment 10.1), craft
Methyl red indicator	pH test solution	Swimming pool supplies
Naphthalene	Some mothballs	Pharmacy, supermarket
Nickel (II) oxide	Green glaze, nickel oxide	Craft
Nickel (III) oxide	Black glaze, nickel oxide	Craft
Para-amino-benzoic acid	PABA	Health food (see the 'Chemistry of cosmetics—Sunscreens', Chapter 4)
Phosphoric acid	Phosphoric acid	Hardware
Phthalocyanines	Phthalocyanine green, blue	Craft
Polyester resin	Fibre glass resin	Craft, hobby, speciality suppliers
Potassium aluminium sulfate	Alum	Pharmacy, pool chemicals
Potassium chloride	Salt substitute 'No Salt', 50% KCl	Supermarket, health food
Potassium hydrogen tartrate	Cream of tartar	Supermarket
Potassium ferricyanide/ ferrocyanide	Case hardening agents	Speciality hardware
Potassium permanganate	Condy's crystals	Pharmacy
Silica gel	Drying agent in packaging (blue/red)	Parcels
Silicon carbide	Carborundum	Hardware, craft
Silicon oxide	Silica, sand	Craft
Silver chloride	Silver chloride glaze	Craft
Sodium bicarbonate	Baking soda, carb. soda	Supermarket, health
Sodium bisulfite	Sodium metabisulfite	Beer/wine kits
Sodium carbonate	Washing soda, soda ash	Supermarket, craft
Sodium chlorate	Sodium chlorate	Hardware (weed-killer)
Sodium chloride	Table salt	Supermarket
Sodium hydrogen glutamate	Monosodium glutamate	Health food, supermarket, Chinese restaurants
Sodium hydrogen sulfate	Pool acid	Swimming pool supplies
Sodium hydroxide	Caustic soda	Hardware
Sodium hypochlorite	Bleach, chlorine water	Supermarket
Sodium perborate	Non-chlorine bleach	In detergent formulations
Sodium silicate	Water glass (egg preservative)	Pharmacy, craft
Sodium sulfate	Glauber's salt	Pharmacy
Sodium tetraborate	Borax	Supermarket, hardware, craft
Sodium thiosulfate	Photo 'hypo'	Pharmacy, photo supplies
Stannic oxide	Tin oxide	Craft
Starch	Starch	Supermarket
Sulfur	Sulfur	Pharmacy
Sulfuric acid	Battery acid	Garage, pharmacy

TABLE 13.9
continued

Chemical name	Common name	Availability
Tartaric acid	Tartaric acid	Health food, supermarket
Tin	Modern pewter, can plating	
Titanium dioxide	White paint and 'white out' pigments, titanium white	Stationers, hardware, craft. sporting goods
Tungsten	Light bulb filaments	Home, supermarket
Tunsten carbide	Tungsten carbide	Hardware
Urea	Urea	Garden supplies
Zinc	Zinc	Metal casings of used dry cells
Zinc chloride (32% in HCl)	Solder flux	Hardware, craft
Zinc oxide	Zinc oxide	Craft
Zirconium	Zirconium	Photoflash bulbs
Zirconium oxide	Zirconium oxide	Craft

a All products are denatured with chemicals to make them undrinkable.

REFERENCES

1. I. Rae, 'Letter from Monash'. *Chem. Aust.* **55**(11) 1988, 380.
2. E. V. Lassak, 'The Australian eucalyptus oil industry, past and present'. *Chem. Aust.* **55**(11), 1988, 396.
3. I. A. Southwell, 'Australian tea tree: oil of melaleuca, terpinen-4-ol type'. *Chem. Aust.* **55**(11), 1988, 400.

BIBLIOGRAPHY

General
Tooley, P. *Experiments in Applied Chemistry*. John Murray, London, 1975.

Chromatography
Dodds, C. J. 'Food colouring in Smarties'. *School Science Review* **53**(184), 1972, 589; Hillman, R. A. H. *ibid.* **56**(197), 1975, 763.
Youatt, J. B. *Education in Chemistry* **14**(1), 1977, 29.

Dipstick chemistry
Freeland, P. W. 'Some applications of glucose-sensitive reagent strips in biology teaching', *School Science Review* **55**(190), 1973, 91.
Freeland, P. W. 'Some applications of reagent strips in soil experiments', *School Science Review* **56**(197), 1974, 38.
Freeland, P. W. 'Determination of glucose produced by embryo- and endosperm-halves of germinating barley grains', *School Science Review* **56**(194), 1974, 88.
Freeland, P. W. 'Test and indicator papers for the semi-quantitative determination of ions and various compounds', *School Science Review* **59**(206), 1977, 59.

Extraction
Laswick, J. A. and Laswick, P. H. 'Caffeine and benzoic acid in soft drinks', *J. Chem. Ed.* **49**(10), 1972, 708.

Infra-red experiments

Fox, R. H. and Scheutzman, H. I. 'The infrared identification of microscopic samples of man-made fibres'. *J. Forensic Sciences* **1**, 1968, 397.

Gove, R. C., Hannah R. W., Patlacini, S. C., Porro, T. J. (Perkin Elmer Corp.). 'Infrared and ultraviolet spectra of seventy-six pesticides'. *J.A.O.A.C.* **54**, 1971, 1040.

Haslam, J., Willis, H. A. and Squirrell, D. C. M. *The identification and analysis of plastics,* 2nd edn. Iliffe Books, London, 1972; reprinted Heyden, 1981.

Hausdoff, H. H. 'Infrared applications to the analysis of cosmetics and essential oils'. *J. Cosmetic Chemists* **4**, 1953, 251.

Krause, A., Lange, A., Eyrin, M. *Plastics analysis guide: chemical and instrumental methods,* Hanser Publ., Munich, 1983.

Murphy, J. E. and Schwemer, W. C. 'Infrared analysis of emulsion polishes'. *Anal. Chem.* **30**, 1958, 116.

Perkin Elmer Infrared Applications Studies: Polymer Plastics Industry, 3 pp., 1968; *Paints and coatings*, 3 pp. (reflectance methods), 1968; *Packaging and containers*, 3 pp. (plastic films), 1968; *Waxes*, 3 pp. 1968; *Analysis of diesel and petroleum base lubricants*, 3 pp.; *Drugs*, 34 pp., 1972 (a very well prepared publication with a flow diagram system of identification of about 60 drugs).

Sammul, O. R., Brannon, W. L. and Hayden, A. L. (US FDA). 'Infrared spectra of compounds of pharmaceutical interest. *J.A.O.A.C.* **47**, 1964, 918–91.

Selinger, B. 'A near I.R. colorimeter'. *Education in Chemistry* **5**, 1968, 61.

Stavinoha, L. L. and Wright, B. R. *Spectrometric analysis of used oils.* Society of Automotive Engineers, New York, 1969.

Textiles and plastics

Identification of textile materials, 6th edn. Textile Institute, Manchester, 1970.

Ridley, A. and Williams, D. *Simple experiments in textile science.* Heinemann, 1974.

Water

Duckworth, R. B., Oswin, C. R., Johnson, D. S., Lamb, J., Aithin, A., Ranken, M. D. and Goodall, J. B. 'The roles of water in food', seven articles in *Chem. in Indust.,* 18 December 1976, 1039–60.

Experiment 13.2

Talburt, W. F. and Smith, O. *Potato processing.* AVI Westport, Connecticut, 1959.

Williams, A. E. *Potato crisps.* Food Trade Press, London, 1951.

Experiment 13.4

Allen, R. L. M. *Colour chemistry.* Nelson, London, 1971.

Hallas, G. 'Colour chemistry, parts I–IV', *School Science Review* **56**(196), 1975, 507; **56**(197), 1975, 699; **57**(198), 1975, 38; **57**(199), 1975, 635.

Experiment 13.6

Gyorgy, R., and Pearson, W. N. (eds). *The vitamins*, vol. 7, 2nd edn. Academic Press, New York, 1967, pp. 27–51.

Sebrell, W. H., and Harris, R. S. (eds). *The vitamins*, vol. 1, 2nd edn. Academic Press, New York, 1967, pp. 338–359.

Experiment 13.16

Amerine, M. A., and Ough, C. S. *Wine and must analysis.* Wiley Interscience, New York, 1974.

Fowles, G. W. A. 'Sulfur dioxide and tannin'. *Education in Chemistry* **15**, 1978, 89 (as expanded by Melbourne State (Teachers) College).

Haddad, P. R., Sterns, M., and Wardlaw, J. 'Analysis of wine: an undergraduate experiment'. *Education in Chemistry* **15**, 1978, 87.

Rankine, B. C. 'Chemical aspects of wine control'. *Chem. Aust.* **46**, 1979, 225.

Simpson R. F. 'Some important aroma components of white wine', *Food Tech. Aust.* **31**, 1979, 516.

Somers, T. C. 'In search of quality in red wines', *Food Tech. Aust.* **27**, 1975, 49. (The chemical connection in red wine is lucidly explored.)

Vogel, A. I. *A textbook of quantitative chemical analysis*, 3rd edn. Longman, 1961, pp. 304–309, 370.

Volger, A., and Kunkely, H., 'Photochemistry and beer'. *J. Chem. Ed.* **59**(1), 1982, 25.

Williams, P. J., 'Brandy and grape spirits: distillation, composition and flavour', *Chem. Aust.*, (Proceedings) 1976, 179.

Appendixes

APPENDIX I
WATER QUALITY IN PRIVATE SWIMMING POOLS

It is recommended that interested readers consult the Australian Standard *Private swimming pools—water quality*, issued as AS 3633.

This appendix covers the following topics:

A. Pollutants

B. Water treatment

C. Definition of terms

A. Pollutants

SOURCES OF POLLUTION

Bathers are the chief source of nitrogenous compounds in swimming pools. Quite large quantities of both ammonia nitrogen and organic nitrogen are introduced into the water via perspiration and urine. The amounts of each may vary, but it would be reasonable to assume that children are responsible for the greatest proportion of urine in the pool, while high ambient temperatures would increase the rate of perspiration.

PERSPIRATION

An active swimmer may lose up to 1 L of perspiration per hour in pool water at 24°C and an ambient air temperature of 38°C. Perspiration contains sodium chloride, calcium and magnesium salts, and nitrogenous compounds (which consist of large amounts of organic nitrogen, ammonia nitrogen, urea, creatinine and amino acids).

The pH of perspiration is in the range of 4.0 to 6.8. However, the nitrogen content will vary according to diet.

URINE

The chemical composition of urine is more complex than that of perspiration and causes more difficulty in treatment because of its high nitrogen content. Urine is composed of urea, creatinine, uric acid, hippuric acid and inorganic salts.

Ammonia and urea (from perspiration and urine) are the main products which affect the chlorination process. Ammonia nitrogen reacts with chlorine within minutes to form chloramines. However, urea must first go through a hydrolysis reaction before it can combine with chlorine and this can be accelerated by the presence of certain enzymes. This hydrolysis takes three to four hours under normal conditions.

ORGANIC NITROGEN
The presence of organic nitrogen in pool water seriously interferes with the chlorination process.

Whereas chlorine reacts with ammonia in a matter of minutes at pH levels of 7 to 8 with complete destruction of ammonia and 75 to 80% loss of nitrogen, the reaction of chlorine with some organic nitrogen compounds may take days before reaching completion. The persistent residuals formed with chlorine read in the test as dichloramines although they are much more complex compounds.

BACTERIA
Both humans and animals can pollute pool water with bacteria. The presence of organic matter will promote bacterial growth, whereas sanitising chemicals such as chlorine will destroy or inactivate the bacteria.

Human pathogenic micro-organisms that can be found in swimming pools include coliforms (*Escherichia coli*), *Staphylococcus*, *Streptococcus*, *Salmonella* and *Neisseria* species. *E. coli* is used as an indicator of fecal pollution from human or animal sources. *Staphylococci* and *Streptococci* are used as indicators of pollution originating from the nose, throat, skin and mouth, and are more resistant to chlorine than coliforms.

Pseudomonas aeruginosa is an opportunistic bacterial pathogen that can cause eye, ear and skin infections in pools that are inadequately disinfected.

Another commonly occurring bacterium, *Mycobacterium marinum*, is often found on wet swimming pool surrounds and this causes skin granulomatoses.

VIRUSES
Both humans and animals can also pollute pool water with viruses, which can then infect other swimmers.

Many viruses can be transmitted from one swimmer to another via pool water. Most of them, especially the enteroviruses, are more resistant to chlorine than *E. coli*, or the many adenoviruses that can be transmitted. Three are particularly important with respect to swimming pools. Two cause pharyngoconjunctival fever, which is an upper respiratory tract infection with fever, very severe conjunctivitis and often abdominal pain. The other type causes keratoconjunctivitis, which is a fairly severe conjunctivitis.

Because of the large volume of water in a typical swimming pool, and the often low virus concentration, virus detection is both difficult and expensive. However, there are methods (e.g. the swab method) that have been used with some success in ascertaining the presence of viruses in swimming pools. However, these methods require trained personnel and sophisticated laboratory facilities for testing.

AMOEBAE
Acanthamoeba species and *Naegleria fowlerii*, both of which occur widely in nature, can produce a usually fatal meningoencephalitis. These amoebae invade the swimmer through the cribriform plate (in the nasal cavity) and then migrate to the brain.

ALGAE
Algae are microscopic plants. Two varieties are found in pools: those that float freely in the water and the more persistent variety which imbeds itself into pores and crevices in concrete and between tiles. Sunlight, carbon dioxide, mineral matter, nitrogenous compounds (or atmospheric nitrogen) and other nutrients are essential for algae to grow.

Algal growths are not only objectionable because of the obvious discolouring of the water (i.e. turbidity), but also because they are slimy, and contribute to accidents in and around a swimming pool. Algae will also harbour and foster bacterial growth and retard the action of chlorine. Their reaction with chlorine may even give rise to odour problems.

Heavy algal growths will increase the chlorine demand to a point where the ordinary levels of free chlorine will not kill them off. It is then necessary to shock dose the pool by maintaining a free chlorine level in excess of 10 mg/L overnight, when the pool is not in use. Usually the next morning the algae will brush off quite readily. If not, the shock dose should be repeated until all the algae are killed.

The presence of algae in a swimming pool is an indication that a regular free chlorine residual (during the swimming season) or sufficient algicide (during the non-swimming season) is not being maintained in the water.

CORROSION

Corrosion in swimming pools is caused by prevailing acidic conditions, i.e. when the pH of the water is less than 7.0. Such conditions can cause deterioration of structural concrete and cement rendering, which may cause tiles to lift. Metal fittings including pumps, ladders, underwater light fittings and heat exchangers will also corrode.

Chemically unbalanced water will also etch the cement finish or the grouting between tiles. However, if the pH of the water is above 7.0, metal fittings are less affected by such conditions.

The design of fresh water pools may *not* be suitable for conversion to salt water use. This is because the high concentration of salt required for chlorine generation, or that naturally present in sea water pools, may result in the salt penetrating the concrete shell of the pool and attacking the reinforcing steel, which results in weakening and extensive staining of the shell.

TOTAL DISSOLVED SOLIDS (TDS)

Excessive concentrations of total dissolved solids in a fresh water pool (i.e. a non-salt chlorinated pool) may indicate that the pool water is 'stale' (affected by nitrogen, ammonia, amines or other organic contaminants). Accordingly, the pool water may require excessive amounts of sanitiser.

To reduce the concentration of TDS it may be necessary to totally or partially empty the pool and refill with fresh water.

Source: Standards Australia postal ballot draft.

B. Water treatment

WATER SANITISING PRODUCTS

The dose of sanitiser required to achieve and maintain the recommended minimum levels is dependent on a number of factors such as the number of bathers, water temperature, sunlight conditions, pollution from leaves, grass and debris.

Chlorination

Calcium hypochlorite

Calcium hypochlorite is commonly available as white to off-white granules which contain 650 g/kg of available chlorine or in tablet form which contains 700 g/kg of available chlorine. The granular form should preferably dissolve in water prior to dosing or it may be added directly to the pool water. However, the tablet form *must* be placed in a floating dispenser, to allow for slow release of the available chlorine, before it is added to the pool water.

> **CAUTION**
> GRANULAR calcium hypochlorite MUST NOT be used in a fixed or floating dispenser.

If stored in accordance with the instructions, calcium hypochlorite will have a storage life of several years.

When this product is added to pool water, the calcium hardness of the water will be increased, resulting in a tendency towards scale formation. Also, because the product contains a small amount of free alkali, the pH of the water will be increased slightly. Moreover, when this product is dissolved in water, its stability to ultraviolet (UV) light and temperatures in excess of 26°C is poor. However, the stability to UV light can be substantially improved by use of a pool stabiliser (i.e. isocyanuric acid).

Sodium hypochlorite

Sodium hypochlorite is commonly available as a yellow solution containing 80 to 125 g/L of available chlorine. As a solution, this product is convenient to dispense but it is relatively bulky to store. Although it contains an alkali stabiliser, the product is still not very stable, and even if stored in a cool area away from direct sunlight its shelf life is only four to eight weeks. Since the alkali stabiliser is caustic soda, the addition of sodium hypochlorite to pool water will slightly increase the pH of the water. This product does not affect water hardness. As with other sanitisers that do not include a chlorine stabiliser, the stability of the available chlorine in this product to ultraviolet (UV) light and temperatures in excess of 26°C is poor, although the UV stability of the product can be substantially improved by use of a chlorine stabiliser (i.e. isocyanuric acid).

Stabilised chlorine

With this type of pool treatment the available chlorine is released in the water to react with bacteria and organic matter. At the same time, a chlorine stabiliser (i.e. isocyanuric acid) is released, but is not consumed.

The most common stabilised chlorine compounds are as follows:

a. Sodium dichloroisocyanurate (a chlorinated isocyanurate salt) which is available in granular form, containing up to 630 g/kg of available chlorine. This product is soluble and, because its pH is almost neutral, it has no effect on either the pH or total alkalinity of the pool water.

b. Trichloroisocyanuric acid (chlorinated cyanuric acid), which is available in tablet form, containing up to 900 g/kg of available chlorine. As this product is an acid, the total alkalinity should be at least 150 mg/L to maintain the pool water at optimum pH.

Both sodium dichloroisocyanurate and trichloroisocyanuric acid are stable compounds and as such have long storage lives. The recommended concentration of isocyanuric acid is limited to between 30 mg/L and 50 mg/L, because at higher concentrations isocyanuric acid *retards* the activity of chlorine in killing bacteria. If these compounds are used as the source of chlorine, regular tests for the concentration of cyanuric acid will be necessary. If it is found that the concentration of cyanuric acid is above 50 mg/L, the use of chlorinated cyanurates should cease and calcium hypochlorite or sodium hypochlorite should be used for chlorination. Alternatively, a proportion of the pool water can be drained off and refilled with fresh water. For example, if the concentration of isocyanuric acid is found to be

100 mg/L, then 1/2 of the volume of pool water should be drained off and refilled with fresh water.

Electrolytic chlorine generator (salt chlorinator)

Equipment is available for the generation of sodium hypochlorite solutions by means of the electrolysis of salt (sodium chloride) solution. The electrolytic chlorinator generates sodium hypochlorite either by

a. the addition of sodium chloride directly to the pool and direct electrolysis of the pool water; or
b. electrolysis of a sodium chloride solution to give relatively concentrated sodium hypochlorite solution, which is stored and metered into the pool as required.

In method (a), fouling of the electrodes necessitates regular electrode maintenance. Corrosion problems will also occur with some metal pool fittings, including some stainless steels, as a result of the high pool chloride levels, necessitating installation of special corrosion-resistant fittings.

In method (b), electrode-fouling problems are minimised and corrosion problems associated with high pool chloride levels are avoided. The storage of sodium hypochlorite solution allows high chlorine dosage rates for either superchlorination or use during periods of high bather loads, using relatively low-capacity electrolysis systems.

To achieve and maintain the chlorine concentration, salt chlorinating equipment must be of the appropriate capacity for the size of pool. Also, the equipment (especially the electrodes) must be regularly serviced and salt concentrations maintained in accordance with the manfacturer's instructions.

CAUTION
Some pool constructions may NOT be suitable
for salt chlorination

Salt chlorinators should *not* be used in a concrete pool that has a shell thickness less than 175 mm, *unless* the shell is specially treated to prevent salt penetrations.

Polyhexamethylene biguanide and hydrogen peroxide system (e.g. BAQUACIL)

Commercially available polyhexamethylene biguanide (e.g. BAQUACIL) is a pale blue solution that comprises 20% active ingredients. When added to swimming pool water in sufficient quantity to give a concentration of 25 to 50 mg/L, the biguanide will keep the water clear, clean and hygienic for relatively long periods of time (testing is necessary only once a week).

Hydrogen peroxide must be added on a regular basis to assist in the control of algae. The commercially available system (BAQUACIL) provides a 30% strength hydrogen peroxide solution, which is to be added to the pool every fortnight to give a concentration of 100 mg/L. There is no need to test for hydrogen peroxide provided the pool is dosed in accordance with the manufacturer's instructions.

In this system, the biguanide, unlike chlorine, is not consumed by organic material (which is always present in pool water because of leaves, etc., as well as body wastes), nor does it lose much of its effectiveness in the presence of such organic matter. Further, the compounds do not produce an odour.

This system is *not* compatible with chlorine. If the product is added to pool water containing chlorine, a heavy brown precipitate forms which is a complex between both products. Both the polyhexamethylene biguanide and chlorine are rendered inactive.

Polyhexamethylene biguanide is a very effective flocculant, which not only places additional strain on the filter system but is removed from pool water together with the flocculated solids. Accordingly, the concentration of the biguanide needs to be checked regularly.

Bromine

Bromine is commercially available as sodium or potassium bromide, used in conjunction with sodium hypochlorite or potassium persulfate which is supplied as a liquid, or as bromochlorodimethylhydantoin (BCDMH), supplied in tablet form. Bromide-based compounds, from which bromine is generated by adding sodium hypochlorite, are available cially. Their use and effect on water quality is essentially the same as those for sodium hypochlorite used alone. Although mainly used for hot water spas, BCDMH may be used for the sanitisation of private swimming pools. It is formulated as tablets and has a long shelf life. As BCDMH is acidic (pH 4.5), it reduces the total alkalinity of pool water, but has no effect on calcium hardness. Bromine is more rapidly destroyed than chlorine in sunlight and is only suitable for indoor use. Unlike chlorine, bromine cannot be stabilised to ultraviolet light.

Liquid bromine is *not recommended* for private swimming pools, as it is highly toxic and corrosive.

Ozone generators

Ozone is a sky-blue coloured gas that can be generated by two different methods. In the corona discharge method, air is passed between two electrodes, across which a high voltage passes. In the photochemical method, air is passed over a short wavelength ultraviolet lamp. In both cases, the oxygen in the air is converted to ozone, which is relatively unstable and decomposes to re-form oxygen. Ozone is water soluble and a very effective bactericide. However, as ozone is also very toxic, it must not come into contact with bathers and therefore must be removed from the water before it is returned to the pool. Accordingly, the tubes are to be regularly cleaned and maintained and the charcoal filter must be regularly renewed. However, as ozone destroys micro-organisms in the pool water, the consumption of supplementary sanitiser is substantially reduced.

Ionic water purifiers: copper–silver electrolysis

This system is *not recommended*. In this system an electric current is discharged from copper and silver electrodes through an electrolyte medium of comparatively fresh pool water (i.e. no salt deliberately added). The current flows only while the recirculating pump is in operation.

It is claimed that the copper acts as the algicide and the silver acts as the bactericide. However, it is generally believed that this system is *not* as effective as the other sanitisers mentioned above.

Chlorine gas

Although chlorine gas is an effective sanitiser and it is used extensively for sanitising public swimming pools, this product is *not recommended* for private swimming pools because of the hazards associated with its use and storage.

ALGICIDE PRODUCTS

General

When a pool is to be used for bathing, an algicide may be used but *only* together with an appropriate sanitiser. When a pool is not being used for bathing (e.g. during winter) an algicide may be used for algal control without a sanitiser.

Quaternary ammonium compounds (quats)

These compounds are primarily algicides. Quats complement chlorine because they are effective on the walls and floor of the pool, while chlorine is more effective in the body of water.

This algicide is usually applied at weekly to monthly intervals to maintain a concentration of about 2.5 mg/L in the pool water. However, care should be taken to ensure that excessive amounts of the product are not added. Test kits are not currently available for determining the concentration of these compounds. They do not decompose readily and the main loss occurs through splash-out and filtration.

To prevent the water appearing hazy, quats should *not* be added for at least 48 hours after superchlorinating the pool. During normal dosing there is no need for a time delay between the application of the quats and the chlorinated compound. In practice, there is no problem with chlorinating a pool already containing a quat.

Polymeric cationic compounds

Polymeric cationic compounds are not strictly quaternary ammonium compounds, although they have similar and sometimes better algicidal properties than quats. The dosage rate and use is similar to those of quats and they should be used in conjunction with an approved sanitiser during the swimming season. These compounds do not foam nor do they turn the pool water hazy.

Sodium perborate tetrahydrate

Sodium perborate tetrahydrate is a white free-flowing crystalline powder. This product is a strong oxidising agent, which destroys algae and chloramines. Accordingly, during the bathing season this product must be used in conjunction with an appropriate sanitiser. When added to pool water this product will increase the pH of the water. Although this product is not regarded as particularly hazardous, it should not be mixed with other pool chemicals, especially combustible materials such as pool chlorines in granular or tablet form.

Potassium monopersulfate

This product is a white, granular, free-flowing powder that is a unique triple-salt, comprising two moles of potassium monopersulfate ($KHSO_5$), one mole of potassium bisulfate ($KHSO_4$), and one mole of potassium sulfate (K_2SO_4). Its solution in water is strongly acid (pH of 1% solution is 2.3). It is odourless, and a powerful oxidising agent. Its active oxygen content is 4.5% to 5%. It can convert sodium chloride (salt) to sodium hypochlorite (the ingredient of liquid chlorine) in water. Its solution is not very stable, between pH 7 and 8. It is capable of destroying chloramines, and it is beneficial in pools using salt water chlorinators or in spas using bromine. It is not compatible with chlorine compounds or copper-based algicides.

Copper-based compounds

Because of their varying nature and strengths, these compounds are intended to be applied to pools at intervals ranging from several weeks to 6 monthly. If used correctly, these compounds are effective algicides which neither foam nor turn pool water hazy.

Copper concentration in pool water must *not* be allowed to exceed 1 mg/L. Test kits for measuring copper concentrations are available, but only in the more professional packs. Copper in water acts as a catalyst to reduce the activity of chlorine and bromine.

CAUTION

Copper-based compounds in which the manufacturer specifies pool water copper concentrations greater than 1 mg/L are *not* recommended.

Excessive copper concentrations may stain hair and swimwear as well as discolour pool surfaces (particularly aged marble pools in which the water has a pH that is more than 7.8 and the calcium hardness is high).

STABILISER

Isocyanuric acid
In outdoor pools, chlorine is rapidly destroyed by sunlight. When isocyanuric acid is added to the water, the chlorine is protected from destruction by sunlight. This results in considerable savings of chlorine and extended periods of correct sanitisation. However, as stabilised chlorine is not as effective as a sanitiser as unstabilised chlorine, the free chlorine concentration of a stabilised pool has to be increased to maintain the same disinfection rate.

The use of isocyanuric acid is *not recommended* for indoor pools because there is no direct sunlight to destroy chlorine in the pool and the stabiliser inhibits the effectiveness of chlorine.

Isocyanuric acid may be added to the pool water as free isocyanuric acid, or in combination with chlorine as trichloroisocyanuric acid tablets, or as sodium dichloroisocyanurate granules.

Isocyanuric acid should be added at the start of the season to a concentration of 50 mg/L, and then tested monthly and replenished if required. This makes up for any isocyanuric acid losses during backwashing or from splash-out. A good indicator for the need to test is when chlorine retention becomes difficult.

Isocyanuric acid dissolves slowly and to a limited extent in water, and therefore cannot be added to pools as a highly concentrated solution.

The product may be added to the balance tank or to the top of the filters by tipping the calculated weight into a skimmer, hair and lint strainer, or by sprinkling directly into the pool. However, the latter method may cause fading or bleaching of some pool interiors. Filters should be backwashed before the addition of the product and then run without backwashing for two to three days to ensure complete dissolution of the solid, before further backwashing. During this procedure, the water is suitable for swimming.

Where cyanuric acid is regularly added in the form of stabilised chlorine compounds its concentration will gradually build up in the pool. High concentrations of isocyanuric acid significantly inhibit the action of chlorine as an effective sanitiser. Therefore, when the chlorinated isocyanurates are used as the chlorine source, regular tests for isocyanuric acid are necessary.

SUPPLEMENTARY POOL CHEMICALS

Dry acid (sodium bisulfate)
Dry acid is a white crystalline or granular powder and contains the equivalent of 350 g/kg of

sulfuric acid. This product is used to lower the pH and the total alkalinity of pool water and is incompatible with chlorine compounds when concentrated. This product must be dissolved in water before it is added to the pool.

Hydrochloric acid

Hydrochloric acid is a colourless to greenish-yellow fuming liquid and contains 345 g/L of hydrochloric acid. The fumes are very corrosive to metal and mucous tissue, and care must be taken to avoid contact with eyes and skin. This product is used to lower the pH and the total alkalinity of pool water. Hydrochloric acid is incompatible with chlorine compounds when concentrated, as it will release toxic chlorine gas.

CAUTION

Hydrochloric acid is to be substantially diluted in water *before* it is added to the pool.

Sodium bicarbonate

Sodium bicarbonate is a white powder containing 990 g/kg of sodium bicarbonate. Although this product is alkaline, it is not corrosive to body tissue. This product increases the total alkalinity of pool water but, because of its buffering effect, it only slowly increases pH. Because the modest rise in pH is permanent, sodium bicarbonate is a favoured treatment for pH control of private pools.

Soda ash (sodium carbonate anhydrous)

Soda ash is a white powder that contains 990 g/kg of sodium carbonate. This product is a stronger alkali than sodium bicarbonate. It can cause irritation to body tissue and will cause severe eye irritation or damage. It can also be corrosive to brass and galvanised fittings. When added to pool water, this product causes a large increase in pH and total alkalinity.

Initially the pH of the pool water will rise with the addition of sodium carbonate. However, the pH will usually fall again over a period, to a level slightly above the original level. Successive sodium carbonate additions will gradually increase the alkalinity until a permanent pH rise is achieved. Because of its large effect on pH at normal pool alkalinities, it is not usually suitable for increasing total alkalinity alone. Sodium carbonate may cause a cloudy or milky precipitate to appear on the pool walls and floor because of its reaction with calcium in the pool water.

Sodium thiosulfate

Sodium thiosulfate is a dechlorinator and as such is used to neutralise chlorine in water *only* if it is desired to sanitise the pool with a product (e.g. polymeric biguanide) that is not compatible with chlorine.

Alum (aluminium sulfate)

Alum or aluminium sulfate is a water soluble salt that produces acid when dissolved in water. It is used in conjunction with an alkali (e.g. soda ash) to flocculate turbidity in water, especially when the pool is filled for the first time. Flocculation with this product is most effective if the pH of the pool water is between 7 and 8. Accordingly, the pH of the water should be checked and adjusted after using alum. Avoid overdosing the pool with alum as the floc may disperse and hence not be removable.

Polymeric flocculants

Polymeric flocculants are water-soluble organic polymers carrying positive charges that trap and flocculate particles causing water cloudiness. The organic flocculants are removed together with the flocculated solids.

Nitric acid

This product is *not recommended* for use in private swimming pools because it is very corrosive to the skin and eyes, and its fumes are toxic. Nitric acid is also incompatible with chlorine and bromine treatments.

Sulfuric acid

This product is *not recommended* for use in private swimming pools because it is very corrosive to the skin and eyes, and its fumes are toxic. Sulfuric acid is also very dangerous to handle as it reacts violently with water, evolving great quantities of heat.

Sulfamic acid

This product is *not recommended* for use in private swimming pools because it contains nitrogen and will therefore form chloramines in pool water, where chlorine (bromine) is used.

Sodium metabisulfite

This product is *not recommended* for use in private swimming pools because it is not compatible with sanitisers. However, it may be used by pool professionals to remove localised rust stains from the pool floor or walls when the pool is drained. This product must be neutralised before the pool is refilled and re-sanitised.

C. Definition of terms

Acid demand   A measurement of the amount of acid that needs to be added to the pool water to lower the pH and total alkalinity to acceptable levels.

Acidic   Water with a pH between 0 and 7.

Algicide   A chemical that is capable of killing algae.

Alkaline   Water with a pH between 7 and 14.

Alkalinity, total   A measure of the total amount of dissolved alkaline compounds in the pool water. Total alkalinity is a measurement of the resistance of the pool water to a change in pH. For example, if the pool water has high pH and low total alkalinity, the addition of a small amount of acid will lower the pH sharply.

Bactericide   A chemical that is capable of killing bacteria and preferably other microorganisms such as viruses and amoebae.

Bromine, combined   Bromine that has combined with ammonium compounds or organic matter containing nitrogen to form bromamines. Ammonium compounds and organic matter containing nitrogen are normally transmitted to pool water by body wastes (e.g. perspiration) and by organic contamination (e.g. leaves). Combined bromine (i.e. bromamine) is a more effective disinfecting agent than combined chlorine (i.e. chloramine).

Bromine, free   Bromine that is not combined with ammonia or organic matter containing nitrogen is free to kill bacteria and algae and to destroy organic contamination introduced into the pool water.

Bromine, total   The sum of combined bromine and free bromine.

Chlorine, combined   Chlorine that has combined with ammonium compounds or organic matter containing nitrogen to form chloramines. Ammonium compounds and organic matter containing nitrogen are normally transmitted to pool water by body wastes (e.g. perspiration) and organic contamination (e.g. leaves). Bactericidal properties of combined chlorine (i.e. chloramines) are only approximately one hundredth that of a similar level of free chlorine in water.

Chlorine, free   Chlorine that is not combined with ammonia but is free to kill bacteria and algae and to destroy organic contamination introduced into the pool water. Free chlorine is also known as 'free available chlorine' and 'free residual chlorine'.

Chlorine, total   The sum of combined chlorine and free chlorine.

Hardness, calcium   A measure of the amount of dissolved calcium compounds in the water.

pH   A scale (ranging from 0 to 14) that indicates the amount of acid or alkali present in the water. Water with a pH of 7 is neutral. Since pH is logarithmic, a solution with a pH of 5 is *ten* times more acidic than a solution with a pH of 6, whilst a solution with a pH of 4 is *one hundred* times more acidic than a solution with a pH of 6.

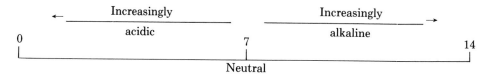

Sanitiser   A compound or system which, when applied as instructed to a swimming pool, kills algae and harmful micro-organisms in order to achieve and maintain an effective level of disinfection in the pool water, and in this way renders the pool suitable for swimming.

Shock dose   The addition to pool water of five to seven times the normal daily dose of chlorine (without stabiliser) for treatment of excessive algae growth, discoloured water, odour, or to restore the pool after it has been neglected for an extended period of time.

Solids, total dissolved   A measure of the total amount of dissolved compounds in the pool water.

Stabiliser   A compound added to pool water to reduce the rate of loss of chlorine due to sunlight (i.e. ultraviolet light). An example of a stabiliser is isocyanuric acid (also known as cyanuric acid).

Superchlorination   The addition to pool water of two to four times the normal daily dose of chlorine for destruction of combined chlorine (i.e. chloramines) and for treatment of impurities introduced by prolonged use, increased number of bathers or debris resulting from inclement weather.

Source: Standards Australia AS 3633.

APPENDIX II

POLYMERS

A. Sources of polymer materials

Material	Source
acrylonitrile-butadiene copolymer	3
acrylonitrile-butadiene styrene (ABS)	5,6
acrylonitrile-styrene-butyl acrylate (ASA)	2
cellulose acetate butyrate (CAB)	6
chlorinated polyethylene (CPE)	6
EPS	5
ethylene vinyl acetate (EVA)	6
ionomer	6
melamine-formaldehyde resin	9
phenol-formaldehyde resin	5

Material	Source
polyacrylic (inc. PMMA)	3,7
amide (e.g. nylon 6,66)	2,6,7
butadiene (high cis) elastomer	1
butadiene-styrene elastomer	1
butadiene-styrene resins	1
butylene terephthalate (PBT)	2
carbonate	6
epichlorahydrin	3
ester (thermoplastic)	7
ester (thermoset)	5
ethylene, LD	6,11
ethylene, LLD	6,11
ethylene, HD	4,6,11
oxymethylene (polyacetal)	2,4
phenylene oxide (e.g. Naryl®)	6
propylene co-polymer	4,6
propylene homopolymer (PP)	4,6
styrene (PS)	2,5
urethane-prepolymer	12
urethane-thermoplastic	2,3
vinyl acetate (PVA)	6
vinyl chloride (PVC)	3,6
vinyl chloride copolymer	10
styrene-acrylonitrile (SAN)	2,5
urea-formaldehyde resin	9
polymer alloys	8
Other details	
Literature available	1,2,3,4,5,10,11,12
Specifications available	1,2,3,6,10,11
Confidentiality	8

1. *Australian Synthetic Rubber Co. Ltd:* Chief Chemist, Maidstone St, Altona, Vic. 3018.
2. *BASF Australia Ltd:* Technical Sales Representative, 120 Taren Point Rd, Taren Point, NSW 2229.
3. *B.F. Goodrich Chemical Ltd:* Marketing Services Manager, 14 Queens Rd, Melbourne, Vic. 3004.
4. *Hoechst Aust. Ltd:* Manager, Plastics Technical Service Section, PO Box 4300, Melbourne, Vic. 3001.
5. *Monsanto Aust. Ltd:* Manager, Technology, Sommerville Rd, West Footscray, Vic. 3012.
6. *Nylex Corporation Ltd:* Analytical Services Manager, PO Box 68, Mentone, Vic. 3194.
7. *Plastrol Trading Co.:* Marketing Manager, 11B Lathlan St, Waterloo, NSW 2019.
8. *Polymer Alloys Pty Ltd:* Marketing Director, 8 Monomeeth Dr., Mitcham, Vic. 3021.
9. *The Swift Watts Winter Co.:* Business Manager, 149 Milton St, Ashfield, NSW 2131.
10. *Stauffer Aust. Ltd:* Product Manager, 6 Grand Ave, Camellia, NSW 2142.
11. *Union Carbide Aust. Ltd:* Manufacturing Technology Manager, Maidstone St, Altona, Vic. 3018.
12. *Vinroyal Aust. Pty Ltd:* Sales Manager, Urethane Elastomer, PO Box 217, Granville, NSW 2142.

B. Some trade names and trivial names of polymers

ABS	Generic name for plastics from acrylonitrile, butadiene and styrene units
Acetate	Generic name for fibres from cellulose (2½) acetate
Acrilan	Trade name for fibres from poly(acrylonitrile) (Chemstrand)
Acryl	Generic name for fibres with at least 85% acrylonitrile units
Alcantara	Synthetic suede from polyester fibres in a polyurethane matrix
Araldite	Epoxy resins (Ciba-Geigy)
Aramide	Generic name for fibres from aromatic polyamides
Bakelite	Thermoset from phenol and formaldehyde (Bakelite Corp.)
Balata	Natural trans-1,4-polyisoprene
Cellophane	Films from regenerated cellulose (Kalle)
Celluloid	Cellulose nitrate plasticised with camphor
Dacron	Fibres from polyethylene terephthalate (DuPont)
Diolen	Fibres from poly(ethylene terephthalate) (Vereinigte Glanzstoff)
Dralon	Fibres from poly(acrylonitrile) (Bayer)
Grilen	Fibre from poly(ethylene terephthalate) (Ems Chemistry)
Guttapercha	Natural trans-1.4-polyisoprene
Hostalen	Polyethylene thermoplastic (Hoechst)
Hostalen PP	Polypropylene thermoplastic (Hoechst)
Hostalit	Poly(vinyl chloride) thermoplastic (Hoechst)
Hostaphan	Films from poly(ethylene terephthalate) (Kalle)
Igelit	Poly(vinyl chloride) thermoplastic (BASF)
Lucite	Poly(methyl methacrylate) thermoplastic (DuPont)
Lupolen	Polyethylene thermoplastic (BASF)
Luran	Thermoplastic from styrene and acrylonitrile (BASF)
Lustron	Thermoplastic polystyrene (Monsanto)
Marlex	Thermoplastic polyethylene (Phillips Chemical)
Moltopren	Cellular polymer from polyurethanes (Bayer)
Mylar	Films from polyethylene terephthalate (DuPont)
Natural rubber	cis-1.4-Polyisoprene
Neoprene	Elastomeric polymers and copolymers from chloroprene (DuPont)
Nylon	Generic name for polyamides
Orlon	Fibre from polyacrylonitrile (DuPont)
Perlon	Fibre from polycaprolactam (Bayer)
Plexiglas	Thermoplastic poly(methyl methacrylate) (Rohm and Haas)
Polyester	Generic name for fibres from polyesters with at least 85% terephthalic acid and ethylene glycol units
Polyester, unsaturated	Thermosets from maleic acid/ethylene glycol polymers (or similar compounds which are crosslinked with e.g. polystyrene
Qiana	Polyamide from trans,trans-diaminodicyclohexylmethane and dodecanedicarboxylic acid or sebacic acid (fibre from DuPont)
Rayon	Generic name for fibres from regenerated cellulose

Saran	Films and fibres from polymers with at least 80% vinylidene chloride units (Dow Chemical)
Silicone	Generic name for polymers with a siloxane chain
Spandex	Generic name for elastic fibres from polymers with at least 85% segmented polyurethane
Styrofoam	Cellular plastic from polystyrene (Dow Chemical)
Styron	Polystyrene thermoplastic (Dow Chemical)
Styropor	Cellular thermoplastic from polystyrene (BASF)
Teflon	Thermoplastic poly(tetrafluoroethylene) (or other fluorinated polymers if letters follow the word Teflon) (DuPont)
Terital	Fibres from poly(ethylene terephthalate) (Montedison)
Terlenka	Fibres from poly(ethylene terephthalate) (AKU)
Terylene	Fibres from poly(ethylene terephthalate) (ICI)
Trevira	Fibres from poly(ethylene terephthalate (Hoechst)
Triacetate	Generic name for fibres from cellulose triacetate
Tricel	Fibres from cellulose triacetate (British Celanese)
Ultrasuede	Synthetic suede (see Alcantara)
Vestolen	Various polyamides (Huels)
Vestolit	Poly(vinyl chloride) (Huels)
Viscose	Generic name for fibres from regenerated cellulose
Zytel	Various aliphatic polyamides (DuPont)

Source: H.-G. Elias, *Mega molecules*, Springer-Verlag, Heidelberg, 1987, pp. 190–191.

APPENDIX III

BURNING BEHAVIOUR OF FIBRES

Class of fibre and generic name	Some trade names or common names[a]	Notes
Natural cellulose Cotton Linen		All burn readily unless given adequate flame-retardant treatments.
Protein (animal) Wool Silk Mohair Hair and fur		Usually difficult to ignite, burn slowly and in heavier weight cloths tend to extinguish. Very light weight cloths may burn strongly.

Class of fibre and generic name	Some trade names or common names[a]	Notes
Synthetic		These fibres exhibit a variety of flammability properties. Some burn, some melt and drip away, and yet others do not support combustion.
Cupro-viscose modal	Evlan, Rayon, Cupresa, Vincel	Regenerated cellulose fibres burn readily unless given an adequate flame-retardant treatment.
Acetate rayon (including triacetate)	Arnel Dicel, Tricel Filcel, Textella	Acetate fibres burn before melting. They ignite and burn readily and generally drip molten polymer.
Flame-retardant regenerated cellulose fibres		The fibre is made with flame-retardant compounds incorporated in the resin. It burns slowly, and is more difficult to ignite than other regenerated cellulose fibres.
Acrylic	Acrilan, Beslon, Cashmilon, Courtelle, Creslan, Crylor, Exlan, Dolan, Dralon, Orlon, Zefran, Vonnel, Filcryl, Nomelle, Exlan	Acrylic fibres generally burn before melting, although some ignite only with difficulty. Once ignited acrylic fibres burn strongly and may drip molten polymer.
Chloro-fibres— poly(vinyl chloride), poly(vinylidene chloride) and co-polymers	Clevyl, Geon, Movil, PeCe, Teviron, Thermovyl, Tygan, Valren, Vinyon, Rhovyl	These fibres generally do not burn under normal conditions. On application of a flame they shrink away. If ignition does occur, burning ceases on removal of the flame.
Modacrylic	Kanekalon, Teklan, Verel, Vonnel	Fibres are difficult to ignite. Flammability properties very similar to those of chloro-fibres.
Polyamide	Antron, Bri-Nylon, Caprolan, Celon, Enkalon, Nylon 6, Nylon 11 (Rilsan), Nylon 66, Nylon 610, Perlon, Amilan, Antron, Cantrece, Delfion, Dorix, Fluflon, Promilan	These fibres melt when heat is applied and generally drip away from the flame. Because of this they may not burn. When blended with some fibres (e.g. cotton, rayon, wool) they are unable to drip away from the flame and will burn, often quite fiercely. Polyamide fabric may also burn when sewn into garments with cotton thread; the threads may act as a support to prevent the burning fibre from dripping away.
Polyester	Dacron, Diolen, Fortrel, Kodel, Tergal, Terlenka, Terylene, Teteron, Trevira, Vycon, Crimplene	Very similar flammability characteristics to polyamide fibres.
Fire-retardant polyester		The fibre is made with flame retardant compounds incorporated in the resin. They have reduced burning characteristics in blends.

Class of fibre and generic name	Some trade names or common names[a]	Notes
Polyethylene and polypropylene	Courlene, Drylene, Herculon, Marvess, Meraklon, Nymplex, Polycrest, Pylen or Pylene, Reevon, Spunstron, Ulstron	These fibres melt when ignited and drip molten drops while continuing to burn.
Vinylal—poly(vinyl alcohol)	Mewlon, Vinylon, Kuralon	Similar flammability characteristics to polyethylene and polypropylene fibres but burn somewhat less readily.
Other fibres of various types	Durette, Kynol, Nomex, PBI, Kevlar, Cordelan	These fibres do not burn under normal conditions.

[a] The list of trade names is not intended to be all-inclusive but covers most fibres that are available in Australia.

Source: Appendix D from AS1249, 1983.

APPENDIX IV

THALIDOMIDE: A TRAGEDY OF MALPRACTICE AND DECEPTION

The following book review was written by Yvonne Preston and published in the *Financial Review*.

Thalidomide and the Power of the Drug Companies by Henning Sjöström and Robert Nilsson, published by Penguin.

In October 1960 two German doctors, members of the staff of the Institute of Human Genetics in Munster, exhibited two grossly deformed infants at the annual meeting of paediatricians in Kassel.

The babies' arms were so short that the hands seemed to project almost directly from the shoulders. The legs, less affected, showed similar distortions. An abnormality of the blood vessels disfigured the faces of both children and one of them also suffered from a constriction of the small intestine. The two physicians had never before seen such a combination of malformations; not surprisingly, since this type of congenital malformation, on the research evidence then available, occurred no more frequently than one in four million births.

The German paediatricians concluded that the gross malformation picture represented a new clinical syndrome. Its cause has subsequently and tragically become a household word—thalidomide.

The tragedy of thalidomide, seen as a story of an innocent case of a tranquilliser drug turning out to have monstrous side effects on the unborn, is hard enough for the mind to comprehend and accept.

It is almost impossible to adjust the mind to take in the possibility that the whole tragedy has far more sinister overtones; that thalidomide was marketed with an eye to profit and tragic side effects a long way second; and that it was recognised as being a dangerous drug, for the damage it could do to the nervous system, when it was put on the market; and even that the threat to the foetus was recognised for some time before it was withdrawn.

Yet this is how the authors of *Thalidomide and the Power of the Drug Companies* see the disaster. Far from the tragedy which occurred despite the best efforts of doctors and scientists to ensure the safety of the drug, [it is] a tragedy which should never have occurred if regulations had been tight enough. If clinical testing had been thorough enough, if repeated warnings from physicians had been heeded and if, and this above all, the greed for profit had not so easily taken precedence over almost every other consideration in the marketing of thalidomide-containing drugs, notably on the part of the German company which first produced it.

Chemie Grünenthal was formed as a subsidiary of a German soap, detergent and cosmetics firm, in 1946, and began to produce antibiotics, which had a ready sale in epidemic-fearful post-war Germany.

Henning Sjöström, a Swedish lawyer who advised the prosecution in several international thalidomide law suits, and Robert Nilsson, a Swedish research chemist, the authors of this shocking indictment of the company, argue that it frequently revealed itself to be a less than ethical operation, both in its marketing tactics and its lack of concern for the side-effects of its products, long before thalidomide shot it into the unwelcome headlines.

In the mid-fifties Chemie Grünenthal entered a new area of pharmaceutical preparations, moving from antibiotics to the highly profitable field of sedatives and hypnotics.

In 1958 Grünenthal opened a massive publicity campaign for Contergan, the brand name for its thalidomide-containing tranquilliser. Soon it was launched on the international market, sold by licensees in 11 European, 7 African and 17 Asiatic countries and 11 countries in the Western hemisphere, under 37 different brand names.

By 1961, the year when Contergan was finally withdrawn from the German market, total sales of thalidomide had reached a value of DM12.4 million, which, added to export income of at least 25 per cent of that total, amounted to a highly profitable market for Grünenthal.

The drug was sold without prescription in Germany. The special publicity for thalidomide, masterminded for all countries by the manufacturers themselves, made great play of the product as 'completely innocuous', 'atoxic'. A British advertisement emphasised the safety of the drug with a picture of a small child taking a bottle from a medicine shelf.

By 1959, with the sales explosively increasing with the growing exploitation of markets—a liquid form of Contergan for children became West Germany's babysitter—adverse reports of severe side effects began to appear. Doctors and pharmacists commented on severe constipation, dizziness, decrease in blood pressure.

As early as September, 1959, the use of Contergan was dropped in a German hospital because of severe cases of purpura, or local haemorrhage of the skin. Grünenthal's Swiss affiliate relayed to the home company that Swiss doctors had reported tremor of the hands following the use of Softenon—the Swiss brand name.

'Once and never again. This is a horrible drug', announced a Swiss physician—in 1959!

Soon reports of irreversible side effects began to filter through to Grünenthal. A case of polyneuritis, a permanent impairment of the nervous system, was reported. Grünenthal commented that this had never been observed before, though later evidence showed that the company's own clinical trials of thalidomide in 1956 had revealed the possibility of polyneuritis.

Worse followed—reports of disturbances of the gait and sensibility in fingers and toes, involuntary twitching of the facial muscles, double vision, even epileptic seizures.

None of this mounting body of evidence inspired any reaction in Grünenthal, beyond concern for their sales, which continued to rocket. Dr Heinrich Muckter, who developed thalidomide and directed its sales campaigns, wrote in April, 1960, 'Everything must be done to avoid prescription enforcement, since already a substantial amount of our turnover comes from over-the-counter sales'.

In the first four months of 1960, again despite the mounting body of adverse evidence against thalidomide, a quarter of a million leaflets were distributed extolling the drug as 'completely harmless, even for infants', 'non-toxic' and 'harmless even over long periods of use'.

And in the background the company went to enormous lengths to suppress critical articles; it hired private detectives to report on the lives of physicians who displayed a hostile attitude to Contergan; it continued its policy of minimising the risks associated with the drug by advertising the rarity of adverse effects and the drug's therapeutic effectiveness; and it persuaded cooperative physicians to report favourably on the drug—one of whom testified at the later prosecution of Chemie Grünenthal that he had observed a case of Contergan polyneuritis the day after his article praising the drug's virtues had been published, but could not prevent the company from continuing to use his article for promotion purposes.

This story of intrigue, deception, callousness and greed is enough on its own. It is merely the curtain-raiser to the nightmare which was to come. Read further and, accepting the accuracy of these two Swedes' laboriously and magnificently researched account, you cannnot but feel sick in the stomach.

It would surely not have been too much to have hoped that a mere hint of possible foetal damage of the magnitude of that which began to be revealed at the end of 1960 would have prompted a major reaction from the manufacturers.

But no. International requests for information about the passage of the drug from mother to embryo were treated non-committally—as they had to be in the light of Grünenthal's failure to include any trials on pregnant animals prior to launching their drug in a fanfare of publicity and recommendations for use by pregnant women.

Cases of extremely rare phocomelia [Gk: a monster having limbs so short as to suggest the flappers of a seal] appeared more and more frequently.

But leaflets to the effect that 'Contergan is a safe drug' continued to pour out and even when the mounting pressure of public opinion finally forced the withdrawal of the drug from the German market, the company was active in promoting the idea of resuming production as soon as possible.

The book goes much further than a straight indictment of what its authors clearly see as a company without a shred of integrity. Its record of the slowly grinding, badly oiled wheels of drug control agencies in Japan, Sweden, America, Canada, as well as Germany, show clearly that the machinery for drug control was inadequate and that there were major deficiencies in provisions for the development of new drugs. Even now pre-clinical animal tests are recommended only indirectly in Germany and this in a country with possibly 6700 cases of phocomelia.

The book names names; quotes damning evidence from the pens of scientists and doctors; and quotes extensively from the profit-oriented comments of Grünenthal from the company's own documents—most of which had to be seized in police raids.

The book spares nobody. It comments that Dr William McBride notified the Australian representatives of Distillers, the British distributors of Distaval, of his suspicions of thalidomide malformations, but his observations never reached head office.

The Swedish Medical Board did not finally advise the Swedish distributors, Astra, to include a warning for polyneuritis in their thalidomide brochures until two weeks after Chemie Grünenthal had actually been forced to withdraw the drug completely from the German market.

Richardson Merrell, United States licensees for thalidomide, distributed 2 528 412 thalidomide tablets for 'experimental purposes' before Kevadon, the US thalidomide brand, was released and approved by the FDA, but the FDA itself was unable, in the event, to track them all down.

They had been given to 20 000 patients in containers bearing no more than the directions for use.

The totally inadequate material available before the introduction of thalidomide in various countries, the faults of which must have been quite obvious to Chemie Grünenthal and its licensees, shows, says the book, that 'thalidomide was introduced according to the method of Russian roulette'.

The large number of trade names under which thalidomide was sold meant that a teratogenic action as the result of Contergan reported in a Stockholm newspaper was not recognised by Swedish doctors as Neurosedyn—the Swedish brand name for thalidomide.

And to round the whole horrifying episode off is the story of the extensive prosecution of the German company at Aachen, a case which was ultimately abandoned, but in which bitter wrangling and accusation, character assassination and high-powered drug company public relations activity went on, while outside the courtroom three thalidomide victims played while their mothers followed the case, and were daily reminders to the protagonists of what had been achieved in the name of profit.

The book raises a thousand different questions. It strongly suggests that the mysteries of science may be placing too much power in the hands of those who are out for profits.

It questions the role of commercial interests in drug production; the expenditure of enormous sums on drug publicity; the use of marketing brand names rather than generic terms; the emphasis of much pharmaceutical research, inevitably in a competitive market; on producing minor variations of a competitor's product rather than concentrating on new and beneficial drug developments.

It argues a case for some form of State-owned drug company, if only as a means to break up an unsound non-competitive situation.

It pursues the increasingly pressing question of the need to evaluate much more carefully the therapeutic advantages of each new drug against the side-effects that the drug may impose.

The tragedy of thalidomide is that its function, simply as a tranquilliser and sleeping agent, was by no means of such a life-saving nature as to justify ignoring even the early revealed and relatively minor side-effects, let alone the ultimate and monstrous side-effects.

It argues that pharmaceutical companies still shy away from accepting strict liability on the grounds that the drug industry is working uniquely for the benefit of mankind and is therefore not comparable to other industries—a spurious argument since this, like any other type of industry in the West, exists for profit.

It argues above all, and most tragically of all, that the industry and medical authorities cannot legitimately argue of thalidomide that 'nobody ever thought of such a possibility' and 'this catastrophe was unavoidable'.

The shocking evidence of this book is that the 'possibility was always there; that the catastrophe was not inevitable'.

The knowledge of large-scale screening of drugs for teratogenic activity was available within the field of experimental embryology but, with some exceptions in the USA, no use was made of these methods within the pharmaceutical industry.

It was well known, even before the tragedy of multiple malformations woke the world to the problem, that drugs which may be innocuous to adults may cause severe damage to the young underdeveloped child, and most particularly to the unborn child.

Medical science, like nuclear science, can work miracles or wreak havoc.

'Sometimes man does not seem to learn even from disaster.'

See also *Suffer the children: the story of thalidomide* by the *Sunday Times*' Insight Team (André Deutsch, London, 1979).

APPENDIX V

CHILD-POISONING

The problem of accidental poisoning in children is a serious one. The figures in the following tables are from the Adelaide Children's Hospital for 1986. Earlier statistics are given in previous editions of *Chemistry in marketplace*.

To place the problem in perspective, a national survey conducted in 1981 using 1980 statistics gives a breakdown of the serious injuries to children requiring admission to hospital, as shown in Table V.1.

TABLE V.1
Categories of child injury, 1980

Type of accident	Number	Percentage
Falls	237	31.5
Poisoning	223	29.6
Burns and scalds	154	20.5
Road	78	10.4
Cuts	18	2.4
Other	43	5.6

Source: National Safety Council.

The age breakdown in Table V.4 indicates that poisoning is largely a problem of the 1 to 4 age group.

Hospital statistics can give only an indication of the problem because there is no legislation to compel doctors and injury clinics to report their treatments. The Adelaide Children's Hospital is one of three major hospitals for children and the percentage of cases collected depends on city demography.

TELEPHONE CALLS

Of 9857 telephone calls about poisons received by the hospital in 1986, 7245 concerned human exposure, 255 the exposure of animals and 2357 were general enquiries that did not involve exposure (such as requests for statistics, and general information on poisonous plants and venomous creatures).

TABLE V.2
Human exposure to poisons, 1986: categories of telephone calls, by type of poison

Type of poison	Number of calls	Percentage
Drugs	2653	37
Household products	1193	16
Miscellaneous (includes plants, paints, foreign bodies)	852	12
Chemicals	583	8
Bites and stings	428	6
Insecticides	408	5
Other	1128	16
TOTAL	7245	100

CHILDREN TREATED

A total of 730 children (429 boys and 301 girls) were treated at the hospital in 1986, and 163 of these were admitted. The majority of the children treated were aged 1 to 3 years. Children in this age group accounted for 424 out of the total 730 children treated (58%). Accidental poisoning (497 cases, 68%) and bites and stings (180 cases, 25%) were the most common kinds of poisoning amongst the children treated.

TABLE V.3
Categories and modes of child-poisoning, 1986

Categories	Number of children treated
Drugs	222
Bites/stings	180
Chemicals	37
Children [sic]	5
Cosmetics	15
Fertilisers	1
Fungicides	—
Gases	3
Handyman	6
Herbicides	—
Household	43
Insecticides	21
Miscellaneous	185
Rodenticides	5
Veterinary	4
Unclassified	3
TOTAL	730

Mode of poisoning:	
Suicidal	8
Deliberate misuse	41
Accidental	497
Bites/stings	183
Other	1
TOTALS	730

TABLE V.4
Child-poisoning by age group, 1986

Age	Male	Female
Under 1 year	25	15
1 year	95	92
2 years	113	58
3 years	37	29
4 years	25	9
5–9 years	71	41
10–18 years	63	57
TOTALS	429	301

TABLE V.5
Types of drugs taken by children treated, 1986

Type of drug	Number of children treated	Percentage
Analgesics (e.g. Panadol, Tempra, Dymadon)	34	15
Respiratory and common cold preparations (e.g. Demazin, Dimetapp, Sudafed)	25	11
Alcohol	20	9
Tranquillisers (e.g. Valium, Serepax)	13	6
Skin (e.g. creams, ointments)	13	6
Allergy (antihistamines, e.g. Phenergan, Polaramine)	12	5
Cardiovascular drugs	10	5
Ear, nose and throat preparations (nose drops, inhalant solutions)	10	5
Other	85	38

APPENDIX VI

TOP 50 GENERIC DRUGS

This appendix lists the top 50 generic drugs sold under the Pharmaceutical Benefits Scheme in the year ended 30 June 1986, by number of prescriptions (not including hospitals).

Position	Generic code(s)	Drug	Example of proprietary name	Example of drug or treatment	Prescriptions	Percentage of total prescriptions	Sales $
1	532 538	Amoxycillin	Amoxil	Penicillin antibiotic (particularly for children)	5 132 530	5.66	32 978 218
2	711	Levonorgestrel with ethinyloestradiol	Microgynon	Contraceptive	3 886 393	4.28	21 770 587
3	940	Trimethoprim with sulphamethoxazole	Bactrim	Enhanclo sulpha (urinary tract infection)	3 175 447	3.50	18 358 538
4	162	Salbutamol	Ventolin	Anti-asthmatic	2 908 851	3.21	18 613 534
5	256	Hydrochlorothiazide with amiloride	Modoretic	Potassium-saving diuretic	2 285 297	2.52	11 511 252
6	200	Diclofenac sodium	Voltaren	Joint pains	2 146 083	2.36	18 389 205
7	380	Naproxen	Naprosyn	Joint pains	1 698 519	1.87	13 372 024
8	796	Metoprolol	Betaloc	β_1-blocker (high blood pressure)	1 683 957	1.86	17 318 417
9	680	Frusemide	Lasix	Diuretic	1 614 971	1.78	8 830 940
10	835	Doxycycline	Vibramycin	Antibiotic STD clinics (Penicillin resistant strains only)	1 610 892	1.78	10 351 591
11	149	Atenolol	Tenormin	β_1-blocker	1 507 140	1.66	15 665 946
12	432	Oxazepam	Serepax	tranquilliser	1 480 055	1.63	6 543 945
13	404	Paracetamol	Panamax	Pain or fever	1 390 933	1.53	5 578 173
14	665	Theophylline	Nuelin	Bronchial asthma	1 355 058	1.49	8 009 071
15	803	Allopurinol	Zyloprim	Gout	1 265 982	1.40	9 070 670
16	112	Aluminium hydroxide with magnesium hydroxide	Mylanta	Acid stomach	1 256 695	1.38	6 265 134

Position	Generic code(s)	Drug	Example of proprietary name	Example of drug or treatment	Prescriptions	Percentage of total prescriptions	Sales $
17	859	Nitrazepam	Mogadon	Sleeping pill	1 148 981	1.27	4 706 377
18	363	Methyldopa	Aldomet	High blood pressure	1 100 215	1.21	11 577 122
19	465	Potassium chloride	Slow-K	Potassium replacement	1 076 289	1.19	5 109 693
20	221	Digoxin	Lanoxin	Cardiac failure	1 055 753	1.16	4 633 999
21	788	Propanolol	Inderal	β-blocker (migraine)	1 009 245	1.11	9 378 746
22	698	Indomethacin	Indocid	Anti-inflammation	1 000 296	1.10	6 592 198
23	565	Tetracycline with nystatin	Mysteclin V	Antibiotic (e.g. for acne)	970 343	1.07	6 488 805
24	562	Prazosin hydrochloride	Minipress	High blood pressure	946 469	1.04	11 350 596
25	361	Ibuprofen	Brufen	Anti-inflammation	927 058	1.02	5 431 166
26	783	Diazepam	Valium	Tranquilliser	921 582	1.02	4 043 431
27	345	Doxepin	Sinequan	Tricyclic anti-depressant	920 142	1.01	4 941 880
28	521	Sulindac	Clinoril	Anti-inflammation	900 658	0.99	6 924 786
29	692	Amitriptyline	Tryptanol	Tricyclic anti-depressant	899 530	0.99	4 501 049
30	103 196	Aspirin; aspirin (buffered)		Pain and fever	880 475	0.97	4 261 345
31	207	Diflunisal	Dolobid	Anti-inflammatory (bone pain)	868 226	0.96	7 385 956
32	858	Temazepam	Normison	Sleeping pill	866 887	0.96	3 557 379
33	255	Erythromycin	Erythrocin	Antibiotic	857 358	0.94	5 345 532
34	197	Codeine phosphate with paracetamol	Panadeine	Pain and fever	836 954	0.92	3 605 902
35	419	Verapamil hydrochloride chloride	Isoptin	High blood pressure, angina	824 533	0.91	10 236 572
36	340 341	Clotrimazole	Canesten	Anti-fungal	823 054	0.91	4 961 852
37	175	Chlorothiazide	Chlotride	Diuretic	775 409	0.85	4 344 082

38	554	Beclomethasone deipropionate	Becotide	Bronchial asthma	711 973	0.78	7 427 415
39	670 927 929	Prochlorperazine	Stemetil	Anti-nauseant (navican)	680 003	0.75	2 669 978
40	962 805	Betamethasone valerate	Celestone	anti-inflammatory (steroid)	676 934	0.75	3 185 052
41	416	Phenoxymethyl-penicillin	Abbocillin V	penicillin (oral)	621 722	0.69	3 432 658
42	799 810	Timolol	Timoptol	β-blocker (eye drops)	572 396	0.63	5 696 761
43	872	Isosorbide dinitrate	Isordil	Angina	568 337	0.63	2 617 602
44	509 511	Metoclopramide hydrochloride	Maxolon	Anti-emetic	531 599	0.59	2 856 199
45	306	Hydrochlorothiazide with triamterene	Amizide	Diuretic and lowering blood pressure	527 445	0.58	2 832 447
46	311	Erythromycin stearate	E-Mycin	Antibiotic	499 789	0.55	3 099 255
47	821	Sodium citrotartrate	Citravescent	urinary infection (makes urine alkaline)	482 941	0.53	2 422 938
48	292 298 728 915	Hydrocortisone	Cortef	Skin condition (steroid)	482 941	0.53	2 422 938
49	193	Amiloride	Midamor	Diuretic (potassium sparing)	478 695	0.52	2 630 023
50	347 518	Terbutaline sulphate	Bricanyl	Bronchospasms	478 606	0.52	2 985 022
TOTALS					61 331 951	67.59	396 362 478

Note: Apparent minor errors are due to rounding. Cost includes patient contribution. Only items listed in the top 50 for 1985 are numbered. Figures used in this table are extracted from the 90 738 032 prescriptions fully processed during 1985/86 for a cost of $648 980 793. Actual prescriptions paid for by the government during 1985/86 totalled 119 841 936 for a cost of $839 566 038 including patient contribution.

APPENDIX VII

TOP 50 PRESCRIPTION AND OVER-THE-COUNTER (OTC) PRODUCTS, 1986

Prescription	Use	OTC	Use
Voltaren	joint pains	S26	baby food
Ventolin	asthma	Panadol	pain and fever
Tenormin	high blood pressure	Panadeine	pain and fever
Amoxil	antibiotic	Codral Cold & Flu	pain and fever
Tagamet	ulcers	Benadryl Expectorant	decongestant
Capoten	high blood pressure	Orthoxicol Cold & Flu	decongestant
Betaloc	high blood pressure	Cepacol	gargle
Minipress	high blood pressure	Combantrin	worms
Naprosyn	joint pains	Berocca	multivitamin
Isoptin	heart disease	Vicks Sinex	decongestant
Aldomet	high blood pressure	Pro Sobee	Baby food
Adalat	angina	Aspro	pain and fever
Zantac	ulcers	Actifed CC Sugarfree	decongestant
Insulin CSL-Novo	diabetes	Infasoy	baby food
Zyloprim	gout	Betadine	antiseptic (skin)
Inderal	migraine	Disprin	pain and fever
Moduretic	diuretic	Medi Slim	diet
Mylanta	ulcers	Enfamil	infant formula
Becotide	asthma	Strepsils	sore throat
Dolobid	joint pain	Sustagen	hospital formula
Lasix	diuretic	Oil of Ulan	cosmetic
Bactrim	sulpha drug + trimetho	Modifast	diet
Intal	asthmatics	Colgate	toothpaste
Triquilar	contraceptive	Myadec	vitamin
Timoptol	eye drops	Vita Valu Vit C	vitamin
Clinoril	anti-inflammation	Hydrocare	contact lenses
Moxacin	antibiotic	Veganin	headache
Indocid	anti-inflammation	Supradyn	vitamin
Septrin	sulfa drug	Visine	eyedrops
Theo-Dur	asthma	Enfalac	baby food
Visken	high blood pressure	J & J Baby	baby food
Renitec	high blood pressure	Codral	pain, cough
Serepax	tranquilliser	Tinaderm	tinea
Epilim	epilepsy	Orthoxicol Cough Supp.	coughs
Transderm-Nitro	angina	Isomil	infant formula
Mersyndol	pain killer	Vermox	worms
Demazin	antihistamine	Macro B High Potency	vitamin
Sudafed	decongestant	Bonjela	mouth ulcers
Brufen	joint pain	Vaseline Intensive Care	skin lotion
Tegretol	anti-convulsant	Sensodyne	toothpaste
Triphasil	contraceptive	Aim	toothpaste
Microgynon 30	contraceptive	Savlon	soap
Vibra	antibiotic	Topex	acne
Mysteclin V	acne	Pentavite	child vitamins
Daonil	diabetes	SMA	baby food
Magadon	sleeping pill	Steradent	false teeth
Canesten	anti-fungal	Selsun	dandruff
Slow-K	potassium replacement	Vita Valu Vit E	vitamin
Gaviscon	indigestion	Vicks Vapour Rub	decongestant
Questran	reduction of bile	Orthoxicol Expectorant	decongestant

This table represents national totals and includes non-pharmaceutical benefits and hospital prescriptions. Prescription products are those requiring a doctor's prescription (S4) or those which may be written on a prescription (S2, S3). OTC products include other therapeutic agents and some retail lines. Products were selected by sales value.

Source: Australian Pharmaceutical Index, Intercontinental Medical Statistics (Australasia) Pty Ltd

APPENDIX VIII

THE ENTROPY GAME

The following game was developed to illustrate the statistical approach to entropy, discussed at the beginning of Chapter 10. Perhaps not quite the chemist's answer to Dungeons and Dragons, but Counters and Grids ™ not only introduces you to the Einstein solid and statistical thermodynamics, but also to social theory, organisational structure, political prioritisation and epidemiology. It displays in one *connected* swoop, the main probability distributions of interest to chemists. (The program of the Entropy Game is available from the author. Send a Macintosh disk plus $35 to: Dr B Selinger, Chemistry Department, The Faculties, Australian National University, GPO Box 4, Canberra, ACT 2601.)

You proceed as follows: set up a grid, say 6 × 6, and on it place any number of counters, say 108 (to give a ratio of three counters per grid). You can place them on randomly or evenly, or all on one square, or choose your own unique scheme. You then plot out a histogram of the number of squares with 0, 1, 2 etc. counters versus the number of counters 0, 1, 2 etc. For random placement, you place a counter on each square as specified by the throw of a pair of dice (6 sided for a 6 × 6 grid, or 4, 8, 10 sided etc. for other grids), to specify a particular square, like in a city map reference. One die defines the horizontal coordinate and the other the vertical coordinate. The result of one attempt is seen in Figure VIII.1.

2	2	1	3	4	6
1	4	3	4	2	2
1	3	2	4	1	1
3	6	6	2	4	1
5	6	1	4	2	5
2	4	4	1	1	5

Fig. VIII.1 Number of counters on each square (random placement)

Now throw the dice again but use the result in the following manner. On the first throw of the dice, you pick up a counter, if there is one, from the square selected by the throw. If you hit a blank square you throw again. On the next throw you place the counter down again, on the square selected by the throw. You repeat this hundreds or thousands of times and note the changing shape of the histogram along the way. We programmed the game for the Macintosh, but it can be run on the tiniest of computers. What happens is very interesting. No matter how you start you end up with the same result, a fluctuating, approximately exponentially falling distribution. Let us start with 108 counters placed randomly on a 6 × 6 grid. You usually obtain a histogram with a hump in the middle, as in Figure VIII.2 (approaches normal for a high ratio of counters to squares).

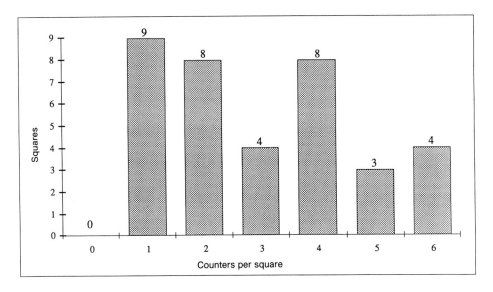

Fig. VIII.2   Number of squares with 0,1,2 etc. counters. 108 counters on 36 squares.
$W = 1.825 \times 10^{23}$; variance = 2.857

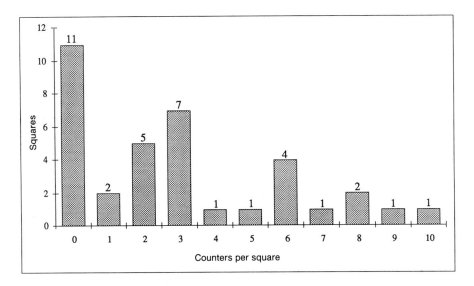

Fig. VIII.3   Histogram after 1000 iterations. 108 counters on 36 squares. $W = 1.6 \times 10^{26}$;
variance = 8.686

The *random* experiment of dice throwing (say 1000 times) causes the histogram to change shape (see Fig. VIII.3) to an exponential distribution.

This becomes more obvious if you smooth out the fluctuations and display the *average* of a number of distributions (say 500), after the 1000 iterations (see Fig. VIII.4).

The explanation for the changes can be seen in Table VIII.1 if we calculate the number of possibilities for each arrangement.[1] Consider a starting position that is uniform and calculate the possibilities of each consequent position. N is the number of squares while the

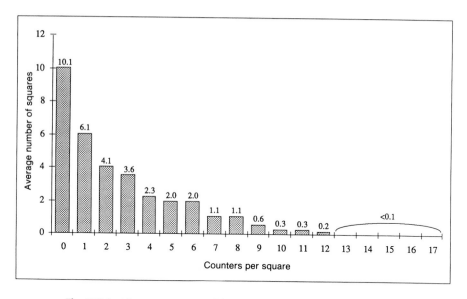

Fig. VIII.4 Histogram averaged for 500 iterations, after 1000 iterations, variance = 10.747

n_0, n_1 etc. are the number of squares with 0, 1 etc. counters. W is the number of possibilities of distributing counters amongst squares without changing the histogram. The sample variance is given by:

$$S^2 = \frac{\Sigma(y_i\text{-}y)}{(N\text{-}1)} = \frac{[\Sigma y_i^2\text{-}(\Sigma y_i)^2/N]}{(N\text{-}1)}$$

where y_i is the sample value and y is the mean.

TABLE VIII.1
Number of squares with 0,1,2 ... counters q

Number of throws	n_0	n_1	n_2	n_3	n_4	n_5	n_6	$W = \dfrac{N!}{n_0!\,n_1!\,n_2!...}$	Variance
uniform	0	36	0	0	0	0	0	1	0
5	5	26	5	0	0	0	0	6×10^{10}	0.29
10	13	12	9	2	0	0	0	2×10^{17}	0.57
20	18	9	4	2	2	1	0	1.67×10^{18}	1.83
40	16	9	7	3	1	0	0	1.62×10^{18}	1.26
60	15	11	6	3	1	0	0	1.65×10^{18}	1.17
80	15	10	8	2	1	0	0	9.72×10^{17}	1.14
100	19	9	3	2	1	1	1	7×10^{17}	2.29
All on one	0	0	0	$0 ... n_{36} = 1$				36	36

Thus for

n_0	n_1	n_2	n_3	n_4
16	9	7	3	1

$$S^2 = [\{(16 \times 0^2) + (9 \times 1^2) + (7 \times 2^2) + (3 \times 3^2) + (1 \times 4^2)\} - 36]/35 = 1.26$$

The maximum number of possibilities means keeping the product of the numbers on the bottom line of the expression for W_0 as small as possible, i.e. keeping the numbers themselves as small as possible, subject to two constraints:

1. $(n_1 \times 1) + (n_2 \times 2) + (n_3 \times 3) + \ldots = $ no. of counters
2. $n_0 + n_1 + n_2 + n_3 + \ldots \qquad\qquad = $ no. of squares

Note the following:
1. Very few changes in the original distribution cause a rapid increase in W.
2. W_{max} is the value of W for the most probable configuration and is of the order of 10^{18}.
3. The distribution can fluctuate at equilibrium provided the value of W stays near to W_{max}.

W_{total} is the sum of all the possibilities for all the configurations and is given by[2]:

$$W_{total} = \frac{(N + q - 1)!}{(N - 1)! \, q!} = \text{sum of } W \text{ over all configurations.}$$

For $N = 36$, $q = 36$, $W_{max} = 1 \times 10^{18}$, $W_{total} = 2 \times 10^{20}$.

Some idea of the magnitude of the number of possibilities can be obtained when you consider the number of possibilities for setting Rubik's Cube. Hofstadter[3] estimates this as 4.3×10^{19} (43 252 003 274 489 856 000, to be precise). You can imagine that throwing a dice to select a move will never bring a random cube configuration back to the single START.

This, then, is the rationale of the Second Law; systems drift towards configurations that are made up from the greatest number of possibilities or *in*distinguishable microstates.

There are some other interesting observations to make about the distributions. When you start placing counters at random one at a time on the squares, you produce a *binomial* distribution. For 36 squares and 36 counters (using the usual statistical notation), $n = 36$ and $p = 1/36$. The mean is $np = 1$, and the expected variance is $np(1 - p)$. Under these conditions, the binomial is approximately *Poisson* where the mean is μ and equals the expected variance, because $np(1 - p) \to np$. When, in addition, μ is small, the Poisson distribution becomes a geometric distribution. On the other hand keeping p the same but increasing n to give a *large np* (≥ 20), we find that the binomial $\to$ Poisson, tends to a *normal* or bell-shaped error distribution (still with equal mean and variance). With 108 counters on 36 squares we are on the way towards this limit.

When we start the game with 108 on 36, we quickly see that the random starting distribution does not have the largest number of possibilities W, and thus the distribution changes in shape with play. Seeing that we started with a *random* placement, to say that the drive is now dominated by a *tendency to randomness* is unhelpful, unless we explain how the constraints have changed when we changed the rule. We started by *depositing* the counters binomially. This fixed the expected variance at $np(1 - p)$ ($= 3$). However, as the game proceeds, the variance is no longer constrained, but increases and another distribution takes over.

For *aficionados* you can actually derive the equilibrium to which this distribution tends. For i counters per square the probability distribution $\Pi(i)$ is given by:

$$\Pi(i) = \frac{(N + q - i - 2) \, \mathbf{C} \, (N - 1)}{(N + q - 1) \, \mathbf{C} \, (N)}$$

Where $\mathbf{C}$ is the combination symbol.

In the limit as $N \to \infty$, $q \to \infty$ (or taking an average of a large number of runs), the distribution becomes a *geometric* distribution. The expected equilibrium probability $\Pi(i)$, for this limit is now given[4] as:

$$\Pi(i) = \{1/(1 + \theta)\} \times \{\theta/(1 + \theta)\}^i$$
$$\text{where } \theta = q/N$$

$$\text{The variance} = \{\theta/(1 + \theta)\} \div \{1/(1 + \theta)^2\}$$

Thus for 36 counters on 36 squares the equilibrium distribution $\Pi(i)$ is given by $(1/2) \times (1/2)^i$, while for 108 counters it is $1/4 \times (3/4)^i$. (The expected equilibrium variances are 2 and 12 respectively. Try some experiments to show that this is so.) These geometric distributions are discrete *exponential* in the limit of large (or averaged) samples.

This game illustrates the fundamental concept of the Second Law of Thermodynamics, the understanding of which the author, C. P. Snow, declared in his Rede lecture[5] as *the* test of scientific literacy, the equivalent to being able to appreciate a play by Shakespeare. Interestingly enough, he later drew back from this test. 'This law', he later said, 'is one of the greatest depth and generality; it has its own sombre beauty; like all major scientific laws it evokes reverence. There is of course no value in the non-scientist knowing it by the rubric in an encyclopaedia. It needs understanding, which cannot be attained unless one has learned some of the language of physics. That understanding ought to be part of the common twentieth century culture. Nevertheless, I wish I had chosen another example.' He goes on to say that he would now have chosen molecular biology, but adds, 'the ideas in this branch of science are not as physically deep or of such universal significance as those of the Second Law. The Second Law is a generalisation which covers the cosmos'.

Counters and grids is a game that provides a small-scale simulation of the probability interpretation of entropy, the Einstein solid, Bose–Einstein statistics with the Boltzmann distribution as limit. In fact the Boltzmann limit is found for any large-scale distribution for which the only constraint, or sole information, is the constant value of the mean.

Examples include the intensity of light passing through an absorbing solution (constant absorbance); first order kinetics (constant rate constant); radioactive or fluorescence decay (constant lifetime); radial electron density in an *s* orbital (fixed Bohr radius) etc.

However what is incredibly interesting is that the thinking behind the Second Law is by no means limited to these physical examples.

CONSUMER EXAMPLES

Just call the counters *money* and the squares *people*. You then discover a truism. In a laissez-faire economy in which money is exchanged freely between individuals for goods and services without any restrictions (in particular redistribution via taxation and social benefits; the model fits authoritarian economies better than democratic ones), most people end up with few dollars and a few end up with most of the dollars. Even if you increase the mean number of counters per square from less than one to greater than one, the final distribution does not change significantly in shape, so people's relativities hardly change, even though in absolute terms they are better off.

A closer model is where hundreds of people enter a casino with, say, an equal modest stake and at the end of the evening leave with their pockets emptied or filled. Their group winnings follow very closely the exponential distribution. It is extremely important to note that we cannot predict who will be rich and who will be poor. The more possibilities for distributing the money *between* people without affecting the *overall* distribution, the more probable that distribution.

Even more interesting is that even where we ourselves control the result, as for example in the way the organisers deliberately distribute prizes in lotteries, there are few large prizes, moderate numbers of moderate prizes and many small prizes. With a fixed amount of prize money and number of prizes, the exponential distribution is the most *natural* way to do things.

You say that you're not into gambling. Well you are, you know! As a policy holder in any insurance scheme (car, health, house etc.), you contribute a small premium to protect yourself against the *certainty* that natural events will dictate that a few people will need large payouts and more will need smaller ones. The small premium monies will redistribute themselves as claims, roughly along the lines of the exponential distribution.

From the probability of occurrence of individual letters in this text to the size of cities in a country or oil wells in the world or stars in the galaxy, the distribution as determined by the Second Law of Thermodynamics (with minor reservations) gives the gist of the answer. Further examples can be seen in the loss of control of a private company when it distributes shares to a large public (look at the share distribution of shareholders in a large company).

If we move slightly laterally to the random exchange of power and influence in an organisation, then its distribution follows the same pattern. The number of people at any level decreases with the height of the level. Hierarchies behave just like the real atmosphere and for the same reason; they become rarer the higher you go.

The thinking behind the Second Law has other everyday analogues. For example, what is the fundamental reason for it being so much harder to park a car in a tight spot than to drive it out again afterwards? (Consider the number of possibilities for being parked compared to being 'unparked'.)

One could stretch the argument and say that the reason for lack of support for teaching innovation is a consequence of the Second Law. It is much more natural to distribute funds to provide a small spectrum of more spectacular items to fill annual reports and for parliamentarians to make speeches about, than to uniformly improve the whole field. Laying foundation stones for high tech hospitals beats small and unnoticed public health improvements across the board. For the same reason, it is very difficult to replace the concept of a large centralised power generation system (be it coal-fired or nuclear) with many small, widely distributed solar units. Finally, no lottery would attract custom if, for the same total amount of prize money, it offered only equal, relatively small prizes and no outstanding ones.

In regard to quantum theory, Einstein once said 'God does not play dice'. Einstein was probably wrong. Anyway in thermodynamics he definitely does play dice. And so do we, all the time.

EPIDEMIOLOGY

For our last example, we return to the random distribution with which we set up the game in Figures VIII.1 and VIII.2 and study it in more detail. We use the distribution to study epidemiology—incidence of disease in the community.[6] Leukaemia (which is a combination of related diseases) is responsible for about one in 40 deaths in the West, so it is rather rare. However it is the main cancer found in children. While there are several postulated causes of leukaemia, there have been suggestions that childhood leukaemia, in particular, may also be caused by low levels of radioactivity, such as those emitted by nuclear installations. A UK TV program in 1983 revealed an apparent cluster of four children with leukaemia close to the nuclear reprocessing plant at Sellafield, formerly Windscale. The statistical occurrence of clusters when the expected value is quite low can be illustrated with our random distribution.

Choose a 12×12 grid and enter 72 counters at random. The average number of counters per square is 0.50. However notice the occupancy of the grids. In the particular throw seen in Figure VIII.5, there are 94 squares with no counters, 35 with one counter, 9 with two counters, 5 with three counters and none with four counters. Although the average number of counters per grid is 0.5, one grid contains 4 counters, eight times the average. The expected number is given by the Poisson distribution:

$$N(x) = \frac{\mu^x \exp[-\mu]}{x!}$$

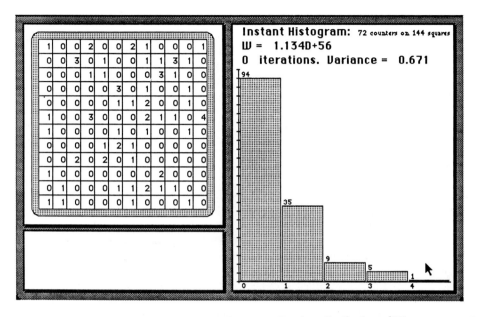

Fig. VIII.5 Random distribution of 72 counters on 144 squares

where the mean $\mu = 0.5$, and $x =$ number of counters per square. Thus $N(x)$ is the number of squares with x counters.

The Poisson distribution predicts, on average, 87 squares with zero counters, 44 with one counter, 11 with two counters, 2 with three counters and 0.02 with four counters. Our result is in good agreement with the expected (as can be shown by a statistical test of goodness of fit). We note that four counters on one grid would be expected on average every $1/0.02 =$ five throws.

Although on average two-thirds of the grids are empty, this is clearly not the probability of obtaining empty squares, since it is impossible to cover 144 squares with 72 counters. The probability of finding one or more empty grids is 100%. The frequency with which multiple occupancy occurs is $[9 + 5 + 1]/144$ or 10%, and yet the probability that at least one square has multiple occupancy is:

$$1 - (144/144) \times 143/144) \times (132/144) \times \ldots (73/144).$$

which is over 99.9%.

For this section you need to have access to a computer, because you need to throw the dice 72 times (or your choice) each time. However you can play a similar game by packing the bottom of a shallow dish (such as a petri dish) with, say, 400 white beads and 50 (i.e. one-eighth) coloured beads. Pour the balls from one dish to another and then let them settle flat. If the balls are fairly close packed, then the distribution will look fairly even and each ball will be surrounded by about six others (see Fig. VIII.6 and also the section and plates on models for atom packing in Chapter 6). However the random distribution is far from even in colour, and will on average reveal several colour clusters. A random distribution is not uniform.

If you divide the space up into smaller regions, you can produce a greater or smaller number of regions with clusters, depending on how you gerrymander the boundaries. In fairness, you should mark out the divisions before you shake the balls, not afterwards!

Producing 'honest electorates' is a central problem in epidemiology. Read the *New Scientist* article[6] for more on leukaemia and radiation.

The game can also be used to illustrate another interesting occurrence in everyday life. When you are waiting to cross at a busy intersection, eventually you get a break. Assume that the traffic is random, i.e. there are no obvious traffic lights or other obstacles that are bunching up the traffic. The 10-second interval you need to cross the intersection can be made to be equivalent to say four adjacent empty squares. In our example there are, on average, half as many counters as squares, so on average every second square should be occupied. On this basis you should never be able to cross the intersection! However moving along a row at a time in Figure VIII.5 shows that several clusters of four empty squares occur, and so you can cross the intersection.

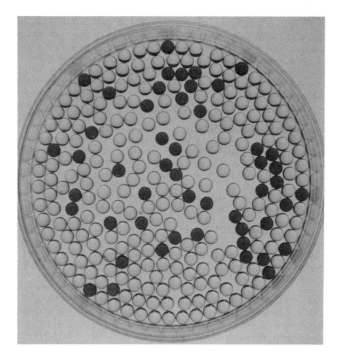

Fig. VIII.6

References
1. 'The teaching of thermodynamics: a teaching problem and an opportunity', *School Science Review* **57**(210), 1976, 654; 'A picture of shuffling quanta', in *Nuffield Advanced Science Chemistry, Teachers' Guide I*, Topics 1–11, pp. 75–79, Longman, Harlow, Essex, 1984.
2. H. A. Bent, *The Second Law*, Oxford University Press, New York, 1965, ch. 21.
3. D. Hofstadter, *Metamagical themas: questing for the essence of mind and pattern*, Penguin, 1985, p. 305.
4. B. Selinger, and R. Sutherland 'Counters and grids', *J. Chem. Ed.* June 1989.
5. C. P. Snow, *The two cultures and a second look*, Cambridge University Press, Cambridge, 1969.
6. D. Taylor and D. Wilke, 'Drawing the line with leukemia', *New Scientist*, 21 July 1988, 53.

APPENDIX IX

AUSTRALIA'S SPACE HEATERS PERFORMING POORLY

This winter millions of Australians are spending their evenings in rooms warmed by conventional space heaters—and most of us are not experiencing a satisfactory degree of thermal comfort. In other words our heaters are not doing the job we expect of them!

Even with oil, gas or electric space heaters turned to 'high' many people will find that their feet are too cold for comfort. Some householders may decide—on the basis of this discomfort—to insulate their ceilings, and although this may reduce their fuel bills it will not solve their cold feet problems.

What 'everybody knows'—but it still took quite some work to prove to them—is that our feet get cold if the air near the floor is cold. Medical science has proved that once our feet get cold our bodies switch off the flow of warm blood to that region in an effort to conserve heat for the vital organs contained in the trunk. Frostbite is an extreme example of this 'switching off'. Because of this bodily reaction we can say that the temperature of air a few centimetres from the ground largely determines the comfort of a person in the normal sitting position. It is not uncommon for people in a heated room to feel 'stuffy' and nod off to sleep. Work in the UK has described this 'stuffiness' as a condition caused by a temperature gradient—hot near the head, and cold near the floor—and attributed sleepiness to this variation.

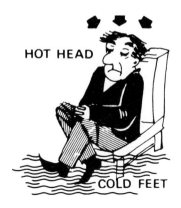

HOT HEAD

COLD FEET

The aims of effective room heating must include the elimination of this temperature gradient. If we are to overcome cold feet and sleepiness our heating systems need to warm the air near the floor and maintain an even temperature from floor level to ceiling. Most popular space heaters do not achieve this—they pour out warm air at too great a height above floor level and we get a high temperature at head height or higher, and an unacceptably low temperature at feet level.

In an effort to determine how to achieve optimum room heating DBR [Division of Building Research, CSIRO] scientists set up a test room with an area of 20 m^2—the size of a large living room. They experimented with heated air being blown into the room at different heights above floor level, and rated each against a scale of zero (poor heating) to 100 (excellent). This score was achieved by taking the temperature rise recorded near the floor and dividing it by the average temperature rise recorded in the room. The result was then multiplied by 100 to give a workable figure.

The diagrams in Figure IX.1 show that the best performance was achieved when air was blown into the room at a height of 0.3 m above floor level. It can be assumed that a similar result would have been achieved at any height below 0.3 m. In these experiments, the incoming air was kept at a temperature of 17°C above that of room temperature. The figures appearing in boxes (in diagrams A–D) indicate the performance change which occurred when the incoming air was 30°C above that of room temperature. This increased temperature always reduced heating effectiveness—thereby indicating that there must be an adequate volume of circulating air to keep the incoming temperature within about 17°C of the room air. More precisely, this means about 50 litres per second per kW.

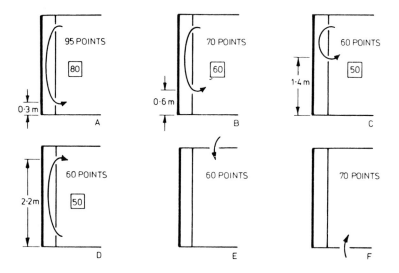

Fig. IX.1

Tests were also done with convection 'stoves' of both fan and fanless types (65 points); hot water radiators of the stand-up variety (60 points); and hot water baseboard radiators (80 points when placed opposite a window, and 90 points when placed under a window).

The results indicate that few methods are well suited to room heating, and that much present equipment is unsatisfactory. The only reasonable possibilities appear to be:

Heating of the actual flooring material (not tested). This generally has to be installed during house construction. Heating systems installed in a concrete floor can pose another problem, that of very slow response to change.

Convection 'stoves' using oil, gas or solid fuel with forced air flowing in the opposite direction to normal so that air intake is at say waist level, and the forced air outlet is almost at floor level. (Such units may not be available commercially.)

Hot water baseboard radiators.

Fan-assisted heaters blowing air horizontally near the floor.

Radiant heaters (gas, electric, fuel, oil) facing the people to be kept warm.

While conventional space heaters continue to pour out warm air at the wrong height above floor level we will not solve our hot head and cold feet problem.

Source: Rebuild, June 1977, CSIRO Division of Building Research.

APPENDIX X

OCTANE RATINGS

TABLE X.1
Octane values of hydrocarbons. Octane values vary widely with hydrocarbon branching,
unsaturation, and size.

Hydrocarbon	Structure	Unleaded research octane number (RON)
n-Octane		−19.0
n-Heptane		0.0
n-Hexane		24.8
n-Propylcyclopentane		31.2
Octene-2, cis-isomer		56.2
n-Pentane		61.7
Isopropylcyclohexane		62.8
2,4-Dimethylhexane		65.2
Octane-4, trans-isomer		73.3
Isopropylcyclopentane		81.1
Cyclohexane		83.0
Pentene-1		90.9
Hexene-2, trans-isomer		92.7
n-Butane		93.8
Propane		97.1

Butene-1		97.4
2,2,4-Trimethylpentane (iso-octane)		*100.0*
Cyclopentane		101.3
Propylene		102.5
2,4,4-Trimethylpentene-1		102.5
Benzene, technical grade		105.8
1,4-Diethylbenzene		106.0
Ethylbenzene		107.4
o-Xylene		107.4
Isopropylbenzene (cumene)		113.1
p-Xylene		116.4
m-Xylene		117.5
Toluene, technical grade		117.8
Toluene, chemically pure		120.1

Source: 'Gasoline', *Chemical and Engineering News*, 9 November 1970, p. 52.

TABLE X.2
Octane number requirements of automobile engines. Octane
number requirements of automobile engines increase as
compression ratio increases.

Engine compression ratio	Typical research octane no. required for knock-free operation
4:1	60
5:1	73
6:1	81
7:1	87
8:1	91
9:1	95
10:1	98
11:1	100
12:1	102

Source: 'Gasoline', *Chemical and Engineering News*, 9 November 1970, p. 52.

APPENDIX XI

FOOD LAW IN AUSTRALIA

In Australia food law is a State matter, although there is a move towards a federal (i.e. national) Food Law. In order to achieve some measure of uniformity the NH & MRC set up a Food Standards Committee (in 1953) which coordinates the activities of the States (and Territories) and produces a list of approved food standards for incorporation into State (and Territory) legislation. Reference in this chapter is made to these standards. They were produced by the Commonwealth Department of Health in the form of a book, which also gave the approximate current status of the law in the various States (and Territories). *The NH & MRC Food Standards have no legal significance until such time as they are incorporated into the legislation of the States and Territories of Australia.*

In June 1982 the Model Food Act with Model Food Standard Regulations was established by the NH & MRC in the hope that it would be adopted by the State Governments. This uniformity in food law throughout Australia would benefit both manufacturers and consumers.

ADOPTION OF THE FOOD STANDARDS CODE
In April 1987 the Food Standards Code replaced the Model Food Standards Regulations. Responsibility for the secretariat for food standards went from the Department of Health to the Federal Bureau of Consumer Affairs, currently in the Department of the Attorney General. The hope that the States would adopt these directly or by reference went along also.

- *Queensland* (very progressive). Food Act 1981. Has adopted the code (Food Standard Regulation 1987) by reference, automatically updated as amended by the NH & MRC.
- *Victoria* (progressive). Food Act 1984. Has adopted the code, but by reference, not automatically.
- *Western Australia* (independently progressive). The code has been incorporated specifically into the Health 'Food Standards' General Regulations 1987, but with

redrafting to meet the cultural needs of Western Australian legal draftspersons, and to allow for peculiarities, such as the level of milk fats in Western Australian cows.

- *South Australia* (progressive but slow). Regulations are being drafted to provide for the code to be adopted by reference.
- *Tasmania* (very progressive). Public Health (Food Standards) Regulations 1986: automatic adoption by reference.
- *Northern Territory* (progressive but very slow). Food Act 1986, yet to commence. Regulations to be drafted to adopt code by reference.
- *Australian Capital Territory. No food law since its establishment in 1901.* (Attorney Generals are always too busy with federal matters). The promise is to have a law something like that of one of the States. It could never be decided whether to adopt the Victorian or NSW model, so the ACT has received nothing. This is the best place in Australia for post-dated food, illegal imports etc.
- *New South Wales* (very conservative). After all, what is wrong with the Pure Food Act 1908; its centennial is only 20 years off. The Act has been suitably amended and the regulations occasionally include some of the NH & MRC food standards, but rewritten, presumably to suit NSW's cows.

NH & MRC principles for the evaluation of food additives—May 1977

1. The use of a food additive is justified only when it serves one or more of the following purposes:
   a. to maintain or improve the nutritional quality of a food;
   b. to improve the palatability, storage life or appearance of a food;
   c. to render a food more appetising;
   d. to provide aids in producing, manufacturing, packing, processing, preparing, treating, packaging, transporting, holding, or storing foods;
   e. as a preservative only when necessary because there is no alternative practicable means of preservation of the food.
2. The use of a food additive is *not* justified:
   a. if the proposed level of use constitutes a hazard to the health of the consumer;
   b. if it causes an appreciable reduction in the nutritive value of a food;
   c. if it disguises the faulty or inferior qualities of a product or the use of processing and handling techniques which are not permitted;
   d. if it deceives the purchaser or consumer;
   e. if the desired effect can be obtained by another method of processing which is economically and technologically feasible.
3. Approval for the use of a food additive should be based on anticipated intake in relation to consumption patterns of the community. Special regard should be given to vulnerable groups with special diets such as infants and the elderly.
4. Where it is necessary to use a food additive for any purpose, the purpose must be specific and in the best interests of the consumer. It must be established that there are no alternative means of achieving the purpose more consistent with the best interests of the consumer.
5. Approval for the use of a food additive shall not be general but shall be limited to specific foods for specific purposes under specific conditions, unless otherwise determined.
6. A food additive must be:
   a. used in the minimum amount necessary to effect the intended purpose under good manufacturing practice;
   b. acceptable at the level approved on toxicological grounds.
7. A food additive must be in conformity with an acceptable standard of purity.
8. a. All food additives proposed for use shall have had adequate toxicological evaluation;

  b. Permitted food additives are subject to continuing observation and are re-appraised in the light of changing conditions of use and new scientific information.
9. Incidental food additives shall not exceed the lowest levels that are technologically feasible.

An applicant wishing to introduce a new food additive must make a detailed submission to the NH & MRC Food Science and Technology sub-Committee. The information must include (*inter alia*) the following:

1. The proposed minimum and maximum level of use.
2. The limits of the probable daily intake of the additive in the diet.
3. Any evidence of rejection of the additive by any statutory body or authority.
4. Details of the precise chemical structure and physical details of the additive.
5. Details of the nature and amounts of impurities present.
6. The advantages that will accrue to the consumer from the use of the additive.
7. Analytical methods that can be used for verification.
8. Method of manufacture.
9. Results of pharmacological and toxicological investigations including acute, short-term and long-term (chronic) toxicity studies. Chronic toxicity data should be given for at least two species, one of which should be the dog, and carried out over the major portion of the life span of the experimental animal. Chronic toxicity experiments should aim to give the data needed to establish a 'no effect' level.
10. Reports of any physiological effects and any abnormal reactions, including carcino-genesis, teratogenesis in pregnant species, sensitivity, tolerance, or idiosyncrasy in response to additive.
11. Biochemical information on the possible mode of action if available; metabolic studies to show rate, extent and mode of elimination.

Note: Details of any reports which could bias an evaluation of the safety of the additive should *not* be omitted. The information supplied should be attested to by a statutory declaration.

APPENDIX XII

FOOD ADDITIVES IN AUSTRALIA

SYNTHETIC FOOD COLOURS IN AUSTRALIA (NH & MRC, 1983)

The food colours fall into different chemical groups. All the reds, oranges and yellows are *azo dyes* having the —N=N— linkage, except erythrosine, which is a *xanthene*. The green, blue and violet are *triphenylene* dyes, while the browns and black are azo dyes again. Indigo carmine is a different type again. Where there is a sensitivity to azo dyes, there is often no reaction to food green and blue (unless the colour is a mixture).

The *Colour index* (3rd edn, Soc. Dyers and Colourists. Bradford, England. Vols 1-4, 1971. Vols 5-6, 1975.) lists an estimated 38 000 commercial colourants involving 7000-8000 different chemical structures. This system identifies each product by a generic name (e.g. acid yellow 23) which describes the application type and colour. If the chemical structure is known, a five digit constitution number is allocated (in this case CI 19140). Some cross-referencing to commercial brands is provided. *Chemical Abstracts* issues a registered serial number to all chemicals, including colourants. The legislatures of various countries issuing lists of allowed additives have their own coding systems as well; for example, the EEC and USA (FD & C number).

The *Chemical Abstracts* Service (CAS) Registry number allows the easiest search of chemical literature.

Red shades

CI 16035   Allura red AC
CI food red 17
FD & C red no. 40
CAS reg. no. 2956–17–6

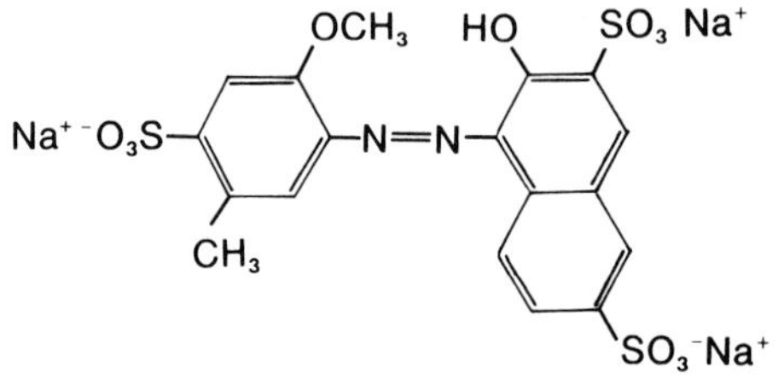

CI 16185   Amaranth
CI acid red 27
CI food red 9
FD & C red no. 2
E123
CAS reg. no. 915–67–3

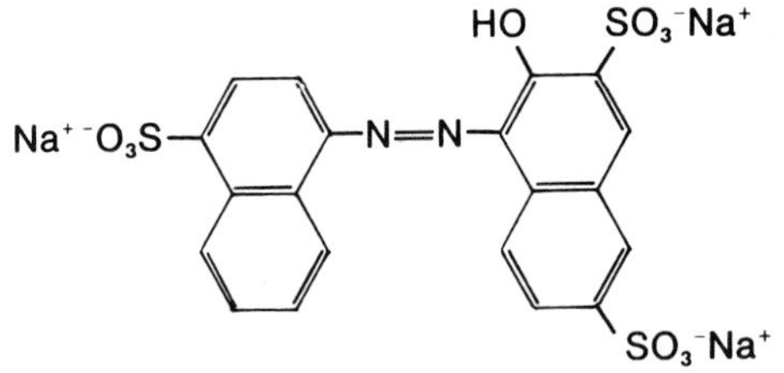

CI 16255   Brilliant scarlet 4R
CI acid red 18
CI food red 7
E124
CAS reg. no. 2611–82–7

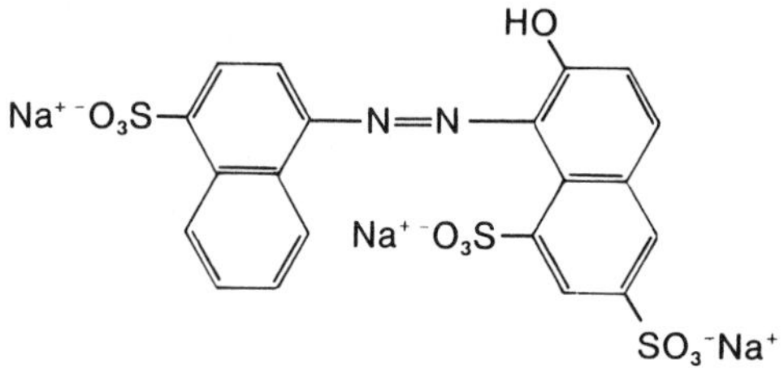

CI 14720   Carmoisine
CI acid red 14
CI food red 3
CAS reg. no. 3567–69–9

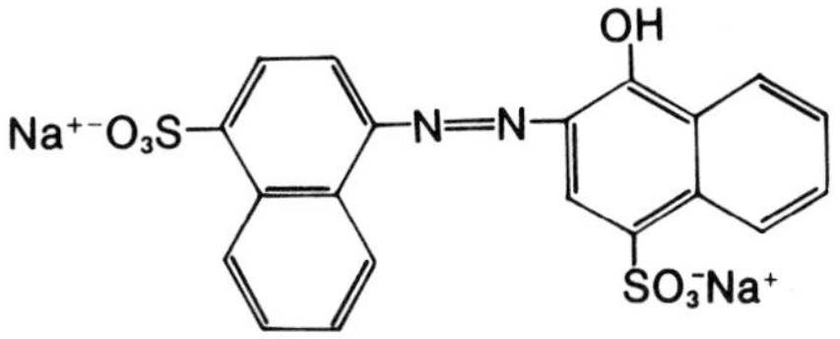

CI 14780   Chlorazol pink Y
(Hexacol red FB)
CI direct red 45
CI food red 13
CAS reg. no. 2150–33–6
Deleted by NH & MRC 1980

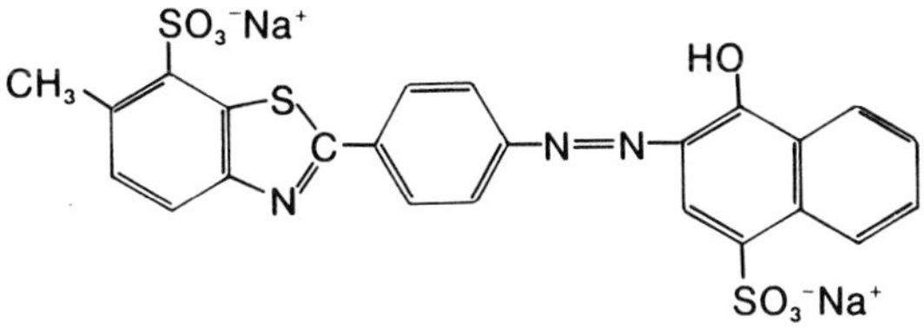

CI 45430   Erythrosine
(tetraiodinate fluorescein)
CI acid red 51
CI food red 14
E127
CAS reg. no. 568–63–8

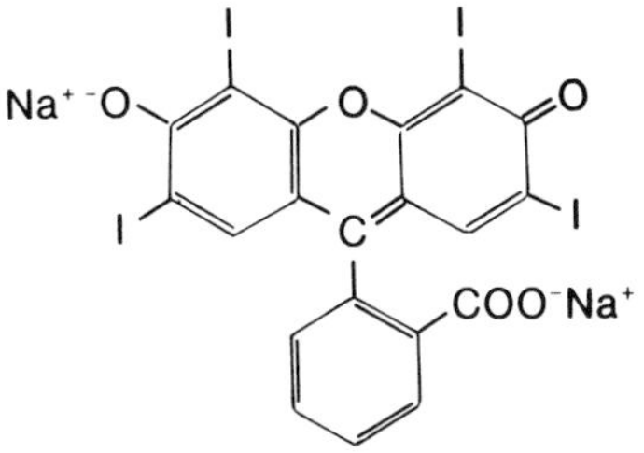

Orange shade

CI 15980   Orange GGN
CI food orange 2
Delisted in the EEC (E111)
CAS reg. no. 2347–72–0
Deleted by NH & MRC 1978

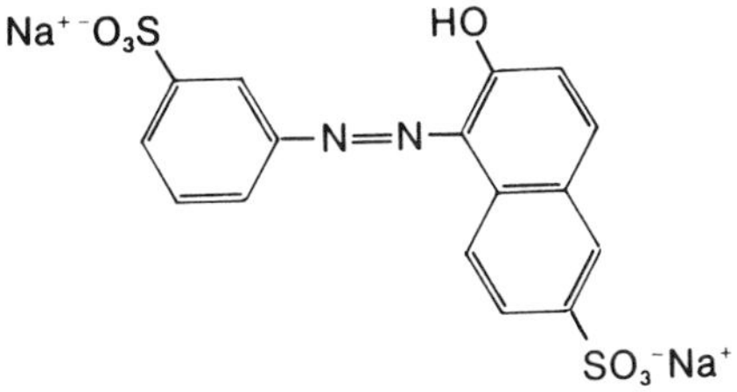

Yellow shades

CI 15985   Sunset yellow FCF
CI food yellow 3
FD & C yellow 6
E110
CAS reg. no. 2783–94–0

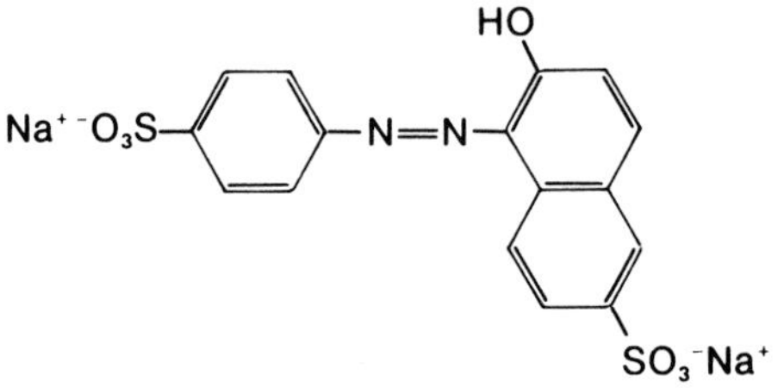

CI 19140   Tartrazine
CI acid yellow 23
CI food yellow 4
FC & C yellow 5
E102
CAS reg. no. 1934–21–60
Rejected by FAO/WHO

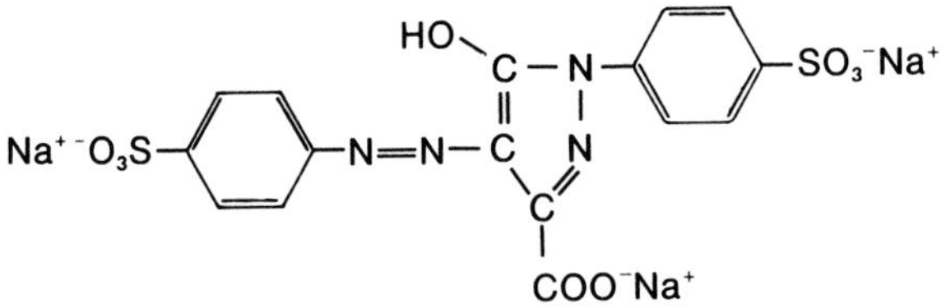

CI 18965   Yellow 2G
CI acid yellow 17
CI food yellow 5
Delisted in the EEC
CAS reg. no. 6359–98–4

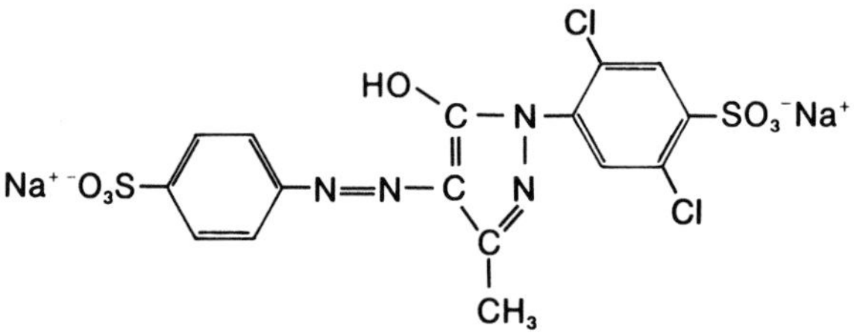

Green shade

CI 44090   Green S
CI acid green 50
CI food green 4
E142
CAS reg. no. 3087–16–9

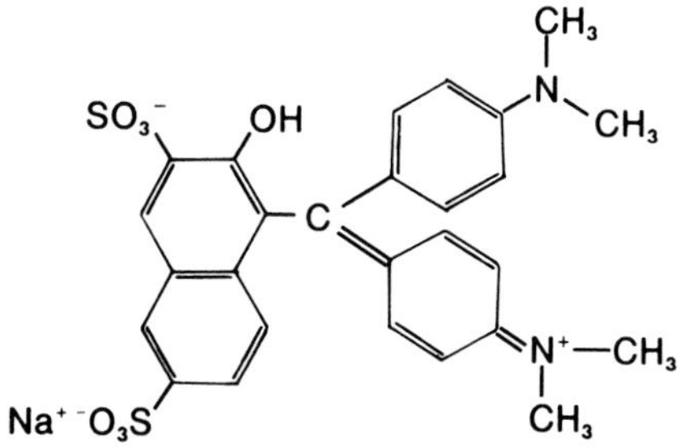

Blue shades

CI 42090   Brilliant blue FCF
CI acid blue 9
CI food blue 2
CI pigment blue 24
Delisted in the EEC
CAS reg. no. 2650–18–2

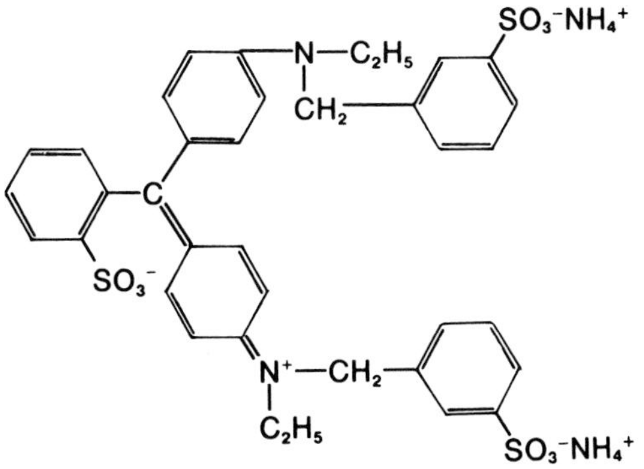

CI 73015   Indigo carmine
CI acid blue 74
CI food blue 1
FD & C blue 2
E132
CAS reg. no. 860–22–0

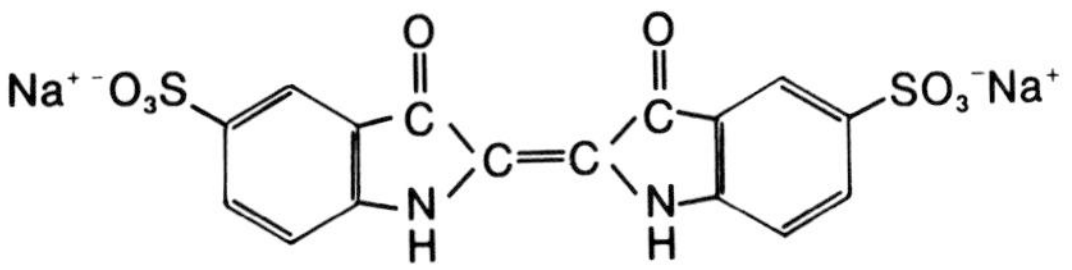

Violet shade

CI 42580   Sodium salt of 4, 4′-di(dimethylamino)-4″-di(*p*-sulfo-benzylamino) triphenyl-methanol anhydride.
CI acid violet 21 (*Colour Index*, 3rd edn, 1971)
CAS reg. no. 5905–37–3
Deleted by NH & MRC 1980
Note: CI 42581, CI acid violet 17, is *para* substituted and is the compound depicted here. Acid violet 21 is *meta* substituted.

Brown shade

Chocolate brown FB
The product of coupling diazotised naphthionic acid (l-naphthylamine-4-sulfonic acid) with a mixture of morin and maclurin.
CI food brown 2
Not allowed in USA or Canada
Delisted in the EEC
No CAS reg. no.

Fig. XII.1 Chocolate brown, anyone? Morin and Maclurin are found in the wood of the Dyer's mulberry, *Chlorophora tinctoria*. Morin is also found in the wood of the osage orange tree, found in the United States and as an ornamental tree in Australia. They are reacted with 1-naphthylamine-4-sulfonic acid to give chocolate brown FB, a colour allowed in Australia presumably because it is needed. However, this colour was not available commercially! It was on the list in 1975, but not in 1979. It was formally removed (after the second edition of this book) in 1980

CI 20285

    Chocolate brown HT
    CI food brown 3
    Not allowed in USA or Canada
    Delisted in the EEC
    CAS reg. no. 4553–89–3

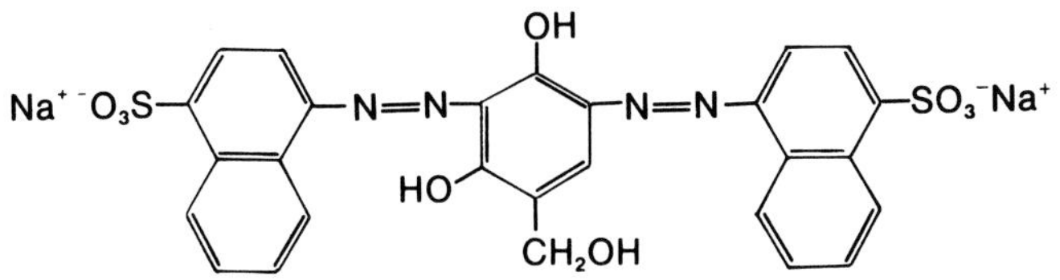

Black shade

CI 28440   Brilliant black BN
                CI food black 1
                (bluish-violet in water)
                E151
                CAS reg. no. 2519–30–4

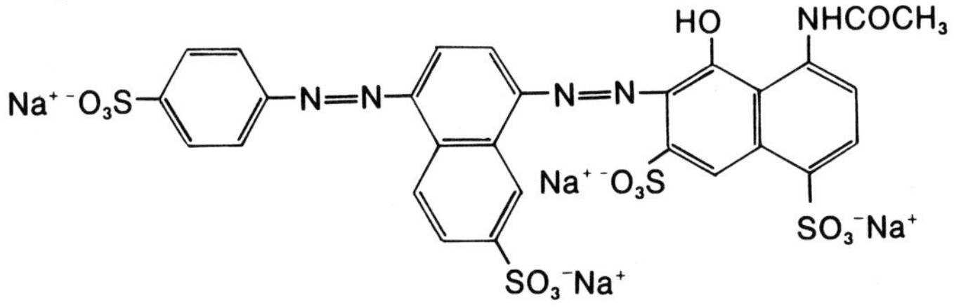

TABLE XII.1
History of approved synthetic colours inAustralia

CI code	1955[a]	CI code	1967[b]	Since 1967[c]
16185	Amaranth	16186	Amaranth	
16255	Brilliant scarlet 4R	16255	Brilliant scarlet 4R	
14720	Carmoisine	14720	Carmoisine	
14780	Chlorazol pink Y	14780	Chlorazol pink Y	deleted 1980
45430	Erythrosine	45430	Erythrosine	
		16045	Fast red E	deleted 1975
45435	Rose Bengale	45435	Rose Bengale	deleted 1975
14815	Scarlet GN	14815	Scarlet GN	deleted 1975
14700	Ponceau SX		removed *c.* 1965	16035 Allura red added 1975
45170	Rhodamine B		removed *c.* 1961	
15980	Orange GGN	15980	Orange GGN	deleted 1978
13015	Acid yellow G	13015	Acid yellow G (kond)	deleted 1975
15985	Sunset yellow FCF	15985	Sunset yellow FCF	
19140	Tartrazine	19140	Tartrazine	
		13011	Yellow RFS[d]	deleted 1968
		18965	Yellow 2G	
		14330	Yellow RY	deleted 1968
		44090	Green S	

TABLE XII.1
Continued

CI code	1955[a]	CI code	1967[b]	Since 1967[c]
42095	Light green SF		removed c. 1961	
42090	Brilliant blue FCF	42090	Brilliant blue FCF	
73015	Indigo carmine	73015	Indigo carmine	
42045	Patent blue		—	
			Violet BNP[d]	deleted 1980
42650	Acid violet 5BN		—	
			Brown FK[d]	deleted 1974
			Chocolate brown FB	deleted 1980
		20285	Chocolate brown HT	
	Chocolate brown NS		—	
20220	Thiazine brown R			
28440	Brilliant black BN	28440	Brilliant black BN	
35445	Black 5410		removed c. 1961	

[a] Before 1955 some 40 coal-tar dyes were approved as food colours.

[b] Now called synthetic colouring substances.

[c] In 1970 a standard for purity of food colours was adopted by the NH & MRC (BS3210, 1960). This was replaced by FCC (International Food Colour Codex) and FAO/WHO specifications in 1983. The old standard allowed about 15% unidentified impurities in the colour (see *Chemistry in the marketplace*, 2nd edn). Consumer representatives on the Food Standards Committee (NH & MRC) since 1974.

[d] Added c. 1961.

TABLE XII.2
Food additive code numbers, numerical order. EEC code numbers adopted by NH & MRC

Code no.	Prescribed name	Code no.	Prescribed name
100	Curcumin	160(f)	Ethyl ester of beta-apo-8'-carotenoic acid
101	Riboflavin	161	Xanthophylls
102	Tartrazine	161(g)	Canthaxanthine
107	Yellow 2G	162	Beetroot red, betanin
110	Sunset yellow FCF	163	Anthocyanins
120	Cochineal, carminic acid	170	Calcium carbonate
122	Carmoisine	171	Titanium dioxide
123	Amaranth	172	Iron oxides and hydroxides
124	Brilliant scarlet 4R	200	Sorbic acid
127	Erythrosine	201	Sodium sorbate
132	Indigo carmine	202	Potassium sorbate
133	Brilliant blue FCF	203	Calcium sorbate
140	Chlorophylls	210	Benzoic acid
142	Green S	211	Sodium benzoate
150	Caramel	212	Potassium benzoate
151	Brilliant black BN	213	Calcium benzoate
153	Carbo medicinalis vegetalis (charcoal)	220	Sulfur dioxide
155	Chocolate brown HT	221	Sodium sulfite
160	Carotenoids	222	Sodium bisulfite
160(a)	Carotene, alpha-, beta-, gamma-	223	Sodium metabisulfite
		224	Potassium metabisulfite
160(b)	Annatto (bixin, norbixin)	234	Nisin
160(e)	Beta-apo-8' carotenal	249	Potassium nitrite

TABLE XII.2
Continued

Code no.	Prescribed name	Code no.	Prescribed name
250	Sodium nitrite	401	Sodium alginate
251	Sodium nitrate	402	Potassium alginate
252	Potassium nitrate	403	Ammonium alginate
260	Acetic acid	404	Calcium alginate
261	Potassium acetate	405	Propylene glycol alginate
262	Sodium acetates	406	Agar
263	Calcium acetate	407	Carrageenan
270	Lactic acid	410	Locust bean gum
280	Propionic acid	412	Guar gum
281	Sodium propionate	413	Tragacanth
282	Calcium propionate	414	Acacia
283	Potassium propionate	415	Xanthan gum
290	Carbon dioxide	416	Karaya gum
296	Malic acid	420	Sorbitol
297	Fumaric acid	421	Mannitol
300	Ascorbic acid	422	Glycerol
301	Sodium ascorbate	433	Polyoxyethylene (20) sorbitan mono-oleate
306	Tocopherol-rich extracts of natural origin	435	Polyoxyethylene (20) sorbitan monostearate
307	Synthetic alpha-tocopherol	436	Polyoxyethylene (20) sorbitan tristearate
308	Synthetic gamma-tocopherol		
309	Synthetic delta-tocopherol	440(a)	Pectin
310	Propyl gallate	442	Ammonium phosphatides
311	Octyl gallate	450	Sodium and potassium polyphosphates
312	Dodecyl gallate		
320	Butylated hydroxyanisole (BHA)	460	Microcrystalline cellulose, powdered cellulose
321	Butylated hydroxytoluene (BHT)	461	Methylcellulose
322	Lecithins	464	Hydroxypropylmethylcellulose
325	Sodium lactate	465	Ethylmethylcellulose
326	Potassium lactate	466	Carboxymethylcellulose
327	Calcium lactate	471	Mono- and di-glycerides of fatty acids
330	Citric acid	472(e)	
331	Sodium citrates	473	Mono-and diacetyltartaric acid esters of mono- and diglycerides of fatty acids
332	Potassium citrates		
333	Calcium citrates		
334	Tartaric acid	473	Sucrose esters of fatty acids
335	Sodium tartrates	475	Polyglycerol esters of fatty acids
336	Potassium tartrates		
337	Sodium potassium tartrate	476	Polyglycerol polyricinoleate
339	Sodium orthophosphates	481	Sodium stearoyl-2-lactylate
340	Potassium orthophosphates	482	Calcium stearoyl-2-lactylate
341	Calcium orthophosphates	491	Sorbitan monostearate
350	Sodium malates	500	Sodium carbonates
351	Potassium malates	501	Potassium carbonates
352	Calcium malates	503	Ammonium carbonates
353	Metatartaric acid	504	Magnesium carbonate
354	Calcium tartrate	508	Potassium chloride
355	Adipic acid	509	Calcium chloride
363	Succinic acid	529	Calcium oxide
380	Tri-ammonium citrate	536	Potassium ferrocyanide
400	Alginic acid	541	Sodium aluminium phosphate

TABLE XII.2
Continued

Code no.	Prescribed name	Code no.	Prescribed name
551	Silicon dioxide	900	Dimethylpolysiloxane
553(b)	Talc	901	Beeswaxes
554	Sodium aluminium silicate	903	Carnauba wax
558	Bentonite	904	Shellac
559	Kaolins	905	Paraffins
570	Stearic acid	909	Stearic acid
572	Magnesium stearate	920	L-Cysteine and its hydrochlorides
575	Glucono-delta-lactone		
621	Monosodium glutamate	924	Potassium bromate
627	Sodium guanylate	925	Chlorine
631	Sodium inosinate	926	Chlorine dioxide
637	Ethyl maltol		

Note: These schedules are not yet complete. They will be completed when the codex Committee on Food Additives finalises its International Code Numbering System for Food Additives.

Source: NH & MRC.

APPENDIX XIII

LEAD POTTERY GLAZES

The following is a reprint from *Some facts about lead glazes for workshop and studio potters*, a pamphlet issued by the National Health and Medical Research Council in 1975.

APPROVED BY THE NH & MRC, 77TH SESSION, NOVEMBER 1973

Introduction

For centuries, lead compounds have been included as constituents in pottery glazes because of the many advantages that accrue to both the potter and the consumer from their use. Lead compounds readily dissolve other essential ingredients such as alumina and silica to form a glaze with a high gloss and brilliance. Lead glazes have a relatively low melting point and a wide softening range. The low surface tension and viscosity of the melt allow minor imperfections on the surface of the clay body to be covered and a ready release of trapped air and good healing of the surface. Generally, there is sufficient reaction between the molten glaze and the underlying clay to form an intermediate layer which relieves stresses and offers a high resistance to crazing and devitrification. The affinity of lead glazes for colouring agents permits the development of a wide range of colours, many of which are difficult to attain by other means. It is often claimed that lead cannot be replaced by any other material that will provide comparable aesthetic effects with equivalent ease and effort.

Lead is, however, a toxic metal. When lead glazes are applied to surfaces used in contact with food and beverage, lead may be leached out by acids present in the food and beverage at levels which are toxicologically unsafe for human consumption. This hazard, primarily due to insufficiencies in glaze formulation, application and firing, has been recognised for a long time. Research, however, into formulation, application and firing procedures has enabled [safe] lead glazes to be produced.

Ceramic glaze

A ceramic glaze is a thin glossy coating fused onto the clayware body. An analogy exists between a glaze and a mixture of rock components fused together at a high temperature, much the same way as nature produces molten lava in an erupting volcano which then coats the earth with a glossy substance. Volcanic rocks can be pulverised and used as glazes. However, the major difference between a lava and an artificial glaze is that, in the latter, a flux is added.

Glazes are formulated from basic compounds such as alumina and silica with derivatives of barium, boron, calcium, lead, lithium, magnesium, potassium, sodium, strontium and zinc. In addition, the formulations may include colouring agents containing antimony, cadmium, chromium, cobalt, copper, iron, manganese, nickel or selenium, as well as opacifiers such as tin, titanium and zirconium. Consequently, the chemistry of glazes is extremely complex, being further complicated by the reaction between the glaze and the clay body during the firing operation.

Flux

A flux is an additive which permits the basic components of a glaze to fuse together at a lower temperature to form a homogeneous mass. The more commonly used fluxes are compounds of boron, calcium, lead, potassium and sodium.

Frit

A frit is a pre-formulated glaze. Selected raw materials are carefully proportioned, mixed and fused in a high-temperature furnace to form a glass. The glass mass is then milled to a fine powder. The frit is evenly dispersed in water and applied to the surface of the shaped piece. During firing the particles re-melt to form a thin layer of finished glaze. By the use of frits it is possible for the potter to exercise rigid control over the formulation of the glaze.

Frits are generally designed to be either lead-free or to contain a high percentage of lead. In the latter case, lead is chemically bound with the other constituents so that the level of leachable lead from the finished surface is negligible, even under the most stringent conditions. Lead leachability depends on different formulation parameters and is not necessarily related to the actual content of lead in the glaze. Frits are designed to be safe when used by themselves or with compatible on-glaze decorations and other accessories. Their indiscriminate use with incompatible materials can lead to increased lead leachability. Such practices are to be strongly discouraged. Not all manufactured lead frits are safe. In some cases lead frits are merely fluxes. Many commercial glazes are designed exclusively for the decoration of artware such as tiles, sculptures and architectural ceramics. The normal use of such wares does not present a health hazard. Care should, however, be exercised to ensure that such commercial glazes are not used on wares that could contain food and beverage.

Formulation

Instead of purchasing a commercial frit, the potter may elect to prepare glazes from his own basic constituents. A multitude of published recipes are available for this purpose. In the selection of a glaze recipe only those proven reliable by laboratory examination of the finished article should be considered.

It is frequent practice to reduce the melting point of a glaze by the addition of a greater amount of flux. In so doing, the potter should realise that he may present serious health hazards both to himself and the consumer. Some potters also, for economic reasons, pool excess glazes and then use the haphazard mixture. This practice is to be strongly discouraged because of the uncontrolled imbalance of formulation that inevitably results.

With a full understanding of the physical chemistry of glazes and proper care to formulation, safe glazes can be obtained. A very wide freedom of choice is permissible when glazes are intended for ornamental ware alone.

Colouring agents

Problems sometimes occur when colouring agents are introduced. Although a lead frit or a balanced formulation may be safe, the incorporation of oxides or carbonates of copper or cobalt, and to a lesser extent those of nickel and of other metals, can cause a release of lead from the silicate matrix. Regardless of the care taken in firing, research has shown that the association of copper or cobalt with a lead glaze triggers off chemical interactions which predispose to a marked increase in lead leachability. In general, any glaze which finishes with a blue or green colour should not be applied to a surface of a utensil intended to contain food or beverage. In some instances, black surfaces can also be unacceptable.

Leadless glazes

A wide range of leadless glazes with well-balanced formulations are available. When doubt prevails on the reliability of a lead glaze intended for domestic ware, preference should be given to the use of a leadless glaze from a reliable source. Both leadless frits and many leadless formulations have been developed which provide excellent results with an attractive and serviceable finish. Leadless glazes are available which mature at temperatures above 1200°C and below 1200°C respectively for stoneware and ceramicware. In some instances where a high percentage of alkali is present, advanced knowledge and skill are required to avoid subsequent deterioration of the surfaces. Some formulations are intended entirely for art glazes on ornamental ware.

Kiln operation

Adequate glaze maturation requires due care to firing conditions, particularly kiln temperature, firing time and the kiln atmosphere. At temperatures below 1080°C glaze constituents may not react sufficiently with each other or with the clay body to provide adequate maturation. Sufficient time is also required to allow the products of reaction between the glaze and clay body to migrate through the glaze network, thereby strengthening and making the glaze more impervious. For adequate maturation, a heavy glaze application (2 mm or more) requires considerably longer firing time than that of a thin layer. The correct firing schedule for temperature and time can be conveniently achieved by the use of pyrometric cones.

Oxidising conditions in the kiln atmosphere are essential. Without proper controls, products of combustion in gas, oil and wood fired kilns may result in reducing conditions. These conditions favour the reduction of metallic compounds to a form in which they can no longer be securely bound to silica, alumina and other adjuncts.

An improperly fired glaze cannot be made safe by refiring, washing with acids or baking in an oven.

Occupational hygiene

Unless handled with due care and with the use of proper equipment, lead glazes are hazardous to the health of the potter. Good housekeeping is important. The workshop should be vacuumed and mopped regularly. Any spill of material should be immediately damp sponged and, if dust appears, the workshop vacuumed thoroughly. Dust of any kind is to be avoided. All operations which disperse dust and fumes should be controlled by forced exhaust ventilation. A dust respirator complying with the specification of Australian standard AS Z18 should be used when dry glazes are mixed or ground. The spraying of glazes should only be done in a well-ventilated booth exhausted to the exterior.

As lead volatilises during the firing procedure, the atmosphere around the kiln may, under certain circumstances, constitute a health hazard. The kiln should therefore be located in an area where children or adults are not unwittingly exposed to lead fumes. The kiln should also be carefully fitted with a hood exhausting to the exterior.

Personal hygiene

Extreme care should be taken to avoid transferring lead from the hands to mouth. If gloves are not worn when handling glazes, hands and fingernails should be thoroughly scrubbed upon finishing work. Food, drink and tobacco should not be brought into the workshop area. Changes of protective clothing should be provided for use in the workshop; but never worn elsewhere. Children should not be allowed to enter a workshop unless supervised by a responsible person.

RECOMMENDATIONS OF THE NH & MRC, 73RD SESSION, OCTOBER 1971

Lead hazards from pottery glazes

Council noted that, following cases of poisoning by lead leached from pottery glazes, Australian and overseas investigations have shown that a number of pottery food and drink utensils are available from which lead may be leached by acid foods such as fruit juice, soft drinks, wines, cider, vinegar, sauerkraut and tomatoes and the use of such utensils may therefore constitute a threat to human health. Council also noted that the implications inherent in the results of these investigations have led to a strengthening of overseas regulations concerned with pottery glazes on food and drink utensils.

Council considered that pottery utensils with glazes which release 7 ppm or more of lead, as determined by the ASTM method, C-555-71, 'Estimation of lead extracted from glazed ceramic surfaces', are unsafe for use as human food or drink containers. Council therefore recommended that:

(i) legislation be enacted by the Commonwealth and the States to prohibit the sale of pottery food and drink utensils which may, by the release of lead, be hazardous to human health;

(ii) glazing formulations containing lead which are available in Australia should be labelled 'WARNING this glazing material contains lead';

(iii) amateur or handicraft potters should not apply glazes bearing lead to the insides of food and drink utensils unless they are able to ensure that the techniques they employ preclude the subsequent release of unsafe amounts of lead from the glaze; and

(iv) acidic foods and beverages should not be stored in pottery containers unless the containers are known not to release significant amounts of lead.

75TH SESSION, MAY 1973

Teaching of pottery crafts in schools

Council recommended that in the teaching of pottery crafts in primary and secondary schools lead compounds should not be used in the making of utensils that could be used as containers for food and beverages.

APPENDIX XIV

CONSUMER INFORMATION

SOURCES OF CONSUMER INFORMATION

Consumer organisation publications

Australia

Canberra Consumer (quarterly), Canberra Consumers Inc., GPO Box 591, Canberra, ACT 2601.

Choice (monthly), *Consuming Interest* (quarterly), Australian Consumers' Association, 57 Carrington Road, Marrickville, NSW 2204.

Consumer Comment (quarterly), Consumers' Association of Victoria, PO Box 339, Mt. Waverley, Vic. 3149.

Consumer Views (monthly), Australian Federation of Consumer Organizations, PO Box 75, Manuka, ACT 2603.

Malaysia

Consumer Currents (monthly), International Organization of Consumers' Unions, PO Box 1045, 10830, Penang, Malaysia.

The Netherlands

IOCU Newsletter (10 issues per year), International Organization of Consumers' Unions, Emmastraat 9, 2595EG, The Hague, The Netherlands.

New Zealand

Consumer (11 issues per year), Consumers' Institute, Private Bag, Te Aro, Wellington 1, New Zealand.

United Kingdom

Which?(monthly), Consumers' Association, Ltd, PO Box 44, Hertford, SG14 1SH UK.

United States of America

Consumer Reports (monthly), Consumers' Union, PO Box 53029, Boulder, Colorado 80322–3029, USA.

Government consumer interest publications—Australia*

ACT Consumer Affairs Council/Bureau, *Annual Report*. GPO Box 158, Canberra, ACT 2601.

NSW Department of Consumer Affairs, *Annual Report*. PO Box 468, Darlinghurst, NSW 2010.

Consumer Affairs Council (NT), *Annual Report* and *Consumer Newsletter* (quarterly). GPO Box 4344, Darwin, NT 5794.

Queensland Consumer Affairs Council and Bureau, *Annual Report*. PO Box 252, North Quay, Brisbane, Qld 4000.

Department of Public and Consumer Affairs, South Australia, *Annual Report*, GPO Box 1268, Adelaide, SA 5001.

Consumer Affairs Council of Tasmania, *Annual Report*. GPO Box 1320, N. Hobart, Tas. 7001.

Ministry of Consumer Affairs (Victoria), *Annual Report*. 500 Bourke Street, Melbourne, Vic. 3000.

Department of Consumer Affairs, WA, *Annual Report*, PO Box 779, West Perth, WA 6005.

National Consumer Affairs Advisory Council, *Annual Report*, GPO Box 1967, Canberra, ACT 2601.

Note: In the USA, readers are referred to the Annual Reports of the Consumer Product Safety Commission.

Law Reform Commission

Community law reform for the ACT, Report 28, 1985.

Debt recovery and insolvency, Report 36, 1987.

General insolvency inquiry, 1985.

Insurance agents and brokers, Report 16, 1980.
Insurance contracts, Report 20, 1982.
Privacy: credit records, Research Paper no. 9, 1980.
Privacy and intrusions, 1980.
Privacy and personal information, 1980.
Reform of evidence law, 1980.
Standing in public interest litigation, Report 27, 1985.

Periodicals and series of the International Organization of Consumer Unions (IOCU)

Periodicals

IOCU Newsletter. Covers news and activities of IOCU and its member organisations (10 issues a year). Free to IOCU members.

CEN-ter Folder. Consumer education newsletter (quarterly).

Consumer Currents. A digest of consumer news (10 issues a year).

Consumer Interpol Focus. Each issue features a single safety problem to provide consumer groups with information and suggestions on what they can do (quarterly). Free to Consumer Interpol correspondents and IOCU members.

HAI News. Newsletter of Health Action International, an informal network of groups working on pharmaceuticals issues (bimonthly).

La Voz Del Consumidor. Produced in Spanish by the Association Mexicana de Estudios para la Defensa del Consumidor (AMEDEC) for IOCU (quarterly). From AMEDEC, Amores 109-Bis A, Col. de Valle, 03100, Mexico DF.

Series and other publications

Anabolic steroids: availability and marketing. The survey looked at the marketing of anabolic steroids in 12 countries and how they were labelled for use.

Comparative testing guide. This provides guidelines to consumer organisations on carrying out comparative tests of consumer goods. 1977.

Consumercraft no. 1: consumer action in developing countries. A selection of papers and reports on the consumer movement and issues of special interest to developing countries. 1980.

Consumercraft no. 2: ideas for consumer action. Selection of papers to provide strategies and resources for projects in newer consumer organisations. 1981.

Consumercraft no. 3: appropriate products. Contains papers from an IOCU seminar held in collaboration with the International Labour Organisation and the International Development Research Centre. 1982.

Consumers in a shrinking world. Proceedings of the 10th IOCU World Congress held in The Hague, 22–26 June 1981. 1982.

The energy crisis. Papers and speeches submitted to the 9th IOCU World Congress, Energy Crisis Seminar, held in London in 1978.

The food crisis. Papers and speeches submitted to the 9th IOCU World Congress, Food Crisis Seminar, held in London in 1978. 1979.

Forty-four problem drugs: a consumer action and research kit on pharmaceutics. 1981.

The global trade in hazardous products: underhand but over the counter. This press pack is based on the concerns and work of Consumer Interpol, a major IOCU program dealing with the international trade in harmful products. The press pack contains two news features, three case studies in addition to a background reader about Consumer Interpol. Each story is accompanied by line drawings, cartoons and photos. 1983.

Health, safety and the consumer. Proceedings of the IOCU seminar held in Ranzan, Japan, 6–9 April 1983.

IOCU directory 1983. Lists international, national and principal local consumer organisations in some 50 countries.

The law and the consumer. Proceedings of the 1980 IOCU seminar held in Hong Kong.

Multinationals and the consumer interest. An introductory paper on the subject by the former president of IOCU, Peter Goldman. 1974.

The pesticide handbook: profiles for action. Tailored for groups involved in action against the misuse and overuse of chemical pesticides. A handy reference containing data profiles on 44 problem pesticides; directory of groups involved in pesticide issues, a bibliography and selected papers on the issue. 1984.

Prescription for change. This guide is a recipe book with ideas to achieve the political will for change on health policy. 1983.

Protecting tomorrow's world today. A memento to the 10th anniversary of the IOCU office in Penang. It recalls its beginnings, its branching out and its flowering as a centre for many of the global activities of IOCU. 1983.

Readings on appropriate products. Collection for the 1982 IOCU seminar on the subject held in Penang.

Readings on consumer education in food and nutrition. Collection for the 1978 IOCU Seminar, held in the Philippines on this subject.

Readings on consumer testing and research. Collection of already published materials. 1976.

Readings on law and the consumer. Collection for the 1980 IOCU seminar on this topic held in Hong Kong.

Survey work: how to conduct user and opinion surveys. 1974.

Thesaurus of consumer terms. Part 1: classified display. An arrangement of consumer topics into 23 subject classes to which there is an alphabetical index. 1979.

Thesaurus of consumer terms. Part 2: alphabetically structured display. A listing of terms showing hierarchical and associative relationships. 1982.

UN draft guidelines on consumer protection: IOCU comments. A booklet written to help IOCU members formulate a position and to promote the adoption of the guidelines. 1983.

World Consumer Rights Day. This folder of materials is meant to provide inspiration and ideas to consumer groups for action around the World Consumer Rights Day, March 15. Annual.

A world in crisis: the consumer response. Proceedings of the 9th IOCU World Congress held in London, 9–14 July 1978. 1979.

These publications are available from IOCU, Emmastraat 9, 2595EG, The Hague, The Netherlands.

POPULAR CONSUMER TEXTS

The following is a selection of books only. Readers should also refer to the previous edition of *Chemistry in the marketplace* and check the subject indexes of catalogues in their local libraries. The lists are in chronological order.

Advertising

1. *Blood, brains and beer*, D. Ogilvy, Hamilton, Melbourne, 1978.
2. *Children and television*, Senate Standing Committee on Education and the Arts, Australian Government Publishing Service (AGPS), Canberra, 1978.
3. *Soft soap hard sell*, R. R. Walker, Hutchinson, Melbourne, 1979.
4. *Hucksters in the classroom*, S. Harty, Centre for the Study of Responsive Law, Washington, DC, 1979.
5. *Self-regulation in Australian advertising*, Australian Advertising Industry Council (AAIC), Sydney, 1979, 1981, 1982.

6. *The positive case for marketing children's products to children*, AAIC, Sydney, 1979.
7. *The media and business*, H. Simons and J. Califano, Vintage, New York, 1979.
8. *Your choice?* A. Williams, Longman, London, 1980.
9. *Advertising. The people have their say*, AAIC, Sydney, 1980.
10. *The role of advertising in Australia*, AAIC, Sydney, 1981.
11. *Advertising and the Trade Practices Act*, W. Pengilley, Commercial Clearing House (CCH), Sydney, 1981.
12. *Advertising and selling*, Trade Practices Commission (TPC), AGPS, Canberra, 1981.
13. *Alcohol and advertising*, Australian Broadcasting Advisory Committee (ABAC), Sydney, 1982.
14. *Adman and Eve*, Equal Opportunities Commission, UK, 1982.
15. *Advertising and women's portrayal*, J. Wood, Office of Women's Affairs, Canberra, 1982.
16. *Fair exposure*, Office of the Status of Women, AGPS, Canberra, 1983.
17. *A look at advertising*, Department of Public and Consumer Affairs, Adelaide, 1983.
18. *The structure and procedures of advertising self-regulation*, AAIC, Sydney, 1983.
19. *Advertising. The people have their say.* Report 2, AAIC, Sydney, 1983.
20. *Deposits and loans. Their advertising and the Trade Practices Act*, TPC, Canberra, 1986.

Business
1. *The chairman as God*, G. Tempel, Blond, London, 1970.
2. *The corporate oligarch*, D. Finn, Simon & Schuster, New York, 1970.
3. *Small is beautiful*, E. F. Schumacher, Sphere Books, London, 1970.
4. *The consumer and corporate accountability*, R. Nader (ed.), Harcourt Brace, New York, 1973.
5. *Social responsibility and the business predicament*, Brookings Institution, Washington, 1974.
6. *Power Inc.*, Morton Mintz and Jerry Cohen, Bantam Books, New York, 1974.
7. *Multinationals and the consumer interest*, P. Goldman, IOCU, The Hague, 1974.
8. *What every corporate communicator should know about his hostile audience*, Clemenger, Sydney, 1975.
9. *The social responsibility of corporations*, P.J. Dunstan, Committee for Economic Development of Australia (CEDA), Sydney, 1976.
10. *Consumerism: the corporate response*, Clemenger, Sydney, 1977.
11. *Corruption in business, facts on file*, New York, 1977.
12. *The social audit consumer handbook*, C. Medawar, Macmillan, London, 1978.
13. *The corporate responsibility*, Wally Olins, Mayflower Books, New York, 1978.
14. *Shoppers' rights*, TPC, AGPS, Canberra, 1979.
15. *The corporate dilemma*, Clemenger, Sydney, 1980.
16. *Corporate crime*, M. Clinard and P. Yeager, Free Press, New York, 1980.
17. *Tools of power*, Andersen, O'Donnell and Parloff, Viking Press, New York, 1980.
18. *The ugly face of Australian business*, Timothy Hall, Harper & Row, Sydney, 1980.
19. *Micro-electronics and retailing*, Office of Fair Trading, London, 1982.
20. *Consumers in business*, Colin Adamson, National Consumer Council, London, 1982.
21. *Freedom of the air*, National Consumer Council, London, 1983.
22. *The impact of publicity on corporate offenders*, B. Fisse and J. Braithwaite, State University of New York, 1983.
23. *The corporate alchemists*, L.N. Davis, Temple Smith, London, 1984.
24. *Corporate responsibility in the pharmaceutical industry*, Lars Broch, IOCU, The Hague, 1984.
25. *The world alcohol industry with special reference to Australia, New Zealand and the*

Pacific Islands, J. Cavanagh, Transnational Corporations Research Project (TNCP), Sydney, 1985.

26. *Interim compendium of self-regulation schemes in Australia*, TPC, Canberra, 1985.
27. *Air transport and the consumer*, National Consumer Council, London, 1986.
28. *Consumers, transnational corporations and development*, E. Wheelwright (ed.), Transnational Corporations Research Project, Sydney, 1986.
29. *Of manners gentle*, P. Grabosky and J. Braithwaite, Oxford, Melbourne, 1986.
30. *Consumers and international trade*, Joyce Blow, IOCU/Bureau Européen des Unions de Consommateurs (BEUC), Brussels, 1986.
31. *Review of business regulations—1*. Business Regulation Review Unit (BRRU), Canberra, January 1986.
32. *Review of business regulations—2*. BRRU, Canberra, May 1986.
33. *Review of business regulations—3. Concerning industrial chemicals*, BRRU, Canberra, 1986.
34. *Review of business regulations—4. Packaging and labelling of food products*, BRRU, Canberra, 1986.
35. *Review of business regulations—5. Major changes in business regulation during 1986*, BRRU, Canberra, March 1987.
36. *Review of business regulations—7. Transport of dangerous goods*, BRRU, Canberra, 1987.
37. *Review of business regulations—9. Australian food standards regulations*, BRRU, Canberra, 1987.
38. *Consumers and trade policy—action guide*, IOCU/BEUC, The Hague, 1987.

Consumerism—general

1. *Consumer complaint handling in America: final report*, Technical Assistance Research Programs Inc. (TARP), Washington DC, 1979.
2. *Consumer complaint handling in America: summary of findings and recommendations*, TARP, Washington DC, 1979.
3. *A nation of guinea pigs*, M. S. Shapo, Free Press, New York, 1979.
4. *In the consumer interest*, E. E. Carpenter, Consumers' Institute, New Zealand, 1980.
5. *Conducting the consumer survey—a primer for volunteers*, G. and M. Mitchell, Virginia Polytechnic Institute, 1980.
6. *Consumer action in developing countries*, IOCU, Consumercraft, Penang, 1980.
7. *Gobbledygook*, Tom Vernon, National Consumer Council, London, 1980.
8. 'Criminal law and consumer protection', W. B. Fisse. In *Consumer protection law and theory*, A. J. Duggan and L. W. Darvall, Law Book Co., Sydney, 1980.
9. 'Consumer redress and the legal system', A. J. Duggan. In *Consumer protection law and theory*, A. J. Duggan and L. W. Darvall, Law Book Co., Sydney, 1980.
10. *You the consumer*, R. Maclean and A. Lawrence, McGraw-Hill, Sydney, 1980.
11. *Consumer protection and information policy*, European Economic Community (EEC), Brussels, 1981.
12. *Ideas for consumer action*, IOCU, Consumercraft, Penang, 1981.
13. *Perspectives on productivity: Australia*, Sentry Insurance, Sydney 1981 (?)
14. *When consumers complain*, A. Best, Columbia University Press, New York, 1981.
15. *Australian consumers' handbook*, Tony Blackie, Macmillan, Melbourne, 1981.
16. *How to stick up for yourself*, R. Smith and A. Stokes, Fontana, Melbourne, 1981.
17. *Background reader for 'Consumers in a shrinking world'*, IOCU, The Hague, 1981.
18. *The electronic revolution*, Clemenger, Sydney, 1981.
19. *Consumer concerns survey*, National Consumer Council, London, 1981.
20. *Faulty goods* [consumer concerns], National Consumer Council, London, 1981.
21. *Bureaucracies* [consumer concerns], National Consumer Council, London, 1981.

22. *It's your money they're after*, Dick Smithies, Allen and Unwin, Auckland, 1982.
23. *Our cheque is in the post*, C. Ward, Pan, London, 1982.
24. *How to talk to a company and get action*, Coca Cola Co., Atlanta, USA, 1982.
25. *Appropriate products*, IOCU, Penang, 1982.
26. *Measuring the grapevine—consumer response and word-of-mouth*, Coca Cola Co., Atlanta, USA, 1982.
27. *Consumer representation in the European communities*, EEC, Luxembourg, 1983.
28. *Small print*, M. Cutts and C. Maher, National Consumer Council, London, 1983.
29. *Towards a consumer rights charter*, IOCU/ACA, 1983.
30. *Freedom of the air*, National Consumer Council, London, 1983.
31. *Consumer initiatives*, S. Harty, Copesthetic, Washington, 1983.
32. *The information society*, National Consumer Council, London/IOCU, 1984.
33. *The consumer protection portfolio*, Consumer Information Documentation Centre (CIDOC), c/o IOCU, Penang, 1984.
34. *Protecting tomorrow's world today*, IOCU, Penang, 1984.
35. *Giving a voice to the world's consumers*, IOCU, Penang, 1984 (?).
36. *Generating power*, Wayne Ellwood, IOCU, Penang, 1984.
37. *Roche versus Adams*, S. Adams, Jonathan Cape, London, 1984, and Fontana, London, 1985.
38. *Beyond the stereotypes*, Clemenger (auth. & publ.), Sydney, 1984.
39. *Plain words for consumers*, National Consumer Council, London, 1984.
40. *Plain English for lawyers*, National Consumer Council, London, 1984.
41. *The European Community and consumers*, EEC, Brussels, 1985.
42. *Guidelines for consumer protection*, United Nations, New York, 1986.
43. *Australia and the United Nations guidelines for consumer protection*, National Consumer Affairs Advisory Council (NCAAC), Canberra, 1986.
44. *Consumer guarantees*, Office of Fair Trading, London, 1986.
45. *Consumers, transnational corporations and development*, E. Wheelwright (ed.), Transnational Corporations Research Project, Sydney, 1986.
46. *Consumer dissatisfaction*, Office of Fair Trading, London, 1986.
47. *Pitfalls in prices surveys*, Prices Surveillance Authority, Sydney, 1986.
48. *Background reader for 'Consumer policy 2000'*, IOCU, The Hague, 1986.
49. *Background reader for 'Consumers and the economic crisis'*, IOCU, The Hague, 1986.
50. *Background reader for 12th IOCU World Congress*, IOCU, The Hague, 1987.
51. *Survey of consumer opinion in Australia*, TPC, Canberra, 1987.
52. *Till they have faces: women as consumers*, T. Wells and G. S. Foo, IOCU/Isis, 1987.

Consumer safety

1. *Consumer health and product hazards*, S. S. Epstein (ed.), MIT Press, Boston, 1974.
2. *Consumer safety, a consultative document* Her Majesty's Stationery Office (HMSO), London, 1976.
3. *Childhood accident project report*, Parts 1–3, S. D. McIntosh, Health Commission of NSW, Sydney, 1976–77.
4. *Children and accidents*, L. Gonski, Child Safety Centre, Royal Alexandra Hospital for Children, Camperdown, NSW, 1979.
5. *A nation of guinea pigs*, M. S. Shapo, Free Press, New York, 1979.
6. *Analysis of domestic accidents to children*, Consumer Safety Unit, London, 1979 and 1982.
7. *Health, safety and environmental regulation: how effective?* American Enterprise Institute for Public Policy Research, Washington DC, 1980.
8. *Causes of death*, Australian Bureau of Statistics, Canberra, annual publication.
9. 'Consumer product safety', M. J. Vernon. In *Consumer protection law and theory*, A. J. Duggan and L. W. Darvall (eds), Law Book Company, Sydney, 1980.

10. *Personal factors in domestic accidents*, Consumer Safety Unit, London, 1980.
11. *Lives at stake*, L. Pringle, Macmillan, New York, 1980.
12. *Child accident prevention*, H. Hayes, Child Accident Prevention Foundation of Australia (CAPFA), Melbourne, 1981.
13. *Safety in childhood*, J. Pearn and J. Nixon, CAPFA, Melbourne, 1981.
14. *An analysis of child accident statistics*, P. J. O'Connor, CAPFA, Melbourne, 1982.
15. *An analysis of Australian child accident statistics*, P. J. O'Connor, CAPFA, Melbourne, 1982.
16. *Towards an effective community policy on the safety of consumer products*, BEUC, Brussels, 1984.
17. *Product safety, the consumer and consumer Interpol*, M. J. Vernon, Proceedings, 11th IOCU World Congress, Bangkok, IOCU, 1984.
18. *Deadly neglect: regulating the manufacture of therapeutic goods*, Australian Federation of Consumer Organisations, Canberra, 1984.
19. *Bicycle helmet safety, Report of the House of Representatives Standing Committee on Road Safety*, AGPS, Canberra, November 1985.
20. *Product safety. Guide to the operation of the Trade Practices Act*, Office of Consumer Affairs, Canberra, 1986.
21. *Hazard assessment and the process of standards setting*. A report to the (Australian) National Consumer Affairs Advisory Council, M. J. Vernon, Canberra, 1986.
22. *Child safety and standards—general guidelines*, ISO Guide 50, ISO Geneva, 1987.
23. *Product safety*, NCAAC, Canberra, 1987.
24. *Accidental poisoning in childhood*, Proprietary Association of Great Britain, London, July 1987.

Cosmetics

1. *The price of beauty*, R. Simon, Longman, London, 1971.
2. *Understanding allergies*, Which? Books, London, 1986.

Drugs

1. *Ailments and remedies*, Consumers' Association, London, 1965.
2. *The drugs you take*, S. Bradshaw, Hutchinson, London, 1966.
3. *Licit and illicit drugs*, E. Brecher, Consumers' Union, New York, 1972.
4. *A dictionary of drugs*, R. B. Fisher and G. A. Christie, Paladin, London, 1975.
5. *Clioquinol: availability and instructions for use*, IOCU, The Hague, 1975.
6. *The selection of essential drugs*, WHO, Geneva. 1979.
7. *The use and abuse of medication*, Senate Hansard, 10 August 1979.
8. *The pill book*, H. Silverman and G. Simon, Bantam, New York, 1979.
9. *The A-Z of Australian family medicines*, R. Spencer, Butterworths, Sydney, 1980.
10. *Drug disinformation*, Charles Medawar, Social Audit, London, 1980.
11. *Pills that don't work*, S. Wolfe and C. Coley, Farrar, Straus, Giroux, New York, 1980.
12. *Forty-four problem drugs*, IOCU, Penang, 1981.
13. *International NGO Seminar on Pharmaceuticals, Geneva, Report*, IOCU, 1981.
14. *The WHO and the pharmaceutical industry*, Health Action International (HAI), The Hague, 1982.
15. *A draft international code on pharmaceuticals*, HAI, 1982.
16. *Prescription for change*, V. Beardshaw, HAI, The Hague, 1983.
17. [Council of Europe] *Report on the sale of European pharmaceutical products in the countries of the Third World*, HAI, 1983.
18. *The wrong kind of medicine*, Charles Medawar, Consumers' Association, London, 1984.

19. *Drugs and world health*, C. Medawar, IOCU, The Hague, 1984.
20. *The rules governing medicaments in the European Community*, EEC, Brussels, 1984.
21. *Standard for the uniform scheduling of drugs and poisons*, NH & MRC, AGPS, Canberra, 1987.
22. *Review of drug evaluation procedures*, Public Service Board, Canberra, 1987.

Food
1. *Food quality and safety: a century of progress*, HMSO, London, 1975.
2. *Eat your heart out*, Jim Hightower, Vintage Books, New York, 1976.
3. *The food crisis*, papers at 9th IOCU World Congress, London, 1978.
4. *Insult or injury?* Charles Medawar, Social Audit, London, 1979.
5. *Environmental contaminants in food*, US Congress, Washington DC, 1979.
6. *Food and the consumer*, A. Kramer, AVI Publishing, USA, 1980.
7. *Consuming passions*, P. Farb and G. Armelagos, Houghton Mifflin, Boston, 1980.
8. *Processed Foods and the Consumer*, Council of Australian Food Technology Associations (CAFTA), Sydney, 1981.
9. *Processed food: a pain in the belly*, ACA, Sydney, 1982.
10. *A pauper's guide to fast food*, AFCO, Canberra, 1984.
11. *The infant feeding portfolio*, CIDOC, Penang, 1984.
12. *The jam industry in Australia*, TPC, Canberra, 1984.
13. *Food law in Australia*, M. W. and R. J. Gerkens, Law Book Company, Sydney, 1985.
14. *Review of business regulations—9. Australian Food Standards Regulations, BRRU, 1987.*
15. *Food standards code*, NH & MRC, AGPS, Canberra, 1987.
16. *Food irradiation: the facts*, T. Webb and T. Lang, Thorsons, Wellingborough, 1987.
17. *The market basket (noxious substances) survey 1985*, NH & MRC, Canberra, 1987.

Food additives
1. *Sowing the wind*, H. Wellford, Bantam Books, New York, 1973.
2. *Natural poisons in natural foods*, A. H. Wertheim, Lyle Stewart, USA, 1974.
3. *Eating may be hazardous to your health*, J. Verrett and J. Carper, Doubleday, New York, 1975.
4. 'Pour boycotter les colorants', *Que choisir?* Paris, undated.
5. *Eater's digest*, M. F. Jacobson, Doubleday, USA, 1976.
6. *Food quality and safety*, HMSO, London, 1976.
7. *Food quality in Australia*, Australian Academy of Science, Canberra, 1974.
8. *Why additives?* British Nutrition Foundation, London, 1977.
9. *Environmental contaminants in food*, US Congress, Washington DC, 1979.
10. *Food additives and the consumer*, European Commission (EC), Brussels, 1980.
11. *Report on mercury in fish and fish products*, Department of Primary Industry, AGPS, Canberra, 1980.

Pesticides
1. *The problems of persistent chemicals*, OECD, Paris, undated.
2. *Agricultural chemicals*, Manufacturing Chemists' Association, Washington DC, undated.
3. *Silent spring*, R. Carson, Penguin, London, 1965.
4. *Since silent spring*, F. Graham, Houghton Mifflin, Boston, 197о.
5. *The chemical feast*, J. S. Turner, Grossman, New York, 1970.
6. *Herbicides, pesticides and human health*, Parliamentary Paper 15/1978, AGPS, Canberra, 1978.

7. *The pesticide conspiracy*, R. Van den Bosch, Doubleday and Co., 1978.
8. *Pesticides and the Australian environment—public safety and consumer protection*, J. T. Snelson, AGPS, Canberra 1979.
9. *Circle of poison*, D. Weir and M. Schapiro, IFDP, San Francisco, 1981.
10. *Hazardous chemicals*, Report of the House of Representatives Standing Committee on Environment and Conservation, AGPS, Canberra, 1982.
11. *Withholding periods, maximum residue limits and poison schedules for agricultural and veterinary chemicals*, 4th edn., Department of Primary Industry (Pesticide Section) Australia, AGPS, Canberra, 1982.
12. *Consolidated list of products whose consumption, and/or sale have been banned, withdrawn, severely restricted or not approved by governments*, first issue, revised, United Nations, New York, 1984 and second issue, 1986.
13. *Getting tough*, World Resources Institute, Washington DC, 1984.
14. *The pesticide portfolio*, CIDOC, Penang, 1984.
15. *The dirty dozen*, IOCU, Penang, 1985.
16. *Pesticides don't know when to stop killing*, PAN, San Francisco, 1985.
17. *The pesticide poisoning report*, IOCU, Penang, 1985.
18. *A better mousetrap*, World Resources Institute, Washington DC, 1985.
19. *The pesticide handbook*, 2nd edn, IOCU, Penang 1986.
20. *Problem pesticides, pesticide problems*, G. Goldenman and S. Rengam, IOCU/PAN, 1987.

Recipes and trade names
1. *The chemical formulary*, H. Bennet, Chemical Publishing Co., New York, 1933- (vol. 24, 1982).
2. *Chemical synonyms and trade names*, W. Gardiner, Technical Press, London, 1971.
3. *The condensed chemical dictionary*, 10th edn, G. G. Hawley, Van Nostrand Reinhold, New York, 1981.

Safety with chemicals
1. *The analytical toxicology of industrial inorganic poisons*, M. B. Jacobs, Wiley Interscience, New York, 1967.
2. *Cancer causing chemicals*, N. I. Sax, Van Nostrand Reinhold, New York, 1981.
3. *Chemical hazards in the workplace*, N. H. Proctor and R. C. Hughes, Lippincott, Philadelphia, Pennsylvania, 1978.
4. *The clinical toxicology of commercial compounds*, M. N. Gleason *et al.*, Williams & Wilkins, Baltimore, Md, 1969.
5. *Dangerous chemicals: emergency first aid guide*, A. Houston, Walters Samson (UK), 1983.
6. *Dangerous properties of industrial materials*, 6th edn, N.I. Sax, Van Nostrand Reinhold, New York, 1984.
7. *Deposition of toxic drugs and chemicals in man*, 2nd edn, R.C. Baselt, Biomedical Publications, Davis, Ca, 1982.
8. *Encyclopedia of occupational health and safety*, 3rd edn, 2 vols, L. Parmeggiani (ed.), ILO, Geneva, 1983.
9. *Guideboook to Australian occupational health and safety laws*, A. Merritt, CCH, Sydney, 1983.
10. *Guide to safety in the chemical laboratory*, 2nd edn, Manufacturing Chemists Association, Van Nostrand Reinhold, New York, 1972.
11. *Handbook of laboratory safety*, 2nd edn, N.V. Steer (ed.), Chemical Rubber Co., Cleveland, Oh., 1971.

12. *Hazards in the chemical laboratory*, 3rd edn, L. Bretherich (ed.), Royal Society of Chemistry, London, 1981.
13. *International register of potentially toxic chemicals*, United Nations Environment Programme, Geneva, Switzerland, 1983.
14. *Long-term hazards from environmental chemicals*, A Royal Society Discussion, The Royal Society, London, 1979.
15. *The Merck index*, 9th edn, M. Windholz, Merck & Co., Rahway, NJ, 1977.
16. *Occupational health and safety concepts: chemical and processing hazards*, G.R.C. Atherley, Applied Science, London, 1978.
17. *Patty's industrial hygiene and toxicology*, 3 vols, 3rd edn, G.D. Clayton and F.E. Clayton, Wiley Interscience, New York, 1981.
18. *Registry of toxic effects of chemical substances*, 2 vols, R. J. Lewis, Sr, and R. L. Tathen, US Department of Health and Human Services, National Institute for Occupational Health and Safety, Ohio, USA, 1980.
19. *Report on the occupational safety and health management review committee*, D. P. Craig, CSIRO, Melbourne, 1983.
20. *Toxic and hazardous: industrial chemical safety manual*, 2nd edn, International Technical Information Institute, Tokyo, Japan, 1979.

Standards

Readers are referred to *TAS—the Australian standard* (published monthly) and the *Annual catalogue of Standards Australia publications*.

TAS can be obtained free from Standards Australia, PO Box 458, North Sydney, NSW 2059.

Draft standards are identified as in the following example:

> DR 82110 Microwave oven leakage detectors for household use

and, when finally published, the standard is referred to as:

> AS 2889—1987 Microwave oven leakage detectors for household use.

A period of three months is allowed for public comment on a draft standard. The committee then considers all the comments received and a postal ballot of committee members determines the final content of the standard. The standard is then published and sold, and manufacturers can apply for the Standards Australia (SA) mark if their products comply and are subjected to the SA inspection procedures.

CONSUMER STANDARDS

Consumer standards published during 1986-87 by committees on which consumers were represented through the Australian Federation of Consumer Organizations (AFCO) are as follows:

AS 1430	Household refrigerators and freezers
AS 1900	Children's flotation toys and swimming aids
AS 1926	Fences and gates for private swimming pools and Amdt. 1
AS 1927	Pedal bicycles for normal road use: safety requirements, Amdt. 2
AS 1957	Care labelling of clothing, textiles, furnishings, bedding etc.
AS 2002	Solar hot water heaters—installation
AS 2063	Lightweight protective helmets (for use in pedal cycling, horse riding and

other activities requiring similar protection): Part 1—Basic performance requirements: Part 2—Helmets for pedal cyclists

AS 2070   Plastics materials for food contact use: Part 7—Poly(vinylidene chloride) (PVDC)

AS 2450   Textiles—natural and man-made fibres: generic names

AS 2462   Cellulosic fibre thermal insulation, Amdt. 2

AS 2479   Down and/or feather filling materials

AS 2512   Methods of testing protective helmets

    2512.4   Determination of penetration resistance

    2512.5   Determination of strength of retention system and its attachment points

    2512.7   Determination of stability

AS 2575   Energy consumption labelling of household appliances

AS 2604   Sunscreen products: evaluation and classification

AS 2615   Trolley jacks

AS 2622   Textile products: fibre content labelling

AS 2690   Motor vehicle sale contracts: Part 1—Contract for the sale of a used motor vehicle

AS 2693   Vehicle jacks

AS 2803   Hinged security screen doors

AS 2818   Guide to swimming pool safety

AS 2869   Tampons: menstrual

AS 2877   Methods of test for fuel consumption of motor vehicles

AS 2889   Microwave oven leakage detectors for household use

AS 2895   Performance of household electrical appliances: microwave ovens

AS 2898   Radar speed detection: Part 1—Functional requirements and definitions; Part 2—Operational procedures

AS 2899   Public information symbol signs: Part 2—Water safety signs

AS 2914   Textile floor coverings: informative labelling

AS 2984   Solar water heaters: outdoor test method

AS 3172   Electric cooking appliances

AS 3179   Small self-contained air conditioners

AS 3181   Electrically operated projectors for household use

AS 3184   Electric dishwashing machines

AS 3185   Electric rotary clothes dryers

AS 3300   Household and similar electrical appliances

Consumer standards committees

The following were the consumer standards committees of Standards Australia, December 1987. (The author has served or is serving on the committees responsible for CS/2, CS/42, CS/53, CS/64, CH/3 and FT/8.)

CS/1   Small powered boats

CS/2   Consumer standards for household detergents

CS/3   Safety requirements for children's furniture

CS/4   Care labelling of textiles

CS/5   Playground equipment

CS/6   Garden incinerators

CS/7   Guards for domestic heating appliances

CS/8   Bicycle lamps and reflectors

CS/9   Contraceptive devices

CS/10   Pedal bicycles

CS/11	Swimming pool covers
CS/12	Household soaps
CS/14	Safety helmets for sport and recreation
CS/15	Petrol for motor vehicles
CS/16	Quality of school and college wear
CS/17	Swimming pool fences
CS/18	Safety of children's toys
CS/19	Child-resistant medicine cupboards
CS/20	Prams and strollers
CS/21	Swimming aids
CS/22	Contracts for consumer transactions
CS/23	Security screen doors
CS/26	Labelling of household chemicals
CS/27	Fuel consumption
CS/28	Solar water heaters
CS/29	Walking track signs
CS/30	Precious metals for jewellery
CS/31	Chemical carpet cleaners
CS/32	Sanitisers
CS/33	Power lawn-mowers and edge-trimmers
CS/34	Safety of private swimming pools
CS/35	Continental quilts
CS/36	Termite treatment in existing buildings
CS/37	Garden soils
CS/39	Slide fasteners
CS/40	Babies' dummies
CS/41	Safety ladders for above-ground pools
CS/42	Sunscreen agents
CS/43	Vehicle stands and vehicle ramps
CS/44	Labelling of clothing and other textiles
CS/45	Size colour coding of clothing
CS/46	Sizing of infants' and children' clothing
CS/47	Inflatable pleasure boats
CS/48	Automotive diesel fuel
CS/49	First-aid kits
CS/50	Water-resistance of watches
CS/51	Yachtsmen's safety harnesses and lines
CS/52	Portable LP gas appliances
CS/53	Sunglasses
CS/54	Bed sheets
CS/55	Car jacks and trolley jacks
CS/56	Kerosene room heaters
CS/57	Corrosion protection of automobiles
CS/58	Sizing of women's wear
CS/59	Spa pools
CS/60	Buoyancy aids
CS/61	Sanitisation of private swimming-pools
CS/62	Solid fuel burning appliances
CS/63	Swimming-pool contracts
CS/64	Solaria
CS/65	Tampons
CS/66	Trampolines

CS/67	Men's industrial clothing
CS/68	Radar speed detection
CS/69	Saddles tack and harness
CS/70	Vehicle rustproofing contracts
CS/71	Spa baths
CS/73	Mattresses
CS/74	Plastics for cooking applications
CS/77	Blood alcohol testing devices
CS/78	Automatic transmission repair terminology

Other committees dealing with consumer standards are as follows:

AU/8	Adult seat belts
AU/11	Safety glass for land transport
AU/16	Passenger car tyres
AU/22	Child car restraints
BD/58	Thermal insulation of houses
CH/3	Paints
CH/17	Adhesives
EL/1/15	Adequate electrical insulation
EL/2	Electrical approvals standards
EL/5	Accumulators
EL/5/3	Household battery chargers
EL/5/4	Automotive batteries
EL/15	Household electrical appliances
EL/22	Dry cells and batteries
EN/2	Energy labelling
FT/8	Plastics for food contact
MD/9	Personal medical information
ME/23	Household refrigerators
MS/3/1	Water safety symbols
PK/13	Aerosol containers
PL/13	Plastics, garden hose
TX/9	Carpets
TX/13	Burning behaviour of textiles
TX/18	Quality requirements for textile furnishings

OTHER PUBLICATIONS

Australian Bureau of Statistics (ABS)

ABS publishes *Catalogue of Publications, Australia* which includes (still) a large number of free publications. In addition, information for researchers can be obtained directly from the Bureau, PO Box 10, Belconnen, ACT 2616. Each year ABS publishes *Year Book Australia*, which contains a wide range of statistical information on the economy and social conditions of Australia, as well as descriptive matter dealing with Australia's history, geography, physiography, climate and meteorology, government, defence, repatriation services and international relations.

Commonwealth Department of Health

The following publications can be obtained from the Commonwealth Department of Health, PO Box 100, Woden, ACT 2606.

Addiction at work
Alcohol in Australia
Applications and costs of modern technology
Australian Government Rehabilitation Service
Chemotherapy with antibiotics
Chiropractic, osteopathy, homoeopathy inquiry
Community health services
Consultative arrangements report
Developing child
Directory of services for disabled
Disabled persons at work
Discussion paper—praying for health care
Drug problems in Australia
Early intervention programs
Efficiency and administration of hospitals, 3 vols
Handbook on health manpower
Evaluating hospitals
Health reference paper
Family planning and health care
Food consumption patterns
Guidelines on residue trials

The National Health and Medical Research Council (NH & MRC) is part of the Commonwealth Department of Health and, amongst other things, publishes *Food Standards* and issues reports.

Commonwealth Scientific and Industrial Research Organisation (CSIRO)

The CSIRO, Division of Building Research, publications can be obtained from PO Box 56, Highett, Vic. 3190. Examples (listed by catalogue code number) are:

10–2	*Termites and ant caps*
10–18	*Mould growth in houses*
10–19	*Ridding roofs of lichen growth*
10–20	*Insulating your home*
10–25	*Dealing with slippery concrete*
10–26	*Cleaning areas of oil-stained concrete*
10–38	*Keeping cool in summer—windows*
10–63	*Keeping outside noise outside*
10–85	*Hints on curing a smoky fireplace*

This series also includes information on a variety of timbers.

The CSIRO, Division of Food Research, publications can be obtained from PO Box 52, North Ryde, NSW 2113. The following are available:

A guide to dairy products
Australian crustaceans: lobsters, crabs and freshwater crayfish
Australian molluscs: oysters, scallops, mussels, pipis, abalone, squid, octopus and cuttlefish
Citrus juices: how to preserve your own
Fish from Australian waters

Handling and storage of fresh fruit and vegetables in the home
Handling food in the home (also in Greek and Italian)
History of food preservation
How to handle chicken—fresh or frozen
The nutritional value of processed foods
Prawns—fresh and frozen
Seafoods: hints for buying, home freezing and preparation
Storage and market diseases of fruit
Storage life of foods: frozen, canned and dry

The Division of Food Research also distributes *Meat and the home freezer*, a publication of the Australian Meat and Livestock Corporation.

National Building Technology Centre (NBTC)

The National Building Technology Centre's publications are available from AGPS, GPO Box 84, Canberra, ACT 2601, or direct from PO Box 30, Chatswood, NSW 2057. A catalogue of publications is also available. A popular and very cheap series is *Notes on the science of building*, illustrated pamphlets dealing with items of practical interest to those engaged in the building industry and to students. Many of them are also of use to the consumer. Examples are:

Carpets
Cleaning of brickwork
Cooling a home
Fire hazards in the home
House design for hot climates
Houses exposed to bushfires
Insulating a house
Internal ceramic tiling
Paints

Industries Assistance Commission (IAC)

The IAC has forwarded over 300 reports to the Commonwealth Government since it took over from the Tariff Board in 1974. Some of the more recent ones of interest are:

Abattoir and meat processing industry
Asbestos
Brassieres
Canning fruit
Certain pigment dyestuffs and colour lakes
Copper and certain copper products
Dental materials
Ethylene, diethylene and triethylene glycols
Fatty acids and certain aliphatic acids, their salts and esters
Floor and wall coverings
Fluorescent and filament lamps
Fruit and fruit products
Gelatin
Harvesting and processing of fish, crustacea and molluscs
Lawn-mowers, certain engines and parts
Musical instruments and parts and accessories therefor
Orange and tangerine juices
Perfumery, cosmetics and toilet preparations

Phosphatic and nitrogenous fertilisers
Polymeric plasticisers and certain polyester polyals
Polyunsaturated margarine
Preparation extracts and juices of meat
Primary batteries
Sugar industry
Tableware and certain other goods of pottery
Tanned and finished leather, dressed fur
Textile clothing and footwear
Tin ores and concentrates
Tobacco industry (Australian)
Wheat industry

These reports (many of which are referenced in this book) are a very mixed bag but the occasional gems of background information and statistics give one an insight seldom available from other sources. A report of the Tariff Board (the precursor of the IAC) on dumping of phenoxy herbicide precursor chemicals was the *sine qua non* of the research on 2,4,5-T and dioxin described in Chapter 5. The reports are available from AGPS, GPO Box 84, Canberra, ACT 2601.

Patent, Trademark and Design Office (PTDO)
The patent literature is an excellent source of technical information but care must be taken because many companies file patents to lead competitors to think that their development plans are moving in a different direction.

Publications of the PTDO include *Patent literature: a source of technical information* and *The Australian Official Journal of Patents, Trade Marks and Designs*. A complete list of publications, including prices, and information to assist users of the PTDO are available from PO Box 200, Woden, ACT 2606.

Trade Practices Commission (TPC)
Current publications of the TPC are as follows, and are available from TPC, PO Box 19, Belconnen, ACT 2616.

Leaflets
> *Deposits and loans: advertising checklist* (May 1986)
> *Consumer product standards bans and recalls* (June 1986)
> *The Trade Practices Commission* (Aug. 1986)
> *The Trade Practices Act and labelling the origin of goods* (Oct. 1986)
> *Refusal to deal* (May 1987)
> *A fair go for consumers* (Sept. 1987)
> *Consumers and refunds* (Sept. 1987)
> *Business and refunds* (Sept. 1987)

Guides
> *Objectives, policies and priorities in relation to restrictive trade practices* (Jan. 1986)
> *Guide to amendments to the Trade Practices Act* (May 1986)
> *Restatement of future directions of TPC consumer protection work* (May 1986)
> *Conferences on recalls and bans* (June 1986)
> *Guidelines for the merger provisions of the Trade Practices Act 1974* (Oct. 1986)
> *Summary of the Trade Practices Act 1974* (Dec. 1986)
> *Representative applications under the Trade Practices Act—*
> *explanation and guiding principles* (Dec. 1986)

A look at the Trade Practices Commission for high school students (March 1987)
Unconscionable conduct (March 1987)
Trade Practices Act teaching package (for TAFE colleges) (copies available on loan from the Commission's Canberra office) (July 1987)

The above publications are available free from Trade Practices Commission offices.

Guidelines

Small business and the Trade Practices Act (1983)
Insurance and the Trade Practices Act (1985)
Deposits and loans: their advertising and the Trade Practices Act (1986)

Reports

Packaging and labelling laws in Australia—June 1977 (limited quantities available free of charge from the Commission's Canberra office)
Interim compendium of self-regulation schemes in Australia—June 1985 (limited quantities available free of charge from the Commission's Canberra office)
Survey of consumer opinion in Australia—January 1987 (available from Commonwealth Government bookshops at $19.95)
Trade Practices Commission Annual Report 1986–87 (available from Commonwealth Government bookshops at $9.95)

Journals

Trade Practices Commission Bulletin (published every two months and available for a $10 annual subscription fee).

Each year legislative action is taken by the Commonwealth Government under the *Trade Practices Act* and Customs (Prohibited Import) Regulations to ban the sale of unsafe goods and to declare consumer product safety standards. Such items are often listed in the daily press and in *Choice*. Details are available from the Trade Practices Commission.

Glossary

This glossary contains brief definitions of chemical terms used in the text, together with derivations when these are considered of interest to the reader. The abbreviations used in the text are also included. Readers should also consult the index in order to obtain further information contained in the body of the text.

Today there are a number of general dictionaries available that contain chemical and other scientific terms (e.g. *The Macquarie Dictionary*, in Australia) and many readers will find the definitions they contain helpful. For the specialist, as well as the general reader with a keen interest in chemistry, a number of chemical dictionaries are available.

ACA Australian Consumers' Association. The largest consumer organisation in Australia, with over 200 000 subscribers to its monthly magazine, *Choice*. It is a completely independent, non-profit, non-party-political organisation. No advertising or subsidies are accepted.

acid Originally a sour material [L *acidus*, sour]. For more details see Chapter 1.

Act Act of parliament; law passed by parliament. Regulations are made under the Act and these regulate the activities controlled by the Act. The regulations are easy to introduce or amend; the Act is not easy to change. If different States have different Acts covering the same area, then the corresponding regulations may have to be worded differently.

ADI Acceptable daily intake, see MTDI.

aerobic Name coined by Pasteur in 1863 to denote bacterial processes occurring only in the presence of oxygen; opposite is anaerobic. [Gk *aero-* (*aer* air), + *bios* life.]

AFCO The Australian Federation of Consumer Organizations is primarily a federal lobby allied or competing with manufacturing, agricultural and union lobbies in Canberra. It is funded by the Federal government.

aflatoxins A very toxic group of substances, formed by the mould, *Aspergillus flavus*.

AGAL Australian Government Analytical Laboratories.

agar (agar-agar) Seaweed of various kinds; a gelatinous substance obtained from red seaweed (Rhodophyceae); used in bacterial cultures and in food. [From Malay.]

AGPS Australian Government Publishing Service. The equivalent in the UK is Her Majesty's Stationery Office (HMSO); in the USA, Superintendent of Documents, US Government Printing Office.

alcohols Organic compounds with the functional group –OH (not when attached to an aromatic ring; they then become phenols).

aldehyde A compound containing the group —CHO. Coined by Liebig in 1837.

aliphatic   Term introduced by Hjelt, *c.* 1860, to distinguish carbon compounds found in fats from those found in aromatic substances; i.e. open-chain hydrocarbons and derivatives as distinct from derivatives of benzene. [Gk *aleiphar*, oil, fat.]

aliquot   A portion taken for analysis and which is a known fraction of the whole sample. [From L *aliquot*, some, so many.]

alkali   A water-soluble base yielding a caustic solution, i.e. pH > 7. [Arabic *alqili*, the roasted—product of roasting marine plants; hence any substance having properties similar to those of roasted calcium carbonate, i.e. CaO from $Ca(OH)_2$ in water.]

alkaloid   Organic nitrogen bases occurring in plants and having powerful action on animals; e.g. nicotine, morphine, quinine, strychnine. [From Gk, used by Dumas in 1835.]

alkylolamide   The product from the reaction of the amine part of an alkylolamine with an organic acid. An alkylolamine is an alkylol (fatty alcohol) in which one of the hydrogens (on carbon) has been replaced by an amine group.

allergen   A substance, often a protein, that elicits an allergic reaction.

alloy   An intimate association (which may be a compound, solution or mixture) of two or more metals which has metallic properties. [L *ad*, to, + *ligare*, bind.]

alum   Name given to certain double salts which crystallise readily as octahedra; e.g. $KAl(SO_4)_2 \cdot 12H_2O$, common alum. [L *alumen*, bitter salt.]

amide   The group $-CONH_2$ made by replacement of a hydrogen atom in ammonia or an amine by an acid radical; hence am(ammonia) and -ide, coined by Weiz. (See *alkylolamide*.)

amine   The group $-NH_2$ made by replacement of one or more hydrogens in ammonia by alkyl or other hydrocarbon radicals.

amino acid   An organic acid with an amino group attached. These acids are components of proteins.

amorphous   Without crystalline, or regular, structure. [Gk *amorphos*, without shape.]

amphibole   A group of asbestos and asbestos-like minerals of variable composition; Haüy (1743-1822). [L *amphibolos*, ambiguous.]

amphoteric   Capable of acting both as an acid or a base. [Gk *amphoteros*—comparative of *ampho*, both.]

amyl   The group $-C_5H_{11}$. [Gk *amylon*, L *amylum*, starch.] An alcohol obtained originally from starch (amyl alcohol); coined by Balard 1844.

anaerobe   A microbe that thrives only in an oxygen-deficient environment.

angstrom, ÅH   A unit of measurement for wavelength, equal to 10^{-10} metres. Convenient for molecular dimensions but being phased out for SI units. [Named after A.J. Ångstrom, 1814-74, the Swedish physicist.]

antioxidant   Substance which inhibits oxidation (loosely, the reaction with oxygen) of another material to which it has been added.

arachidic   Arachidic acid, $C_{19}H_{39}COOH$, found in peanuts (ground nuts, genus *Arachis*).

argon   [Gk *argos*, inactive.] A colourless, odourless, chemically inactive, monatomic gaseous element used in incandescent light bulbs and for welding.

aromatic   Having planar ring type groups usually composed of carbon atoms (e.g. benzene and naphthalene) which have alternating double and single bonds (these in fact *don't* alternate but are smeared around the ring uniformly as an average of one and one half bonds); in contrast to aliphatic.

aryl The term used to refer to a derivative of an aromatic group such as benzene or naphthalene (*aromatic* + *yl*).

asbestos A group of fibrous silicate minerals. [Gk, unquenchable.]

aspirin Synthetic acetylsalicylic acid ['A' for acetyl and L *spiraea*, plant source of salycilate.]; coined by Dreser in 1899 for the *synthetic* but nature identical substance found in *Spiraea ulmaria* (willow tree).

atropine Poisonous alkaloid found in deadly nightshade (belladonna). Used medicinally; e.g. to widen pupils for eye examination. [L *Atropa*, the belladonna genus, from Gk *Atropos*, inflexible, one of the Fates in Greek mythology.]

azeotrope A mixture of liquids that boils at constant temperature. [Gk *a* without, *zein* boil + *tropos* turning, change.] *See also* eutectic.

barium [Gk *barys*, heavy.] An element related to calcium but of greater atomic mass.

becquerel Unit of radioactive activity, symbol, Bq, corresponding to one nuclear transition per second.

bi-acid A substance that has two acid groups instead of the more usual single acid group.

bifunctional A molecule with two groups attached that can react.

billion The value 10^9, one thousand million, is used in this book.

biodegradable The property of a complex chemical compound to be able to be broken down into simpler components under naturally occurring biological processes—such as those which form part of the normal life-cycle in a river or soil.

biological half-life The mean time required for half of a quantity of specified material in a living organism to be biologically eliminated.

borax sodium borate decahydrate. [Arabic, Persian.]

branched-hard The branched-chain compounds are unable to be broken down at a reasonable rate in the environment.

bromine A halogen, intermediate between chlorine (gas) and iodine (solid); a dense, fuming liquid at room temperature. [Gk *bromos*, stink.]

buffer A mixture of substances which tend to hinder large changes in acid or basic properties of a solution. Used in a more general sense outside chemistry.

burette A graduated glass tube open at one end and fitted with a tap at the other. [F *burette* a small carafe; hence carburettor?]

butyric acid C_3H_7COOH [L *butyrum*, butter.]. The acid from rancid butter.

C & EN *Chemical and Engineering News*, published by American Chemical Society.

carbamate Ester of carbamic acid; from carb(onic) and am(ide).

carboxylic Refers to the organic acids having functional group –COOH.

carcinogen An agent capable of inducing cancer.

casein The phosphoprotein of milk and cheese. [L *caseus*, cheese.]

catalyst An agent that speeds up a chemical reaction without itself being used up in the process. Enzymes are catalysts for biological reactions, whereas the transition metals (Co, Ni, Pt etc.) are common chemical reaction catalysts. [Gk *cata*, down, + *lysis*, a loosening, setting free; coined by Berzelius 1835.]

catechol (1,2-dihydroxybenzene) From dry distillation of catechu from wood of certain trees. [Malay *cachu*.]

catecholamines (biogenic amines) A series of biologically active amines; e.g. dopamine (3-hydroxytyramine), noradrenaline, adrenaline.

catenation   Atoms joining in a chain. [L *catena*, chain.]

caustic   Very alkaline—capable of dissolving skin, fat etc. to form soap.

chemistry   From alchemy, the art of transmutation of elements. [Arabic *al-kimiya*.]

Chemistry and Industry   Published by the Society of Chemical Industry, UK.

Chemistry in Britain   Journal published by the UK Royal Society of Chemistry and sent to members.

Chem Tech   Published by the American Chemical Society.

chiral   Any geometrical figure which like a hand, cannot be brought into coincidence with its mirror image. [Gk *chei*, hand.] Coined by Lord Kelvin, 1894.

chloramphenicol   An antibiotic, natural and synthetic. One of the very few natural products containing nitro group [from chlor-amide-phenyl-nitro-glycol].

chlorine   A yellowish green gas. Pool chlorine is a compound of chlorine and calcium hydroxide. It is approximately 70% calcium oxychloride (*see* Swimming pools, Chapter 5). [Gk *chloros*, pale green.]

Choice   Monthly magazine of Australian Consumers' Association.

cholesterol   A fat-like molecule (chemically not a fat, but an alcohol) with a structure on which all the steroids (e.g. sex hormones, bile acids and cortisone) are based. Produced by the body. [Gk *chole*, gall, bile, + *stereos*, solid.]

chromatography   Literally, the separation of colours. Now a general technique of separation of chemicals based on the difference in the strength of adsorption onto a solid. [From Gk *chroma*, colour, + *graphe*, writing.]

chrysotile   Chief asbestos mineral, a magnesium silicate. [Gk *chrysos*, gold; iron impurities give it a yellowish colour.]

cis   On this side of [L]; cf. *trans*, on opposite sides.

cobalt   [Gk *kobalt*, an evil sprite, a goblin, a diluent of silver ores. The ore could not be made to yield a useful metal.]

collagen   A protein in animal connective tissue yielding gelatin on boiling. [Gk *colla*, glue, + gennan, to generate]

colligative   Properties of a solution which depend on the number, but not nature, of dissolved particles. [L *colligare*, to bind together]

compounding   A polymer is formed from monomer units and is sometimes called a resin. In order to make a useful plastic material, the resin must be mixed with other materials or *compounded*.

conjugated   Alternating double and single bonds as in polyunsaturated.

continuous phase/outer phase   The first liquid which surrounds droplets of a second liquid (being the discontinuous phase) in an emulsion.

co-polymer   A polymer formed from linking two (or more) different monomer types.

copper   [L *cuprum*, from the island of Cyprus.] The copper ores of Cyprus were the Romans, main source of the metal.

corrosion   Corrosion is an example of oxidation caused by oxygen or other oxidising agents.

crocidolite   An asbestos mineral (blue). [From Gk *crocydos*, a nap of woollen cloth, + *lithos*, stone.]

CSIRO   Commonwealth Scientific and Industrial Research Organisation. This has many divisions (such as Food Research, Entomology, Plant Industry, Building Research, Human Nutrition, Chemical Physics) engaged in research for Australian industry.

curie The former unit of radioactivity, based on one gram of radium. Symbol Ci.

denature The tertiary structure of a protein collapses on 'denaturing' by heating, agitation or exposure to air.

detergent [L *detergere*, to cleanse, wipe away.] Synthetic surfactants, not including soaps (which are the sodium salts of natural fatty acids).

diastereoisomer Isomeric molecules with some asymmetric centres the same and others different. [From Gk *dia*, through; *stereos*, solid; *isos*, *equal*; *meros*, part.] Not an easy concept!

dioxan The cyclic ether 1,4-dioxan is a solvent miscible with water. Not to be confused with dioxin.

dioxins e.g. 2,3,7,8-tetrachlorodibenzo-*p*-dioxin (2,3,7,8-TCDD). Exhibits powerful biological activity in most species at very low concentrations.

effective half-life (of radioactive contaminants in biological systems) The time required for the activity of a radionuclide in a living organism to fall to half its original value as a result of both biological elimination and radioactive decay.

elastomer A polymer material with elastic properties, namely the ability to return to the original dimensions after distortion.

electrolysis Electricity is carried through a solution by ions. At the electrodes, these (or other) ions are neutralised and discharged as neutral atoms or molecules. In a car battery, electrolysis occurs during the charging process. When the battery is operating in reverse, the chemical reactions provide electricity. The process is called an electrochemical reaction. [From Gk *electron*, amber, + *lysis*, setting free.]

emollient A substance that softens and soothes the skin.

emulsions The suspension of one liquid as fine droplets in another with which it does not mix. Hence also emulsifiers. See also *glycerides*.

enantiomer One of a pair of molecules which are optical isomers of each other (see *chiral*).

enzyme A biological molecule which can promote or catalyse a particular reaction (to the exclusion of others). [From Gk *en*, in, + *zyme*, yeast. Coined by Kuhne, 1878.]

eosin Tetrabromofluorescein. [Gk *eos*, dawn]

epidemiology The study of disease in populations, with the ultimate aim of establishing cause and effect.

epoxy Oxygen directly linked to two adjacent bonded carbon atoms forming a triangle. [Gk *epi*, beside.]

equilibrium In chemistry this has a very specific meaning and refers to reactions in which the forward and reverse rates are matched so that the composition of the mixture appears unchanging in time. [L *aequus*, equal, + *libra*, balance.]

ergot A disease of rye caused by a fungus which causes bread made from diseased rye to become poisonous. The word also refers to the toxic substance, which is also used medicinally.

erythema Surface inflammation of the skin. [Gk *eruthema*, be red.]

esters Combination of (organic) acids and alcohols. The carboxylic esters with short chains are often pleasant smelling. Fats and oils belong within the classification of esters. [G *essig*, vinegar, + *ather*, ether, coined by Gmelin, 1848.]

ether Compound in which two hydrocarbon groups are linked by one oxygen. [L *aether*, Gk *aither*, pure upper air.]

eutectic   A mixture of two or more substances at the composition yielding a lowest (local) melting point [Gk *eu*, easily, + *tekein*, to melt.] Contrast *dystectic*, a composition of maximum melting point. See also *azeotrope*.

excise   Duty or tax levied on goods produced or sold within the country. Customs duties are levied on imported goods.

'factored'   From factor, an agent buying and selling on commission. Process of hiding import and export of goods. Compare 'laundered' for money.

fatty   Having a long chain of carbon atoms, usually 10–18 members. These chains are the backbone of the fatty acids in fats.

FAO   Food and Agricultural Organisation (of the UN).

FDA   Food and Drug Administration (USA).

flammable   Easily set on fire. Used now officially in place of the word 'inflammable' (which is philologically more correct) because of the confusion whereby 'inflammable' was thought to mean non-flammable.

flocculate   To coagulate in fluffy lumps. [L *flocculus*, a little flock of wool.]

fluorescence   The rapid emission of light at longer wavelengths than that which is absorbed; e.g. adsorption of ultraviolet light can yield blue fluorescence.

fluorine   [L *fluere*, to flow.]

flux   A substance added to lower the melting temperature in metallurgy (and soldering). [L *fluere*, to flow.] Hence also *flux* as a measure of particles flowing through a surface of 1 m per second.

Food Technology in Australia   Published by the Council of Australian Food Technologists Association, Inc. (CAFTA).

gallium   [L *gallia*, France.]

galvanise   To cover metal by electrodeposition. Incorrectly used when referring to covering steel with zinc by other means. [After Luigi Galvani (1737–1798), who investigated electric effects on frogs' legs—jerking their muscles—hence galvanise into action.]

gelatin   See *collagen*—protein from animal tissues; used as a glue. [L *gelare*, to freeze, set solid; hence Italian *gelato*.]

glycerides   Esters of the tri-alcohol glycerol; sometimes called triglycerides (fats). Mono-glycerides are made synthetically and used as emulsifiers. [Gk *glyceros*, sweet.]

glycosides   Compounds of sugars with non-sugars.

gray   Unit of absorbed dose of radiation corresponding to one joule per kilogram of matter. Symbol Gy.

hafnium   [L *Hafnia*, Copenhagen.]

half-life   The length of time for a substance to drop in concentration to half its original level. Strictly useful only for exponential decay (which never reaches zero) but used more generally.

hazardous   Hazard depends on the *chance* of contact or injection (of a toxic material). Flammable materials become hazardous if there is an unacceptable chance that they will be exposed to fire.

heavy metal   The metals with higher atomic mass tend to form compounds which are more poisonous (e.g. Hg, Cd, Pb etc.).

histamine   An amine found in human tissue and released upon injury, causing local swelling and allergic reaction.

HMSO   Her Majesty's Stationery Office (UK)

humectants   Additives for keeping a product moist, or a product for keeping something else (e.g. skin) moist.

hydrogen   [Gk *hydros*, water, + *genes*, forming. Compare G *Wasserstoff*, stuff of water.]

hydrogen bond   A special bond in which hydrogen already bonded to an electronegative atom, such as oxygen, nitrogen, or sulfur, forms an additional 'extra' attraction for a second such atom close by. Hydrogen bonds account for the unique structure of water and determine the overall (secondary structure of proteins (enzymes) and nucleic acids. The double helix of DNA is held together by hydrogen bonds. [Gk *hydro*, water, + *geinomai*, I produce.]

hygroscopic   Materials which absorb water from the air. This property depends on how much moisture is in the air. Hydroscopic has a different meaning.

hyper-   Prefix meaning high. Poor use of language because of the similarity in sound of the word 'hypo-' which has the opposite meaning. Hyperkinetic = overactive; hypertension = high blood-pressure.

hypo-   Prefix meaning low (cf. *hyper*). These prefixes are generally restricted to medical terms. Photographic 'hypo' (sodium thiosulfate) has one sulphur atom with a low valency of two.

-ide   Suffix used for inorganic compounds containing two elements.

imine   As $-NH_2$ is an amine, $=NH$ became an imine. A compound of $-NH_2$ is amino (amide), so a compound of $=NH$ became an imino (imide). Logical!

indole   A low-melting solid with a faecal smell but used in low concentrations in perfumes. [L *ind(igo)* + *ol(eum)*, oil from indigo.]

Industries Assistance Commission (IAC)   Was known as the Tariff Board (Australia). A Commonwealth statutory body which holds inquiries and makes reports on the level of protection it believes necessary for industries.

infra-red   The region of the sun's radiation which lies beyond the red colour as seen in a rainbow. It is not visible but it is sensed as heat. Just as materials absorb different colours from white light, so they also absorb different sections of infra-red radiation.

-ine   As a suffix it can indicate (a) organic base; (b) amino acid; or (c) halogen element. Very versatile!

inhibit   Slow down a chemical reaction by blocking a part of the mechanism. [L inhibitus curb, restrain.]

initiator   A substance used to start a polymerisation reaction.

inorganic   Chemistry of elements other than those with carbon as the most important constituent.

iodophors   A group of compounds that release iodine slowly in solution to act as local disinfectants.

ionise   When hydrochloric acid gas is dissolved in water, it breaks up into ions; i.e. it ionises. Ions can come together and form an *un*-ionised compound (nothing to do with industrial associations!). [Gk *ion* (ienai), to go, coined by Faraday, 1834.]

ion/ionic   Chemical entity carrying electric charge (positive or negative). Ions occur in equal number of oppositely charged members either close together, as in solid salt ($Na^+ Cl^-$), or free to wander, as in a salt solution in water.

interstice   Space between packed atoms or ions in a crystal. [L *inter* between, *store* to stand]

isotopes   Atoms with the same atomic number and hence almost identical chemistry, but with different atomic mass, hence often having different radioactive properties.

-ium   Suffix normally used for metals or groups supposedly having properties of metals; e.g. *ammonium*. Wrong use in *helium*.

IUPAC   International Union of Pure and Applied Chemistry. Sets international chemical standards.

JAOAC *Journal of the Association of Official Agricultural Chemists* (USA). Seems no point in including if origin uncertain and derivation can be given.

K   Symbol for potassium. [L *kalium*, Arabic *quili*, charred ashes of the glasswort plant, used in glass making.] K is also the symbol for kelvin, absolute temperature.

ketone   Group $C=O$. [L *acetum*, vinegar, with Gk suffix *-one*, a female patronymic used in chemistry to denote a weaker derivative.]

labile   Unstable, liable to change to another form or to move away.

lanolin   Wool fat (the palmitate and stearate esters of cholesterol). [L *lan(a)*, wool, + *ol(eum)*, oil, fat.]

lanthanum   A rare metal. [Gk *lanthanum*, to escape notice.]

latex   An emulsion of rubber globules (in water), extended to include globules of synthetic materials.

lattice   Regular three-dimensional array of atoms or ions in a crystal. [Compare *lath* (OE *laett*), piece of sawn or split timber in form of thin strip used for support for slates, plaster or trellis, etc.]

lecithin   A biological fat and cell wall component with phosphate polar lead group.

linoleic acid   An unsaturated fatty acid, $C_{17}H_{31}COOH$. [L *linum*, flax, + *ol(eum)*, oil; hence also linoleum.]

litmus   Blue colouring matter from certain lichens used as an acid/base indicator. [Old Norse *litmose*, lichen for dyeing.]

logarithm   [Gk *logos*, word, ratio, + *arithmos*, number; thence mod. L *logarithmus*, coined by J. Napier, 1614.]

lyophilic   Solvent-loving; refers to colloid sized molecules (surfactants, proteins, etc.). [Gk *lysis*, a loosening, dissolution, + *philos*, loving.] Hence also lipophilic, fat soluble; *hydrophilic*, water soluble and then opposites (next entry).

lyophobic   Solvent-hating. [Gk *phobia*, fear of.]

-lysis   Breaking down, decomposition. [Gk *lysis*, a loosening, dissolution.] Hence hydrolysis, breaking down *by* water (hydro); pyrolysis, breaking down *by* heat (pyro), and haemolysis, splitting *of* blood corpuscles (haem).

M, mega   Power 10^6, million times.

malic, maleic   Organic acids. [L *malum*, apple.]

malt   Malt is the grain of barley which has been caused to sprout by being kept moist. [Indo-European *mel*, Gk *mill*, L *molere* to grind.]

margarine   Butter substitute. [Gk *margarites*, a pearl, hence 'margaric' coined by Chevreul, 1813, for a solid fat obtained from lard with a pearly appearance.] Today *margaric acid* refers to an *un*natural fatty acid, $C_{16}H_{33}COOH$.

mercury   Mercury was the messenger of the gods in Roman mythology. Symbol Hg [Gk *hydragyros*, L *hydrargyrus* water-silver, quicksilver.]

meta-   [Gk *meta*, changed (in form, etc.), next to, between.] Hence meta substituted benzene, metabisulfite, metabolism, metastable form.

methylene Group —CH_2—. [Gk *methy*, + wine, *hyle*, wood. Methyl alcohol was originally made by distilling wood.]

microcosmic salt Sodium ammonium hydrogen phosphate, a crystalline salt obtained from evaporation of urine at a time when humans were regarded as the centre (microcosm) of the universe.

miscible Two liquids are miscible when they mix completely in all proportions. Often liquids are partially miscible. [L *miscere*, to mix.]

mole/moles/mols/mol Bakers have their dozen (= 13) and so do chemists. The chemist's one is called a mole and contains 6.023×10^{23} single units (atoms, molecules, electrons, etc.). This number is chosen because this number of atoms and molecules always weighs in grams an amount equal to the atomic (molecular) mass. Thus a mole of carbon weighs 12 g and a mole of oxygen weighs 16 g. A mole of carbon monoxide (CO) weighs 28 g. [Coined by Ostwald 1853–1932.]

molybdenum [Gk *molybdos*, lead.] Originally found in small amounts in lead ores.

morphine An alkaloid, $C_{17}H_{19}O_3N$. [Gk *Morpheus*, god of dreams, son of sleep.]

MTDI Maximum tolerable daily intake, originally called acceptable daily intake (ADI). The amount of a substance that can be ingested per day for a lifetime without ill effects. This is based on a study of the toxicity of the substance (its poisonous nature, including the possible formation of tumours). It does *not* take into account any pharmacological effects of the substance, its ability to act as a drug.

mutagen An agent capable of causing mutations in the genetic material, which can affect either the organism or its offspring, depending on which cells are affected.

mycotoxin Poison produced by a mould.

nanometre, nm 10^{-9} or one billionth of a metre. SI prefix.

NBSL National Biological Standards Laboratory (Australia).

New Scientist UK science weekly. Available at larger newsagents or by subscription, and now printed in Australia.

NH & MRC National Health and Medical Research Council (Australia).

nickel [G *kupfernickel*, coppernickel or bedevilled copper; cf. 'Old Nick', the devil.] The ore yielded no copper (cf. names cobalt and wolfram) where copper was also expected.

nicotine Jean Nicot introduced tobacco into France in 1560. [Mod. L (*herba*) *Nicotiana*, herb of Nicot = tobacco.]

nitrogen [L *nitrium*, Gk *nitron*, native soda. Compare G *Stickstoff*, suffocating material.]

octane C_8H_{18}. [Gk octo, eight.]

oleic acid A fatty acid. [L *oleum*, oil; Gk *elaion*, olive oil.]

organic Originally referred to chemicals produced by living organisms. Today it simply refers to the chemistry of compounds containing carbon. [Gk *organor*, instrument, bodily organ.]

organochlorine Compounds which are generally composed only of carbon and hydrogen, to which chlorine has been added. These compounds are often biologically very active but not easy to break down; hence they are persistent.

ortho- Regular form. [Gk *ortho*.]

osmosis Selective passage of solvent molecule, but not solute molecule, through a semipermeable membrane. [Mod. L.]

oxalic acid The simplest organic acid with two carboxylic acid groups—poisonous.

oxidation Originally reaction with oxygen as in burning. Now generalised and can include reaction with chlorine or other oxidation agents. [Gk *oxys eidos*, species; coined by de Morveau, 1790.]

oxidation number A method of bookkeeping in balancing reactions in which oxidation and reduction are simultaneously occurring.

oxygen [From Gk *oxys*, acid, + *genes*, forming; an erroneous name given by Lavoisier, who thought oxygen was essential in acids. Compare G *Sauerstoff*, stuff of acid (Sauer)]

para- Alongside, beyond, near, contrary to. [Gk *para*.] Very flexible prefix in chemistry.

parenteral (Drugs) absorbed into the body by a route other than by the intestinal tract.

peptide Molecule formed when two amino acids are joined; hence peptide bond, polypeptide. A protein is a large polypeptide.

perspex A transparent polymer. [L *perspicere*, to look through.] Lucite, Plexiglas.

petroleum fraction A fraction of oil selected in a refinery on the basis of boiling point.

pH Measure of acidity and basicity of water solution. See Chapter 1.

phase A term introduced by Willard Gibbs in his treatise on thermodynamics (1876–78). [Gk *phasis*, mod. L *plasis*, appearance (of a star).] In chemistry, materials in a different thermodynamic state or composition but not in quantity or shape.

phenols Aromatic groups such as benzene and naphthalene with the functional group –OH attached.

phocomelia [Gk *phoke*, seal.] Having limbs looking like the flappers of seals.

photodynamic Caused by light.

pK$_a$ A measure of the degree to which an acid or base will dissociate in water.

plasticiser This is an additive which makes a plastic material more flexible or less rigid.

plastisols A plastic formulation that has a high degree of fluid behaviour, generally due to a large amount of added plasticiser.

platinum [Sp. *platina*, silver.] The original source of platinum was in silver ores.

poison Substance that, when introduced into or absorbed by a living organism, destroys life or impairs health. Used anthropomorphically by chemists when referring to catalysts, which when 'poisoned' can no longer function.

polymerisation The process by which single units (monomers) are joined together, like linked paper-clips, to form a giant molecule.

polymorphism Occurrence of a substance in a number of solid forms.

ppm Parts per million (by mass), equivalent to a grain of sugar in a cup of tea (very approximate). Now called milligrams per kilogram, or given as 0.0001%.

PRACI Proceedings of the Royal Australian Chemical Institute. Renamed *Chemistry in Australia* in July 1977.

precipitate/precipitation To fall out of solution as a sediment. [L *praecipitatus*, thrown down headlong.]

prostaglandins Biological chemicals isolated most readily from sperm and prostate glands of sheep, but very widespread in animals. These compounds are hormones with widely differing functions.

protein Essential constituents of life. [Gk *proteion*, the first place, the chief rank; coined by G.J. Mulder, 1838.]

proton Hydrogen atom without its electron; i.e. a hydrogen ion. Hence protonation, which is adding a proton.

Q Q is the ratio of the fusion power generated to the electrical power input. The lower the Q value at which a reaction can operate, the easier the process is to manage. Care: there are uses and definitions of the symbol Q for other purposes.

quantum mechanics The modern theory of chemistry is based on a mathematical theory which suggests that the mechanical behaviour of very small things such as electrons and atoms is different from the mechanics of Newton, which described the behaviour of everyday size objects. [L *quantus*, how much, how many; coined by Planck, 1900.]

racemic A one-to-one mixture of left-handed and right-handed chiral forms of the same molecule. Most chemical reactions produce products as such mixtures, whereas biological reactions generally produce one or the other form only. [L *racemus*, cluster of grapes.]

radiology The study and application of x-rays, gamma rays, etc., in the diagnosis and treatment of disease.

radium [L *radius*, a ray.]

radon From radium.

reduction The opposite to oxidation. Reduction and oxidation go together. When a material is oxidised by oxygen for example, the oxygen is reduced to water. Such processes are known as redox reactions.

refractive index A measure of the ability of a material to bend a ray of light.

rhodium [Gk *rhodios*, roselike.] The first compounds of this metal to be made were rose pink.

R, R' The R designates an undefined organic group, in our case generally a hydrocarbon chain. The R' just says it is not necessarily the same as R.

SA Standards Australia, formerly Standards Association of Australia (SAA).

salicylic acid *o*-hydroxy benzoic acid. Acetylsalicylic acid = aspirin (analgesic). Methyl salicylate = oil of wintergreen (used as a liniment for sore muscles). Phenyl salicylates (salol) = used in sunscreens and as a stabiliser in plastics. Salicylanilide = compound of salicylic acid and aniline derivatives, used as antiseptic in soap.

saponify To make soap from fat. [L *sapo*, soap, + *facere*, to make, do.]

Sci Quest A journal once called Chemistry published by the American Chemical Society.

sequester To take out of circulation, to tie up metal ions so that they don't interfere (by precipitating soaps, etc.). [L *sequestare*, to commit for safe keeping.]

sequestering agent A chemical which ties up metallic ions in solution.

sievert Unit of radiation dose equivalent corresponding to the absorption of one joule in one kilogram of biological matter, taking into account the quality factor and other modifying factors. Symbol Sv.

SI units Système international d'Unités. The international system of units based on a restricted metric system. It replaces older c.g.s. (centigrade gram second) and m.k.s. (metre kilogram second) schemes, and British units.

silicon [L *silex*, flint.] Flint, like quartz and sand, is silicon dioxide, also called silica.

solute Dissolved material.

solvent Dissolving liquid.

spatula A spoon-like instrument. [L *spatha*, a broad tool or weapon.]

spectroscopy The separation and analysis of light into its component parts and study of the interaction of materials with the different regions of light. Now applied over the whole range of electromagnetic radiation from radio waves to gamma and x-rays. The most powerful and most common source of information on the structure of molecules.

spontaneous   In chemistry this has the specialised meaning of referring to a process which can occur on its own without further input. However, it may occur so slowly as to be unmeasurable. A mixture of LP gas and air will burn *spontaneously*, but to do so requires a catalyst or a match. The direction in which a reaction goes *spontaneously* is the natural direction it will go, when it goes. [Latin L *spontaneus*, of one's own free will; coined by Hobbes, 1656.]

stereospecific   As applied to polymers, it is the ability to cause a polymer to be formed in a single geometry rather than as a mixture of structures. This often means the polymer can pack together more efficiently to give a material of greater order and higher density. See also *racemic*.

stoich(e)iometry   Ratio of atoms in a molecule or a reaction. [Gk *stoicheion*, element, + *metria*, process of measuring; coined by Richter, 1792.]

strontium   From Strontian, a town in Scotland, where it was first discovered, ore strontianite.

substrate   A basis on which something else is placed; a starting material.

sulfonation   The addition of the function group $-SO_3H$ to a molecule.

sulfonic acid/sulfonate   Organic compound with the functional group $-SO_3H/-SO_3^-$.

surface tension   The force of attraction for itself that gives a liquid such as water an apparent skin which contracts so as to form drops rather than sheets (on waxy surfaces). On some surfaces (e.g. clean glass) the attraction of the water is greater for the glass than for itself and so the water wets the glass. It is also called surface (free) energy.

surfactant   A molecule attracted to the surface of water and capable of changing the properties of the surface generally by lowering the surface tension.

synergist   A substance which itself has no activity but increases the activity of another substance when it is added to it.

Tappi.   Journal of the Technical Association of the Pulp and Paper Industry.

tautomer   When an atom (often hydrogen) moves from one point in a molecule to another, the new and original molecules form a tautomeric pair.

technetium   [Gk *technitos*, artificial.] This element does not occur at all in nature. It is a fission product of uranium, and hence artificial.

teratogen   An agent capable of inducing fetal abnormalities (monstrosities). [Gk *teras*, monster.]

thermodynamics   The study of whether chemical reactions are feasible in theory. No information on how fast and hence how practical the process may be is obtained.

thermoplastic   Applied to a polymer that melts on heating. Its structure consists of independent long molecules.

thermosetting   Applied to a polymer that is sometimes made by heating but once formed does not melt on heating, charring instead. Its structure has cross-linking from one chain to the next.

thio-   Sulfur. [Gk *theion*, brimstone.] Used in names such as sodium thiosulfate where one sulfur atom replaces an oxygen atom; hence two different words are needed.

thixotropy   The property of becoming less viscous on stirring. [Gk *thixis*, a touch, + *tropos*, a turning, change.]

threshold   (Of poisons) the maximum level of intake that produces no clinically detectable effect—no response dose. This is not necessarily the level below which no damage is done.

titrate   Determining strength (concentration) of a solution by finding the volume of a standard solution with which it reacts. [F *titre*, to give title to, to determine strength.]

toxicity   The *property* of poisons.

toxin   Originally used for poisons secreted by microbes.

trade mark   A device or word legally registered to distinguish a manufacturer's or trader's goods. It is spelt with a capital letter to distinguish it from a common name; e.g. Coke produced by the Coca Cola Company. Both coca and cola are common names.

trans   On opposite sides. [L] See *cis*.

tungsten   [From Swed. *tung*, heavy, + *sten*, stone.] Symbol W comes from *wolfram*, an earlier name, see below.

urticaria   Nettle-rash, but used to describe other skin allergies often made worse by exposure to sunlight.

valence   The combining power of an atom (see Chapter 1). [L *valentia*, strength, capacity.]

vermiculite   Group of mica minerals so called because when slowly heated they open into long worm-like structures. [L *vermiculus*, diminutive of *vermis*, worm; hence vermin.]

viscosity   Resistance to change in shape or form of a material—'internal friction'.

vitamin   [L *vita*, life, + G *Amin*, amine; coined by C. Funk, 1913.]

vitreous   [L *vitrum*, glass.]

volatile   Readily forming a gas.

v/v   Volume/volume.

'wetting'   The covering of a solid by a liquid with a thin film. The contact angle the liquid makes on the solid is small. See also *surface tension*.

WHO   World Health Organisation (United Nations).

wolfram   Old name for tungsten. Agricola (1556) called a mineral now known as *wolframite* 'wolf foam' because it seemed to eat up tin in the same way as a wolf eats sheep.

w/v   Weight/volume.

w/w   Weight/weight.

xenon   [Gk *xenos*, a stranger.] When Ramsay was investigating the inert gases, this gas turned up unexpectedly.

xylene   Dimethyl benzenes [Gk *xylon* wood] from destructive distillation of wood.

yttrium   [From Ytterby, a village 15 km north of Gothenburg, Sweden's second largest city. The four elements yttrium, erbium, ytterbium and terbium were discovered at Ytterby.]

zeolite   A group of hydrated aluminosilicates (natural minerals or synthetic) [Gk *zein* to boil and *lithos* stone; coined by A.F. Cronstedt, 1756, for a mineral species which appeared to boil when heated in a blowpipe.]

Original Source: S.C. Bevan, S.J. Gregg, A. Rosseinsky, *Concise Etymological Dictionary of Chemistry*, Applied Science Publishers, London, 1976.

Further reading: M.P. Crosland, *Historical studies in the language of chemistry*, Heinemann, London, 1962.

Index

In this index prefixes such as *beta* and *ortho* are ignored, and entries are filed under the main part of the name. A letter-by-letter filing arrangement has been used. Sub-entries are arranged in strict alphabetical order, according to the first word of the entry, even when this is a less significant word (preposition or conjunction) such as *in*, *of* or *and*.

5 6 7 8 9 0
E F G H I J